# Springer Texts in Statistics

*Advisors:*
George Casella    Stephen Fienberg    Ingram Olkin

**Springer**
*New York*
*Berlin*
*Heidelberg*
*Barcelona*
*Hong Kong*
*London*
*Milan*
*Paris*
*Singapore*
*Tokyo*

# Springer Texts in Statistics

*Alfred:* Elements of Statistics for the Life and Social Sciences

*Berger:* An Introduction to Probability and Stochastic Processes

*Bilodeau and Brenner:* Theory of Multivariate Statistics

*Blom:* Probability and Statistics: Theory and Applications

*Brockwell and Davis:* An Introduction to Times Series and Forecasting

*Chow and Teicher:* Probability Theory: Independence, Interchangeability, Martingales, Third Edition

*Christensen:* Plane Answers to Complex Questions: The Theory of Linear Models, Second Edition

*Christensen:* Linear Models for Multivariate, Time Series, and Spatial Data

*Christensen:* Log-Linear Models and Logistic Regression, Second Edition

*Creighton:* A First Course in Probability Models and Statistical Inference

*Dean and Voss:* Design and Analysis of Experiments

*du Toit, Steyn, and Stumpf:* Graphical Exploratory Data Analysis

*Durrett:* Essentials of Stochastic Processes

*Edwards:* Introduction to Graphical Modelling

*Finkelstein and Levin:* Statistics for Lawyers

*Flury:* A First Course in Multivariate Statistics

*Jobson:* Applied Multivariate Data Analysis, Volume I: Regression and Experimental Design

*Jobson:* Applied Multivariate Data Analysis, Volume II: Categorical and Multivariate Methods

*Kalbfleisch:* Probability and Statistical Inference, Volume I: Probability, Second Edition

*Kalbfleisch:* Probability and Statistical Inference, Volume II: Statistical Inference, Second Edition

*Karr:* Probability

*Keyfitz:* Applied Mathematical Demography, Second Edition

*Kiefer:* Introduction to Statistical Inference

*Kokoska and Nevison:* Statistical Tables and Formulae

*Kulkarni:* Modeling, Analysis, Design, and Control of Stochastic Systems

*Lehmann:* Elements of Large-Sample Theory

*Lehmann:* Testing Statistical Hypotheses, Second Edition

*Lehmann and Casella:* Theory of Point Estimation, Second Edition

*Lindman:* Analysis of Variance in Experimental Design

*Lindsey:* Applying Generalized Linear Models

*Madansky:* Prescriptions for Working Statisticians

*McPherson:* Statistics in Scientific Investigation: Its Basis, Application, and Interpretation

*Mueller:* Basic Principles of Structural Equation Modeling

*Nguyen and Rogers:* Fundamentals of Mathematical Statistics: Volume I: Probability for Statistics

*Nguyen and Rogers:* Fundamentals of Mathematical Statistics: Volume II: Statistical Inference

*(continued after index)*

Christian P. Robert    George Casella

# Monte Carlo
# Statistical Methods

With 65 Figures

 Springer

Christian P. Robert
CREST-INSEE
Laboratoire de Statistique
75675 Paris Cedex 14
France

Dept. de Mathematique
UFR des Sciences
Universite de Rouen
76821 Mont Saint Aignan cedex
France

George Casella
Biometrics Unit
Cornell University
Ithaca, NY 14853-7801
USA

Library of Congress Cataloging-in-Publication Data
Robert, Christian P., 1961–
    Monte Carlo statistical methods / Christian P. Robert, George
Casella.
        p.      cm. — (Springer texts in statistics)
    Includes bibliographical references (p.      -     ) and index.
    ISBN 0-387-98707-X (alk. paper)
    1. Mathematical statistics.   2. Monte Carlo method.   I. Casella,
George.   II. Title   III. Series.
QA276.R575    1999
519.5—dc21                                              98-55413

Printed on acid-free paper.

Production managed by Allan Abrams; manufacturing supervised by Jeffrey Taub.
Photocomposed pages prepared from the authors' TeX files.
Printed and bound by Maple-Vail Book Manufacturing Group, York PA.
Printed in the United States of America.

9 8 7 6 5 4 3 2 1

ISBN 0-387-98707-X  Springer-Verlag  New York  Berlin  Heidelberg    SPIN 10707599

*In memory of our colleague and dear friend, Costas Goutis (1962-1996)*

*To Benjamin, Joachim, Rachel and Sarah, our favorite random generators!*

# Preface

Monte Carlo statistical methods, particularly those based on Markov chains, have now matured to be part of the standard set of techniques used by statisticians. This book is intended to bring these techniques into the classroom, being (we hope) a self-contained logical development of the subject, with all concepts being explained in detail, and all theorems, etc. having detailed proofs. There is also an abundance of examples and problems, relating the concepts with statistical practice and enhancing primarily the application of simulation techniques to statistical problems of various difficulties.

This is a textbook intended for a second-year graduate course. We do not assume that the reader has any familiarity with Monte Carlo techniques (such as random variable generation) or with any Markov chain theory. We do assume that the reader has had a first course in statistical theory at the level of *Statistical Inference* by Casella and Berger (1990). Unfortunately, a few times throughout the book a somewhat more advanced notion is needed. We have kept these incidents to a minimum and have posted warnings when they occur. While this is a book on simulation, whose actual implementation must be processed through a computer, no requirement is made on programming skills or computing abilities: algorithms are presented in a program-like format but in plain text rather than in a specific programming language. (Most of the examples in the book were actually implemented in C, with the S-Plus graphical interface.)

Chapters 1–3 are introductory. Chapter 1 is a review of various statistical methodologies and of corresponding computational problems. Chapters 2 and 3 contain the basics of random variable generation and Monte Carlo integration. Chapter 4, which is certainly the most theoretical in the book, is an introduction to Markov chains, covering enough theory to allow the reader to understand the workings and evaluate the performance of Markov chain Monte Carlo (MCMC) methods. Section 4.1 is provided for the reader who already is familiar with Markov chains, but needs a refresher, especially in the application of Markov chain theory to Monte Carlo calculations. Chapter 5 covers optimization and provides the first application of Markov chains to simulation methods. Chapters 6 and 7 cover the heart of MCMC methodology, the Metropolis–Hastings algorithm and the Gibbs sampler. Finally, Chapter 8 presents the state-of-the-art methods for monitoring convergence of the MCMC methods and Chapter 9 shows how these methods apply to some statistical settings which cannot be processed otherwise, namely the missing data models.

Each chapter concludes with a section of notes that serve to enhance the discussion in the chapters, describe alternate or more advanced methods, and point the reader to further work that has been done, as well as to current research trends in the area. The level and rigor of the notes are variable, with some of the material being advanced.

The book can be used at several levels and can be presented in several ways. For example, Chapters 1–3 and most of Chapter 5 cover standard simulation theory, and hence serve as a basic introduction to this topic. Chapters 6–9 are totally concerned with MCMC methodology. A one-semester course, assuming no familiarity with random variable generation or Markov chain theory could be based on Chapters 1–7, with some illustrations from Chapters 8 and 9. For instance, after a quick introduction with examples from Chapter 1 or §3.1, and a description of Accept–Reject techniques of §2.3, the course could cover Monte Carlo integration (§3.2, §3.3 [except §3.3.3], §3.4, §3.7), Markov chain theory through either §4.1 or §4.2–§4.8 (while adapting the depth to the mathematical level of the audience), mention stochastic optimization via §5.3, and describe Metropolis–Hastings and Gibbs algorithms as in Chapters 6 and 7 (except §6.5, §7.1.5, and §7.2.4). Dpending on the time left, the course could conclude with some diagnostic methods of Chapter 8 (for instance, those implemented in CODA) and/or some models of Chapter 9 (for instance, the mixture models of §9.3 and §9.4). Alternatively, a more advanced audience could cover Chapter 4 and Chapters 6–9 in one semester and have a thorough introduction to MCMC theory and methods.

Much of the material in this book had its original incarnation as the French monograph *Méthodes de Monte Carlo par Chaînes de Markov* by Christian Robert (Paris: Economica, 1996), which has been tested for several years on graduate audiences (in France, Québec, and even Norway). Nonetheless, it constitutes a major revision of the French text, with the inclusion of problems, notes, and the updating of current techniques, to keep up with the advances that took place in the past 2 years (like Langevin diffusions, perfect sampling, and various types of monitoring).

Throughout the preparation of this book, and its predecessor, we were fortunate to have colleagues who provided help. Sometimes this was in the form of conversations or references (thanks to Steve Brooks and Sid Chib!), and a few people actually agreed to read through the manuscript. Our colleague and friend, Costas Goutis, provided many helpful comments and criticisms, mostly on the French version, but these are still felt in this version. We are also grateful to Brad Carlin, Dan Fink, Jim Hobert, and Krishanu Maulik for detailed reading of parts of the manuscript, to our historian Walter Piegorsch, and to Richard Tweedie, who taught from the manuscript and provided many helpful suggestions, and to his students, Alex Trindade, Sandy Thompson, Nicole Benton, Sarah Streett, and Sue Taylor. Virginie Braïdo, Arnaud Doucet, Jean-Louis Foulley, Ana Justel, Anne Philippe, Sandrine Micaleff, and Randall Douc pointed out typos and mistakes in the French version, but should not be held responsible for those remaining! Part of Chapter 8 has a lot of common with a "reviewww" written by Christian Robert with Chantal Guihenneuc–Jouyaux and Kerrie Mengersen for the Valencia Bayesian meeting (and the Internet!). The input of the French working group "MC Cube," whose focus is on convergence diagnostics, can also be felt in several places of this book. Wally Gilks and David Spiegelhalter granted us permission to use their graph (Figure 2.3.1) and examples as Problems 7.44–7.55, for which we are grateful. Agostino Nobile kindly provided the data on which Figures 7.3.2 and 7.3.2 are based. Finally, Arnoldo Frigessi (from Roma) made the daring move of teaching (in English) from the French version in Olso, Norway; not only providing us with very helpful feedback but also contributing to making the European Union more of a reality!

Christian P. Robert
George Casella

December 1998

# Contents

List of Tables      xvii

List of Figures      xix

**1  Introduction**      **1**
  1.1  Statistical Models      1
  1.2  Likelihood Methods      5
  1.3  Bayesian Methods      11
  1.4  Deterministic Numerical Methods      17
  1.5  Problems      19
  1.6  Notes      31

**2  Random Variable Generation**      **35**
  2.1  Basic Methods      35
    2.1.1  Introduction      35
    2.1.2  The *Kiss* Generator      39
    2.1.3  Beyond Uniform Distributions      43
  2.2  Transformation Methods      45
  2.3  Accept–Reject Methods      49
    2.3.1  General Principle      49
    2.3.2  Envelope Accept–Reject Methods      53
    2.3.3  Log-Concave Densities      56
  2.4  Problems      60
  2.5  Notes      69

**3  Monte Carlo Integration**      **71**
  3.1  Introduction      71
  3.2  Classical Monte Carlo Integration      74
  3.3  Importance Sampling      80
    3.3.1  Principles      80
    3.3.2  Finite Variance Estimators      83
    3.3.3  Comparing Importance Sampling with Accept–Reject      92
  3.4  Riemann Approximations      96
  3.5  Laplace Approximations      103

3.6   The Saddlepoint Approximation                       104
   3.6.1   An Edgeworth Derivation                 105
   3.6.2   Tail Areas                              108
3.7   Acceleration Methods                                109
   3.7.1   Antithetic Variables                    109
   3.7.2   Control Variates                        113
   3.7.3   Conditional Expectations                116
3.8   Problems                                            119
3.9   Notes                                               131

**4   Markov Chains**                                     **139**
4.1   Essentials for MCMC                                 140
4.2   Basic Notions                                       141
4.3   Irreducibility, Atoms, and Small Sets              145
   4.3.1   Irreducibility                          145
   4.3.2   Atoms and Small Sets                    147
   4.3.3   Cycles and Aperiodicity                 150
4.4   Transience and Recurrence                           151
   4.4.1   Classification of Irreducible Chains    151
   4.4.2   Criteria for Recurrence                 153
   4.4.3   Harris Recurrence                       154
4.5   Invariant Measures                                  155
4.6   Ergodicity and Convergence                          159
   4.6.1   Ergodicity                              159
   4.6.2   Geometric Convergence                   164
   4.6.3   Uniform Ergodicity                      166
4.7   Limit Theorems                                      167
   4.7.1   Ergodic Theorems                        168
   4.7.2   Central Limit Theorems                  170
4.8   Covariance in Markov Chains                         172
4.9   Problems                                            174
4.10  Notes                                               185

**5   Monte Carlo Optimization**                          **193**
5.1   Introduction                                        193
5.2   Stochastic Exploration                              194
   5.2.1   A Basic Solution                        194
   5.2.2   Gradient Methods                        196
   5.2.3   Simulated Annealing                     199
   5.2.4   Prior Feedback                          202
5.3   Stochastic Approximation                            206
   5.3.1   Missing Data Models and Demarginalization  207
   5.3.2   Monte Carlo Approximation               208
   5.3.3   The EM Algorithm                        213
   5.3.4   Monte Carlo EM                          216
5.4   Problems                                            218

5.5  Notes                                                       227

**6  The Metropolis–Hastings Algorithm                          231**
6.1  Monte Carlo Methods Based on Markov Chains                 231
6.2  The Metropolis–Hastings algorithm                          233
   6.2.1  Definition                             233
   6.2.2  Convergence Properties                  236
6.3  A Collection of Metropolis–Hastings Algorithms            238
   6.3.1  The Independent Case                    239
   6.3.2  Random Walks                            245
   6.3.3  ARMS: A General Metropolis–Hastings Algorithm   249
6.4  Optimization and Control                                   251
   6.4.1  Optimizing the Acceptance Rate          252
   6.4.2  Conditioning and Accelerations          255
6.5  Further Topics                                             259
   6.5.1  Reversible Jumps                         259
   6.5.2  Langevin Algorithms                      264
6.6  Problems                                                   268
6.7  Notes                                                      279

**7  The Gibbs Sampler                                          285**
7.1  General Principles                                         285
   7.1.1  Definition                              285
   7.1.2  Completion                              288
   7.1.3  Convergence Properties                  293
   7.1.4  Gibbs Sampling and Metropolis–Hastings  296
   7.1.5  The Hammersley–Clifford Theorem         297
   7.1.6  Hierarchical Structures                 299
7.2  The Two-Stage Gibbs Sampler                               303
   7.2.1  Dual Probability Structures             303
   7.2.2  Reversible and Interleaving Chains      308
   7.2.3  Monotone Covariance and Rao–Blackwellization   310
   7.2.4  The Duality Principle                    315
7.3  Hybrid Gibbs Samplers                                      317
   7.3.1  Comparison with Metropolis–Hastings Algorithms   317
   7.3.2  Mixtures and Cycles                      319
   7.3.3  Metropolizing the Gibbs Sampler         322
   7.3.4  Reparameterization                      326
7.4  Improper Priors                                            328
7.5  Problems                                                   332
7.6  Notes                                                      348

**8  Diagnosing Convergence                                     363**
8.1  Stopping the Chain                                         363
   8.1.1  Convergence Criteria                    363
   8.1.2  Multiple Chains                          366

| | | | |
|---|---|---|---|
| | 8.1.3 | Conclusions | 367 |
| 8.2 | Monitoring Convergence to the Stationary Distribution | | 368 |
| | 8.2.1 | Graphical Methods | 368 |
| | 8.2.2 | Nonparametric Tests of Stationarity | 370 |
| | 8.2.3 | Renewal Methods | 371 |
| | 8.2.4 | Distance Evaluations | 375 |
| 8.3 | Monitoring Convergence of Averages | | 380 |
| | 8.3.1 | Graphical Methods | 380 |
| | 8.3.2 | Multiple Estimates | 381 |
| | 8.3.3 | Renewal Theory | 388 |
| | 8.3.4 | Within and Between Variances | 394 |
| 8.4 | Simultaneous Monitoring | | 397 |
| | 8.4.1 | Binary Control | 397 |
| | 8.4.2 | Valid Discretization | 401 |
| 8.5 | Problems | | 402 |
| 8.6 | Notes | | 407 |

**9 Implementation in Missing Data Models** — **415**

| | | | |
|---|---|---|---|
| 9.1 | Introduction | | 415 |
| 9.2 | First Examples | | 416 |
| | 9.2.1 | Discrete Data Models | 416 |
| | 9.2.2 | Data Missing at Random | 419 |
| 9.3 | Finite Mixtures of Distributions | | 422 |
| 9.4 | A Reparameterization of Mixtures | | 425 |
| 9.5 | Extensions | | 429 |
| | 9.5.1 | Hidden Markov Chains | 430 |
| | 9.5.2 | Changepoint models | 432 |
| | 9.5.3 | Stochastic Volatility | 435 |
| 9.6 | Problems | | 438 |
| 9.7 | Notes | | 448 |

**A Probability Distributions** — **451**

**B Notation** — **455**

| | | |
|---|---|---|
| B.1 | Mathematical | 455 |
| B.2 | Probability | 455 |
| B.3 | Distributions | 456 |
| B.4 | Markov Chains | 457 |
| B.5 | Statistics | 457 |
| B.6 | Algorithms | 458 |

**C References** — **459**

**Subject Index** — **488**

**Author Index** — **495**

# List of Tables

1.6.1 Some conjugate families     31

3.2.1 Evaluation of some normal quantiles     78
3.3.1 Comparison of instrumental distributions     90
3.3.2 Comparison between Monte Carlo and importance sampling estimators     96
3.5.1 Laplace approximation of a Gamma integral     104
3.6.1 Saddlepoint approximation of a noncentral chi squared integral     109

5.2.1 Stochastic gradient runs     199
5.2.2 Simulated annealing runs     202
5.2.3 Sequence of Bayes estimators for the gamma distribution     205
5.2.4 Average grades of first year students     206
5.2.5 Maximum likelihood estimates of mean grades     206
5.3.1 Estimation result for the factor ARCH model (5.3.10)     213
5.4.1 Tree swallow movements     225
5.4.2 Selected batting average data     226

6.3.1 Monte Carlo saddlepoint approximation of a noncentral chi squared integral     244
6.3.2 Estimators of the mean and the variance of a normal distribution using a random walk Metropolis–Hastings algorithm     246
6.4.1 Estimation in the inverse Gaussian distribution $\mathcal{IN}(\theta_1, \theta_2)$ by the Metropolis–Hastings algorithm     254
6.4.2 Decrease of squared error risk associated with Rao–Blackwellization     257
6.4.3 Improvement in quadratic risk from Rao–Blackwellization     258
6.6.1 Braking distances.     275
6.6.2 Galaxy data     276
6.7.1 Performance of the Metropolis–Hastings algorithm [A.28]     282

7.1.1 Convergence of estimated quantiles simulated from the slice sampler [A.33]     291

7.1.2 Failures of pumps in a nuclear plant   302

7.2.1 Frequencies of passage for 360 consecutive observations.   305

7.3.1 Interquantile ranges for the Gibbs sampling and the modification of Liu   326

7.5.1 Occurrences of clinical mastitis in dairy herds   334

8.3.1 Estimation of the asymptotic variance $\gamma_h^2$ for $h(\theta) = \theta$ and renewal control   392

8.3.2 Estimators of $\gamma_h^2$ for $h(x) = x$, obtained by renewal in $i_0$.   393

8.3.3 Approximations by Gibbs sampling of posterior expectations and evaluation of variances by the renewal method   394

8.4.1 Minimum test sample sizes for the diagnostic of Raftery and Lewis   399

8.4.2 Parameters of the binary control method for three different parameters of the grouped multinomial model   400

8.4.3 Evolution of initializing and convergence times   401

9.2.1 Observation of two characteristics of the habitat of 164 lizards   417

9.2.2 Average incomes and numbers of responses/nonresponses to a survey on the income by age, sex, and marital status   419

9.6.1 Independent observations of $Z = (X, Y) \sim \mathcal{N}_2(0, \Sigma)$ with missing data   441

9.6.2 Yearly number of mining accidents in England, $1851 - 1962$   445

# List of Figures

1.2.1 Cauchy likelihood    10

2.1.1 Plot of chaotic pairs    39
2.1.2 Representation of the line $y = 69069x$ mod 1 by uniform sampling    41
2.1.3 Plots of pairs from the *Kiss* generator    43
2.3.1 Lower and upper envelopes of a log-concave density    58

3.2.1 Convergence of the Bayes estimator of $||\theta||^2$    77
3.2.2 Empirical cdf's of log-likelihoods    80
3.3.1 Approximations squared error risks, exponential and lognormal    83
3.3.2 Convergence of three estimators of $\mathbb{E}_f[|X/(1 - X)|^{1/2}]$    86
3.3.3 Convergence of four estimators of $\mathbb{E}_f[X^5 \mathbb{I}_{X \geq 2.1}]$    87
3.3.4 Convergence of four estimators of $\mathbb{E}_f[h_3(X)]$    88
3.3.5 Convergence of four estimators of $\mathbb{E}_f[h_5(X)]$    91
3.3.6 Convergence of estimators of $\mathbb{E}[X/(1 + X)]$    95
3.3.7 Convergence of the estimators of $\mathbb{E}[X/(1 + X)]$    96
3.4.1 Convergence of the estimators of $\mathbb{E}[X \log(X)]$    101
3.4.2 Convergence of estimators of $\mathbb{E}_\nu[(1 + e^X)\mathbb{I}_{X \leq 0}]$    102
3.4.3 Convergence of estimators of $\mathbb{E}_\nu[(1 + e^X)\mathbb{I}_{X \leq 0}]$    102
3.6.1 Student's $t$ saddlepoint approximation    106
3.7.1 Approximate risks of truncated James–Stein estimators    111
3.7.2 Comparison of an antithetic and standard iid estimate    113
3.7.3 Convergence of estimators of $\mathbb{E}[\exp(-X^2)]$    117
3.7.4 Comparisons of an Accept–Reject and importance sampling estimator    119
3.9.1 Graphs of the variance coefficients    136

5.2.1 Representation of the function $h(x, y)$ of Example 5.2.1    196
5.2.2 Stochastic gradient paths    198
5.2.3 Simulated annealing sequence    203
5.3.1 Estimated slices of the log-likelihood ratio for the factor ARCH model (5.3.10)    212

6.3.1 Convergence of Accept–Reject and Metropolis–Hastings
      estimators for the algorithms [A.26] and [A.27]                          243
6.3.2 Convergence of Accept–Reject and Metropolis–Hastings
      estimators for the algorithms [A.26] and [A.27]                          243
6.3.3 Histograms of three samples produced by the
      algorithm [A.28]                                                         246
6.3.4 90% confidence envelopes of the means produced by the
      random walk Metropolis–Hastings algorithm [A.24]                         248
6.5.1 Linear regression with reversible jumps                                  262
6.5.2 Convergence of the empirical averages for the Langevin
      random walk and stochastic gradient Metropolis–Hastings
      algorithms                                                               268
6.5.3 Convergence of the empirical averages for the Langevin
      Metropolis–Hastings and iid simulation algorithms                        269

7.1.1 Comparison of the Box–Muller algorithm and the
      slice sampler                                                            292
7.2.1 Evolution of the estimator $\delta_{rb}$ of [A.35]                       306
7.3.1 Successive moves of a Gibbs chain                                        319
7.3.2 Gibbs chain for the probit model                                        322
7.3.3 Hybrid chain for the probit model                                       322
7.3.4 Comparison of the Gibbs sampling and of the modification
      of Liu                                                                   326
7.4.1 Iterations of the chains of a divergent Gibbs sampling
      algorithm                                                                330
7.4.2 Evolution of $(\beta^{(t)})$ for a random effects model                 331
7.6.1 Cross-representation of a quasi-random sequence                         360

8.2.1 Witch's hat distribution                                                 369
8.2.2 Evolution of the chain $(\theta_1^{(t)})$ around the mode of a witch's
      hat distribution                                                         370
8.2.3 Plot of successive Kolmogorov–Smirnov statistics                        372
8.2.4 Plot of renewal probabilities for the pump failure data                 375
8.2.5 Stationarity of the indicator $J_t$ for $m = 50$ parallel chains
      converging to the witch's hat distribution                              376
8.2.6 Evolutions of convergence indicators for the model of Gaver
      and O'Muircheartaigh                                                     379
8.2.7 Convergence of empirical averages for the algorithm [[A.30]]
      and the Metropolis–Hastings algorithm                                   380
8.3.1 Evolution of the $D_t^i$ criterion for the chain produced by
      [[A.30]]                                                                 381
8.3.2 Evolution of the CUSUM criterion $D_t^i$ for the chain produced
      by algorithm [[A.43]]                                                    381
8.3.3 Comparison of the density (8.3.6) and of the histogram from
      a sample of 20,000 points simulated by Gibbs sampling                   385
8.3.4 Convergence of four estimators of the expectation under
      (8.3.6)                                                                  386

8.3.5 Convergence of four estimators of $\mathbb{E}[(X^{(t)})^{0.8}]$ after elimination of the first $200,000$ iterations — 387

8.3.6 Evolutions of $R_T$ and $W_T$ for the posterior distribution (8.3.6) — 395

8.3.7 Evolutions of $R_T$ and $W_T$ for the witch's hat distribution — 396

8.4.1 Convergence of the mean $\hat{\varrho}_t$ for a chain generated from $[A.30]$ — 401

8.4.2 Discretization of a continuous Markov chain, based on three small sets — 402

9.4.1 Evolution of the estimation of the density (9.4.1) — 428

9.4.2 Convergence of estimators of the parameters of the mixture (9.4.5) — 429

9.5.1 Simulated sample of a switching $AR$ model — 431

9.5.2 The number of moves of a lamb fetus during 240 successive 5-second periods — 432

9.5.3 Simulated sample of a stochastic volatility process — 435

9.5.4 Gelman and Rubin's (1992) shrink factors for a stochastic volatility model — 436

9.5.5 Geweke's (1992) convergence indicators for a stochastic volatility model — 437

9.5.6 Successive values of the minimum $p$-value evaluating the stationarity of a sample — 437

9.5.7 Allocation map and average versus true allocation for a stochastic volatility model — 438

# CHAPTER 1

# Introduction

There must be, he thought, some key, some crack in this mystery he could use to achieve an answer.

—P.C. Doherty, *Crown in Darkness*

Until the advent of powerful and accessible computing methods, the experimenter was often confronted with a difficult choice. Either describe an accurate model of a phenomenon, which would usually preclude the computation of explicit answers, or choose a standard model which would allow this computation, but may not be a close representation of a realistic model. This dilemma is present in many branches of statistical applications, for example, in electrical engineering, aeronautics, biology, networks, and astronomy. To use realistic models, the researchers in these disciplines have often developed original approaches for model fitting that are customized for their own problems. (This is particularly true of physicists, the originators of Markov chain Monte Carlo methods.) Traditional methods of analysis, such as the usual numerical analysis techniques, are not well adapted for such settings.

In this introductory chapter, we examine some of the statistical models and procedures that contributed to the development of simulation-based inference. The first section of this chapter looks at some statistical models, and the remaining sections examine different statistical methods. Throughout these sections, we describe many of the computational difficulties associated with the methods. The final section of the chapter contains a discussion of deterministic numerical analysis techniques.

## 1.1 Statistical Models

In a purely statistical setup, computational difficulties occur at both the level of *probabilistic modeling* of the inferred phenomenon and at the level of *statistical inference* on this model (estimation, prediction, tests, variable selection, etc.). In the first case, a detailed representation of the causes of the phenomenon, such as accounting for potential explanatory variables linked to the phenomenon, can lead to a probabilistic structure that is too complex to allow for a parametric representation of the model. Moreover, there may be no provision for getting closed-form estimates of quantities of interest. A frequent setup with this type of complexity is *expert systems* (in medicine, physics, finance, etc.) or, more generally, *graph structures*. See Pearl 1988, Robert[1] (1991), Spiegelhalter *et al.* (1993), or Lauritzen (1996) for examples of complex expert systems.[2]

Another situation where model complexity prohibits an explicit representation appears in econometrics (and in other areas) for structures of *latent* (or *missing*) variable models. Given a "simple" model, aggregation or removal of some components of this model may sometimes produce such involved structures that simulation is really the only way to draw an inference. In these situations, an often used method for estimation is the EM algorithm (Dempster *et al.* 1977), which is described in Chapter 3. In the following example, we illustrate a common missing data situation. Chapter 9 will study a series of models where simulation methods are necessary.

**Example 1.1.1 Censored data models.** *Censored data models* are missing data models where densities are not sampled directly. To obtain estimates and make inferences in such models usually requires involved computations and precludes analytical answers.

In a typical simple statistical model, we would observe random variables[3] (rv's) $Y_1, \ldots, Y_n$, drawn independently from a population with distribution $f(y|\theta)$. The distribution of the sample would then be given by the product $\prod_{i=1}^{n} f(y_i|\theta)$. Inference about $\theta$ would be based on this distribution.

In many studies, particularly in medical statistics, we have to deal with *censored* random variables; that is, rather than observing $Y_1$, we may observe $\min\{Y_1, \overline{u}\}$, where $\overline{u}$ is a constant. For example, if $Y_1$ is the survival time of a patient receiving a particular treatment and $\overline{u}$ is the length of the study being done (say $\overline{u} = 5$ years), then if the patient survives longer than 5 years, we do not observe the survival time, but rather the censored value $\overline{u}$. This modification leads to a more difficult evaluation of the sample density.

Barring cases where the censoring phenomenon can be ignored (see Chapter 9), several types of censoring can be categorized by their relation with an underlying (unobserved) model, $Y_i \sim f(y_i|\theta)$:

(i) Given random variables $Y_i$, which are, for instance, times of observation or concentrations, the actual observations are $Y_i^* = \min\{Y_i, \overline{u}\}$, where $\overline{u}$ is the maximal observation duration, the smallest measurable concentration rate, or some other truncation point.

(ii) The original variables $Y_i$ are kept in the sample with probability $\rho(y_i)$ and the number of censored variables is either known or unknown.

(iii) The variables $Y_i$ are associated with auxiliary variables $X_i \sim g$ such

---

[1] Claudine, not Christian!

[2] Section 7.6.6 also gives a brief introduction to graphical models in connection with Gibbs sampling.

[3] Throughout the book we will use uppercase Roman letters for random variables and lowercase Roman letters for their realized values. Thus, we would observe $X = x$, where the random variable $X$ produces the observation $x$. (For aesthetic purposes this distinction is sometimes lost with Greek letter random variables.)

that $y_i^* = h(y_i, x_i)$ is the observation. Typically, $h(y_i, x_i) = \min(y_i, x_i)$. The fact that truncation occurred, namely the variable $\mathbb{I}_{Y_i > X_i}$, may be either known or unknown.

As a particular example, if $X \sim \mathcal{N}(\theta, \sigma^2)$ and $Y \sim \mathcal{N}(\mu, \rho^2)$, the variable $Z = X \wedge Y = \min(X, Y)$ is distributed as

$$
\left[ 1 - \Phi\left( \frac{z - \theta}{\sigma} \right) \right] \rho^{-1} \varphi\left( \frac{z - \mu}{\rho} \right)
$$

$$
(1.1.1) \qquad + \left[ 1 - \Phi\left( \frac{z - \mu}{\rho} \right) \right] \sigma^{-1} \varphi\left( \frac{z - \theta}{\sigma} \right) ,
$$

where $\varphi$ is the density of the normal $\mathcal{N}(0,1)$ distribution and $\Phi$ is the corresponding cdf, which is not easy to compute. Similarly, if $X$ has a Weibull distribution with two parameters, $\mathcal{W}e(\alpha, \beta)$, and density

$$
f(x) = \alpha \beta x^{\alpha - 1} \exp(-\beta x^\alpha)
$$

on $\mathbb{R}^+$, the observation of the censored variable $Z = X \wedge \omega$, where $\omega$ is constant, has the density

$$
(1.1.2) \quad f(z) = \alpha \beta z^{\alpha - 1} e^{-\beta z^\alpha} \, \mathbb{I}_{z \leq \omega} + \left( \int_\omega^\infty \alpha \beta x^{\alpha - 1} e^{-\beta x^\alpha} dx \right) \delta_\omega(z) ,
$$

where $\delta_a(\cdot)$ is the Dirac mass at $a$. In this case, the weight of the Dirac mass, $P(X \geq \omega)$, can be explicitly computed (Problem 1.3).

The distributions (1.1.1) and (1.1.2) appear naturally in quality control applications. There, testing of a product may be of a duration $\omega$, where the quantity of interest is time to failure. If the product is still functioning at the end of the experiment, the observation on failure time is censored. Similarly, in a longitudinal study of a disease, some patients may leave the study either due to other causes of death or by simply being lost to follow up. ‖

In some cases, the additive form of a density, while formally explicit, prohibits the computation of the density of a sample $(X_1, \ldots, X_n)$ for $n$ large. (Here, "explicit" has the restrictive meaning that "it can be computed in a reasonable time.")

**Example 1.1.2 Mixture models.**  Models of *mixtures of distributions* are based on the assumption that the observations $X_i$ are generated from one of $k$ elementary distributions $f_j$ with probability $p_j$, the overall density being

$$
X \sim p_1 f_1(x) + \ldots + p_k f_k(x) .
$$

If we observe a sample of independent random variables $(X_1, \ldots, X_n)$, the sample density is

$$
\prod_{i=1}^n \{ p_1 f_1(x_i) + \cdots + p_k f_k(x_i) \} .
$$

Expanding this product shows that it involves $k^n$ elementary terms, which is prohibitive to compute in large samples. While the computation of standard moments like the mean or the variance of these distributions is feasible in many setups (and thus the derivation of moment estimators, see Problem 1.5), the representation of the likelihood (and therefore the analytical computation of maximum likelihood or Bayes estimates) is generally impossible for mixtures. ‖

Finally, we look at a particularly important example in the processing of temporal (or time series) data where the likelihood cannot be written explicitly.

**Example 1.1.3 Moving average model.** An $MA(q)$ model describes variables $(X_t)$ that can be modeled as $(t = 0, \ldots, n)$

(1.1.3)
$$X_t = \varepsilon_t + \sum_{j=1}^{q} \beta_j \varepsilon_{t-j} \, ,$$

where for $i = -q, -(q-1), \ldots$, the $\varepsilon_i$'s are iid random variables $\varepsilon_i \sim \mathcal{N}(0, \sigma^2)$ and for $j = 1, \ldots, q$, the $\beta_j$'s are unknown parameters. If the sample consists of the observation $(X_0, \ldots, X_n)$, where $n > q$, the sample density is (Problem 1.4)

(1.1.4)
$$
\int_{\mathbb{R}^q} \sigma^{-(n+q)} \prod_{i=1}^{q} \varphi\left(\frac{\varepsilon_{-i}}{\sigma}\right) \varphi\left(\frac{x_0 - \sum_{i=1}^{q} \beta_i \varepsilon_{-i}}{\sigma}\right)
$$
$$
\times \varphi\left(\frac{x_1 - \beta_1 \hat{\varepsilon}_o - \sum_{i=2}^{q} \beta_i \varepsilon_{1-i}}{\sigma}\right) \cdots
$$
$$
\times \varphi\left(\frac{x_n - \sum_{i=1}^{q} \beta_i \hat{\varepsilon}_{n-i}}{\sigma}\right) d\varepsilon_{-1} \cdots d\varepsilon_{-q} \, ,
$$

with

$$\hat{\varepsilon}_0 = x_0 - \sum_{i=1}^{q} \beta_i \varepsilon_{-i} \, ,$$

$$\hat{\varepsilon}_1 = x_1 - \sum_{i=2}^{q} \beta_i \varepsilon_{1-i} - \beta_1 \hat{\varepsilon}_0 \, ,$$

$$\vdots$$

$$\hat{\varepsilon}_n = x_n - \sum_{i=1}^{q} \beta_i \hat{\varepsilon}_{n-i} \, .$$

The iterative definition of the $\hat{\varepsilon}_i$'s is a real obstacle to an explicit integration in (1.1.4) and hinders statistical inference in these models. Note that for $i = -q, -(q-1), \ldots, -1$ the perturbations $\varepsilon_{-i}$ can be interpreted as missing data (see Chapter 9). ‖

Before the introduction of simulation-based inference, computational difficulties encountered in the modeling of a problem often forced the use of "standard" models and "standard" distributions. One course would be to use models based on *exponential families*, defined below by (1.2.4) (see Brown 1986, Robert 1994a, or Lehmann and Casella 1998), which enjoy numerous regularity properties (see Note 1.6.1). Another course was to abandon parametric representations for nonparametric approaches which are, by definition, robust against modeling errors.

We also note that the reduction to simple, perhaps non-realistic, distributions (necessitated by computational limitations) does not necessarily eliminate the issue of nonexplicit expressions, whatever the statistical technique. Our major focus is the application of simulation-based techniques to provide solutions and inference for a more realistic set of models and, hence, circumvent the problems associated with the need for explicit or computationally simple answers.

## 1.2 Likelihood Methods

The statistical techniques that we will be most concerned with are *maximum likelihood* and *Bayesian* methods, and the inferences that can be drawn from their use. In their implementation, these approaches are customarily associated with specific mathematical computations, the former with maximization problems—and thus to an *implicit* definition of estimators as solutions of maximization problems— the later with integration problems—and thus to a (formally) *explicit* representation of estimators as an integral. (See Berger 1985, Casella and Berger 1990, Robert 1994a, or Lehmann and Casella 1998 for an introduction to these techniques.)

The method of maximum likelihood estimation is quite a popular technique for deriving estimators. Starting from an iid sample $X_1, \ldots, X_n$ from a population with density $f(x|\theta_1, \ldots, \theta_k)$, the *likelihood function* is

$$L(\boldsymbol{\theta}|\mathbf{x}) = L(\theta_1, \ldots, \theta_k | x_1, \ldots, x_n)$$

(1.2.1)
$$= \prod_{i=1}^{n} f(x_i | \theta_1, \ldots, \theta_k).$$

More generally, when the $X_i$'s are not iid, the likelihood is defined as the joint density $f(x_1, \ldots, x_n | \boldsymbol{\theta})$ taken as a function of $\boldsymbol{\theta}$. The value of $\boldsymbol{\theta}$, say $\hat{\boldsymbol{\theta}}$, which is the parameter value at which $L(\theta|\mathbf{x})$ attains its maximum as a function of $\boldsymbol{\theta}$, with $\mathbf{x}$ held fixed, is known as a *maximum likelihood estimator (MLE)*. Notice that, by its construction, the range of the MLE coincides with the range of the parameter. The justifications of the maximum likelihood method are primarily asymptotic, in the sense that the MLE is converging almost surely to the true value of the parameter, under fairly general conditions (see Lehmann and Casella 1998 and Problem 1.15), although it can also be interpreted as being at the fringe of the Bayesian paradigm (see, e.g., Berger and Wolpert 1988).

**Example 1.2.1 Gamma MLE.** A maximum likelihood estimator is typically calculated by maximizing the logarithm of the likelihood function (1.2.1). Suppose $X_1, \ldots, X_n$ are iid observations from the gamma density $f(x|\alpha, \beta) = \frac{1}{\Gamma(\alpha)\beta^\alpha} x^{\alpha-1} e^{-x/\beta}$, where we assume that $\alpha$ is known. The *log likelihood* is

$$\log L(\alpha, \beta|x_1, \ldots, x_n)$$

$$= \log \prod_{i=1}^{n} f(x_i|\alpha, \beta)$$

$$= \log \prod_{i=1}^{n} \frac{1}{\Gamma(\alpha)\beta^\alpha} x_i^{\alpha-1} e^{-x_i/\beta}$$

$$= -n \log \Gamma(\alpha) - n\alpha \log \beta + (\alpha - 1) \sum_{i=1}^{n} \log x_i - \sum_{i=1}^{n} x_i/\beta,$$

where we use the fact that the log of the product is the sum of the logs, and have done some simplifying algebra. Solving $\frac{\partial}{\partial \beta} \log L(\alpha, \beta|x_1, \ldots, x_n) = 0$ is straightforward and yields the MLE of $\beta$, $\hat{\beta} = \sum_{i=1}^{n} x_i/(n\alpha)$.

Suppose now that $\alpha$ was also unknown, and we additionally had to solve

$$\frac{\partial}{\partial \alpha} \log L(\alpha, \beta|x_1, \ldots, x_n) = 0.$$

This results in a particularly nasty equation, involving some difficult computations (such as the derivative of the gamma function, the *digamma function*). An explicit solution is no longer possible. ‖

Calculation of maximum likelihood estimators can sometimes be implemented through the minimization of a sum of squared residuals, which is the basis of the *method of least squares*.

**Example 1.2.2 Least squares estimators.** Estimation by *least squares* can be traced back to Legendre (1805) and Gauss (1810) (see Stigler 1986). In the particular case of linear regression, we observe $(x_i, y_i)$, $i = 1, \ldots, n$, where

(1.2.2) $$Y_i = a + bx_i + \varepsilon_i, \quad i = 1, \ldots, n,$$

and the variables $\varepsilon_i$'s represent errors. The parameter $(a, b)$ is estimated by minimizing the distance

(1.2.3) $$\sum_{i=1}^{n} (y_i - ax_i - b)^2$$

in $(a, b)$, yielding the least squares estimates. If we add more structure to the error term, in particular that $\varepsilon_i \sim \mathcal{N}(0, \sigma^2)$, independent (equivalently, $Y_i|x_i \sim \mathcal{N}(ax_i + b, \sigma^2)$), the log-likelihood function for $(a, b)$ is proportional to

$$\log(\sigma^{-n}) - \sum_{i=1}^{n} (y_i - ax_i - b)^2/2\sigma^2,$$

and it follows that the maximum likelihood estimates of $a$ and $b$ are identical to the least squares estimates.

If, in (1.2.3), we assume $\mathbb{E}(\varepsilon_i) = 0$, or, equivalently, that the linear relationship $\mathbb{E}[Y|x] = ax + b$ holds, minimization of (1.2.3) is equivalent, from a computational point of view, to imposing a normality assumption on $Y$ conditionally on $x$ and applying maximum likelihood. In this latter case, the additional estimator of $\sigma^2$ is consistent if the normal approximation is asymptotically valid. (See Gouriéroux and Monfort 1996 for the related theory of *pseudo-likelihood*.)                                                                    ‖

Although somewhat obvious, this formal equivalence between the optimization of a function depending on the observations and the maximization of a likelihood associated with the observations has a nontrivial outcome and applies in many other cases. For example, in the case where the parameters are constrained, Robertson *et al.* (1988) consider a $p \times q$ table of random variables $Y_{ij}$ with means $\theta_{ij}$, where the means are increasing in $i$ and $j$. Estimation of the $\theta_{ij}$'s by minimizing the sum of the $(y_{ij} - \theta_{ij})^2$'s is possible through the (numerical) algorithm called *"pool-adjacent-violators,"* developed by Robertson *et al.* (1988) to solve this specific problem. (See Problems 1.22 and 1.23.) An alternative is to use an algorithm based on simulation and a representation using a normal likelihood (see §5.2.4).

In the context of exponential families, that is, distributions with density

$$(1.2.4) \qquad f(x) = h(x)\, e^{\theta \cdot x - \psi(\theta)}, \qquad \theta, x \in \mathbb{R}^k,$$

the approach by maximum likelihood is (formally) straightforward. The maximum likelihood estimator of $\theta$ is the solution of

$$(1.2.5) \qquad x = \nabla \psi \{\hat{\theta}(x)\},$$

which also is the equation yielding a method of moments estimator, since $\mathbb{E}_\theta[X] = \nabla \psi(\theta)$. The function $\psi$ is the log-Laplace transform, or *cumulant generating function* of $h$; that is, $\psi(t) = \log \mathbb{E}[\exp\{th(X)\}]$, where we recognize the right side as the log *moment generating function* of $h$.

**Example 1.2.3 Normal MLE.** In the setup of the normal $\mathcal{N}(\mu, \sigma^2)$ distribution, the density can be written as in (1.2.4), since

$$f(y|\mu, \sigma) \propto \sigma^{-1} e^{-(\mu - y)^2 / 2\sigma^2}$$
$$= \sigma^{-1} e^{(\mu/\sigma^2)\, y - (1/2\sigma^2)\, y^2 - \mu^2 / 2\sigma^2}.$$

The so-called *natural parameters* are then $\theta_1 = \mu/\sigma^2$ and $\theta_2 = -1/2\sigma^2$, with $\psi(\theta) = -\theta_1^2/4\theta_2 + \log(-\theta_2/2)/2$. While there is no MLE for a single observation from $\mathcal{N}(\mu, \sigma^2)$, equation (1.2.5) leads to

$$(1.2.6) \qquad \begin{aligned} -n\frac{\theta_1}{2\theta_2} &= \sum_{i_1}^n y_i = n\bar{y}, \\ n\frac{\theta_1^2}{4\theta_2^2} - \frac{n}{2\theta_2} &= \sum_{i_1}^n y_i^2 = n(s^2 + \bar{y}^2), \end{aligned}$$

in the case of $n$ iid observations $y_1, \ldots, y_n$, that is, to the regular MLE, $(\hat{\mu}, \hat{\sigma}^2) = (\bar{y}, s^2)$, where $s^2 = \sum(y_i - \bar{y})^2/n$.                    ‖

Unfortunately, there are many settings where $\psi$ cannot be computed explicitly. Even if it could be done, it may still be the case that the solution of (1.2.5) is not explicit, or there are constraints on $\theta$ such that the maximum of (1.2.4) is not a solution of (1.2.5). This last situation occurs in the estimation of the table of $\theta_{ij}$'s in the discussion above.

**Example 1.2.4 Beta MLE.** The beta $\mathcal{B}e(\alpha, \beta)$ distribution is a particular case of exponential family since its density,

$$f(y|\alpha, \beta) = \frac{\Gamma(\alpha + \beta)}{\Gamma(\alpha)\,\Gamma(\beta)}\, y^{\alpha-1}(1 - y)^{\beta-1}, \qquad 0 \le y \le 1,$$

can be written as (1.2.4), with $\theta = (\alpha, \beta)$ and $x = (\log y, \log(1 - y))$. Equation (1.2.5) becomes

$$(1.2.7) \qquad \begin{aligned} \log y &= \Psi(\alpha) - \Psi(\alpha + \beta), \\ \log(1 - y) &= \Psi(\beta) - \Psi(\alpha + \beta), \end{aligned}$$

where $\Psi(z) = d\log\Gamma(z)/dz$ denotes the *digamma function* (see Abramowitz and Stegun 1964). There is no explicit solution to (1.2.7). As in Example 1.2.3, although it may seem absurd to estimate both parameters of the $\mathcal{B}e(\alpha, \beta)$ distribution from a single observation, $Y$, the formal computing problem at the core of this example remains valid for a sample $Y_1, \ldots, Y_n$ since (1.2.7) is then replaced by

$$\frac{1}{n}\sum_i \log y_i = \Psi(\alpha) - \Psi(\alpha + \beta),$$

$$\frac{1}{n}\sum_i \log(1 - y_i) = \Psi(\beta) - \Psi(\alpha + \beta). \qquad \|$$

When the parameter of interest $\lambda$ is not a one-to-one function of $\theta$, that is, when there are *nuisance* parameters, the maximum likelihood estimator of $\lambda$ is still well defined. If the parameter vector is of the form $\theta = (\lambda, \psi)$, where $\psi$ is a nuisance parameter, a typical approach is to calculate the full MLE $\hat{\theta} = (\hat{\lambda}, \hat{\psi})$ and use the resulting $\hat{\lambda}$ to estimate $\lambda$. In principle, this does not require more complex calculations, although the distribution of the maximum likelihood estimator of $\lambda$, $\hat{\lambda}$, may be quite involved. Many other options exist, such as *conditional, marginal, or profile* likelihood (see, for example, Barndorff-Nielsen and Cox 1994).

**Example 1.2.5 Noncentrality parameter.** If $X \sim \mathcal{N}_p(\theta, I_p)$ and if $\lambda = \|\theta\|^2$ is the parameter of interest, the nuisance parameters are the angles $\Psi$ in the polar representation of $\theta$ (see Problem 1.24) and the maximum likelihood estimator of $\lambda$ is $\hat{\lambda}(x) = \|x\|^2$, which has a constant bias equal to $p$. Surprisingly, an observation $Y = \|X\|^2$ which has a noncentral chi squared distribution, $\chi_p^2(\lambda)$ (see Appendix A), leads to a maximum likelihood estimator of $\lambda$ which differs[4] from $Y$, since it is the solution of

---

[4] This phenomenon is not paradoxical, as $Y = \|X\|^2$ is not a sufficient statistic in the original problem.

the implicit equation

$$(1.2.8) \qquad \sqrt{\lambda}\, I_{(p-1)/2}\left(\sqrt{\lambda y}\right) = \sqrt{y}\, I_{p/2}\left(\sqrt{\lambda y}\right), \qquad y > p,$$

where $I_\nu$ is the *modified Bessel function* (see Problem 1.13)

$$I_\nu(t) = \frac{(z/2)^\nu}{\sqrt{\pi}\,\Gamma(\nu + \frac{1}{2})} \int_0^\pi e^{t\cos(\theta)} \sin^{2\nu}(\theta)\,d\theta$$

$$= \left(\frac{t}{2}\right)^\nu \sum_{k=0}^\infty \frac{(z/2)^{2k}}{k!\,\Gamma(\nu + k + 1)}.$$

So even in the favorable context of exponential families, we are not necessarily free from computational problems, since the resolution of (1.2.8) requires us first to evaluate the special functions $I_{p/2}$ and $I_{(p-1)/2}$. Note also that the maximum likelihood estimator is not a solution of (1.2.8) when $y < p$ (see Problems 1.25 and 1.26). ‖

When we leave the exponential family setup, we face increasingly challenging difficulties in using maximum likelihood techniques. One reason for this is the lack of a *sufficient statistic of fixed dimension* outside exponential families, barring the exception of a few families such as uniform or Pareto distributions whose support depends on $\theta$ (Robert 1994a, Section 3.2). This result, known as the *Pitman–Koopman Lemma* (see Lehmann and Casella 1998, Theorem 1.6.18), implies that, outside exponential families, the complexity of the likelihood increases quite rapidly with the number of observations, $n$ and, thus, that its maximization is delicate, even in the simplest cases.

**Example 1.2.6 Student's $t$ distribution.** Modeling random perturbations using normally distributed errors is often (correctly) criticized as being too restrictive and a reasonable alternative is the Student's $t$ distribution, denoted by $\mathcal{T}(p, \theta, \sigma)$, which is often more "robust" against possible modeling errors (and others). The density of $\mathcal{T}(p, \theta, \sigma)$ is proportional to

$$(1.2.9) \qquad\qquad \sigma^{-1}\left(1 + \frac{(x - \theta)^2}{p\sigma^2}\right)^{-(p+1)/2}.$$

Typically, $p$ is known and the parameters $\theta$ and $\sigma$ are unknown. Based on an iid sample $(X_1, \ldots, X_n)$ from (1.2.9), the likelihood is proportional to a power of the product

$$\sigma^{n\frac{p+1}{2}} \prod_{i=1}^n \left(1 + \frac{(x_i - \theta)^2}{p\sigma^2}\right).$$

When $\sigma$ is known, for some configurations of the sample, this polynomial of degree $2n$ may have $n$ local minima, each of which needs to be calculated to determine the global maximum, the maximum likelihood estimator (see also Problem 1.14). Figure 1.2.1 illustrates this multiplicity of modes of the likelihood from a Cauchy distribution $\mathcal{C}(\theta, 1)$ $(p = 1)$ when $n = 3$ and $X_1 = 0$, $X_2 = 5$, and $X_3 = 9$. ‖

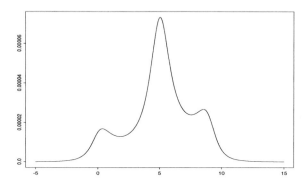

Figure 1.2.1. *Likelihood of the sample* $(0, 5, 9)$ *from the distribution* $\mathcal{C}(\theta, 1)$.

**Example 1.2.7 (Continuation of Example 1.1.2)** In the special case of a mixture of two normal distributions,

$$p\mathcal{N}(\mu, \tau^2) + (1 - p)\mathcal{N}(\theta, \sigma^2) \, ,$$

an iid sample $(X_1, \ldots, X_n)$ results in a likelihood function proportional to

$$(1.2.10) \qquad \prod_{i=1}^{n} \left[ p\tau^{-1}\varphi\left(\frac{x_i - \mu}{\tau}\right) + (1 - p)\,\sigma^{-1}\,\varphi\left(\frac{x_i - \theta}{\sigma}\right) \right]$$

containing $2^n$ terms. Standard maximization techniques often fail to find the global maximum because of multimodality of the likelihood function, and specific algorithms must be devised (to obtain the global maximum with high probability).

The problem is actually another order of magnitude more difficult, since the likelihood is unbounded here. The expansion of the product (1.2.10) contains the terms

$$p^n \tau^{-n} \prod_{i=1}^{n} \varphi\left(\frac{x_i - \mu}{\tau}\right)$$

$$+ p^{n-1}(1 - p)\,\tau^{-n+1}\sigma^{-1}\,\varphi\left(\frac{x_1 - \theta}{\sigma}\right) \prod_{i=2}^{n} \varphi\left(\frac{x_i - \mu}{\tau}\right) + \ldots \, .$$

This expression is unbounded in $\sigma$ (let $\sigma$ go to 0 when $\theta = x_1$). However, this difficulty with the likelihood function does not preclude us from using the maximum likelihood approach in this context, since Redner and Walker (1984) have shown that there exist solutions to the *likelihood equations*, that is, local maxima of (1.2.10), which have acceptable properties. (Similar problems occur in the context of linear regression with "errors in variables." See Casella and Berger 1990, Chapter 12, for an introduction.)                    ∥

In addition to the difficulties associated with optimization problems, likelihood-related approaches may also face settings where the likelihood function is only expressible as an integral (for example, the censored data models of Example 1.1.1). Similar computational problems arise in the determination of the *power* of a testing procedure in the Neyman–Pearson approach (see Lehmann 1986, Casella and Berger 1990, Robert 1994a, or Gouriéroux and Monfort 1996).

For example, inference based on a likelihood ratio statistic requires computation of quantities such as

$$P_\theta(L(\theta|X)/L(\theta_0|X) \leq k) \,,$$

with fixed $\theta_0$ and $k$, where $L(\theta|x)$ represents the likelihood based on observing $X = x$. Outside of the more standard (simple) settings, this probability cannot be explicitly computed because dealing with the distribution of test statistics under the alternative hypothesis may be quite difficult. A particularly delicate case is the *Behrens–Fisher problem*, where the above probability is difficult to evaluate even under the null hypothesis (see Lehmann 1986, Lee 1989, or Problem 1.52). (Note that likelihood ratio tests cannot be rigorously classified as a likelihood-related approach, since they violate the *Likelihood Principle* (Berger and Wolpert 1988, Robert 1994a), but the latter does not provide a testing theory *per se.*)

## 1.3 Bayesian Methods

Whereas the difficulties related to maximum likelihood methods are mainly *optimization* problems (multiple modes, solution of likelihood equations, links between likelihood equations and global modes, etc.), the Bayesian approach more often results in *integration* problems. In the Bayesian paradigm, information brought by the data $x$, a realization of $X \sim f(x|\theta)$, is combined with prior information that is specified in a *prior distribution* with density $\pi(\theta)$ and summarized in a probability distribution, $\pi(\theta|x)$, called the *posterior distribution*. This is derived from the *joint* distribution $f(x|\theta)\pi(\theta)$, according to *Bayes formula*

$$(1.3.1) \qquad \pi(\theta|x) = \frac{f(x|\theta)\pi(\theta)}{\int f(x|\theta)\pi(\theta)d\theta},$$

where $m(x) = \int f(x|\theta)\pi(\theta)d\theta$ is the *marginal density* of $X$ (see Berger 1985, Bernardo and Smith 1994 or Robert 1994a, for more details, in particular about the philosophical foundations of this inferential approach).

For the estimation of a particular parameter $h(\theta)$, the decision-theoretic approach to statistical inference (see, e.g., Berger 1985) requires the specification of a loss function $L(\delta, \theta)$, which represents the loss incurred by estimating $h(\theta)$ with $\delta$. The Bayesian version of this approach leads to the minimization of the *Bayes risk*,

$$\int \int L(\delta, \theta) f(x|\theta) \pi(\theta) dx d\theta \, ,$$

that is, the loss integrated against both $X$ and $\theta$. A straightforward inversion of the order of integration (Fubini's theorem) leads to choosing the estimator $\delta$ that minimizes (for each $x$) the *posterior loss*,

(1.3.2)          $$\mathbb{E}[L(\delta, \theta)|x] = \int L(\delta, \theta) \, \pi(\theta|x) \, d\theta \, .$$

In the particular case of the quadratic loss

$$L(\delta, \theta) = \|h(\theta) - \delta\|^2 \, ,$$

the Bayes estimator of $h(\theta)$ is $\delta^\pi(x) = \mathbb{E}^\pi[h(\theta)|x]$. (See Problem 1.27.)

Some of the difficulties related to the computation of $\delta^\pi(x)$ are, first, that $\pi(\theta|x)$ is not generally available in closed form and, second, that in many cases the integration of $h(\theta)$ according to $\pi(\theta|x)$ cannot be done analytically. Loss functions $L(\delta, \theta)$ other than the quadratic loss function are usually even more difficult to deal with.

The computational drawback of the Bayesian approach has been so great that, for a long time, the favored types of priors in a Bayesian modeling were those allowing explicit computations, namely *conjugate priors*. These are prior distributions for which the corresponding posterior distributions are themselves members of the original prior family, the Bayesian updating being accomplished through updating of parameters. (See Note 1.6.1 and Robert 1994a for a discussion of the link between conjugate priors and exponential families.)

**Example 1.3.1 Binomial Bayes estimator.** For an observation $X$ from the binomial distribution $\mathcal{B}(n, p)$, a family of conjugate priors is the family of beta distributions $\mathcal{B}e(a, b)$. To find the Bayes estimator of $p$ under squared error loss, we can find the minimizer of the Bayes risk, that is,

$$\min_\delta \int_0^1 \sum_{x=1}^n [p - \delta(x)]^2 \binom{n}{x} \frac{\Gamma(a+b)}{\Gamma(a)\Gamma(b)} p^{x+a-1} (1-p)^{n-x+b-1} dp.$$

Equivalently, we can work with the posterior expected loss (1.3.2) and find the estimator that yields

$$\min_\delta \frac{\Gamma(a+b+n)}{\Gamma(a+x)\Gamma(n-x+b)} \int_0^1 [p - \delta(x)]^2 p^{x+a-1} (1-p)^{n-x+b-1} dp \, ,$$

where we note that the posterior distribution of $p$ (given $x$) is $\mathcal{B}e(x+a, n-x+b)$. The solution is easily obtained through differentiation, and the Bayes estimator $\delta^\pi$ is the posterior mean

$$\delta^\pi(x) = \frac{\Gamma(a+b+n)}{\Gamma(a+x)\Gamma(n-x+b)} \int_0^1 p \, p^{x+a-1} (1-p)^{n-x+b-1} dp = \frac{x+a}{a+b+n}.$$

The use of squared error loss results in the Bayes estimator being the mean of the posterior distribution, which usually simplifies calculations. If, instead, we had specified a *absolute error loss* $|p - \delta(x)|$ or had used a nonconjugate prior, the calculations can become somewhat more involved (see Problem 1.27).                                                                           ‖

The use of conjugate priors for computational reasons implies a restriction on the modeling of the available prior information and may be detrimental to the usefulness of the Bayesian approach as a method of statistical inference. This is because it perpetuates an impression both of subjective "manipulation" of the background (prior information) and of formal expansions unrelated to reality. The considerable advances of Bayesian decision theory have often highlighted the negative features of modeling using only conjugate priors. For example, Bayes estimators are the optimal estimators for the three main classes of optimality (admissibility, minimaxity, invariance), but those based on conjugate priors only partially enjoy these properties (see Berger 1985, Section 4.7, or Robert 1994a, Chapter 6).

**Example 1.3.2 (Continuation of Example 1.2.5).** For the estimation of $\lambda = \|\theta\|^2$, a *reference prior*[5] on $\theta$ is $\pi(\theta) = \|\theta\|^{-(p-1)}$ (see Berger *et al.* 1998), with corresponding posterior distribution

$$(1.3.3) \qquad\qquad \pi(\theta|x) \propto \frac{e^{-\|x-\theta\|^2/2}}{\|\theta\|^{p-1}} \; .$$

The normalizing constant corresponding to $\pi(\theta|x)$ is not easily obtainable and the Bayes estimator of $\lambda$, the posterior mean

$$(1.3.4) \qquad\qquad \frac{\int_{\mathbb{R}^p} \|\theta\|^{2-p} \, e^{-\|x-\theta\|^2/2} \, d\theta}{\int_{\mathbb{R}^p} \|\theta\|^{1-p} \, e^{-\|x-\theta\|^2/2} \, d\theta} \; ,$$

cannot be explicitly computed. (See Problem 1.26.)                                 ‖

The computation of the normalizing constant of $\pi(\theta|x)$ is not just a formality. Although the derivation of a posterior distribution is generally done through proportionality relations, that is, using *Bayes Theorem* in the form

$$\pi(\theta|x) \propto \pi(\theta) \, f(x|\theta) \; ,$$

it is sometimes necessary to know the posterior distribution or, equivalently, the marginal distribution, exactly. For example, this is the case in the Bayesian comparison of (statistical) models. If $\mathcal{M}_1, \mathcal{M}_2, \ldots, \mathcal{M}_k$ are possible models for the observation $X$, with densities $f_j(\cdot|\theta_j)$, if the associated parameters $\theta_1, \theta_2, \ldots, \theta_k$ are *a priori* distributed from $\pi_1, \pi_2, \ldots,$

---

[5] A reference prior is a prior distribution which is derived from maximizing a distance measure between the prior and the posterior distributions. When there is no nuisance parameter in the model, the standard reference prior is Jeffreys' (1961) prior (see Bernardo and Smith 1994, Robert 1994a, and Note 1.6.1).

$\pi_k$, and if these models have the prior weights $p_1, p_2, \ldots, p_k$, the posterior probability that $X$ originates from model $\mathcal{M}_j$ is

(1.3.5)
$$\frac{p_j \int f_j(x|\theta_j)\,\pi_j(\theta_j)\,d\theta_j}{\sum_{i=1}^{k} p_i \int f_i(x|\theta_i)\,\pi_i(\theta_i)\,d\theta_i}.$$

In particular, the comparison of two models $\mathcal{M}_1$ and $\mathcal{M}_2$ is often implemented through the *Bayes factor*

$$B^\pi(x) = \frac{\int f_1(x|\theta_1)\,\pi_1(\theta_1)\,d\theta_1}{\int f_2(x|\theta_2)\,\pi_2(\theta_2)\,d\theta_2}.$$

for which the proportionality constant is quite important (see Kass and Raftery 1995 and Goutis and Robert 1998 for different perspectives on Bayes factors).

**Example 1.3.3 Logistic regression.** A useful regression model for binary $(0 - 1)$ responses is the *logit model*, where the distribution of $Y$ conditional on the explanatory (or dependent) variables $x \in \mathrm{I\!R}^p$ is modeled by the relation

(1.3.6)
$$P(Y = 1) = p = \frac{\exp(x^t\beta)}{1 + \exp(x^t\beta)}.$$

Equivalently, the *logit* transform of $p$, $\mathrm{logit}(p) = \log[p/(1 - p)]$, satisfies the linear relationship $\mathrm{logit}(p) = x^t\beta$.

A classical illustration of the model comparison procedure above occurs in the selection of the explanatory variables in a regression. Suppose that the vector of explanatory variables is $\mathbf{x} = (x^1, x^2, \ldots, x^p)$ and we use the notation $x^{(-j)} = (x^1, x^2, \ldots, x^{j-1}, x^{j+1}, \ldots, x^p)$. If, for instance, one considers whether the first explanatory variable, $x^1$, must be excluded, the models to be compared are $\mathcal{M}_F$ and $\mathcal{M}_R$, $\mathcal{M}_F$ being associated with the full model (1.3.6) and $\mathcal{M}_R$ with the reduced model

$$P(Y = 1) = \frac{\exp(x^{(-1)t}\gamma)}{1 + \exp(x^{(-1)t}\gamma)}, \qquad \gamma \in \mathrm{I\!R}^{p-1}.$$

When the parameters $\beta$ and $\gamma$ are normally distributed *a priori*, $\beta \sim \mathcal{N}_p(\mu, \Sigma)$ and $\gamma \sim \mathcal{N}_{p-1}(\zeta, \Delta)$, the Bayes factor for comparing $\mathcal{M}_F$ and $\mathcal{M}_R$, based on a sample $((x_1, y_1), \ldots, (x_n, y_n))$, is

$$B^\pi(x) = \frac{|\Sigma|^{-1/2} \displaystyle\int_{\mathrm{I\!R}^p} \prod_{i=1}^{n} \frac{e^{y_i \mathbf{x}_i^t \beta}}{1 + e^{\mathbf{x}_i^t \beta}}\, e^{-(\beta-\mu)^t\,\Sigma^{-1}(\beta-\mu)/2}\, d\beta}{\sqrt{2\pi}\,|\Delta|^{-1/2} \displaystyle\int_{\mathrm{I\!R}^{p-1}} \prod_{i=1}^{n} \frac{e^{y_i \mathbf{x}_i^{(-1)t}\gamma}}{1 + e^{\mathbf{x}_i^{(-1)t}\gamma}}\, e^{-(\gamma-\zeta)^t\,\Delta^{-1}(\gamma-\zeta)/2}\, d\gamma}.$$

An explicit calculation of this quantity is quite difficult.                    ‖

The computational problems encountered in the Bayesian approach are not limited to the computation of integrals or normalizing constants. For instance, the determination of confidence regions (also called *credible regions*) with *highest posterior density*,

$$C^\pi(x) = \{\theta; \pi(\theta|x) \geq k\},$$

requires the solution of the equation $\pi(\theta|x) = k$ for the value of $k$ that satisfies

$$P(\theta \in C^\pi(x)|x) = P(\pi(\theta|x) \geq k|x) = \gamma,$$

where $\gamma$ is a predetermined confidence level.

**Example 1.3.4 Bayes credible regions.** For iid observations $X_1, \ldots,$ $X_n$ from a normal distribution $\mathcal{N}(\theta, \sigma^2)$, and a prior distribution $\theta \sim \mathcal{N}(0, \tau^2)$, the posterior density of $\theta$ is normal with mean and variance

$$\delta^\pi = \frac{n\tau^2}{n\tau^2 + \sigma^2}\bar{x} \quad \text{and} \quad \frac{n\tau^2\sigma^2}{n\tau^2 + \sigma^2},$$

respectively. If we assume that $\sigma^2$ and $\tau^2$ are known, the highest posterior density region is

$$\left\{\theta : \sqrt{\frac{n\tau^2 + \sigma^2}{2\pi n\tau^2}} \exp\left[-\frac{n\tau^2 + \sigma^2}{2n\tau^2}(\theta - \delta^\pi)^2\right] \geq k\right\}.$$

Since the posterior distribution is symmetric and unimodal, this set is equivalent to (Problem 1.29)

$$\{\theta; \delta^\pi - k' \leq \theta \leq \delta^\pi + k'\},$$

for some constant $k'$ that is chosen to yield a specified posterior probability. Since the posterior distribution of $\theta$ is normal, this can be done by hand using a normal probability table.

For the situation of Example 1.3.1, the posterior distribution of $p$, $\pi(p|x, a, b)$, was found to be $\mathcal{Be}(x + a, n - x + b)$, which is not necessarily symmetric. To find the 90% highest posterior density region for $p$, we must find limits $l(x)$ and $u(x)$ that satisfy

$$\int_{l(x)}^{u(x)} \pi(p|x, a, b)dp = .9 \quad \text{and} \quad \pi(l(x)|x, a, b) = \pi(u(x)|x, a, b).$$

This cannot be solved analytically.                                        ‖

Computation of a confidence region can be quite delicate when $\pi(\theta|x)$ is not explicit. In particular, when the confidence region involves only one component of a vector parameter, calculation of $\pi(\theta|x)$ requires the integration of the joint distribution over all the other parameters. Note that knowledge of the normalizing factor is of minor importance in this setup. (See Robert 1994a, Chapter 9, for other examples.)

**Example 1.3.5 Cauchy confidence regions.** Consider $X_1, \ldots, X_n$ an iid sample from the Cauchy distribution $\mathcal{C}(\theta, \sigma)$, with associated prior distribution $\pi(\theta, \sigma) = \sigma^{-1}$. The confidence region on $\theta$ is then based on

$$\pi(\theta|x_1, \ldots, x_n) \propto \int_0^\infty \sigma^{-n-1} \prod_{i=1}^n \left[ 1 + \left( \frac{x_i - \theta}{\sigma} \right)^2 \right]^{-1} d\sigma \,,$$

an integral which cannot be evaluated explicitly. Similar computational problems occur with likelihood estimation in this model. One method for obtaining a likelihood confidence interval for $\theta$ is to use the *profile likelihood*

$$\ell^P(\theta|x_1, \ldots, x_n) = \max_\sigma \ell(\theta, \sigma|x_1, \ldots, x_n)$$

and consider the region $\{\theta : \ell^P(\theta|x_1, \ldots, x_n) \geq k\}$. Explicit computation is also difficult here.                    ‖

**Example 1.3.6 Linear calibration.** In a standard regression model, $Y = \alpha + \beta x + \varepsilon$, there is interest in estimating or predicting features of $Y$ from knowledge of $x$. In *linear calibration models* (see Osborne 1991 for an introduction and review of these models), the interest is in determining values of $x$ from observed responses $y$. For example, in a chemical experiment, one may want to relate the precise but expensive measure $y$ to the less precise but inexpensive measure $x$. A simplified version of this problem can be put into the framework of observing the independent random variables

$$Y \sim \mathcal{N}_p(\beta, \sigma^2 I_p), \ Z \sim \mathcal{N}_p(x_0 \beta, \sigma^2 I_p), \ S \sim \sigma^2 \chi_q^2 \,,$$

with $x_0 \in \mathbb{R}$, $\beta \in \mathbb{R}^p$. The parameter of interest is now $x_0$ and this problem is equivalent to Fieller's (1954) problem (see, e.g., Lehmann and Casella 1998).

A reference prior on $(x_0, \beta, \sigma)$ is given in Kubokawa and Robert (1994), and yields the joint posterior distribution

(1.3.7)
$$\pi(x_0, \beta, \sigma^2|y, z, s) \propto \sigma^{-(3p+q)-\frac{1}{2}} \exp\{-(s + \|y - \beta\|^2 + \|z - x_0\beta\|^2)/2\sigma^2\} (1 + x_0^2)^{-1/2} \,.$$

This can be analytically integrated to obtain the marginal posterior distribution of $X_0$ to be

$$\pi(x_0|y, z, s) \propto \frac{(1 + x_0^2)^{(p+q-1)/2}}{\left\{ \left( x_0 - \frac{y^t z}{s + \|y\|^2} \right)^2 + \frac{\|z\|^2 + s}{\|y\|^2 + s} - \frac{(y^t z)^2}{(s + \|y\|^2)^2} \right\}^{(2p+q)/2}} \,.$$

(1.3.8)

However, the computation of the posterior mean, the Bayes estimate of $x_0$, is not feasible analytically; neither is the determination of the confidence region $\{\pi(x_0|\mathcal{D}) \geq k\}$. Nonetheless, it is desirable to determine this confidence region since alternative solutions, for example the *Fieller-Creasy* interval, suffer from defects such as having infinite length with positive probability (see Gleser and Hwang 1987, Casella and Berger 1990, Ghosh *et al.* 1995, or Philippe and Robert 1998a).                    ‖

## 1.4 Deterministic Numerical Methods

The previous examples illustrated the need for techniques, in both the construction of complex models and estimation of parameters, that go beyond the standard analytical approaches. However, before starting to describe simulation methods, which is the purpose of this book, we should recall that there exists a well-developed alternative approach for integration and optimization, based on *numerical* methods. We refer the reader to classical textbooks on numerical analysis (see, for instance, Fletcher 1980 or Evans 1993) for a description of these methods, which are generally efficient and can deal with most of the above examples (see also Kennedy and Gentle 1980 and Thisted 1988 for presentations in statistical settings).

We briefly recall here the more standard approaches to optimization and integration problems, both for comparison purposes and for future use. When the goal is to solve an equation of the form $f(x) = 0$, the most common approach is to use a *Newton–Raphson* algorithm, which produces a sequence $x_n$ such that

$$x_{n+1} = x_n - \left( \left. \frac{\partial f}{\partial x} \right|_{x=x_n} \right)^{-1} f(x_n)$$

until it stabilizes around a solution of $f(x) = 0$. (Note that $\frac{\partial f}{\partial x}$ is a matrix in multidimensional settings.) Optimization problems associated with smooth functions $F$ are then based on this technique, using the equation $\nabla F(x) = 0$, where $\nabla F$ denotes the *gradient* of $F$, that is, the vector of derivatives of $F$. (When the optimization involves a constraint $G(x) = 0$, $F$ is replaced by a Lagrangian form $F(x) - \lambda G(x)$, where $\lambda$ is used to satisfy the constraint.) The corresponding techniques are then the *gradient methods*, where the sequence $x_n$ is such that

$$(1.4.1) \qquad x_{n+1} = x_n - \left( \nabla \nabla^t F \right)^{-1} (x_n) \nabla F(x_n) \, ,$$

where $\nabla \nabla^t F$ denotes the matrix of second derivatives of $F$. Numerous variants can be found in the literature, among which one can mention the *steepest descent* method, where each iteration results in a unidimensional optimizing problem for $F(x_n + td_n)$ ($t \in \mathbb{R}$), $d_n$ being an acceptable direction, namely such that

$$\left. \frac{d^2 F}{dt^2} (x_n + td_n) \right|_{t=0}$$

is of the proper sign. The direction $d_n$ is often chosen as $\nabla F$ or as the smoothed version of (1.4.1),

$$\left[ \nabla \nabla^t F(x_n) + \lambda I \right]^{-1} \nabla F(x_n),$$

in the *Levenberg–Marquardt version*. Other versions are available that do not require differentiation of the function $F$.

Turning to integration, the numerical computation of an integral

$$I = \int_a^b h(x)dx$$

can be done by simple *Riemann integration* (see §3.4), or by improved techniques such as the *trapezoidal rule*

$$\hat{I} = \frac{1}{2}\sum_{i=1}^{n-1}(x_{i+1} - x_i)(h(x_i) + h(x_{i+1}))\,,$$

where the $x_i$'s constitute an ordered partition of $[a, b]$, or *Simpson's rule*, whose formula is

$$\tilde{I} = \frac{\delta}{3}\left\{ f(a) + 4\sum_{i=1}^{n} h(x_{2i-1}) + 2\sum_{i=1}^{n} h(x_{2i}) + f(b) \right\}$$

in the case of equally spaced samples with $(x_{i+1}-x_i) = \delta$. Other approaches involve orthogonal polynomials (Gram–Charlier, Legendre, etc.), as illustrated by Smith and Naylor (1982) for statistical problems, or splines (see Wahba 1981 for a statistical connection). However, these methods may not work well in high dimensions, as stressed by Thisted (1988).

Comparison between the approaches, simulation versus numerical analysis, is delicate because both approaches can provide well-suited tools for most problems (possibly needing a preliminary study) and the distinction between these two groups of techniques can be vague. So, rather than addressing the issue of a general comparison, we will focus on the requirements of each approach and the objective conditions for their implementation in a statistical setup.

By nature, standard numerical methods do not take into account the probabilistic aspects of the problem; that is, the fact that many of the functions involved in the computations are related to probability densities. Therefore, a numerical integration method may consider regions of a space which have zero (or low) probability under the distribution of the model, a phenomenon which usually does not appear in a simulation experiment.[6] Similarly, the occurrence of local modes of a likelihood will often cause more problems for a deterministic gradient method than for a simulation method that explores high-density regions. (But multimodality must first be identified for these efficient methods to apply, as in Oh 1989 or Oh and Berger 1993.)

On the other hand, simulation methods very rarely take into account the specific analytical form of the functions involved in the integration or optimization. For instance, because of the randomness induced by the simulation, a gradient method yields a much faster determination of the mode of a unimodal density. For small dimensions, integration by Riemann sums or

---

[6] Simulation methods using a distribution other than the distribution of interest, such as importance sampling (§2.2.2) or Metropolis–Hastings algorithms (Chapter 6), may suffer from such a drawback.

by quadrature converges faster than the mean of a simulated sample. More-over, existing scientific software (for instance, Gauss, Maple, Mathematica, Matlab) and scientific libraries like IMSL often provide highly efficient nu-merical procedures, whereas simulation is, at best, implemented through pseudo-random generators for the more common distributions. However, software like BUGS (see Note 7.6.4) are progressively bridging the gap.

Therefore, it is often reasonable to use a numerical approach when deal-ing with regular functions in small dimensions and in a given single prob-lem. On the other hand, when the statistician needs to study the details of a likelihood surface or posterior distribution, or needs to simultaneously estimate several features of these functions, or when the distributions are highly multimodal (see Examples 1.1.2 and 1.2.5), it is preferable to use a simulation-based approach. Such an approach captures (if only approx-imately through the generated sample) the different characteristics of the density and thus allows, at little cost, extensions of the inferential scope to, perhaps, another test or estimator.

However, given the dependence on specific problem characteristics, it is fruitless to advocate the superiority of one method over the other, say of the simulation-based approach over numerical methods. Rather, it seems more reasonable to justify the use of simulation-based methods by the statistician in terms of *expertise*. The intuition acquired by a statistician in his or her every-day processing of random models can be directly exploited in the implementation of simulation techniques (in particular in the evaluation of the variation of the proposed estimators or of the stationarity of the resulting output), while purely numerical techniques rely on less familiar branches of mathematics. Finally, note that many desirable approaches are those which efficiently combine both perspectives, as in the case of *simulated annealing* (see §5.2.3) or of Riemann sums (see §3.4).

## 1.5 Problems

**1.1** In the situation of Example 1.1.1, establish that the densities are indeed (1.1.1) and (1.1.2).

**1.2** In Example 1.1.1, the distribution of the random variable $Z = \min(X, Y)$ was of interest. Derive the distribution of $Z$ in the following case of *informative censoring*, where $Y \sim \mathcal{N}(\theta, \sigma^2)$ and $X \sim \mathcal{N}(\theta, \theta^2 \sigma^2)$. Pay attention to the identifiability issues.

**1.3** In Example 1.1.1, show that the integral

$$\int_\omega^\infty \alpha \beta x^{\alpha-1} e^{-\beta x^\alpha} dx$$

can be explicitly calculated. (*Hint:* Use a change of variables.)

**1.4** For the model (1.1.3), show that the density of $(X_0, \ldots, X_n)$ is given by (1.1.4).

**1.5** In the setup of Example 1.1.2, derive the moment estimator of the weights $(p_1, \ldots, p_k)$ when the densities $f_j$ are known.

**1.6** In the setup of Example 1.2.3, show that the likelihood equations are given by (1.2.6) and that their solution is the standard $(\bar{y}, s^2)$ statistic.

**1.7** (Pearson 1894) For a mixture of two normal distributions,

$$\pi \mathcal{N}(\mu_1, \sigma_1^2) + (1 - \pi)\mathcal{N}(\mu_2, \sigma_2^2) \,,$$

give the five first moments in terms of $m_s$ and $k_s$, the $s$th sample central moment and sample cumulant, respectively. Show that the resolution of the moment equations leads to a ninth-degree polynomial (*Note:* Historically, Pearson 1894 was the first to estimate the parameters of a mixture of two normal distributions.)

**1.8** (Titterington *et al.* 1985) In the case of a mixture of two exponential distributions

$$\pi \mathcal{E}xp(1) + (1 - \pi)\mathcal{E}xp(2) \,,$$

show that $\mathbb{E}[X^s] = \{\pi + (1 - \pi)2^{-s}\}\Gamma(s + 1)$. Deduce the best (in $s$) moment estimator based on $t_s(x) = x^s/\Gamma(s + 1)$.

**1.9** Give the moment estimator for a mixture of $k$ Poisson distributions, based on $t_s(x) = x(x - 1) \cdots (x - s + 1)$. (*Note:* Pearson 1915 and Gumbel 1940 proposed partial solutions in this setup. See Titterington *et al.* 1985, pp. 80–81, for details.)

**1.10** The *Weibull distribution* $\mathcal{We}(\alpha, c)$ is widely used in engineering and reliability. Its density is given by

$$f(x|\alpha, c) = c\alpha^{-1}(x/\alpha)^{c-1}e^{-(x/\alpha)^c}.$$

(a) Show that when $c$ is known, this model is equivalent to a Gamma model.

(b) Give the likelihood equations in $\alpha$ and $c$ and show that they do not allow for explicit solutions.

(c) Consider an iid sample $X_1, \ldots, X_n$ from $\mathcal{We}(\alpha, c)$ censored from the right in $y_0$. Give the corresponding likelihood function when $\alpha$ and $c$ are unknown and show that there is no explicit maximum likelihood estimator in this case either.

**1.11** (Continuation of Problem 1.10) Show that the cdf of the Weibull distribution $\mathcal{We}(\alpha, \beta)$ can be written explicitly, and show that the scale parameter $\alpha$ determines the behavior of the hazard rate $h(t) = \frac{f(t)}{1-F(t)}$, where $f$ and $F$ are the density and the cdf, respectively.

**1.12** (Continuation of Problem 1.10) The following sample gives the times (in days) at which carcinoma was diagnosed in rats exposed to a carcinogen:

$$143, \, 164, \, 188, \, 188, \, 190, \, 192, \, 206, \, 209, \, 213, \, 216, \, 220,$$
$$227, \, 230, \, 234, \, 246, \, 265, \, 304, \, 216^*, \, 244^*,$$

where the observations with an asterisk are censored (see Pike 1966 for details). Fit a three parameter Weibull $\mathcal{We}(\alpha, \beta, \gamma)$ distribution to this dataset, where $\gamma$ is a translation parameter, for (a) $\gamma = 100$ and $\alpha = 3$, (b) $\gamma = 100$ and $\alpha$ unknown and (c) $\gamma$ and $\alpha$ unknown. (Treat the asterisked observations as ordinary data here. See Problem 5.23 for a method of dealing with the censoring.)

**1.13** (Robert 1990) The modified Bessel function (see Abramowitz and Stegun 1964) $I_\nu$ ($\nu \geq 0$) is a solution to the differential equation $z^2 f'' + z f' - (z^2 + \nu^2) f(z) = 0$ and can be represented by the power series

$$I_\nu(z) = \left(\frac{z}{2}\right)^\nu \sum_{k=0}^{\infty} \frac{(z/2)^{2k}}{k! \, \Gamma(\nu + k + 1)}.$$

(a) Show that the above series converges in $\mathbb{R}$.

(b) Expanding

$$\int_0^\pi e^{z \cos(\theta)} \sin^{2\nu}(\theta) \, d\theta$$

in a power series, show that $I_\nu$ can be written as

$$I_\nu(z) = \frac{(z/2)^\nu}{\pi^{1/2} \Gamma(\nu + \frac{1}{2})} \int_0^\pi e^{z \cos(\theta)} \sin^{2\nu}(\theta) \, d\theta.$$

(c) Establish the following recurrence formulas:

$$\begin{cases} I_{\nu+1}(z) = I_{\nu-1}(z) - (2\nu/z) I_\nu(z), \\ I_\nu'(z) = I_{\nu-1}(z) - (\nu/z) I_\nu(z). \end{cases}$$

(d) From part (b), use an integration by parts to establish that for $z > 0$,

$$I_{\nu+1}(z) \leq I_\nu(z).$$

(e) Starting from the power series representation of $I_\nu$, show that $t^{-\nu} I_\nu(t)$ is increasing in $t$. If we define $r_\nu$ as

$$r_\nu(t) = \frac{I_{\nu+1}(t)}{I_\nu(t)},$$

show that $r_\nu$ is increasing, concave, and that $r_\nu(t)/t$ is decreasing.

(f) Show that

$$\lim_{t \to \infty} r_\nu(t) = 1, \qquad \lim_{t \to \infty} \frac{r_\nu(t)}{t} = \frac{1}{2(\nu + 1)},$$

and that

$$r_\nu'(t) = 1 - \frac{2\nu + 1}{t} r_\nu(t) - r_\nu^2(t).$$

(g) Show that, using a modified Bessel function, the density of the noncentral chi squared distribution with noncentrality parameter $\lambda$ and $\nu$ degrees of freedom can be written as

$$p_{\lambda,\nu}(x) = \frac{1}{2} \left(\frac{x}{\lambda}\right)^{\frac{\nu-2}{4}} I_{\frac{\nu-2}{2}}(\sqrt{\lambda x}) e^{-\frac{x+\lambda}{2}}.$$

**1.14** Let $X_1, X_2, \ldots, X_n$ be iid with density $f(x|\theta, \sigma)$, the Cauchy distribution $\mathcal{C}(\theta, \sigma)$, and let $L(\theta, \sigma | \mathbf{x}) = \prod_{i=1}^n f(x_i | \theta, \sigma)$ be the likelihood function.

(a) If $\sigma$ is known, show that a solution to the likelihood equation $\frac{d}{d\theta} L(\theta, \sigma | \mathbf{x}) = 0$ is the root of a $2n - 1$ degree polynomial. Hence, finding *the* likelihood estimator can be challenging.

(b) For $n = 3$, if both $\theta$ and $\sigma$ are unknown, find the maximum likelihood estimates and show that they are unique.

(c) For $n \geq 3$, if both $\theta$ and $\sigma$ are unknown, show that the likelihood is unimodal.

(*Note:* See Copas 1975 and Ferguson 1978 for details.)

**1.15** Show that if $\mathbf{X}_n = (X_1, \ldots, X_n)$ is an iid random sample, then as $n \to \infty$, the log-likelihood $\log L(\theta|\mathbf{X}_n)$ converges in probability to the quantity $\mathbb{E}_{\theta_0}[\log f(X_1|\theta)]$, where $\theta_0$ denotes the true value of the parameter. Deduce that the limiting loglikelihood is maximized by $\theta_0$. (*Hint:* Use the Law of Large Numbers and Jensen's inequality.)

**1.16** Show that (1.3.7) implies (1.3.8).

**1.17** Consider a Bernoulli random variable $Y \sim \mathcal{B}([1 + e^\theta]^{-1})$.

(a) If $y = 1$, show that the maximum likelihood estimator of $\theta$ is $\infty$.

(b) Show that the same problem occurs when $Y_1, Y_2 \sim \mathcal{B}([1 + e^\theta]^{-1})$ and $y_1 = y_2 = 0$ or $y_1 = y_2 = 1$. Give the maximum likelihood estimator in the other cases.

**1.18** Consider $n$ observations $x_1, \ldots, x_n$ from $\mathcal{B}(k, p)$ where both $k$ and $p$ are unknown.

(a) Show that the maximum likelihood estimator of $k$, $\hat{k}$, satisfies

$$(\hat{k}(1 - \hat{p}))^n \geq \prod_{i=1}^n (\hat{k} - x_i) \quad \text{and} \quad ((\hat{k} + 1)(1 - \hat{p}))^n < \prod_{i=1}^n (\hat{k} + 1 - x_i),$$

where $\hat{p}$ is the maximum likelihood estimator of $p$.

(b) If the sample is $16, 18, 22, 25, 27$, show that $\hat{k} = 99$.

(c) If the sample is $16, 18, 22, 25, 28$, show that $\hat{k} = 190$. Discuss the stability of the maximum likelihood estimator.

(*Note:* Olkin *et al.* 1981 were one of the first to investigate the stability of the MLE for the binomial parameter $n$; see also Carroll and Lombard 1985, Casella 1986, and Hall 1994.)

**1.19** (Basu 1988) An urn contains 1000 tickets; 20 are tagged $\theta$ and 980 are tagged $100\theta$. A ticket is drawn at random with tag $x$.

(a) Give the maximum likelihood estimator of $\theta$, denoted by $\delta(x)$, and show that $P(\delta(x) = \theta) = 0.98$.

(b) Suppose now there are 20 tickets tagged $\theta$ and 980 tagged $a_i\theta$ $(i \leq 980)$, such that $a_i \in [10, 10.1]$ and $a_i \neq a_j$ $(i \neq j)$. Give the new maximum likelihood estimator, $\delta'$, and show that $P(\delta'(x) < 100\theta) = 0.02$. Discuss the appeal of maximum likelihood estimation in this case.

**1.20** (Romano and Siegel 1986) Given

$$f(x) = \frac{1}{x} \exp\left[-50\left(\frac{1}{x} - 1\right)^2\right] \qquad (x > 0),$$

show that $f$ is integrable and that there exist $a, b > 0$ such that

$$\int_0^b af(x)dx = 1 \quad \text{and} \quad \int_1^b af(x)dx = 0.99.$$

For the distribution with density

$$p(y|\theta) = a\theta^{-1} f(y\theta^{-1}) \mathbb{I}_{[0, b\theta]}(y),$$

give the maximum likelihood estimator, $\delta(y)$, and show that $P(\delta(y) > 100\theta) = 0.99$.

**1.21** (Romano and Siegel 1986)  Consider $X_1, X_2,$ and $X_3$ iid $\mathcal{N}(\theta, \sigma^2)$.

(a) Give the maximum likelihood estimator of $\sigma^2$ if $(x_1, x_2, x_3) = (9, 10, 11)$ or if $(x_1, x_2, x_3) = (29, 30, 31)$.

(b) Given three additional observations $x_4, x_5$ and $x_6$, give the maximum likelihood estimator if $(x_1, \ldots, x_6) = (9, 10, 11, 29, 30, 31)$. Does this result contradict the Likelihood Principle?

**1.22** (Robertson *et al.* 1988)  For a sample $X_1, \ldots, X_n$, and a function $f$ on $\mathcal{X}$, the isotonic regression of $f$ with weights $\omega_i$ is the solution of the minimization in $g$ of

$$\sum_{i=1}^{n} \omega_i (g(x_i) - f(x_i))^2,$$

under the constraint $g(x_1) \leq \cdots \leq g(x_n)$.

(a) Show that a solution to this problem is obtained by the *pool-adjacent-violators* algorithm:

**Algorithm A.1 –Pool-adjacent-violators–**
If $f$ is not isotonic, find $i$ such that
$f(x_{i-1}) > f(x_i)$, replace $f(x_{i-1})$ and $f(x_i)$ by

$$f^*(x_i) = f^*(x_{i-1}) = \frac{\omega_i f(x_i) + \omega_{i-1} f(x_{i-1})}{\omega_i + \omega_{i-1}},$$

and repeat until the constraint is satisfied. Take $g = f^*$.

(b) Apply this algorithm to the case $n = 4$, $f(x_1) = 23$, $f(x_2) = 27$, $f(x_3) = 25$, and $f(x_4) = 28$, when the weights are all equal.

**1.23** (Continuation of Problem 1.22)  The simple *tree ordering* is obtained when one compares treatment effects with a control state. The isotonic regression is then obtained under the constraint $g(x_i) \geq g(x_1)$ for $i = 2, \ldots, n$.

(a) Show that the following provides the isotonic regression $g^*$:

**Algorithm A.2 –Tree ordering–**
If $f$ is not isotonic, assume w.l.o.g.
that the $f(x_i)$'s are in increasing order $(i \geq 2)$.
Find the smallest $j$ such that

$$A_j = \frac{\omega_1 f(x_1) + \cdots + \omega_j f(x_j)}{\omega_1 + \cdots \omega_j} < f(x_{j+1}),$$

take $g^*(x_1) = A_j = g^*(x_2) = \cdots = g^*(x_j)$, $g^*(x_{j+1}) = f(x_{j+1})$, ....

(b) Apply this algorithm to the case where $n = 5$, $f(x_1) = 18$, $f(x_2) = 17$, $f(x_3) = 12$, $f(x_4) = 21$, $f(x_5) = 16$, with $\omega_1 = \omega_2 = \omega_5 = 1$ and $\omega_3 = \omega_4 = 3$.

**1.24** In the setting of Example 1.2.5, it was seen that a polar transformation could be useful. Given a vector $\theta = (\theta_1, \ldots, \theta_p) \in \mathbb{R}^p$, show that the following representation

$$\begin{cases} \theta_1 = \rho \cos \varphi_1, \\ \theta_2 = \rho \sin \varphi_1 \cos \varphi_2, \\ \vdots \\ \theta_{p-1} = \rho \sin \varphi_1 \cdots \cos \varphi_{p-1}, \\ \theta_p = \rho \sin \varphi_1 \cdots \sin \varphi_{p-1}, \end{cases}$$

provides a one-to-one correspondence between the parameterization $(\theta_1, \ldots, \theta_p)$ and the so-called *polar parameterization* $(\rho, \varphi_1, \ldots, \varphi_{p-1})$ when $\rho > 0$ and $\varphi_1, \ldots, \varphi_{p-2} \in [0, 2\pi)$, $\varphi_{p-1} \in [0, \pi)$.

**1.25** (Saxena and Alam 1982) In the setting of Example 1.2.5, use the series expansion of $I_\nu$ in Problem 1.13 to show that the maximum likelihood estimator $\hat{\lambda}$ satisfies

$$\hat{\lambda}(z) = z - p + 0.5 + O(1/z).$$

Show that $\hat{\lambda}$ is dominated by $(z - p)^+$ under quadratic loss.

**1.26** For the setup of Example 1.2.5, where $X \sim \mathcal{N}_p(\theta, I_p)$:

(a) Show that the maximum likelihood estimator of $\lambda = \|\theta\|^2$ is $\hat{\lambda}(x) = \|x\|^2$ and that it has a constant bias equal to $p$.

(b) If we observe $Y = \|X\|^2$, distributed as a noncentral chi squared random variable (see Appendix A), show that the MLE of $\lambda$ is the solution in (1.2.8). Discuss what happens if $y < p$.

(c) In part (a), if the reference prior $\pi(\theta) = \|\theta\|^{-(p-1)}$ is used, show that the posterior distribution is given by (1.3.3), with posterior mean (1.3.4).

**1.27** Suppose that $X \sim f(x|\theta)$, with prior distribution $\pi(\theta)$, and interest is in the estimation of the parameter $h(\theta)$.

(a) Using the loss function $L(\delta, h(\theta))$, show that the estimator that minimizes the Bayes risk

$$\int \int L(\delta, h(\theta)) f(x|\theta) \pi(\theta) dx d\theta$$

is given by the estimator $\delta$ that minimizes (for each $x$)

$$\int L(\delta, \theta) \, \pi(\theta|x) \, d\theta \ .$$

(b) For $L(\delta, \theta) = \|h(\theta) - \delta\|^2$ , show that the Bayes estimator of $h(\theta)$ is $\delta^\pi(x) = \mathbb{E}^\pi[h(\theta)|x]$.

(c) For $L(\delta, \theta) = |h(\theta) - \delta|$ , show that the Bayes estimator of $h(\theta)$ is the median of the posterior distribution.

**1.28** For each of the following cases, give the posterior and marginal distributions.

(a) $X|\sigma \sim \mathcal{N}(0, \sigma^2)$, $1/\sigma^2 \sim \mathcal{G}(1, 2)$;

(b) $X|\lambda \sim \mathcal{P}(\lambda)$, $\lambda \sim \mathcal{G}(2, 1)$;

(c) $X|p \sim \mathcal{N}eg(10, p)$, $p \sim \mathcal{B}e(1/2, 1/2)$.

**1.29** Let $f(x)$ be a unimodal density, and for a given value of $\alpha$, let the interval $[a, b]$ satisfy $\int_a^b f = 1 - \alpha$.

(a) Show that the shortest interval satisfying the probability constraint is given by $f(a) = f(b)$, where $a$ and $b$ are on each side of the mode of $f$.

(b) Show that if $f$ is symmetric, then the shortest interval satisfies $a = -b$.

(c) Find the 90% highest posterior credible regions for the posterior distributions of Problem 1.28

**1.30** (Bauwens 1991) Consider $X_1, \ldots, X_n$ iid $\mathcal{N}(\theta, \sigma^2)$ with prior

$$\pi(\theta, \sigma^2) = \sigma^{-2(\alpha+1)} \exp(-s_0^2/2\sigma^2).$$

(a) Compute the posterior distribution $\pi(\theta, \sigma^2 | x_1, \ldots, x_n)$ and show that it only depends on $\bar{x}$ and $s^2 = \sum_{i=1}^{n}(x_i - \bar{x})^2$.

(b) Derive the posterior expectation $\mathbb{E}^{\pi}[\sigma^2 | x_1, \ldots, x_n]$ and show that its behavior when $\alpha$ and $s_0$ both converge to 0 depends on the limit of $(s_0^2/\alpha) - 1$.

**1.31** (Raiffa and Schlaifer 1961) Consider a $\mathcal{B}e(\alpha m, (1 - m)\alpha)$ prior on $p \in [0, 1]$. Show that if $m$ is held fixed and $\alpha$ approaches 0, the prior distribution converges to a two-point mass distribution with weight $m$ on $p = 1$ and $(1-m)$ on $p = 0$. Discuss the drawbacks of such a setting.

**1.32**

(a) Show that, if the prior distribution is improper, the marginal distribution is also improper.

(b) Show that if the prior $\pi(\theta)$ is improper and the sample space $\mathcal{X}$ is finite, the posterior distribution $\pi(\theta | x)$ is not defined for some value of $x$.

(c) Consider $X_1, \ldots, X_n$ distributed according to $\mathcal{N}(\theta_j, 1)$, with $\theta_j \sim \mathcal{N}(\mu, \sigma^2)$ $(1 \le j \le n)$ and $\pi(\mu, \sigma^2) = \sigma^{-2}$. Show that the posterior distribution $\pi(\mu, \sigma^2 | x_1, \ldots, x_n)$ is not defined.

**1.33** Assuming that $\pi(\theta) = 1$ is an acceptable prior for real parameters, show that this generalized prior leads to $\pi(\sigma) = 1/\sigma$ if $\sigma \in \mathbb{R}^+$ and to $\pi(\varrho) = 1/\varrho(1-\varrho)$ if $\varrho \in [0, 1]$ by considering the "natural" transformations $\theta = \log(\sigma)$ and $\theta = \log(\varrho/(1 - \varrho))$.

**1.34** For each of the following situations, exhibit a conjugate family for the given distribution:

(a) $X \sim \mathcal{G}(\theta, \beta)$; that is, $f_{\beta}(x|\theta) = \beta^{\theta} x^{\theta-1} e^{-\beta x}/\Gamma(\theta)$.

(b) $X \sim \mathcal{B}e(1, \theta), \theta \in \mathbb{N}$.

**1.35** Show that, if $X \sim \mathcal{B}e(\theta_1, \theta_2)$, there exist conjugate priors on $\theta = (\theta_1, \theta_2)$ but that they do not lead to tractable posterior quantities, except for the computation of $\mathbb{E}^{\pi}[\theta_1/(\theta_1 + \theta_2)|x]$.

**1.36** Show that under the specification $\theta_j \sim \pi_j$ and $X \sim f_j(x|\theta_j)$ for model $\mathcal{M}_j$, the posterior probability of model $\mathcal{M}_j$ is given by (1.3.5).

**1.37** This problem examines the connection among Bayes estimation, sufficiency, and the likelihood function. Let $(X_1, X_2, \ldots, X_n)$ have density $f(x_1, \ldots, x_n|\theta)$ and likelihood function $L(\theta|\mathbf{x}) = L(\theta|x_1, \ldots, x_n) = f(x_1, \ldots, x_n|\theta)$. Show that

(a) $f(x_1, \ldots, x_n|\theta) = L(\theta|\mathbf{x}) = g(\mathbf{t}|\theta)h(\mathbf{x})$, where $T(x) = \mathbf{t}$ is sufficient for $\theta$ and the function $h(\cdot)$ does not depend on $\theta$;

(b) for any prior distribution $\pi(\theta)$, the posterior distribution is

$$\pi(\theta|\mathbf{x}) = \frac{g(\mathbf{t}|\theta)\pi(\theta)}{\int g(\mathbf{t}|\theta')\pi(\theta') \, d\theta'}.$$

Thus, the likelihood function and the Bayesian posterior are functions of a minimal sufficient statistic.

**1.38** Consider $X_1, \ldots, X_n \sim \mathcal{N}(\mu + \nu, \sigma^2)$, with $\pi(\mu, \nu, \sigma) \propto 1/\sigma$.

(a) Show that the posterior distribution is not defined for every $n$.

(b) Extend this result to overparameterized models with improper priors.

**1.39** Consider estimation in the *linear model*

$$Y = b_1 X_1 + b_2 X_2 + \epsilon,$$

under the constraint $0 \leq b_1, b_2 \leq 1$, for a sample $(Y_1, X_{11}, X_{21}), \ldots, (Y_n, X_{1n}, X_{2n})$ when the errors $\epsilon_i$ are independent and distributed according to $\mathcal{N}(0, 1)$. A noninformative prior is

$$\pi(b_1, b_2) = \mathbb{I}_{[0,1]}(b_1) \mathbb{I}_{[0,1]}(b_2).$$

(a) Show that the posterior means are given by $(i = 1, 2)$

$$\mathbb{E}^\pi[b_i | y_1, \ldots, y_n] = \frac{\int_0^1 \int_0^1 b_i \prod_{j=1}^n \varphi(y_j - b_1 X_{1j} - b_2 X_{2j}) \, db_1 \, db_2}{\int_0^1 \int_0^1 \prod_{j=1}^n \varphi(y_j - b_1 X_{1j} - b_2 X_{2j}) \, db_1 \, db_2},$$

where $\varphi$ is the density of the standard normal distribution.

(b) Show that an equivalent expression is

$$\delta_i^\pi(y_1, \ldots, y_n) = \frac{\mathbb{E}^\pi\left[b_i \mathbb{I}_{[0,1]^2}(b_1, b_2) | y_1, \ldots, y_n\right]}{P^\pi\left[(b_1, b_2) \in [0, 1]^2 | y_1, \ldots, y_n\right]},$$

where the right-hand term is computed under the distribution

$$\begin{pmatrix} b_1 \\ b_2 \end{pmatrix} \sim \mathcal{N}_2\left(\begin{pmatrix} \hat{b}_1 \\ \hat{b}_2 \end{pmatrix}, (X^t X)^{-1}\right),$$

with $(\hat{b}_1, \hat{b}_2)$ the unconstrained least squares estimator of $(b_1, b_2)$ and

$$X = \begin{pmatrix} X_{11} & X_{21} \\ \vdots & \vdots \\ X_{1n} & X_{2n} \end{pmatrix}.$$

(c) Show that the posterior means cannot be written explicitly, except in the case where $(X^t X)$ is diagonal.

**1.40** (Berger 1985) Consider the hierarchical model

$$X | \theta \sim \mathcal{N}_p(\theta, \sigma_1^2 I_p),$$
$$\theta | \xi \sim \mathcal{N}_p(\xi \mathbf{1}, \sigma_\pi^2 I_p),$$
$$\xi \sim \mathcal{N}(\xi_0, \tau^2),$$

where $\mathbf{1} = (1, \ldots, 1)^t \in \mathbb{R}^p$.

(a) Show that

$$\delta(x | \xi, \sigma_\pi) = x - \frac{\sigma_1^2}{\sigma_1^2 + \sigma_\pi^2}(x - \xi \mathbf{1}),$$

$$\pi_2(\xi, \sigma_\pi^2 | x) \propto (\sigma_1^2 + \sigma_\pi^2)^{-p/2} \exp\left\{-\frac{\|x - \xi \mathbf{1}\|^2}{2(\sigma_1^2 + \sigma_\pi^2)}\right\} e^{-(\xi - \xi_0)^2 / 2\tau^2} \pi_2(\sigma_\pi^2)$$

$$\propto \frac{\pi_2(\sigma_\pi^2)}{(\sigma_1^2 + \sigma_\pi^2)^{p/2}} \exp\left\{-\frac{p(\bar{x} - \xi)^2 + s^2}{2(\sigma_1^2 + \sigma_\pi^2)} - \frac{(\xi - \xi_0)^2}{2\tau^2}\right\}$$

with $s^2 = \sum_i (x_i - \bar{x})^2$. Deduce that $\pi_2(\xi | \sigma_\pi^2, x)$ is the normal distribution $\mathcal{N}(\mu(x, \sigma_\pi^2), V_\pi(\sigma_\pi^2))$.

(b) Show that

$$\delta^\pi(x) = \mathbb{E}^{\pi_2(\sigma_\pi^2|x)} \left[ x - \frac{\sigma_1^2}{\sigma_1^2 + \sigma_\pi^2} (x - \bar{x}\mathbf{1}) - \frac{\sigma_1^2 + \sigma_\pi^2}{\sigma_1^2 + \sigma_\pi^2 + p\tau^2} (\bar{x} - \xi_0)\mathbf{1} \right]$$

and

$$\pi_2(\sigma_\pi^2|x) \propto \frac{\tau \exp -\frac{1}{2} \left[ \dfrac{s^2}{\sigma_1^2 + \sigma_\pi^2} + \dfrac{p(\bar{x} - \xi_0)^2}{p\tau^2 + \sigma_1^2 + \sigma_\pi^2} \right]}{(\sigma_1^2 + \sigma_\pi^2)^{(p-1)/2}(\sigma_1^2 + \sigma_\pi^2 + p\tau^2)^{1/2}} \, \pi_2(\sigma_\pi^2).$$

(c) Deduce the representation

$$\delta^\pi(x) = x - \mathbb{E}^{\pi_2(\sigma_\pi^2|x)} \left[ \frac{\sigma_1^2}{\sigma_1^2 + \sigma_\pi^2} \right] (x - \bar{x}\mathbf{1})$$

$$- \mathbb{E}^{\pi_2(\sigma_\pi^2|x)} \left[ \frac{\sigma_1^2 + \sigma_\pi^2}{\sigma_1^2 + \sigma_\pi^2 + p\tau^2} \right] (\bar{x} - \xi_0)\mathbf{1}.$$

and discuss the appeal of this expression from an integration point of view.

**1.41** A classical linear regression can be written as $Y \sim \mathcal{N}_p(X\beta, \sigma^2 I_p)$ with $X$ a $p \times q$ matrix and $\beta \in \mathbb{R}^q$.

(a) When $X$ is known, give the natural parameterization of this exponential family and derive the conjugate priors on $(\beta, \sigma^2)$.

(b) Generalize to $\mathcal{N}_p(X\beta, \Sigma)$.

**1.42** [7] An *autoregressive model* $AR(1)$ connects the random variables in a sample $X_1, \ldots, X_n$ through the relation $X_{t+1} = \varrho X_t + \epsilon_t$, where $\epsilon_t \sim \mathcal{N}(0, \sigma^2)$ is independent of $X_t$.

(a) Show that the $X_t$'s induce a Markov chain and derive a stationarity condition on $\varrho$. Under this condition, what is the stationary distribution of the chain?

(b) Give the covariance matrix of $(X_1, \ldots, X_n)$.

(c) If $x_0$ is a (fixed) starting value for the chain, express the likelihood function and derive a conjugate prior on $(\varrho, \sigma^2)$. (*Hint:* Note that $X_t|x_{t-1} \sim \mathcal{N}(\varrho x_{t-1}, \sigma^2)$.)

**1.43** (Continuation of Problem 1.42) Consider a so-called *state-space* model

$$\begin{cases} Y_t = GX_t + W_t, \\ X_{t+1} = FX_t + V_t, \end{cases}$$

where $Y_t, W_t \in \mathbb{R}^w$, $X_t, V_t \in \mathbb{R}^v$, $W_t \sim \mathcal{N}_w(0, R)$, and $V_t \sim \mathcal{N}_v(0, Q)$.

(a) Show that the $MA(q)$ model of Example 1.1.3 also fits in this representation.

(b) Show that the best linear predictor of $X_{t+1}$ in terms of $Y_0, Y_1, \ldots, Y_t$ is

$$\hat{X}_1 = \mathbb{E}[X_1|Y_0], \qquad \hat{X}_{t+1} = F\hat{X}_t + \Theta_t \Delta_t^{-1}(Y_t - G\hat{X}_t),$$

---

[7] This problem requires material that will be covered in Chapter 4. It is put here for those already familiar with Markov chains.

where

$$\Delta_t = G\Omega_t G' + R,$$
$$\Theta_t = F\Omega_t G',$$
$$\Omega_{t+1} = F\Omega_t F' + Q - \Theta_t \Delta_t^{-1} \Theta_t',$$

and $\Omega_1$ is the covariance matrix of $X_1$ given $Y_0$. (*Note:* This is the *Kalman filter;* see Brockwell and Davis 1996, pp. 265–266 for a complete proof.)

(c) Express the $AR(1)$ model of Problem 1.42 in the state-space form and apply the results of part (c) to find the best linear predictor.

*Note:* The next seven problems involve properties of the exponential family, conjugate prior distributions, and Jeffreys prior distributions. Brown (1986) is a book-length introduction to exponential families, and shorter introductions can be found in Casella and Berger (1990, Section 3.3), Robert (1994a, Section 3.2), or Lehmann and Casella (1998, Section 1.5). For conjugate and Jeffreys priors, in addition to Note 1.6.1, see Berger (1985, Section 3.3) or Robert (1994a, Section 3.4).

**1.44** Show that the Jeffreys prior is invariant under one-to-one reparameterization; that is, that the Jeffreys priors derived for two different parameterizations of the same model agree by the Jacobian transformation rule.

**1.45** Consider $\mathbf{x} = (x_{ij})$ and $\Sigma = (\sigma_{ij})$ symmetric positive-definite $m \times m$ matrices. The *Wishart* distribution, $\mathcal{W}_m(\alpha, \Sigma)$, is defined by the density

$$p_{\alpha,\Sigma}(\mathbf{x}) = \frac{|\mathbf{x}|^{\frac{\alpha-(m+1)}{2}} \exp(-(\mathrm{tr}(\Sigma^{-1}\mathbf{x})/2)}{\Gamma_m(\alpha)|\Sigma|^{\alpha/2}},$$

with $\mathrm{tr}(A)$ the trace of $A$ and

$$\Gamma_m(\alpha) = 2^{\alpha m/2} \pi^{m(m-1)/4} \prod_{i=1}^{m} \Gamma\left(\frac{\alpha-i+1}{2}\right).$$

(a) Show that this distribution belongs to the exponential family. Give its natural representation and derive the mean of $\mathcal{W}_m(\alpha, \Sigma)$.

(b) Show that, if $Z_1, \ldots, Z_n \sim \mathcal{N}_m(0, \Sigma)$,

$$\sum_{i=1}^{n} Z_i Z_i' \sim \mathcal{W}_m(n, \Sigma).$$

**1.46** Consider $X \sim \mathcal{N}(\theta, \theta)$ with $\theta > 0$.

(a) Indicate whether the distribution of $X$ belongs to an exponential family and derive the conjugate priors on $\theta$.

(b) Determine the Jeffreys prior $\pi^J(\theta)$.

**1.47** Show that a Student's $t$ distribution $\mathcal{T}_p(\nu, \theta, \tau^2)$ does not allow for a conjugate family, apart from $\mathcal{F}_0$, the (trivial) family that contains all distributions.

**1.48** (Robert 1991) The generalized inverse normal distribution $\mathcal{IN}(\alpha, \mu, \tau)$ has the density

$$K(\alpha, \mu, \tau)|y|^{-\alpha} \exp\left\{-\left(\frac{1}{y} - \mu\right)^2 / 2\tau^2\right\},$$

with $\alpha > 0$, $\mu \in \mathbb{R}$, and $\tau > 0$.

(a) Show that this density is well defined and that the normalizing factor is

$$K(\alpha,\mu,\tau)^{-1} = \tau^{\alpha-1} e^{-\mu^2/2\tau^2} 2^{(\alpha-1)/2} \, \Gamma\left(\frac{\alpha-1}{2}\right) \, {}_1F_1\left(\frac{\alpha-1}{2};1/2;\frac{\mu^2}{2\tau^2}\right),$$

where $_1F_1$ is the *confluent hypergeometric function*

$$_1F_1(a;b;z) = \sum_{k=0}^{\infty} \frac{\Gamma(a+k)}{\Gamma(b+k)} \frac{\Gamma(b)}{\Gamma(a)} \frac{z^k}{k!}$$

(see Abramowitz and Stegun 1964).

(b) If $X \sim \mathcal{N}(\mu,\tau^2)$, show that the distribution of $1/X$ is in the $\mathcal{IN}(\alpha,\mu,\tau)$ family.

(c) Deduce that the mean of $Y \sim \mathcal{IN}(\alpha,\mu,\tau)$ is defined for $\alpha > 2$ and is

$$\mathbb{E}_{\alpha,\mu,\tau}[Y] = \frac{\mu}{\tau^2} \frac{{}_1F_1(\frac{\alpha-1}{2};3/2;\mu^2/2\tau^2)}{{}_1F_1(\frac{\alpha-1}{2};1/2;\mu^2/2\tau^2)}.$$

(d) Show that $\theta \sim \mathcal{IN}(\alpha,\mu,\tau)$ constitutes a conjugate family for the multiplicative model $X \sim \mathcal{N}(\theta,\theta^2)$.

**1.49** Consider a population divided into $k$ categories (or *cells*) with probability $p_i$ for an individual to belong to the $i$th cell ($1 \le i \le n$). A sequence $(\pi_k)$ of prior distributions on $p^k = (p_1,\ldots,p_k)$, $k \in \mathbb{N}$, is called *coherent* if any grouping of cells into $m$ categories leads to the prior $\pi_m$ for the transformed probabilities.

(a) Determine coherence conditions on the sequence $(\pi_k)$.

(b) In the particular case when $\pi_k$ is a Dirichlet distribution $\mathcal{D}_k(\alpha_1,\ldots,\alpha_k)$, express these conditions in terms of the $\alpha_k$'s.

(c) Does the Jeffreys prior induce a coherent sequence?

(d) What about $\pi_k(p^k) \propto \prod_i p_i^{-1/k}$, proposed by Perk (1947)?

**1.50** Recall the situation of Example 1.2.5 (see also Example 1.3.2), where $X \sim \mathcal{N}(\theta,I_p)$.

(a) For the prior $\pi(\lambda) = 1/\sqrt{\lambda}$, show that the Bayes estimator of $\lambda = ||\theta||^2$ under quadratic loss can be written as

$$\delta^\pi(x) = \frac{{}_1F_1(3/2;p/2;||x||^2/2)}{{}_1F_1(1/2;p/2;||x||^2/2)},$$

where $_1F_1$ is the confluent hypergeometric function.

(b) Using the series development of $_1F_1$ derive an asymptotic expansion of $\delta^\pi$ (for $||x||^2 \to +\infty$) and compare it with $\delta_0(x) = ||x||^2 - p$.

(c) Compare the risk behavior of the estimators of part (b) under the weighted quadratic loss

$$L(\delta,\theta) = \frac{(||\theta||^2 - \delta)^2}{2||\theta||^2 + p}.$$

**1.51** (Dawid *et al.* 1973) Consider $n$ random variables $X_1,\ldots,X_n$, such that the first $\xi$ of these variables has an $\mathcal{E}xp(\eta)$ distribution and the $n-\xi$ other have a $\mathcal{E}xp(c\eta)$ distribution, where $c$ is known and $\xi$ takes its values in $\{1,2,\ldots,n-1\}$.

(a) Give the shape of the posterior distribution of $\xi$ when $\pi(\xi, \eta) = \pi(\xi)$ and show that it only depends on $z = (z_2, \ldots, z_n)$, with $z_i = x_i/x_1$.

(b) Show that the distribution of $Z$, $f(z|\xi)$, only depends on $\xi$.

(c) Show that the posterior distribution $\pi(\xi|x)$ cannot be written as a posterior distribution for $Z \sim f(z|\xi)$, whatever $\pi(\xi)$, although it only depends on $z$. How do you explain this phenomenon?

(d) Show that the paradox of part (c) does not occur when $\pi(\xi, \eta) = \pi(\xi)\eta^{-1}$.

**1.52** A famous problem in classical Statistics is the *Behrens–Fisher* problem. It comes from the simple setting of two normal populations with unknown means and variances, because there is no optimal test[8] to compare the means through the test of the hypothesis $H_0 : \theta = \mu$. Consider $X_1, \ldots, X_n$ a sample from $\mathcal{N}(\theta, \sigma^2)$ and $Y_1, \ldots, Y_m$ a sample from $\mathcal{N}(\mu, \tau^2)$ where $\theta, \mu, \tau$ and $\sigma$ are unknown.

(a) Explain why a reasonable test should be based on the pivotal quantity

$$T = \frac{(\theta - \mu) - (\bar{x} - \bar{y})}{\sqrt{s_x^2/n + s_y^2/m}}$$

with $\bar{x} = \sum_i x_i/n$, $\bar{y} = \sum_j y_j/m$, $s_x^2 = \sum_i (x_i - \bar{x})^2/n - 1$, and $s_y^2 = \sum_j (y_j - \bar{y})^2/m - 1$.

(b) Show that the distribution of $T$ depends on $\sigma/\tau$ even when $\theta = \mu$ and *is not* a Student $t$ distribution.

(c) Give the posterior distribution of $T$ when $\pi(\theta, \mu, \sigma, \tau) = 1/(\sigma^2 \tau^2)$ and show that it depends only on $(s_X/\sqrt{n})(s_Y/\sqrt{m})$.

(d) Study the power of the test based on large values of $T$ for different configurations of $(\theta, \sigma, \mu, \tau)$. (Consider, in particular, the cases when $\sigma = \tau$ and when $\sigma/\tau$ goes to 0.)

(*Note:* See Robinson 1982 for a detailed survey of the different issues related to this problem.)

**1.53** (Smith and Makov 1978)  Consider

$$X \sim f(x|p) = \sum_{i=1}^{k} p_i f_i(x),$$

where $p_i > 0$, $\sum_i p_i = 1$, and the densities $f_i$ are known. The prior $\pi(p)$ is a Dirichlet distribution $\mathcal{D}(\alpha_1, \ldots, \alpha_k)$.

(a) Explain why the computing time could get prohibitive as the sample size increases.

(b) A sequential alternative which approximates the Bayes estimator is to replace $\pi(p|x_1, \ldots, x_n)$ by $\mathcal{D}(\alpha_1^{(n)}, \ldots, \alpha_k^{(n)})$, with

$$\alpha_1^{(n)} = \alpha_1^{(n-1)} + P(Z_{n1} = 1|x_n), \ldots, \alpha_k^{(n)} = \alpha_k^{(n-1)} + P(Z_{nk} = 1|x_n),$$

and $Z_{ni}$ $(1 \leq i \leq k)$ is the component indicator vector of $X_n$. Justify this approximation and compare with the updating of $\pi(\theta|x_1, \ldots, x_{n-1})$ when $x_n$ is observed.

---

[8] The theory of optimal tests will sometimes lead to Uniformly Most Powerful (UMP) tests. See Lehmann (1986) for a complete development.

(c) Examine the performances of the approximation in part (b) for a mixture of two normal distributions $\mathcal{N}(0,1)$ and $\mathcal{N}(2,1)$ when $p = 0.1, 0.25$, and 0.5.

(d) If $\pi_i^n = P(Z_{ni} = 1|x_n)$, show that

$$\hat{p}_i^{(n)}(x_n) = \hat{p}_i^{(n-1)}(x_{n-1}) - a_{n-1}\{\hat{p}_i^{(n-1)} - \pi_i^n\},$$

where $\hat{p}_i^{(n)}$ is the quasi-Bayesian approximation of $\mathbb{E}^\pi(p_i|x_1,\ldots,x_n)$.

## 1.6 Notes

### 1.6.1 Prior Distributions

#### (i) Conjugate Priors

When prior information about the model is quite limited, the prior distribution is often chosen from a parametric family. Families $\mathcal{F}$ that are *closed under sampling* (that is, such that, for every prior $\pi \in \mathcal{F}$, the posterior distribution $\pi(\theta|x)$ also belongs to $\mathcal{F}$) are of particular interest, for both parsimony and invariance motivations. These families are called *conjugate* families. Most often, the main motivation for using conjugate priors is their tractability; however, such choices may constrain the subjective input.

For reasons related to the *Pitman–Koopman Lemma* (see the discussion following Example 1.2.5), conjugate priors can only be found in exponential families. In fact, if the sampling density is of the form

(1.6.1)          $$f(x|\theta) = C(\theta)h(x)\exp\{R(\theta) \cdot T(x)\},$$

which include many common continuous and discrete distributions (see Brown 1986), a conjugate family for $f(x|\theta)$ is given by

$$\pi(\theta|\mu, \lambda) = K(\mu, \lambda)\,e^{\theta \cdot \mu - \lambda\psi(\theta)},$$

since the posterior distribution is $\pi(\theta|\mu + x, \lambda + 1)$. Table 1.6.1 presents some standard conjugate families.

| Distribution | Sample Density | Prior Density |
|---|---|---|
| Normal | $\mathcal{N}(\theta, \sigma^2)$ | $\theta \sim \mathcal{N}(\mu, \tau^2)$ |
| Normal | $\mathcal{N}(\mu, 1/\theta)$ | $\theta \sim \mathcal{G}(\alpha, \beta)$ |
| Poisson | $\mathcal{P}(\theta)$ | $\theta \sim \mathcal{G}(\alpha, \beta)$ |
| Gamma | $\mathcal{G}(\nu, \theta)$ | $\theta \sim \mathcal{G}(\alpha, \beta)$ |
| Binomial | $\mathcal{B}(n, \theta)$ | $\theta \sim \mathcal{B}e(\alpha, \beta)$ |
| Multinomial | $\mathcal{M}_k(\theta_1, \ldots, \theta_k)$ | $\theta_1, \ldots, \theta_k \sim \mathcal{D}(\alpha_1, \ldots, \alpha_k)$ |

Table 1.6.1. *Some conjugate families of distributions.*

Extensions of (1.6.1) which allow for parameter-dependent support enjoy most properties of the exponential families. In particular, they extend the applicability of conjugate prior analysis to other types of distributions like the uniform or the Pareto distribution. (See Robert 1994a, Section 3.2.2.)

Another justification of conjugate priors, found in Diaconis and Ylvisaker (1979), is that some Bayes estimators are then linear. If $\xi(\theta) = \mathbb{E}_\theta[x]$, which is equal to $\nabla\psi(\theta)$, the prior mean of $\xi(\theta)$ for the prior $\pi(\theta|\mu,\lambda)$ is $x_0/\lambda$ and if $x_1,\ldots,x_n$ are iid $f(x|\theta)$,

$$\mathbb{E}^\pi[\xi(\theta)|x_1,\ldots,x_n] = \frac{x_0+n\bar{x}}{\lambda+n}.$$

For more details, see Bernardo and Smith (1994, Section 5.2).

*(ii) Noninformative Priors*

If there is no strong prior information, a Bayesian analysis may proceed with a "noninformative" prior; that is, a prior distribution which attempts to impart no information about the parameter of interest. A classic noninformative prior is the *Jeffreys prior* (Jeffreys 1961). For a sampling density $f(x|\theta)$, this prior has a density that is proportional to $\sqrt{|I(\theta)|}$, where $|I(\theta)|$ is the determinant of the information matrix,

$$I(\theta) = E_\theta\left[\frac{\partial}{\partial\theta}\log f(X|\theta)\right]^2 = -E_\theta\left[\frac{\partial^2}{\partial\theta^2}\log f(X|\theta)\right].$$

For further details see Berger (1985), Bernardo and Smith (1994), Robert (1994a), or Lehmann and Casella (1998).

*(iii) Reference Priors*

An alternative approach to constructing a "noninformative" prior is that of *reference priors* (Bernardo 1979; Berger and Bernardo 1989, 1992). We start with Kullback–Leibler information, $K[f,g]$, also known as Kullback–Leibler *information for discrimination* between two densities. For densities $f$ and $g$, it is given by

$$K[f,g] = \int \log\left[\frac{f(t)}{g(t)}\right] f(t)\,dt.$$

The interpretation is that as $K[f,g]$ gets larger, it is easier to discriminate between the densities $f$ and $g$.

A reference prior can be thought of as the density $\pi(\cdot)$ that maximizes (asymptotically) the expected Kullback-Leibler information (also known as Shannon information)

$$\int K[\pi(\theta|x),\pi(\theta)]m_\pi(x)\,dx,$$

where $m_\pi(x) = \int f(x|\theta)\pi(\theta)\,d\theta$ is the marginal distribution.

Further details are in Bernardo and Smith (1994) and Robert (1994a, Section 3.4), and there are approximations due to Clarke and Barron (1990) and Clarke and Wasserman (1993).

*1.6.2 Method of Moments*

The computing bottleneck created by the need for explicit solutions has led to alternative approaches to inference (see, for instance, Gouriéroux and Monfort 1996). Such approaches rely on the use of *linear* structures of dependence or may involve solving implicit equations for *method of moments* or minimization of generalized distances (for *M-estimators*). Approaches using minimal distances can, in general, be reformulated as maximizations of formal likelihoods, as illustrated in Example 1.2.2, whereas the method of moments can sometimes be expressed as a difference equation.

These interpretations are somewhat rare and, moreover, the method of moments is generally suboptimal when compared to Bayesian or maximum likelihood approaches. These latter two methods are more efficient in using the information contained in the distribution of the observations, according to the *Likelihood Principle* (see Berger and Wolpert 1988, Robert 1994a, or Problem 1.37). However, the moment estimators are still of interest as starting values for iterative methods aiming at maximizing the likelihood, since they will converge in many setups. For instance, in the case of Normal mixtures, although the likelihood is not bounded (see Example 1.2.7) and therefore there is no maximum likelihood estimator, it can be shown that the solution of the likelihood equations which is closer to the moment estimator is a convergent estimator (see Lehmann and Casella 1998, Section 6.4).

### 1.6.3 Bootstrap Methods

*Bootstrap* (or *resampling*) techniques are a collection of computationally intensive methods that are based on resampling from the observed data. They were first introduced by Efron (1979a) and are described more fully in Efron (1982), Efron and Tibshirani (1994), or Hjorth (1994). (See also Hall 1992 and Barbe and Bertail 1994 for more theoretical treatments.) Although these methods do not call for, in principle, a simulation-based implementation, in many cases where their use is particularly important, intensive simulation is required. The basic idea of the bootstrap[9] is to evaluate the properties of an arbitrary estimator $\hat{\theta}(x_1, \ldots, x_n)$ through the empirical cdf of the sample $X_1, \ldots, X_n$,

$$F_n(x) = \frac{1}{n} \sum_{i=1}^{n} \mathbb{I}_{X_i \leq x},$$

instead of the theoretical cdf $F$. More precisely, if an estimate of $\theta(F) = \int h(x)dF(x)$ is desired, an obvious candidate is $\theta(F_n) = \int h(x)dF_n(x)$. When the $X_i$'s are iid, the *Glivenko–Cantelli Theorem* (see Billingsley 1995) guarantees the *sup-norm* convergence of $F_n$ to $F$, and hence guarantees that $\theta(F_n)$ is a consistent estimator of $\theta(F)$. The bootstrap provides a somewhat "automatic" method of computing $\theta(F_n)$, by resampling the data.

It has become common to denote a bootstrap sample with a superscript "⋆", so we can draw bootstrap samples

$$\mathbf{X}^{*i} = (X_1^*, \ldots, X_n^*) \sim F_n, \quad \text{iid.}$$

(Note that the $X_i^*$'s are equal to one of the $x_j$'s and that a same value $x_j$ can appear several times in $\mathbf{X}^{*i}$.) Based on drawing $\mathbf{X}^{*1}, \ldots, \mathbf{X}^{*B}$, $\theta(F_n)$ can be approximated by the bootstrap estimator

(1.6.2)
$$\hat{\theta}(F_n) \approx \frac{1}{B} \sum_{i=1}^{B} h(\mathbf{x}^{*i})$$

with the approximation becoming more accurate as $B$ increases.

If $\hat{\theta}$ is an arbitrary estimator of $\theta(F)$, the bias, the variance, or even the error distribution, of $\hat{\theta}$,

$$\mathbb{E}_F[\hat{\theta} - \theta(F)], \quad \text{var}_F(\hat{\theta}), \quad \text{and} \quad \mathrm{P}_F(\hat{\theta} - \theta(F) \leq u),$$

---

[9] This name comes from the French novel *L'Histoire Comique Contenant les Estats and Empires de la Lune* (1657) by Cyrano de Bergerac, where the hero reaches the moon by pulling on his own... bootstraps!

can then be approximated by replacing $F$ with $F_n$. For example, the bootstrap estimator of the bias will thus be

$$b_n = \mathbb{E}_{F_n}[\hat{\theta}(F_n) - \theta(F_n)]$$
$$\approx \frac{1}{B}\sum_{i=1}^{B}\{\hat{\theta}(F_{ni}^*) - \hat{\theta}(F_n)\}\,,$$

where $\hat{\theta}(F_{ni}^*)$ is constructed as in (1.6.2), and a *confidence interval* $[\hat{\theta} - \beta, \hat{\theta} - \alpha]$ on $\theta$ can be constructed by imposing the constraint

$$P_{F_n}(\alpha \le \hat{\theta}(F_n) - \theta(F_n) \le \beta) = c$$

on $(\alpha, \beta)$, where $c$ is the desired confidence level. There is a huge body of literature, not directly related to the purpose of this book, in which the authors establish different optimality properties of the bootstrap estimates in terms of bias and convergence (see Hall 1992, Efron and Tibshirani 1994, or Lehmann 1998).

Although direct computation of $\hat{\theta}$ is possible in some particular cases, most setups require simulation to approximate the distribution of $\hat{\theta} - \theta(F_n)$. Indeed, the distribution of $(X_1^*, \ldots, X_n^*)$ has a discrete support $\{x_1, \ldots, x_n\}^n$, but the cardinality of this support, $n^n$, increases much too quickly to permit an exhaustive processing of the points of the support even for samples of average size. There are some algorithms (such as those based on *Gray Codes*; see Diaconis and Holmes 1994) which may allow for exhaustive processing in larger samples.

# CHAPTER 2

# Random Variable Generation

"Have you any thought," resumed Valentin, "of a tool with which it could be done?"

"Speaking within modern probabilities, I really haven't," said the doctor.

—G.K. Chesterton, *The Innocence of Father Brown*

The methods developed in this book mostly rely on the possibility of producing (with a computer) a supposedly endless flow of random variables (usually iid) for well-known distributions. Such a simulation is, in turn, based on the production of uniform random variables. We thus provide in this chapter some basic methodology for doing this. In particular, we present a uniform random number generator and illustrate methods for using these uniform random variables to produce random variables from both standard and nonstandard distributions.

## 2.1 Basic Methods

### 2.1.1 Introduction

Methods of simulation are based on the production of random variables, often independent random variables, that are distributed according to a distribution $f$ that is not necessarily explicitly known (see, for example, Examples 1.1.1, 1.1.2, and 1.1.3). The type of random variable production is formalized below in the definition of a *pseudo-random number generator*. In this chapter, we concentrate on the generation of random variables that are uniform on the interval $[0, 1]$, because the uniform distribution $\mathcal{U}_{[0,1]}$ provides the basic probabilistic representation of randomness. In fact, in describing the structure of a space of random variables, it is always possible to represent the generic probability triple $(\Omega, \mathcal{F}, P)$ (where $\Omega$ represents the whole space, $\mathcal{F}$ represents a $\sigma$-algebra on $\Omega$, and $P$ is a probability measure) as $([0, 1], \mathcal{B}, \mathcal{U}_{[0,1]})$ (where $\mathcal{B}$ are the Borel sets on $[0, 1]$) and therefore equate the variability of $\omega \in \Omega$ with that of a uniform variable in $[0, 1]$ (see, for instance, Billingsley 1995, Section 2). The random variables $X$ are then functions from $[0, 1]$ to $\mathcal{X}$, transformed by the *generalized inverse*.

**Definition 2.1.1** For an increasing function $F$ on $\mathbb{R}$, the *generalized inverse* of $F$, $F^-$, is the function defined by

$$(2.1.1) \qquad F^-(u) = \inf\{x; \ F(x) \geq u\} \ .$$

We then have the following lemma, sometimes known as the *probability integral transform*, which gives us a representation of any random variable as a transform of a uniform random variable.

**Lemma 2.1.2** *If* $U \sim \mathcal{U}_{[0,1]}$, *then the random variable* $F^-(U)$ *has the distribution* $F$.

*Proof.* For all $u \in [0,1]$ and for all $x \in F^-([0,1])$, the generalized inverse satisfies

$$F(F^-(u)) \geq u \quad \text{and} \quad F^-(F(x)) \leq x .$$

Therefore,

$$\{(u,x): F^-(u) \leq x\} = \{(u,x): F(x) \geq u\}$$

and

$$P(F^-(U) \leq x) = P(U \leq F(x)) = F(x) . \qquad \square\square$$

Thus, formally, in order to generate a random variable $X \sim F$, it suffices to generate $U$ according to $\mathcal{U}_{[0,1]}$ and then make the transformation $x = F^-(u)$. The generation of uniform random variables is therefore a key determinant in the behavior of simulation methods for other probability distributions, since those distributions can be represented as a deterministic transformation of uniform random variables. (Although, in practice, we often use methods other than that of Lemma 2.1.2, this basic representation is usually a good way to think about things. Note also that Lemma 2.1.2 implies that a bad choice of a uniform random number generator can invalidate the resulting simulation procedure.)

Before presenting a reasonable uniform random number generator, we first digress a bit to discuss what we mean by a "bad" random number generator. The logical paradox[1] associated with the generation of "random numbers" is the problem of producing a *deterministic* sequence of values in $[0,1]$ which imitates a sequence of *iid* uniform random variables $\mathcal{U}_{[0,1]}$. (Techniques based on the physical imitation of a "random draw" using, for example, the internal clock of the machine have been ruled out. This is because, first, there is no guarantee on the *uniform* nature of numbers thus produced and, second, there is no reproducibility of such samples.) However, we really do not want to enter here into the philosophical debate on the notion of "random," and whether it is, indeed, possible to "reproduce randomness" (see, for example, Chaitin 1982, 1988).

For our purposes, there are methods that use a fully deterministic process to produce a random sequence in the following sense: Having generated $(X_1,\ldots,X_n)$, knowledge of $X_n$ [or of $(X_1,\ldots,X_n)$] imparts no discernible knowledge of the value of $X_{n+1}$. Of course, given the initial value $X_0$, the

[1] Von Neumann (1951) summarizes this problem very clearly by writing *"Any one who considers arithmetical methods of reproducing random digits is, of course, in a state of sin. As has been pointed out several times, there is no such thing as a random number—there are only methods of producing random numbers, and a strict arithmetic procedure of course is not such a method."*

sample $(X_1, \ldots, X_n)$ is always the same. Thus, the "pseudo-randomness" produced by these techniques is limited since two samples $(X_1, \ldots, X_n)$ and $(Y_1, \ldots, Y_n)$ produced by the algorithm will not be independent, nor identically distributed, nor comparable in any probabilistic sense. This limitation should not be forgotten: The validity of a random number generator is based on a single sample $X_1, \ldots, X_n$ when $n$ tends to $+\infty$ and not on replications $(X_{11}, \ldots, X_{1n}), (X_{21}, \ldots, X_{2n}), \ldots (X_{k1}, \ldots, X_{kn})$, where $n$ is fixed and $k$ tends to infinity. In fact, the distribution of these $n$-tuples depends only on the manner in which the initial values $X_{r1}$ $(1 \leq r \leq k)$ were generated.

With these limitations in mind, we can now introduce the following operational definition, which avoids the difficulties of the philosophical distinction between a deterministic algorithm and the reproduction of a random phenomenon.

**Definition 2.1.3** *A uniform pseudo-random number generator* is an algorithm which, starting from an initial value $u_0$ and a transformation $D$, produces a sequence $(u_i) = (D^i(u_0))$ of values in $[0, 1]$. For all $n$, the values $(u_1, \ldots, u_n)$ reproduce the behavior of an iid sample $(V_1, \ldots, V_n)$ of uniform random variables when compared through a usual set of tests.

This definition is clearly restricted to *testable* aspects of the random variable generation, which are connected through the deterministic transformation $u_i = D(u_{i-1})$. Thus, the validity of the algorithm consists in the verification that the sequence $U_1, \ldots, U_n$ leads to acceptance of the hypothesis

$$H_0 : U_1, \ldots, U_n \quad \text{are iid} \quad \mathcal{U}_{[0,1]}.$$

The set of tests used is generally of some consequence. There are classical tests of uniformity, such as the Kolmogorov–Smirnov test. Many generators will be deemed adequate under such examination. In addition, and perhaps more importantly, one can use methods of *time series* to determine the degree of correlation between between $U_i$ and $(U_{i-1}, \ldots, U_{i-k})$, by using an $ARMA(p, q)$ model, for instance. One can use nonparametric tests, like those of Lehmann (1975) or Randles and Wolfe (1979), applying them on arbitrary decimals of $U_i$. Marsaglia[2] has assembled a set of tests called *Die Hard*. See also Martin-Löef (1966) for a more mathematical treatment of this subject, including the notion of random sequences and a corresponding formal test of randomness.

Definition 2.1.3 is therefore *functional*: An algorithm that generates uniform numbers is acceptable if it is not rejected by a set of tests. This methodology is not without problems, however. Consider, for example, particular applications that might demand a large number of iterations, as the theory of large deviations (Bucklew 1990), or particle physics, where

---

[2] These tests are now available as a freeware on the site
http://stat.fsu.edu/~geo/diehard.html

algorithms resistant to standard tests may exhibit fatal faults. In particular, algorithms having hidden periodicities (see below) or which are not uniform for the smaller digits may be difficult to detect. Ferrenberg et al. (1992) show, for instance, that an algorithm of Wolff (1989), reputed to be "good," results in systematic biases in the processing of Ising models (see Example 5.2.5), due to long-term correlations in the generated sequence.

The notion that a deterministic system can imitate a random phenomenon may also suggest the use of *chaotic* models to create random number generators. These models, which result in complex deterministic structures (see Bergé et al. 1984, Gleick 1987 or Ruelle 1987) are based on dynamic systems of the form $X_{n+1} = D(X_n)$ which are very sensitive to the initial condition $X_0$.

**Example 2.1.4 The logistic function.** The logistic function $D_\alpha(x) = \alpha x(1 - x)$ produces, for some values of $\alpha \in [3.57, 4.00]$, chaotic configurations. In particular, the value $\alpha = 4.00$ yields a sequence $(X_n)$ in $[0, 1]$ that, theoretically, has the same behavior as a sequence of random numbers (or random variables) distributed according to the *arcsine distribution* with density $1/\pi\sqrt{x(1 - x)}$. In a similar manner, the "tent" function,"

$$D(x) = \begin{cases} 2x & \text{if } x \leq 1/2, \\ 2(1 - x) & \text{if } x > 1/2, \end{cases}$$

produces a sequence $(X_n)$ that tends to $\mathcal{U}_{[0,1]}$ (see Problem 2.1).     ‖

Although the limit distribution (also called the stationary distribution) associated with a dynamic system $X_{n+1} = D(X_n)$ is sometimes defined and known, the chaotic features of the system do not guarantee acceptable behavior (in the probabilistic sense) of the associated generator. In particular, the second generator of Example 2.1.4 can have disastrous behavior. Given the finite representation of real numbers in the computer, the sequence $(X_n)$ sometimes will converge to a fixed value. (For instance, the tent function progressively eliminates the last decimals of $X_n$.) Moreover, even when these functions give a good approximation of randomness in the unit square $[0, 1] \times [0, 1]$ (see Example 2.1.5), the hypothesis of randomness is rejected by many standard tests. Classic examples from the theory of chaotic functions do not lead to acceptable pseudo-random number generators.

**Example 2.1.5 (Continuation of Example 2.1.4)** Figure 2.1.1 illustrates the properties of the generator based on the logistic function $D_\alpha$. The histogram of the transformed variables $Y_n = 0.5 + \arcsin(X_n)/\pi$, of a sample of successive values $X_{n+1} = D_\alpha(X_n)$ fits the uniform density extremely well. Moreover, while the plots of $(Y_n, Y_{n+1})$ and $(Y_n, Y_{n+10})$ do not display characteristics of uniformity, Figure 2.1.1 shows that the sample of $(Y_n, Y_{n+100})$ satisfactorily fills the unit square. However, the 100 calls to $D_\alpha$ between two generations are excessive in terms of computing time.    ‖

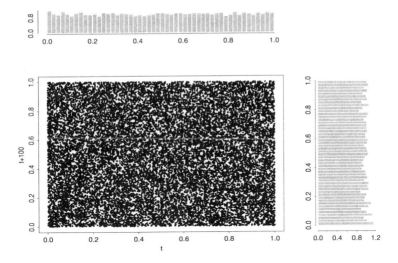

Figure 2.1.1. *Plot of the sample* $(y_n, y_{n+100})$ $(n = 1, \ldots, 9899)$ *for the sequence* $x_{n+1} = 4x_n(1 - x_n)$ *and* $y_n = F(x_n)$, *along with the (marginal) histograms of* $y_n$ *(on top) and* $y_{n+100}$ *(right margin).*

We have presented in this introduction some necessary basic notions to now describe a very good pseudo-random number generator, the algorithm *Kiss*[3] of Marsaglia and Zaman (1993). To keep our presentation simple, we only present a single generator, instead of a catalog of usual generators. For those, the books of Knuth (1981), Rubinstein (1981), Ripley (1987), and Fishman (1996) are excellent sources.

### 2.1.2 The Kiss Generator

As we have remarked above, the finite representation of real numbers in a computer can radically modify the behavior of a dynamic system. Preferred generators are those that take into account the specifics of this representation and provide a uniform sequence. It is important to note that such a sequence does not really take values in the interval $[0, 1]$ but rather on the integers $\{0, 1, \ldots, M\}$, where $M$ is the largest integer accepted by the computer. One manner of characterizing the performance of these integer generators is through the notion of *period*.

**Definition 2.1.6** The *period*, $T_0$, of a generator is the smallest integer $T$ such that $u_{i+T} = u_i$ for every $i$; that is, such that $D^T$ is equal to the identity function.

---

[3] The name is an acronym of the saying *Keep it simple, stupid!*, and not reflective of more romantic notions. After all, this is a Statistics text!

The period is a very important parameter, having direct impact on the usefulness of a random number generator. If the number of needed generations exceeds the period of a generator, there may be noncontrollable artifacts in the sequence (cyclic phenomena, false orderings, etc.). Unfortunately, a generator of the form $X_{n+1} = f(X_n)$ has a period no greater than $M + 1$, for obvious reasons. In order to overcome this bound, a generator must utilize several sequences $X_n^i$ simultaneously (which is a characteristic of *Kiss*) or must involve $X_{n-1}, X_{n-2}, \ldots$ in addition to $X_n$, or must use other methods such as *start-up tables*, that is, using an auxiliary table of random digits to restart the generator.

*Kiss* simultaneously uses two generation techniques, namely *congruential* generation and *shift register* generation.

**Definition 2.1.7** A *congruential generator* on $\{0, 1, \ldots, M\}$ is defined by the function

$$D(x) = (ax + b) \bmod (M + 1).$$

The period and, more generally, the performance of congruential generators depend heavily on the choice of $(a, b)$ (see Ripley 1987). When transforming the above generator into a generator on $[0, 1]$, with $\tilde{D}(x) = (ax + b)/(M + 1) \bmod 1$, the graph of $D$ should range throughout $[0, 1]^2$, and a choice of the constant $a \notin \mathbb{Q}$ would yield a "recovery " of $[0, 1]^2$; that is, an infinite sequence of points should fill the space.

Although ideal, the choice of an irrational $a$ is impossible (since $a$ needs be specified with a finite number of digits). With $a$ rational, a congruential generator will produce pairs $(x_n, D(x_n))$ that lie on parallel lines. Figure 2.1.2 illustrates this phenomenon for $a = 69069$, representing the sequence $(3k \ 10^{-4}, D(3k))$ for $k = 1, 2, \ldots, 333$. It is thus important to select $a$ in such a way as to maximize the number of parallel segments in $[0, 1]^2$ (see Problem 2.2).

Most commercial generators use congruential methods, with perhaps the most disastrous choice of $(a, b)$ being that of the old (and notorious) procedure *RANDU* (see Ripley 1987 and Problem 2.5). Even when the choice of $(a, b)$ assures the acceptance of the generator by standard tests, non-uniform behavior will be observed in the last digits of the real numbers produced by this method, due to round-up errors.

The second technique employed by *Kiss* is based on the (theoretical) independence between the $k$ binary components of $X_n \sim \mathcal{U}_{\{0,1,\ldots,M\}}$ (where $M = 2^k - 1$) and is called a *shift register* generator.

**Definition 2.1.8** For a given $k \times k$ matrix $T$, whose entries are either 0 or 1, the associated *shift register generator* is given by the transformation

$$x_{n+1} = Tx_n,$$

where $x_n$ is represented as a vector of binary coordinates $e_{ni}$, that is to say,

$$x_n = \sum_{i=0}^{k-1} e_{ni} 2^i.$$

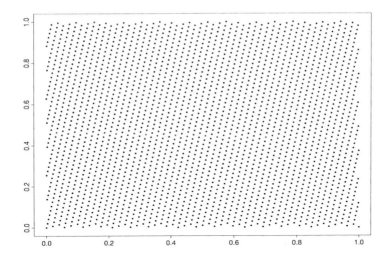

Figure 2.1.2. *Representation of the line* $y = 69069x \bmod 1$ *by uniform sampling with sampling step* $3 \ 10^{-4}$.

with $e_{ni}$ equal to 0 or 1.

This second class of generators is motivated by both the internal (computer-dependent) representation of numbers as sequences of *bits* and the speed of manipulation of these elementary algebraic operations. Since the computation of $Tx_n$ is done modulo 2, each addition is then the equivalent of a logical OR. Moreover, as the matrix $T$ only contains 0 and 1 entries, multiplication by $T$ amounts to shifting the content of coordinates, which gives the technique its name.

For instance, if the $i$th line of $T$ contains a 1 in the $i$th and $j$th positions uniquely, the $i$th coordinate of $x_{n+1}$, will be obtained by

$$e_{(n+1)i} = (e_{ni} + e_{nj}) \mod 2$$
$$= e_{ni} \vee e_{nj} - e_{ni} \wedge e_{nj} \, ,$$

where $a \wedge b = \min(a, b)$ and $a \vee b = \max(a, b)$. This is a comparison of the $i$th coordinate of $x_n$ and the coordinate corresponding to a *shift* of $(j - i)$. There also exist sufficient conditions on $T$ for the associated generator to have period $2^k$ (see Ripley 1987).

The generators used by *Kiss* are based on the matrices

$$T_L = \begin{pmatrix} 1 & 1 & & & \\ & \ddots & & 0 & \\ & & \ddots & 1 \\ 0 & & & 1 \end{pmatrix} \quad \text{and} \quad T_R = \begin{pmatrix} 1 & & & \\ 1 & \ddots & 0 & \\ & & \ddots & \\ 0 & & 1 & 1 \end{pmatrix},$$

whose entries are 1 on the main diagonal and on the first upper diagonal and first lower diagonal, respectively, the other elements being 0. They are related to the right and left shift matrices,

$$R(e_1, \ldots, e_k)^t = (0, e_1, \ldots, e_{k-1})^t,$$
$$L(e_1, \ldots, e_k)^t = (e_2, e_3, \ldots, e_k, 0)^t,$$

since $T_R = (I + R)$ and $T_L = (I + L)$.

To generate a sequence of integers $X_1, X_2, \ldots$, the *Kiss* algorithm generates three sequences of integers. First, the algorithm uses a congruential generator to obtain

$$I_{n+1} = (69069 \times I_n + 23606797) \pmod{2^{32}},$$

and then two shift register generators of the form,

$$J_{n+1} = (I + L^{15})(I + R^{17}) J_n \pmod{2^{32}},$$
$$K_{n+1} = (I + L^{13})(I + R^{18}) K_n \pmod{2^{31}}.$$

These are then combined to produce

$$X_{n+1} = (I_{n+1} + J_{n+1} + K_{n+1}) \mod 2^{32}.$$

Formally, this algorithm is not of the type specified in Definition 2.1.3, since it uses three parallel chains of integers. However, this feature yields advantages over algorithms based on a single dynamic system $X_{n+1} = f(X_n)$ since the period of *Kiss* is of order $2^{95}$, which is almost $(2^{32})^3$. In fact, the (usual) congruential generator $I_n$ has a maximal period of $2^{32}$, the generator of $K_n$ has a period of $2^{31} - 1$ and that of $J_n$ a period of $2^{32} - 2^{21} - 2^{11} + 1$ for almost all initial values $J_0$ (see Marsaglia and Zaman 1993 for more details). The *Kiss* generator has been successfully tested on the different criteria of *Die Hard*, including tests on random subsets of *bits*. Figure 2.1.3 presents plots of $(X_n, X_{n+1})$, $(X_n, X_{n+2})$, $(X_n, X_{n+5})$ and $(X_n, X_{n+10})$ for $n = 1, \ldots, 5000$, where the sequence $(X_n)$ has been generated by *Kiss*, without exhibiting any nonuniform feature on the square $[0, 1]^2$. A version of this algorithm in the programming language C is given below.

### Algorithm A.3 The Kiss Algorithm.

```
long int kiss (i,j,k)
unsigned long *i,*j,*k
{
*j = *j ∧ (*j << 17);                              [A.3]
*k = (*k ∧ (*k << 18)) & 0X7FFFFFFF ;
return  ((*i = 69069 * (*i) + 23606797) +
(*j ∧ = (*j >> 15)) + (*k ∧ = (*k >> 13)) ;
}
```

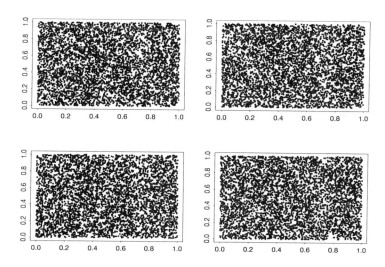

Figure 2.1.3. *Plots of pairs* $(X_t, X_{t+1})$, $(X_t, X_{t+2})$, $(X_t, X_{t+5})$ *and* $(X_t, X_{t+10})$ *for a sample of* 5000 *generations from Kiss.*

(see Marsaglia and Zaman 1993 for a Fortran version of *Kiss*). Note that some care must be exercised in the use of this program as a generator on $[0, 1]$, since it implies dividing by the largest integer available on the computer and may sometimes result in uniform generation on $[-1, 1]$.

### 2.1.3 Beyond Uniform Distributions

As mentioned above, from a theoretical point of view an operational version of any probability space $(\Omega, \mathcal{A}, P)$ can be created from the uniform distribution $\mathcal{U}_{[0,1]}$ and Lemma 2.1.2. Thus, the generation of any sequence of random variables can be formally implemented through the uniform generator *Kiss*. In practice, however, this approach only applies when the cumulative distribution functions are "explicitly" available, in the sense that there exists an algorithm allowing the computation of $F^-(u)$ in acceptable time. In particular, for distributions with explicit forms of $F^-$ (for instance, the exponential, double-exponential, or Weibull distributions; see Problem 2.10 for other examples), Lemma 2.1.2 can be implemented. However, this situation only covers a small number of cases, and in this section we present alternative techniques for generating nonuniform random variables. Some of these methods are rather case-specific and are difficult to generalize as they rely on properties of the distribution under consideration and its relation with other probability distributions. Other methods, for example the accept–reject method (see §2.3), are more general and do not use any strong analytic property of the densities.

**Example 2.1.9 Normal random variables.** Although $\Phi$, the cumulative distribution function of the normal distribution cannot be expressed explicitly, since

$$\Phi(x) = \frac{1}{\sqrt{2\pi}} \int_{-\infty}^{x} \exp\{-z^2/2\}dz,$$

there exist approximations of $\Phi$ and of $\Phi^{-1}$ up to an arbitrary precision. For instance, Abramowitz and Stegun (1964) give the approximation

$$\Phi(x) \simeq 1 - \varphi(x)\left[b_1 t + b_2 t^2 + b_3 t^3 + b_4 t^4 + b_5 t^5\right] \qquad (x > 0),$$

where $\varphi$ denotes the normal density, $t = (1 + px)^{-1}$ and

$$p = 0.2316419, \quad b_1 = 0.31938, \quad b_2 = -0.35656,$$
$$b_3 = 1.78148, \quad b_4 = -1.82125, \quad b_5 = 1.33027.$$

Similarly, we also have the approximation

$$\Phi^{-1}(\alpha) \simeq t - \frac{a_0 + a_1 t}{1 + b_1 t + b_2 t^2},$$

where $t^2 = \log(\alpha^{-2})$ and

$$a_0 = 2.30753, \quad a_1 = 0.27061, \quad b_1 = 0.99229, \quad b_2 = 0.04481.$$

These two approximations are exact up to an error of order $10^{-8}$, the error being absolute. If no other fast simulation method was available, this approximation could be used in settings which do not require much precision in the tails of $\mathcal{N}(0,1)$. (However, as shown in Example 2.2.2, there exists an exact and much faster algorithm.)      ‖

Devroye (1985) presents a more comprehensive (one could say almost exhaustive!) treatment of the methods of random variable generation than the one presented here, in particular looking at refinements of existing algorithms in order to achieve uniformly optimal performances. (We strongly urge the reader to consult this book for a better insight on the implications of this goal in terms of probabilistic and algorithmic complexity.)

Some refinements of the simulation techniques introduced in this chapter will be explored in Chapter 5, where we consider ways to accelerate Monte Carlo methods. At this point, we note that the concepts of "optimal" and "efficient" algorithms are particularly difficult to formalize. We can naively compare two algorithms, $[B_1]$ and $[B_2]$ say, in terms of time of computation, for instance through the average generation time of one observation. However, such a comparison depends on many subjective factors like the quality of the programming, the particular programming language used to implement the method, and the particular machine on which the program runs. More importantly, it does not take into account the conception and programming (and debugging) times, nor does it incorporate the specific use of the sample produced, partly because a quantification of these factors is generally impossible. For instance, some algorithms have a decreasing efficiency when the sample size increases. The reduction of the efficiency of a given algorithm to its average computation time is therefore misleading and we only use this type of measurement in settings where $[B_1]$ and $[B_2]$ are already of the same complexity. Devroye (1985) also notes that the simplicity of algorithms should be accounted for in their evaluation, since complex algorithms facilitate programming errors and, therefore, may lead to important time losses.[4]

A last remark to bring this introduction to its end is that simulation of the standard distributions presented here is accomplished quite efficiently by many statistical programming packages (for instance, Gauss, Mathematica, Matlab, Splus). When the generators from these general-purpose packages are easily accessible (in terms of programming), it is probably preferable to use such a generator rather than to write one's own, as the time required to implement and test the algorithm may offset any gain in computational efficiency! However, if a generation technique will get extensive use or if there are particular features of a problem that can be exploited, the creation of a personal library of random variable generators can accelerate analyses and even improve results, especially if the setting involves "extreme" cases (sample size, parameter values, correlation structure, rare events) for which the usual generators are poorly adapted. The investment represented by the creation and validation of such a personal library must therefore be weighed against the potential benefits.

## 2.2 Transformation Methods

When a distribution $f$ is linked in a relatively simple way to another distribution that is easy to simulate, this relationship can often be exploited to construct an algorithm to simulate variables from $f$. We begin with an illustration of some distributions that are simple to generate.

**Example 2.2.1 Exponential variable generation.** If $U \sim \mathcal{U}_{[0,1]}$, the random variable $X = -\log U$ has distribution

$$P(X \leq x) = P(-\log U \leq x) = P(U \leq 1 - e^{-x}) = 1 - e^{-x},$$

the exponential distribution. Similarly, if $U \sim \mathcal{U}_{[0,1]}$, then $X = -\log U/\lambda$ is distributed from the exponential $\mathcal{E}xp(\lambda)$ distribution.

As an illustration of the random variables that can be generated starting from an exponential distribution, consider the following cases: If the $U_i$'s are iid $\mathcal{U}_{[0,1]}$ random variables, then

$$Y = -2 \sum_{j=1}^{\nu} \log(U_j) \sim \chi_{2\nu}^2 \ ,$$

(2.2.1)
$$Y = -\beta \sum_{j=1}^{a} \log(U_j) \sim \mathcal{G}a(a, \beta) \ ,$$

---

[4] In fact, in numerous settings, the time required by a simulation is overwhelmingly dedicated to programming. This is, at least, the case for the authors themselves!!!

$$Y = \frac{\sum_{j=1}^{a} \log(U_j)}{\sum_{j=1}^{a+b} \log(U_j)} \sim \mathcal{B}e(a, b).$$

Other derivations are possible (see Problem 2.11).      ‖

This transformation is quite simple to use and, hence, will often be a favorite. However, there are limits to the usefulness of this transformation, both in scope of variables that can be generated and in efficiency of generation. For example, as we will see, there are more efficient algorithms for gamma and beta random variables. Also, we cannot use it to generate gamma random variables with a non-integer shape parameter. For instance, we cannot get a $\chi_1^2$ variable, which would, in turn, get us a $\mathcal{N}(0,1)$ variable. For that, we look at the following example of the *Box–Muller* algorithm (1958) for the generation of $\mathcal{N}(0,1)$ variables.

**Example 2.2.2 Normal variable generation.** If $r$ and $\theta$ are the polar coordinates of $(X_1, X_2)$, then $r^2 = X_1^2 + X_2^2 \sim \chi_2^2$, which is also the $\mathcal{E}xp(1/2)$ distribution, whereas $\theta$ follows a uniform distribution on $[0, 2\pi]$, since the distribution of $(X_1, X_2)$ is rotation invariant (see Problem 2.12). If $U_1$ and $U_2$ are iid $\mathcal{U}_{[0,1]}$, the variables $X_1$ and $X_2$ defined by

$$X_1 = \sqrt{-2\log(U_1)} \, \cos(2\pi U_2) \,, \qquad X_2 = \sqrt{-2\log(U_1)} \, \sin(2\pi U_2) \,,$$

are then iid $\mathcal{N}(0,1)$. The *Box–Muller algorithm* is therefore

**Algorithm A.4 –Box-Muller–**

1 Generate $U_1, U_2$ iid $\mathcal{U}_{[0,1]}$ ;

2 Define                                                              [A.4]

$$\begin{cases} x_1 = \sqrt{-2\log(u_1)} \cos(2\pi u_2) \,, \\ x_2 = \sqrt{-2\log(u_1)} \sin(2\pi u_2) \,; \end{cases}$$

3 Take $x_1$ and $x_2$ as two independent draws from $\mathcal{N}(0,1)$.

In comparison with algorithms based on the Central Limit Theorem, this algorithm is exact, producing two normal random variables from two uniform random variables, the only drawback (in speed) being the necessity of calculating functions such as log, cos, and sin. If this is a concern, Devroye (1985) gives faster alternatives that avoid the use of these functions (see also Problems 2.13 and 2.14).      ‖

**Example 2.2.3 Poisson generation.** The Poisson distribution is connected to the exponential distribution through the Poisson process; that is, if $N \sim \mathcal{P}(\lambda)$ and $X_i \sim \mathcal{E}xp(\lambda)$, $i \in \mathbb{N}^*$, then

$$P_\lambda(N = k) = P_\lambda(X_1 + \cdots + X_k \leq 1 < X_1 + \cdots + X_{k+1}) \,.$$

Thus, the Poisson distribution can be simulated by generating exponential random variables until their sum exceeds 1. This method is simple, but is really practical only for smaller values of $\lambda$. On average, the number of exponential variables required is $\lambda$, and this could be prohibitive for large values of $\lambda$. In these settings, Devroye (1981) proposed a method whose computation time is uniformly bounded (in $\lambda$) and we will see another approach, suitable for large $\lambda$'s, in Example 2.3.9. Note also that a generator of Poisson random variables can produce negative binomial random variables since, when $Y \sim \mathcal{G}a(n, (1-p)/p)$ and $X|y \sim \mathcal{P}(y)$, $X \sim \mathcal{N}eg(n, p)$. (See Problem 2.19.)                                                                    ∥

**Example 2.2.4 Beta generation.** Consider $U_1, \ldots, U_n$ an iid sample from $\mathcal{U}_{[0,1]}$. If $U_{(1)} \leq \cdots \leq U_{(n)}$ denotes the ordered sample, that is, the *order statistics* of the original sample, $U_{(i)}$ is distributed as $\mathcal{B}e(i, n-i+1)$ and the vector of the differences $(U_{(i_1)}, U_{(i_2)} - U_{(i_1)}, \ldots, U_{(i_k)} - U_{(i_{k-1})}, 1 - U_{(i_k)})$ has a Dirichlet distribution $\mathcal{D}(i_1, i_2 - i_1, \ldots, n - i_k + 1)$ (see Problem 2.22). However, even though these probabilistic properties allow the direct generation of beta and Dirichlet random variables from uniform random variables, they do not yield efficient algorithms. The calculation of the order statistics can, indeed, be quite time-consuming since it requires sorting the original sample. Moreover, it only applies for integer parameters in the beta distribution.

The following result allows for an alternative generation of beta random variables from uniform random variables: Jöhnk's Theorem (see Jöhnk 1964 or Devroye 1985) states that if $U$ and $V$ are iid $\mathcal{U}_{[0,1]}$, the distribution of

$$\frac{U^{1/\alpha}}{U^{1/\alpha} + V^{1/\beta}},$$

conditional on $U^{1/\alpha} + V^{1/\beta} \leq 1$, is the $\mathcal{B}e(\alpha, \beta)$ distribution. However, given the constraint on $U^{1/\alpha} + V^{1/\beta}$, this result does not provide a good algorithm to generate $\mathcal{B}e(\alpha, \beta)$ random variables for large values of $\alpha$ and $\beta$. ∥

**Example 2.2.5 Gamma generation.** Given a generator of beta random variables, we can derive a generator of gamma random variables $\mathcal{G}a(\alpha, 1)$ ($\alpha < 1$) the following way: If $Y \sim \mathcal{B}e(\alpha, 1 - \alpha)$ and $Z \sim \mathcal{E}xp(1)$, then $X = YZ \sim \mathcal{G}a(\alpha, 1)$. Indeed, by making the transformation $x = yz, w = z$ and integrating the joint density, we find

$$f(x) = \frac{\Gamma(1)}{\Gamma(\alpha)\Gamma(1-\alpha)} \int_x^\infty \left(\frac{x}{w}\right)^{\alpha-1} \left(1 - \frac{x}{w}\right)^{-\alpha} w^{-1}e^{-w}dw$$

$$(2.2.2) \qquad = \frac{1}{\Gamma(\alpha)} x^{\alpha-1}e^{-x}.$$

Alternatively, if we can start with a gamma random variable, a more efficient generator for $\mathcal{G}a(\alpha, 1)$ ($\alpha < 1$) can be constructed: If $Y \sim \mathcal{G}a(\alpha+1, 1)$ and $U \sim \mathcal{U}_{[0,1]}$, independent, then $X = YU^{1/\alpha}$ is distributed according to

$\mathcal{G}a(\alpha, 1)$, since

$$(2.2.3) \qquad f(x) \propto \int_x^\infty w^\alpha e^{-w} \left(\frac{x}{w}\right)^{\alpha-1} w^{-1} dw = x^{\alpha-1} e^{-x} \ .$$

(See Stuart 1962 or Problem 2.24).                                            ‖

The representation of a probability density as in (2.2.2) is a particular case of a *mixture distribution*. Not only does such a representation induce relatively efficient simulation methods, but it is also related to methods in Chapter 7. The principle of a mixture representation is to represent a density $f$ as the marginal of another distribution, in the form

$$(2.2.4) \qquad f(x) = \int_{\mathcal{Y}} g(x, y)\, dy \qquad \text{or} \qquad f(x) = \sum_{i \in \mathcal{Y}} p_i\, f_i(x) \ ,$$

depending on whether $\mathcal{Y}$ is continuous or discrete. For instance, if the joint distribution $g(x, y)$ is simple to simulate, then the variable $X$ can be obtained as a component of the generated $(X, Y)$. Alternatively, if the component distributions $f_i(x)$ can be easily generated, $X$ can be obtained by first choosing $f_i$ with probability $p_i$ and then generating an observation from $f_i$.

**Example 2.2.6 Student's $t$ generation.** A useful form of (2.2.4) is

$$(2.2.5) \qquad f(x) = \int_{\mathcal{Y}} g(x, y)\, dy = \int_{\mathcal{Y}} h_1(x|y) h_2(y)\, dy \ ,$$

where $h_1$ and $h_2$ are the conditional and marginal densities of $X|Y = y$ and $Y$, respectively. For example, we can write Student's $t$ density with $\nu$ degrees of freedom in this form, where

$$X|y \sim \mathcal{N}(0, \nu/y) \quad \text{and} \quad Y \sim \chi_\nu^2.$$

Such a representation is also useful for discrete distributions. In Example 2.2.3, we noted an alternate representation for the negative binomial distribution. If $X$ is negative binomial, $X \sim \mathcal{N}eg(n, p)$, then $P(X = x)$ can be written as (2.2.5) with

$$X|y \sim \mathcal{P}(y) \quad \text{and} \quad Y \sim \mathcal{G}(n, \beta),$$

where $\beta = (1 - p)/p$. Note that the discreteness of the negative binomial distribution does not result in a discrete mixture representation of the probability. The mixture is continuous, as the distribution of $Y$ is itself continuous.                                            ‖

**Example 2.2.7 Noncentral chi squared generation.** The noncentral chi squared distribution, $\chi_p^2(\lambda)$, also allows for a mixture representation, since it can be written as a sum of central chi squared densities. In fact, it is of the form (2.2.5) with $h_1$ the density of a $\chi_{p+2K}^2$ distribution and $h_2$ the density of $\mathcal{P}(\lambda/2)$. However, this representation is not as efficient as the algorithm obtained by generating $Z \sim \chi^2_{p-1}$ and $Y \sim \mathcal{N}(\sqrt{\lambda}, 1)$, and using the fact that $Z + Y^2 \sim \chi^2_p(\lambda)$. Note that the noncentral chi squared distribution does not have an explicit form for its density function. It is either represented as an infinite mixture (see (3.6.9)) or by using modified Bessel functions (see Problem 1.13).  ||

In addition to the above two examples, other distributions can be represented as mixtures (see, for instance, Gleser 1989). In many cases this representation can be exploited to produce algorithms for random variable generation (see Problems 2.28–2.30 and Note 2.5.2).

## 2.3 Accept–Reject Methods

### 2.3.1 General Principle

There are many distributions from which it is difficult, or even impossible, to directly simulate. Moreover, in some cases, we are not even able to represent the distribution in a usable form, such as a transformation or a mixture. In such settings, it is impossible to exploit direct probabilistic properties to derive a simulation method. We thus turn to another class of methods that only requires us to know the functional form of the density $f$ of interest up to a multiplicative constant; no deep analytical study of $f$ is necessary. The key to this method is to use a simpler (simulationwise) density $g$ from which the simulation is actually done. For a given density $g$—called the *instrumental density*— there are thus many densities $f$— called the *target densities*—which can be simulated this way.

We refer to the following algorithm as the *Accept–Reject method*. Given a density of interest $f$, the first requirement is the determination of a density $g$ and a constant $M$ such that

$$(2.3.1) \qquad\qquad f(x) \leq Mg(x)$$

on the support of $f$. The algorithm is based on the following result:

**Lemma 2.3.1** *The algorithm*

**Algorithm A.5 –Accept–Reject Method–**

1. Generate $X \sim g$, $U \sim \mathcal{U}_{[0,1]}$ ;
2. Accept $Y = X$ if $U \leq f(X)/Mg(X)$ ;            [A.5]
3. Return to 1. otherwise.

*produces a variable $Y$ distributed according to $f$.*

*Proof.* The distribution function of $Y$ is given by

$$P(Y \leq y) = P\left(X \leq y | U \leq \frac{f(X)}{Mg(X)}\right) = \frac{P\left(X \leq y, U \leq \frac{f(X)}{Mg(X)}\right)}{P\left(U \leq \frac{f(X)}{Mg(X)}\right)}.$$

Now, writing out the probability integral yields

$$P(Y \leq y) = \frac{\int_{-\infty}^{y} \int_{0}^{f(x)/Mg(x)} du \; g(x) dx}{\int_{-\infty}^{\infty} \int_{0}^{f(x)/Mg(x)} du \; g(x) dx} = \frac{\frac{1}{M} \int_{-\infty}^{y} f(x) dx}{\frac{1}{M} \int_{-\infty}^{\infty} f(x) dx},$$

and this last expression is equal to $\int_{-\infty}^{y} f(x) dx$, completing the proof.    □□

This lemma has two consequences. First, it provides a generic method to simulate from any density $f$ that is known *up to a multiplicative factor*; that is, the normalizing constant of $f$ need not be known, since the method only requires input of the ratio $f/M$, which does not depend on the normalizing constant. This property is particularly important in Bayesian calculations. There, a quantity of interest is the posterior distribution, defined according to Bayes Theorem by

$$\pi(\theta|x) \propto \pi(\theta) \; f(x|\theta) .$$

Thus, the posterior density $\pi(\theta|x)$ is easily specified up to a normalizing constant and, to use the Accept–Reject algorithm, this constant need not be calculated. (Obviously, there remains the task to find a density $g$ satisfying the constraint (2.3.1).) Note also that the bound (2.3.1) need not to be tight, in the sense that the algorithm [A.5] remains valid when $M$ is replaced with $2M$. (See Problem 2.34.)

A second consequence of the lemma is that the probability of acceptance in the algorithm [A.5] is exactly $1/M$, when evaluated for the properly normalized densities, and the expected number of trials until a variable is accepted is $M$ (see Problem 2.36). Thus, a comparison between different Accept–Reject algorithms based on different instrumental densities $g_1, g_2, \ldots$ can be undertaken through the comparison of the respective bounds $M_1, M_2, \ldots$ (as long as the corresponding densities $g_1, g_2, \ldots$ are correctly normalized). In particular, a first method of optimizing the choice of $g$ in $g_1, g_2, \ldots$ is to find the smallest bound $M_i$. However, this first and rudimentary comparison technique has some limitations, which we will see later in this chapter.

In cases where $f$ and $g$ are normalized so they are both probability densities, the constant $M$ is necessarily larger that 1. Therefore, the size of $M$, and thus the efficiency of [A.5], becomes a function of how closely $g$ can imitate $f$, especially in the tails of the distribution. Note that for $f/g$ to remain bounded, it is necessary for $g$ to have tails thicker than those of $f$. It is therefore impossible for instance to use [A.5] to simulate a Cauchy distribution $f$ using a normal distribution $g$; however, the reverse works quite well. (See Example 2.3.2.) Interestingly enough, the opposite case when $g/f$ is bounded can also be processed by a tailored Markov chain Monte Carlo algorithm derived from Doukhan *et al.* (1994) (see Problems 6.2 and 6.3).

**Example 2.3.2 Normal generation from a Cauchy.** When $f(x) = \exp(-x^2/2)$ and $g(x) = 1/(1 + x^2)$, densities of the normal and Cauchy distributions, respectively (ignoring the normalization constants), the ratio

$$\frac{f(x)}{g(x)} = (1 + x^2)\, e^{-x^2/2}$$

is bounded by $2/\sqrt{e}$, which is attained at $x = \pm 1$. When taking the normalization constants into account, the probability of acceptance is then $\sqrt{e/2\pi} = 0.66$, and this implies that, on the average, one out of every three simulated Cauchy variables is rejected. (The mean number of trials to success is $1/.66 = 1.52$; see Problem 2.36.) Replacing $g$ by a Cauchy density with scale parameter $\sigma$,

$$g_\sigma(x) = 1/\{\pi\sigma(1 + x^2/\sigma^2)\}\,,$$

the bound on $f/g_\sigma$ is $2\pi^{-1}\sigma^{-1}\exp\{\sigma^2/2 - 1\}$ for $\sigma^2 < 2$. This bound is minimized by $\sigma^2 = 1$, which shows that $\mathcal{C}(0,1)$ is the best choice among the Cauchy distributions for simulating a $\mathcal{N}(0,1)$ distribution. ‖

The above example shows that an optimization of the Accept–Reject algorithm is possible in the following limited setting: We choose the instrumental density $g$ in a parametric family, and then determine the value of the parameter which minimizes the bound $M$. A similar comparison between two parametric families is much more delicate since it is then necessary to take into account the computation time of one generation from $g$ in [A.5]. In fact, pushing the reasoning to the limit, if $g = f$ and if we simulate $X \sim f$ by numerical inversion of the distribution function, we formally achieve the minimal bound $M = 1$, but this does not guarantee that we have an efficient algorithm, as can be seen in the case of the normal distribution.

**Example 2.3.3 (Continuation of Example 2.1.9)** Consider generating a $\mathcal{N}(0,1)$ by [A.5] using a double-exponential distribution $\mathcal{L}(\alpha)$, with density $g(x|\alpha) = (\alpha/2)\exp(-\alpha|x|)$. It is then straightforward to show that

$$\frac{f(x)}{g(x|\alpha)} \leq \sqrt{\frac{2}{\pi}}\alpha^{-1}e^{-\alpha^2/2}$$

and that the minimum of this bound (in $\alpha$) is attained for $\alpha = 1$. The probability of acceptance is then $\sqrt{\pi/2e} = .76$, which shows that to produce one normal random variable, this Accept-Reject algorithm requires on the average $1/.76 \approx 1.3$ uniform variables, to be compared with the fixed single uniform required by the Box–Muller algorithm. ‖

A real advantage of the Accept–Reject algorithm is illustrated in the following example.

**Example 2.3.4 (Continuation of Example 2.2.5)** We saw in Example 2.2.1 that if $\alpha \in \mathbb{N}$, the gamma distribution $\mathcal{G}a(\alpha, \beta)$ can be represented as the sum of $\alpha$ exponential random variables $\epsilon_i \sim \mathcal{E}xp(\beta)$, which are very easy to simulate, since $\epsilon_i = -\log(U_i)/\beta$, with $U_i \sim \mathcal{U}([0,1])$. In more general cases (for example when $\alpha \notin \mathbb{N}$), this representation does not hold. A possible approach is then to use the Accept–Reject algorithm with instrumental distribution $\mathcal{G}a(a, b)$, with $a = [\alpha]$ ($\alpha \geq 0$). (Without loss of generality, suppose $\beta = 1$.) The ratio $f/g_b$ is $b^{-a} x^{\alpha-a} \exp\{-(1-b)x\}$, up to a normalizing constant, yielding the bound

$$M = b^{-a} \left( \frac{\alpha - a}{(1-b)e} \right)^{\alpha - a}$$

for $b < 1$. Since the maximum of $b^{-a}(1-b)^{\alpha-a}$ is attained at $b = a/\alpha$, the optimal choice of $b$ for simulating $\mathcal{G}a(\alpha, 1)$ is $b = a/\alpha$, which gives the same mean for $\mathcal{G}a(\alpha, 1)$ and $\mathcal{G}a(a, b)$. (See Problem 2.40.)          ‖

It may also happen that the complexity of the optimization is very expensive in terms of analysis or of computing time. In the first case, the construction of the optimal algorithm should still be undertaken when the algorithm is to be subjected to intensive use. In the second case, it is most often preferable to explore the use of another family of instrumental distributions $g$.

**Example 2.3.5 Truncated normal distributions.** *Truncated normal distributions* appear in many contexts, such as in the discussion after Example 1.2.2. When constraints $x \geq \underline{\mu}$ produce densities proportional to

$$e^{-(x-\mu)^2/2\sigma^2} \; \mathbb{I}_{x \geq \underline{\mu}}$$

for a bound $\underline{\mu}$ large compared with $\mu$, there are alternatives which are far superior to the naïve method in which a $\mathcal{N}(\mu, \sigma^2)$ distribution is simulated until the generated value is larger than $\underline{\mu}$. (This approach requires an average number of $1/\Phi((\mu - \underline{\mu})/\sigma)$ simulations from $\mathcal{N}(\mu, \sigma^2)$ for one acceptance.) Consider, without loss of generality, the case $\mu = 0$ and $\sigma = 1$. A potential instrumental distribution is the translated exponential distribution, $\mathcal{E}xp(\alpha, \underline{\mu})$, with density

$$g_\alpha(z) = \alpha e^{-\alpha(z-\underline{\mu})} \; \mathbb{I}_{z \geq \underline{\mu}} \; .$$

The ratio $f/g_\alpha(z) = e^{-\alpha(z-\underline{\mu})} \, e^{-z^2/2}$ is then bounded by $\exp(\alpha^2/2 - \alpha\underline{\mu})$ if $\alpha > \underline{\mu}$ and by $\exp(-\underline{\mu}^2/2)$ otherwise. The corresponding (upper) bound is

$$\begin{cases} 1/\alpha \; \exp(\alpha^2/2 - \alpha\underline{\mu}) & \text{if } \alpha > \underline{\mu}, \\ 1/\alpha \; \exp(-\underline{\mu}^2/2) & \text{otherwise.} \end{cases}$$

The first expression is minimized by

$$(2.3.2) \qquad \alpha^* = \underline{\mu} + \frac{\sqrt{\underline{\mu}^2 + 4}}{2} \; ,$$

whereas $\tilde{\alpha} = \underline{\mu}$ minimizes the second bound. The optimal choice of $\alpha$ is therefore (2.3.2), which requires the computation of the square root of $\underline{\mu}^2 + 4$. Robert (1995a) proposes a similar algorithm for the case where the normal distribution is restricted to the interval $[\underline{\mu}, \overline{\mu}]$. For some values of $[\underline{\mu}, \overline{\mu}]$, the optimal algorithm is associated with a value of $\alpha$, which is a solution to an implicit equation. (See also Geweke 1991 for a similar resolution of this simulation problem and Marsaglia 1964 for an earlier solution.)                                                    ‖

One criticism of the Accept–Reject algorithm is that it generates "useless" simulations when rejecting. We will see in Chapter 3 how the method of importance sampling (§3.3) can be used to bypass this problem and also how both methods can be compared.

### 2.3.2 Envelope Accept–Reject Methods

In numerous settings, the distribution associated with the density $f$ is difficult to simulate because of the complexity of the function $f$ itself, which may require substantial computing time at each evaluation. In the setup of Example 1.2.6 for instance, if a Bayesian approach is taken with $\theta$ distributed (a posteriori) as

$$(2.3.3) \qquad \prod_{i=1}^{n} \left[ 1 + \frac{(x_i - \theta)^2}{p\sigma^2} \right]^{-\frac{p+1}{2}} ,$$

where $\sigma$ is known, each single evaluation of $\pi(\theta|x)$ involves the computation of $n$ terms in the product. It turns out that an acceleration of the simulation of densities such as (2.3.3) can be accomplished by an algorithm that is "one step beyond" the Accept–Reject algorithm. This algorithm is an *envelope* algorithm and relies on the evaluation of a simpler function $g_l$ which bounds the target density $f$ from below. The algorithm is based on the following elementary lemma:

**Lemma 2.3.6** *If there exist a density $g_m$, a function $g_l$ and a constant $M$ such that*

$$g_l(x) \leq f(x) \leq M g_m(x) ,$$

*then the algorithm*

**Algorithm A.6 –Envelope Accept–Reject–**

1. Generate $X \sim g_m(x)$, $U \sim \mathcal{U}_{[0,1]}$;
2. Accept $X$ if $U \leq g_l(X)/M g_m(X)$;                                    [A.6]
3. otherwise, accept $X$ if $U \leq f(X)/M g_m(X)$

*produces random variables that are distributed according to $f$.*

By the construction of a lower envelope on $f$, based on the function $g_l$, the number of evaluations of $f$ is potentially decreased by a factor

$$\frac{1}{M} \int g_l(x) \, dx,$$

which is the probability that $f$ is not evaluated. This method is called the *squeeze principle* by Marsaglia (1977) and the ARS algorithm presented in

§2.3.3 is based upon it. A possible way of deriving the bounds $g_l$ and $Mg_m$ is to use a Taylor expansion of $f(x)$.

**Example 2.3.7 (Continuation of Example 2.3.2)** It follows from the Taylor series expansion of $\exp(-x^2/2)$ that $\exp(-x^2/2) \geq 1 - (x^2/2)$, and hence

$$\left(1 - \frac{x^2}{2}\right) \leq f(x),$$

which can be interpreted as a lower bound for the simulation of $\mathcal{N}(0,1)$. This bound is obviously useless when $|X| < \sqrt{2}$, an event which occurs with probability 0.61 for $X \sim \mathcal{C}(0,1)$.                                    ‖

**Example 2.3.8 (Continuation of Example 2.3.4)** An efficient algorithm to generate from a $\mathcal{G}a(\alpha, 1)$ ($\alpha > 1$) distribution has been proposed by Best (1978). It relies on the instrumental distribution $\mathcal{T}(2, 0, 1)$, with explicit density

$$g(x) = \frac{1}{2\sqrt{2}} \left(1 + \frac{x^2}{2}\right)^{-3/2}$$

and cdf

$$(2.3.4) \qquad G(x) = \frac{1}{2}\left[1 + \frac{x/\sqrt{2}}{\sqrt{1 + x^2/2}}\right],$$

which can thus be directly simulated by inversion. If $f_\alpha$ denotes the $\mathcal{G}a(\alpha, 1)$ density and $g_\alpha$ denotes the density of $\mathcal{T}(2, \alpha - 1, \sqrt{(12\alpha - 3)/8})$, one can show (see Devroye 1985) that

$$\frac{f_\alpha(x)}{g_\alpha(x)} \leq \frac{\sqrt{12\alpha - 3}}{\Gamma(\alpha)}\left(\frac{\alpha - 1}{e}\right)^{\alpha - 1} = c_\alpha\,,$$

with, moreover, $c_\alpha < e\sqrt{6/\pi}$, a bound which ensures the applicability of the algorithm in the sense that it does not go to infinity with some values of $\alpha$.

It is also possible to propose a lower envelope which avoids the computation of logarithms. Given the cdf (2.3.4), the simulation of $\mathcal{T}(2, 0, 3(4\alpha - 1)/8)$ corresponds to the transformation

$$Y = \frac{\sqrt{c}(U - 1/2)}{\sqrt{U(1 - U)}}\,,$$

of the uniform variable $U$, where $c = (12\alpha - 3)/4$. Introducing $X = Y + \alpha - 1$, the ratio $f_\alpha(X)/c_\alpha g_\alpha(X)$ can be written

$$(2.3.5) \qquad \left(\frac{X}{\alpha - 1}\right)^{\alpha - 1} e^{-Y}\left(1 + \frac{4Y^2}{12\alpha - 3}\right)^{3/2}$$

and $Y$ is accepted if $V \sim \mathcal{U}_{[0,1]}$ is less than this ratio. Let $w = u(1 - u)$, $z = 64v^2 w^3$, and $b = \alpha - 1$. It then follows that $v$ is less than (2.3.5) whenever

$$(2.3.6) \qquad 2\left(b \log\left(\frac{x}{b}\right) - y\right) \geq \log(z).$$

Moreover, the number of calls to the logarithmic function in (2.3.6) can be dramatically reduced through use of the bounds

$$2\{b \log(x/b) - y\} \geq 1 - 2y^2/x \qquad \text{and} \qquad \log(z) \leq z - 1.$$

Best's (1978) algorithm to generate a $\mathcal{G}a(\alpha, 1)$ $(\alpha > 1)$ distribution then becomes

**Algorithm A.7 –Best's Gamma Simulation–**

0. Define $b = \alpha - 1$, $c = (12\alpha - 3)/4$;

1. Generate $u, v$ iid $\mathcal{U}_{[0,1]}$ and define

$$w = u(1 - u), \quad y = \sqrt{\frac{c}{w}}\left(u - \frac{1}{2}\right), \quad x = b + y \ ;$$

2. If $x > 0$, take $z = 64v^2 w^3$ and accept $x$ when                    [A.7]

$$z \leq 1 - 2y^2/x$$

   or when

$$2\left(b \log\ (x/b) - y\right) \geq \log(z) \ ;$$

3. Otherwise, start from 1.

An alternative based on *Burr's distribution,* a distribution with density

$$g(x) = \lambda\mu \ \frac{x^{\lambda-1}}{(\mu + x^\lambda)^2} \ , \qquad x > 0 \ ,$$

has been developed by Cheng (1977) and Cheng and Feast (1979) and enjoys similar performances to those of Best's algorithm [A.7] (see Devroye 1985 and Problem 2.25).                    ‖

**Example 2.3.9 (Continuation of Example 2.2.3)** As indicated in Example 2.2.3, the simulation of the Poisson distribution $\mathcal{P}(\lambda)$ using a Poisson process and exponential variables can be rather inefficient. Here, we describe a simpler alternative of Atkinson (1979), who uses the relationship between the Poisson $\mathcal{P}(\lambda)$ distribution and the *logistic distribution*. The logistic distribution has density and distribution function

$$f(x) = \frac{1}{\beta} \frac{\exp\{-(x - \alpha)/\beta\}}{[1 + \exp\{-(x - \alpha)/\beta\}]^2} \qquad \text{and} \qquad F(x) = \frac{1}{1 + \exp\{-(x - \alpha)/\beta\}}$$

and is therefore analytically invertible.

To better relate the continuous and discrete distributions, we consider $N = \lfloor x + 0.5 \rfloor$, the integer part of $x + 0.5$. Also, the range of the logistic distribution is $(-\infty, \infty)$, but to better match it with the Poisson, we restrict the range to $[-1/2, \infty)$. Thus, the random variable $N$ has distribution function

$$P(N = n) = \frac{1}{1 + e^{-(n+0.5-\alpha)/\beta}} - \frac{1}{1 + e^{-(n-0.5-\alpha)/\beta}}$$

if $x > 1/2$ and

$$P(N = n) = \left( \frac{1}{1 + e^{-(n+0.5-\alpha)/\beta}} - \frac{1}{1 + e^{-(n-0.5-\alpha)/\beta}} \right) \frac{1 + e^{-(0.5+\alpha)/\beta}}{e^{-(0.5+\alpha)/\beta}}$$

if $-1/2 < x \leq 1/2$, and the ratio of the densities is

(2.3.7) $$\lambda^n / P(N = n) \, e^\lambda \, n! \, .$$

Although it is difficult to compute a bound on (2.3.7) and, hence, to optimize it in $(\alpha, \beta)$, Atkinson (1979) proposed the choice $\alpha = \lambda$ and $\beta = \pi/\sqrt{3\lambda}$. This identifies the two first moments of $X$ with those of $\mathcal{P}(\lambda)$. For this choice of $\alpha$ and $\beta$, analytic optimization of the bound on (2.3.7) remains impossible, but numerical maximization and interpolation yields the bound $c = 0.767 - 3.36/\lambda$. The resulting algorithm is then

**Algorithm A.8 –Atkinson's Poisson Simulation–**
To generate $N \sim \mathcal{P}(\lambda)$:

0. Define

$$\beta = \pi/\sqrt{3\lambda}, \quad \alpha = \lambda\beta \quad \text{and} \quad k = \log c - \lambda - \log \beta;$$

1. Generate $U_1 \sim \mathcal{U}_{[0,1]}$ and calculate

$$x = \{\alpha - \log\{(1 - u_1)/u_1\}\}/\beta$$

     until $X > -0.5$ ;

2. Define $N = [X + 0.5]$ and generate $U_2 \sim \mathcal{U}_{[0,1]}$;

3. Accept $N$ if                                   [A.8]

$$\alpha - \beta x + \log \left( u_2 / \{1 + \exp(\alpha - \beta x)\}^2 \right) \leq k + N \log \lambda - \log N! \, .$$

Although the resulting simulation is exact, this algorithm is based on a number of approximations, both through the choice of $(\alpha, \beta)$ and in the computation of the majorization bounds and the density ratios. Moreover, note that it requires the computation of factorials, $N!$, which may be quite time-consuming. Therefore, although [A.8] usually has a reasonable efficiency, more complex algorithms such as those of Devroye (1985) may be preferable.      ‖

### 2.3.3 Log-Concave Densities

The particular case of *log-concave densities* (that is, densities whose logarithm is concave) allows the construction of a generic algorithm that can be quite efficient.

**Example 2.3.10 Log-concave densities.** Recall the exponential family (1.2.4)

$$f(x) = h(x) \, e^{\theta \cdot x - \psi(\theta)}, \qquad \theta, x \in \mathbb{R}^k.$$

This density is log-concave if

$$\frac{\partial^2}{\partial x^2} \log f(x) = \frac{\partial^2}{\partial x^2} \log h(x) = \frac{h(x)h''(x) - [h'(x)]^2}{h^2(x)} < 0,$$

which will often be the case for the exponential family. For example, if $X \sim \mathcal{N}(\theta, 1)$, then $h(x) \propto \exp\{-x^2/2\}$ and $\partial^2 \log h(x)/\partial x^2 = -1$. See Problem 2.42 for other examples of log-concave densities.          ‖

Devroye (1985) describes some algorithms that take advantage of the log-concavity of the density, but, here, we present a universal method. The algorithm, which was proposed by Gilks (1992) and Gilks and Wild (1992), is based on the construction of an envelope and the derivation of a corresponding Accept–Reject algorithm. The method is called *adaptive rejection sampling* (ARS) and it provides a sequential evaluation of lower and upper envelopes of the density $f$ when $h = \log f$ is concave.

Let $\mathcal{S}_n$ be a set of points $x_i, i = 0, 1, \ldots, n + 1$, in the support of $f$ such that $h(x_i) = \log f(x_i)$ is known up to the same constant. Given the concavity of $h$, the line $L_{i,i+1}$ through $(x_i, h(x_i))$ and $(x_{i+1}, h(x_{i+1}))$ is below the graph of $h$ in $[x_i, x_{i+1}]$ and is above this graph outside this interval (see Figure 2.3.1). For $x \in [x_i, x_{i+1}]$, if we define

$$\overline{h}_n(x) = \min\{L_{i-1,i}(x), L_{i+1,i+2}(x)\} \quad \text{and} \quad \underline{h}_n(x) = L_{i,i+1}(x),$$

the envelopes are

(2.3.8)                    $$\underline{h}_n(x) \leq h(x) \leq \overline{h}_n(x)$$

uniformly on the support of $f$. (We define

$$\underline{h}_n(x) = -\infty \quad \text{and} \quad \overline{h}_n(x) = \min(L_{0,1}(x), L_{n,n+1}(x))$$

on $[x_0, x_{n+1}]^c$.) Therefore, for $\underline{f}_n(x) = \exp \underline{h}_n(x)$ and $\overline{f}_n(x) = \exp \overline{h}_n(x)$, (2.3.8) implies that

$$\underline{f}_n(x) \leq f(x) \leq \overline{f}_n(x) = \varpi_n \, g_n(x) \,,$$

where $\varpi_n$ is the normalized constant of $f_n$; that is, $g_n$ is a density. The ARS algorithm to generate an observation from $f$ is thus

**Algorithm A.9 –ARS Algorithm–**

0. Initialize $n$ and $\mathcal{S}_n$.

1. Generate $X \sim g_n(x)$, $U \sim \mathcal{U}_{[0,1]}$.

2. If $U \leq \underline{f}_n(X)/\varpi_n \, g_n(X)$, accept $X$;                    [A.9]
   otherwise, if $U \leq f(X)/\varpi_n \, g_n(X)$, accept $X$
   and update $\mathcal{S}_n$ to $\mathcal{S}_{n+1} = \mathcal{S}_n \cup \{X\}$.

An interesting feature of this algorithm is that the set $\mathcal{S}_n$ is only updated when $f(x)$ has been previously computed. As the algorithm produces variables $X \sim f(x)$, the two envelopes $\underline{f}_n$ and $\overline{f}_n$ become increasingly accurate and, therefore, we progressively reduce the number of evaluations of $f$. Note that in the initialization of $\mathcal{S}_n$, a necessary condition is that $\varpi_n < +\infty$ (i.e.,

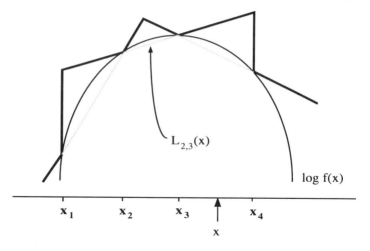

Figure 2.3.1. *Lower and upper envelopes of a log-concave density (Source: Gilks et al., 1995).*

that $g_n$ is actually a probability density). To achieve this requirement, $L_{0,1}$ needs to have positive slope if the support of $f$ is not bounded on the left and $L_{n,n+1}$ needs to have a negative slope if the support of $f$ is not bounded on the right.

Step 1 of [A.9] relies on simulations from $g_n$, which can be represented in the form

$$\varpi_n^{-1}\left\{\sum_{i=0}^{r_n} e^{\alpha_i x+\beta_i}\, \mathbb{I}_{[x_i,x_{i+1}]}(x) + e^{\alpha_{-1}x+\beta_{-1}}\mathbb{I}_{[-\infty,x_0]}(x) \right.$$
$$\left. e^{\alpha_{r_n+1}x+\beta_{r_n+1}}\, \mathbb{I}_{[x_{n+1},+\infty]}(x)\right\},$$

where $y = \alpha_i x + \beta_i$ is the equation of the segment of line corresponding to $g_n$ on $[x_i, x_{i+1}]$, $r_n$ denotes the number of segments, and

$$\varpi_n = \int_{-\infty}^{x_0} e^{\alpha_{-1}x+\beta_{-1}}dx + \sum_{i=0}^{n}\int_{x_i}^{x_{i+1}} e^{\alpha_i x+\beta_i}dx + \int_{x_{n+1}}^{+\infty} e^{\alpha_{r_n+1}x+\beta_{r_n+1}}dx$$
$$= \frac{e^{\alpha_{-1}x_0+\beta_{-1}}}{\alpha_{-1}} + \sum_{i=0}^{n} e^{\beta_i}\,\frac{e^{\alpha_i x_{i+1}}-e^{\alpha_i x_i}}{\alpha_i} - \frac{e^{\alpha_{r_n+1}x_{n+1}}}{\alpha_{r_n+1}},$$

when supp $f = \mathbb{R}$. This representation as a sequence suggests the following algorithm for the simulation from $g_n$:

**Algorithm A.10 –Supplemental ARS Algorithm–**

1. Select the interval $[x_i, x_{i+1}]$ with probability

$$e^{\beta_i}\,\frac{e^{\alpha_i x_{i+1}} - e^{\alpha_i x_i}}{\varpi_n\,\alpha_i}.\qquad\qquad [A.10]$$

2. Generate $U \sim \mathcal{U}_{[0,1]}$ and take

$$X = \alpha_i^{-1}\,\log[e^{\alpha_i x_i} + U(e^{\alpha_i x_{i+1}} - e^{\alpha_i x_i})].$$

The ARS algorithm is not optimal in the sense that it is often possible to devise a better specialized algorithm for a given log-concave density. However, although Gilks and Wild (1992) do not provide theoretical evaluations of simulation speeds, they mention reasonable performances in the cases they consider. Note that, in contrast to the previous algorithms, the function $g_n$ is updated during the iterations and, therefore, the average computation time for one generation from $f$ decreases with $n$. This feature makes the comparison with other approaches quite delicate.

The major advantage of [A.9] compared with alternatives is its *universality*. For densities $f$ that are only known through their functional form, the ARS algorithm yields an automatic Accept–Reject algorithm that only requires checking $f$ for log-concavity. Moreover, the set of log-concave densities is wide; see Problems 2.42 and 2.43. The ARS algorithm thus allows for the generation of samples from distributions that are rarely simulated, without requiring the development of case-specific Accept–Reject algorithms.

**Example 2.3.11 Capture–recapture models.** In a heterogeneous *capture–recapture model* (see Seber 1982, 1992 or Dupuis 1995b), animals are captured at time $i$ with probability $p_i$, the size $N$ of the population being unknown. The corresponding likelihood is therefore

$$L(p_1,\ldots,p_I|N,n_1,\ldots,n_I) = \frac{N!}{(N-r)!} \prod_{i=1}^{I} p_i^{n_i}(1-p_i)^{N-n_i},$$

where $I$ is the number of captures, $n_i$ is the number of captured animals during the $i$th capture, and $r$ is the total number of different captured animals. If $N$ is a priori distributed as a $\mathcal{P}(\lambda)$ variable and the $p_i$'s are from a *normal logistic* model,

$$\alpha_i = \log\left(\frac{p_i}{1-p_i}\right) \sim \mathcal{N}(\mu_i,\sigma^2),$$

as in George and Robert (1992), the posterior distribution satisfies

$$\pi(\alpha_i|N,n_1,\ldots,n_I) \propto \exp\left\{\alpha_i n_i - \frac{1}{2\sigma^2}(\alpha_i-\mu_i)^2\right\}\Big/(1+e^{\alpha_i})^N.$$

If this conditional distribution must be simulated (for reasons which will be made clearer in Chapter 7), the ARS algorithm can be implemented. In fact,

(2.3.9)          $\alpha_i n_i - \dfrac{1}{2\sigma^2}(\alpha_i-\mu_i)^2 - N\,\log(1+e^{\alpha_i})$

is concave in $\alpha_i$, as can be shown by computing the second derivative.   ‖

The above example also illustrates that checking for log-concavity of a Bayesian posterior distribution is straightforward, as $\log \pi(\theta|x) = \log \pi(\theta) + \log f(x|\theta) + c$, where $c$ is a constant. This implies that the log-concavity of $\pi(\theta)$ and of $f(x|\theta)$ (in $\theta$) are sufficient to conclude the log-concavity of $\pi(\theta|x)$.

**Example 2.3.12 Poisson regression.** Consider a sample $(Y_1, x_1), \ldots,$ $(Y_n, x_n)$ of integer-valued data $Y_i$ with explanatory variable $x_i$, where $Y_i$ and $x_i$ are connected via a Poisson distribution,

$$Y_i | x_i \sim \mathcal{P}(\exp\{a + bx_i\}) .$$

If the prior distribution of $(a, b)$ is a normal distribution $\mathcal{N}(0, \sigma^2) \times \mathcal{N}(0, \tau^2)$, the posterior distribution of $(a, b)$ is given by

$$\pi(a, b | \mathbf{x}, \mathbf{y}) \propto \exp\left\{ a \sum_i y_i + b \sum_i y_i x_i - e^a \sum_i e^{x_i b} \right\} e^{-a^2/2\sigma^2} \, e^{-b^2/2\tau^2} .$$

We will see in Chapter 7 that it is often of interest to simulate successively the (full) conditional distributions $\pi(a, |\mathbf{x}, \mathbf{y}, b)$ and $\pi(b | \mathbf{x}, \mathbf{y}, a)$. Since

$$\log \pi(a | \mathbf{x}, \mathbf{y}, b) = a \sum_i y_i - e^a \sum_i e^{x_i b} - a^2/2\sigma^2 ,$$

$$\log \pi(b | \mathbf{x}, \mathbf{y}, a) = b \sum_i y_i x_i - e^a \sum_i e^{x_i b} - b^2/2\tau^2 ,$$

and

$$\frac{\partial^2}{\partial a^2} \log \pi(a | \mathbf{x}, \mathbf{y}, b) = - \sum_i e^{x_i b} \, e^a - \sigma^{-2} < 0 ,$$

$$\frac{\partial^2}{\partial b^2} \log \pi(b | \mathbf{x}, \mathbf{y}, a) = -e^a \sum_i x_i^2 \, e^{x_i b} - \tau^{-2} < 0 ,$$

the ARS algorithm directly applies for both conditional distributions.    ‖

## 2.4 Problems

**2.1** Referring to Example 2.1.4, show the following:

(a) The *arcsine distribution*, with density $f(x) = 1/\pi\sqrt{x(1-x)}$, is invariant under the transform $y = 1 - x$, that is, $f(x) = f(y)$.

(b) The uniform distribution $\mathcal{U}_{[0,1]}$ is invariant under the "tent" transform,

$$D(x) = \begin{cases} 2x & \text{if } x \le 1/2 \\ 2(1 - x) & \text{if } x > 1/2. \end{cases}$$

(c) Check these two properties via a computer experiment by plotting the histograms of the successive iterates.

**2.2** This problem will look into one of the failings of congruential generators, the production of parallel lines of output. Consider a congruential generator $D(x) = ax \bmod 1$, that is, the output is the fractional part of $ax$.

(a) For $k = 1, 2, \ldots, 333$, plot the pairs $(k*0.003, D(k*0.003))$ for $a = 5, 20, 50$. What can you conclude about the parallel lines?

(b) Show that each line has slope $a$ and the lines repeat at intervals of $1/a$ (hence, larger values of $a$ will increase the number of lines). (*Hint:* Let $x = \frac{i}{a} + \delta$, for $i = 1, \ldots, a$ and $0 < \delta < \frac{1}{a}$. For this $x$, show that $D(x) = a\delta$, regardless of the value of $i$.)

**2.3** If $M$ is the largest integer accepted by a computer, show that the period of a generator of the form $x_{n+1} = f(x_n)$ is at most $M + 1$.

**2.4** If $D_1$ and $D_2$ are two random generators with period $M$, indicate ways of combining both generators to produce a generator of period $M \times M$. Discuss the advantage of the resulting generator.

**2.5** The RANDU generator, once popular on IBM machines, is based on the recursion

$$X_{n+1} = 65539 X_n \bmod 2^{31} .$$

Illustrate the undesirable behavior of this generator with a computer experiment. (*Hint:* Show that $X_{t+2} = (6X_t - 9X_{t-1}) \bmod 2^{31}$.)

**2.6**(a) Test the properties of the NAG congruential generator, based on

$$X_{n+1} = 13^{13} X_n \bmod 2^{59} .$$

(b) Test the properties of the IMSL congruential generator, based on

$$X_{n+1} = (2^7 + 1) X_n + 1 \bmod 2^{35} .$$

(c) Examine the properties of the generator based on the Fibonacci series

$$X_{n+1} = (X_n + X_{n-1}) \bmod 2^{32} .$$

(d) Examine whether the decimals of $\pi$ constitute a valid random generator and discuss the implementation issues. (*Note:* See Dodge 1996 for a proposal in this direction.)

**2.7** (Continuation of Problem 2.6) For one of the generators above, test its dependence on the computer used for the simulation by comparing two outputs from two different computers initialized with the same starting value.

**2.8** Even when the choice of $(a, b)$ ensures the acceptance of the congruential generator by standard tests, nonuniform behavior will be observed in the last digits of the real numbers produced by this method. Demonstrate this using the generators

$$X_{n+1} = 16807 X_n \bmod 2^{31} \quad \text{and} \quad X_{n+1} = 69069 X_n + 1 \bmod 2^{32} .$$

(*Hint:* See L'Ecuyer and Cordeau 1996 for a complete study.)

**2.9** Show that a single uniform variable $U \sim \mathcal{U}([0, 1])$ can formally generate a finite sequence of random variables with finite support, plus a continuous random variable with invertible cdf. (*Hint:* Break the unit interval into a partition with appropriate weights and iterate the method.)

**2.10** For each of the following distributions, calculate the explicit form of the distribution function and show how to implement its generation using the *Kiss* algorithm: (a) exponential; (b) double exponential; (c) Weibull; (d) Pareto; (e) Cauchy; (f) extreme value; (g) arcsine.

**2.11** Referring to Example 2.2.1:

(a) Show that if $U \sim \mathcal{U}_{[0,1]}$, then $X = -\log U / \lambda \sim \mathcal{E}xp(\lambda)$.

(b) Verify the distributions in (2.2.1).

(c) Show how to generate an $\mathcal{F}_{m,n}$ random variable, where both $m$ and $n$ are even integers.

(d) Show that if $U \sim \mathcal{U}_{[0,1]}$, then $X = \log \frac{u}{1-u}$ is a Logistic$(0, 1)$ random variable. Show also how to generate a Logistic$(\mu, \beta)$ random variable.

**2.12** Establish the properties of the *Box–Muller algorithm* of Example 2.2.2. If $U_1$ and $U_2$ are iid $\mathcal{U}_{[0,1]}$, show that:

(a) The transforms

$$X_1 = \sqrt{-2\log(U_1)}\,\cos(2\pi U_2)\,, \qquad X_2 = \sqrt{-2\log(U_1)}\,\sin(2\pi U_2)\,,$$

are iid $\mathcal{N}(0,1)$.

(b) The polar coordinates are distributed as

$$r^2 = X_1^2 + X_2^2 \sim \chi_2^2\,,$$
$$\theta = \arctan\frac{X_1}{X_2} \sim \mathcal{U}[0,2\pi].$$

(c) Establish that $-2\log r^2 \sim \mathcal{U}[0,1]$, and so $r^2$ and $\theta$ can be simulated directly.

**2.13** (Continuation of Problem 2.12)

(a) Show that a faster version of the Box–Muller algorithm is

**Algorithm A.11 –Box-Muller (2)–**

1. Generate                                                                                      [A.11]

$$U_1, U_2 \sim \mathcal{U}([-1,1])$$

   until $S = U_1^2 + U_2^2 \leq 1$.
2. Define $Z = \sqrt{-2\log(S)/S}$ and take

$$X_1 = Z\,U_1, \qquad X_2 = Z\,U_2.$$

(*Hint:* Show that $(U_1, U_2)$ is uniform on the unit sphere and that $X_1$ and $X_2$ are independent.)

(b) Give the average number of generations in 1. and compare with the original Box–Muller algorithm [A.4] on a small experiment.

(c) Examine the effect of not constraining $(U_1, U_2)$ to the unit circle.

**2.14** Show that the following version of the Box–Muller algorithm produces one normal variable and compare the execution time with both versions [A.4] and [A.11]:

**Algorithm A.12 –Box–Muller (3)–**

1. Generate

$$Y_1, Y_2 \sim \mathcal{E}xp(1)$$

   until $Y_2 > (1 - Y_1)^2/2$.                                                                 [A.12]
2. Generate $U \sim \mathcal{U}([0,1])$ and take

$$X = \begin{cases} Y_1 & \text{if } U < 0.5 \\ -Y_1 & \text{if } U > 0.5. \end{cases}$$

**2.15** Detail the properties of the algorithm based on the Central Limit Theorem for a sequence of uniforms $U_1, \ldots, U_n$ for $n = 12$, $n = 48$ and $n = 96$. (Consider, in particular, the range of the generated variable.)

**2.16** Compare the performances of the polar Box–Muller algorithm [A.11] with those of the algorithm of Example 2.1.9 based on high order approximations of the normal cdf, in terms of running time and of precision.

**2.17** For the generation of a Student's $t$ distribution, $T(\nu, 0, 1)$, Kinderman *et al.* (1977) provide an alternative to the generation of a normal random variable and a chi squared random variable.

**Algorithm A.13 –Student's $t$–**

1. Generate $U_1, U_2 \sim \mathcal{U}([0, 1])$.
2. If $U_1 < 0.5$, $X = 1/(4U_1 - 1)$ and $V = X^{-2}U_2$;
   otherwise, $X = 4U_1 - 3$ and $V = U_2$.                    [A.13]
3. If $V < 1 - (|X|/2)$ or $V < (1 + (X^2/\nu))^{-(\nu+1)/2}$, take $X$;
   otherwise, repeat.

Validate this algorithm and compare it with the algorithm of Example 2.2.6.

**2.18** For $\alpha \in [0, 1]$, show that the algorithm

**Algorithm A.14**
Generate
$$U \sim \mathcal{U}([0, 1])$$
until $U < \alpha$.

produces a simulation from $\mathcal{U}([0, \alpha])$ and compare it with the genuine transform $\alpha U$ for values of $\alpha$ close to 1.

**2.19** Recall the algorithm of Example 2.2.3.

(a) Show that if $N \sim \mathcal{P}(\lambda)$ and $X_i \sim \mathcal{E}xp(\lambda)$, $i \in \mathbb{N}^*$, independent, then
$$P_\lambda(N = k) = P_\lambda(X_1 + \cdots + X_k \leq 1 < X_1 + \cdots + X_{k+1}).$$

(b) Use the results of part (a) to justify that the following algorithm simulates a Poisson $\mathcal{P}(\lambda)$ random variable:

**Algorithm A.15 –Poisson simulation–**

$p = 1$, $N = 0$, $c = e^{-\lambda}$.
1. Repeat                                                      [A.15]
   $N = N + 1$
   generate $U_i$
   update $p = pU_i$
   until $p < c$.
2. Take $X = N - 1$

**2.20** Consider the naïve algorithm for generating Poisson $\mathcal{P}(\lambda)$ variables based on the simple comparison of $U \sim \mathcal{U}([0, 1])$ with $\exp(-\lambda)$, $\exp(-\lambda)\lambda + \exp(-\lambda)$, etc. Evaluate the improvements brought by a preliminary ranking of the probabilities and by the storage of the most frequent probabilities in a table.

**2.21** Establish the validity of Knuth's (1981) $\mathcal{B}(n, p)$ generator:

**Algorithm A.16 –Binomial–**

Define $k = n$, $\theta = p$ and $x = 0$.
1. Repeat
   $i = [1 + k\theta]$
   $V \sim \mathcal{B}e(i, k + 1 - i)$
   if $\theta > V$, $\theta = \theta/V$ and $k = i - 1$;
      otherwise, $x = x + i$, $\theta = (\theta - V)/(1 - V)$ and $k = k - i$
   until $k \leq K$.                                           [A.16]

2. For $i = 1, 2, \ldots, k$,

     generate $U_i$

     if $U_i < p,\ x = x + 1$.

3. Take $x$.

**2.22** Establish the claims of Example 2.2.4: If $U_1, \ldots, U_n$ is an iid sample from $\mathcal{U}_{[0,1]}$ and $U_{(1)} \leq \cdots \leq U_{(n)}$ are the corresponding order statistics, show that

(a) $U_{(i)} \sim \mathcal{B}e(i, n - i + 1)$;

(b) $(U_{(i_1)}, U_{(i_2)} - U_{(i_1)}, \ldots, U_{(i_k)} - U_{(i_{k-1})}, 1 - U_{(i_k)}) \sim \mathcal{D}(i_1, i_2 - i_1, \ldots, n - i_k + 1)$;

(c) if $U$ and $V$ are iid $\mathcal{U}_{[0,1]}$, the distribution of

$$\frac{U^{1/\alpha}}{U^{1/\alpha} + V^{1/\beta}},$$

conditional on $U^{1/\alpha} + V^{1/\beta} \leq 1$, is the $\mathcal{B}e(\alpha, \beta)$ distribution.

**2.23** Show that the order statistics of the uniform sample of Example 2.2.4 can be directly generated via the *Renyi representation* $u_{(i)} = \sum_{j=1}^{i} \nu_j / \sum_{j=1}^{n} \nu_j$, where the $\nu_j$'s are iid $\mathcal{E}xp(1)$.

**2.24** There are (at least) two ways to establish (2.2.3) of Example 2.2.5:

(a) Make the transformation $x = yu^{1/\alpha}$, $w = y$, and integrate out $w$.

(b) Make the transformation $x = yz$, $w = z$, and integrate out $w$.

**2.25** For $\alpha > 1$, the Cheng and Feast's (1979) algorithm for a Gamma $\mathcal{G}a(\alpha, 1)$ distribution is:

**Algorithm A.17 –Cheng and Feast's Gamma–**

Define $c_1 = \alpha - 1$, $c_2 = (\alpha - (1/6\alpha))/c_1$, $c_3 = 2/c_1$, $c_4 = 1 + c_3$, and $c_5 = 1/\sqrt{\alpha}$.

1. Repeat
generate $U_1, U_2$ and take $U_1 = U_2 + c_5(1 - 1.86U_1)$ if $\alpha > 2.5$
until $0 < U_1 < 1$.                                                   [A.17]

2. $W = c_2 U_2 / U_1$.

3. If $c_3 U_1 + W + W^{-1} \leq c_4$ or $c_3 \log U_1 - \log W + W \leq 1$, take $c_1 W$;
otherwise, repeat.

Show that this algorithm produces variables generated from $\mathcal{G}a(\alpha, 1)$.

**2.26** Ahrens and Dieter (1974) propose the following algorithm to generate a Gamma $\mathcal{G}a(\alpha, 1)$ distribution:

**Algorithm A.18 –Ahrens and Dieter's Gamma–**

1. Generate $U_0, U_1$.

2. If $U_0 > e/(e + \alpha)$, $x = -\log\{(\alpha + e)(1 - U_0)/\alpha e\}$ and $y = x^{\alpha-1}$;
otherwise, $x = \{(\alpha + e)U_0/e\}^{1/\alpha}$ and $y = e^{-x}$.           [A.18]

3. If $U_1 < y$, take $x$;
otherwise, repeat.

Show that this algorithm produces variables generated from $\mathcal{G}a(\alpha, 1)$. Compare with Problem 2.25.

**2.27** To generate the beta distribution $\mathcal{B}e(\alpha, \beta)$ we can use the following representation:

(a) Show that, if $Y_1 \sim \mathcal{G}a(\alpha, 1)$, $Y_2 \sim \mathcal{G}a(\beta, 1)$, then

$$X = \frac{Y_1}{Y_1 + Y_2} \sim \mathcal{B}e(\alpha, \beta) \ .$$

(b) Use part (a) to construct an algorithm to generate a beta random variable.

(c) Compare this algorithm with the method given in Problem 2.22 for different values of $(\alpha, \beta)$.

(d) Compare this algorithm with an Accept–Reject algorithm based on (i) the uniform distribution; (ii) the truncated normal distribution (when $\alpha \geq 1$ and $\beta \geq 1$).

(*Note:* See Schmeiser and Shalaby 1980 for an alternative Accept–Reject algorithm to generate beta rv's.)

**2.28**(a) Show that Student's $t$ density can be written in the form (2.2.5), where $h_1(x|y)$ is the density of $\mathcal{N}(0, \nu/y)$ and $h_2(y)$ is the density of $\chi_\nu^2$.

(b) Show that Fisher's $\mathcal{F}_{m,\nu}$ density can be written in the form (2.2.5), with $h_1(x|y)$ the density of $\mathcal{G}a(m/2, \nu/m)$ and $h_2(y)$ the density of $\chi_\nu^2$.

**2.29** Show that a negative binomial random variable is a mixture of Poisson distributions; that is, if $Y \sim \mathcal{G}a(n, (1-p)/p)$ and $X|y \sim \mathcal{P}(y)$, then $X \sim \mathcal{N}eg(n, p)$. Use this fact to write an algorithm for simulating negative binomial random variables.

**2.30** The noncentral chi squared distribution, $\chi_p^2(\lambda)$, can be defined by a mixture representation (2.2.5), where $h_1(x|K)$ is the density of $\chi_{p+2K}^2$ and $h_2(k)$ is the density of $\mathcal{P}(\lambda/2)$. Show that it can also be expressed as the sum of a $\chi_{p-1}^2$ random variable and of the square of a standard normal variable. Compare the two algorithms which can be derived from these representations. Discuss whether a direct approach via an Accept–Reject algorithm is at all feasible.

**2.31** (Walker 1997) Show that the Weibull distribution, $\mathcal{W}e(\alpha, \beta)$, with density

$$f(x|\alpha, \beta) = \beta \alpha x^{\alpha-1} \exp\left(-\beta x^\alpha\right),$$

can be represented as a mixture of $X \sim \mathcal{B}e(\alpha, \omega^{1/\alpha})$ by $\omega \sim \mathcal{G}a(2, \beta)$. Examine whether this representation is helpful from a simulation point of view.

**2.32** This problem looks at properties of the Wishart distribution and contains some advanced matrix algebra concepts.

(a) The *Wishart* distribution $\mathcal{W}_p(n, \Sigma)$ is a distribution on the space of $p \times p$ symmetric matrices, which is defined as the distribution of the sum of $p$ matrices of the form $X_i X_i^t$, with the $X_i$'s iid from $\mathcal{N}_p(0, \Sigma)$. Adapt the method of Example 2.30 to derive a fast simulation method.

(b) Consider the *Choleski decomposition* of a symmetric covariance matrix, $\Sigma = AA^t$, where $A$ is a lower triangular matrix. Deduce an implementation of the simulation of the standard simulation for the multivariate normal $\mathcal{N}_p(0, \Sigma)$ distribution. (*Hint:* Give an algorithm for the Choleski decomposition.)

(c) Compare the method of part (b) with a stepwise naïve implementation which simulates $X_1$, $X_2|X_1$, etc.

**2.33** An application of the mixture representation can be used to establish the following result (see Note 2.5.2):

**Lemma 2.4.1** *If*

$$f(x) = \frac{f_1(x) - \varepsilon f_2(x)}{1 - \varepsilon}$$

*where $f_1$ and $f_2$ are probability densities such that $f_1(x) \geq \varepsilon f_2(x)$, the algorithm*

Generate

$$(X, U) \sim f_1(x) \, \mathbb{I}_{[0,1]}(u)$$

until $U \geq \varepsilon f_2(X)/f_1(X)$

*produces a variable $X$ distributed according to $f$.*

(a) Show that the distribution of $X$ satisfies

$$P(X \leq x_0) = \int_{-\infty}^{x_0} \left(1 - \frac{\varepsilon f_2(x)}{f_1(x)}\right) f_1(x)\, dx \sum_{i=0}^{\infty} \varepsilon^i \ .$$

(b) Evaluate the integral in (a) to complete the proof.

**2.34** For the Accept–Reject algorithm [A.5], with $f$ and $g$ properly normalized,

(a) Show that the probability of accepting a random variable is

$$P\left(U < \frac{f(X)}{Mg(X)}\right) = \frac{1}{M}.$$

(b) Show that $M \geq 1$.

(c) Show that the bound $M$ does not have to be tight; that is, there may be $M' < M$ such that $f(x) \leq M'g(x)$. Give an example where it makes sense to use $M$ instead of $M'$.

(d) When the bound $M$ is too tight (i.e., when $f(x) > Mg(x)$ on a non-negligible part of the support of $f$), show that the algorithm [A.5] does not produce a generation from $f$. Give the resulting distribution.

(e) When the bound is not tight, show that there is a way, using Lemma 2.4.1, to recycle part of the rejected random variables (*Note:* See Casella and Robert 1998 for details.)

**2.35** When generating $\mathcal{G}a(\alpha, \beta)$, compare the Accept–Reject algorithms based on $\mathcal{G}a(\lceil \alpha \rceil, \beta')$ and on $\mathcal{G}a(\lfloor \alpha \rfloor, \beta'')$, where $\lfloor \alpha \rfloor$ denotes the integer part of $\alpha$ (i.e., the largest integer smaller than $\alpha$) and $\lceil \alpha \rceil$ is the smallest integer larger than $\alpha$. (*Hint:* Optimize first in $\beta'$ and $\beta''$.)

**2.36** For the Accept–Reject algorithm [A.5], let $N$ be the number of trials until the $k$th random variable is accepted. Show that, for the normalized densities, $N$ has the negative binomial distribution $\mathcal{N}eg(k, p)$, where $p = 1/M$. Deduce that the expected number of trials until $k$ random variables are obtained is $kM$.

**2.37** For the situation of Example 2.3.2:

(a) Show that for the normalized densities, the probability of acceptance is $\sqrt{e/2\pi} = 0.66$.

(b) If the instrumental density is $g_\sigma$ instead of $g$, show that the bound on $f/g_\sigma$ is $2\sigma^{-1} \exp\{\sigma^2/2 - 1\}$ and is minimized by $\sigma^2 = 1$.

**2.38** Given a normal distribution $\mathcal{N}(0,1)$ restricted to $\mathbb{R}^+$, construct an Accept–Reject algorithm based on $\mathcal{E}xp(\lambda)$ and optimize in $\lambda$.

**2.39** For the generation of a Cauchy random variable, compare the inversion method with one based on the generation of the normal pair of the polar Box–Muller method (Problem 2.13).

(a) Show that, if $X_1$ and $X_2$ are iid normal, $Y = X_1/X_2$ is distributed as a Cauchy random variable.

(b) Show that the Cauchy distribution function is $F(x) = \tan^{-1}(x)/\pi$, so the inversion method is easily implemented.

(c) Is one of the two algorithms superior?

**2.40** For the Accept–Reject algorithm of the $\mathcal{G}a(n,1)$ distribution, based on the $\mathcal{E}xp(\lambda)$ distribution, determine the optimal value of $\lambda$.

**2.41** This problem looks at a generalization of Example 2.3.4.

(a) If the target distribution of an Accept–Reject algorithm is the gamma distribution $\mathcal{G}a(\alpha, \beta)$, where $\alpha$ is not necessarily an integer, show that the instrumental distribution $\mathcal{G}a(a, b)$ is associated with the ratio

$$\frac{f(x)}{g(x)} = \frac{\Gamma(a)}{\Gamma(\alpha)} \frac{\beta^\alpha}{b^a} x^{\alpha-a} e^{-(\beta-b)x}.$$

(b) Why do we need $a < \alpha$ and $b < \beta$?

(c) For $a = [\alpha]$, show that the bound is maximized (in $x$) at $x = (\alpha-a)/(\beta-b)$.

(d) For $a = [\alpha]$, find the optimal choice of $b$.

(e) Compare with $a' = [\alpha] - 1$, when $\alpha > 2$.

**2.42** As mentioned in §2.3.3, many densities are log-concave.

(a) Show that the so-called *natural* exponential family,

$$dP_\theta(x) = \exp\{x \cdot \theta - \psi(\theta)\}d\nu(x)$$

is log-concave.

(b) Show that the logistic distribution of (2.3.9) is log-concave.

(c) Show that the *Gumbel* distribution

$$f(x) = \frac{k^k}{(k-1)!} \exp\left\{-kx - ke^{-x}\right\}, \quad k \in \mathbb{N}^*,$$

is log-concave (Gumbel 1958).

(d) Show that the *generalized inverse Gaussian* distribution,

$$f(x) \propto x^\alpha \, e^{-\beta x - \alpha/x}, \qquad x > 0,\ \alpha > 0,\ \beta > 0,$$

is log-concave.

**2.43** (George *et al.* 1993) For the natural exponential family, the conjugate prior measure is defined as

$$d\pi(\theta|x_0, n_0) \propto \exp\{x_0 \cdot \theta - n_0 \psi(\theta)\}d\theta,$$

with $n_0 > 0$. (See Brown 1986, Chapter 1, for properties of exponential families.)

(a) Show that

$$\varphi(x_0, n_0) = \log \int_\Theta \exp\{x_0 \cdot \theta - n_0 \psi(\theta)\} d\theta$$

is convex.

(b) Show that the so-called *conjugate likelihood distribution*

$$L(x_0, n_0 | \theta_1, \dots, \theta_p) \propto \exp\left\{ x_0 \cdot \sum_{i=1}^p \theta_i - n_0 \sum_{i=1}^p \psi(\theta) - p\varphi(x_0, n_0) \right\}$$

is log-concave in $(x_0, n_0)$.

(c) Deduce that the ARS algorithm applies in hierarchical Bayesian models with conjugate priors on the natural parameters and log-concave hyperpriors on $(x_0, n_0)$.

(d) Apply the ARS algorithm to the case

$$X_i | \theta_i \sim \mathcal{P}(\theta_i t_i), \qquad \theta_i \sim \mathcal{G}a(\alpha, \beta), \qquad i = 1, \dots, n,$$

with fixed $\alpha$ and $\beta \sim \mathcal{G}a(0.1, 1)$.

**2.44** The right-truncated gamma distribution $\mathcal{TG}(a, b, t)$ is defined as the restriction of the Gamma distribution $\mathcal{G}a(a, b)$ to the interval $(0, t)$.

(a) Show that we can consider $t = 1$ without loss of generality.

(b) Give the density $f$ of $\mathcal{TG}(a, b, 1)$ and show that it can be expressed as the following mixture of beta $\mathcal{B}e(a, k + 1)$ densities:

$$f(x) = \sum_{k=0}^\infty \frac{b^a e^{-b}}{\gamma(a, b)} \frac{b^k}{k!} x^{a-1}(1 - x)^k,$$

where $\gamma(a, b) = \int_0^1 x^{a-1} e^{-x}$.

(c) If $f$ is replaced with $g_n$ which is the series truncated at term $k = n$, show that the acceptance probability of the Accept–Reject algorithm based on $(g_n, f)$ is

$$\frac{1 - \dfrac{\gamma(n + 1, b)}{n!}}{1 - \dfrac{\gamma(a + n + 1, b)\Gamma(a)}{\Gamma(a + n + 1)\gamma(a, b)}}$$

(d) Evaluate this probability for different values of $(a, b)$.

(e) Give an Accept–Reject algorithm based on the pair $(g_n, f)$ and a computable bound. (*Note:* See Philippe 1997c for a complete resolution of the problem.)

**2.45** (Continuation of Problem 2.44) Consider $\mathcal{TG}^+(a, b)$, the Gamma distribution restricted to $(1, \infty)$.

(a) Give the density $f$ of $\mathcal{TG}^+(a, b)$.

(b) When $a \in \mathbb{N}$, show that $\mathcal{TG}^+(a, b)$ can be expressed as a mixture of gamma distributions translated by 1,

$$f(x) = \sum_{k=1}^a \frac{(a - 1)! b^a e^{-b(x-1)} (x - 1)^{a-1}}{\Gamma(a, b)(a - k)!(k - 1)!},$$

where $\Gamma(a, b) = \Gamma(a) - \gamma(a, b)$.

(c) For an arbitrary $a > 0$, give an Accept–Reject algorithm based on $\mathcal{TG}^+(\alpha, \beta)$ with $\alpha \in \mathbb{N}$ and $\alpha > a$. Optimize in $(\alpha, \beta)$.

(d) Derive the optimal choice of $(\alpha, \beta)$ when $\alpha \in \mathbb{N}$ and $\alpha < a$, and show that there is no explicit solution in $\beta$.

(e) For $\alpha = [a]$ and $\beta = [a]b/a$ when $b < a$ and $b - a + [a]$ otherwise, show that the acceptance probability is always larger than $e/4$. (*Note:* See Philippe 1997c for a full proof.)

**2.46** Show that the quasi-Monte Carlo methods discussed in Note 2.5.1 lead to standard Riemann integration for the equidistributed sequences in dimension 1.

**2.47** Establish (2.5.1) and show that the divergence $D(x_1, \ldots, x_n)$ leads to the Kolmogorov–Smirnov test in nonparametric Statistics.

## 2.5 Notes

### 2.5.1 Quasi-Monte Carlo Methods

Quasi-Monte Carlo methods were proposed in the 1950s to overcome some drawbacks of regular Monte Carlo methods by replacing probabilistic bounds on the errors with deterministic bounds. The idea at the core of quasi-Monte Carlo methods is to substitute the randomly (or pseudo-randomly) generated sequences used in regular Monte Carlo methods with a deterministic sequence $(x_n)$ in order to minimize the so-called *divergence*,

$$D(x_1, \ldots, x_n) = \sup_u \left| \frac{1}{n} \sum_{i=1}^{n} \mathbb{I}_{[0,u]}(x_i) - u \right|.$$

This is also the *Kolmogorov–Smirnov distance* between the empirical cdf and that of the uniform distribution, used in nonparametric tests. For fixed $n$, the solution is obviously $x_i = (2i-1)/2n$ in dimension 1, but the goal here is to get a *low-discrepancy* sequence $(x_n)$ which provides small values of $D(x_1, \ldots, x_n)$ for all $n$'s (i.e., such that $x_1, \ldots, x_{n-1}$ do not depend on $n$) and can thus be updated sequentially.

As shown in Niederreiter (1992), there exist such sequences, which ensure a divergence rate of order $O(n^{-1} \log(n)^{d-1})$, where $d$ is the dimension of the integration space.[5] Since it can be shown that the divergence is related to the overall approximation error by

(2.5.1)
$$\left| \frac{1}{n} \sum_{i=1}^{n} h(x_i) - \int f(x)dx \right| \leq V(h)D(x_1, \ldots, x_n)$$

(see Niederreiter 1992), where $V(h)$ is the total variation of $h$,

$$V(f) = \lim_{N \to \infty} \sup_{x_0 = 1 \leq \cdots \leq x_N = 1} \sum_{j=1}^{N} |h(x_j) - h(x_{j-1})|,$$

---

[5] The notation $O(1/n)$ denotes a function that satisfies

$$\lim_{n \to \infty} nO(1/n) = C,$$

with $C$ a positive constant.

the gain over standard Monte Carlo methods can be substantial since standard methods lead to order $O(n^{-1/2})$ errors (see §3.4). The advantage over standard integration techniques such as Riemann sums is also important when the dimension $d$ increases since the latter are of order $n^{4/d}$ (see Yakowitz *et al.* 1978).

The true comparison with regular Monte Carlo methods is however more delicate than a simple assessment of the order of convergence. Construction of these sequences, although independent from $h$, can be quite involved (see Fang and Wang 1994 for examples). More importantly, the construction requires that the functions to be integrated have bounded support, which can be a hindrance in practice. See Niederreiter (1992) for extensions in optimization setups. Note 7.6.7 also expands on the difficulties of an extension to Markov Chain Monte Carlo methods.

*2.5.2 Mixture Representations*

Mixture representations, such as those used in Examples 2.2.6 and 2.2.7, can be extended (theoretically, at least) to many other distributions. For instance, a random variable $X$ (and its associated distribution) is called *infinitely divisible* if for every $n$ there exist iid random variables $X_1', \cdots, X_n'$ such that $X \sim X_1' + \cdots + X_n'$ (see Feller 1971, Section XVII.3 or Billingsley 1995, Section 28). It turns out that most distributions that are infinitely divisible can be represented as mixtures of Poisson distributions, the noncentral $\chi_p^2(\lambda)$ distribution being a particular case of this phenomenon. (However, this theoretical representation does not necessarily guarantee that infinitely divisible distributions are always easy to simulate.)

We also note that if the finite mixture

$$\sum_{i=1}^{k} p_i \, f_i(x)$$

can result in a decomposition of $f(x)$ into simple components (for instance, uniform distributions on intervals) and a last residual term with a small weight, the following approximation applies: We can use a trapezoidal approximation of $f$ on intervals $[a_i, b_i]$, the weight $p_i$ being of the order of $\int_{a_i}^{b_i} f(x)dx$. Devroye (1985) details the applications of this method in the case where $f$ is a polynomial on $[0, 1]$.

CHAPTER 3

# Monte Carlo Integration

Cadfael had heard the words without hearing them and enlightenment fell on him so dazzlingly that he stumbled on the threshold.
—Ellis Peter, *The Heretic's Apprentice*

## 3.1 Introduction

Two major classes of numerical problems that arise in statistical inference are *optimization* problems and *integration* problems. (An associated problem, that of *implicit equations*, can often be reformulated as an optimization problem.) Although optimization is generally associated with the likelihood approach, and integration with the Bayesian approach, these are not strict classifications, as shown by Examples 1.2.2 and 1.3.5, and Examples 3.1.1, 3.1.2 and 3.1.3, respectively.

Examples 1.1.1–1.3.5 have also shown that it is not always possible to derive explicit probabilistic models and even less possible to analytically compute the estimators associated with a given paradigm (maximum likelihood, Bayes, method of moments, etc.). Moreover, other statistical methods, such as *bootstrap* methods (see Note 1.6.3), although unrelated to the Bayesian approach, may involve the integration of the empirical cdf. Similarly, alternatives to standard likelihood, such as *marginal* likelihood, may require the integration of the nuisance parameters (Barndorff-Nielsen and Cox 1994).

Note in addition that Bayes estimators are not always posterior expectations. In general, the Bayes estimate under the loss function $L(\theta, \delta)$ and the prior $\pi$ is the solution of the minimization program

$$(3.1.1) \qquad \min_{\delta} \int_{\Theta} L(\theta, \delta)\, \pi(\theta)\, f(x|\theta)\, d\theta\,.$$

Only when the loss function is the quadratic function $\|\theta - \delta\|^2$ will the Bayes estimator be a posterior expectation. While some other loss functions lead to general solutions $\delta^\pi(x)$ of (3.1.1) in terms of $\pi(\theta|x)$ (see, for instance, Robert 1994a, 1996c for the case of *intrinsic losses*), a specific setup where the loss function is constructed by the decision-maker almost always precludes analytical integration of (3.1.1). This necessitates an approximate solution of (3.1.1) either by numerical methods or by simulation.

Thus, whatever the type of statistical inference, we are led to consider numerical solutions. The previous chapter has illustrated a number of methods for the generation of random variables with any given distribution and, hence, provides a basis for the construction of solutions to our statistical problems. Thus, just as the search for a stationary state in a dynamical system in physics or in economics can require one or several simulations of successive states of the system, statistical inference on complex models will often require the use of simulation techniques. (See, for instance, Bauwens 1984, Bauwens and Richard 1985, and Gouriéroux and Monfort 1995 for illustrations in econometrics.) We now look at a number of examples illustrating these situations before embarking on a description of simulation-based integration methods.

**Example 3.1.1** $L_1$ **loss.** For $\theta \in \mathbb{R}$ and $L(\theta, \delta) = |\theta - \delta|$, the Bayes estimator associated with $\pi$ is the posterior median of $\pi(\theta|x)$, $\delta^{\pi}(x)$, which is the solution to the equation

$$(3.1.2) \qquad \int_{\theta \leq \delta^{\pi}(x)} \pi(\theta) \, f(x|\theta) \, d\theta = \int_{\theta \geq \delta^{\pi}(x)} \pi(\theta) \, f(x|\theta) \, d\theta \ .$$

In the setup of Example 1.2.4, that is, when $\lambda = \|\theta\|^2$ and $X \sim \mathcal{N}_p(\theta, I_p)$, this equation is quite complex, since, when using the reference prior of Example 1.3.2,

$$\pi(\lambda|x) \propto \lambda^{p-1/2} \int e^{-\|x-\theta\|^2/2} \prod_{i=1}^{p-2} \sin(\varphi_i)^{p-i-1} \, d\varphi_1 \ldots d\varphi_{p-1} \ ,$$

where $\lambda, \varphi_1, \ldots, \varphi_{p-1}$ are the polar coordinates of $\theta$, that is, $\theta_1 = \lambda \cos(\varphi_1)$, $\theta_2 = \lambda \sin(\varphi_1) \cos(\varphi_2)$, .... (See Problem 1.24.)                $\|$

**Example 3.1.2 Piecewise linear and quadratic loss functions.** Consider a loss function which is piecewise quadratic,

$$(3.1.3) \quad L(\theta, \delta) = w_i(\theta - \delta)^2 \quad \text{when} \quad \theta - \delta \in [a_i, a_{i+1}), \quad w_i > 0.$$

Differentiating the posterior expectation (3.1.3) shows that the associated Bayes estimator satisfies

$$\sum_i w_i \int_{a_i}^{a_{i+1}} (\theta - \delta^{\pi}(x)) \, \pi(\theta|x) \, d\theta = 0 \ ,$$

that is,

$$\delta^{\pi}(x) = \frac{\sum_i w_i \int_{a_i}^{a_{i+1}} \theta \, \pi(\theta) \, f(x|\theta) \, d\theta}{\sum_i w_i \int_{a_i}^{a_{i+1}} \pi(\theta) \, f(x|\theta) \, d\theta} \ .$$

Although formally explicit, the computation of $\delta^{\pi}(x)$ requires the computation of the posterior means restricted to the intervals $[a_i, a_{i+1})$ and of the posterior probabilities of these intervals.

Similarly, consider a piecewise linear loss function,

$$L(\theta, \delta) = w_i|\theta - \delta| \quad \text{if} \quad \theta - \delta \in [a_i, a_{i+1}),$$

or Huber's (1972) loss function,

$$L(\theta, \delta) = \begin{cases} \rho(\theta - \delta)^2 & \text{if } |\theta - \delta| < c \\ 2\rho c\{|\theta - \delta| - c/2\} & \text{otherwise,} \end{cases}$$

where $\rho$ and $c$ are specified constants. Although a specific type of prior distribution leads to explicit formulas, most priors result only in integral forms of $\delta^\pi$. Some of these may be quite complex.          ‖

Inference based on *classical decision theory* evaluates the performance of estimators (maximum likelihood estimator, best unbiased estimator, moment estimator, etc.) through the loss imposed by the decision-maker or by the setting. Estimators are then compared through their expected losses, also called risks. In most cases, it is impossible to obtain an analytical evaluation of the risk of a given estimator, or even to establish that a new estimator (uniformly) dominates a standard estimator.

It may seem that the topic of *James–Stein* estimation is an exception to this observation, given the abundant literature on the topic. In fact, for some families of distributions (such as exponential or spherically symmetric) and some types of loss functions (such as quadratic or concave), it is possible to analytically establish domination results over the maximum likelihood estimator or unbiased estimators (see Robert 1994a, Chapter 2, or Lehmann and Casella 1998, Chapter 5). Nonetheless, in these situations, estimators such as *empirical Bayes estimators*, which are quite attractive in practice, will rarely allow for analytic expressions. This makes their evaluation under a given loss problematic.

Given a sampling distribution $f(x|\theta)$ and a conjugate prior distribution $\pi(\theta|\lambda, \mu)$, the empirical Bayes method estimates the *hyperparameters* $\lambda$ and $\mu$ from the *marginal distribution*

$$m(x|\lambda, \mu) = \int f(x|\theta) \, \pi(\theta|\lambda, \mu) \, d\theta$$

by maximum likelihood. The estimated distribution $\pi(\theta|\hat{\lambda}, \hat{\mu})$ is often used as in a standard Bayesian approach (that is, without taking into account the effect of the substitution) to derive a point estimator. See Searle *et al.* (1992, Chapter 9) or Carlin and Louis (1996) for a more detailed discussion on this approach. (We note that this approach is sometimes called *parametric* empirical Bayes, as opposed to the *nonparametric* empirical Bayes approach developed by Robbins. See Robbins 1964, 1983 or Maritz and Lwin 1989 for details.) The following example illustrates some difficulties encountered in evaluating empirical Bayes estimators (see also Example 3.7.1).

**Example 3.1.3 Empirical Bayes estimator.** Let $X$ have the distribution $X \sim \mathcal{N}_p(\theta, I_p)$ and let $\theta \sim \mathcal{N}_p(\mu, \lambda I_p)$, the corresponding conjugate prior. The hyperparameter $\mu$ is often specified, and here we take $\mu = 0$. In the empirical Bayes approach, the scale hyperparameter $\lambda$ is replaced by the maximum likelihood estimator, $\hat{\lambda}$, based on the marginal distribution $X \sim \mathcal{N}_p(0, (\lambda + 1)I_p)$. This leads to the maximum likelihood estimator $\hat{\lambda} = (\|x\|^2/p - 1)^+$. Since the posterior distribution of $\theta$ given $\lambda$ is $\mathcal{N}_p(\lambda x/(\lambda + 1), \lambda I_p/(\lambda + 1))$, the empirical Bayes inference may be based on the pseudo-posterior $\mathcal{N}_p(\hat{\lambda}x/(\hat{\lambda} + 1), \hat{\lambda}I_p/(\hat{\lambda} + 1))$. If, for instance, $\|\theta\|^2$ is the quantity of interest, and if it is evaluated under a quadratic loss, the empirical Bayes estimator is

$$
\begin{aligned}
\delta^{eb}(x) &= \left(\frac{\hat{\lambda}}{\hat{\lambda}+1}\right)^2 \|x\|^2 + \frac{\hat{\lambda}p}{\hat{\lambda}+1} \\
&= \left[\left(1 - \frac{p}{\|x\|^2}\right)^+\right]^2 \|x\|^2 + p\left(1 - \frac{p}{\|x\|^2}\right)^+ \\
&= (\|x\|^2 - p)^+ .
\end{aligned}
$$

This estimator dominates both the best unbiased estimator, $\|x\|^2 - p$, and the maximum likelihood estimator based on $\|x\|^2 \sim \chi_p^2(\|\theta\|^2)$ (see Saxena and Alam 1982 and Example 1.2.5). However, since the proof of this second domination result is quite involved, one might first check for domination through a simulation experiment that evaluates the risk function,

$$
R(\theta, \delta) = \mathbb{E}_\theta[(\|\theta\|^2 - \delta)^2] ,
$$

for the three estimators. This quadratic risk is often normalized by $1/(2\|\theta\|^2 + p)$ (which does not affect domination results but ensures the existence of a minimax estimator; see Robert 1994a). ‖

A general solution to the different computational problems contained in the previous examples and in those of §1.1 is to use simulation, of either the true or approximate distributions to calculate the quantities of interest. In the setup of Decision Theory, whether it is classical or Bayesian, this solution is natural, since risks and Bayes estimators involve integrals with respect to probability distributions. We will see in Chapter 5 why this solution also applies in the case of maximum likelihood estimation. Note that the possibility of producing an almost infinite number of random variables distributed according to a given distribution gives us access to the use of *frequentist* and *asymptotic* results much more easily than in usual inferential settings (see Serfling 1980 or Lehmann and Casella 1998, Chapter 6) where the sample size is most often fixed. One can, therefore, apply probabilistic results such as the Law of Large Numbers or the Central Limit Theorem, since they allow for an assessment of the convergence of simulation methods (which is equivalent to the deterministic bounds used by numerical approaches.)

### 3.2 Classical Monte Carlo Integration

Before applying our simulation techniques to more practical problems, we first need to develop their properties in some detail. This is more easily accomplished by looking at the generic problem of evaluating the integral

$$(3.2.1) \qquad \mathbb{E}_f[h(X)] = \int_{\mathcal{X}} h(x)\, f(x)\, dx \, .$$

Based on previous developments, it is natural to propose using a sample $(X_1, \ldots, X_m)$ generated from the density $f$ to approximate (3.2.1) by the empirical average[1]

$$\overline{h}_m = \frac{1}{m} \sum_{j=1}^{m} h(x_j) \, ,$$

since $\overline{h}_m$ converges almost surely to $\mathbb{E}_f[h(X)]$ by the Strong Law of Large Numbers. Moreover, when $h^2$ has a finite expectation under $f$, the speed of convergence of $\overline{h}_m$ can be assessed since the variance

$$\mathrm{var}(\overline{h}_m) = \frac{1}{m} \int_{\mathcal{X}} (h(x) - \mathbb{E}_f[h(X)])^2 \, f(x) dx$$

can also be estimated from the sample $(X_1, \ldots, X_m)$ through

$$v_m = \frac{1}{m^2} \sum_{j=1}^{m} [h(x_j) - \overline{h}_m]^2 \, .$$

For $m$ large,

$$\frac{\overline{h}_m - \mathbb{E}_f[h(X)]}{\sqrt{v_m}}$$

is therefore approximately distributed as a $\mathcal{N}(0,1)$ variable, and this leads to the construction of a convergence test and of confidence bounds on the approximation of $\mathbb{E}_f[h(X)]$.

**Example 3.2.1 Cauchy prior.** For the problem of estimating a normal mean, it is sometimes the case that a *robust* prior is desired (see, for example, Berger 1985, Section 4.7). A degree of robustness can be achieved with a Cauchy prior, so we have the model

$$(3.2.2) \qquad X \sim \mathcal{N}(\theta, 1), \quad \theta \sim \mathcal{C}(0, 1).$$

Under squared error loss, the posterior mean is

$$(3.2.3) \qquad \delta^\pi(x) = \frac{\displaystyle\int_{-\infty}^{\infty} \frac{\theta}{1+\theta^2} e^{-(x-\theta)^2/2} d\theta}{\displaystyle\int_{-\infty}^{\infty} \frac{1}{1+\theta^2} e^{-(x-\theta)^2/2} d\theta}$$

---

[1] This approach is often referred to as the *Monte Carlo method*, following Metropolis and Ulam (1949). We will meet Metropolis again in Chapters 5 and 6, with the simulated annealing and MCMC methods.

From the form of $\delta^\pi$ we see that we can simulate iid variables $\theta_1, \cdots, \theta_m \sim \mathcal{N}(x, 1)$ and calculate

$$\hat{\delta}^\pi_m(x) = \frac{\sum_{i=1}^m \dfrac{\theta_i}{1 + \theta_i^2}}{\sum_{i=1}^m \dfrac{1}{1 + \theta_i^2}} .$$

The Law of Large Numbers implies that $\hat{\delta}^\pi_m(x)$ goes to $\delta^\pi(x)$ as $m$ goes to $\infty$ (see Problem 3.1 for details).          ‖

**Example 3.2.2 (Continuation of Example 3.1.3)**  If the quadratic loss of Example 3.1.3 is normalized by $1/(2\|\theta\|^2 + p)$, the resulting Bayes estimator is

$$\delta^\pi(x) = \mathbb{E}^\pi \left[ \frac{\|\theta\|^2}{2\|\theta\|^2 + p} \,\Big|\, x, \lambda \right] \Big/ \mathbb{E}^\pi \left[ \frac{1}{2\|\theta\|^2 + p} \,\Big|\, x, \lambda \right] ,$$

which does not have an explicit form.

Consider the evaluation of $\delta^\pi$ for $p = 3$ and $x = (0.1, 1.2, -0.7)$. We can write $\delta^\pi$ as

$$\delta^\pi(x) = \frac{1}{2} \left\{ \mathbb{E}^\pi [(2\|\theta\|^2 + 3)^{-1}|x]^{-1} - 3 \right\} ,$$

requiring the computation of

$$\mathbb{E}^\pi [(2\|\theta\|^2 + 3)^{-1}|x] = \int_{\mathbb{R}^3} (2\|\theta\|^2 + 3)^{-1} \, \pi(\theta|x) \, d\theta .$$

If $\pi(\theta)$ is the noninformative prior distribution proportional to $\|\theta\|^{-2}$ (see Example 1.3.2), we would need a sample $(\theta_1, \ldots, \theta_m)$ from the posterior distribution

(3.2.4)          $$\pi(\theta|x) \propto \|\theta\|^{-2} \exp\left\{-\|x - \theta\|^2/2\right\} .$$

The simulation of (3.2.4) can be done by representing $\theta$ in polar coordinates $(\rho, \varphi_1, \varphi_2)$ $(\rho > 0, \varphi_1 \in [0, 2\pi], \varphi_2 \in [-\pi/2, \pi/2])$, with $\theta = (\rho\cos\varphi_1, \rho\sin\varphi_1\cos\varphi_2, \rho\sin\varphi_1\sin\varphi_2)$, which yields

$$\pi(\rho, \varphi_1, \varphi_2|x) \propto \exp\left\{\rho x \cdot (\theta/\rho) - \rho^2/2\right\} \sin(\varphi_1) .$$

If we denote $\xi = \theta/\rho$, which only depends on $(\varphi_1, \varphi_2)$, then $\rho|\varphi_1, \varphi_2 \sim \mathcal{N}(x \cdot \xi, 1)$ truncated to $\mathbb{R}^+$. The integration of $\rho$ then leads to

$$\pi(\varphi_1, \varphi_2|x) \propto \Phi(-x \cdot \xi) \exp\{(x \cdot \xi)^2/2\} \sin(\varphi_1),$$

where $x \cdot \xi = x_1 \cos(\varphi_1) + x_2 \sin(\varphi_1) \cos(\varphi_2) + x_3 \sin(\varphi_1) \sin(\varphi_2)$. Unfortunately, the marginal distribution of $(\varphi_1, \varphi_2)$ is not directly available since it involves the cdf of the normal distribution, $\Phi$.

For simulation purposes, we can modify the polar coordinates in order to remove the positivity constraint on $\rho$. The alternative constraint becomes $\varphi_1 \in [-\pi/2, \pi/2]$, which ensures identifiability for the model. Since $\rho$ now varies in $\mathbb{R}$, the marginal distribution of $(\varphi_1, \varphi_2)$ is

$$\pi(\varphi_1, \varphi_2|x) \propto \exp\{(x \cdot \xi)^2/2\} \sin(\varphi_1),$$

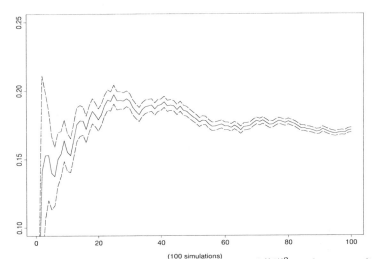

Figure 3.2.1.   *Convergence of the Bayes estimator of $||\theta||^2$ under normalized quadratic loss for the reference prior $\pi(\theta) = ||\theta||^{-2}$ and the observation $x = (0.1, 1.2, -0.7)$. The envelope provides a nominal 95% confidence interval on $||\theta||^2$.*

which can be simulated by an Accept–Reject algorithm using the instrumental function $\sin(\varphi_1)\exp\{||x||^2/2\}$. One can, therefore, simulate $(\varphi_1, \varphi_2)$ based on a uniform distribution on the half-unit sphere. The algorithm corresponding to this decomposition is the following:

### Algorithm A.19 –Polar Simulation–

1. Simulate $(\varphi_1, \varphi_2)$ from the uniform distribution on the half-unit sphere and $U$ from $\mathcal{U}_{[0,1]}$ until

$$U \le \exp\{x \cdot \xi - ||x||^2/2\} \ . \qquad [A.19]$$

2. Generate $\rho$ from the normal distribution

$$\mathcal{N}(x_1\cos(\varphi_1) + x_2\sin(\varphi_1)\cos(\varphi_2) + x_3\sin(\varphi_1)\sin(\varphi_2), 1) \ .$$

The sample resulting from $[A.19]$ provides a subsample $(\rho_1, \ldots, \rho_m)$ in step 2 and an approximation of $\mathbb{E}^\pi[(2\rho^2 + 3)^{-1}|X]$,

$$T_m = \frac{1}{m} \sum_{j=1}^m (2\rho_j^2 + 3)^{-1} \ .$$

Figure 3.2.1 gives a realization of a sequence of $T_m$, the envelope being constructed from the normal approximation through the 95% confidence interval $T_m \pm 1.96\sqrt{v_m}$ .     $\parallel$

The approach followed in the above example can be successfully utilized in many cases, even though it is often possible to achieve greater efficiency

| $n$ | 0.0 | 0.67 | 0.84 | 1.28 | 1.65 | 2.32 | 2.58 | 3.09 | 3.72 |
|---|---|---|---|---|---|---|---|---|---|
| $10^2$ | 0.485 | 0.74 | 0.77 | 0.9 | 0.945 | 0.985 | 0.995 | 1 | 1 |
| $10^3$ | 0.4925 | 0.7455 | 0.801 | 0.902 | 0.9425 | 0.9885 | 0.9955 | 0.9985 | 1 |
| $10^4$ | 0.4962 | 0.7425 | 0.7941 | 0.9 | 0.9498 | 0.9896 | 0.995 | 0.999 | 0.9999 |
| $10^5$ | 0.4995 | 0.7489 | 0.7993 | 0.9003 | 0.9498 | 0.9898 | 0.995 | 0.9989 | 0.9999 |
| $10^6$ | 0.5001 | 0.7497 | 0.8 | 0.9002 | 0.9502 | 0.99 | 0.995 | 0.999 | 0.9999 |
| $10^7$ | 0.5002 | 0.7499 | 0.8 | 0.9001 | 0.9501 | 0.99 | 0.995 | 0.999 | 0.9999 |
| $10^8$ | 0.5 | 0.75 | 0.8 | 0.9 | 0.95 | 0.99 | 0.995 | 0.999 | 0.9999 |

Table 3.2.1. *Evaluation of some normal quantiles by a regular Monte Carlo experiment based on n replications of a normal generation. The last line gives the exact values.*

through numerical methods (Riemann quadrature, Simpson method, etc.) in dimension 1 or 2. The scope of application of this Monte Carlo integration method is obviously not only limited to the Bayesian paradigm since, similar to Example 3.2.2, the performances of complex procedures can be measured in any setting where the distributions involved in the model can be simulated. We mentioned in §3.1 the potential of this approach in evaluating estimators based on a decision-theoretic formulation. The same applies for testing, when the level of significance of a test, and its power function, cannot be easily computed, and simulation thus can provide a useful improvement over asymptotic approximations when explicit computations are impossible.

**Example 3.2.3 Normal cdf.** Since the normal cdf cannot be written in an explicit form, a possible way to construct normal distribution tables is to use simulation. Consider the generation of a sample of size $n$, $(x_1, \ldots, x_n)$, based on the Box–Muller algorithm $[A_4]$ of Example 2.2.2.

The approximation of

$$\Phi(t) = \int_{-\infty}^{t} \frac{1}{\sqrt{2\pi}} e^{-y^2/2} dy$$

by the Monte Carlo method is thus

$$\hat{\Phi}(t) = \frac{1}{n} \sum_{i=1}^{n} \mathbb{I}_{x_i \leq t},$$

with (exact) variance $\Phi(t)(1 - \Phi(t))/n$ (as the variables $\mathbb{I}_{x_i \leq t}$ are independent Bernoulli with success probability $\Phi(t)$). For values of $t$ around $t = 0$, the variance is thus approximately $1/4n$, and to achieve a precision of four decimals, the approximation requires on average $n = (\sqrt{2}\ 10^4)^2$ simulations, that is, 200 million iterations. Table 3.2.1 gives the evolution of this approximation for several values of $t$ and shows an accurate evaluation for 100 million iterations. Note that greater accuracy is achieved in the tails and that more efficient simulations methods could be used, as in Example 3.3.1. ‖

Many tests are based on an asymptotic normality assumption as, for instance, the *likelihood ratio test*. Given $H_0$, a null hypothesis corresponding to $r$ independent constraints on the parameter $\theta \in \mathbb{R}^k$, denote by $\hat{\theta}$ and $\hat{\theta}^0$ the unconstrained and constrained (under $H_0$) maximum likelihood estimators of $\theta$, respectively. The likelihood ratio $\ell(\hat{\theta}|x)/\ell(\hat{\theta}^0|x)$ then satisfies

$$(3.2.5) \quad \log[\ell(\hat{\theta}|x)/\ell(\hat{\theta}^0|x)] = 2 \{\log \ell(\hat{\theta}|x) - \log \ell(\hat{\theta}^0|x)\} \xrightarrow{\mathcal{L}} \chi_r^2 ,$$

when the number of observations goes to infinity (see Lehmann 1986, Section 8.8, or Gouriéroux and Monfort 1996). However, the convergence in (3.2.5) is only satisfied under regularity constraints on the likelihood function (see Lehmann and Casella 1998, Chapter 6, for a full development); hence, the asymptotics may not apply.

**Example 3.2.4 Testing the number of components.** A situation where the standard regularity conditions do not apply is that of the normal mixture (see Example 1.1.2)

$$p \, \mathcal{N}(\mu, 1) + (1 - p) \, \mathcal{N}(\mu + \theta, 1) ,$$

where the constraint $\theta > 0$ ensures *identifiability*. A test on the existence of a mixture cannot be easily represented in a hypothesis test since $H_0$ : $p = 0$ effectively eliminates the mixture and results in the identifiability problem related with $\mathcal{N}(\mu + \theta, 1)$. (The inability to estimate the nuisance parameter $p$ under $H_0$ results in the likelihood not satisfying the necessary regularity conditions; see Davies 1977. However, see Lehmann and Casella 1998, Section 6.6 for mixtures where it is possible to construct efficient estimators.)

A slightly different formulation of the problem will allow a solution, however. If the identifiability constraint is taken to be $p \geq 1/2$ instead of $\theta > 0$, then $H_0$ can be represented as

$$H_0 : \quad p = 1 \quad \text{or} \quad \theta = 0 .$$

We therefore want to determine the limiting distribution of (3.2.5) under this hypothesis and under a local alternative. Figure 3.2.2 represents the empirical cdf of $2 \{\log \ell(\hat{p}, \hat{\mu}, \hat{\theta}|x) - \log \ell(\hat{\mu}^0|x)\}$ and compares it with the $\chi_2^2$ cdf, where $\hat{p}, \hat{\mu}, \hat{\theta}$, and $\hat{\mu}^0$ are the respective MLEs for 1000 simulations of a normal $\mathcal{N}(0, 1)$ sample of size 100 (Since $(\hat{p}, \hat{\mu}, \hat{\theta})$ cannot be computed analytically, we use the EM algorithm of §5.3.3.) The poor agreement between the asymptotic approximation and the empirical cdf is quite obvious. Figure 3.2.2 also shows how the $\chi_2^2$ approximation is improved if the limit (3.2.5) is replaced by an equally weighted mixture of a Dirac mass at 0 and a $\chi_2^2$ distribution.

Note that the resulting sample of the log-likelihood ratios can also be used for inferential purposes, for instance to derive an exact test via the estimation of the quantiles of the distribution of (3.2.5) under $H_0$ or to evaluate the power of a standard test.                    ‖

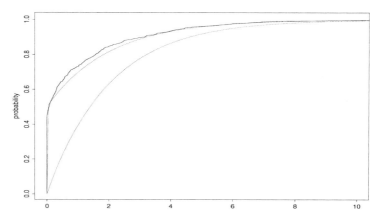

Figure 3.2.2. *Empirical cdf of a sample of log-likelihood ratios for the test of presence of a Gaussian mixture (solid lines) and comparison with the cdf of a $\chi_2^2$ distribution (dotted lines) and with the cdf of a .5 − .5 mixture of a $\chi_2^2$ distribution and of a Dirac mass at 0 (dashed lines) (based on 1000 simulations of a normal $\mathcal{N}(0,1)$ sample of size 100).*

It may seem that the method proposed above is sufficient to approximate integrals like (3.2.1) in a controlled way. However, while the straightforward Monte Carlo method indeed provides good approximations of (3.2.1) in most regular cases, there exist more efficient alternatives which not only avoid a direct simulation from $f$ but also can be used repeatedly for several integrals of the form (3.2.1). The repeated use can be either for a family of functions $h$ or a family of densities $f$. In particular, the usefulness of this flexibility is quite evident in Bayesian analyses of *robustness*, of *sensitivity* (see Berger 1990, 1994), or for the computation of power functions of specific tests (see Lehmann 1986, or Gouriéroux and Monfort 1996).

## 3.3 Importance Sampling

### 3.3.1 Principles

The method we now study is called *importance sampling* because it is based on so-called *importance functions*, and although it would be more accurate to call it "weighted sampling," we will follow common usage. We start this section with a somewhat unusual example, borrowed from Ripley (1987), which shows that it may actually pay to generate from a distribution other than the distribution $f$ of interest. (See Note 3.9.1 for a global approach to the approximation of tail probabilities by *large deviation* techniques.)

**Example 3.3.1 Cauchy tail probability.** Suppose that the quantity of interest is the probability, $p$, that a Cauchy $\mathcal{C}(0,1)$ variable is larger than

2, that is,

$$p = \int_2^{+\infty} \frac{1}{\pi(1+x^2)} \, dx \; .$$

When $p$ is evaluated through the empirical average

$$\hat{p}_1 = \frac{1}{m} \sum_{j=1}^m \mathbb{I}_{X_j > 2}$$

of an iid sample $X_1, \ldots, X_m \sim \mathcal{C}(0,1)$, the variance of this estimator is $p(1-p)/m$ (equal to $0.127/m$ since $p = 0.15$). This variance can be reduced by taking into account the symmetric nature of $\mathcal{C}(0,1)$, since the average

$$\hat{p}_2 = \frac{1}{2m} \sum_{j=1}^m \mathbb{I}_{|x_j| > 2}$$

has variance $p(1 - 2p)/2m$ equal to $0.052/m$.

The (relative) inefficiency of these methods is due to the generation of values outside the domain of interest, $[2, +\infty)$, which are, in some sense, irrelevant for the approximation of $p$. If $p$ is written as

$$p = \frac{1}{2} - \int_0^2 \frac{1}{\pi(1+x^2)} \, dx \; ,$$

the integral above can be considered to be the expectation of $h(X) = 2/\pi(1 + X^2)$, where $X \sim \mathcal{U}_{[0,2]}$. An alternative method of evaluation for $p$ is therefore

$$\hat{p}_3 = \frac{1}{2} - \frac{1}{m} \sum_{j=1}^m h(U_j)$$

for $U_j \sim \mathcal{U}_{[0,2]}$. The variance of $\hat{p}_3$ is $(\mathbb{E}[h^2] - \mathbb{E}[h]^2)/m$ and an integration by parts shows that it is equal to $0.0092/m$. Moreover, since $p$ can be written as

$$p = \int_0^{1/2} \frac{y^{-2}}{\pi(1 + y^{-2})} \, dy \; ,$$

this integral can also be seen as the expectation of $\frac{1}{4} h(Y) = 1/2\pi(1 + Y^2)$ against the uniform distribution on $[0, 1/2]$ and another evaluation of $p$ is

$$\hat{p}_4 = \frac{1}{4m} \sum_{j=1}^m h(Y_j)$$

when $Y_j \sim \mathcal{U}_{[0,1/2]}$. The same integration by parts shows that the variance of $\hat{p}_4$ is then $0.95 \; 10^{-4}/m$.

Compared with $\hat{p}_1$, the reduction in variance brought by $\hat{p}_4$ is of order $10^{-3}$, which implies, in particular, that this evaluation requires $\sqrt{1000} \approx 33$ times fewer simulations than $\hat{p}_1$ to achieve the same precision.                    ‖

The evaluation of (3.2.1) based on simulation from $f$ is therefore not necessarily optimal and Theorem 3.3.4 shows that this choice is, in fact, always suboptimal. Note also that the integral (3.2.1) can be represented in an infinite number of ways by triplets $(\mathcal{X}, h, f)$. Therefore, the search for an optimal estimator should encompass all these possible representations (as in Example 3.3.1). As a side remark, we should stress that the very notion of "optimality" of a representation is quite difficult to define precisely. Indeed, as already noted in Chapter 2, the comparison of simulation methods cannot be equated with the comparison of the variances of the resulting estimators. Conception and computation times should also be taken into account. At another level, note that the optimal method proposed in Theorem 3.3.4 depends on the function $h$ involved in (3.2.1). Therefore, it cannot be considered as optimal when several integrals related to $f$ are simultaneously evaluated. In such cases, which often occur in Bayesian analysis, only generic methods can be compared(that is to say, those which are independent of $h$).

The principal alternative to direct sampling from $f$ for the evaluation of (3.2.1) is to use importance sampling, defined as follows:

**Definition 3.3.2** The method of *importance sampling* is an evaluation of (3.2.1) based on generating a sample $X_1, \ldots, X_n$ from a given distribution $g$ and approximating

$$(3.3.1) \qquad \mathbb{E}_f[h(X)] \approx \frac{1}{m} \sum_{j=1}^{m} \frac{f(X_j)}{g(X_j)} \, h(X_j) \, .$$

This method is based on the alternative representation of (3.2.1):

$$\mathbb{E}_f[h(X)] = \int_{\mathcal{X}} h(x) \, \frac{f(x)}{g(x)} \, g(x) \, dx \, ,$$

and the estimator (3.3.1) converges to (3.2.1) for the same reason the regular Monte Carlo estimator $\overline{h}_m$ converges, whatever the choice of the distribution $g$ (as long as supp$(g) \supset$ supp$(f)$). This method is therefore of considerable interest since it puts very little restriction on the choice of the instrumental distribution $g$, which can be chosen from distributions that are easy to simulate. Moreover, the same sample (generated from $g$) can be used repeatedly, not only for different functions $h$ but also for different densities $f$, a feature which is quite attractive for robustness and Bayesian sensitivity analyses.

**Example 3.3.3 Exponential and log-normal comparison.** Consider $X$ as an estimator of $\lambda$, when $X \sim \mathcal{E}xp(1/\lambda)$ or when $X \sim \mathcal{LN}(0, \sigma^2)$ (with $e^{\sigma^2/2} = \lambda$, see Problem 3.5). If the goal is to compare the performances of this estimator under both distributions for the scaled squared error loss

$$L(\lambda, \delta) = (\delta - \lambda)^2/\lambda^2,$$

a *single* sample from $\mathcal{LN}(0, \sigma^2)$, $X_1, \ldots, X_T$, can be used for both purposes, the risks being evaluated by

$$\hat{R}_1 = \frac{1}{T\lambda^2} \sum_{t=1}^{T} X_t e^{-X_t/\lambda} \lambda^{-1} e^{\log(X_t)^2/2\sigma^2} \sqrt{2\pi\sigma}(X_t - \lambda)^2$$

in the exponential case and by

$$\hat{R}_2 = \frac{1}{T\lambda^2} \sum_{t=1}^{T} (X_t - \lambda)^2$$

in the log-normal case. In addition, the scale nature of the parameterization allows a single sample $(Y_1^0, \ldots, Y_T^0)$ from $\mathcal{N}(0, 1)$ to be used for all $\sigma$'s, with $X_t = \exp(\sigma Y_t^0)$.

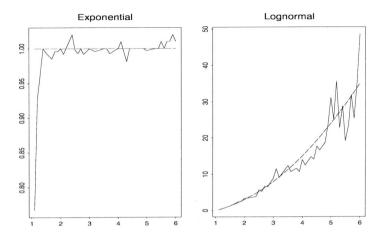

Figure 3.3.1. *Graph of approximate scaled squared error risks of X vs.* $\lambda$ *for an exponential and a log-normal observation, compared with the theoretical values (dashes) for* $\lambda \in [1, 6]$ *(10,000 simulations).*

The comparison of these evaluations is given in Figure 3.3.1 for $T = 10,000$, each point corresponding to a sample of size $T$ simulated from $\mathcal{LN}(0, \sigma^2)$ by the above transformation. The exact values are given by 1 and $(\lambda+1)(\lambda-1)$, respectively. Note that implementing importance sampling in the opposite way offers little appeal since the weights $\exp\{-\log(X_t)^2/2\sigma^2\} \times \exp(\lambda X_t)/X_t$ have infinite variance (see below). The graph of the risk in the exponential case is then more stable than for the original sample from the log-normal distribution. ‖

### 3.3.2 Finite Variance Estimators

Although the distribution $g$ can be almost any density for the estimator (3.3.1) to converge, there are obviously some choices that are better than others, and it is natural to try to compare different distributions $g$ for the evaluation of (3.2.1). First, note that, while (3.3.1) does converge almost surely to (3.2.1), its variance is finite only when the expectation

$$\mathbb{E}_g\left[h^2(X)\frac{f^2(X)}{g^2(X)}\right] = \mathbb{E}_f\left[h^2(X)\frac{f(X)}{g(X)}\right] = \int_{\mathcal{X}} h^2(x)\,\frac{f^2(x)}{g(x)}\,dx < \infty .$$

Thus:

- Instrumental distributions with tails lighter than those of $f$ (that is, those with unbounded ratios $f/g$) are not appropriate for importance sampling. In fact, in these cases, the variances of the corresponding estimators (3.3.1) will be infinite for many functions $h$.

- If the ratio $f/g$ is unbounded, the weights $f(x_j)/g(x_j)$ will vary widely, giving too much importance to a few values $x_j$.

Distributions $g$ with thicker tails than $f$ ensure that the ratio $f/g$ does not cause the divergence of $\mathbb{E}_f[h^2 f/g]$. In particular, Geweke (1989) mentions two types of sufficient conditions:

(a) $f(x)/g(x) < M \;\; \forall x \in \mathcal{X}$ and $\mathrm{var}_f(h) < \infty$ ;

(b) $\mathcal{X}$ is compact, $f(x) < F$ and $g(x) > \varepsilon \;\; \forall x \in \mathcal{X}$.

These conditions are quite restrictive. In particular, $f/g < M$ implies that the Accept–Reject algorithm [$A.5$] also applies. (A comparison between both approaches is given in §3.3.3.)

Among the distributions $g$ leading to finite variances for the estimator (3.3.1), it is, in fact, possible to exhibit the optimal distribution corresponding to a given function $h$ and a fixed distribution $f$, as stated by the following result of Rubinstein (1981); see also Geweke (1989).

**Theorem 3.3.4** *The choice of $g$ that minimizes the variance of the estimator (3.3.1) is*

$$g^*(x) = \frac{|h(x)|\, f(x)}{\int_{\mathcal{X}} |h(z)|\, f(z)\, dz} .$$

*Proof.* First note that

$$\mathrm{var}\left[\frac{h(X)f(X)}{g(X)}\right] = \mathbb{E}_g\left[\frac{h^2(X)f^2(X)}{g^2(X)}\right] - \left(\mathbb{E}_g\left[\frac{h(X)f(X)}{g(X)}\right]\right)^2 ,$$

and the second term does not depend on $g$. So, to minimize variance, we only need minimize the first term. From Jensen's inequality it follows that

$$\mathbb{E}_g\left[\frac{h^2(X)f^2(X)}{g^2(X)}\right] \geq \left(\mathbb{E}_g\left[\frac{|h(X)|f(X)}{g(X)}\right]\right)^2 = \left(\int |h(x)|f(x)dx\right)^2 ,$$

which provides a lower bound that is independent of the choice of $g$. It is straightforward to verify that this lower bound is attained by choosing $g = g^*$.          □□

This optimality is rather formal since, when $h(x) > 0$, the optimal choice of $g^*(x)$ requires the knowledge of $\int h(x)f(x)dx$, the integral of interest! A practical alternative taking advantage of Theorem 3.3.4 is to use

$$(3.3.2) \qquad \frac{\sum_{j=1}^{m} h(x_j)\, f(x_j)/g(x_j)}{\sum_{j=1}^{m} f(x_j)/g(x_j)} ,$$

where $f$ and $g$ are known up to constants, since (3.3.2) also converges to $\int h(x)f(x)dx$ by the Strong Law of Large Numbers. However, the optimality of Theorem 3.3.4 does not transfer to (3.3.2), which is biased. From a general point of view, Casella and Robert (1998) have nonetheless shown that the weighted estimator (3.3.2) may perform better (when evaluated under squared error loss) in some settings. (See also Van Dijk and Kloeck 1984.)

From a practical point of view, Theorem 3.3.4 suggests looking for distributions $g$ for which $|h|f/g$ is almost constant with finite variance. It is important to note that although the finite variance constraint is not necessary for the convergence of (3.3.1) and of (3.3.2), importance sampling performs quite poorly when

$$(3.3.3) \qquad \int \frac{f^2(x)}{g(x)}\, dx = +\infty ,$$

whether in terms of behavior of the estimator (high-amplitude jumps, instability of the path of the average, slow convergence) or of comparison with direct Monte Carlo methods. Distributions $g$ such that (3.3.3) occurs are therefore not recommended.

The next two examples show that importance sampling methods can bring considerable improvement of naïve Monte Carlo estimates when implemented with care. However, they can encounter disastrous performances and produce extremely poor estimates when the variance conditions are not met.

**Example 3.3.5 Student's $t$ distribution.** Consider $X \sim T(\nu, \theta, \sigma^2)$, with density

$$f(x) = \frac{\Gamma((\nu+1)/2)}{\sigma\sqrt{\nu\pi}\,\Gamma(\nu/2)} \left(1 + \frac{(x-\theta)^2}{\nu\sigma^2}\right)^{-(\nu+1)/2} .$$

Without loss of generality, we take $\theta = 0$ and $\sigma = 1$. If the quantities of interest are $\mathbb{E}_f[h_i(X)]$ $(i = 1, 2, 3)$ with

$$h_1(x) = \sqrt{\left|\frac{x}{1-x}\right|}, \qquad h_2(x) = x^5 \mathbb{I}_{[2.1,\infty[}(x), \qquad h_3(x) = \frac{x^5}{1+(x-3)^2}\, \mathbb{I}_{x\geq 0} ,$$

it is possible to generate directly from $f$, since $f$ is the ratio of a $\mathcal{N}(0,1)$ variable and the square root of a $\mathcal{G}a(\nu/2, \nu/2)$ variable (a $\chi^2$ random variable with $\nu$ degrees of freedom). If $\nu$ is large, this method can be costly; importance sampling alternatives are to use a Cauchy $\mathcal{C}(0,1)$ distribution and a normal $\mathcal{N}(0, \nu/(\nu-2))$ distribution (scaled so that the variance is the

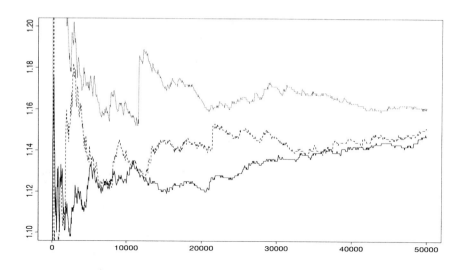

Figure 3.3.2. *Convergence of three estimators of* $\mathbb{E}_f[|X/(1-X)|^{1/2}]$ *for* $\nu = 12$: *sampling from* $f$ *(solid lines), importance sampling with a Cauchy instrumental distribution (dashes) and importance sampling with normal importance distribution (dots). The final values are respectively* 1.14, 1.14, *and* 1.16, *to compare with an exact value of* 1.13.

same as $\mathcal{T}(\nu, \theta, \sigma^2)$). The choice of the normal distribution is not expected to be optimal, as the ratio

$$\frac{f^2(x)}{g(x)} \propto \frac{e^{x^2(\nu-2)/2\nu}}{[1 + x^2/\nu]^{(\nu+1)}}$$

does not have a finite integral. However, this will give us an opportunity to study the performance of importance sampling in such a situation. On the other hand, the $\mathcal{C}(0,1)$ distribution has larger tails than $f$ and ensures that the variance of $f/g$ is finite.

Figure 3.3.2 illustrates the performances of the three corresponding estimators for the function $h_1$ when $\nu = 12$. The estimator constructed from the normal distribution converges, but with large jumps. Moreover, it differs significantly from the two other estimators after 50,000 iterations. On the other hand, the estimators derived from the true distribution and the importance sampling estimator associated with the Cauchy distribution have similar patterns, even though the latter has some jumps.

Perhaps, a more sophisticated approach is to note that both $h_2$ and $h_3$ have restricted supports and we could benefit by having the instrumental distributions take this information into account. In the case of $h_2$, a uniform distribution on $[0, 1/2.1]$ is reasonable, since the expectation $\mathbb{E}_f[h_2(X)]$ can

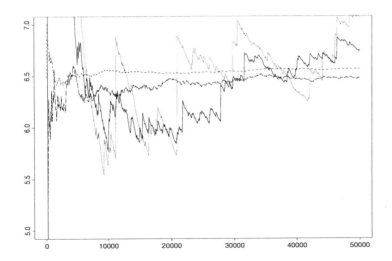

Figure 3.3.3. *Convergence of four estimators of* $\mathbb{E}_f[X^5\mathbb{I}_{X \geq 2.1}]$ *for* $\nu = 12$: *Sampling from* $f$ *(solid lines), importance sampling with Cauchy instrumental distribution (short dashes), importance sampling with uniform* $\mathcal{U}([0, 1/2.1])$ *instrumental distribution (long dashes) and importance sampling with normal instrumental distribution (dots). The final values are respectively* 6.75, 6.48, 6.57, *and* 7.06, *for an exact value of* 6.54.

be written as

$$\int_0^{1/2.1} u^{-7} f(1/u) \, du = \frac{1}{2.1} \int_0^{1/2.1} 2.1 \, u^{-7} f(1/u) \, du \, .$$

The corresponding importance sampling estimator is then

$$\delta_2 = \frac{1}{2.1m} \sum_{j=1}^{m} U_j^{-7} f(1/U_j) \, ,$$

where the $U_j$'s are iid $\mathcal{U}([0, 1/2.1])$. Figure 3.3.3 shows the improvement brought by this choice, with the estimator $\delta_2$ converging to the true value after only a few hundred iterations. The importance sampling estimator associated with the Cauchy distribution is also quite stable, but it requires more iterations to achieve the same precision. Both of the other estimators (which are based on the true distribution and the normal distribution, respectively) fluctuate around the exact value with high-amplitude jumps.

In the case of $h_3$, a reasonable candidate for the instrumental distribution is $g(x) = \exp(x)\mathbb{I}_x \geq 0$, leading to the estimation of

$$\mathbb{E}_f[h_3(X)] = \int_0^{\infty} \frac{x^5}{1 + (x-3)^2} \, f(x) \, dx$$

$$= \int_0^\infty \frac{x^5 e^x}{1 + (x-3)^2} \, f(x) \, e^{-x} \, dx$$

by

(3.3.4)                    $$\frac{1}{m} \sum_{j=1}^m h_3(X_j) \, w(X_j) \,,$$

where the $X_j$'s are iid $\mathcal{E}xp(1)$ and $w(x) = f(x) \exp(x)$. Figure 3.3.4 shows that, although this weight does not have a finite expectation under $\mathcal{T}(\nu, 0, 1)$, the estimator (3.3.4) provides a good approximation of $\mathbb{E}_f[h_3(X)]$, having the same order of precision as the estimation provided by the exact simulation, and greater stability. The estimator based on the Cauchy distribution is, as in the other case, stable, but its bias is, again, slow to vanish, and the estimator associated with the normal distribution once more displays large fluctuations which considerably hinder its convergence.                    ‖

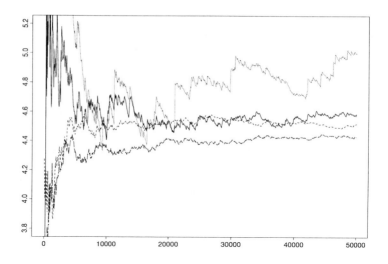

Figure 3.3.4.  *Convergence of four estimators of* $\mathbb{E}_f[h_3(X)]$*: Sampling from f (solid lines), importance sampling with Cauchy instrumental distribution (short dashes), with normal instrumental distribution (dots), and with exponential instrumental distribution (long dashes). The final values are respectively 4.58, 4.42, 4.99, and 4.52, for a true value of 4.64.*

**Example 3.3.6 Transition matrix estimation.**  Consider a Markov chain with two states, 1 and 2, whose transition matrix is

$$T = \begin{pmatrix} p_1 & 1 - p_1 \\ 1 - p_2 & p_2 \end{pmatrix} ,$$

that is,

$$P(X_{t+1} = 1|X_t = 1) = 1 - P(X_{t+1} = 2|X_t = 1) = p_1,$$
$$P(X_{t+1} = 2|X_t = 2) = 1 - P(X_{t+1} = 1|X_t = 2) = p_2 .$$

Assume, in addition, that the constraint $p_1 + p_2 < 1$ holds (see Geweke 1989 for a motivation related to continuous time processes). If the sample is $X_1, \ldots, X_m$ and the prior distribution is

$$\pi(p_1, p_2) = 2 \, \mathbb{I}_{p_1 + p_2 < 1} \, ,$$

the posterior distribution of $(p_1, p_2)$ is

$$\pi(p_1, p_2|m_{11}, m_{12}, m_{21}, m_{22}) \propto p_1^{m_{11}} (1 - p_1)^{m_{12}} (1 - p_2)^{m_{21}} p_2^{m_{22}} \, \mathbb{I}_{p_1 + p_2 < 1} \, ,$$

where $m_{ij}$ is the number of passages from $i$ to $j$, that is,

$$m_{ij} = \sum_{t=2}^{m} \mathbb{I}_{x_t = i} \mathbb{I}_{x_{t+1} = j},$$

and it follows that $\mathcal{D} = (m_{11}, \ldots, m_{22})$ is a sufficient statistic.

Suppose now that the quantities of interest are the posterior expectations of the probabilities and the associated odds:

$$h_1(p_1, p_2) = p_1, \quad h_2(p_1, p_2) = p_2, \quad h_3(p_1, p_2) = \frac{p_1}{1 - p_1}$$

and

$$h_4(p_1, p_2) = \frac{p_2}{1 - p_2}, \quad h_5(p_1, p_2) = \log\left(\frac{p_1(1 - p_2)}{p_2(1 - p_1)}\right) \, ,$$

respectively.

The distribution $\pi(p_1, p_2|\mathcal{D})$ is the restriction of the product of two distributions $\mathcal{B}e(m_{11} + 1, m_{12} + 1)$ and $\mathcal{B}e(m_{22} + 1, m_{21} + 1)$ to the simplex $\{(p_1, p_2) : p_1 + p_2 < 1\}$. So a reasonable first approach is to simulate these two distributions until the sum of two realizations is less than 1. Unfortunately, this naïve strategy is rather inefficient since, for the given data $(m_{11}, m_{12}, m_{21}, m_{22}) = (68, 28, 17, 4)$ we have $P^\pi(p_1 + p_2 < 1|\mathcal{D}) = 0.21$ (Geweke 1989). The importance sampling alternatives are to simulate distributions which are restricted to the simplex.

A solution inspired from the shape of $\pi(p_1, p_2|\mathcal{D})$ is a Dirichlet distribution $\mathcal{D}(m_{11} + 1, m_{22} + 1, m_{12} + m_{21} + 1)$, with density

$$\pi_1(p_1, p_2|\mathcal{D}) \propto p_1^{m_{11}} p_2^{m_{22}} (1 - p_1 - p_2)^{m_{12} + m_{21}} \, .$$

However, the ratio $\pi(p_1, p_2|\mathcal{D})/\pi_1(p_1, p_2|\mathcal{D})$ is not bounded and the expectation of this ratio under $\pi(p_1, p_2|\mathcal{D})$ is infinite. Geweke's (1989) proposal is to use the normal approximation to the binomial distribution, that is,

$$\pi_2(p_1, p_2|\mathcal{D}) \propto \exp\{-(m_{11} + m_{12})(p_1 - \hat{p}_1)^2/2\,\hat{p}_1(1 - \hat{p}_1)\}$$
$$\times \exp\{-(m_{21} + m_{22})(p_2 - \hat{p}_2)^2/2\,\hat{p}_2(1 - \hat{p}_2)\} \, \mathbb{I}_{p_1 + p_2 \leq 1} \, ,$$

where $\hat{p}_i$ is the maximum likelihood estimator of $p_i$, that is, $m_{ii}/(m_{ii} + m_{i(3-i)})$. An efficient way to simulate $\pi_2$ is then to simulate $p_1$ from the

| Distribution | $h_1$ | $h_2$ | $h_3$ | $h_4$ | $h_5$ |
|:---:|:---:|:---:|:---:|:---:|:---:|
| $\pi_1$ | 0.748 | 0.139 | 3.184 | 0.163 | 2.957 |
| $\pi_2$ | 0.689 | 0.210 | 2.319 | 0.283 | 2.211 |
| $\pi_3$ | 0.697 | 0.189 | 2.379 | 0.241 | 2.358 |
| $\pi$ | 0.697 | 0.189 | 2.373 | 0.240 | 2.358 |

Table 3.3.1. *Comparison of the evaluations of $\mathbb{E}_f[h_j]$ for the estimators (3.3.2) corresponding to three instrumental distributions $\pi_i$ and to the true distribution $\pi$ (10,000 simulations).*

normal distribution $\mathcal{N}(\hat{p}_1, \hat{p}_1(1-\hat{p}_1)/(m_{12}+m_{11}))$ restricted to $[0, 1]$, then $p_2$ from the normal distribution $\mathcal{N}(\hat{p}_2, \hat{p}_2(1-\hat{p}_2)/(m_{21}+m_{22}))$ restricted to $[0, 1-p_1]$, using the method proposed by Geweke (1991) and Robert (1995a). The ratio $\pi/\pi_2$ then has a finite expectation under $\pi$, since $(p_1, p_2)$ is restricted to $\{(p_1, p_2) : p_1 + p_2 < 1\}$.

Another possibility is to keep the distribution $\mathcal{B}(m_{11} + 1, m_{12} + 1)$ as the marginal distribution on $p_1$ and to modify the conditional distribution $p_2^{m_{22}}(1-p_2)^{m_{21}} \, \mathbb{I}_{p_2 < 1-p_1}$ into

$$\pi_3(p_2|p_1, \mathcal{D}) = \frac{2}{(1-p_1)^2} \, p_2 \, \mathbb{I}_{p_2 < 1-p_1} \, .$$

The ratio $w(p_1, p_2) \propto p_2^{m_{22}-1}(1-p_2)^{m_{21}}(1-p_1)^2$ is then bounded in $(p_1, p_2)$.

Table 3.3.1 provides the estimators of the posterior expectations of the functions $h_j$ evaluated for the true distribution $\pi$ (simulated the naïve way, that is, until $p_1 + p_2 < 1$) and for the three instrumental distributions $\pi_1$, $\pi_2$, and $\pi_3$. The distribution $\pi_3$ is clearly preferable to the two other instrumental distributions since it provides the same estimation as the true distribution, at a lower computational cost. Note that $\pi_1$ does worse in all cases.

Figure 3.3.5 describes the evolution of the estimators (3.3.2) of $\mathbb{E}[h_5]$ as $m$ increases for the three instrumental distributions considered. Similarly to Table 3.3.1, it shows the improvement brought by the distribution $\pi_3$ upon the alternative distributions, since the precision is of the same order as the true distribution, for a significantly lower simulation cost. The jumps in the graphs of the estimators associated with $\pi_2$ and, especially, with $\pi_1$ are characteristic of importance sampling estimators with infinite variance. ‖

We therefore see that importance sampling cannot be applied blindly. Rather, care must be taken in choosing an instrumental density as the almost sure convergence of (3.3.1) is only formal (in the sense that it may require an enormous number of simulations to produce an accurate approximation of the quantity of interest). These words of caution are meant to make the user aware of the problems that might be encountered

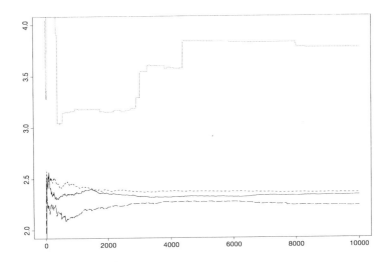

Figure 3.3.5. *Convergence of four estimators of* $\mathbb{E}_f[h_3(X)]$ *for the true distribution $\pi$ (solid lines) and for the instrumental distributions $\pi_1$ (dots), $\pi_2$ (long dashes), and $\pi_3$ (short dashes). The final values are 2.373, 3.184, 2.319, and 2.379, respectively.*

if importance sampling is used when $\mathbb{E}_f[\|f(X)/g(X)\|]$ is infinite. (When $\mathbb{E}_f[f(X)/g(X)]$ is finite, the stakes are not so high, as convergence is more easily attained.) If the issue of finiteness of the variance is ignored, and not detected, it may result in strong biases. For example, it can happen that the obvious divergence behavior of the previous examples does not occur. Thus, other measures, such as monitoring of the range of the weights $f(X_i)/g(X_i)$ (which are of mean 1 in all cases), can help to detect convergence problems. (See also Note 3.9.3.)

The finiteness of the ratio $\mathbb{E}_f[f(X)/g(X)]$ can be achieved by substituting a mixture distribution for the density $g$,

$$(3.3.5) \qquad \rho g(x) + (1 - \rho)\ell(x),$$

where $\rho$ is close to 1 and $\ell$ is chosen for its heavy tails (for instance, a Cauchy or a Pareto distribution). From an operational point of view, this means that the observations are generated with probability $\rho$ from $g$ and with probability $1 - \rho$ from $\ell$. However, the mixture ($g$ versus $\ell$) does not play a role in the computation of the importance weights; that is, by construction, the estimator integrates out the uniform variable used to decide between $g$ and $\ell$. (We discuss in detail such a marginalization perspective in §3.7.3, where uniform variables involved in the simulation are integrated out in the estimator.) Obviously, (3.3.5) replaces $g(x)$ in the weights of (3.3.1) or (3.3.2), which can then ensures a finite variance for integrable functions $h^2$. Hesterberg (1998) studies the performances of this approach, called a *defensive mixture.*

### 3.3.3 Comparing Importance Sampling with Accept–Reject

Theorem[2] 3.3.4 formally solves the problem of comparing Accept–Reject and importance sampling methods, since with the exception of the constant functions $h(x) = h_0$, the optimal density $g^*$ is always different from $f$. However, a more realistic comparison should also take account of the fact that Theorem 3.3.4 is of limited applicability in a practical setup, as it prescribes an instrumental density that depends on the function $h$ of interest. This may not only result in a considerable increase of the computation time for every new function $h$ (especially if the resulting instrumental density is not easy to generate from), but it also eliminates the possibility of reusing the generated sample to estimate a number of different quantities, as in Example 3.3.6. Now, when the Accept–Reject method is implemented with a density $g$ satisfying $f(x) \leq Mg(x)$ for a constant $1 < M < \infty$, the density $g$ can serve as the instrumental density for importance sampling. A positive feature is that $f/g$ is bounded, thus ensuring finiteness of the variance for the corresponding importance sampling estimators. Bear in mind, though, that in the Accept–Reject method the resulting sample, $X_1, \ldots, X_n$, is a subsample of $Y_1, \ldots, Y_t$, where the $Y_i$'s are simulated from $g$ and where $t$ is the (random) number of simulations from $g$ required for produce the $n$ variables from $f$.

To undertake a comparison of estimation using Accept–Reject and estimation using importance sampling, it is reasonable to start with the two traditional estimators

$$(3.3.6) \quad \delta_1 = \frac{1}{n} \sum_{i=1}^{n} h(X_i) \quad \text{and} \quad \delta_2 = \frac{1}{t} \sum_{j=1}^{t} h(Y_j) \frac{f(Y_j)}{g(Y_j)} .$$

These estimators correspond to the straightforward utilization of the sample produced by Accept–Reject and to an importance sampling estimation derived from the overall sample, that is, to a recycling of the variables rejected by algorithm [A.5].[3] If the ratio $f/g$ is only known up to a constant, $\delta_2$ can be replaced by

$$\delta_3 = \sum_{j=1}^{t} h(Y_j) \frac{f(Y_j)}{g(Y_j)} \bigg/ \sum_{j=1}^{t} \frac{f(Y_j)}{g(Y_j)} .$$

If we write $\delta_2$ in the more explicit form

$$\delta_2 = \frac{n}{t} \left\{ \frac{1}{n} \sum_{i=1}^{n} h(X_i) \frac{f(X_i)}{g(X_i)} + \frac{t-n}{n} \frac{1}{t-n} \sum_{i=1}^{t-n} h(Z_i) \frac{f(Z_i)}{g(Z_i)} \right\} ,$$

---

[2] This section contains more specialized material and may be omitted on a first reading.
[3] This obviously assumes a relatively tight control on the simulation methods rather than the use of a pseudo-random generation software, which only delivers the accepted variables.

where $\{Y_1, \ldots, Y_t\} = \{X_1, \ldots, X_n\} \cup \{Z_1, \ldots, Z_{t-n}\}$ (the $Z_i$'s being the variables rejected by the Accept–Reject algorithm [A.5]), one might argue that, based on sample size, the variance of $\delta_2$ is smaller than that of the estimator

$$\frac{1}{n} \sum_{i=1}^{n} h(X_i) \frac{f(X_i)}{g(X_i)} .$$

If we could apply Theorem 3.3.4, we could then conclude that this latter estimator dominates $\delta_1$ (for an appropriate choice of $g$) and, hence, that it is better to recycle the $Z_i$'s than to discard them. Unfortunately, this reasoning is flawed since $t$ is a random variable, being the *stopping rule* of the Accept–Reject algorithm. The distribution of $t$ is therefore a negative binomial distribution, $\mathcal{N}eg(n, 1/M)$ (see Problem 2.36 ) so $(Y_1, \ldots, Y_t)$ is not an iid sample from $g$. (Note that the $Y_j$'s corresponding to the $X_i$'s, including $Y_t$, have distribution $f$, whereas the others do not.)

The comparison between $\delta_1$ and $\delta_2$ can be reduced to comparing $\delta_1 = f(y_t)$ and $\delta_2$ for $t \sim \mathcal{G}eo(1/M)$ and $n = 1$. However, even with this simplification, the comparison is quite involved (see Problem 3.41 for details), so a general comparison of the bias and variance of $\delta_2$ with $\text{var}_f(h(X))$ is difficult (Casella and Robert 1998).

While the estimator $\delta_2$ is based on an incorrect representation of the distribution of $(Y_1, \ldots, Y_t)$, a reasonable alternative based on the correct distribution of the sample is

$$(3.3.7) \qquad \delta_4 = \frac{n}{t} \delta_1 + \frac{1}{t} \sum_{j=1}^{t-n} h(Z_j) \frac{(M-1)f(Z_j)}{Mg(Z_j) - f(Z_j)} ,$$

where the $Z_j$'s are the elements of $(Y_1, \ldots, Y_t)$ that have been rejected. This estimator is also unbiased and the comparison with $\delta_1$ can also be studied in the case $n = 1$; that is, through the comparison of the variances of $h(X_1)$ and of $\delta_4$, which now can be written in the form

$$\delta_4 = \frac{1}{t} h(X_1) + (1 - \rho) \frac{1}{t} \sum_{j=1}^{t-1} h(Z_j) \left( \frac{g(Z_j)}{f(Z_j)} - \rho \right)^{-1} .$$

Assuming again that $\mathbb{E}_f[h(X)] = 0$, the variance of $\delta_4$ is

$$\text{var}(\delta_4) = \mathbb{E}\left[ \frac{t-1}{t^2} \int h^2(x) \frac{f^2(x)(M-1)}{Mg(x) - f(x)} \, dx + \frac{1}{t^2} \mathbb{E}_f[h^2(X)] \right] ,$$

which is again too case-specific (that is, too dependent on $f$, $g$, and $h$) to allow for a general comparison.

The marginal distribution of the $Z_i$'s from the Accept–Reject algorithm is $(M g - f)/(M - 1)$, and the importance sampling estimator $\delta_5$ associated with this instrumental distribution is

$$\delta_5 = \frac{1}{t-n} \sum_{j=1}^{t-n} \frac{(M-1)f(Z_j)}{Mg(Z_j) - f(Z_j)} h(Z_j),$$

which allows us to write $\delta_4$ as

$$\delta_4 = \frac{n}{t}\delta_1 + \frac{t-n}{t}\delta_5,$$

a weighted average of the usual Monte Carlo estimator and of $\delta_5$.

According to Theorem 3.3.4, the instrumental distribution can be chosen such that the variance of $\delta_5$ is lower than the variance of $\delta_1$. Since this estimator is unbiased, $\delta_4$ will dominate $\delta_1$ for an appropriate choice of $g$. This domination result is of course as formal as Theorem 3.3.4, but it indicates that, for a fixed $g$, there exist functions $h$ such that $\delta_4$ improves on $\delta_1$.

If $f$ is only known up to the constant of integration (hence, $f$ and $M$ are not properly scaled), $\delta_4$ can replaced by

$$\delta_6 = \frac{n}{t}\delta_1 + \frac{t-n}{t}\sum_{j=1}^{t-n} \frac{h(Z_j)f(Z_j)}{Mg(Z_j) - f(z_j)}$$

(3.3.8)
$$\bigg/ \sum_{j=1}^{t-n} \frac{f(Z_j)}{Mg(Z_j) - f(Z_j)}\ .$$

Although the above domination of $\delta_1$ by $\delta_4$ does not extend to $\delta_6$, nonetheless, $\delta_6$ correctly estimates constant functions while being asymptotically equivalent to $\delta_4$. See Casella and Robert (1998) for additional domination results of $\delta_1$ by weighted estimators.

**Example 3.3.7 Gamma simulation.** For illustrative purposes, consider the simulation of $\mathcal{G}a(\alpha,\beta)$ from the instrumental distribution $\mathcal{G}a(a,b)$, with $a = [\alpha]$ and $b = a\beta/\alpha$. (This choice of $b$ is justified in Example 2.3.4 as maximizing the acceptance probability in an Accept–Reject scheme.) The ratio $f/g$ is therefore

$$w(x) = \frac{\Gamma(a)}{\Gamma(\alpha)} \frac{\beta^\alpha}{b^a} x^{\alpha-a} e^{(b-\beta)x}\ ,$$

which is bounded by

$$M = \frac{\Gamma(a)}{\Gamma(\alpha)} \frac{\beta^\alpha}{b^a} \left(\frac{\alpha-a}{\beta-b}\right)^{\alpha-a} e^{-(\alpha-a)}$$

(3.3.9)
$$= \frac{\Gamma(a)}{\Gamma(\alpha)} \exp\{\alpha(\log(\alpha) - 1) - a(\log(a) - 1)\}\ .$$

Since the ratio $\Gamma(a)/\Gamma(\alpha)$ is bounded from above by 1, an approximate bound that can be used in the simulation is

$$M' = \exp\{a(\log(a) - 1) - \alpha(\log(\alpha) - 1)\}\ ,$$

with $M'/M = 1 + \varepsilon = \Gamma(\alpha)/\Gamma([\alpha])$. In this particular setup, the estimator $\delta_4$ is available since $f/g$ and $M$ are explicitly known. In order to assess the effect of the approximation (3.3.8), we also compute the estimator $\delta_6$ for the following functions of interest:

$$h_1(x) = x^3, \quad h_2(x) = x \log x, \quad \text{and} \quad h_3(x) = \frac{x}{1+x}.$$

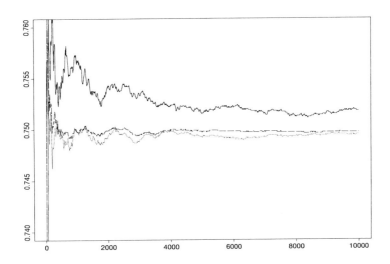

Figure 3.3.6. *Convergence of the estimators of* $\mathbb{E}[X/(1+X)]$, $\delta_1$ *(solid lines), $\delta_4$ (dots) and $\delta_6$ (dashes), for $\alpha = 3.7$ and $\beta = 1$. The final values are respectively 0.7518, 0.7495, and 0.7497, for a true value of the expectation equal to 0.7497.*

Figure 3.3.6 describes the convergence of the three estimators of $h_3$ in $m$ for $\alpha = 3.7$ and $\beta = 1$ (which yields an Accept–Reject acceptance probability of $1/M = .10$). Both estimators $\delta_4$ and $\delta_6$ have more stable graphs than the empirical average $\delta_1$ and they converge much faster to the theoretical expectation 0.7497, $\delta_6$ then being equal to this value after 6000 iterations. For $\alpha = 3.08$ and $\beta = 1$ (which yields an Accept–Reject acceptance probability of $1/M = .78$), Figure 3.3.7 illustrates the change of behavior of the three estimators of $h_3$ since they now converge at similar speeds. Note the proximity of $\delta_4$ and $\delta_1$, $\delta_6$ again being the estimator closest to the theoretical expectation 0.7081 after 10,000 iterations.

Table 3.3.2 provides another evaluation of the three estimators in a case which is *a priori* very favorable to importance sampling, namely for $\alpha = 3.7$. The table exhibits, in most cases, a strong domination of $\delta_4$ and $\delta_6$ over $\delta_1$ and a moderate domination of $\delta_4$ over $\delta_6$.                    ‖

In contrast to the general setup of §3.3, $\delta_4$ (or its approximation $\delta_6$) can always be used in an Accept–Reject sampling setup since this estimator does not require additional simulations. It provides a second evaluation of $\mathbb{E}_f[h]$, which can be compared with the Monte Carlo estimator for the purpose of convergence assessment.

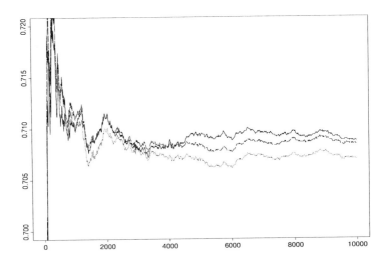

Figure 3.3.7.   *Convergence of estimators of* $\mathbb{E}[X/(1 + X)]$, $\delta_1$ *(solid lines),* $\delta_4$
*(dots) and* $\delta_6$ *(dashes) for* $\alpha = 3.08$ *and* $\beta = 1$. *The final values are respectively*
$0.7087$, $0.7069$, *and* $0.7084$, *for a true value of the expectation equal to* $0.7081$.

| $m$ | 100 | | | 1000 | | | 5000 | | |
|---|---|---|---|---|---|---|---|---|---|
| | $\delta_1$ | $\delta_4$ | $\delta_6$ | $\delta_1$ | $\delta_4$ | $\delta_6$ | $\delta_1$ | $\delta_4$ | $\delta_6$ |
| $h_1$ | 87.3 | 55.9 | 64.2 | 36.5 | 0.044 | 0.047 | 2.02 | 0.54 | 0.64 |
| $h_2$ | 1.6 | 3.3 | 4.4 | 4.0 | 0.00 | 0.00 | 0.17 | 0.00 | 0.00 |
| $h_3$ | 6.84 | 0.11 | 0.76 | 4.73 | 0.00 | 0.00 | 0.38 | 0.02 | 0.00 |

Table 3.3.2.   *Comparison of the performances of the Monte Carlo estimator* $(\delta_1)$
*with two importance sampling estimators* $(\delta_4$ *and* $\delta_6)$ *under squared error loss*
*after* $m$ *iterations for* $\alpha = 3.7$ *and* $\beta = 1$. *The squared error loss is multiplied by*
$10^2$ *for the estimation of* $\mathbb{E}[h_2(X)]$ *and by* $10^5$ *for the estimation of* $\mathbb{E}[h_3(X)]$.
*The squared errors are actually the difference from the theoretical values* $(99.123,$
$5.3185,$ *and* $0.7497,$ *respectively) and the three estimators are based on the same*
*unique sample, which explains the lack of monotonicity (in* $m$) *of the errors.*
*(Source: Casella and Robert 1998.)*

## 3.4 Riemann Approximations

In approximating an integral like (3.2.5), the simulation-based approach is
based on a *probabilistic* convergence result for the empirical average

$$\frac{1}{m} \sum_{i=1}^{m} h(X_i) \, ,$$

when the $X_i$'s are simulated according to $f$. Numerical integration is based
on the *analytical* definition of the integral, namely as a limit of Riemann

sums. In fact, for every sequence $(a_{i,n})_i$ $(0 \leq i \leq n)$ such that $a_{0,n} = a$, $a_{n,n} = b$, and $a_{i,n} - a_{i,n-1}$ converges to 0 (in $n$), the (Riemann) sum

$$\sum_{i=0}^{n-1} h(a_{i,n}) \, f(a_{i,n})(a_{i+1,n} - a_{i,n}) \,,$$

converges to the integral

$$I = \int_a^b h(x) \, f(x) dx$$

as $n$ goes to infinity. When $\mathcal{X}$ has dimension greater than 1, the same approximation applies with a grid on the domain $\mathcal{X}$ (see Rudin 1976 for more details.)

When these approaches are put together, the result is a Riemann sum with random steps, with the $a_{i,n}$'s simulated from $f$ (or from an instrumental distribution $g$). This method was first introduced by Yakowitz *et al.* (1978) as *weighted Monte Carlo integration* for uniform distributions on $[0, 1]^d$. In a more general setup, we call this approach *simulation by Riemann sums* or *Riemannian simulation*, following Philippe (1997a,b), although it is truly an integration method.

**Definition 3.4.1** The method of *simulation by Riemann sums* approximates the integral

$$I = \int h(x) \, f(x) dx$$

by

(3.4.1) $$\sum_{i=0}^{m-1} h(X_{(i)}) \, f(X_{(i)})(X_{(i+1)} - X_{(i)})$$

where $X_0, \ldots, X_m$ are iid random variables from $f$ and $X_{(0)} \leq \cdots \leq X_{(m)}$ are the order statistics associated with this sample.

Suppose first that the integral $I$ can be written

$$I = \int_0^1 h(x) \, dx,$$

and that $h$ is a differentiable function. We can then establish the following result about the validity of the Riemannian approximation:

**Proposition 3.4.2** *Let* $U = (U_0, U_1, \ldots, U_m)$ *be an ordered sample from* $\mathcal{U}_{[0,1]}$*. If the derivative* $h'$ *is bounded on* $[0, 1]$*, the estimator*

$$\delta(U) = \sum_{i=0}^{m-1} h(U_i)(U_{i+1} - U_i) + h(0)U_0 + h(U_m)(1 - U_m)$$

*has a variance of order* $O(m^{-2})$*.*

*Proof.* If we define $U_{-1} = 0$ and $U_{m+1} = 1$, then $\delta$ can be written

$$\delta = \sum_{i=-1}^{m} h(U_i)(U_{i+1} - U_i) = \sum_{i=-1}^{m} \int_{U_i}^{U_{i+1}} h(U_i)du,$$

and thus the difference $(I - \delta)$ can be written as

$$\sum_{i=-1}^{m} \int_{U_i}^{U_{i+1}} (h(u) - h(U_i))\, du .$$

The first-order expansion

$$h(u) = h(U_i) + h'(\zeta)(u - U_i), \qquad \zeta \in [U_i, u] ,$$

implies that

$$|h(u) - h(U_i)| < c(u - U_i), \quad \text{with} \quad c = \sup_{[0,1]} |h'(X)|,$$

and, thus,

$$\text{var}(\delta) = \mathbb{E}[(I - \delta)^2] < c^2\, \mathbb{E}\left[ \left( \sum_{i=-1}^{m} (U_{i+1} - U_i)^2 \right)^2 \right]$$
$$= c^2 \left\{ (m + 2)\, \mathbb{E}[Z_i^4] + (m + 1)(m + 2)\, \mathbb{E}[Z_i^2 Z_j^2] \right\} ,$$

where $Z_i = U_{i+1} - U_i$. Since the $U_i$'s are the order statistics of a uniform distribution, the variables $Z_i$ are jointly distributed according to a Dirichlet distribution $\mathcal{D}_m(1, \ldots, 1)$, with $Z_i \sim \mathcal{B}e(1, m)$ and $(Z_i, Z_j, 1 - Z_i - Z_j) \sim \mathcal{D}_3(1, 1, m - 1)$. Therefore,

$$\mathbb{E}[Z_i^4] = m \int_0^1 z^4 (1 - z)^{m-1}\, dz = \frac{24\, m!}{(m + 4)!}$$

and

$$\mathbb{E}[Z_i^2 Z_j^2] = m(m - 1) \int z_1^2 z_2^2 (1 - z_1 - z_2)^{m-2}\, dz_1 dz_2$$
$$= \frac{(2!)^2}{(m + 4) \ldots (m + 1)} = \frac{4\, m!}{(m + 4)!} ,$$

which yields

$$\text{var}(\delta) \leq c \left\{ \frac{24}{(m + 4)(m + 3)(m + 1)} + \frac{4}{(m + 4)(m + 3)} \right\}$$
$$= O(m^{-2}) . \qquad \qquad \square\square$$

Yakowitz *et al.* (1978) improve on the order of the variance by symmetrizing $\delta$ into

$$\tilde{\delta} = \sum_{i=-1}^{m} (U_{i+1} - U_i) \frac{h(U_i) + h(U_{i+1})}{2} .$$

When the second derivative of $h$ is bounded, the error of $\tilde{\delta}$ is then of order $O(m^{-4})$.

Even when the additional assumption on the second derivative is not satisfied, the practical improvement brought by Riemann sums (when compared with regular Monte Carlo integration) is substantial since the magnitude of the variance decreases from $m^{-1}$ to $m^{-2}$. Unfortunately, this dominance fails to extend to the case of multidimensional integrals, a phenomenon that is related to the so-called "curse of dimensionality"; that is, the subefficiency of numerical methods compared with simulation algorithms for dimensions $d$ larger than 4 since the error is then of order $O(m^{-4/d})$ (see Yakowitz et al. 1978). The intuitive reason behind this phenomenon is that a numerical approach like the Riemann sum method basically covers the whole space with a grid. When the dimension of the space increases, the number of points on the grid necessary to obtain a given precision increases too, which means, in practice, a much larger number of iterations for the same precision.

The result of Proposition 3.4.2 holds for an arbitrary density, due to the property that the integral $I$ can also be written as

$$(3.4.2) \qquad \int_0^1 H(x)\, dx \,,$$

where $H(x) = h(F^-(x))$, and $F^-$ is the generalized inverse of $F$, cdf of $f$ (see Lemma 2.1.2). Although this is a formal representation when $F^-$ is not available in closed form and cannot be used for simulation purposes in most cases (see §2.1.3), (3.4.2) is central to this extension of Proposition 3.4.2. Indeed, since

$$X_{(i+1)} - X_{(i)} = F^-(U_{i+1}) - F^-(U_i),$$

where the $X_{(i)}$'s are the order statistics of a sample from $F$ and the $U_i$'s are the order statistics of a sample from $\mathcal{U}_{[0,1]}$,

$$\sum_{i=0}^{m-1} h(X_{(i)}) f(X_{(i)})(X_{(i+1)} - X_{(i)})$$

$$= \sum_{i=0}^{m-1} H(U_i) f(F^-(U_i))(F^-(U_{i+1}) - F^-(U_i))$$

$$\simeq \sum_{i=0}^{m-1} H(U_i)(U_{i+1} - U_i) \,,$$

given that $(F^-)'(u) = 1/f(F^-(u))$. Since the remainder is negligible in the first-order expansion of $F^-(U_{i+1}) - F^-(U_i)$, the above Riemann sum can then be expressed in terms of uniform variables and Proposition 3.4.2 does apply in this setup, since the extreme terms $h(0)U_0$ and $h(U_m)(1 - U_m)$ are again of order $m^{-2}$ (variancewise). (See Philippe 1997a,b, for more details on the convergence of (3.4.1) to $I$ under some conditions on the function $h$.)

The above results imply that the Riemann sums integration method will perform well in unidimensional setups when the density $f$ is known. It, thus, provides an efficient alternative to standard Monte Carlo integration in this setting, since it does not require additional computations (although it requires keeping track of and storing all the $X_{(i)}$'s). Also, as the convergence is of a higher order, there is no difficulty in implementing the method. When $f$ is known only up to a constant (that is, $f_0(x) \propto f(x)$), (3.4.1) can be replaced by

$$(3.4.3) \qquad \sum_{i=0}^{m-1} h(X_{(i)}) \, f_0(X_{(i)}) (X_{(i+1)} - X_{(i)})$$

$$\Big/ \sum_{i=0}^{m-1} f_0(X_{(i)}) (X_{(i+1)} - X_{(i)}) \,,$$

since both the numerator and the denominator almost surely converge. This approach thus provides, in addition, an efficient estimation method for the normalizing constant via the denominator in (3.4.3). Note also that when $f$ is entirely known, the denominator converges to 1, which can be used as a convergence assessment device (see Philippe 1997a).

**Example 3.4.3 (Continuation of Example 3.3.7)** When $X \sim \mathcal{G}a(3.7, 1)$, assume that $h_2(x) = x \log(x)$ is the function of interest. A sample $X_1, \dots, X_m$ from $\mathcal{G}a(3.7, 1)$ can easily be produced by the algorithm [A.7] of Chapter 2 and we compare the empirical mean, $\delta_{1m}$, with the Riemann sum estimator

$$\delta_{2m} = \frac{1}{\Gamma(3.7)} \sum_{i=1}^{m-1} h_2(X_{(i)}) \, X_{(i)}^{2.7} e^{-X_{(i)}} \, (X_{(i+1)} - X_{(i)}) \,,$$

which uses the known normalizing constant. Figure 3.4.1 clearly illustrates the difference of convergence speed for both estimators and the much greater stability of $\delta_{2m}$, which is close to the theoretical value 5.3185 after 3000 iterations. ‖

If the original simulation is done by importance sampling (that is, if the sample $X_1, \dots, X_m$ is generated from an instrumental distribution $g$), since the integral $I$ can also be written

$$I = \int h(x) \frac{f(x)}{g(x)} g(x) dx,$$

the Riemann sum estimator (3.4.1) remains unchanged. Although it has similar convergence properties, the boundedness conditions on $h$ are less explicit and, thus, more difficult to check. As in the original case, it is possible to establish an equivalent to Theorem 3.3.4, namely to show that $g(x) \propto |h(x)| f(x)$ is optimal (in terms of variance) (see Philippe 1997c), with the additional advantage that the normalizing constant does not need to be known, since $g$ does not appear in (3.4.1).

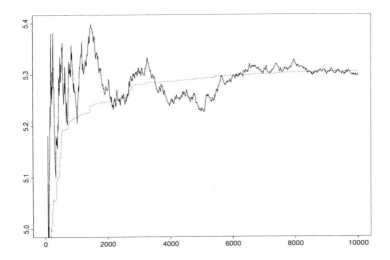

Figure 3.4.1. *Convergence of estimators of* $\mathbb{E}[X \log(X)]$, $\delta_{1m}$ *(solid lines) and* $\delta_{2m}$ *(dots) for* $\alpha = 3.7$ *and* $\beta = 1$. *The final values are* 5.3007 *and* 5.3057, *respectively, for a true value of* 5.31847.

**Example 3.4.4 (Continuation of Example 3.3.5)** If $\mathcal{T}(\nu, 0, 1)$ is simulated by importance sampling from the normal instrumental distribution $\mathcal{N}(0, \nu/(\nu - 2))$, the difference between the two distributions is mainly visible in the tails. This makes the importance sampling estimator $\delta_{1m}$ very unstable (see Figures 3.3.2, 3.3.3, and 3.3.4). Figure 3.4.2 compares this estimator to the Riemann sum estimator

$$\delta_{2m} = \sum_{i=1}^{m-1} h_4(X_{(i)}) \frac{\Gamma((\nu+1)/2)}{\sqrt{\nu\pi}\,\Gamma(\nu/2)} \left[1 + X_{(i)}^2/\nu\right]^{-\frac{\nu+1}{2}} (X_{(i+1)} - X_{(i)})$$

and its normalized version

$$\delta_{3m} = \frac{\sum_{i=0}^{m-1} h_4(X_{(i)}) \left[1 + X_{(i)}^2/\nu\right]^{-\frac{\nu+1}{2}} (X_{(i+1)} - X_{(i)})}{\sum_{i=0}^{m-1} \left[1 + X_{(i)}^2/\nu\right]^{-\frac{\nu+1}{2}} (X_{(i+1)} - X_{(i)})},$$

for $h_4(X) = (1 + e^X)\,\mathbb{I}_{X \leq 0}$ and $\nu = 2.3$.

We can again note the stability of the approximations by Riemann sums, the difference between $\delta_{2m}$ and $\delta_{3m}$ mainly due to the bias introduced by the approximation of the normalizing constant in $\delta_{3m}$. For the given sample, note that $\delta_{2m}$ dominates the other estimators.

If, instead, the instrumental distribution is chosen to be the Cauchy distribution $\mathcal{C}(0, 1)$, the importance sampling estimator is much better behaved. Figure 3.4.3 shows that the speed of convergence of the associated estimator is much faster than with the normal instrumental distribution.

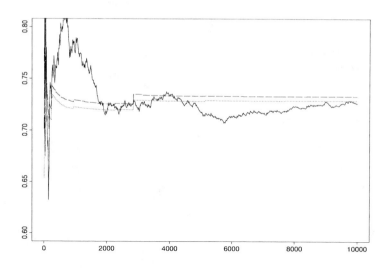

Figure 3.4.2.  *Convergence of estimators of* $\mathbb{E}_\nu[(1+e^X)\mathbb{I}_{X\leq 0}]$, $\delta_{1m}$ *(solid lines),* $\delta_{2m}$ *(dots), and* $\delta_{3m}$ *(dashes), for a normal instrumental distribution and* $\nu = 2.3$. *The final values are respectively* 0.7262, 0.7287, *and* 0.7329, *for a true value of* 0.7307.

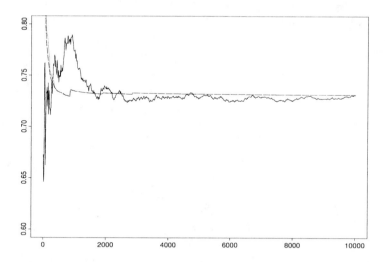

Figure 3.4.3.  *Convergence of estimators of* $\mathbb{E}_\nu[(1+e^X)\mathbb{I}_{X\leq 0}]$, $\delta_{1m}$ *(solid lines),* $\delta_{2m}$ *(dots), and* $\delta_{3m}$ *(dashes) for a Cauchy instrumental distribution and* $\nu = 2.3$. *The two Riemann sum approximations are virtually equal except for the beginning simulations. The final values are respectively* 0.7325, 0.7314, *and* 0.7314, *and the true value is* 0.7307.

Although $\delta_{2m}$ and $\delta_{3m}$ have the same representation for $\mathcal{N}(0, \nu/(\nu-2))$ and $\mathcal{C}(0,1)$, the corresponding samples differ, and these estimators exhibit an even higher stability in this case, giving a good approximation of $\mathbb{E}[h_4(X)]$ after only a few hundred iterations. Both estimators are actually identical almost from the start, a fact which indicates how fast the denominator of $\delta_{3m}$ converges to the normalizing constant.  ∥

## 3.5 Laplace Approximations

As an alternative to simulation of integrals, we can also attempt analytic approximations. One of the oldest and most useful approximations is the integral Laplace approximation. It is based on the following argument: Suppose that we are interested in evaluating the integral

$$(3.5.1) \qquad \int_A f(x|\theta)dx$$

for a fixed value of $\theta$. (The function $f$ needs to be nonnegative and integrable; see Tierney and Kadane 1986 and Tierney *et al.* 1989 for extensions.). Write $f(x|\theta) = \exp\{nh(x|\theta)\}$, where $n$ is the sample size or another parameter which can go to infinity, and use a Taylor series expansion of $h(x|\theta)$ about a point $x_0$ to obtain

$$h(x|\theta) \approx h(x_0|\theta) + (x - x_0)h'(x_0|\theta) + \frac{(x - x_0)^2}{2!}h''(x_0|\theta)$$

$$(3.5.2) \qquad + \frac{(x - x_0)^3}{3!}h'''(x_0|\theta) + R_n(x) \,,$$

where we write

$$h'(x_0|\theta) = \left.\frac{\partial h(x|\theta)}{\partial x}\right|_{x=x_0} \,,$$

and similarly for the other terms, while the remainder $R_n(x)$ satisfies

$$\lim_{x \to x_0} R_n(x)/(x - x_0)^3 = 0.$$

Now choose $x_0 = \hat{x}_\theta$, the value that maximizes $h(x|\theta)$ for the given value of $\theta$. Then, the linear term in (3.5.2) is zero and we have the approximation

$$\int_A e^{nh(x|\theta)}dx \simeq e^{nh(\hat{x}_\theta|\theta)} \int_A e^{n\frac{(x-\hat{x}_\theta)^2}{2}h''(\hat{x}_\theta|\theta)} e^{n\frac{(x-\hat{x}_\theta)^3}{3!}h'''(\hat{x}_\theta|\theta)}dx,$$

which is valid within a neighborhood of $\hat{x}_\theta$. (See Schervish 1995, Section 7.4.3, for detailed conditions.) Note the importance of choosing the point $x_0$ to be a maximum.

The cubic term in the exponent is now expanded in a series around $\hat{x}_\theta$ to yield

$$\int_A e^{nh(x|\theta)}dx \simeq e^{nh(\hat{x}_\theta|\theta)} \int_A e^{n\frac{(x-\hat{x}_\theta)^2}{2}h''(\hat{x}_\theta|\theta)}$$

| Interval | Approximation | Exact |
|----------|---------------|-------|
| $(7,9)$ | 0.193351 | 0.193341 |
| $(6,10)$ | 0.375046 | 0.37477 |
| $(2,14)$ | 0.848559 | 0.823349 |
| $(15.987, \infty)$ | 0.0224544 | 0.100005 |

Table 3.5.1. *Laplace approximation of a Gamma integral for* $\alpha = 5$ *and* $\beta = 2$.

$$(3.5.3) \quad \times \left[ 1 + n \frac{(x - \hat{x}_\theta)^3}{3!} h'''(\hat{x}_\theta | \theta) + n^2 \frac{(x - \hat{x}_\theta)^6}{2!(3!)^2} [h'''(\hat{x}_\theta | \theta)]^2 + R_n \right] dx,$$

where $R_n$ again denotes a remainder term (see Problem 3.18).

We call the integral approximations in (3.5.3) the *first-order* approximation if it excludes the last three terms in the right-hand side (including $R_n$), and the *second-order* approximation if it excludes the last two terms; finally the *third-order* approximation includes all the terms except $R_n$.

Since the above integrand is the kernel of a normal density with mean $\hat{x}_\theta$ and variance $-1/n \, h''(\hat{x}_\theta | \theta)$, we can evaluate these expressions further. More precisely, letting $\Phi(\cdot)$ denote the standard normal cdf, and taking $A = [a, b]$, we can evaluate the integral in the first-order approximation to obtain (see Problem 3.20)

$$\int_a^b e^{nh(x|\theta)} dx \simeq e^{nh(\hat{x}_\theta | \theta)} \sqrt{\frac{2\pi}{-nh''(\hat{x}_\theta | \theta)}}$$

$$(3.5.4) \qquad \times \left\{ \Phi[\sqrt{-nh''(\hat{x}_\theta | \theta)}(b - \hat{x}_\theta)] - \Phi[\sqrt{-nh''(\hat{x}_\theta | \theta)}(a - \hat{x}_\theta)] \right\}.$$

**Example 3.5.1 Gamma approximation.** As a simple illustration of the Laplace approximation, consider estimating a gamma integral, say

$$(3.5.5) \qquad \qquad \int_a^b \frac{x^{\alpha - 1}}{\Gamma(\alpha)\beta^\alpha} e^{-x/\beta} dx.$$

Applying (3.5.2) with $\hat{x}_\theta = (\alpha - 1)\beta$ (the mode of the density) yields the Laplace approximation.

For $\alpha = 5$ and $\beta = 2$, $\hat{x}_\theta = 8$, and the approximation will be best in that area. In Table 3.5.1 we see that although the approximation is reasonable in the central region of the density, it becomes quite unacceptable in the tails.                                                                                        ‖

## 3.6 The Saddlepoint Approximation

The saddlepoint approximation, in contrast to the Laplace approximation, is mainly a technique for approximating a function rather than an integral, although it naturally leads to an integral approximation. Here, we give some basic background to help understand this technique, some of which is adapted from Goutis and Casella (1999). For more thorough introductions to the saddlepoint approximation, see the review articles by Reid (1988, 1991), or the books by Field and Ronchetti (1990), Kolassa (1994), or Jensen (1995).

Starting from (3.5.1), suppose we would like to evaluate

$$(3.6.1) \qquad g(\theta) = \int_A f(x|\theta)dx$$

for a range of values of $\theta$. A natural thing to do is, for each $\theta$, to approximate $g(\theta)$ by a Laplace-type approximation of the integral. Using the first-order Laplace approximation, we find that for $A = [a, b]$, $g(\theta)$ is approximated by (3.5.4). This is a basic idea of a *saddlepoint approximation*, that for each value of $\theta$, $\hat{x}_\theta$ (the *saddlepoint*) is chosen in some optimal way.[4] We gain greater accuracy at the expense of increased computing effort, as now the function $h(x|\theta)$ must be maximized for each value of $\theta$ for which we evaluate $g(\theta)$.

**Example 3.6.1 Student's $t$ saddlepoint.** We can write the density of Student's $t$, $f_\nu(t)$, in the form

$$(3.6.2) \qquad f_\nu(t) = \int_0^\infty \left[ \frac{\sqrt{x}e^{\frac{-t^2 x}{2\nu}}}{\sqrt{2\pi\nu}} \right] \left[ \frac{x^{(\nu/2)-1}e^{-x/2}}{\Gamma(\nu/2)2^{\nu/2}} \right] dx,$$

that is, a variance mixture of normals. For each value of $t$, we apply the approximation (3.5.4), with the $h$ function being a transform of the integrand of (3.6.2), as in the representation $f(x|\theta) = \exp\{nh(x|\theta)\}$. This yields the saddlepoint $\hat{x}(t) = \frac{(-1+\nu)\nu}{t^2+\nu}$. For $\nu = 6$, the approximation is shown in Figure 3.6.1 and yields remarkable accuracy. When the approximation is renormalized (that is, divided by its integral on $\mathbb{R}$), it is virtually exact. ‖

### 3.6.1 An Edgeworth Derivation

From (3.6.1), we see that one way to derive the saddlepoint approximation is to start from an integral representation of the function of interest. This was Daniels' (1954) technique (see also Daniels 1980, 1983), where he started from the integral representation of a density using the characteristic function (the so-called *inversion formula*). This naturally led to analysis in the complex plane. There is an alternate derivation, that Reid (1988) calls "a more statistical version," based on Edgeworth expansions. In many ways, it is similar to the original Daniels' derivation (as it involves inversion of an integral), but it has the advantage of helping us make our statements about the "order of the approximation" somewhat clearer. Moreover, it provides a more precise picture of how the greater accuracy is achieved.

---

[4] The saddlepoint approximation got its name because its original derivation (Daniels 1954) used a complex analysis argument, and the point $\hat{x}_\theta$ is a saddlepoint in the complex plane.

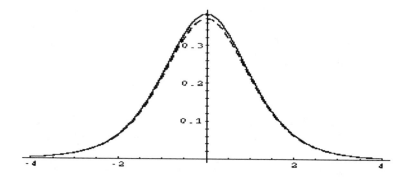

Figure 3.6.1. *Student's t saddlepoint approximation for $\nu = 6$. The dashed line is the saddlepoint approximation based on (3.5.4) for each t, the solid line is both the exact density function and the renormalized saddlepoint.*

An *Edgeworth expansion* of a distribution is accomplished by first expanding the cumulant generating function, that is,

$$(3.6.3) \qquad K(\tau) = \log\left(\mathbb{E}[e^{\tau X}]\right),$$

in a Taylor series around 0, then inverting (see Feller 1971, Chapter XVI, or Stuart and Ord 1987, Chapter 6 and Sections 11.130–11.16).

Let $X_1, X_2, \dots, X_n$ be iid with density $f$, mean $\mu$, and variance $\sigma^2$. A useful form of an Edgeworth expansion is given by Hall (1992, Equation 2.17), and can be written

$$P\left(\frac{\sqrt{n}(\bar{X} - \mu)}{\sigma} \leq w\right) =$$

$$(3.6.4) \qquad \Phi(w) + \varphi(w)\left[\frac{-1}{6\sqrt{n}}\kappa(w^2 - 1) + O(1/n)\right],$$

where $\Phi$ and $\varphi$ are respectively the distribution and density function of a standard normal and $\kappa = \mathbb{E}(X_1 - \mu)^3$ is the *skewness*. For the Edgeworth expansion, the term in $O(1/n)$ is of the form $p(w)/n$, where $p$ is a polynomial. Since it is multiplied by $\varphi(w)$, it follow that its derivatives (in $w$) maintain the same order of approximation. Thus, (3.6.4) can be differentiated to obtain a density approximation with the same order of accuracy. If we do that, and then make the transformation $x = \sigma w + \mu$, we obtain the approximation to the density of $\bar{X}$ to be

$$f_{\bar{X}}(x) = \frac{\sqrt{n}}{\sigma}\varphi\left(\frac{x - \mu}{\sigma/\sqrt{n}}\right)$$

$$(3.6.5) \qquad \times \left[1 + \frac{\kappa}{6\sqrt{n}}\left\{\left(\frac{x - \mu}{\sigma/\sqrt{n}}\right)^3 - 3\left(\frac{x - \mu}{\sigma/\sqrt{n}}\right)\right\} + O(1/n)\right].$$

Ignoring the term with braces produces the usual normal approximation, which is accurate to $O(1/\sqrt{n})$. If we are using (3.6.5) for values of $x$ near $\mu$, then the value of the expression in braces is close to zero, and the approximation will then be accurate to $O(1/n)$. The trick of the saddlepoint approximation is to make this always be the case.

To do so, we use a strategy similar to that used in Example 3.6.1; that is, we use a family of densities so that, for each $x$, we can choose a density from the family to cancel the term in braces in (3.6.5). One method of creating such a family is through a technique known as *exponential tilting* (see Efron 1981, Stuart and Ord 1987, Section 11.13, or Reid 1988). Starting with a density $f$ of interest, we create the exponential family

$$\mathcal{F} = \{f(\cdot|\tau); f(x|\tau) = \exp[\tau x - K(\tau)]f(x)\} \ ,$$

where $K(\tau)$ is the cumulant generating function of $f$ given in (3.6.3). It immediately follows that if $X_1, X_2, \ldots, X_n$ are iid from $f(x|\tau)$, the density of $\bar{X}$ is

(3.6.6) $$f_{\bar{X}}(x|\tau) = \exp\{n[\tau x - K(\tau)]\} f_{\bar{X}}(x),$$

where $f_{\bar{X}}(x)$ is the density of the average of an iid sample from $f$ (see Problem 3.22). Let $\mu_\tau$, $\sigma_\tau^2$, and $\kappa_\tau$ denote the mean, variance and skewness of $f(\cdot|\tau)$, respectively, and apply (3.6.5) to $f_{\bar{X}}(x|\tau)$ to obtain an Edgeworth expansion for $f_{\bar{X}}(x|\tau)$. A family of Edgeworth expansions for $f_{\bar{X}}(x)$ is then obtained by inversion, that is,

$$f_{\bar{X}}(x) = \exp\{-n[\tau x - K(\tau)]\} \frac{\sqrt{n}}{\sigma_\tau} \varphi \left( \frac{x - \mu_\tau}{\sigma_\tau/\sqrt{n}} \right)$$

(3.6.7) $$\times \left[ 1 + \frac{\kappa_\tau}{6\sqrt{n}} \left\{ \left( \frac{x - \mu_\tau}{\sigma_\tau/\sqrt{n}} \right)^3 - 3 \left( \frac{x - \mu_\tau}{\sigma_\tau/\sqrt{n}} \right) \right\} + O(1/n) \right].$$

Now, the parameter $\tau$ is free for us to choose in (3.6.7). Therefore, given $x$, we can choose $\tau$ so that $\mu_\tau = x$; that is, we equate the mean of $f_{\bar{X}}(x|\tau)$ with the point $x$. This choice cancels the middle term in the square brackets in (3.6.7), thereby improving the order of the approximation. Recalling that $K(\tau)$ is the cumulant generating function, we can equivalently choose $\tau$ so that $K'(\tau) = x$, which is the saddlepoint equation. Denoting this value by $\hat{\tau} = \hat{\tau}(x)$ and noting that $\sigma_\tau = K''(\tau)$, we get the saddlepoint approximation

$$f_{\bar{X}}(x) = \frac{\sqrt{n}}{\sigma_{\hat{\tau}}} \varphi(0) \exp\{n[K(\hat{\tau}) - \hat{\tau}x]\} [1 + O(1/n)]$$

(3.6.8) $$\approx \left( \frac{n}{2\pi K''(\hat{\tau}(x))} \right)^{1/2} \exp\{n[K(\hat{\tau}(x)) - \hat{\tau}(x)x]\}.$$

**Example 3.6.2 (Continuation of Example 2.2.7)** The noncentral chi squared distribution can be represented as a mixture of central chi squared distributions. This would suggest a saddlepoint approximation using the technique of Example 3.6.1. This can be done, but here is another attack presented by Hougaard (1988). The noncentral chi squared density $\chi_p^2(\lambda)$

has no closed form and is usually written

(3.6.9) $$f(x|\lambda) = \sum_{k=0}^{\infty} \frac{x^{p/2+k-1}e^{-x/2}}{\Gamma(p/2+k)2^{p/2+k}} \frac{\lambda^k e^{-\lambda}}{k!},$$

where $p$ is the degrees of freedom and $\lambda$ is the noncentrality parameter. It turns out that calculation of the moment generating function is simple, and it can be expressed in closed form as

(3.6.10) $$\phi_X(t) = \frac{e^{2\lambda t/(1-2t)}}{(1-2t)^{p/2}}.$$

Solving the saddlepoint equation $\partial \log \phi_X(t)/\partial t = x$ yields the saddlepoint

(3.6.11) $$\hat{t}(x) = \frac{-p + 2x - \sqrt{p^2 + 8\lambda x}}{4x}$$

and applying (3.6.8) yields the approximate density. Here, the saddlepoint and the renormalized saddlepoint approximations are remarkably accurate, with the renormalized density being virtually exact (see Problem 3.23). ‖

### 3.6.2 Tail Areas

There is another use of the saddlepoint approximation that is, perhaps, even more important than the approximation of a density function, namely the use of the saddlepoint to approximate the tail area of a distribution. From (3.6.8), we have the approximation

$$P(\bar{X} > a) = \int_a^{\infty} \left( \frac{n}{2\pi K_X''(\hat{\tau}(x))} \right)^{1/2} \exp\left\{ n\left[ K_X\left(\hat{\tau}(x)\right) - \hat{\tau}(x)x\right]\right\} dx$$

$$= \int_{\hat{\tau}(a)}^{1/2} \left( \frac{n}{2\pi} \right)^{1/2} [K_X''(t)]^{1/2} \exp\left\{ n\left[ K_X(t) - tK_X'(t)\right]\right\} dt,$$

where we make the transformation $K_X'(t) = x$ and $\hat{\tau}(a)$ satisfies

$$K_X'(\hat{\tau}(a)) = a .$$

This transformation was noted by Daniels (1983, 1987), and allows the evaluation of the integral with only one saddlepoint evaluation. However, it requires an integration, most likely through a numerical method.

**Example 3.6.3 Tail area approximation.** In Example 2.1.9, we saw an approximation to normal tail areas based on a Taylor series approximation. If we apply the saddlepoint tail area approximation, we find that it is exact (which in some sense is not helpful). To examine the accuracy of the saddlepoint tail area, we return to the noncentral chi squared. Table 3.6.1 compares the tail areas calculated by integrating the exact density and using the regular and renormalized saddlepoints. As can be seen, the accuracy is quite impressive.                    ‖

| Interval | Approximation | Renormalized approximation | Exact |
|----------|---------------|----------------------------|-------|
| $(36.225, \infty)$ | 0.1012 | 0.0996 | 0.10 |
| $(40.542, \infty)$ | 0.0505 | 0.0497 | 0.05 |
| $(49.333, \infty)$ | 0.0101 | 0.0099 | 0.01 |

Table 3.6.1. *Saddlepoint approximation of a noncentral chi squared tail probability for $p = 6$ and $\lambda = 9$.*

## 3.7 Acceleration Methods

Although the wealth of methods proposed above seems to require a comparison between these different methods, we do not expect there to be any clear-cut domination (as was the case with the comparison between Accept–Reject and importance sampling in §3.3.3). Instead, we look at more global acceleration strategies, which are independent of the simulation setup.

The acceleration methods described below can be used not only in a single implementation but also as a control device to assess the convergence of a simulation algorithm, following the argument of *parallel estimators* (see also §8.3.2). For example, if $\delta_{1m}, \ldots, \delta_{pm}$ are $p$ convergent estimators of a same quantity $I$, a stopping rule for convergence is that $\delta_{1m}, \ldots, \delta_{pm}$ are identical or, given a minimum precision requirement $\varepsilon$, that

$$\max_{1 \leq i < j \leq p} |\delta_{im} - \delta_{jm}| < \varepsilon.$$

### 3.7.1 Antithetic Variables

Although the usual simulation methods lead to iid samples (or quasi-iid, see Chapter 2), it may actually be preferable to generate samples of correlated variables when estimating an integral $I$, as they may reduce the variance of the corresponding estimator.

A first setting where the generation of independent samples is less desirable corresponds to the comparison of two quantities which are close in value. If

$$(3.7.1) \qquad I_1 = \int g_1(x) f_1(x) dx \quad \text{and} \quad I_2 = \int g_2(x) f_2(x) dx$$

are two such quantities, where $\delta_1$ estimates $I_1$ and $\delta_2$ estimates $I_2$, independently of $\delta_1$, the variance of $(\delta_1 - \delta_2)$, is then $\text{var}(\delta_1) + \text{var}(\delta_2)$, which may be too large to support a fine enough analysis on the difference $I_1 - I_2$. However, if $\delta_1$ and $\delta_2$ are positively correlated, the variance is reduced by a factor $-2 \, \text{cov}(\delta_1, \delta_2)$, which may greatly improve the analysis of the difference.

A convincing illustration of the improvement brought by correlated samples is the comparison of (regular) estimators via simulation. Given a density $f(x|\theta)$ and a loss function $L(\delta, \theta)$, two estimators $\delta_1$ and $\delta_2$ are evaluated through their risk functions, $R(\delta_1, \theta) = \mathbb{E}[L(\delta_1, \theta)]$ and $R(\delta_2, \theta)$. In general, these risk functions are not available analytically, but they may be approximated, for instance, by a regular Monte Carlo method,

$$\hat{R}(\delta_1, \theta) = \frac{1}{m} \sum_{i=1}^{m} L(\delta_1(X_i), \theta), \qquad \hat{R}(\delta_2, \theta) = \frac{1}{m} \sum_{i=1}^{m} L(\delta_2(Y_i), \theta),$$

the $X_i$'s and $Y_i$'s being simulated from $f(\cdot|\theta)$. Positive correlation between $L(\delta_1(X_i), \theta)$ and $L(\delta_2(Y_i), \theta)$ then reduces the variability of the approximation of $R(\delta_1, \theta) - R(\delta_2, \theta)$.

Before we continue with the development in this section, we pause to make two elementary remarks that should be observed in any simulation comparison.

(i) First, the *same sample* $(X_1, \ldots, X_m)$ should be used in the evaluation of $R(\delta_1, \theta)$ and of $R(\delta_2, \theta)$. This repeated use of a single sample greatly improves the precision of the estimated difference $R(\delta_1, \theta) - R(\delta_2, \theta)$, as shown by the comparison of the variances of $\hat{R}(\delta_1, \theta) - \hat{R}(\delta_2, \theta)$ and of

$$\frac{1}{m} \sum_{i=1}^{m} \{L(\delta_1(X_i), \theta) - L(\delta_2(X_i), \theta)\} \; .$$

(ii) Second, *the same sample* should be used for the comparison of risks *for every value of* $\theta$. Although this sounds like an absurd recommendation since the sample $(X_1, \ldots, X_m)$ is usually generated from a distribution depending on $\theta$, it is often the case that the same uniform sample can be used for the generation of the $X_i$'s for every value of $\theta$. Also, in many cases, there exists a transformation $M_\theta$ on $\mathcal{X}$ such that if $X^0 \sim f(X|\theta_0)$, $M_\theta X^0 \sim f(X|\theta)$. A single sample $(X_1^0, \ldots, X_m^0)$ from $f(X|\theta_0)$ is then sufficient to produce a sample from $f(X|\theta)$ by the transform $M_\theta$. (This second remark is somewhat tangential for the theme of this section; however, it brings significant improvement in the practical implementation of Monte Carlo methods.)

The variance reduction associated with the conservation of the underlying uniform sample is obvious in the graphs of the resulting risk functions, which then miss the irregular peaks of graphs obtained with independent samples and allow for an easier comparison of estimators. See, for instance, the graphs in Figure 3.3.1, which are based on samples generated independently for each value of $\lambda$. By comparison, an evaluation based on a single sample corresponding to $\lambda = 1$ would give a constant risk in the exponential case.

**Example 3.7.1 James-Stein estimation.** In the case $X \sim \mathcal{N}_p(\theta, I_p)$, the transform is the location shift $M_\theta X = X + \theta - \theta_0$. When studying *truncated James–Stein estimators*

$$\delta_a(X) = \left(1 - \frac{a}{\|x\|^2}\right)^+ X, \qquad 0 < a < 2(p-2)$$

(see Robert 1994a for a motivation), the squared error risk of $\delta_a$ can be computed "explicitly," but the resulting expression involves several special functions (Robert 1988) and the approximation of the risks by simulation is much more helpful in comparing these estimators. Figure 3.7.1 illustrates this comparison in the case $p = 5$ and exhibits a crossing phenomenon for the risk functions in the same region; however, as shown by the inset, the crossing point for the risks of $\delta_a$ and $\delta_c$ depends on $(a, c)$.    ‖

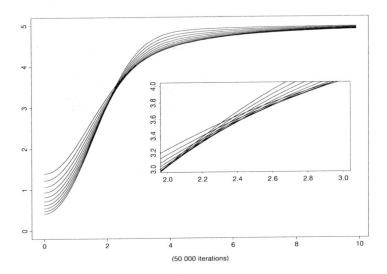

(50 000 iterations)

Figure 3.7.1. *Approximate squared error risks of truncated James–Stein estimators for a normal distribution $\mathcal{N}_5(\theta, I_5)$, as a function of $\|\theta\|$. The inset gives a magnification of the intersection zone for the risk functions.*

In a more general setup, creating a strong enough correlation between $\delta_1$ and $\delta_2$ is rarely so simple, and the quest for correlation can result in an increase in the conception and simulation times, which may even have a negative overall effect on the efficiency of the analysis. Indeed, to use the same uniform sample for the generation of variables distributed from $f_1$ and $f_2$ in (3.7.1) is only possible when there exists a simple transformation from $f_1$ to $f_2$. For instance, if $f_1$ or $f_2$ must be simulated by Accept–Reject methods, the use of a random number of uniform variables prevents the use of a common sample.[5]

The method of *antithetic variables* is based on the same idea that higher efficiency can be brought about by correlation. Given two samples $(X_1, \ldots, X_m)$ and $(Y_1, \ldots, Y_m)$ from $f$ used for the estimation of

$$I = \int_{\mathbb{R}} h(x)f(x)dx \ ,$$

[5] The same difficulty occurs in Chapter 8 with coupling methods.

the estimator

(3.7.2) $$\frac{1}{2m} \sum_{i=1}^{m} [h(X_i) + h(Y_i)]$$

is more efficient than an estimator based on an iid sample of size $2m$ if the variables $h(X_i)$ and $h(Y_i)$ are *negatively correlated*. In this setting, the $Y_i$'s are the *antithetic variables*, and it remains to develop a method for generating these variables in an optimal (or, at least, useful) way. However, the correlation between $h(X_i)$ and $h(Y_i)$ depends both on the pair $(X_i, Y_i)$ and on the function $h$. (For instance, if $h$ is even, $X_i$ has mean 0, and $X_i = -Y_i$, $X_i$ and $Y_i$ are negatively correlated, but $h(X_i) = h(Y_i)$.) A solution proposed in Rubinstein (1981) is to use the uniform variables $U_i$ to generate the $X_i$'s and the variables $1 - U_i$ to generate the $Y_i$'s. The argument goes as follows: If $H = h{\circ}F^-$, $X_i = F^-(U_i)$, and $Y_i = F^-(1-U_i)$, then $h(X_i)$ and $h(Y_i)$ are negatively correlated *when $H$ is a monotone function*. Again, such a constraint is often difficult to verify and, moreover, the technique only applies for direct transforms of uniform variables, thus excluding the Accept–Reject methods.

Geweke (1988) proposed the implementation of an inversion at the level of the $X_i$'s by taking $Y_i = 2\mu - X_i$ when $f$ *is symmetric around $\mu$*. With some additional conditions on the function $h$, the improvement brought by

$$\frac{1}{2m} \sum_{i=1}^{m} [h(X_i) + h(2\mu - X_i)]$$

upon

$$\frac{1}{2m} \sum_{i=1}^{2m} h(X_i)$$

is quite substantial for large sample sizes $m$. Empirical extensions of this approach can then be used in cases where $f$ is not symmetric, by replacing $\mu$ with the mode of $f$ or the median of the associated distribution. Moreover, if $f$ is unknown or, more importantly, if $\mu$ is unknown, $\mu$ can be estimated from a first sample (but caution is advised!).

**Example 3.7.2 (Continuation of Example 3.1.3)**  Assume, for the sake of illustration, that the noncentral chi squared variables $Z_i = ||X_i||^2$ are simulated from normal random variables $X_i \sim \mathcal{N}_p(\theta, I_p)$. We can create negative correlation by using $Y_i = 2\theta - X_i$, which has a correlation of $-1$ with $X_i$, to produce a second sample, $Z_i' = ||Y_i||^2$. However, the negative correlation does not necessarily transfer to the pairs $(h(Z_i), h(Z_i'))$. Figure 3.7.2 illustrates the behavior of (3.7.2) for

$$h_1(z) = z \qquad \text{and} \qquad h_2(z) = \mathbb{I}_{z<||\theta||^2+p},$$

when $m = 500$ and $p = 4$, compared to an estimator based on an iid sample of size $2m$. As shown by the graphs in Figure 3.7.2, although the correlation between $h(Z_i)$ and $h(Z_i')$ is actually positive for small values of $||\theta||^2$, the improvement brought by (3.7.2) over the standard average is quite impressive in the case of $h_1$. The setting is less clear for $h_2$, but the variance of the terms of (3.7.2) is much smaller than its independent counterpart.                                                                    ‖

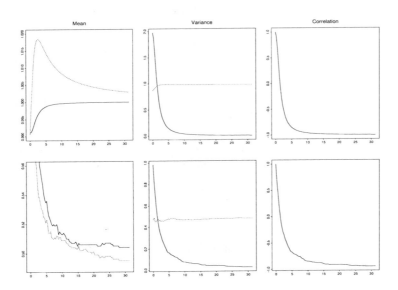

Figure 3.7.2. *Average of the antithetic estimator (3.7.2) (solid lines) against the average of an standard iid estimate (dots) for the estimation of* $\mathbb{E}[h_1(Z)]$ *(upper left) and* $\mathbb{E}[h_2(Z)]$ *(lower left), along with the empirical variance of* $h_1(X_i) + h_1(Y_i)$ *(Upper center) and* $h_2(X_i) + h_2(Y_i)$ *(Lower center), and the correlation between* $h_1(Z_i)$ *and* $h_1(Z_i')$ *(upper right) and between* $h_2(Z_i)$ *and* $h_2(Z_i')$ *(lower right), for* $m = 500$ *and* $p = 4$. *The horizontal axis is scaled in terms of* $||\theta||$ *and the values in the upper left graph are divided by the true expectation,* $||\theta||^2 + p$, *and the values in the upper central graph are divided by* $8||\theta||^2 + 4p$.

### 3.7.2 Control Variates

In some settings, there exist functions $h_0$ whose mean under $f$ is known. For instance, if $f$ is symmetric around $\mu$, the mean of $h_0(X) = \mathbb{I}_{X \geq \mu}$ is $1/2$. We also saw a more general example in the case of Riemann sums with known density $f$, with a convergent estimator of 1. This additional information can reduce the variance of an estimator of $I = \int h(x)f(x)dx$ in the following way. If $\delta_1$ is an estimator of $I$ and $\delta_3$ an unbiased estimator of $\mathbb{E}_f[h_0(X)]$, consider the weighted estimator $\delta_2 = \delta_1 + \beta(\delta_3 - \mathbb{E}_f[h_0(X)])$. The estimators $\delta_1$ and $\delta_2$ have the same mean and

$$\text{var}(\delta_2) = \text{var}(\delta_1) + \beta^2 \text{var}(\delta_3) + 2\beta \text{cov}(\delta_1, \delta_3) .$$

For the optimal choice

$$\beta^* = -\frac{\text{cov}(\delta_1, \delta_3)}{\text{var}(\delta_3)} ,$$

we have

$$\text{var}(\delta_2) = (1 - \rho_{13}^2)\, \text{var}(\delta_1)\,,$$

$\rho_{13}^2$ being the correlation coefficient between $\delta_1$ and $\delta_3$, so the control variate strategy will result in decreased variance. In particular, if

$$\delta_1 = \frac{1}{m} \sum_{i=1}^{m} h(X_i) \qquad \text{and} \qquad \delta_3 = \frac{1}{m} \sum_{i=1}^{m} h_0(X_i)\,,$$

the control variate estimator is

$$\delta_2 = \frac{1}{m} \sum_{i=1}^{m} (h(X_i) + \beta^* h_0(X_i)) - \beta^* \mathbb{E}_f[h_0(X)]\,,$$

with $\beta^* = -\text{cov}(h(X), h_0(X))/\text{var}(h_0(X))$. Note that this construction is only formal since it requires the computation of $\beta^*$. An incorrect choice of $\beta$ may lead to an increased variance; that is, $\text{var}(\delta_2) > \text{var}(\delta_1)$. (However, in practice, the sign of $\beta^*$ can be evaluated by simulation. More generally, functions with known expectations can be used as side controls in convergence diagnoses.)

**Example 3.7.3 Control variate integration.** Let $X \sim f$, and suppose that we want to evaluate

$$P(X > a) = \int_a^\infty f(x)dx.$$

The natural place to start is with $\delta_1 = \frac{1}{n} \sum_{i=1}^n \mathbb{I}(X_i > a)$, where the $X_i$'s are iid from $f$.

Suppose now that $f$ is symmetric or, more, generally, that for some parameter $\mu$ we know the value of $P(X > \mu)$ (where we assume that $a > \mu$). We can then take $\delta_3 = \frac{1}{n} \sum_{i=1}^n \mathbb{I}(X_i > \mu)$ and form the control variate estimator

$$\delta_2 = \frac{1}{n} \sum_{i=1}^n \mathbb{I}(X_i > a) + \beta \left( \frac{1}{n} \sum_{i=1}^n \mathbb{I}(X_i > \mu) - P(X > \mu) \right).$$

Since $\text{var}(\delta_2) = \text{var}(\delta_1) + \beta^2 \text{var}(\delta_3) + 2\beta \text{cov}(\delta_1, \delta_3)$ and

$$\text{cov}(\delta_1, \delta_3) = \frac{1}{n} P(X > a)[1 - P(X > \mu)]\,,$$

(3.7.3) $$\qquad \text{var}(\delta_3) = \frac{1}{n} P(X > \mu)[1 - P(X > \mu)]\,,$$

it follows that $\delta_2$ will be an improvement over $\delta_1$ if

$$\beta < 0 \quad \text{and} \quad |\beta| < 2 \frac{\text{cov}(\delta_1, \delta_3)}{\text{var}(\delta_3)} = 2 \frac{P(X > a)}{P(X > \mu)}.$$

If $P(X > \mu) = 1/2$ and we have some idea of the value of $P(X > a)$, we can choose an appropriate value for $\beta$ (see Problem 3.28). ‖

**Example 3.7.4 Logistic regression.** Consider the logistic regression model introduced in Example 1.3.3,

$$P(Y_i = 1) = \exp(x_i^t \beta)/\{1 + \exp(x_i^t \beta)\}.$$

The likelihood associated with a sample $((x_1, Y_1), \ldots, (x_n, Y_n))$ can be written

$$\exp\left(\beta^t \sum_i Y_i x_i\right) \prod_{i=1}^{n} \{1 + \exp(x_i^t \beta)\}^{-1}.$$

When $\pi(\beta)$ is a *conjugate prior* (see Note 1.6.1),

$$(3.7.4) \quad \pi(\beta|\zeta, \lambda) \propto \exp(\beta^t \zeta) \prod_{i=1}^{n} \{1 + \exp(x_i^t \beta)\}^{-\lambda}, \qquad \lambda > 0,$$

the posterior distribution of $\beta$ is of the same form, with $(\sum_i Y_i x_i + \zeta, \lambda + 1)$ replacing $(\zeta, \lambda)$.

The expectation $\mathbb{E}^{\pi}[\beta | \sum_i Y_i x_i + \zeta, \lambda + 1]$ is derived from variables $\beta_j$ $(1 \le j \le m)$ generated from (3.7.4). Since the logistic distribution is in an exponential family, the following holds (see Brown 1986 or Robert 1994a, Section 3.2.3):

$$\mathbb{E}_\beta\left[\sum_i Y_i x_i\right] = n\nabla\psi(\beta)$$

and

$$\mathbb{E}^{\pi}\left[\nabla\psi(\beta) \middle| \sum_i Y_i x_i + \zeta, \lambda + 1\right] = \frac{\sum_i Y_i x_i + \zeta}{n(\lambda + 1)}.$$

Therefore, the posterior expectation of the function

$$n\nabla\psi(\beta) = \sum_{i=1}^{n} \frac{\exp(x_i^t \beta)}{1 + \exp(x_i^t \beta)} x_i$$

is known and equal to $(\sum_i Y_i x_i + \zeta)/(\lambda + 1)$ under the prior distribution $\pi(\beta|\zeta, \lambda)$. Unfortunately, a control variate version of

$$\delta_1 = \frac{1}{m} \sum_{j=1}^{m} \beta_j$$

is not available since the optimal constant $\beta^*$ (or even its sign) cannot be evaluated. Thus, the fact that the posterior mean of $\nabla\psi(\beta)$ is known does not help us to establish a control variate estimator. This information can be used in a more informal way to study convergence of $\delta_1$ (see, for instance, Robert 1993).     ‖

In conclusion, the technique of control variates is only manageable in very specific cases: the control function $h$ must be available, as well as the optimal weight $\beta^*$.

*3.7.3 Conditional Expectations*

Another approach to reduce the variance of an estimator is to use the conditioning inequality

$$\text{var}(\mathbb{E}[\delta(X)|Y]) \le \text{var}(\delta(X)),$$

sometimes called *Rao–Blackwellization* (Gelfand and Smith 1990; Liu *et al.* 1994; Casella and Robert 1996) because the inequality is associated with the Rao–Blackwell Theorem (Lehmann and Casella 1998), although the conditioning is not always in terms of sufficient statistics.

In a simulation context, if $\delta(X)$ is an estimator of $I = \mathbb{E}_f[h(X)]$ and if $X$ can be simulated from the joint distribution $\tilde{f}(x, y)$, such that

$$\int \tilde{f}(x, y) dy = f(x),$$

the estimator $\delta^*(Y) = \mathbb{E}_{\tilde{f}}[\delta(X)|Y]$ dominates $\delta$ in terms of variance (and of squared error loss, since the bias is the same). Obviously, this result only applies in settings where $\delta^*(Y)$ can be explicitly computed.

**Example 3.7.5 Student's $t$ expectation.** Consider the expectation of $h(x) = \exp(-x^2)$ when $X \sim T(\nu, \mu, \sigma^2)$. The Student's $t$ distribution can be simulated as a mixture of a normal distribution and of a gamma distribution by Dickey's decomposition (1968),

$$X|y \sim \mathcal{N}(\mu, \sigma^2 y) \qquad \text{and} \qquad Y^{-1} \sim \mathcal{G}a(\nu/2, \nu/2).$$

Therefore, the empirical average

$$\delta_m = \frac{1}{m} \sum_{j=1}^{m} \exp(-X_j^2)$$

can be improved upon when the $X_j$ are parts of the sample $((X_1, Y_1), \ldots, (X_m, Y_m))$, since

$$\delta_m^* = \frac{1}{m} \sum_{j=1}^{m} \mathbb{E}[\exp(-X^2)|Y_j]$$

$$= \frac{1}{m} \sum_{j=1}^{m} \frac{1}{\sqrt{2\sigma^2 Y_j + 1}}$$

is the conditional expectation when $\mu = 0$. Figure 3.7.3 provides an illustration of the difference of the convergences of $\delta_m$ and $\delta_m^*$ to $\mathbb{E}_g[\exp(-X^2)]$ for $(\nu, \mu, \sigma) = (4.6, 0, 1)$. For $\delta_m$ to have the same precision as $\delta_m^*$ requires 10 times as many simulations.                                              ‖

Unfortunately, this conditioning method seems about as applicable as the previous methods since it involves a particular type of simulation (joint variables) and requires functions which are sufficiently regular for the conditional expectations to be explicit.

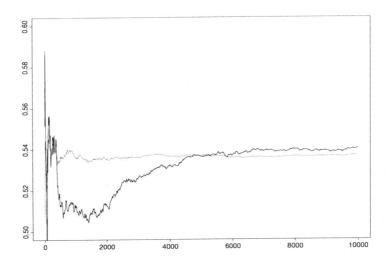

Figure 3.7.3. *Convergence of the estimators of* $\mathbb{E}[\exp(-X^2)]$, $\delta_m$ *(solid lines) and* $\delta_m^*$ *(dots) for* $(\nu, \mu, \sigma) = (4.6, 0, 1)$. *The final values are* 0.5405 *and* 0.5369, *respectively, for a true value equal to* 0.5373.

We conclude this section with a demonstration of a specific situation where Rao–Blackwellization is always possible.[6] This is in the general setup of Accept–Reject methods, which are not always amenable to other acceleration techniques of §3.7.1 and §3.7.2. (Casella and Robert 1996 distinguish between *parametric* Rao–Blackwellization and *nonparametric* Rao–Blackwellization, the parametric version being more restrictive and only used in specific setups such as Gibbs sampling. See §6.4.2.)

Consider an Accept–Reject method based on the instrumental distribution $g$. If the original sample produced by the algorithm is $(X_1, \ldots, X_m)$, it can be associated with two iid samples, $(U_1, \ldots, U_N)$ and $(Y_1, \ldots, Y_N)$, with corresponding distributions $\mathcal{U}_{[0,1]}$ and $g$; $N$ is then the stopping time associated with the acceptance of $m$ variables $Y_j$. An estimator of $\mathbb{E}_f[h]$ based on $(X_1, \ldots, X_m)$ can therefore be written

$$\delta_1 = \frac{1}{m} \sum_{i=1}^{m} h(X_i)$$

$$= \frac{1}{m} \sum_{j=1}^{N} h(Y_j) \, \mathbb{I}_{U_j \leq w_j}$$

with $w_j = f(Y_j)/Mg(Y_j)$. A reduction of the variance of $\delta_1$ can be obtained

---

[6] This part contains rather specialized material, which will not be used again in the book. It can be omitted at first reading.

by integrating out the $U_i$'s, which leads to the estimator

$$\delta_2 = \frac{1}{m} \sum_{j=1}^{N} \mathbb{E}[\mathbb{I}_{U_j \leq w_j} | N, Y_1, \ldots, Y_N] \, h(Y_j) = \frac{1}{m} \sum_{i=1}^{N} \rho_i h(Y_i),$$

where, for $i = 1, \ldots, n-1$, $\rho_i$ satisfies

$$\rho_i = P(U_i \leq w_i | N = n, Y_1, \ldots, Y_n)$$

(3.7.5)
$$= w_i \frac{\sum_{(i_1, \ldots, i_{m-2})} \prod_{j=1}^{m-2} w_{i_j} \prod_{j=m-1}^{n-2} (1 - w_{i_j})}{\sum_{(i_1, \ldots, i_{m-1})} \prod_{j=1}^{m-1} w_{i_j} \prod_{j=m}^{n-1} (1 - w_{i_j})},$$

while $\rho_n = 1$. The numerator sum is over all subsets of $\{1, \ldots, i-1, i+1, \ldots, n-1\}$ of size $m-2$, and the denominator sum is over all subsets of size $m-1$. The resulting estimator $\delta_2$ is an average over all the possible permutations of the realized sample, the permutations being weighted by their probabilities. The Rao–Blackwellized estimator is then a function only of $(N, Y_{(1)}, \ldots, Y_{(N-1)}, Y_N)$, where $Y_{(1)}, \ldots, Y_{(N-1)}$ are the order statistics.

Although the computation of the $\rho_i$'s may appear formidable, a recurrence relation of order $n^2$ can be used to calculate the estimator. Define $(k \leq m < n)$

$$S_k(m) = \sum_{(i_1, \ldots, i_k)} \prod_{j=1}^{k} w_{i_j} \prod_{j=k+1}^{m} (1 - w_{i_j}),$$

with $\{i_1, \ldots, i_m\} = \{1, \ldots, m\}$, $S_k(m) = 0$ for $k > m$, and $S_k^i(i) = S_k(i-1)$. Then, we can recursively calculate

(3.7.6)
$$S_k(m) = w_m S_{k-1}(m-1) + (1 - w_m) S_k(m-1),$$
$$S_k^i(m) = w_m S_{k-1}^i(m-1) + (1 - w_m) S_k^i(m-1)$$

and note that weight $\rho_i$ of (3.7.5) is given by

$$\rho_i = w_i \, S_{t-2}^i(n-1)/S_{t-1}(n-1) \qquad (i < n).$$

If the random nature of $N$ is ignored in taking the conditional expectation, this leads to the importance sampling estimator,

$$\delta_3 = \sum_{j=1}^{N} w_j \, h(Y_j) \bigg/ \sum_{j=1}^{N} w_j \,,$$

which does not necessarily improve upon $\delta_1$ (see §3.3.3).

Casella and Robert (1996) establish the following proposition, showing that $\delta_2$ can be computed and dominates $\delta_1$. The proof is left to Problem 3.32.

**Proposition 3.7.6** *The estimator $\delta_2 = \frac{1}{m} \sum_{j=1}^{N} \rho_j \, h(y_j)$ dominates the estimator $\delta_1$ under quadratic loss.*

The computation of the weights $\rho_i$ is obviously more costly than the derivation of the weights of the importance sampling estimator $\delta_3$ or of the corrected estimator of §3.3.3. However, the recursive formula (3.7.6)

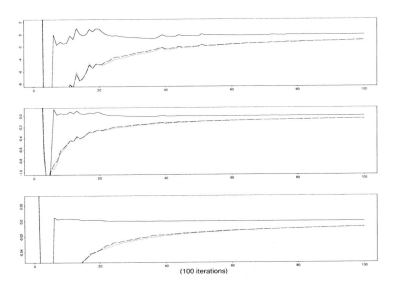

Figure 3.7.4. *Comparisons of the errors* $\delta - \mathbb{E}[h_i(X)]$ *of the Accept–Reject estimator* $\delta_1$ *(long dashes), of the importance sampling estimator* $\delta_3$ *(dots), and of the conditional version of* $\delta_1$, $\delta_2$ *(solid lines), for* $h_1(x) = x^3$ *(top),* $h_2(x) = x \log(x)$ *(middle), and* $h_3(x) = x/(1+x)$ *(bottom) and* $\alpha = 3.7$, $\beta = 1$. *The final errors are respectively* $-0.998$, $-0.982$, *and* $-0.077$ *(top),* $-0.053$, $-0.053$, *and* $-0.001$ *(middle), and* $-0.0075$, $-0.0074$, *and* $-0.00003$ *(bottom).*

leads to an overall simplification of the computation of the coefficients $\rho_i$. Nonetheless, the increase in computing time can go as high as seven times (Casella and Robert 1996), but the corresponding variance decrease is even greater (80%).

**Example 3.7.7 (Continuation of Example 3.3.7)** If we repeat the simulation of $\mathcal{G}a(\alpha, \beta)$ from the $\mathcal{G}a(a, b)$ distribution, with $a \in \mathbb{N}$, it is of interest to compare the estimator obtained by Rao–Blackwellization, $\delta_2$, with the standard estimator based on the Accept–Reject sample and with the biased importance sampling estimator $\delta_3$. Figure 3.7.4 illustrates the substantial improvement brought by conditioning, since $\delta_2$ (uniformly) dominates both alternatives for the observed simulations. Note also the strong similarity between $\delta_1$ and the importance sampling estimator, which does not bring any noticeable improvement in this setup.  ‖

For further discussion, estimators and examples see Casella (1996) and Casella and Robert (1996).

## 3.8 Problems

**3.1** For the situation of Example 3.2.1:

(a) Show that $\hat{\delta}^\pi_m(x) \to \delta^\pi(x)$ as $m \to \infty$.

(b) Show that the Central Limit Theorem can be applied to $\hat{\delta}_m^\pi(x)$.

(c) Generate random variables $\theta_1, \ldots, \theta_m \sim \mathcal{N}(x, 1)$ and calculate $\hat{\delta}_m^\pi(x)$ for $x = 0, 1, 4$. Use the Central Limit Theorem to construct a measure of accuracy of your calculation.

**3.2**(a) If $X \sim \mathcal{N}(0, \sigma^2)$, show that

$$\mathbb{E}[e^{-X^2}] = \frac{1}{\sqrt{2\sigma^2 + 1}} \ .$$

(b) Generalize to the case $X \sim \mathcal{N}(\mu, \sigma^2)$.

**3.3** For the situation of Example 3.3.3, recreate Figure 3.3.1 using the following simulation strategies with a sample size of 10, 000:

(a) For each value of $\lambda$, simulate a sample from the $\mathcal{E}xp(1/\lambda)$ distribution and a separate sample from the log-normal $\mathcal{LN}(0, 2 \log \lambda)$ distribution. Plot the resulting risk functions.

(b) For each value of $\lambda$, simulate a sample from the $\mathcal{E}xp(1/\lambda)$ distribution and then transform it into a sample from the $\mathcal{LN}(0, 2 \log \lambda)$ distribution. Plot the resulting risk functions.

(c) Simulate a sample from the $\mathcal{E}xp(1)$ distribution. For each value of $\lambda$, transform it into a sample from $\mathcal{E}xp(1/\lambda)$, and then transform it into a sample from the $\mathcal{LN}(0, 2 \log \lambda)$ distribution. Plot the resulting risk functions.

(d) Compare and comment on the accuracy of the plots.

**3.4** Compare (in a simulation experiment) the performances of the regular Monte Carlo estimator of

$$\int_1^2 \frac{e^{-x^2/2}}{\sqrt{2\pi}} \, dx$$

with those of an estimator based on an optimal choice of instrumental distribution.

**3.5** In the setup of Example 3.3.3, give the two first moments of the log-normal distribution $\mathcal{LN}(\mu, \sigma^2)$.

**3.6** In the setup of Example 3.3.5, examine whether or not the different estimators of the expectations $\mathbb{E}_f[h_i(X)]$ have finite variances.

**3.7** Establish the equality (3.3.9) using the representation $b = \beta a/\alpha$.

**3.8** (Ó Ruanaidh and Fitzgerald 1996) For simulating random variables from the density $f(x) = \exp\{-\sqrt{x}\}[\sin(x)]^2$, $0 < x < \infty$, compare the following choices of instrumental densities in terms of variance conditions and speed of simulation:

$$g_1(x) = \tfrac{1}{2} e^{-|x|}, \qquad g_2(x) = \tfrac{1}{2\sqrt{2}} \operatorname{sech}^2(x/\sqrt{2}),$$

$$g_3(x) = \tfrac{1}{2\pi} \tfrac{1}{1+x^2/4}, \qquad g_4(x) = \tfrac{1}{\sqrt{2\pi}} e^{-x/2} \ .$$

**3.9** In this chapter, the importance sampling method is developed for an iid sample $(Y_1, \ldots, Y_n)$ from $g$.

(a) Show that the importance sampling estimator is still unbiased if the $Y_i$'s are correlated while being marginally distributed from $g$.

(b) Show that the importance sampling estimator can be extended to the case when $Y_i$ is generated from a conditional distribution $q(y_i|Y_{i-1})$.

(c) Implement a scheme based on an iid sample $(Y_1, Y_3, \ldots, Y_{2n-1})$ and a secondary sample $(Y_2, Y_4, \ldots, Y_{2n})$ such that $Y_{2i} \sim q(y_{2i}|Y_{2i-1})$. Show that the covariance

$$\text{cov}\left(h(Y_{2i-1})\frac{f(Y_{2i-1})}{g(Y_{2i-1})}, h(Y_{2i})\frac{f(Y_{2i})}{q(Y_{2i}|Y_{2i-1})}\right)$$

is null. Generalize.

**3.10** For a sample $(Y_1, \ldots, Y_h)$ from $g$, the weights $\omega_i$ are defined as

$$\omega_i = \frac{f(Y_i)/g(Y_i)}{\sum_{j=1}^{h} f(Y_j)/g(Y_j)}.$$

Show that the following algorithm (Rubin 1987) produces a sample from $f$ such that the empirical average

$$\frac{1}{M}\sum_{m=1}^{M} h(X_m)$$

is asymptotically equivalent to the importance sampling estimator based on $(Y_1, \ldots, Y_N)$:

For $m = 1, \ldots, M$,
take $X_m = Y_i$ with probability $\omega_i$

(*Note:* This is the SIR algorithm.)

**3.11** (Smith and Gelfand 1992) When evaluating an integral based on a posterior distribution

$$\pi(\theta|x) \propto \pi(\theta)\ell(\theta|x),$$

where $\pi$ is the prior distribution and $\ell$ the likelihood function, show that the prior distribution can always be used as instrumental distribution.

(a) Show that the variance is finite when the likelihood is bounded.

(b) Compare with choosing $\ell(\theta|x)$ as instrumental distribution when the likelihood is proportional to a density. (*Hint:* Consider the case of exponential families.)

(c) Discuss the drawbacks of this (these) choice(s) in specific settings.

(d) Show that a mixture between both instrumental distributions can ease some of the drawbacks.

**3.12** Consider

$$f(x) = \frac{\Gamma(2)}{\sqrt{3\pi}\Gamma(3/2)}\left(1 + x^2/3\right)^{-2}, \qquad g(x) = \frac{1}{\pi}\left(1 + x^2\right)^{-1}.$$

(a) Show that

$$f(x)/g(x) \leq 3\sqrt{3}/4.$$

(b) Show that

$$\delta^{IS} = \frac{1}{n+t-1}\sum_{i=1}^{n+t-1}\frac{f(Z_i)}{g(Z_i)}h(Z_i)$$

is unbiased.

(c) For the functions

$$h_1(y) = y, \qquad h_2(y) = |y|^{0.1}, \qquad \text{and} \qquad h_3(y) = \mathbb{I}_{y>1.96},$$

compare the usual Monte Carlo approximation $\delta^{AR}$ (based on sample size $n$) with the importance sampling estimator (based on sample size $n + t$)

$$\delta^{IS} = \frac{1}{n+t-1} \sum_{i=1}^{n+t-1} \frac{f(z_i)}{g(z_i)} h(z_i)$$

and the rescaled importance sampling estimator

$$\delta^{ISW} = \sum_{i=1}^{n+t-1} \frac{f(z_i)}{g(z_i)} h(z_i) \bigg/ \sum_{i=1}^{n+t-1} \frac{f(z_i)}{g(z_i)}.$$

(d) Repeat the experiment with a Student $\mathcal{T}_7$ distribution for $f$.

**3.13** Monte Carlo marginalization is a technique for calculating a marginal density when simulating from a joint density. Let $(X_i, Y_i) \sim f_{XY}(x, y)$, independent, and the corresponding marginal distribution $f_X(x) = \int f_{XY}(x, y)dy$.

(a) Let $w(x)$ be an arbitrary density. Show that

$$\lim_n \frac{1}{n} \sum_{i=1}^{n} \frac{f_{XY}(x^*, y_i)w(x_i)}{f_{XY}(x_i, y_i)} = \int \int \frac{f_{XY}(x^*, y)w(x)}{f_{XY}(x, y)} f_{XY}(x, y)dxdy = f_X(x^*)$$

and so we have a Monte Carlo estimate of $f_X$, the marginal distribution of $X$, from only knowing the form of the joint distribution.

(b) Let $X|Y = y \sim \mathcal{G}a(y, 1)$ and $Y \sim \mathcal{E}xp(1)$. Use the technique of part (a) to plot the marginal density of $X$. Compare it to the exact marginal.

(c) Choosing $w(x) = f_{X|Y}(x|y)$ works to produce the marginal distribution, and it is optimal. In the spirit of Theorem 3.3.4, can you prove this?

**3.14** (Evans and Swartz 1995) Devise and implement a simulation experiment to approximate the probability $P(Z \in (0, \infty)^6)$ when $Z \sim \mathcal{N}_6(0, \Sigma)$ and

$$\Sigma^{-1/2} = \text{diag}(0, 1, 2, 3, 4, 5) + e \cdot e^t,$$

with $e^t = (1, 1, 1, 1, 1, 1)$:

(a) when using the $\Sigma^{-1/2}$ transform of a $\mathcal{N}_6(0, I_6)$ random variables;

(b) when using the Choleski decomposition of $\Sigma$;

(c) when using a distribution restricted to $(0, \infty)^6$ and importance sampling.

**3.15** (Chen and Shao 1997) As mentioned, normalizing constants are superfluous in Bayesian inference except in the case when several models are considered at once (as in the computation of Bayes factors). In such cases, where $\pi_1(\theta) = \tilde{\pi}_1(\theta)/c_1$ and $\pi_2(\theta) = \tilde{\pi}_2(\theta)/c_2$, and only $\tilde{\pi}_1$ and $\tilde{\pi}_2$ are known, the quantity to approximate is $\varrho = c_1/c_2$ or $\xi = \log(c_1/c_2)$.

(a) Show that the ratio $\varrho$ can be approximated by

$$\frac{1}{n} \sum_{i=1}^{n} \frac{\tilde{\pi}_1(\theta_i)}{\tilde{\pi}_2(\theta_i)}, \qquad \theta_1, \dots, \theta_n \sim \pi_2.$$

(*Hint:* Use an importance sampling argument.)

(b) Show that

$$\frac{\int \tilde{\pi}_1(\theta)\alpha(\theta)\pi_2(\theta)d\theta}{\int \tilde{\pi}_2(\theta)\alpha(\theta)\pi_1(\theta)d\theta} = \frac{c_1}{c_2}$$

holds for every function $\alpha(\theta)$ such that both integrals are finite.

(c) Deduce that

$$\frac{\frac{1}{n_2}\sum_{i=1}^{n_2}\tilde{\pi}_1(\theta_{2i})\alpha(\theta_{2i})}{\frac{1}{n_1}\sum_{i=1}^{n_1}\tilde{\pi}_2(\theta_{1i})\alpha(\theta_{1i})},$$

with $\theta_{1i} \sim \pi_1$ and $\theta_{2i} \sim \pi_2$, is a convergent estimator of $\varrho$.

(d) Show that part (b) covers the case of the Newton and Raftery (1994) representation

$$\frac{c_1}{c_2} = \frac{\mathbb{E}^{\pi_2}\left[\tilde{\pi}_2(\theta)^{-1}\right]}{\mathbb{E}^{\pi_1}\left[\tilde{\pi}_1(\theta)^{-1}\right]}.$$

(e) Show that the optimal choice (in terms of mean square error) of $\alpha$ in part (c) is

$$\alpha(\theta) = c\frac{n_1 + n_2}{n_1\pi_1(\theta) + n_2\pi_2(\theta)}$$

where $c$ is a constant. (*Note:* See Meng and Wong 1996.)

**3.16** (Continuation of Problem 3.15) When the priors $\pi_1$ and $\pi_2$ belong to a parameterized family (that is, $\pi_i(\theta) = \pi(\theta|\lambda_i)$), the corresponding constants are denoted by $c(\lambda_i)$.

(a) Verify the identity

$$-\log\left(\frac{c(\lambda_1)}{c(\lambda_2)}\right) = \mathbb{E}\left[\frac{U(\theta,\lambda)}{\pi(\lambda)}\right],$$

where

$$U(\theta,\lambda) = \frac{d}{d\lambda}\log(\tilde{\pi}(\theta|\lambda))$$

and $\pi(\lambda)$ is an arbitrary distribution on $\lambda$.

(b) Show that $\xi$ can be estimated with the *bridge estimator* of Gelman and Meng (1998),

$$\hat{\xi} = \frac{1}{n}\sum_{i=1}^{n}\frac{U(\theta_i,\lambda_i)}{\pi(\lambda_i)},$$

when the $(\theta_i, \lambda_i)$'s are simulated from the joint distribution induced by $\pi(\lambda)$ and $\pi(\theta|\lambda_i)$.

(c) Show that the minimum variance estimator of $\xi$ is based on

$$\pi(\lambda) \propto \sqrt{\mathbb{E}_\lambda[U^2(\theta,\lambda)]}$$

and examine whether this solution gives the Jeffreys prior.

**3.17** Consider the evaluation of

$$I = \int_{\mathbb{R}} h(x)f(x)dx .$$

(a) Examine whether the substitution of

$$\sum_{i=1}^{n-1}(X_{(i+1)} - X_{(i)}) \frac{h(X_{(i)}) + h(X_{(i+1)})}{2}$$

into

$$\sum_{i=1}^{n-1}(X_{(i+1)} - X_{(i)}) h(X_{(i)})$$

improves the speed of convergence. (*Hint*: Examine the influence of the remainders

$$\int_{-\infty}^{X_{(1)}} h(x)f(x)dx \qquad \text{and} \qquad \int_{X_{(n)}}^{+\infty} h(x)f(x)dx \ .\Bigg)$$

(b) Show that it is always possible to express I as

$$I = \int_0^1 \tilde{h}(y)\tilde{f}(y)dy \ .$$

(c) Examine whether this new representation allows for improvement when using

$$\sum_{i=0}^{n}(Y_{(i+1)} - Y_{(i)}) \frac{\tilde{h}(Y_{(i)}) + \tilde{h}(Y_{(i+1)})}{2} \ .$$

**3.18** Verify (3.5.3). Show that $h''(\hat{x}_\theta|\theta)$ is necessarily negative.

**3.19** Using the facts that

$$\int y^3 e^{-cy^2/2}dy = \frac{-1}{2c}\left[y^2 + \frac{1}{c}\right]e^{-cy^2/2} \ ,$$

$$\int y^6 e^{-cy^2/2}dy = \frac{-1}{2c}\left[y^5 + \frac{5y^3}{2c} + \frac{15y}{4c}\right]e^{-cy^2/2} + 30\sqrt{\frac{\pi}{c^7}}\Phi(\sqrt{2c}y) \ ,$$

derive expressions similar to (3.5.4) for the second- and third-order approximations.

**3.20** By evaluating the normal integral for the first order approximation from (3.5.3), establish (3.5.4).

**3.21** Referring to Example 3.5.1, derive the Laplace approximation for the gamma density and reproduce Table 3.5.1.

**3.22** Verify Equation (3.6.6). (*Hint*: Calculate the moment generating function (mgf) of an iid sample from $f(x|\tau)$ and then directly calculate the mgf of $f_{\bar{X}}(x|\tau)$.)

**3.23** For the situation of Example 3.6.2:

(a) Verify the mgf in (3.6.10).

(b) Show that the solution to the saddlepoint equation is given by (3.6.11).

(c) Plot the saddlepoint density for $p = 7$ and $n = 1, 5, 20$. Compare your results to the exact density.

**3.24**(a) Use the mixture representation of (3.6.9) and the technique of Example 3.6.1 to derive an alternative saddlepoint approximation to the noncentral chi squared density.

(b) Compare the approximation in part (a) to that of Problem 3.23(c).

**3.25** (Daniels 1983) Show that the renormalized saddlepoint approximation is exact for the normal and gamma densities.

**3.26** The saddlepoint approximation is useful in obtaining approximations to the density of the maximum likelihood estimator, particularly in exponential families. Let $X_1, X_2, \ldots, X_n$, be iid with density

$$f(x|\theta) = \exp\left\{\theta s(x) - K(\theta) - d(x)\right\}.$$

(a) Show that the sufficient statistic is $S = \sum_{i=1}^{n} s(X_i)$, with density

$$f(s|\theta) = \exp\left\{\theta s - nK(\theta) - h(s)\right\}.$$

(b) Since the density in part (a) is in the form of an "exponential tilt" of $\exp\left\{-h(s)\right\}$, use the saddlepoint approximation to verify

$$f(s|\theta) \approx \left[2\pi n\, K''\left(\hat{t}(s)\right)\right]^{-1/2} \exp\left\{\left[\theta - \hat{t}(s)\right]s - n\left[K(\theta) - K\left(\hat{t}(s)\right)\right]\right\},$$

where $\hat{t}(s)$ solves the equation $nK'(t) = s$.

(c) Show that $\hat{t}(s)$ is the maximum likelihood estimator and that its density has a saddlepoint approximation

$$f(\hat{t}|\theta) \approx \left[\frac{nK''(\hat{t})}{2\pi}\right]^{1/2} \exp\left\{(\theta - \hat{t})s(\hat{t}) - n\left(K(\theta) - K(\hat{t})\right)\right\}.$$

(d) Let $X_1, \ldots, X_n$ be iid from the Pareto $\mathcal{P}a(\alpha, \beta)$ distribution with known lower limit $\alpha$. The corresponding density is

$$f(x|\beta) = \frac{\beta \alpha^\beta}{x^{\beta+1}}, \qquad x > \alpha.$$

Show that

$$S = -\sum \log X_i, \qquad K(\beta) = -\log(\beta \alpha^\beta),$$

the saddlepoint is given by

$$\hat{t} = \frac{-n}{s + n\log(\alpha)},$$

and the saddlepoint approximation is

$$f(\hat{\beta}|\beta) \approx \left[\frac{n}{2\pi}\right]^{1/2} \left(\frac{\beta}{\hat{\beta}}\right)^n e^{(1-\beta/\hat{\beta})} \frac{1}{\hat{\beta}}.$$

Show that the renormalized version of this approximation is exact.

**3.27** (Berger *et al.* 1998) For $\Sigma$ a $p \times p$ positive-definite symmetric matrix, consider the distribution

$$\pi(\theta) \propto \frac{\exp\left(-(\theta - \mu)^t \Sigma^{-1}(\theta - \mu)/2\right)}{||\theta||^{p-1}}.$$

(a) Show that the distribution is well defined; that is, that

$$\int_{\mathbb{R}^p} \frac{\exp\left(-(\theta - \mu)^t \Sigma^{-1}(\theta - \mu)/2\right)}{||\theta||^{p-1}} d\theta < \infty.$$

(b) Show that an importance sampling implementation based on the normal instrumental distribution $\mathcal{N}_p(\mu, \Sigma)$ is not satisfactory from both theoretical and practical points of view.

(c) Examine the alternative based on a gamma distribution $\mathcal{G}a(\alpha, \beta)$ on $\eta = \|\theta\|^2$ and a uniform distribution on the angles.

**3.28** For the situation of Example 3.7.3:

(a) Verify (3.7.3)

(b) Verify the conditions on $\beta$ in order for $\delta_2$ to improve on $\delta_1$.

(c) For $f = \mathcal{N}(0, 1)$, find $P(X > a)$ for $a = 3, 5, 7$.

(d) For $f = \mathcal{T}_5$, find $P(X > a)$ for $a = 3, 5, 7$.

(e) For $f = \mathcal{T}_5$, find $a$ such that $P(X > a) = .01, .001, .0001$.

**3.29** A naïve way to implement the antithetic variable scheme is to use both $U$ and $(1 - U)$ in an inversion simulation. Examine empirically whether this method leads to variance reduction for the distributions

$$f_1(x) = \frac{1}{\pi(1+x^2)}, \qquad f_2(x) = \frac{1}{2}e^{-|x|}, \qquad f_3(x) = e^{-x}\mathbb{I}_{x>0},$$
$$f_4(x) = \frac{1}{\sqrt{6}}\frac{1}{\left(1+\frac{x^2}{3}\right)}, \qquad f_5(x) = \frac{2}{x^3}\mathbb{I}_{x>1} ,$$

and different functions of interest (moments, quantiles, etc.)

To calculate the weights for the Rao–Blackwellized estimator of §3.7.3, it is necessary to derive properties of the distribution of the random variables in the Accept-Reject algorithm [A.5]. The following problem is a rather straightforward exercise in distribution theory and is only made complicated by the stopping rule of the Accept-Reject algorithm.

**3.30** From the Accept-Reject Algorithm we get a sequence $Y_1, Y_2, \ldots$ of independent random variables generated from $g$ along with a corresponding sequence $U_1, U_2, \ldots$ of uniform random variables. For a fixed sample size $t$ (i.e. for a fixed number of accepted random variables), the number of generated $Y_i$'s is a random integer $N$.

(a) Show that the joint distribution of $(N, Y_1, \ldots, Y_N, U_1, \ldots, U_N)$ is given by

$$P(N = n, Y_1 \leq y_1, \ldots, Y_n \leq y_n, U_1 \leq u_1, \ldots, U_n \leq u_n)$$
$$= \int_{-\infty}^{y_n} g(t_n)(u_n \wedge w_n)dt_n \int_{-\infty}^{y_1} \cdots \int_{-\infty}^{y_{n-1}} g(t_1) \ldots g(t_{n-1})$$
$$\times \sum_{(i_1, \cdots, i_{t-1})} \prod_{j=1}^{t-1}(w_{i_j} \wedge u_{i_j}) \prod_{j=t}^{n-1}(u_{i_j} - w_{i_j})^+ dt_1 \cdots dt_{n-1},$$

where $w_i = f(y_i)/Mg(y_i)$ and the sum is over all subsets of $\{1, \ldots, n-1\}$ of size $t - 1$.

(b) There is also interest in the joint distribution of $(Y_i, U_i)|N = n$, for any $i = 1, \ldots, n-1$, as we will see in Problem 3.33. Since this distribution is the same for each $i$, we can just derive it for $(Y_1, U_1)$. (Recall that $Y_n \sim f$.) Show that

$$P(N = n, Y_1 \leq y, U_1 \leq u_1)$$
$$= \binom{n-1}{t-1}\left(\frac{1}{M}\right)^{t-1}\left(1 - \frac{1}{M}\right)^{n-t-1}$$
$$\times \left[\frac{t-1}{n-1}(w_1 \wedge u_1)\left(1 - \frac{1}{M}\right) + \frac{n-t}{n-1}(u_1 - w_1)^+\left(\frac{1}{M}\right)\right]\int_{-\infty}^{y} g(t_1)dt_1.$$

(c) Show that part (b) yields the negative binomial marginal distribution of $N$,

$$P(N = n) = \binom{n-1}{t-1} \left(\frac{1}{M}\right)^t \left(1 - \frac{1}{M}\right)^{n-t} ,$$

the marginal distribution of $Y_1$, $m(y)$,

$$m(y) = \frac{t-1}{n-1} f(y) + \frac{n-t}{n-1} \frac{g(y) - \rho f(y)}{1-\rho} ,$$

and

$$P(U_1 \leq w(y)|Y_1 = y, N = n) = \frac{g(y)w(y)M^{\frac{t-1}{n-1}}}{m(y)} .$$

**3.31** Strawderman (1996) adapted the control variate scheme to the Accept–Reject algorithm. When $Y_1, \ldots, Y_N$ is the sample produced by an Accept–Reject algorithm based on $g$, let $m$ denote the density

$$m(y) = \frac{t-1}{n-1} f(y) + \frac{n-t}{n-1} \frac{g(y) - \rho f(y)}{1-\rho} ,$$

when $N = n$ and $\rho = \dfrac{1}{M}$.

(a) Show that

$$I = \int h(x)f(x)dx = \mathbb{E}_N \left[ \mathbb{E}\left[ \frac{h(y)f(y)}{m(y)} \,\Big|\, N \right] \right] ,$$

where $m$ is the marginal density of $Y_i$ (see Problem 3.30).

(b) Show that for any function $c(\cdot)$ and some constant $\beta$,

$$I = \beta \mathbb{E}[c(Y)] + \mathbb{E}\left[ \frac{h(Y)f(Y)}{m(Y)} - \beta c(Y) \right] .$$

(c) Setting $d(y) = h(y)f(y)/m(y)$, show that the optimal choice of $\beta$ is

$$\beta^* = \mathrm{cov}[d(Y), c(Y)]/\mathrm{var}[c(Y)].$$

(d) Examine choices of $c$ for which the optimal $\beta$ can be constructed and, thus, where the control variate method applies.

(*Note:* Strawderman suggests estimating $\beta$ with $\hat{\beta}$, the estimated slope of the regression of $d(y_i)$ on $c(y_i)$, $i = 1, 2, \ldots, n-1$.)

**3.32** For the estimator $\delta_2$ of §3.7.3:

(a) Verify the expression (3.7.5) for $\rho_i$.

(b) Verify the recursion (3.7.6).

(c) Prove Proposition 3.7.6. (*Hint:* Show that $\mathbb{E}[\delta_2] = \mathbb{E}[\delta_1]$ and apply the Rao–Blackwell Theorem.)

**3.33** This problem looks at the performance of a *termwise* Rao–Blackwellized estimator. Casella and Robert (1998) established that such an estimator does not sacrifice much performance over the full Rao–Blackwellized estimator of Proposition 3.7.6. Given a sample $(Y_1, \ldots, Y_N)$ produced by an Accept–Reject algorithm to accept $m$ values, based on $(f, g, M)$:

(a) Show that

$$\frac{1}{n} \sum_{i=1}^{n} \mathbb{E}[\mathbb{I}_{U_i \leq \omega_i} | Y_i] h(Y_i)$$

$$= \frac{1}{n-m} \left( h(Y_n) + \sum_{i=1}^{n-1} b(Y_i) h(Y_i) \right)$$

with

$$b(Y_i) = \left( 1 + \frac{m(g(Y_i) - \rho f(Y_i))}{(n-m-1)(1-\rho)f(Y_i)} \right)^{-1}.$$

(b) If $S_n = \sum_1^{n-1} b(y_i)$, show that

$$\delta = \frac{1}{n-m} \left( h(Y_n) + \frac{n-m-1}{S_n} \sum_{i=1}^{nt-1} b(Y_i) h(Y_i) \right)$$

asymptotically dominates the usual Monte Carlo approximation, conditional on the number of rejected variables $m$ under quadratic loss. (*Hint:* Show that the sum of the weights $S_n$ can be replaced by $(n-m-1)$ in $\delta$ and assume $\mathbb{E}_f[h(X)] = 0$.)

**3.34** If $(Y_1, \ldots, Y_N)$ is the sample produced by an Accept–Reject method based on $(f, g)$, where $M = \sup(f/g)$, $(X_1, \ldots, X_t)$ denotes the accepted subsample and $(Z_1, \ldots, Z_{N-t})$ the rejected subsample.

(a) Show that both

$$\delta_2 = \frac{1}{N-t} \sum_{i=1}^{N-t} h(Z_i) \frac{Mg(Z_i) - f(Z_i)}{M-1}$$

and

$$\delta_1 = \frac{1}{t} \sum_{i=1}^{t} h(X_i)$$

are unbiased estimators of $I = \mathbb{E}_f[h(X)]$ (when $N > t$).

(b) Show that $\delta_1$ and $\delta_2$ are independent.

(c) Determine the optimal weight $\beta^*$ in $\delta_3 = \beta \delta_1 + (1-\beta)\delta_2$ in terms of variance. (*Note:* $\beta$ may depend on $N$ but not on $(Y_1, \ldots, Y_N)$.)

**3.35** Given a sample $Z_1, \ldots, Z_{n+t}$ produced by an Accept–Reject algorithm to accept $n$ values, based on $(f, g, M)$, show that the distribution of a rejected variable is

$$\left( 1 - \frac{f(z)}{Mg(z)} \right) g(z) = \frac{g(z) - \rho f(z)}{1-\rho},$$

where $\rho = 1/M$, that the marginal distribution of $Z_i$ $(i < n+t)$ is

$$f_m(z) = \frac{n-1}{n+t-1} f(z) + \frac{t}{n+t-1} \frac{g(z) - \rho f(z)}{1-\rho},$$

and that the joint distribution of $(Z_i, Z_j)$ $(1 \leq i \neq j < n+t)$ is

$$\frac{(n-1)(n-2)}{(n+t-1)(n+t-2)} f(z_i) f(z_j)$$

$$+\frac{(n-1)t}{(n+t-1)(n+t-2)}\left\{f(z_i)\frac{g(z_j)-\rho f(z_j)}{1-\rho}\frac{g(z_i)-\rho f(z_i)}{1-\rho}f(z_j)\right\}$$

$$+\frac{n(n-1)}{(n+t-1)(n+t-2)}\frac{g(z_i)-\rho f(z_i)}{1-\rho}\frac{g(z_j)-\rho f(z_j)}{1-\rho}.$$

**3.36** (Continuation of Problem 3.35) If $Z_1,\ldots,Z_{n+t}$ is the sample produced by an Accept–Reject algorithm to generate $n$ values, based on $(f,g,M)$, show that the $Z_i$'s are negatively correlated in the sense that for every square integrable function $h$,

$$\mathrm{cov}(h(Y_i),h(Y_j))=-\mathbb{E}_g[h]^2\mathbb{E}_N\left[\frac{(t-1)(n-t)}{(n-1)^2(n-2)}\right]$$
$$=-\mathbb{E}_g[h]^2\{\rho^t{}_2F_1(t-1,t-1;t-1;1-\rho)-\rho^2\},$$

where ${}_2F_1(a,b;c;z)$ is the confluent hypergeometric function (see Abramowitz and Stegun 1964 or Problem 1.48).

**3.37** For $t\sim\mathcal{G}eo(\rho)$, show that

$$\mathbb{E}[t^{-1}]=-\frac{\rho\log\rho}{1-\rho}\,,\qquad \mathbb{E}[t^{-2}]=\frac{\rho\mathrm{Li}(1-\rho)}{1-\rho}\,,$$

where

(3.8.1)
$$\mathrm{Li}(x)=\sum_{k=1}^{\infty}\frac{x^k}{k^2}$$

is the dilog function (see Abramowitz and Stegun 1964, formula 27.7, who also tabulated the function.)

**3.38** (Continuation of Problem 3.37) If $N\sim\mathcal{N}eg(n,\rho)$, show that

$$\mathbb{E}[(N-1)^{-1}]=\frac{\rho}{n-1}\,,\qquad \mathbb{E}[(N-2)^{-1}]=\frac{\rho(n+\rho-2)}{(n-1)(n-2)}$$

$$\mathbb{E}[((N-1)(N-2))^{-1}]=\frac{\rho^2}{(n-1)(n-2)},$$

$$\mathbb{E}[(N-1)^{-2}]=\frac{\rho^t{}_2F_1(n-1,n-1;n;1-\rho)}{(n-1)^2}\,.$$

**3.39** (Continuation of Problem 3.37) If Li is the dilog function (3.8.1), show that

$$\lim_{\rho\to1}\frac{\rho\mathrm{Li}(1-\rho)}{1-\rho}=1\qquad\text{and}\qquad\lim_{\rho\to1}\log(\rho)\mathrm{Li}(1-\rho)=0\,.$$

Deduce that the domination corresponding to (3.9.7) occurs on an interval of the form $[\rho_0,1]$.

**3.40** Given an Accept–Reject algorithm based on $(f,g,\rho)$, we denote by

$$b(y_j)=\frac{(1-\rho)f(y_j)}{g(y_j)-\rho f(y_j)}$$

the importance sampling weight of the rejected variables $(Y_1,\ldots,Y_t)$, and by $(X_1,\ldots,X_n)$ the accepted variables.

(a) Show that the estimator

$$\delta_1=\frac{n}{n+t}\,\delta^{AR}+\frac{t}{n+t}\,\delta_0,$$

with

$$\delta_0 = \frac{1}{t} \sum_{j=1}^{t} b(Y_j) h(Y_j)$$

and

$$\delta^{AR} = \frac{1}{n} \sum_{i=1}^{n} h(X_i),$$

does not uniformly dominate $\delta^{AR}$. (*Hint:* Consider the constant functions.)

(b) Show that

$$\delta_{2w} = \frac{n}{n+t} \delta^{AR} + \frac{t}{n+t} \sum_{j=1}^{t} \frac{b(Y_j)}{S_t} h(Y_j)$$

is asymptotically equivalent to $\delta_1$ in terms of bias and variance.

(c) Deduce that $\delta_{2w}$ asymptotically dominates $\delta^{AR}$ if (3.9.7) holds.

**3.41** For the Accept–Reject algorithm of §3.3.3:

(a) Show that conditionally on $t$, the joint density of $(Y_1, \ldots, Y_t)$ is indeed

$$\prod_{j=1}^{t-1} \left( \frac{Mg(y_j) - f(y_j)}{M - 1} \right) f(y_t)$$

and the expectation of $\delta_2$ of (3.3.6) is given by

$$\mathbb{E} \left[ \frac{t-1}{t} \left\{ \frac{M}{M-1} \mathbb{E}_f[h(X)] - \frac{1}{M-1} \mathbb{E}_f \left[ h(X) \frac{f(X)}{g(X)} \right] \right\} \right.$$
$$\left. + \frac{1}{t} \mathbb{E}_f \left[ h(X) \frac{f(X)}{g(X)} \right] \right].$$

(b) If we denote the acceptance probability of the Accept–Reject algorithm by by $\rho = 1/M$ and assume $\mathbb{E}_f[h(X)] = 0$, show that the *bias* of $\delta_2$ is

$$\left( \frac{1 - 2\rho}{1 - \rho} \mathbb{E}[t^{-1}] - \frac{\rho}{1 - \rho} \right) \mathbb{E}_f \left[ h(X) \frac{f(X)}{g(X)} \right].$$

(c) Establish that for $t \sim Geo(\rho)$, $\mathbb{E}[t^{-1}] = \rho \log(\rho)/(1 - \rho)$, and that the bias of $\delta_2$ can be written as

$$-\frac{\rho}{1 - \rho} (1 - (1 - 2\rho) \log(\rho)) \mathbb{E}_f \left[ h(X) \frac{f(X)}{g(X)} \right].$$

(d) Show that the variance of $\delta_2$ is

$$\mathbb{E} \left[ \frac{t-1}{t^2} \right] \frac{1}{1-\rho} \mathbb{E}_f \left[ h^2(X) \frac{f(X)}{g(X)} \right]$$
$$+ \mathbb{E} \left[ \frac{1}{t^2} \left\{ 1 - \frac{\rho(t-1)}{1-\rho} \right\} \right] \text{var}_f(h(X)f(X)/g(X)).$$

**3.42** Using the information from Note 3.9.1, for a binomial experiment $X_n \sim \mathcal{B}(n, p)$ with $p = 10^{-6}$, determine the minimum sample size $n$ so that

$$P \left( \left| \frac{X_n}{n} - p \right| \leq \epsilon p \right) > .95$$

when $\epsilon = 10^{-1}, 10^{-2}$, and $10^{-3}$.

**3.43** When random variables $Y_i$ are generated from (3.9.1), show that $J^{(m)}$ is distributed as $\lambda(\theta_0)^{-n}\exp(-n\theta J)$. Deduce that (3.9.2) is unbiased.

**3.44** In the setup of Note 3.9.2, show that, for a function $\xi$,

$$\int_0^t (\xi(s) - \xi(0))d\xi(s) = \frac{1}{2}(\xi(t)^2 - t).$$

**3.45** Show that, in (3.9.4), $b(\cdot)$ is defined by $b(x) = b_0(x) + \frac{1}{2}\sigma(x)\sigma'(x)$ in dimension $\alpha = 1$. Extend to the general case by introducing $\partial\sigma_j$, which is a matrix with the $(i,k)$ element $\partial\sigma_{ij}(x)/\partial x_k$.

**3.46** Show that the solution to (3.9.3) can also be expressed through a *Stratonovich integral,*

$$X(t) = X(0) + \int_0^t b_0(X(s))ds + \int_0^t \sigma(X(s)) \circ d\xi(s),$$

where the second integral is defined as the limit

$$\lim_{\Delta \to 0} \sum_{i=1}^n \frac{\sigma(X(s_i)) + \sigma(X(s_{i-1}))}{2}(\xi(s_i) - \xi(s_{i-1})),$$

where $0 = s_0 < s_1 < \cdots < s_n = t$ and $\max_i(b_i - s_{i-1}) \leq \Delta$. (*Note:* The above integral cannot be defined in the usual sense because the Wiener process has almost surely unbounded variations. See Talay 1995, p. 56.)

**3.47** Show that the solution to the Ornstein–Uhlenbeck equation

(3.8.2)                              $$dX(t) = -X(t)dt + \sqrt{2}d\xi(t)$$

is a stationary $\mathcal{N}(0,1)$ process.

**3.48** Show that if $(X(t))$ is solution to (3.8.2), $Y(t) = \arctan X(t)$ is solution of a SDE *(Stochastic Differential Equation)* with $b(x) = \frac{1}{4}\sin(4x) - \sin(2x)$, $\sigma(x) = \sqrt{2}\cos^2(x)$.

**3.49** Given an integral

$$I = \int_{\mathcal{X}} h(x)f(x)dx$$

to be evaluated by simulation, compare the usual Monte Carlo estimator

$$\delta_a = \frac{1}{np}\sum_{i=1}^{np} h(x_i)$$

based on an iid sample $(X_1, \ldots, X_{np})$ with a stratified estimator (see Note 3.9.5)

$$\delta_a = \sum_{j=1}^p \frac{p_j}{n}\sum_{i=1}^n h(Y_i^p),$$

where $p_j = \int_{\mathcal{X}_j} f(x)dx$, $\mathcal{X} = \bigcup_{j=1}^p \mathcal{X}_j$ and $(Y_1^j, \ldots, Y_n^j)$ is a sample from $f\mathbb{I}_{\mathcal{X}_j}$. Show that $\delta_2$ does not bring any improvement if the $p_j$'s are unknown and must be estimated.

## 3.9 Notes

### 3.9.1 Large Deviations Techniques

When we introduced importance sampling methods in §3.3, we showed in Example 3.3.1 that alternatives to direct sampling were preferable when sampling from the tails of a distribution $f$. When the event $A$ is particularly rare, say $P(A) \leq 10^{-6}$, methods such as importance sampling are needed to get an acceptable approximation (see Problem 3.42). Since the optimal choice given in Theorem 3.3.4 is formal, in the sense that it involves the unknown constant $I$, more practical choices have been proposed in the literature. In particular, Bucklew (1990) indicates how the *theory of large deviations* may help in devising proposal distributions in this purpose.

Briefly, the theory of large deviations is concerned with the approximation of tail probabilities $P(|\bar{X}_n - \mu| > \varepsilon)$ when $\bar{X}_n = (X_1 + \cdots + X_n)/n$ is a mean of iid random variables, $n$ goes to infinity, and $\varepsilon$ is large. (When $\varepsilon$ is small, the normal approximation based on the Central Limit Theorem works well enough.)

If $M(\theta) = \mathbb{E}[\exp(\theta X_1)]$ is the moment generating function of $X_1$ and we define $I(x) = \sup_\theta \{\theta x - \log M(\theta)\}$, the large deviation approximation is

$$\frac{1}{n} \log P(S_n \in F) \approx - \inf_F I(x).$$

This result is sometimes called *Cramer's Theorem* and a simulation device based on this result is called *twisted simulation*.

To evaluate

$$I = P\left( \frac{1}{n} \sum_{i=1}^n h(x_i) \geq 0 \right),$$

when $\mathbb{E}[h(X_1)] < 0$, we use the proposal density

$$(3.9.1) \qquad\qquad t(x) \propto f(x) \exp\{\theta_0 h(x)\}$$

where the parameter $\theta_0$ is chosen such that $\int h(x)f(x)e^{\theta_0 h(x)}\, dx = 0$. (Note the similarity with exponential tilting in saddlepoint approximations in §3.6.1.) The corresponding estimate of $I$ is then based on blocks ($m = 1, \ldots, M$)

$$J^{(m)} = \frac{1}{n} \sum_{i=1}^n h(Y_i^{(m)}),$$

where the $Y_i^{(m)}$ are iid from $t$, as follows:

$$(3.9.2) \qquad\qquad \hat{I} = \frac{1}{M} \sum_{m=1}^M \mathbb{I}_{[0,\infty[}(J^{(m)}) e^{-n\theta_0 J^{(m)}} \lambda(\theta_0)^n,$$

with $\lambda(\theta) = \int f(x)e^{\theta_0 f(x)}\, dx$. The fact that (3.9.2) is unbounded follows from a regular importance sampling argument (Problem 3.43). Bucklew (1990) provides arguments about the fact that the variance of $\hat{I}$ goes to 0 exponentially twice as fast as the regular (direct sampling) estimate.

**Example 3.9.1 Laplace distribution.** Consider $h(x) = x$ and the sampling distribution $f(x) = \frac{1}{2a} \exp\{-|x - \mu|/a\}$, $\mu < 0$. We then have

$$t(x) \propto \exp\{-|x - \mu|/a + \theta_0\},$$

$$\theta_0 = \sqrt{\mu^{-2} + a^{-2}} - \mu^{-1},$$

$$\lambda(\theta_0) = \frac{\mu^2}{2} \exp(-C)a^2 C,$$

with $C = \sqrt{1 + \frac{\mu^2}{a^2}} - 1$. A large deviation computation then shows that (Bucklew 1990, p. 139)

$$\lim_n \frac{1}{n} \log(M \operatorname{var} \hat{I}) = 2 \log \lambda(\theta_0),$$

while the standard average $\bar{I}$ satisfies

$$\lim_n \frac{1}{n} \log(M \operatorname{var} \bar{I}) = \log \lambda(\theta_0) .$$

‖

Obviously, this is not the entire story, Further improvements can be found in the theory, while the computation of $\theta_0$ and $\lambda(\theta_0)$ and simulation from $t(x)$ may become quite intricate in realistic setups.

### 3.9.2 Simulation of Stochastic Differential Equations

Given a differential equation[7] in $\mathbb{R}^d$,

$$dX(t) = b_0(X(t))dt,$$

where $b_0$ is a function from $\mathbb{R}^d$ to $\mathbb{R}^d$, it is often of interest to consider the perturbation of this equation by a random noise, $\xi(t)$,

$$(3.9.3) \qquad \frac{dX(t)}{dt} = b_0(X(t)) + \sigma(X(t))\xi(t),$$

which is called a *stochastic differential equation* (SDE), $\sigma$ being the variance factor taking values in the space of $d \times d$ matrices. Applications of SDEs abound in fluid mechanics, random mechanics, and particle Physics.

The perturbation $\xi$ in (3.9.3) is often chosen to be a *Wiener process*, that is, such that $\xi(t)$ is a Gaussian vector with mean 0, independent components, and correlation

$$\mathbb{E}[\xi_i(t)\xi_j(s)] = \delta_{ij} \min(t, s) ,$$

where $\delta_{ij} = \mathbb{I}_{i=j}$.

The solution of (3.9.3) can also be represented through an *Itô integral*,

$$(3.9.4) \qquad X(t) = X(0) + \int_0^t b(X(s))ds + \int_0^t \sigma(X(s))d\xi(s),$$

where $b$ is derived from $b_0$ and $\sigma$ (see Problems 3.45 and 3.46), and the second integral is the limit

$$\lim_{\Delta \to 0} \sum_{i=1}^n \sigma(X(s_i))(\xi(s_i) - \xi(s_{i-1})),$$

where $0 = s_0 < s_1 < \cdots < s_n = t$ and $\max_i(b_i - s_{i-1}) \leq \Delta$. This limit exists whenever $\sigma(X)$ is square integrable; that is, when

$$\mathbb{E}\left[\int_0^t |\sigma(X(s))|^2 ds\right] < \infty.$$

See, for example, Ikeda and Watanabe (1981) for details.

----

[7] The material in this note is of a more advanced mathematical level than the remainder of the book and it will not be used in the sequel. This description of simulation methods for SDEs borrows heavily from the expository paper of Talay (1995), which presents more details on the use of simulation techniques in this setup.

In this setup, simulations are necessary to produce an approximation of the trajectory of $(X(t))$, given the Wiener process $(\xi(t))$, or to evaluate expectations of the form $\mathbb{E}[h(X(t))]$. It may also be of interest to compute the expectation $\mathbb{E}[h(X)]$ under the stationary distribution of $(X(t))$ when this process is ergodic, a setting we will encounter again in the MCMC method (see Chapters 4 and 7).

A first approximation to the solution of (3.9.4) is based on the discretization

$$X(t) \approx X(0) + b(X(0))t + \sigma(X(0))(\xi(t) - \xi(0)) ,$$

for $t$ small. The sequence $(k = 1, \ldots)$

(3.9.5)        $$\bar{X}_{k+1} = \bar{X}_k + b(\bar{X}_k)\varrho + \sigma(\bar{X}_k)[\xi((k+1)\varrho) - \xi(k\varrho)]$$

is then called the *Euler scheme* with step $\varrho$. While intuitive, this approximation is divergent for the approximation of $(X(t))$. An alternative, called the *Milshtein scheme* (Milshtein 1974), is based on a Taylor expansion of $\sigma(X(t))$. In dimension $d = 1$, this approach can be expressed as

$$
\begin{aligned}
X(t) &\approx X(0) + b(X(0))t + \sigma(X(0))(\xi(t) - \xi(0)) \\
&\quad + \sigma(X(0))\sigma'(X(0)) \int_0^t (\xi(s) - \xi(0))d\xi(s) \\
&= X(0) + b(X(0))t + \sigma(X(0))(\xi(t) - \xi(0)) \\
&\quad + \frac{1}{2}\sigma(X(0))\sigma'(X(0))(\xi(t)^2 - t)
\end{aligned}
$$

(3.9.6)

(see Talay 1995, p. 62, for an extension to multidimensional settings). Defining a sequence $\bar{X}_k$ from (3.9.6) as in (3.9.5) then leads to a convergent approximation under very weak conditions on $b$ and $\sigma$ (Talay 1995, p. 66).

The improvement brought by the Milshtein scheme also appears in the approximation of $\mathbb{E}[h(X(t))]$ since Milshtein (1974) shows that there exists a constant $C$ such that

$$\mathbb{E}\left|h(X(t)) - h\left(\bar{X}_{[t/\varrho]}\right)\right|^2 < C\varrho$$

for the Euler scheme, while

$$\mathbb{E}\left|h(X(t)) - h\left(\bar{X}_{[t/\varrho]}\right)\right|^2 < C\varrho^2$$

for the Milshtein scheme. However, when one uses $N$ independent Wiener processes $\xi_i$ to approximate $\mathbb{E}[h(X(t))]$ by

$$\frac{1}{N}\sum_{i=1}^n h\left(\bar{X}^i_{[t/\varrho]}\right),$$

where $\bar{X}^i_k$ is defined as in (3.9.5) with $\xi_i$ instead of $\xi$, the rate of convergence is the same for both schemes (Talay 1995, p. 69).

When the goal is to approximate

$$\mathbb{E}_\mu[h(X)] = \int h(x)d\mu(x),$$

where $\mu$ is the stationary distribution of $(X(t))$, the efficiency of both Euler and Milshtein schemes is $O(\varrho)$, in the sense that

$$\lim_N \frac{1}{N}\sum_{k=1}^N h(\bar{X}_k) = \int h(X)d\mu(X) + O(\varrho),$$

while more involved schemes achieve an efficiency of $O(\varrho^2)$ (Talay 1995).

### 3.9.3 Monitoring Importance Sampling Convergence

With reference to convergence control for simulation methods, importance sampling methods can be implemented in a *monitored* way; that is, in parallel with other evaluation methods. These can be based on alternative instrumental distributions or other techniques (standard Monte Carlo, Markov chain Monte Carlo, Riemann sums, Rao–Blackwellization, etc.). The respective samples then provide separate evaluations of $\mathbb{E}[h(X)]$, through (3.3.1), (3.3.2), or yet another estimator (as in §3.3.3), and the convergence criterion is to stop when most estimators are close enough (see also §8.3.2). Obviously, this empirical method is not completely foolproof, but it generally prevents pseudo-convergences when the instrumental distributions are sufficiently different. On the other hand, this approach is rather *conservative*, as it is only as fast as the slowest estimator. However, it may also point out instrumental distributions with variance problems. From a computational point of view, an efficient implementation of this control method relies on the use of *parallel programming* in order to weight each distribution more equitably, so that a distribution $f$ of larger variance, compared with another distribution $g$, may compensate this drawback by a lower computation time, thus producing a larger sample in the same time.[8]

### 3.9.4 Accept–Reject with Loose Bounds

For the Accept–Reject algorithm [A.5], some interesting results emerge if we assume that
$$f(x) \leq M\, g(x)/(1+\varepsilon) \; ;$$
that is, that the bound $M$ is not tight. If we assume $\mathbb{E}_f[h(X)] = 0$, the estimator $\delta_4$ of (3.3.7) then satisfies
$$\mathrm{var}(\delta_4) \leq \left( \mathbb{E}\left[ \frac{t-1}{t^2} \right] \frac{M-1}{\varepsilon} + \mathbb{E}[t^{-2}] \right) \mathbb{E}_f[h^2(X)] \,,$$
where
$$\mathbb{E}[t^{-2}] = \frac{\rho}{1-\rho}\mathrm{Li}(1-\rho) \,,$$
and $\mathrm{Li}(x)$ denotes the *dilog* function (3.8.1), which can also be written as
$$\mathrm{Li}(x) = \int_x^1 \frac{\log(u)}{1-u} du$$
(see Problem 3.37). The bound on the variance of $\delta_4$ is thus
$$\mathrm{var}(\delta_4) \leq \left\{ \left( \frac{\rho}{1-\rho} - \epsilon^{-1} \right) \mathrm{Li}(1-\rho) - \epsilon^{-1}\log(\rho) \right\} \mathbb{E}_f[h^2(X)]$$
and $\delta_4$ uniformly dominates the usual Accept–Reject estimator $\delta_1$ of (3.3.6) as long as
$$(3.9.7) \qquad \varepsilon^{-1}\{-\log(\rho) - \mathrm{Li}(1-\rho)\} + \frac{\rho\,\mathrm{Li}(1-\rho)}{1-\rho} < 1 \,.$$

This result rigorously establishes the advantage of recycling the rejected variables for the computation of integrals, since (3.9.7) does not depend on the

---

[8] This feature does not necessarily require a truly parallel implementation, since it can be reproduced by the cyclic allocation of uniform random variables to each of the distributions involved.

function $h$. Note, however, that the assumption $\mathbb{E}_f[h(X)] = 0$ is quite restrictive, because the sum of the weights of $\delta_4$ does not equal 1 and, therefore, $\delta_4$ does not correctly estimate constant functions (except for the constant function $h = 0$). Therefore, $\delta_1$ will dominate $\delta_4$ for constant functions, and a uniform comparison between both estimators is impossible.

Figure 3.9.1 gives the graphs of the left-hand side of (3.9.7) for $\varepsilon = 0.1, 0.2, \ldots,$ 0.9. A surprising aspect of these graphs is that domination (that is, where the curve is less than 1) occurs for larger values of $\rho$, which is somewhat counterintuitive, since smaller values of $\rho$ lead to higher rejection rates, therefore to larger rejected subsamples and to a smaller variance of $\delta_4$ for an adequate choice of density functions $g$. On the other hand, the curves are correctly ordered in $\epsilon$ since larger values of $\epsilon$ lead to wider domination zones.

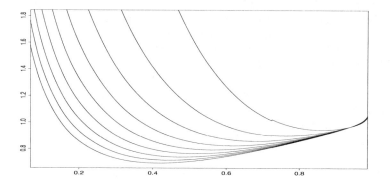

Figure 3.9.1. *Graphs of the variance coefficients of $\delta_4$ for $\epsilon = 0.1, \ldots, 0.9$, the curves decreasing in $\epsilon$. The domination of $\delta_1$ by $\delta_4$ occurs when the variance coefficient is less than 1.*

### 3.9.5 Partitioning

When enough information is available on the function $f$, *stratified sampling* may be used. This technique (see Hammersley and Handscomb 1964 or Rubinstein 1981 for references) decomposes the integration domain $\mathcal{X}$ in a partition $\mathcal{X}_1, \ldots, \mathcal{X}_p$, with separate evaluations of the integrals on each region $\mathcal{X}_i$; that is, the integral (3.2.5) is expressed as

$$\sum_{i=1}^p \int_{\mathcal{X}_i} h(x) f(x) dx = \sum_{i=1}^p \varrho_i \int_{\mathcal{X}_i} h(x) f_i(x) dx$$

$$= \sum_{i=1}^p \varrho_i I_i,$$

where the weights $\varrho_i$ are the probabilities of the regions $\mathcal{X}_i$ and the $f_i$'s are the restrictions of $f$ to these regions. Then samples of size $n_i$ are generated from the $f_i$'s to evaluate each integral $I_i$ separately by a regular estimator $\hat{I}_i$.

The motivation for this approach is that the variance of the resulting estimator, $\varrho_1 \hat{I}_1 + \cdots + \varrho_p \hat{I}_p$, that is,

$$\sum_{i=1}^{p} \varrho_i^2 \frac{1}{n_i} \int_{\mathcal{X}_i} (h(x) - I_i)^2 f_i(x) dx,$$

may be much smaller than the variance of the standard Monte Carlo estimator based on a sample of size $n = n_1 + \cdots + n_p$. The optimal choice of the $n_i$'s in this respect is

$$n_i^* \propto \varrho_i^2 \int_{\mathcal{X}_i} (h(x) - I_i)^2 f_i(x) dx.$$

Thus, if the regions $\mathcal{X}_i$ can be chosen, the variance of the stratified estimator can be reduced by selecting $\mathcal{X}_i$'s with similar variance factors $\int_{\mathcal{X}_i} (h(x) - I_i)^2 f_i(x) dx$.

While this approach may seem far-fetched, because of its requirements on the distribution $f$, note that it can be combined with importance sampling, where the importance function $g$ may be chosen in such a way that its quantiles are well known. Also, it can be iterated, with each step producing an evaluation of the $\varrho_i$ and of the corresponding variance factors, which may help in selecting the partition and the $n_i$'s in the next step (see, however, Problem 3.49).

An extension proposed by McKay *et al.* (1979) introduces stratification on all input dimensions. More precisely, if the domain is represented as the unit hypercube $\mathcal{H}$ in $\mathbb{R}^d$ and the integral (3.2.5) is written as $\int_{\mathcal{H}} h(x) dx$, *Latin hypercube sampling* relies on the simulation of (a) $d$ random permutations $\pi_j$ of $\{1, \ldots, n\}$ and (b) uniform $\mathcal{U}([0,1])$ rv's $U_{i_1, \ldots, i_d, j}$ ($j = 1, \ldots, d$, $1 \leq i_1, \ldots, i_d \leq n$). A sample of $n$ vectors $\mathbf{X}_i$ in the unit hypercube is then produced as $\mathbf{X}(\pi_1(i), \ldots, \pi_d(i))$ with

$$\mathbf{X}(i_1, \ldots, i_d) = (X_1(i_1, \ldots, i_d), \ldots, X_d(i_1, \ldots, i_d)),$$

$$X_j(i_1, \ldots, i_d) = \frac{i_j - U_{i_1, \ldots, i_d, j}}{n}, \qquad 1 \leq j \leq d.$$

The component $X_j(i_1, \ldots, i_d)$ is therefore a point taken at random on the interval $[i_{j-1}/n, i_j/n]$ and the permutations ensure that no uniform variable is taken twice from the same interval, for every dimension. Note that we also need only generate $n \times d$ uniform random variables. (Latin hypercubes are also used in agricultural experiment to ensure that all parcels and all varieties are used, at a minimal cost. See Mead 1988 or Kuehl 1994.) McKay *et al.* (1979) show that when $h$ is a real valued function, the variance of the resulting estimator is substantially reduced, compared with the regular Monte Carlo estimator based on the same number of samples. Asymptotic results about this technique can be found in Stein (1987) and Loh (1996). (It is, however, quite likely that the curse of dimensionality (see §3.4), also applies to this technique.)

### 3.9.6 Saddlepoint Approximations

The presentation in §3.6 only shows that the order of the approximation is $O(1/n)$, not the $O(n^{-3/2})$ that is often claimed. This better error rate is obtained by renormalizing (3.6.8), so that it integrates to 1. Details of this are contained in Daniels' original 1954 paper, Field and Ronchetti (1990, Chapter 3), or Kolassa (1994, Chapter 4). We omit them here, but note that such renormalization can sometimes be quite computer intensive; see, for example, the discussion in Section 3.3 of Field and Ronchetti (1990).

Saddlepoint approximations for tail areas have seen much more development than given here. The work of Lugannani and Rice (1980) produced a very accurate approximation that only requires the evaluation of one saddlepoint and *no* integration. It is derived by further transformations of the saddlepoint approximation; see the discussions by Field and Ronchetti (1990, Section 6.2) or Kolassa (1994, Section 5.3). There are other approaches to tail area approximations; for example, the work of Barndorff-Nielsen (1991) using ancillary statistics or the Bayes-based approximation of DiCiccio and Martin (1993). Wood *et al.* (1993) give generalizations of the Lugannani and Rice formula.

# CHAPTER 4

# Markov Chains

Leaphorn never counted on luck. Instead, he expected order—the natural sequence of behavior, the cause producing the natural effect, the human behaving in the way it was natural for him to behave. He counted on that and on his own ability to sort out the chaos of observed facts and find in them this natural order.

—Tony Hillerman, *The Blessing Way*

In this chapter we introduce fundamental notions of Markov chains and state the results that are needed to establish the convergence of various MCMC algorithms and, more generally, to understand the literature on this topic. Thus, this chapter, along with basic notions of probability theory, will provide enough foundation for the understanding of the following chapters. It is, unfortunately, a necessarily brief and, therefore, incomplete introduction to Markov chains, and we refer the reader to Meyn and Tweedie (1993), on which this chapter is based, for a thorough introduction to Markov chains. Other perspectives can be found in Doob (1953), Chung (1960), Feller (1970, 1971), and Billingsley (1995) for general treatments, and Norris (1997) Nummelin (1984), Revuz (1984), and Resnick (1994) for books entirely dedicated to Markov chains. Given the purely utilitarian goal of this chapter, its style and presentation differ from those of other chapters, especially with regard to the plethora of definitions and theorems and to the rarity of examples and proofs. In order to make the book accessible to those who are more interested in the implementation aspects of MCMC algorithms than in their theoretical foundations, we include a preliminary section that contains the essential facts about Markov chains.

Before formally introducing the notion of a Markov chain, note that we do not deal in this chapter with Markov models in *continuous time* (also called *Markov processes*) since the very nature of simulation leads[1] us to consider only discrete-time stochastic processes, $(X_n)_{n \in \mathbf{N}}$. Indeed, Hastings (1970) notes that the use of pseudo-random generators and the representation of numbers in a computer imply that the Markov chains related with Markov

---

[1] Some Markov chain Monte Carlo algorithms still employ a diffusion representation to speed up convergence to the stationary distribution (see, for instance, §6.5.2, Roberts and Tweedie 1995 or Phillips and Smith 1996).

chain Monte Carlo methods are, in fact, finite state-space Markov chains. However, we also consider arbitrary state-space Markov chains to allow for continuous support distributions and to avoid addressing the problem of approximation of these distributions with discrete support distributions, since such an approximation depends on both material and algorithmic specifics of a given technique (see Roberts *et al.* 1995, for a study of the influence of discretization on the convergence of Markov chains associated with Markov chain Monte Carlo algorithms).

## 4.1 Essentials for MCMC

For those familiar with the properties of Markov chains, this first section provides a brief survey of the properties of Markov chains that are contained in the chapter and are essential for the study of MCMC methods. Starting with §4.2, the theory of Markov chains is developed from first principles.

In the setup of MCMC algorithms, Markov chains are constructed from a *transition kernel* $K$ (Definition 4.2.1), a conditional probability density such that $X_{n+1} \sim K(X_n, X_{n+1})$. A typical example is provided by the *random walk* process, defined as $X_{n+1} = X_n + \epsilon_n$, where $\epsilon_n$ is generated independently of $X_n, X_{n-1}, \ldots$ (see Example 4.10.4). The chains encountered in MCMC settings enjoy a very strong stability property, namely a *stationary probability distribution* exists by construction (Definition 4.5.1); that is, a distribution $\pi$ such that if $X_n \sim \pi$, then $X_{n+1} \sim \pi$, if the kernel $K$ allows for free moves all over the state space. (This freedom is called *irreducibility* in the theory of Markov chains and is formalized in Definition 4.3.1 as the existence of $n \in \mathbb{N}$ such that $P(X_n \in A | X_0) > 0$ for every $A$ such that $\pi(A) > 0$.) This property also ensures that most of the chains involved in MCMC algorithms are *recurrent* (that is, that the average number of visits to an arbitrary set $A$ is infinite (Definition 4.4.5)), or even Harris recurrent (that is, such that the probability of an infinite number of returns to $A$ is 1 (Definition 4.4.8)). Harris recurrence ensures that the chain has the same limiting behavior for *every* starting value instead of *almost* every starting value. (Therefore, this is the Markov chain equivalent of the notion of continuity for functions.)

This latter point is quite important in the context of MCMC algorithms. Since most algorithms are started from some arbitrary point $x_0$, we are in effect starting the algorithm from a set of measure zero (under a continuous dominating measure). Thus, insuring that the chain converges for almost every starting point is not enough, and we need Harris recurrence to guarantee convergence from every starting point.

The *stationary distribution* is also a *limiting distribution* in the sense that the limiting distribution of $X_{n+1}$ is $\pi$ under the total variation norm (see Proposition 4.6.2), notwithstanding the initial value of $X_0$. Stronger forms of convergence are also encountered in MCMC settings, like *geometric* and *uniform* convergences (see Definitions 4.6.8 and 4.6.12). In a simulation setup, a most interesting consequence of this convergence property is that the average

(4.1.1)
$$\frac{1}{N} \sum_{n=1}^{N} h(X_n)$$

converges to the expectation $\mathbb{E}_\pi[h(X)]$ almost surely. When the chain is *reversible* (Definition 4.7.6) (that is, when the transition kernel is symmetric), a Central Limit Theorem also holds for this average.

In Chapter 8, diagnostics will be based on a minorization condition; that is, the existence of a set $C$ such that there also exists $m \in \mathbb{N}$, $\epsilon_m > 0$, and a probability measure $\nu_m$ such that

$$P(X_m \in A | X_0) \geq \epsilon_m \nu_m(A)$$

when $X_0 \in C$. The set $C$ is then called a *small set* (Definition 4.3.7) and visits of the chain to this set can be exploited to create independent batches in the sum (4.1.1), since, with probability $\epsilon_m$, the next value of the Markov chain is generated from the minorizing measure $\nu_m$, which is independent of $X_m$.

## 4.2 Basic Notions

A Markov chain is a sequence of random variables that can be thought of as evolving over time, with probability of a transition depending on the particular set in which the chain is. It therefore seems natural and, in fact, is mathematically somewhat cleaner to define the chain in terms of its *transition kernel*, the function that determines these transitions.

**Definition 4.2.1** A *transition kernel* is a function $K$ defined on $\mathcal{X} \times \mathcal{B}(\mathcal{X})$ such that

(i) $\forall x \in \mathcal{X}$, $K(x, \cdot)$ is a probability measure;

(ii) $\forall A \in \mathcal{B}(\mathcal{X})$, $K(\cdot, A)$ is measurable.

When $\mathcal{X}$ is *discrete*, the transition kernel simply is a (transition) matrix $K$ with elements

$$P_{xy} = P(X_n = y | X_{n-1} = x), \qquad x, y \in \mathcal{X}.$$

In the continuous case, the *kernel* also denotes the conditional density $K(x, x')$ of the transition $K(x, \cdot)$; that is, $P(X \in A | x) = \int_A K(x, x') dx'$.

**Example 4.2.2 Bernoulli–Laplace Model.** Consider $\mathcal{X} = \{0, 1, \ldots, M\}$ and a chain $(X_n)$ such that $X_n$ represents the state, at time $n$, of a tank which contains exactly $M$ particles and is connected to another identical tank. Two types of particles are introduced in the system, and there are $M$ of each type. If $X_n$ denotes the number of particles of the first kind in the first tank at time $n$ and the moves are restricted to a single exchange of particles between both tanks at each instant, the transition matrix is given by (for $0 < x, y < M$)

$$P_{xy} = 0 \quad \text{if} \quad |x - y| > 1,$$

$$P_{xx} = 2\,\frac{x(M-x)}{M^2}\,,\; P_{x(x-1)} = \left(\frac{x}{M}\right)^2,\; P_{x(x+1)} = \left(\frac{M-x}{M}\right)^2$$

and $P_{01} = P_{M(M-1)} = 1$. (This model is the *Bernoulli–Laplace* model; see Feller 1970, Chapter XV.)                                                   ‖

The chain $(X_n)$ is usually defined for $n \in \mathbb{N}$ rather than for $n \in \mathbb{Z}$. Therefore, the distribution of $X_0$, the initial state of the chain, plays an important role. In the discrete case, where the kernel $K$ is a transition matrix, given an initial distribution $\mu = (\omega_1, \omega_2, \ldots)$, the marginal probability distribution of $X_1$ is obtained from the matrix multiplication

$$(4.2.1) \qquad\qquad \mu_1 = \mu K$$

and, by repeated multiplication, $X_n \sim \mu_n = \mu K^n$. Similarly, in the continuous case, if $\mu$ denotes the initial distribution of the chain, namely if

$$(4.2.2) \qquad\qquad X_0 \sim \mu\,,$$

then we let $P_\mu$ denote the probability distribution of $(X_n)$ under condition (4.2.2). When $X_0$ is fixed, in particular for $\mu$ equal to the Dirac mass $\delta_{x_0}$, we use the alternative notation $P_{x_0}$.

**Definition 4.2.3** Given a transition kernel $K$, a sequence $X_0, X_1, \ldots, X_n, \ldots$ of random variables is a *Markov chain*, denoted by $(X_n)$, if, for any $t$, the conditional distribution of $X_t$ given $x_{t-1}, x_{t-2}, \ldots, x_0$ is the same as the distribution of $X_t$ given $x_{t-1}$; that is,

$$P(X_{k+1} \in A | x_0, x_1, x_2, \ldots, x_k) = P(X_{k+1} \in A | x_k)$$
$$(4.2.3) \qquad\qquad\qquad\qquad = \int_A K(x_k, dx)\,.$$

The chain is *time homogeneous*, or simply *homogeneous*, if the distribution of $(X_{t_1}, \ldots, X_{t_k})$ given $x_{t_0}$ is the same as the distribution of $(X_{t_1-t_0}, X_{t_2-t_0}, \ldots, X_{t_k-t_0})$ given $x_0$ for every $k$ and every $(k+1)$-uplet $t_0 \le t_1 \le \cdots \le t_k$.

So, in the case of a Markov chain, if the initial distribution or the initial state is known, the construction of the Markov chain $(X_n)$ is entirely determined by its *transition*, namely by the distribution of $X_n$ conditionally on $x_{n-1}$.

The study of Markov chains is almost always restricted to the time-homogeneous case and we omit this designation in the following. It is, however, important to note here that an incorrect implementation of Markov chain Monte Carlo algorithms can easily produce time-heterogeneous Markov chains for which the standard convergence properties do not apply. (See also the case of the ARMS algorithm in §6.3.3.)

**Example 4.2.4 Simulated Annealing.** The *simulated annealing* algorithm (see §5.2.3 for details) is often implemented in a time-heterogeneous form and studied in time-homogeneous form. Given a finite state-space with size $K$, $\Omega = \{1, 2, \ldots, K\}$, an energy function $E(\cdot)$, and a temperature $T$, the simulated annealing Markov chain $X_0, X_1, \ldots$ is represented by the following transition operator: Conditionally on $X_n$, $Y$ is generated from a fixed probability distribution $(\pi_1, \ldots, \pi_K)$ on $\Omega$ and the new value of the chain is given by

$$X_{n+1} = \begin{cases} Y & \text{with probability } \exp\{(E(Y) - E(X_n))/T\} \wedge 1 \\ X_n & \text{otherwise.} \end{cases}$$

If the temperature $T$ depends on $n$, the chain is time heterogeneous. ‖

**Example 4.2.5** $AR(1)$ **Models.** $AR(1)$ models provide a simple illustration of Markov chains on continuous state-space. If

$$X_n = \theta X_{n-1} + \varepsilon_n , \qquad \theta \in \mathbb{R},$$

with $\varepsilon_n \sim N(0, \sigma^2)$, and if the $\varepsilon_n$'s are independent, $X_n$ is indeed independent from $X_{n-2}, X_{n-3}, \ldots$ conditionally on $X_{n-1}$. The Markovian properties of an $AR(q)$ process can be derived by considering the vector $(X_n, \ldots, X_{n-q+1})$. On the other hand, $ARMA(p, q)$ models do not fit in the Markovian framework (see Problem 4.3). ‖

In the general case, the fact that the kernel $K$ determines the properties of the chain $(X_n)$ can be inferred from the relations

$$P_x(X_1 \in A_1) = K(x, A_1) ,$$

$$P_x((X_1, X_2) \in A_1 \times A_2) = \int_{A_1} K(y_1, A_2) \, K(x, dy_1)$$

$$\cdots$$

$$P_x((X_1, \ldots, X_n) \in A_1 \times \cdots \times A_n) = \int_{A_1} \cdots \int_{A_{n-1}} K(y_{n-1}, A_n)$$
$$\times K(x, dy_1) \cdots K(y_{n-2}, dy_{n-1}) .$$

In particular, the relation $P_x(X_1 \in A_1) = K(x, A_1)$ indicates that $K(x_n, dx_{n+1})$ is a *version* of the conditional distribution of $X_{n+1}$ given $X_n$. However, as we have defined a Markov chain by first specifying this kernel, we do not need to be concerned with different versions of the conditional probabilities. This is why we noted that constructing the Markov chain through the transition kernel was mathematically "cleaner." (Moreover, in the following chapters, we will see that the objects of interest are often these conditional distributions, and it is important that we need not worry about different versions. Nonetheless, the properties of a Markov chain considered in this chapter are independent of the version of the conditional probability chosen.)

If we denote $K^1(x, A) = K(x, A)$, the kernel for $n$ transitions is given by $(n > 1)$

(4.2.4) $$K^n(x, A) = \int_{\mathcal{X}} K^{n-1}(y, A) \, K(x, dy).$$

The following result provides convolution formulas of the type $K^{m+n} = K^m \star K^n$, which are called *Chapman–Kolmogorov equations*.

**Lemma 4.2.6 Chapman–Kolmogorov equations** *For every* $(m, n) \in$ $\mathbb{N}^2$, $x \in \mathcal{X}$, $A \in \mathcal{B}(\mathcal{X})$,

$$K^{m+n}(x, A) = \int_{\mathcal{X}} K^n(y, A) \, K^m(x, dy) .$$

(In a very informal sense, the Chapman–Kolmogorov equations state that to get from $x$ to $A$ in $m+n$ steps, you must pass through some $y$ on the $n$th step.) In the discrete case, Lemma 4.2.6 is simply interpreted as a matrix product and follows directly from (4.2.1). In the general case, we need to consider $K$ as an operator on the space of integrable functions; that is, we define

$$Kh(x) = \int h(y) \, K(x, dy) , \qquad h \in \mathcal{L}_1(\lambda) ,$$

$\lambda$ being the dominating measure of the model. $K^n$ is then the $n$th composition of $P$, namely $K^n = K \circ K^{n-1}$.

**Definition 4.2.7** A *resolvant* associated with the kernel $P$ is a kernel of the form

$$K_\varepsilon(x, A) = (1 - \varepsilon) \sum_{i=0}^{\infty} \varepsilon^i K^i(x, A), \qquad 0 < \epsilon < 1,$$

and the chain with kernel $K_\varepsilon$ is a $K_\varepsilon$-*chain*.

Given an initial distribution $\mu$, we can associate with the kernel $K_\varepsilon$ a chain $(X_n^\varepsilon)$ which formally corresponds to a subchain of the original chain $(X_n)$, where the indices in the subchain are generated from a geometric distribution with parameter $1 - \varepsilon$. Thus, $K_\epsilon$ is indeed a kernel, and we will see that the resulting Markov chain $(X_n^\varepsilon)$ enjoys much stronger regularity. This will be used later to establish many properties of the original chain.

If $\mathbb{E}_\mu[ \, \cdot \, ]$ denotes the expectation associated with the distribution $P_\mu$, the *(weak) Markov property* can be written as the following result, which just rephrases the limited memory properties of a Markov chain:

**Proposition 4.2.8 Weak Markov property** *For every initial distribution* $\mu$ *and every* $(n + 1)$ *sample* $(X_0, \ldots, X_n)$,

$$(4.2.5) \quad \mathbb{E}_\mu[h(X_{n+1}, X_{n+2}, \ldots)|x_0, \ldots, x_n] = \mathbb{E}_{x_n}[h(X_1, X_2, \ldots)],$$

*provided that the expectations exist.*

Note that if $h$ is the indicator function, then this definition is exactly the same as Definition 4.2.3. However, (4.2.5) can be generalized to other classes of functions—hence the terminology "weak"— and it becomes particularly useful with the notion of *stopping time* in the convergence assessment of Markov chain Monte Carlo algorithms in Chapter 8.

**Definition 4.2.9** Consider $A \in \mathcal{B}(\mathcal{X})$. The first $n$ for which the chain enters the set $A$ is denoted by

$$(4.2.6) \qquad\qquad \tau_A = \inf\{n \geq 1; X_n \in A\}$$

and is called the *stopping time* at $A$ with, by convention, $\tau_A = +\infty$ if $X_n \notin A$ for every $n$. More generally, a function $\zeta(x_1, x_2, \ldots)$ is called a *stopping rule* if the set $\{\zeta = n\}$ is measurable for the $\sigma$-algebra induced by $(X_0, \ldots, X_n)$. Associated with the set $A$, we also define

$$(4.2.7) \qquad \eta_A = \sum_{n=1}^{\infty} \mathbb{I}_A(X_n),$$

the *number of passages* of $(X_n)$ in $A$.

Of particular importance are the related quantities $\mathbb{E}_x[\eta_A]$ and $P_x(\tau_A < \infty)$, which are the *average number of passages in $A$* and the *probability of return to $A$ in a finite number of steps*.

We will be mostly concerned with stopping rules of the form given in (4.2.6), which express the fact that $\tau_A$ takes the value $n$ when none of the values of $X_0, X_1, \ldots, X_{n-1}$ are in the given state (or set) $A$, but the $n$th value is. The *strong Markov property* corresponds to the following result, whose proof follows from the weak Markov property and conditioning on $\{\zeta = n\}$:

**Proposition 4.2.10 Strong Markov property** *For every initial distribution $\mu$ and every stopping time $\zeta$ which is almost surely finite,*

$$\mathbb{E}_\mu[h(X_{\zeta+1}, X_{\zeta+2}, \cdots)|x_1, \ldots, x_\zeta] = \mathbb{E}_{x_\zeta}[h(X_1, X_2, \cdots)],$$

*provided the expectations exist.*

We can thus condition on a random number of instants while keeping the fundamental properties of a Markov chain.

**Example 4.2.11 Coin tossing.** In a coin tossing game, player $b$ has a gain of $+1$ if a head appears and player $c$ has a gain of $+1$ if a tail appears (so player $b$ has a "gain" of $-1$ (a loss) if a tail appears). If $X_n$ is the sum of the gains of player $b$ after $n$ rounds of this coin tossing game, the transition matrix $P$ is an infinite dimensional matrix with upper and lower subdiagonals equal to $1/2$. Assume that player $b$ has $B$ dollars and player $c$ has $C$ dollars, and consider the following return times:

$$\tau_1 = \inf\{n; X_n = 0\}, \quad \tau_2 = \inf\{n; X_n < -B\}, \quad \tau_3 = \inf\{n; X_n > C\},$$

which represent respectively the return to null and the ruins of the first and second players, that is to say the first times the fortunes of both players, respectively $B$ and $C$, are spent. The probability of ruin (bankruptcy) for the first player is then $P_0(\tau_2 > \tau_3)$. (Feller 1970, Chapter III, has a detailed analysis of this coin tossing game.)                                                        ‖

## 4.3  Irreducibility, Atoms, and Small Sets

### 4.3.1 Irreducibility

The property of irreducibility is a first measure of the sensitivity of the Markov chain to the initial conditions, $x_0$ or $\mu$. It is crucial in the setup of Markov chain Monte Carlo algorithms, because it leads to a guarantee of convergence, thus avoiding a detailed study of the transition operator, which would otherwise be necessary to specify "acceptable" initial conditions.

In the discrete case, the chain is *irreducible* if all states communicate, namely if

$$P_x(\tau_y < \infty) > 0 , \qquad \forall x, y \in \mathcal{X} ,$$

$\tau_y$ being the first time $y$ is visited, defined in (4.2.6). In many cases, $P_x(\tau_y < \infty)$ is uniformly equal to zero, and it is necessary to introduce an auxiliary measure $\varphi$ on $\mathcal{B}(\mathcal{X})$ to correctly define the notion of irreducibility

**Definition 4.3.1** Given a measure $\varphi$, the Markov chain $(X_n)$ with transition kernel $K(x, y)$ is *$\varphi$-irreducible* if, for every $A \in \mathcal{B}(\mathcal{X})$ with $\varphi(A) > 0$, there exists $n$ such that $K^n(x, A) > 0$ for all $x \in \mathcal{X}$ (equivalently, $P_x(\tau_A < \infty) > 0$). The chain is *strongly $\varphi$-irreducible* if $n = 1$ for all measurable $A$.

**Example 4.3.2 (Continuation of Example 4.2.2)** In the case of the Bernoulli–Laplace model, the (finite) chain is indeed irreducible since it is possible to connect the states $x$ and $y$ in $|x - y|$ steps with probability

$$\prod_{i=x \wedge y}^{x \vee y - 1} \left( \frac{K - i}{K} \right)^2 . \qquad\qquad \|$$

The following result provides equivalent definitions of irreducibility. The proof is left to Problem 4.15, and follows from (4.3.1) and the Chapman–Kolmogorov equations.

**Theorem 4.3.3** *The chain $(X_n)$ is $\varphi$-irreducible if, and only if, for every $x \in \mathcal{X}$ and every $A \in \mathcal{B}(\mathcal{X})$ such that $\varphi(A) > 0$, one of the following properties holds:*

*(i) there exists $n \in \mathbb{N}^*$ such that $K^n(x, A) > 0$;*

*(ii) $\mathbb{E}_x[\eta_A] > 0$;*

*(iii) $K_\epsilon(x, A) > 0$ for every $0 < \epsilon < 1$.*

The introduction of the $K_\varepsilon$-chain then allows for the creation of a strictly positive kernel in the case of a $\varphi$-irreducible chain and this property is used in the following to simplify the proofs. Moreover, the measure $\varphi$ in Definition 4.3.1 plays no crucial role in the sense that irreducibility is an intrinsic property of $(X_n)$ and does not rely on $\varphi$.

The following theorem details the properties of the *maximal irreducibility measure* $\psi$.

**Theorem 4.3.4** *If $(X_n)$ is $\varphi$-irreducible, there exists a probability measure $\psi$ such that:*

*(i) The Markov chain $(X_n)$ is $\psi$-irreducible;*

*(ii) If there exists a measure $\xi$ such that $(X_n)$ is $\xi$-irreducible, then $\xi$ is dominated by $\psi$; that is, $\xi \prec\prec \psi$;*

*(iii) If $\psi(A) = 0$, then $\psi(\{y; P_y(\tau_A < \infty) > 0\}) = 0$;*

*(iv) The measure $\psi$ is equivalent to*

$$(4.3.1) \qquad \psi_0(A) = \int_{\mathcal{X}} K_{1/2}(x, A)\, \varphi(dx), \qquad \forall A \in \mathcal{B}(\mathcal{X}) \,;$$

*that is, $\psi \prec\prec \psi_0$ and $\psi_0 \prec\prec \psi$.*

This result provides a *constructive* method of determining the maximal irreducibility measure $\psi$ through a candidate measure $\varphi$, which still needs to be defined.

**Example 4.3.5 (Continuation of Example 4.2.5)**  When $X_{n+1} = \theta X_n + \varepsilon_{n+1}$ and $\varepsilon_n$ are independent normal variables, the chain is irreducible, the reference measure being the *Lebesgue measure*, $\lambda$. (In fact, $K(x, A) > 0$ for every $x \in \mathbb{R}$ and every $A$ such that $\lambda(A) > 0$.) On the other hand, if $\varepsilon_n$ is uniform on $[-1, 1]$ and $|\theta| > 1$, the chain is not irreducible anymore. For instance, if $\theta > 1$, then

$$X_{n+1} - X_n \geq (\theta - 1)X_n - 1 \geq 0$$

for $X_n \geq 1/(\theta - 1)$. The chain is thus monotonically increasing and obviously cannot visit previous values.    ‖

### 4.3.2 Atoms and Small Sets

In the discrete case, the transition kernel is necessarily atomic in the usual sense; that is, there exist points in the state-space with positive mass. The extension of this notion to the general case by Nummelin (1978) is powerful enough to allow for a control of the chain which is almost as "precise" as in the discrete case.

**Definition 4.3.6** The Markov chain $(X_n)$ has *an atom* $\alpha \in \mathcal{B}(\mathcal{X})$ if there exists a measure $\nu > 0$ such that

$$K(x, A) = \nu(A), \qquad \forall x \in \alpha, \ \forall A \in \mathcal{B}(\mathcal{X}) \,.$$

If $(X_n)$ is $\psi$-irreducible, the atom is *accessible* when $\psi(\alpha) > 0$.

While it trivially applies to every possible value of $X_n$ in the discrete case, this notion is often too strong to be of use in the continuous case since it implies that the transition kernel is *constant* on a set of positive measure. A more powerful generalization is the so-called *minorizing condition*, namely that there exists a set $C \in \mathcal{B}(\mathcal{X})$, a constant $\varepsilon > 0$, and a probability measure $\nu > 0$ such that

$$(4.3.2) \qquad K(x, A) \geq \varepsilon\nu(A), \qquad \forall x \in C, \ \forall A \in \mathcal{B}(\mathcal{X}) \,.$$

The probability measure $\nu$ thus appears as a constant component of the transition kernel on $C$. The minorizing condition (4.3.2) leads to the following notion, which is essential in this chapter and in Chapters 6 and 7 as a technique of proof and as the basis of *renewal theory*.

**Definition 4.3.7** A set $C$ is *small* if there exist $m \in \mathbb{N}^*$ and a measure $\nu_m > 0$ such that

$$(4.3.3) \qquad K^m(x, A) \geq \nu_m(A), \qquad \forall x \in C, \, \forall A \in \mathcal{B}(\mathcal{X}) .$$

**Example 4.3.8 (Continuation of Example 4.3.5)** Since $X_n|x_{n-1} \sim \mathcal{N}(\theta x_{n-1}, \sigma^2)$, the transition kernel is bounded from below by

$$\frac{1}{\sigma\sqrt{2\pi}} \, \exp\left\{ (-x_n^2 + 2\theta x_n \underline{w} - \theta^2 \underline{w}^2 \wedge \overline{w}^2)/2\sigma^2 \right\} \text{ if } x_n > 0,$$

$$\frac{1}{\sigma\sqrt{2\pi}} \, \exp\left\{ (-x_n^2 + \theta x_n \overline{w}\sigma^{-2} - \theta^2 \underline{w}^2 \wedge \overline{w}^2)/2\sigma^2 \right\} \text{ if } x_n < 0,$$

when $x_{n-1} \in [\underline{w}, \overline{w}]$. The set $C = [\underline{w}, \overline{w}]$ is indeed a small set, as the measure $\nu_1$, with density

$$\frac{\exp\{(-x^2 + 2\theta x \underline{w})/2\sigma^2\} \, \mathbb{I}_{x>0} + \exp\{(-x^2 + 2\theta x \overline{w})/2\sigma^2\} \, \mathbb{I}_{x<0}}{\sqrt{2\pi} \, \sigma[\Phi(-\theta \underline{w}/\sigma^2) \, \exp\{\theta^2 \underline{w}^2/2\sigma^2\} + [1 - \Phi(-\theta \underline{w}/\sigma^2)] \, \exp\{\theta^2 \overline{w}^2/2\sigma^2\}]} ,$$

and

$$\varepsilon = \exp\{-\theta^2 \underline{w}^2/2\sigma^2\} \left[ \Phi(-\theta \underline{w}/\sigma^2) \, \exp\{\theta^2 \underline{w}^2/2\sigma^2\} \right.$$
$$\left. + [1 - \Phi(-\theta \underline{w}/\sigma^2)] \, \exp\{\theta^2 \overline{w}^2/2\sigma^2\} \right] ,$$

satisfy (4.3.3) with $m = 1$. ‖

A sufficient condition for $C$ to be small is that (4.3.3) is satisfied by the $K_\varepsilon$-chain in the special case $m = 1$. The following result indicates the connection between small sets and irreducibility.

**Theorem 4.3.9** Let $(X_n)$ be a $\psi$-irreducible chain. For every set $A \in \mathcal{B}(\mathcal{X})$ such that $\psi(A) > 0$, there exist $m \in \mathbb{N}^*$ and a small set $C \subset A$ such that $\nu_m(C) > 0$. Moreover, $\mathcal{X}$ can be decomposed in a denumerable partition of small sets.

The proof of this characterization result is rather involved (see Meyn and Tweedie 1993, pp. 107–109). The decomposition of $\mathcal{X}$ as a denumerable union of small sets is based on an arbitrary small set $C$ and the sequence

$$C_{nm} = \{y; K^n(y, C) > 1/m\}$$

(see Problem 4.22).

Small sets are obviously easier to exhibit than atoms, given the freedom allowed by the minorizing condition (4.3.3). Moreover, they are, in fact, very common since, in addition to Theorem 4.3.9, Meyn and Tweedie (1993, p. 134) show that for sufficiently regular (in a topological sense) Markov chains, every *compact* set is small. Atoms, although a special case of small sets, enjoy stronger stability properties since the transition probability is invariant on $\alpha$. However, *splitting methods* (see below) offer the possibility of extending most of these properties to the general case and it will be used as a technique of proof in the remainder of the chapter.

If the minorizing condition holds for $(X_n)$, there are two ways of deriving a companion Markov chain $(\check{X}_n)$ sharing many properties with $(X_n)$ and possessing an atom $\alpha$. The first method is called *Nummelin's splitting* and constructs a chain made of two copies of $(X_n)$ (see Nummelin 1978 and Meyn and Tweedie 1993, Section 5.1).

A second method, discovered at approximately the same time, is due to Athreya and Ney (1978) and uses a stopping time to create an atom. We prefer to focus on this latter method because it is related to notions of *renewal time*, which are also useful in the control of Markov chain Monte Carlo algorithms (see §8.3.3).

**Definition 4.3.10** A *renewal time* (or *regeneration time*) is a stopping rule $\tau$ with the property that $(X_\tau, X_{\tau+1}, \ldots)$ is independent of $(X_{\tau-1}, X_{\tau-2}, \ldots)$.

For instance, in Example 4.2.11, the returns to zero gain are renewal times. The excursions between two returns to zero are independent and identically distributed (see Feller 1970, Chapter III). More generally, visits to atoms are renewal times, whose features are quite appealing in convergence control for Markov chain Monte Carlo algorithms (see Chapter 8).

If (4.3.3) holds and if the probability $P_x(\tau_C < \infty)$ of a return to $C$ in a finite time is identically equal to 1 on $\mathcal{X}$, Athreya and Ney (1978) modify the transition kernel when $X_n \in C$, by simulating $X_{n+1}$ as

$$
(4.3.4) \quad X_{n+1} \sim \begin{cases} \nu & \text{with probability } \varepsilon \\ \dfrac{K(X_n, \cdot) - \varepsilon \nu(\cdot)}{1 - \varepsilon} & \text{with probability } 1 - \varepsilon; \end{cases}
$$

that is, by simulating $X_{n+1}$ from $\nu$ with probability $\varepsilon$ every time $X_n$ is in $C$. This modification does not change the marginal distribution of $X_{n+1}$ conditionally on $x_n$, since

$$
\epsilon \nu(A) + (1 - \epsilon) \frac{K(x_n, A) - \epsilon \nu(A)}{1 - \epsilon} = K(x_n, A), \qquad \forall A \in \mathcal{B}(\mathcal{X}),
$$

but it produces *renewal times* for each time $j$ such that $X_j \in C$ and $X_{j+1} \sim \nu$.

Now, we clearly see how the renewal times result in independent chains. When $X_{j+1} \sim \nu$, this event is totally independent of any past history, as the current state of the chain has no effect on the measure $\nu$. Note also the key role that is played by the minorization condition. It allows us to create the split chain with the same marginal distribution as the original chain. We denote by $(j > 0)$

$$
\tau_j = \inf\{n > \tau_{j-1}; X_n \in C \text{ and } X_{n+1} \sim \nu\}
$$

the sequence of *renewal times* with $\tau_0 = 0$. Athreya and Ney (1978) introduce the *augmented chain*, also called the *split chain* $\check{X}_n = (X_n, \check{\omega}_n)$, with $\check{\omega}_n = 1$ when $X_n \in C$ and $X_{n+1}$ is generated from $\nu$. It is then easy to show that the set $\check{\alpha} = C \times \{1\}$ *is an atom of the chain* $(\check{X}_n)$, the resulting subchain $(X_n)$ being still a Markov chain with transition kernel $K(x_n, \cdot)$ (see Problem 4.20).

The notion of small set is only useful in finite and discrete settings when the individual probabilities of states are too small to allow for a reasonable rate of renewal. In these cases, small sets are made of collections of states with $\nu$ defined as a minimum. Otherwise, small sets reduced to a single value are also atoms.

### 4.3.3 Cycles and Aperiodicity

The behavior of $(X_n)$ may sometimes be restricted by deterministic constraints on the moves from $X_n$ to $X_{n+1}$. We formalize these constraints here and show in the following chapters that the chains produced by Markov chain Monte Carlo algorithms do not display this behavior and, hence, do not suffer from the associated drawbacks.

In the discrete case, the *period* of a state $\omega \in \mathcal{X}$ is defined as

$$d(\omega) = \text{g.c.d.} \ \{m \geq 1; K^m(\omega, \omega) > 0\} \ ,$$

where we recall that g.c.d. is the greatest common denominator. The value of the period is constant on all states that communicate with $\omega$. In the case of an irreducible chain on a finite space $\mathcal{X}$, the transition matrix can be written (with a possible reordering of the states) as a block matrix

(4.3.5)
$$P = \begin{pmatrix} 0 & D_1 & 0 & \cdots & 0 \\ 0 & 0 & D_2 & & 0 \\ & & & \ddots & \\ D_d & 0 & 0 & & 0 \end{pmatrix} \ ,$$

where the blocks $D_i$ are stochastic matrices. This representation clearly illustrates the forced passage from one group of states to another, with a return to the initial group occurring every $d$th step. If the chain is irreducible (so all states communicate), there is only one value for the period. An irreducible chain is *aperiodic* if it has period 1. The extension to the general case requires the existence of a small set.

**Definition 4.3.11** A $\psi$-irreducible chain $(X_n)$ has a *cycle of length* $d$ if there exists a small set $C$, an associated integer $M$, and a probability distribution $\nu_M$ such that $d$ is the g.c.d. of

$$\{m \geq 1; \ \exists \ \delta_m > 0 \text{ such that } C \text{ is small for } \nu_m \geq \delta_m \nu_M\}.$$

A decomposition like (4.3.5) can be established in general. It is easily shown that the number $d$ is independent of the small set $C$ and that this number intrinsically characterizes the chain $(X_n)$. The *period* of $(X_n)$ is then defined as the largest integer $d$ satisfying Definition 4.3.11 and $(X_n)$ is *aperiodic* if $d = 1$. If there exists a small set $A$ and a minorizing measure $\nu_1$ such that $\nu_1(A) > 0$ (so it is possible to go from $A$ to $A$ in a single step), the chain is said to be *strongly aperiodic*). Note that the $K_\varepsilon$-chain can be used to transform an aperiodic chain into a strongly aperiodic chain.

In discrete setups, if one state $x \in \mathcal{X}$ satisfies $P_{xx} > 0$, the chain $(X_n)$ is aperiodic, although this is not a necessary condition (see Problem 4.37).

**Example 4.3.12 (Continuation of Example 4.3.2)** The Bernoulli-Laplace chain is aperiodic and even strongly aperiodic since the diagonal terms satisfy $P_{xx} > 0$ for every $x \in \{0, \dots, K\}$.                    ‖

When the chain is continuous and the transition kernel has a component which is absolutely continuous with respect to the Lebesgue measure, with density $f(\cdot | x_n)$, a sufficient condition for aperiodicity is that $f(\cdot | x_n)$ is positive in a neighborhood of $x_n$. The chain can then remain in this neighborhood for an arbitrary number of instants before visiting any set $A$. For instance, in Example 4.2.2, $(X_n)$ is strongly aperiodic when $\varepsilon_n$ is distributed according to $\mathcal{U}_{[-1,1]}$ and $|\theta| < 1$ (in order to guarantee irreducibility). The next chapters will demonstrate that Markov chain Monte Carlo algorithms lead to aperiodic chains, possibly via the introduction of additional steps.

## 4.4 Transience and Recurrence

### 4.4.1 Classification of Irreducible Chains

From an algorithmic point of view, a Markov chain must enjoy good *stability* properties to guarantee an acceptable approximation of the simulated model. Indeed, *irreducibility* ensures that every set $A$ will be visited by the Markov chain $(X_n)$, but this property is too weak to ensure that the trajectory of $(X_n)$ will enter $A$ often enough. Consider, for instance, a maximization problem using a random walk on the surface of the function to maximize (see Chapter 5). The convergence to the global maximum cannot be guaranteed without a systematic sweep of this surface. Formalizing this stability of the Markov chain leads to different notions of *recurrence*. In a discrete setup, the *recurrence of a state* is equivalent to a guarantee of a sure return. This notion is thus necessarily satisfied for irreducible chains on a finite space.

**Definition 4.4.1** In a finite state-space $\mathcal{X}$, a state $\omega \in \mathcal{X}$ is *transient* if the average number of visits to $\omega$, $\mathbb{E}_\omega[\eta_\omega]$, is finite and *recurrent* if $\mathbb{E}_\omega[\eta_\omega] = \infty$.

For irreducible chains, the properties of recurrence and transience are properties of the chain, not of a particular state. This fact is easily deduced from the Chapman–Kolmogorov equations. Therefore, if $\eta_A$ denotes the average number of visits defined in (4.2.7), for every $(x, y) \in \mathcal{X}^2$ either $\mathbb{E}_x[\eta_y] < \infty$ in the transient case or $\mathbb{E}_x[\eta_y] = \infty$ in the recurrent case. The chain is then said to be *transient* or *recurrent*, one of the two properties being necessarily satisfied in the irreducible case.

**Example 4.4.2 Branching process.** Consider a population whose individuals reproduce independently of one another. Each individual has $X$ sibling(s), $X \in \mathbb{N}$, distributed according to the distribution with generating function $\phi(s) = \mathbb{E}[s^X]$. If individuals reproduce at fixed instants (thus defining *generations*), the size of the $t$th generation $S_t$ ($t > 1$) is given by

$$S_t = X_1 + \cdots + X_{S_{t-1}},$$

where the $X_i \sim \phi$ are independent. Starting with a single individual at time 0, $S_1 = X_1$, the generating function of $S_t$ is $g_t(s) = \phi^t(s)$, with $\phi^t = \phi \circ \phi^{t-1}$ $(t > 1)$. The chain $(S_t)$ is an example of a *branching process* (see Feller 1971, Chapter XII).

If $\phi$ does not have a constant term (i.e., if $P(X_1 = 0) = 0$), the chain $(S_t)$ is necessarily transient since it is increasing. If $P(X_1 = 0) > 0$, the probability of a return to 0 at time $t$ is $\rho_t = P(S_t = 0) = g_t(0)$, which thus satisfies the recurrence equation $\rho_t = \phi(\rho_{t-1})$. Therefore, there exists a limit $\rho$ different from 1, such that $\rho = \phi(\rho)$, if and only if $\phi'(1) > 1$; namely if $\mathbb{E}[X] > 1$. The chain is thus transient when the average number of siblings per individual is larger than 1. If there exists a restarting mechanism in 0, $S_{t+1}|S_t = 0 \sim \phi$, it is easily shown that when $\phi'(1) > 1$, the number of returns to 0 follows a geometric distribution with parameter $\rho$. If $\phi'(1) \leq 1$, one can show that the chain is recurrent (see Example 4.5.8).            ‖

The treatment of the general (that is to say, non discrete) case is based on chains with atoms, the extension to general chains (with small sets) following from Athreya and Ney's (1978) splitting. We begin by extending the notions of recurrence and transience.

**Definition 4.4.3** A set $A$ is called *recurrent* if $\mathbb{E}_x[\eta_A] = +\infty$ for every $x \in A$. The set $A$ is *uniformly transient* if there exists a constant $M$ such that $\mathbb{E}_x[\eta_A] < M$ for every $x \in A$. It is *transient* if there exists a covering of $\mathcal{X}$ by uniformly transient sets; that is, a countable collection of uniformly transient sets $B_i$ such that

$$A = \bigcup_i B_i.$$

**Theorem 4.4.4** *Let $(X_n)$ be $\psi$-irreducible Markov chain with an accessible atom $\alpha$.*

*(i) If $\alpha$ is recurrent, every set $A$ of $\mathcal{B}(\mathcal{X})$ such that $\psi(A) > 0$ is recurrent.*

*(ii) If $\alpha$ is transient, $\mathcal{X}$ is transient.*

Property (i) is the most relevant in the Markov chain Monte Carlo setup and can be derived from the Chapman–Kolmogorov equations. Property (ii) is more difficult to establish and uses the fact that $P_\alpha(\tau_\alpha < \infty) < 1$ for a transient set when $\mathbb{E}_x[\eta_A]$ is decomposed conditionally on the last visit to $\alpha$ (see Meyn and Tweedie 1993, p. 181, and Problem 4.32).

**Definition 4.4.5** A Markov chain $(X_n)$ is *recurrent* if

(i) there exists a measure $\psi$ such that $(X_n)$ is $\psi$-irreducible, and

(ii) for every $A \in \mathcal{B}(\mathcal{X})$ such that $\psi(A) > 0$, $\mathbb{E}_x[\eta_A] = \infty$ for every $x \in A$.

The chain is *transient* if it is $\psi$-irreducible and if $\mathcal{X}$ is transient.

The classification result of Theorem 4.4.4 can be easily extended to *strongly aperiodic* chains since they satisfy a minorizing condition (4.3.3), thus can be split as in (4.3.2), while the chain $(X_n)$ and its split version $(\check{X}_n)$ (see Problem 4.20) are either both recurrent or both transient. The generalization to an arbitrary irreducible chain follows from the properties of the corresponding $K_\varepsilon$-chain  which is strongly aperiodic, through the relation

(4.4.1)
$$\sum_{n=0}^{\infty} K_\varepsilon^n = \frac{1-\varepsilon}{\varepsilon} \sum_{n=0}^{\infty} K^n ,$$

since
$$\mathbb{E}_x[\eta_A] = \sum_{n=0}^{\infty} K^n(x,A) = \frac{\epsilon}{1-\epsilon} \sum_{n=0}^{\infty} K_\epsilon^n(x,A).$$

This provides us with the following classification result:

**Theorem 4.4.6** *A $\psi$-irreducible chain is either recurrent or transient.*

### 4.4.2 Criteria for Recurrence

The previous results establish a clear dichotomy between transience and recurrence for irreducible Markov chains. Nevertheless, given the requirement of Definition 4.4.5, it is useful to examine simpler criteria for recurrence. By analogy with discrete state-space Markov chains, a first approach is based on small sets.

**Proposition 4.4.7** *A $\psi$-irreducible chain $(X_n)$ is recurrent if there exists a small set $C$ with $\psi(C) > 0$ such that $P_x(\tau_C < \infty) = 1$ for every $x \in C$.*

*Proof.* First, we show that the set $C$ is recurrent. Given $x \in C$, consider $u_n = K^n(x,C)$ and $f_n = P_x(X_n \in C, X_{n-1} \notin C, \ldots, X_1 \notin C)$, which is the probability of first visit to $C$ at the $n$th instant, and define

$$\tilde{U}(s) = 1 + \sum_{n=1}^{\infty} u_n s^n \quad \text{and} \quad Q(s) = \sum_{n=1}^{\infty} f_n s^n.$$

The equation
(4.4.2)
$$u_n = f_n + f_{n-1} u_1 + \cdots + f_1 u_{n-1}$$

describes the relation between the probability of a visit of $C$ at time $n$ and the probabilities of first visit of $C$. This implies

$$\tilde{U}(s) = \frac{1}{1 - Q(s)} ,$$

which connects $\tilde{U}(1) = \mathbb{E}_x[\eta_C] = \infty$ with $Q(1) = P_x(\tau_C < \infty) = 1$. Equation (4.4.2) is, in fact, valid for the split chain $(\check{X}_n)$ (see Problem 4.20), since a visit to $C \times \{0\}$ ensures independence by renewal. Since $\mathbb{E}_x[\eta_C]$, associated with $(X_n)$, is larger than $\mathbb{E}_{\check{x}}[\eta_{C \times \{0\}}]$, associated with $(\check{x}_n)$, and $P_x(\tau_C < \infty)$ for $(X_n)$ is equal to $P_{\check{X}}(\tau_{C \times \{0\}} < \infty)$ for $(\check{X}_n)$, the recurrence can be extended from $(\check{x}_n)$ to $(X_n)$. The recurrence of $(\check{X}_n)$ follows from Theorem 4.4.4, since $C \times \{0\}$ is a recurrent atom for $(\check{X}_n)$. □□

A second method of checking recurrence is based on a generalization of the notions of small sets and minorizing conditions. This generalization involves a *potential function* $V$ and a *drift condition* like (4.10.1) and uses the transition kernel $K(\cdot,\cdot)$ rather than the sequence $K^n$. Note 4.10.1 details this approach, as well as its bearing on the following stability and convergence results.

### 4.4.3 Harris Recurrence

It is actually possible to strengthen the stability properties of a chain $(X_n)$ by requiring not only an infinite average number of visits to every small set but also an infinite number of visits for every path of the Markov chain. Recall that $\eta_A$ is the number of passages of $(X_n)$ in $A$, and we consider $P_x(\eta_A = \infty)$, the probability of visiting $A$ an infinite number of times. The following notion of recurrence was introduced by Harris (1956).

**Definition 4.4.8** A set $A$ is *Harris recurrent* if $P_x(\eta_A = \infty) = 1$ for all $x \in A$. The chain $(X_n)$ is *Harris recurrent* if there exists a measure $\psi$ such that $(X_n)$ is $\psi$-irreducible and for every set $A$ with $\psi(A) > 0$, where $A$ is Harris recurrent.

Recall that recurrence corresponds to $\mathbb{E}_x[\eta_\alpha] = \infty$, a weaker condition than $P_x(\eta_A = \infty) = 1$ (see Problem 4.34). The following proposition expresses Harris recurrence as a condition on $P_x(\tau_A < \infty)$ defined in (4.2.7).

**Proposition 4.4.9** *If for every $A \in \mathcal{B}(\mathcal{X})$, $P_x(\tau_A < \infty) = 1$ for every $x \in A$, then $P_x(\eta_A = \infty) = 1$, for all $x \in \mathcal{X}$, and $(X_n)$ is Harris recurrent.*

*Proof.* The average number of visits to $B$ before a first visit to $A$ is

$$(4.4.3) \qquad U_A(x, B) = \sum_{n=1}^{\infty} P_x(X_n \in B, \tau_A \geq n) \ .$$

Then, $U_A(x, A) = P_x(\tau_A < \infty)$, since, if $B \subset A$, $P_x(X_n \in B, \tau_A \geq n) = P_x(X_n \in B, \tau = n) = P_x(\tau_B = n)$. Similarly, if $\tau_A(k), k > 1$, denotes the time of the $k$th visit to $A$, $\tau_A(k)$ satisfies

$$P_x(\tau_A(2) < \infty) = \int_A P_y(\tau_A < \infty)\, U_A(x, dy) = 1$$

for every $x \in A$ and, by induction,

$$P_x(\tau_A(k + 1) < \infty) = \int_A P_x(\tau_A(k) < \infty)\, U_A(x, dy) = 1.$$

Since $P_x(\eta_A \geq k) = P_x(\tau_A(k) < \infty)$ and

$$P_x(\eta_A = \infty) = \lim_{k \to \infty} P_x(\eta_A \geq k),$$

we deduce that $P_x(\eta_A = \infty) = 1$ for $x \in A$.                    $\Box\Box$

Note that the property of Harris recurrence is only needed when $\mathcal{X}$ is not denumerable. If $\mathcal{X}$ is finite or denumerable, we can indeed show that $\mathbb{E}_x[\eta_x] = \infty$ if and only if $P_x(\tau_x < \infty) = 1$ for every $x \in \mathcal{X}$ through an argument similar to the proof of Proposition 4.4.7. In the general case, it is possible to prove that if $(X_n)$ is Harris recurrent, then $P_x(\eta_B = \infty) = 1$ for every $x \in \mathcal{X}$ and $B \in \mathcal{B}(\mathcal{X})$ such that $\psi(B) > 0$. This property then provides a sufficient condition for Harris recurrence which generalizes Proposition 4.4.7.

**Theorem 4.4.10** *If $(X_n)$ is a $\psi$-irreducible Markov chain with a small set $C$ such that $P_x(\tau_C < \infty) = 1$ for all $x \in \mathcal{X}$, then $(X_n)$ is Harris recurrent.*

Contrast this theorem with Proposition 4.4.7, where $P_x(\tau_C < \infty) = 1$ only for $x \in C$. This theorem also allows us to replace recurrence by Harris recurrence in Theorem 4.10.3. (See Meyn and Tweedie 1993, pp. 204–205 for a discussion of the "almost" Harris recurrence of recurrent chains.) Tierney (1994) and Chan and Geyer (1994) analyze the role of Harris recurrence in the setup of Markov chain Monte Carlo algorithms and note that Harris recurrence holds for most of these algorithms (see Chapters 6 and 7).[2]

## 4.5 Invariant Measures

An increased level of stability for the chain $(X_n)$ is attained if the marginal distribution of $X_n$ is independent of $n$. More formally, this is a requirement for the existence of a probability distribution $\pi$ such that $X_{n+1} \sim \pi$ if $X_n \sim \pi$, and Markov chain Monte Carlo methods are based on the fact that this requirement, which defines a particular kind of recurrence called *positive recurrence*, can be met. The Markov chains constructed from Markov chain Monte Carlo algorithms enjoy this greater stability property (except in very pathological cases; see §7.5). We therefore provide an abridged description of invariant measures and positive recurrence.

**Definition 4.5.1** A $\sigma$-finite measure $\pi$ is *invariant* for the transition kernel $K(\cdot, \cdot)$ (and for the associated chain) if

$$\pi(B) = \int_{\mathcal{X}} K(x, B) \, \pi(dx) , \qquad \forall B \in \mathcal{B}(\mathcal{X}) .$$

When there exists an *invariant probability measure* for a $\psi$-irreducible (hence recurrent by Theorem 4.4.6) chain, the chain is *positive*. Recurrent chains that do not allow for a finite invariant measure are called *null recurrent*.

The invariant distribution is also referred to as *stationary* if $\pi$ is a probability measure, since $X_0 \sim \pi$ implies that $X_n \sim \pi$ for every $n$; thus, the chain is stationary in distribution. (Note that the alternative case when $\pi$ is not finite is more difficult to interpret in terms of behavior of the chain.) It is easy to show that if the chain is irreducible and allows for an $\sigma$-finite invariant measure, this measure is unique, up to a multiplicative

---

[2] Chan and Geyer (1994) particularly stress that *"Harris recurrence essentially says that there is no measure-theoretic pathology (...) The main point about Harris recurrence is that asymptotics do not depend on the starting distribution because of the "split" chain construction."*

factor (see Problem 4.64). The link between positivity and recurrence is given by the following result, which formalizes the intuition that the existence of a invariant measure prevents the probability mass from "escaping to infinity."

**Proposition 4.5.2** *If the chain $(X_n)$ is positive, it is recurrent.*

*Proof.* If $(X_n)$ is transient, there exists a covering of $\mathcal{X}$ by uniformly transient sets, $A_j$, with corresponding bounds

$$\mathbb{E}_x[\eta_{A_j}] \le M_j, \qquad \forall x \in A_j, \ \forall j \in \mathbb{N} .$$

Therefore, by the invariance of $\pi$,

$$\pi(A_j) = \int K(x, A_j)\, \pi(dx) = \int K^n(x, A_j)\, \pi(dx).$$

Therefore, for every $k \in \mathbb{N}$,

$$k\, \pi(A_j) = \sum_{n=0}^{k} \int K^n(x, A_j)\, \pi(dx) \le \int \mathbb{E}_x[\eta_{A_j}]\, \pi(dx) \le M_j,$$

since, from (4.2.7) it follows that $\sum_{n=0}^{k} K^n(x, A_j) \le \mathbb{E}_x[\eta_{A_j}]$. Letting $k$ go to $\infty$ shows that $\pi(A_j) = 0$, for every $j \in \mathbb{N}$, and hence the impossibility of obtaining an invariant probability measure.                    □□

We may, therefore, talk of *positive chains* and of *Harris positive chains*, without the superfluous denomination *recurrent* and *Harris recurrent*. Proposition 4.5.2 is only useful when the positivity of $(X_n)$ can be proved, but, again, the chains produced by Markov chain Monte Carlo methods are, by nature, guaranteed to possess an invariant distribution.

A classical result (see Feller 1970) on irreducible Markov chains with discrete state-space is that the stationary distribution, when it exists, is given by

$$\pi_x = (\mathbb{E}_x[\tau_x])^{-1} , \qquad x \in \mathcal{X},$$

where, from (4.2.6), we can interpret $\mathbb{E}_x[\tau_x]$ as the average number of excursions between two passages in $x$. (It is sometimes called *Kac's Theorem.*) It also follows that $(\mathbb{E}_x[\tau_x]^{-1})$ is the eigenvector associated with the eigenvalue 1 for the transition matrix $\mathbb{P}$ (see Problems 4.12 and 4.65). We now establish this result in the more general case when $(X_n)$ has an atom, $\alpha$.

**Theorem 4.5.3** *Let $(X_n)$ be $\psi$-irreducible with an atom $\alpha$. The chain is positive if and only if $\mathbb{E}_\alpha[\tau_\alpha] < \infty$. In this case, the invariant distribution $\pi$ for $(X_n)$ satisfies*

$$\pi(\alpha) = (\mathbb{E}_\alpha[\tau_\alpha])^{-1} .$$

The notation $\mathbb{E}_\alpha[\,\cdot\,]$ is legitimate in this case since the transition kernel is the same for every $x \in \alpha$ (see Definition 4.3.6). Moreover, Theorem 4.5.3 indicates how positivity is a stability property stronger than recurrence. In fact, the latter corresponds to

$$P_\alpha(\tau_\alpha = \infty) = 0,$$

which is a necessary condition for $\mathbb{E}_\alpha[\tau_\alpha] < \infty$.

*Proof.* If $\mathbb{E}_\alpha[\tau_\alpha] < \infty$, $P_\alpha(\tau_\alpha < \infty) = 1$; thus, $(X_n)$ is recurrent from Proposition 4.4.7. Consider a measure $\pi$ given by

$$(4.5.1) \qquad \pi(A) = \sum_{n=1}^{\infty} P_\alpha(X_n \in A, \tau_\alpha \geq n)$$

as in (4.4.3). This measure is invariant since $\pi(\alpha) = P_\alpha(\tau_\alpha < \infty) = 1$ and

$$\int K(x, A)\pi(dx) = \pi(\alpha)K(\alpha, A) + \int_{\alpha^c} \sum_{n=1}^{\infty} K(x_n, A)\, P_\alpha(\tau_\alpha \geq n, dx_n)$$

$$= K(\alpha, A) + \sum_{n=2}^{\infty} P_\alpha(X_n \in A, \tau_\alpha \geq n) = \pi(A).$$

It is also finite as

$$\pi(\mathcal{X}) = \sum_{n=1}^{\infty} P_\alpha(\tau_\alpha \geq n) = \sum_{n=1}^{\infty} \sum_{m=n}^{\infty} P_\alpha(\tau_\alpha = m)$$

$$= \sum_{m=1}^{\infty} m\, P_\alpha(\tau_\alpha = m) = \mathbb{E}_\alpha[\tau_\alpha] < \infty .$$

Since $\pi$ is invariant when $(X_n)$ is positive, the uniqueness of the invariant distribution implies finiteness of $\pi(\mathcal{X})$, thus of $\mathbb{E}_\alpha[\tau_\alpha]$. Renormalizing $\pi$ to $\pi/\pi(\mathcal{X})$ implies $\pi(\alpha) = (\mathbb{E}_\alpha[\tau_\alpha])^{-1}$. $\qquad \Box\Box$

Following a now "classical" approach, the general case can be treated by splitting $(X_n)$ to $(\check{X}_n)$ (which has an atom) and the invariant measure of $(\check{X}_n)$ induces an invariant measure for $(X_n)$ by marginalization. A converse of Proposition 4.4.7 establishes the generality of invariance for Markov chains (see Meyn and Tweedie 1993, pp. 240–245, for a proof).

**Theorem 4.5.4** *If $(X_n)$ is a recurrent chain, there exists an invariant $\sigma$-finite measure which is unique up to a multiplicative factor.*

**Example 4.5.5 Random walk on $\mathbb{R}$.** Consider the random walk on $\mathbb{R}$, $X_{n+1} = X_n + W_n$, where $W_n$ has a cdf $\Gamma$. Since $K(x, \cdot)$ is the distribution with cdf $\Gamma(y - x)$, the distribution of $X_{n+1}$ is invariant by translation, and this implies that the Lebesgue measure is an invariant measure:

$$\int K(x, A)\lambda(dx) = \int \int_{A-x} \Gamma(dy)\lambda(dx) = \int \Gamma(dy) \int \mathbb{I}_{A-y}(x)\lambda(dx)$$

$$= \lambda(A) .$$

Moreover, the invariance of $\lambda$ and the uniqueness of the invariant measure imply that the chain $(X_n)$ cannot be positive recurrent (in fact, it can be shown that it is null recurrent). $\qquad \|$

**Example 4.5.6 Random walk on $\mathbb{Z}$.** A *random walk* on $\mathbb{Z}$ is defined by

$$X_{n+1} = X_n + W_n,$$

the perturbations $W_n$ being iid with distribution $\gamma_k = P(W_n = k)$, $k \in \mathbb{Z}$. With the same kind of argument as in Example 4.5.5, since the counting measure on $\mathbb{Z}$ is invariant for $(X_n)$, $(X_n)$ cannot be positive. If the distribution of $W_n$ is symmetric, straightforward arguments lead to the conclusion that

$$\sum_{n=1}^{\infty} P_0(X_n = 0) = \infty \,,$$

from which we derive the (null) recurrence of $(X_n)$ (see Feller 1970, Durrett 1991, or Problem 4.28).          ‖

**Example 4.5.7 (Continuation of Example 4.3.12)** Given the quasi-diagonal shape of the transition matrix, it is possible to directly determine the invariant distribution, $\pi = (\pi_0, \ldots, \pi_K)$. In fact, it follows from the equation $P^t \pi = \pi$ that

$$\pi_0 = P_{00}\pi_0 + P_{10}\pi_1,$$
$$\pi_1 = P_{01}\pi_0 + P_{11}\pi_1 + P_{21}\pi_2,$$
$$\vdots$$
$$\pi_K = P_{(K-1)K}\pi_{K-1} + P_{KK}\pi_K \,.$$

Therefore,

$$\pi_1 = \frac{P_{01}}{P_{10}} \pi_0,$$

$$\pi_2 = \frac{P_{01}P_{12}}{P_{21}P_{10}} \pi_0,$$

$$\vdots$$

$$\pi_k = \frac{P_{01} \cdots P_{(k-1)k}}{P_{k(k-1)} \cdots P_{10}} \pi_0 \,.$$

Hence,

$$\pi_k = \binom{K}{k}^2 \pi_0, \qquad k = 0, \ldots, K,$$

and through normalization,

$$\pi_K = \frac{\binom{K}{k}^2}{\binom{2K}{k}} \,,$$

which implies that the hypergeometric distribution $\mathcal{H}(2K, K, 1/2)$ is the invariant distribution for the Bernoulli–Laplace model. Therefore, the chain is positive.          ‖

**Example 4.5.8 (Continuation of Example 4.4.2)** Assume $f'(1) \leq 1$. If there exists an invariant distribution for $(S_t)$, its characteristic function $g$ satisfies

$$(4.5.2) \qquad g(s) = f(s)g(0) + g[f(s)] - g(0) .$$

In the simplest case, that is to say, when the number of siblings of a given individual is distributed according to a Bernoulli distribution $\mathcal{B}(p)$, $f(s) = q + ps$, where $q = 1 - p$, and $g(s)$ is solution of

$$(4.5.3) \qquad g(s) = g(q + ps) + p(s - 1)g(0) .$$

Iteratively substituting (4.5.2) into (4.5.3), we obtain

$$
\begin{aligned}
g(s) &= g[q + p(q + ps)] + p(q + ps - 1)g(0) + p(s - 1)g(0) \\
&= g(q + pq + \cdots + p^{k-1}q + p^k s) + (p + \cdots + p^k)(s - 1)g(0) ,
\end{aligned}
$$

for every $k \in \mathbb{N}$. Letting $k$ go to infinity, we have

$$
\begin{aligned}
g(s) &= g[q/(1 - p)] + [p/(1 - p)](s - 1)g(0) \\
&= 1 + \frac{p}{q} (s - 1)g(0) ,
\end{aligned}
$$

since $q/(1 - p) = 1$ and $g(1) = 1$. Substituting $s = 0$ implies $g(0) = q$ and, hence, $g(s) = 1 + p(s - 1) = q + ps$. The Bernoulli distribution is thus the invariant distribution and the chain is positive. ‖

**Example 4.5.9 (Continuation of Example 4.3.8)** Given that the transition kernel corresponds to the $\mathcal{N}(\theta x_{n-1}, \sigma^2)$ distribution, a normal distribution $\mathcal{N}(\mu, \tau^2)$ is stationary for the $AR(1)$ chain only if

$$\mu = \theta\mu \qquad \text{and} \qquad \tau^2 = \tau^2\theta^2 + \sigma^2 .$$

These conditions imply that $\mu = 0$ and that $\tau^2 = \sigma^2/(1 - \theta^2)$, which can only occur for $|\theta| < 1$. In this case, $\mathcal{N}(0, \sigma^2/(1 - \theta^2))$ is indeed the unique stationary distribution of the $AR(1)$ chain. ‖

## 4.6 Ergodicity and Convergence

### 4.6.1 Ergodicity

Considering the Markov chain $(X_n)$ from a temporal perspective, it is natural (and important) to establish the limiting behavior of $X_n$; that is, *to what is the chain converging?* The existence (and uniqueness) of an invariant distribution $\pi$ makes that distribution a natural candidate for the limiting distribution, and we now turn to finding sufficient conditions on $(X_n)$ for $X_n$ to be asymptotically distributed according to $\pi$. The following theorems are fundamental convergence results for Markov chains and they are at the core of the motivation for Markov chain Monte Carlo algorithms. They are, unfortunately, if not surprisingly, quite difficult to establish and we restrict the proof to the countable case, the extension to the general case being detailed in Meyn and Tweedie (1993, pp. 322–323).

There are many conditions that can be placed on the convergence of $P^n$, the distribution of $X_n$, to $\pi$. Perhaps, the most fundamental and important is that of *ergodicity*, that is, independence of initial conditions.

**Definition 4.6.1** For a Harris positive chain $(X_n)$, with invariant distribution $\pi$, an atom $\alpha$ is *ergodic* if

$$\lim_{n \to \infty} |K^n(\alpha, \alpha) - \pi(\alpha)| = 0 .$$

In the countable case, the existence of an ergodic atom is, in fact, sufficient to establish convergence according to the *total variation norm*,

$$\|\mu_1 - \mu_2\|_{TV} = \sup_A |\mu_1(A) - \mu_2(A)|.$$

**Proposition 4.6.2** *If* $(X_n)$ *is Harris positive on* $\mathcal{X}$, *denumerable, and if there exists an ergodic atom* $\alpha \subset \mathcal{X}$, *then, for every* $x \in \mathcal{X}$,

$$\lim_{n \to \infty} \|K^n(x, \cdot) - \pi\|_{TV} = 0 .$$

*Proof.* The first step follows from a decomposition formula called *"first entrance and last exit"*:

$$
\begin{aligned}
K^n(x, y) = \; & P_x(X_n = y, \tau_\alpha \geq n) \\
& + \sum_{j=1}^{n-1} \left[ \sum_{k=1}^{j} P_x(X_k \in \alpha, \tau_\alpha \geq k) K^{j-k}(\alpha, \alpha) \right] \\
& \times P_\alpha(X_{n-j} = y, \tau_\alpha \geq n - j),
\end{aligned}
$$

(4.6.1)

which relates $K^n(x, y)$ to the last visit to $\alpha$. (See Problem 4.39.) This shows the reduced influence of the initial value $x$, since $P_x(X_n = y, \tau_\alpha \geq n)$ converges to 0 with $n$. The expression (4.5.1) of the invariant measure implies, in addition, that

$$\pi(y) = \pi(\alpha) \sum_{j=1}^{k} P_\alpha(X_j = y, \tau_\alpha \geq j).$$

These two expressions then lead to

$$
\begin{aligned}
\|K^n(x, \cdot) - \pi\|_{TV} = \; & \sum_y |K^n(x, y) - \pi(y)| \\
\leq \; & \sum_y P_x(X_n = y, \tau_\alpha \geq n) \\
& + \sum_y \sum_{j=1}^{n-1} \left| \sum_{k=1}^{j} P_x(X_k \in \alpha, \tau_\alpha = k) K^{j-k}(\alpha, \alpha) - \pi(\alpha) \right| \\
& \times P_\alpha(X_{n-j} = y, \tau_\alpha \geq n - j) \\
& + \sum_y \pi(\alpha) \sum_{j=n-1}^{\infty} P_\alpha(X_j = y, \tau_\alpha \geq j) .
\end{aligned}
$$

The second step in the proof is to show that each term in the above decomposition goes to 0 as $n$ goes to infinity. The first term is actually $P_x(\tau_\alpha \geq n)$ and goes to 0 since the chain is Harris recurrent. The third term is the remainder of the convergent series

$$(4.6.2) \qquad \sum_y \pi(\alpha) \sum_{j=1}^{\infty} P_\alpha(X_j = y, \tau_\alpha \geq j) = \sum_y \pi(y) .$$

The middle term is the sum over the $y$'s of the convolution of the two sequences $a_n = |\sum_{k=1}^{n} P_x(X_k \in \alpha, \tau_\alpha = k)K^{n-k}(\alpha, \alpha) - \pi(\alpha)|$ and $b_n = P_\alpha(X_n = y, \tau_\alpha \geq n)$. The sequence $(a_n)$ is converging to 0 since the atom $\alpha$ is ergodic and the series of the $b_n$'s is convergent, as mentioned. An algebraic argument (see Problem 4.41) then implies that (4.6.2) goes to 0 as $n$ goes to $\infty$.    □□

The decomposition (4.6.1) is quite revealing in that it shows the role of the atom $\alpha$ as the generator of a renewal process. Below, we develop an extension which allows us to deal with the general case using *coupling* techniques. (These techniques are also useful in the assessment of convergence for Markov chain Monte Carlo algorithms; see §8.6.1.) Lindvall (1992) provides an introduction to coupling.

The *coupling* principle uses two chains $(X_n)$ and $(X'_n)$ associated with the same kernel, the "coupling" event taking place when they meet in $\alpha$; that is, at the first time $n_0$ such that $X_{n_0} \in \alpha$ and $X'_{n_0} \in \alpha$. After this instant, the probabilistic properties of $(X_n)$ and $(X'_n)$ are identical and if one of the two chains is stationary, there is no longer any dependence on initial conditions for either chain. Therefore, if we can show that the *coupling time* (that is, the time it takes for both chains to meet), is finite for almost every starting point, the ergodicity of the chain follows.

For a recurrent atom $\alpha$ on a denumerable space $\mathcal{X}$, let $\tau_\alpha(k)$ denote the $k$th visit to $\alpha$ ($k = 1, 2, \ldots$), and let $p = (p(1), p(2), \cdots)$ be the distribution of the *excursion time*,

$$S_k = \tau_\alpha(k+1) - \tau_\alpha(k),$$

between two visits to $\alpha$. If $q = (q(0), q(1), \ldots)$ represents the distribution of $\tau_\alpha(1)$ (which depends on the initial condition, $x_0$ or $\mu$), then the distribution of $\tau_\alpha(n+1)$ is given by the convolution product $q \star p^{n\star}$ (that is, the distribution of the sum of $n$ iid rv's distributed from $p$ and of a variable distributed from $q$), since

$$\tau_\alpha(n+1) = S_n + \cdots + S_1 + \tau_\alpha(1).$$

Thus, consider two sequences $(S_i)$ and $(S'_i)$ such that $S_1, S_2, \ldots$ and $S'_1, S'_2, \ldots$ are iid from $p$ with $S_0 \sim q$ and $S'_0 \sim r$. We introduce the indicator functions

$$Z_q(n) = \sum_{j=0}^{n} \mathbb{I}_{S_1 + \cdots + S_j = n} \quad \text{and} \quad Z_r(n) = \sum_{j=0}^{n} \mathbb{I}_{S'_1 + \cdots + S'_j = n},$$

which correspond to the events that the chains $(X_n)$ and $(X'_n)$ visit $\alpha$ at time $n$. The *coupling time* is then given by

$$T_{qr} = \min \{j; Z_q(j) = Z_r(j) = 1\} ,$$

which satisfies the following lemma, whose proof can be found in Problem 4.47.

**Lemma 4.6.3** *If the mean excursion time satisfies*

$$m_p = \sum_{n=0}^{\infty} np(n) < \infty$$

*and if $p$ is aperiodic (the g.c.d. of the support of $p$ is 1), then the coupling time $T_{pq}$ is almost surely finite, that is,*

$$P(T_{pq} < \infty) = 1 ,$$

*for every $q$.*

If $p$ is aperiodic with finite mean $m_p$, this implies that $Z_p$ satisfies

$$(4.6.3) \qquad \lim_{n \to \infty} |P(Z_q(n) = 1) - m_q^{-1}| = 0 ,$$

as shown in Problem 4.44. The probability of visiting $\alpha$ at time $n$ is thus asymptotically independent of the initial distribution and this result implies that Proposition 4.6.2 holds without imposing constraints in the discrete case.

**Theorem 4.6.4** *For a positive recurrent aperiodic Markov chain on a countable space, for every initial state $x$,*

$$\lim_{n \to \infty} \|K^n(x, \cdot) - \pi\|_{TV} = 0 .$$

*Proof.* Since $(X_n)$ is positive recurrent, $\mathbb{E}_\alpha[\tau_\alpha]$ is finite by Theorem 4.5.3. Therefore, $m_p$ is finite, (4.6.3) holds, and every atom is ergodic. The result follows from Proposition 4.6.2. □□

For general state-spaces $\mathcal{X}$, Harris recurrence is nonetheless necessary in the derivation of the convergence of $K^n$ to $\pi$. (Note that another characterization of Harris recurrence is the convergence of $\|K_x^n - \pi\|_{TV}$ to 0 for *every* value $x$, instead of almost every value.)

**Theorem 4.6.5** *If $(X_n)$ is Harris positive and aperiodic, then*

$$\lim_{n \to \infty} \left\| \int K^n(x, \cdot)\mu(dx) - \pi \right\|_{TV} = 0$$

*for every initial distribution $\mu$.*

This result follows from an extension of the denumerable case to strongly aperiodic Harris positive chains by splitting, since these chains always allow for small sets (see §4.3.3), based on an equivalent to the "first entry – last exit" formula (4.6.1). It is then possible to move to arbitrary chains by the following result.

**Proposition 4.6.6** *If $\pi$ is an invariant distribution for $P$, then*

$$\left\| \int K^n(x,\cdot)\mu(dx) - \pi \right\|_{TV}$$

*is decreasing in $n$.*

*Proof.* First, note the equivalent definition of the norm (Problem 4.56)

$$(4.6.4) \qquad \|\mu\|_{TV} = \frac{1}{2} \sup_{|h|\leq 1} \left| \int h(x)\mu(dx) \right|.$$

We then have

$$2 \left\| \int K^{n+1}(x,\cdot)\mu(dx) - \pi \right\|_{TV}$$

$$= \sup_{|h|\leq 1} \left| \int h(y)K^{n+1}(x,dy)\mu(dx) - \int h(y)\pi(dy) \right|$$

$$= \sup_{|h|\leq 1} \left| \int h(y) \int K^n(x,dw)K(w,dy)\mu(dx) \right.$$

$$\left. - \int h(y) \int K(w,dy)\pi(dw) \right|,$$

since, by definition, $K^{n+1}(x,dy) = \int K^n(x,dw)K(w,dy)$ and by the invariance of $\pi$, $\pi(dy) = \int K(w,dy)\pi(dw)$. Regrouping terms, we can write,

$$2 \left\| \int K^{n+1}(x,\cdot)\mu(dx) - \pi \right\|_{TV}$$

$$= \sup_{|h|\leq 1} \left| \int \left[ \int h(y)K(w,dy) \right] K^n(x,dw)\mu(dx) \right.$$

$$\left. - \int \left[ \int h(y)K(w,dy) \right] \pi(dw) \right|$$

$$\leq \sup_{|h|\leq 1} \left| \int h(w)K^n(x,dw)\mu(dx) - \int h(w)\pi(dw) \right|,$$

where the inequality follows from the fact that the quantity in square brackets is a function with norm less than 1. Hence, monotonicity of the total variation norm is established. □□

Note that the equivalence (4.6.4) also implies the convergence

$$(4.6.5) \qquad \lim_{n\to\infty} |\mathbb{E}_\mu[h(X_n)] - \mathbb{E}^\pi[h(X)]| = 0$$

for every *bounded function* $h$. This equivalence is, in fact, often taken as the defining condition for convergence of distributions (see, for example, Billingsley 1995, Theorem 25.8). We can, however, conclude (4.6.5) from a slightly weaker set of assumptions, where we do not need the full force of Harris recurrence (see Theorem 4.10.11 for an example).

The extension of (4.6.5) to more general functions $h$ is called *h-ergodicity* by Meyn and Tweedie (1993, pp. 342–344).

**Theorem 4.6.7** *Let* $(X_n)$ *be positive, recurrent, and aperiodic.*

*(a) If* $\mathbb{E}^\pi[|h(X)|] = \infty$, $\mathbb{E}_x[|h(X_n)|] \to \infty$ *for every* $x$.

*(b) If* $\int |h(x)|\pi(dx) < \infty$, *then*

$$(4.6.6) \qquad \lim_{n\to\infty} \sup_{|m(x)|\leq|h(x)|} |\mathbb{E}_y[m(X_n)] - \mathbb{E}^\pi[m(X)]| = 0$$

*on all small sets* $C$ *such that*

$$(4.6.7) \qquad \sup_{y\in C} \mathbb{E}_y\left[\sum_{t=0}^{\tau_C-1} h(X_t)\right] < \infty .$$

Similar conditions appear as necessary conditions for the Central Limit Theorem (see (4.7.1) in Theorem 4.7.5). Condition (4.6.7) relates to a coupling argument, in the sense that the influence of the initial condition vanishes "fast enough," as in the proof of Theorem 4.7.4.

### 4.6.2 Geometric Convergence

The convergence (4.6.6) of the expectation of $h(x)$ at time $n$ to the expectation of $h(x)$ under the stationary distribution $\pi$ somehow ensures the proper behavior of the chain $(X_n)$ whatever the initial value $X_0$ (or its distribution). A more precise description of convergence properties involves the study of *the speed of convergence* of $K^n$ to $\pi$. An evaluation of this speed is important for Markov chain Monte Carlo algorithms in the sense that it relates to stopping rules for these algorithms; minimal convergence speed is also a requirement for the application of the Central Limit Theorem.

To study the speed of convergence more closely, we first introduce an extension of the total variation norm, denoted by $\|\cdot\|_h$, which allows for an upper bound other than 1 on the functions. The generalization is defined by

$$\|\mu\|_h = \sup_{|g|\leq h} \left|\int g(x)\mu(dx)\right| .$$

**Definition 4.6.8** A chain $(X_n)$ is *geometrically* $h$*-ergodic*, with $h \geq 1$ on $\mathcal{X}$, if $(X_n)$ is Harris positive, with stationary distribution $\pi$, satisfies $\mathbb{E}^\pi[h] < \infty$, and if there exists $r_h > 1$ such that

$$(4.6.8) \qquad \sum_{n=1}^{\infty} r_h^n \|K^n(x,\cdot) - \pi\|_h < \infty$$

for every $x \in \mathcal{X}$. The case $h = 1$ corresponds to the *geometric ergodicity* of $(X_n)$.

Geometric $h$-ergodicity means that $\|K^n(x,\cdot) - \pi\|_h$ is decreasing at least at a geometric speed since (4.6.8) implies

$$\|K^n(x,\cdot) - \pi\|_h \leq M r_h^{-n}$$

with

$$M = \sum_{n=1}^{\infty} r_h^n \| K^n(x, \cdot) - \pi \|_h \ .$$

If $(X_n)$ has an atom $\alpha$, (4.6.8) implies that for a real number $r > 1$,

$$\mathbb{E}_x \left[ \sum_{n=1}^{\tau_\alpha} h(X_n) r^n \right] < \infty \quad \text{and} \quad \sum_{n=1}^{\infty} |P_\alpha(X_n \in \alpha) - \pi(\alpha)| r^n < \infty.$$

The series associated with $|P_\alpha(X_n \in \alpha) - \pi(\alpha)| \ r^n$ converges outside of the unit circle if the power series associated with $P_\alpha(\tau_\alpha = n)$ converges for values of $|r|$ strictly larger than 1. (The proof of this result, called *Kendall's Theorem*, is based on the renewal equations established in the proof of Proposition 4.4.7.) This equivalence justifies the following definition.

**Definition 4.6.9** An accessible atom $\alpha$ is *geometrically ergodic* if there exists $r > 1$ such that

$$\sum_{n=1}^{\infty} |K^n(\alpha, \alpha) - \pi(\alpha)| \ r^n < \infty$$

and $\alpha$ is a *Kendall atom* if there exists $\kappa > 1$ such that

$$\mathbb{E}_\alpha[\kappa^{\tau_\alpha}] < \infty \ .$$

If $\alpha$ is a Kendall atom, it is thus geometrically ergodic and ensures geometric ergodicity for $(X_n)$:

**Theorem 4.6.10** *If $(X_n)$ is $\psi$-irreducible, with invariant distribution $\pi$, and if there exists a geometrically ergodic atom $\alpha$, then there exist $r > 1$, $\kappa > 1$, and $R < \infty$ such that, for almost every $x \in \mathcal{X}$,*

$$\sum_{n=1}^{\infty} r^n \| K^n(x, \cdot) - \pi \|_{TV} < R \, \mathbb{E}_x \left[ \kappa^{\tau_\alpha} \right] < \infty \ .$$

**Example 4.6.11 Nongeometric returns to 0.** For a chain on $\mathbb{Z}_+$ with transition matrix $\mathbb{P} = (p_{ij})$ such that

$$p_{0j} = \gamma_j, \quad p_{jj} = \beta_j, \quad p_{j0} = 1 - \beta_j, \quad \sum_j \gamma_j = 1,$$

Meyn and Tweedie (1993, p. 361) consider the return time to 0, $\tau_0$, with mean

$$\mathbb{E}_0[\tau_0] = \sum_j \gamma_j \left\{ (1 - \beta_j) + 2\beta_j(1 - \beta_j) + \cdots \right\}$$

$$= \sum_j \gamma_j \left\{ 1 + (1 - \beta_j)^{-1} \right\} .$$

The state 0 is thus an ergodic atom when all the $\gamma_j$'s are positive (yielding irreducibility) and $\sum_j \gamma_j (1 - \beta_j)^{-1} < \infty$. Now, for $r > 0$,

$$\mathbb{E}_0[r^{\tau_0}] = r \sum_j \gamma_j \mathbb{E}_j[r^{\tau_0 - 1}] = r \sum_j \gamma_j \sum_{k=0}^\infty r^k \beta_j^k (1 - \beta_j) \, .$$

For $r > 1$, if $\beta_j \to 1$ as $j \to \infty$, the series in the above expectation always diverges for $j$ large enough. Thus, the chain is not geometrically ergodic. $\|$

### 4.6.3 Uniform Ergodicity

The property of uniform ergodicity is stronger than geometric ergodicity in the sense that the rate of geometric convergence must be uniform over the whole space. It is used in the Central Limit Theorem given in §4.7.

**Definition 4.6.12** The chain $(X_n)$ is *uniformly ergodic* if

$$\lim_{n \to \infty} \sup_{x \in \mathcal{X}} \|K^n(x, \cdot) - \pi\|_{TV} = 0 \, .$$

Uniform ergodicity can be established through one of the following equivalent properties:

**Theorem 4.6.13** *The following conditions are equivalent:*

*(a)* $(X_n)$ *is uniformly ergodic;*

*(b)* *there exist* $R < \infty$ *and* $r > 1$ *such that*

$$\|K^n(x, \cdot) - \pi\|_{TV} < R r^{-n} \, , \qquad \text{for all } x \in \mathcal{X} \, ;$$

*(c)* $(X_n)$ *is aperiodic and* $\mathcal{X}$ *is a small set;*

*(d)* $(X_n)$ *is aperiodic and there exist a small set* $C$ *and a real* $\kappa > 1$ *such that*

$$\sup_{x \in \mathcal{X}} \mathbb{E}_x[\kappa^{\tau_C}] < \infty \, .$$

If the whole space $\mathcal{X}$ is small, there exist a probability distribution, $\varphi$, on $\mathcal{X}$, and constants $\varepsilon < 1$, $\delta > 0$, and $n$ such that, if $\varphi(A) > \varepsilon$ then

$$\inf_{x \in \mathcal{X}} K^n(x, A) > \delta \, .$$

This property is sometimes called *Doeblin's condition*. This requirement shows the strength of the uniform ergodicity and suggests difficulties about the verification. We will still see examples of Markov chain Monte Carlo algorithms which achieve this superior form of ergodicity (see Example 7.1.23). Note, moreover, that in the finite case, uniform ergodicity can be derived from the smallness of $\mathcal{X}$ since the condition

$$P(X_{n+1} = y | X_n = x) \geq \inf_z p_{zy} = \rho_y \quad \text{for every} \quad x, y \in \mathcal{X},$$

leads to the choice of the minorizing measure $\nu$ as

$$\nu(y) = \frac{\rho_y}{\sum_{z \in \mathcal{X}} \rho_z}$$

as long as $\rho_y > 0$ for every $y \in \mathcal{X}$. (If $(X_n)$ is recurrent and aperiodic, this positivity condition can be attained by a subchain $(Y_m) = (X_{nd})$ for $d$ large enough. See Meyn and Tweedie 1993, Chapter 16, for more details.)

## 4.7 Limit Theorems

Although the notions and results introduced in the previous sections are important in justifying Markov chain Monte Carlo algorithms, in the following chapters we will see that this last section is essential to the processing of these algorithms. In fact, the different convergence results (ergodicity) obtained in §4.6 only deal with the probability measure $P_x^n$ (through different norms), which is somewhat of a "snapshot" of the chain $(X_n)$ at time $n$. So, it determines the probabilistic properties of *average* behavior of the chain at a fixed instant. Such properties, even though they provide justification for the simulation methods, are of lesser importance for the control of convergence of a given simulation, where the properties of the *realization* $(x_n)$ of the chain are the only characteristics that truly matter. (Meyn and Tweedie 1993 call this type of properties "sample path" properties.)

We are thus led back to some basic ideas, previously discussed in a statistical setup by Robert (1994a, Chapters 1 and 9); that is, we must consider the difference between *probabilistic analysis*, which describes the average behavior of samples, and *statistical inference*, which must reason by induction from the observed sample. While probabilistic properties can justify or refute some statistical approaches, this does not contradict the fact that statistical analysis must be done *conditional on the observed sample*. Such a consideration can lead to the Bayesian approach in a statistical setup (or at least to consideration of the *Likelihood Principle*; see, e.g., Berger and Wolpert 1988, or Robert 1994a, Section 1.3). In the setup of Markov chains, a conditional analysis can take advantage of convergence properties of $P_x^n$ to $\pi$ only to verify the convergence, to a quantity of interest, of functions of the observed path of the chain. Indeed, the fact that $\|P_x^n - \pi\|$ is close to 0, or even converges geometrically fast to 0 with speed $\rho^n$ ($0 < \rho < 1$), does not bring direct information about the unique available observation from $P_x^n$, namely $X_n$.

The problems in directly applying the classical convergence theorems (Law of Large Numbers, Law of the Iterated Logarithm, Central Limit Theorem, etc.) to the sample $(X_1, \dots, X_n)$ are both due to the Markovian dependence structure between the observations $X_i$ and to the nonstationarity of the sequence. (Only if $X_0 \sim \pi$, the stationary distribution of the chain, will the chain be stationary. Since this is equivalent to integrating over the initial conditions, it eliminates the need for a conditional analysis. Such an occurrence, especially in Markov chain Monte Carlo, is somewhat rare.[3])

---

[3] Nonetheless, there is considerable research in MCMC theory about *perfect simulation*; that is, ways of starting the algorithm with $X_0 \sim \pi$. See Note 8.6.5.

We therefore assume that the chain is started from a point $X_0$ whose distribution is not the stationary distribution of the chain, and thus we deal with non-stationary chains directly. We begin with a detailed presentation of convergence results equivalent to the Law of Large Numbers, which are often called *ergodic theorems*. We then mention in §4.7.2 various versions of the Central Limit Theorem whose assumptions are usually (and unfortunately) difficult to check.

### 4.7.1 Ergodic Theorems

Given observations $X_1, \ldots, X_n$ of a Markov chain, we now examine the limiting behavior of the partial sums

$$S_n(h) = \frac{1}{n} \sum_{i=1}^{n} h(X_i)$$

when $n$ goes to infinity, getting back to the iid case through renewal when $(X_n)$ has an atom. Consider first the notion of *harmonic functions*, which is related to ergodicity for Harris recurrent Markov chains.

**Definition 4.7.1** A measurable function $h$ is *harmonic* for the chain $(X_n)$ if

$$\mathbb{E}[h(X_{n+1})|x_n] = h(x_n).$$

These functions are *invariant* for the transition kernel (in the functional sense) and they characterize Harris recurrence as follows.

**Proposition 4.7.2** *For a positive Markov chain, if the only bounded harmonic functions are the constant functions, the chain is Harris recurrent.*

*Proof.* First, the probability of an infinite number of returns, $Q(x, A) = P_x(\eta_A = \infty)$, as a function of $x$, $h(x)$, is clearly a harmonic function. This is because

$$\mathbb{E}_y[h(X_1)] = \mathbb{E}_y[P_{X_1}(\eta_A = \infty)] = P_y(\eta_A = \infty),$$

and thus, $Q(x, A)$ is constant (in $x$).

The function $Q(x, A)$ describes a *tail event*, an event whose occurrence does not depend on $X_1, X_2, \ldots, X_m$, for any finite $m$. Such events generally obey a $0 - 1$ law, that is, their probabilities of occurrence are either 0 or 1. However, $0 - 1$ laws are typically established in the independence case, and, unfortunately, extensions to cover Markov chains are beyond our scope. (For example, see the *Hewitt–Savage* $0 - 1$ *Law*, in Billingsley 1995, Section 36.) For the sake of our proof, we will just state that $Q(x, A)$ obeys a $0 - 1$ Law and proceed.

If $\pi$ is the invariant measure and $\pi(A) > 0$, the case $Q(x, A) = 0$ is impossible. To see this, suppose that $Q(x, A) = 0$. It then follows that the chain almost surely visits $A$ only a finite number of times and the average

$$\frac{1}{N} \sum_{i=1}^{N} \mathbb{I}_A(X_i)$$

will not converge to $\pi(A)$, contradicting the Law of Large Numbers (see Theorem 4.7.4). Thus, for any $x$, $Q(x, A) = 1$, establishing that the chain is a Harris chain. □□

Proposition 4.7.2 can be interpreted as a continuity property of the transition functional $Kh(x) = \mathbb{E}_x[h(X_1)]$ in the following sense. By induction, a harmonic function $h$ satisfies $h(x) = \mathbb{E}_x[h(X_n)]$ and by virtue of Theorem 4.6.7, $h(x)$ is almost surely equal to $\mathbb{E}^\pi[h(X)]$; that is, it is constant *almost everywhere*. For Harris recurrent chains, Proposition 4.7.2 states that this implies $h(x)$ is constant *everywhere*. (Feller 1971 (pp. 265–267) develops a related approach to ergodicity, where Harris recurrence is replaced by a regularity constraint on the kernel.)

Proposition 4.7.2 will be most useful in establishing Harris recurrence of some Markov chain Monte Carlo algorithms. Interestingly, the behavior of bounded harmonic functions characterizes Harris recurrence, as the converse of Proposition 4.7.2 is true. We state it without its rather difficult proof (see Meyn and Tweedie 1993, p. 415).

**Lemma 4.7.3** *For Harris recurrent Markov chains, the constants are the only bounded harmonic functions.*

A consequence of Lemma 4.7.3 is that if $(X_n)$ is Harris positive with stationary distribution $\pi$ and if $S_n(h)$ converges $\mu_0$-almost surely ($\mu_0$ a.s.) to

$$\int_{\mathcal{X}} h(x) \, \pi(dx) ,$$

for an initial distribution $\mu_0$, this convergence occurs for every initial distribution $\mu$. Indeed, the convergence probability

$$P_x \left( S_N(h) \to \mathbb{E}^\pi[h] \right)$$

is then harmonic. Once again, this shows that Harris recurrence is a superior type of stability in the sense that *almost sure convergence* is replaced by convergence at every point.

The main result of this section, namely the Law of Large Numbers for Markov chains (which is customarily called the *Ergodic Theorem*), guarantees the convergence of $S_n(h)$.

**Theorem 4.7.4 Ergodic Theorem** *If $(X_n)$ has a $\sigma$-finite invariant measure $\pi$, the following two statements are equivalent:*

*(i) If $f, g \in L^1(\pi)$ with $\int g(x)d\pi(x) \neq 0$, then*

$$\lim_{n \to \infty} \frac{S_n(f)}{S_n(g)} = \frac{\int f(x)d\pi(x)}{\int g(x)d\pi(x)} .$$

*(ii) The Markov chain $(X_n)$ is Harris recurrent.*

*Proof.* If (i) holds, take $f$ to be the indicator function of a set $A$ with finite measure and $g$ an arbitrary function with finite and positive integral. If $\pi(A) > 0$,

$$P_x(X \in A \text{ infinitely often}) = 1$$

for every $x \in \mathcal{X}$, which establishes Harris recurrence.

If (ii) holds, we need only to consider the atomic case by a splitting argument. Let $\alpha$ be an atom and $\tau_\alpha(k)$ be the time of the $(k+1)$th visit to $\alpha$. If $\ell_N$ is the number of visits to $\alpha$ at time $N$, we get the bounds

$$\sum_{j=0}^{\ell_N-1} \sum_{n=\tau_\alpha(j)+1}^{\tau_\alpha(j+1)} f(x_n) \qquad \leq \sum_{k=1}^{N} f(x_k)$$

$$\leq \sum_{j=0}^{\ell_N} \sum_{n=\tau_\alpha(j)+1}^{\tau_\alpha(j+1)} f(x_n) + \sum_{k=1}^{\tau_\alpha(0)} f(x_k) \,.$$

The blocks

$$S_j(f) = \sum_{n=\tau_\alpha(j)+1}^{\tau_\alpha(j+1)} f(x_n)$$

are independent and identically distributed. Therefore,

$$\frac{\sum_{i=1}^{n} f(x_i)}{\sum_{i=1}^{n} g(x_i)} \leq \frac{\ell_N}{\ell_N - 1} \; \frac{(\sum_{j=0}^{\ell_N} S_j(f) + \sum_{k=1}^{\tau_\alpha} f(x_k))/\ell_N}{\sum_{j=0}^{\ell_N-1} s_j(g)/(\ell_N - 1)} \,.$$

The theorem then follows by an application of the strong Law of Large Numbers for iid rv's.                                                □□

An important aspect of Theorem 4.7.4 is that $\pi$ does not need to be a probability measure and, therefore, that there can be some type of strong stability even if the chain is null recurrent. In the setup of a Markov chain Monte Carlo algorithm, this result is sometimes invoked to justify the use of improper posterior measures, although we fail to see the relevance of this kind of argument (see §7.4).

### 4.7.2 Central Limit Theorems

There is a natural progression from the Law of Large Numbers to the Central Limit Theorem. Moreover, the proof of Theorem 4.7.4 suggests that there is a direct extension of the Central Limit Theorem for iid variables. Unfortunately this is not the case, as conditions on the finiteness of the variance explicitly involve the atom $\alpha$ of the split chain. Therefore, we provide alternative conditions for the Central Limit Theorem to apply in different settings. The discrete case can be solved directly, as shown by Problems 4.54 and 4.55.

**Proposition 4.7.5** *If $(X_n)$ is a Harris positive chain with an atom $\alpha$ such that*

$$(4.7.1) \qquad \mathbb{E}_\alpha[\tau_\alpha^2] < \infty, \qquad \mathbb{E}_\alpha\left[\left(\sum_{n=1}^{\tau_\alpha} |h(X_n)|\right)^2\right] < \infty$$

*and*

$$\gamma_h^2 = \pi(\alpha)\,\mathbb{E}_\alpha\left[\left(\sum_{n=1}^{\tau_\alpha} \{h(X_n) - \mathbb{E}^\pi[h]\}\right)^2\right] > 0,$$

*the Central Limit Theorem applies; that is,*

$$\frac{1}{\sqrt{N}} \left( \sum_{n=1}^{N} (h(X_n) - \mathbb{E}^{\pi}[h]) \right) \overset{\mathcal{L}}{\rightsquigarrow} \mathcal{N}(0, \gamma_h^2) .$$

*Proof.* Using the same notation as in the proof of Theorem 4.7.4, if $\overline{h}$ denotes $h - \mathbb{E}^{\pi}[h]$, we get

$$\frac{1}{\sqrt{\ell_N}} \sum_{i=1}^{\ell_N} S_i(\overline{h}) \overset{\mathcal{L}}{\rightsquigarrow} \mathcal{N} \left( 0, \mathbb{E}_\alpha \left[ \sum_{n=1}^{\tau_\alpha} \overline{h}(X_n) \right]^2 \right) ,$$

following from the Central Limit Theorem for the independent variables $S_i(\overline{f})$, while $N/\ell_N$ converges a.s. to $\mathbb{E}_\alpha[S_0(1)] = 1/\pi(\alpha)$. Since

$$\left| \sum_{i=1}^{\ell_{N-1}} S_i(\overline{h}) - \sum_{k=1}^{N} \overline{h}(X_k) \right| \leq S_{\ell_N}(|\overline{h}|)$$

and

$$\lim_{n \to \infty} \frac{1}{n} \sum_{j=1}^{n} S_i(|\overline{h}|)^2 = \mathbb{E}_\alpha[S_0(|\overline{h}|)^2] ,$$

we get

$$\limsup_{N \to \infty} \frac{S_{\ell_N}(|\overline{h}|)}{\sqrt{N}} = 0,$$

and the remainder goes to 0 almost surely.   □□

This result indicates that an extension of the Central Limit Theorem to the nonatomic case will be more delicate than for the Ergodic Theorem: Conditions (4.7.1) are indeed expressed in terms of the split chain $(\check{X}_n)$. (See §8.3.3 for an extension to cases when there exists a small set.) In Note 4.10.1, we present some alternative versions of the Central Limit Theorem involving a drift condition. The following theorem avoids the verification of a drift condition, but rather requires the Markov chain to be reversible.

**Definition 4.7.6** A Markov chain $(X_n)$ is *reversible* if the distribution of $X_{n+1}$ conditionally on $X_{n+2} = x$ is the same as the distribution of $X_{n+1}$ conditionally on $X_n = x$.

Thus, for a reversible chain, the direction of time does not matter in the dynamics of the chain (see Problems 4.58 and 4.59). With the assumption of reversibility, the following Central Limit Theorem directly follows from the strict positivity of $\gamma_g$. This was established by Kipnis and Varadhan (1986) using a proof that is beyond our reach.

**Theorem 4.7.7** *If $(X_n)$ is aperiodic, irreducible, and reversible with invariant distribution $\pi$, the Central Limit Theorem applies when*

$$0 < \gamma_g^2 = \mathbb{E}_\pi[\overline{g}^2(X_0)] + 2 \sum_{k=1}^{\infty} \mathbb{E}_\pi[\overline{g}(X_0)\overline{g}(X_k)] < +\infty.$$

The main point here is that even though reversibility is a very restrictive assumption in general, it is often easy to impose in Markov chain Monte Carlo algorithms by introducing additional simulation steps (see Geyer 1992, Tierney 1994, and Green 1995). See also Theorem 4.10.8 for another version of the Central Limit Theorem, which relies on a "drift condition" (see Note 4.10.1) similar to geometric ergodicity.

## 4.8 Covariance in Markov Chains

In many statistical applications, calculation of variances and covariances are of interest. As with typical iid sampling, an application of Chebychev's inequality shows that the convergence of an average of random variables from a Markov chain can be connected to the behavior of the covariances, with a sufficient condition for convergence in probability being that the covariances go to zero. In this section, we look at the behavior of the covariances more closely and see how they are related to the properties of a Markov chain.

In light of the results of §4.6, we want to only look at ergodic Markov chains; in particular, we assume that the Markov chain is Harris positive and aperiodic. When studying convergence properties, from Lemma 4.6.3 we can then assume, without loss of generality, that the Markov chain is stationary. A key assumption in what follows, one that is reasonable from a statistical view, is that the random variables of the chain have finite variance. Thus, let $(X_n)$ be a stationary ergodic Markov chain with mean 0 and finite variance. The variance of the average of the $X_i$'s is

$$\text{var}\left(\frac{\sum_{i=0}^n X_i}{(n+1)}\right) = \frac{1}{(n+1)^2}\sum_{i=0}^n \text{var}(X_i) + \frac{2}{(n+1)^2}\sum_{i=0}^n\sum_{j=i+1}^n \text{cov}(X_i, X_j)$$

$$(4.8.1) \qquad = \frac{1}{n+1}\text{var}(X_0) + \frac{2}{n+1}\sum_{k=1}^n \frac{n-k+1}{n+1}\text{cov}(X_0, X_k),$$

where the last equality follows from the stationarity of the Markov chain. Clearly, the covariance term in (4.8.1) will go to zero if $\sum_{k=1}^n \text{cov}(X_0, X_k)/n$ goes to zero, and a sufficient condition for this is that $\text{cov}(X_0, X_k)$ converges to 0 (Problem 4.40). Hence, by Chebychev's inequality, for any function $h$ with $\text{var}[h(X)] < \infty$,

$$(4.8.2) \qquad \lim_{n\to\infty}\frac{1}{n}\sum_{i=1}^n h(X_i) = \mathbb{E}_f[h(X)]$$

in probability if $\text{cov}(h(X_1), h(X_k))$ converges to 0.

We now want to look at the conditions needed to ensure that in a stationary Markov chain, $\text{cov}(X_0, X_k)$ converges to 0. First, use the Cauchy–Schwarz inequality to write

$$|\text{cov}(X_0, X_k)| = |\mathbb{E}[X_0 X_k]|$$
$$= |\mathbb{E}[X_0\mathbb{E}(X_k|X_0)]|$$
$$(4.8.3) \qquad \leq [\mathbb{E}(X_0^2)]^{1/2}\{\mathbb{E}[\mathbb{E}(X_k|X_0)]^2\}^{1/2}.$$

Since $\mathbb{E}(X_0^2) = \sigma^2$, $\text{cov}(X_0, X_k)$ will go to zero if $\mathbb{E}[\mathbb{E}(X_k|X_0)]^2$ goes to 0.

**Example 4.8.1 (Continuation of Example 4.2.5)** Consider the $AR(1)$ model

$$(4.8.4) \qquad X_k = \theta X_{k-1} + \epsilon_k, \quad k = 0, \ldots, n,$$

when the $\epsilon_k$'s are iid $\mathcal{N}(0, 1)$, $\theta$ is an unknown parameter satisfying $|\theta| < 1$, and $X_0 \sim \mathcal{N}(0, \sigma^2)$. The $X_k$'s all have marginal normal distributions with mean zero. The variance of $X_k$ satisfies $\text{var}(X_k) = \theta^2 \text{var}(X_{k-1}) + 1$ and, $\text{var}(X_k) = \sigma^2$ for all $k$ provided $\sigma^2 = 1/(1 - \theta^2)$. This is the stationary case in which it can be shown that

$$(4.8.5) \qquad \mathbb{E}(X_k|X_0) = \theta^k X_0$$

and, hence, $\mathbb{E}[\mathbb{E}(X_k|X_0)]^2 = \theta^{2k} \sigma^2$, which goes to zero as long as $|\theta| < 1$. Thus, $\text{var}(\bar{X})$ converges to 0. (See Problem 4.71.)                    ∥

Returning to (4.8.3), let $M$ be a positive constant and write

$$\mathbb{E}[\mathbb{E}(X_k|X_0)]^2 = \mathbb{E}[\mathbb{E}(X_k \mathbb{I}_{X_k > M}|X_0) + \mathbb{E}(X_k \mathbb{I}_{X_k \leq M}|X_0)]^2$$
$$(4.8.6) \qquad \leq 2\mathbb{E}[\mathbb{E}(X_k \mathbb{I}_{X_k > M}|X_0)]^2 + 2\mathbb{E}[\mathbb{E}(X_k \mathbb{I}_{X_k \leq M}|X_0)]^2.$$

Examining the two terms on the right side of (4.8.6), the first term can be made arbitrarily small using the fact that $X_k$ has finite variance, while the second term converges to zero as a consequence of Theorem 4.6.5. We formalize this in the following theorem.

**Theorem 4.8.2** *If the Markov chain $(X_n)$ is positive and aperiodic, with $\text{var}(X_n) < \infty$, then $\text{cov}(X_0, X_k)$ converges to 0.*

*Proof.* Again, from Lemma 4.6.3 we assume that the chain is stationary. First, note that

$$\mathbb{E}[\mathbb{E}(X_k \mathbb{I}_{X_k > M}|X_0)]^2 \leq \mathbb{E}[\mathbb{E}(X_k^2 \mathbb{I}_{X_k > M}|X_0)] = \mathbb{E}(X_k^2 \mathbb{I}_{X_k > M}).$$

By stationarity, the last expectation does not depend on $k$ and since $\text{var}(X_k) < \infty$, the expectation goes to zero as $M \to \infty$. Thus, given $\epsilon > 0$, choose $M_1$ (independent of $k$) so that $\mathbb{E}(X_k^2 \mathbb{I}_{X_k > M_1}) < \epsilon/4$.

For the second term in (4.8.6), we can use dominated convergence to write

$$(4.8.7) \lim_{k \to \infty} \mathbb{E}[\mathbb{E}(X_k \mathbb{I}_{X_k \leq M}|X_0)]^2 = \mathbb{E}\left\{ \lim_{k \to \infty} [\mathbb{E}(X_k \mathbb{I}_{X_k \leq M}|X_0)]^2 \right\},$$

and this last expression will go to zero as long as $\mathbb{E}(X_k \mathbb{I}_{X_k \leq M}|X_0)$ goes to 0. Since $X_k \mathbb{I}_{X_k \leq M}$ is bounded, it follows from (4.6.5) that

$$\lim_{k \to \infty} \mathbb{E}(X_k \mathbb{I}_{X_k \leq M}|X_0) = \mathbb{E}(X_k \mathbb{I}_{X_k \leq M})$$

almost everywhere, which is independent of $k$ since the chain is stationary. Moreover, as $\mathbb{E}(X_k) = 0$, we can choose $M_2 > M_1$, and $k$ sufficiently large so that

$$\mathbb{E}[\mathbb{E}(X_k \mathbb{I}_{X_k \leq M}|X_0)]^2 < \epsilon/4 .$$

Thus, the left side of (4.8.6) can be made arbitrarily small, and the theorem is proved.                                                                      □□

## 4.9 Problems

**4.1** Examine whether a Markov chain chain $(X_t)$ may always be represented by the deterministic transform $X_{t+1} = \psi(X_t, \epsilon_t)$, where $(\epsilon_t)$ is a sequence of iid rv's. (*Hint:* Consider that $\epsilon_t$ can be of infinite dimension.)

**4.2** Show that if $(X_n)$ is a time-homogeneous Markov chain, the transition kernel does not depend on $n$. In particular, if the Markov chain has a finite state-space, the transition matrix is constant.

**4.3** Show that an $ARMA(p, q)$ model, defined by

$$X_n = \sum_{i=1}^{p} \alpha_i X_{n-i} + \sum_{j=1}^{q} \beta_j \varepsilon_{n-j} + \varepsilon_n,$$

does not produce a Markov chain. (*Hint:* Examine the relation with an $AR(q)$ process through the decomposition

$$Z_n = \sum_{i=1}^{p} \alpha_i Z_{n-i} + \varepsilon_n, \qquad Y_n = \sum_{j=1}^{q} \beta_j Z_{n-j} + Z_n,$$

since $(Y_n)$ and $(X_n)$ are then identically distributed.)

**4.4** Show that the resolvent kernel of Definition 4.2.7 is truly a kernel.

**4.5** Show that the properties of the resolvent kernel are preserved if the geometric distribution $\mathcal{Geo}(\epsilon)$ is replaced by a Poisson distribution $\mathcal{P}(\lambda)$ with arbitrary parameter $\lambda$.

**4.6** Show that a stopping time is a stopping rule.

**4.7** Prove Proposition 4.2.8.

**4.8** Derive the strong Markov property from the decomposition

$$\mathbb{E}_\mu[h(X_{\zeta+1}, X_{\zeta+2}, \ldots)|x_\zeta, x_{\zeta-1}, \ldots]$$
$$= \sum_{n=1}^{\infty} \mathbb{E}_\mu[h(X_{n+1}, X_{n+2}, \ldots)|x_n, x_{n-1}, \ldots, \zeta = n] P(\zeta = n|x_n, x_{n-1}, \ldots)$$

and from the weak Markov property.

**4.9** Given the transition matrix

$$\mathbb{P} = \begin{pmatrix} 0.0 & 0.4 & 0.6 & 0.0 & 0.0 \\ 0.65 & 0.0 & 0.35 & 0.0 & 0.0 \\ 0.32 & 0.68 & 0.0 & 0.0 & 0.0 \\ 0.0 & 0.0 & 0.0 & 0.12 & 0.88 \\ 0.0 & 0.0 & 0.0 & 0.56 & 0.44 \end{pmatrix},$$

examine whether the corresponding chain is irreducible and aperiodic.

**4.10** Show that irreducibility in the sense of Definition 4.3.1 coincides with the more intuitive notion that two arbitrary states are connected when the Markov chain has a discrete support.

**4.11** Show that a Markov chain on a finite state-space with transition matrix $\mathbb{P}$ is irreducible if and only if there exists $N \in \mathbb{N}$ such that $\mathbb{P}^N$ has no zero entries. (The matrix is then called *regular*.)

**4.12** (Kemeny and Snell 1960) Show that for a regular matrix $\mathbb{P}$:

(a) The sequence $(\mathbb{P}^n)$ converges to a stochastic matrix $A$.

(b) Each row of $A$ is the same probability vector $\pi$.

(c) All components of $\pi$ are positive.

(d) For every probability vector $\mu$, $\mu\mathbb{P}^n$ converges to $\pi$.

(e) $\pi$ satisfies $\pi = \pi\mathbb{P}$.

(*Note:* See Kemeny and Snell 1960, p. 71 for a full proof.)

**4.13** Show that for the measure $\psi$ given by (4.3.1), the chain $(X_n)$ is irreducible in the sense of Definition 4.3.1. Show that for two measures $\varphi_1$ and $\varphi_2$, such that $(X_n)$ is $\varphi_i$-irreducible, the corresponding $\psi_i$'s given by (4.3.1) are equivalent.

**4.14** Let $Y_1, Y_2, \ldots$ be iid rv's concentrated on $\mathbb{N}_+$ and $Y_0$ be another rv also concentrated on $\mathbb{N}_+$. Define

$$Z_n = \sum_{i=0}^{n} Y_i.$$

(a) Show that $(Z_n)$ is a Markov chain. Is it irreducible?

(b) Define the forward recurrence time as

$$V_n^+ = \inf\{Z_m - n; Z_m > n\}.$$

Show that $(V_n^+)$ is also a Markov chain.

(c) If $V_n^+ = k > 1$, show that $V_{n+1}^+ = k - 1$. If $V_n^+ = 1$, show that a renewal occurs at $n + 1$. (*Hint:* Show that $V_{n+1}^+ \sim Y_i$ in the latter case.)

**4.15** Detail the proof of Proposition 4.3.3. In particular, show that the fact that $K_\epsilon$ includes a Dirac mass does not invalidate the irreducibility. (*Hint:* Establish that

$$\mathbb{E}_x[\eta_A] = \sum_n P_x^n(A) > P_x(\tau_A < \infty)\,,$$

$$\lim_{\epsilon \to 1} K_\epsilon(x, A) > P_x(\tau_A < \infty)\,,$$

$$K_\epsilon(x, A) = (1 - \epsilon)\sum_{i=1}^{\infty} \epsilon^i P^i(x, A) > 0$$

imply that there exists $n$ such that $K^n(x, A) > 0$. See Meyn and Tweedie 1993, p. 87.)

**4.16** Prove Theorem 4.3.4. (*Hint:* Start from (iv).)

**4.17** Show that the multiplicative random walk

$$X_{t+1} = X_t \epsilon_t$$

is not irreducible when $\epsilon_t \sim \mathcal{E}xp(1)$ and $x_0 \in \mathbb{R}$. (*Hint:* Show that it produces two irreducible components.)

**4.18** Show that in the setup of Example 4.3.5, the chain is not irreducible when $\epsilon_n$ is uniform on $[-1, 1]$ and $|\theta| > 1$.

**4.19** In the spirit of Definition 4.4.1, we can define a *uniformly transient set* as a set $A$ for which there exists $M < \infty$ with

$$\mathbb{E}_x[\eta_A] \le M, \qquad \forall x \in A\,.$$

Show that *transient* sets are denumerable unions of *uniformly transient* sets.

**4.20** Show that the split chain defined on $\mathcal{X} \times \{0,1\}$ by the following transition kernel:

$$P(\check{X}_{n+1} \in A \times \{0\}|(x_n, 0))$$

$$= \mathbb{I}_C(x_n) \left\{ \frac{P(X_n, A \cap C) - \epsilon\nu(A \cap C)}{1 - \epsilon}(1 - \epsilon) \right.$$

$$\left. + \frac{P(X_n, A \cap C^c) - \epsilon\nu(A \cap C^c)}{1 - \epsilon} \right\}$$

$$+ \mathbb{I}_{C^c}(X_n) \left\{ P(X_n, A \cap C)(1 - \epsilon) + P(X_n, A \cap C^c) \right\},$$

$$P(\check{X}_{n+1} \in A \times \{1\}|(x_n, 0))$$

$$= \mathbb{I}_C(X_n) \frac{P(X_n, A \cap C) - \epsilon\nu(A \cap C)}{1 - \epsilon} \epsilon + \mathbb{I}_{C^c}(X_n) P(X_n, A \cap C)\epsilon,$$

$$P(\check{X}_{n+1} \in A \times \{0\}|(x_n, 1)) = \nu(A \cap C)(1 - \epsilon) + \nu(A \cap C^c),$$

$$P(\check{X}_{n+1} \in A \times \{1\}|(x_n, 1)) = \nu(A \cap C)\epsilon,$$

satisfies

$$P(\check{X}_{n+1} \in A \times \{1\}|\check{x}_n) = \epsilon\nu(A \cap C),$$

$$P(\check{X}_{n+1} \in A \times \{0\}|\check{x}_n) = \nu(A \cap C^c) + (1 - \epsilon)\nu(A \cap C)$$

for every $\check{x}_n \in C \times \{1\}$. Deduce that $C \times \{1\}$ is an atom of the split chain $(\check{X}_n)$.

**4.21** If $C$ is a small set and $B \subset C$, under which conditions on $B$ is $B$ a small set?

**4.22** If $C$ is a small set and $D = \{x; P^m(x, D) > \delta\}$, show that $D$ is a small set for $\delta$ small enough. (*Hint:* Use the Chapman–Kolmogorov equations.)

**4.23** Show that the period $d$ given in Definition 4.3.11 is independent of the selected small set $C$ and that this number characterizes the chain $(X_n)$.

**4.24** Given the transition matrix

$$\mathbb{P} = \begin{pmatrix} 0.0 & 0.4 & 0.6 & 0.0 & 0.0 \\ 0.6 & 0.0 & .35 & 0.0 & 0.05 \\ 0.32 & .68 & 0.0 & 0.0 & 0.0 \\ 0.0 & 0.0 & 0.12 & 0.0 & 0.88 \\ 0.14 & 0.3 & 0.0 & 0.56 & 0.0 \end{pmatrix},$$

show that the corresponding chain is aperiodic, despite the null diagonal.

The *random walk* (Examples 4.5.6 and 4.5.5) is a useful probability model and has been given many colorful interpretations. (A popular one is the description of an inebriated individual whose progress along a street is composed of independent steps in random directions, and a question of interest is to describe where the individual will end up.) Here, we look at a simple version to illustrate a number of the Markov chain concepts.

**4.25** A random walk on the non-negative integers $I = \{0, 1, 2, \ldots\}$ can be constructed in the following way. For $0 < p < 1$, let $Y_0, Y_1, \ldots$ be iid random variables with $P(Y_i = 1) = p$ and $P(Y_i = -1) = 1 - p$, and $X_k = \sum_{i=0}^{k} Y_i$. Then, $(X_n)$ is a Markov chain with transition probabilities

$$P(X_{i+1} = j + 1|X_i = j) = p, \qquad P(X_{i+1} = j - 1|X_i = j) = 1 - p,$$

but we make the exception that $P(X_{i+1} = 1|X_i = 0) = p$ and $P(X_{i+1} = 0|X_i = 0) = 1 - p$.

(a) Show that $(X_n)$ is a Markov chain.

(b) Show that $(X_n)$ is also irreducible.

(c) Show that the invariant distribution of the chain is given by

$$a_k = \left(\frac{p}{1-p}\right)^k a_0, \quad k = 1, 2, \ldots,$$

where $a_k$ is the probability that the chain is at $k$ and $a_0$ is arbitrary. For what values of $p$ and $a_0$ is this a probability distribution?

(d) If $\sum a_k < \infty$, show that the invariant distribution is also the stationary distribution of the chain; that is, the chain is ergodic.

**4.26** If $(X_t)$ is a random walk, $X_{t+1} = X_t + \epsilon_t$, such that $\epsilon_t$ has a moment generating function $f$, defined in a neighborhood of 0, give the moment generating function of $X_{t+1}$, $g_{t+1}$ in terms of $g_t$ and $f$, when $X_0 = 0$. Deduce that there is no invariant distribution with a moment generating function in this case.

Although the property of aperiodicity is important, it is probably less important than properties such as recurrence and irreducibility. It is interesting that Feller (1971, Section XV.5) notes that the classification into periodic and aperiodic states "represents a nuisance." However, this is less true when the random variables are continuous.

**4.27** (Continuation of Problem 4.25)

(a) Using the definition of periodic given here, show that the random walk of Problem 4.25 is periodic with period 2.

(b) Suppose that we modify the random walk of Problem 4.25 by letting $0 < p + q < 1$ and redefining

$$Y_i = \begin{cases} 1 & \text{with probability } p \\ 0 & \text{with probability } 1 - p - q \\ -1 & \text{with probability } q. \end{cases}$$

Show that this random walk is irreducible and aperiodic. Find the invariant distribution, and the conditions on $p$ and $q$ for which the Markov chain is positive recurrent.

**4.28** (Continuation of Problem 4.25) A Markov chain that is not positive recurrent may be either null recurrent or *transient*. In either of these latter two cases, the invariant distribution, if it exists, is not a probability distribution (it does not have a finite integral), and the difference is one of expected return times. For any integer $j$, the probability of returning to $j$ in $k$ steps is $p_{jj}^{(k)} = P(X_{i+k} = j | X_i = j)$, and the expected return time is thus $m_{jj} = \sum_{k=1}^{\infty} k p_{jj}^{(k)}$.

(a) Show that since the Markov chain is irreducible, either $m_{jj} = \infty$ for all $j$ or for no $j$.

(b) An irreducible Markov chain is *transient* if $m_{jj} = \infty$, otherwise it is recurrent. Show that the random walk is positive recurrent if $p < 1/2$ and transient if $p > 1/2$.

(c) Show that the random walk is null recurrent if $p = 1/2$. This is the interesting case where each state will be visited infinitely often, but the expected return time is infinite.

**4.29** Explain why the resolvant chain is necessarily strongly irreducible.

**4.30** Consider a random walk on $\mathbb{R}_+$, defined as

$$X_{n+1} = (X_n + \epsilon)^+.$$

Show that the sets $(0, c)$ are small, provided $P(\epsilon < 0) > 0$.

**4.31** Consider a random walk on $\mathbb{Z}$ with transition probabilities

$$P(Z_t = n + 1 | Z_{t-1} = n) = 1 - P(Z_t = n - 1 | Z_{t-1} = n) \propto n^{-\alpha}$$

and

$$P(Z_t = 1 | Z_{t-1} = 0) = 1 - P(Z_t = -1 | Z_{t-1} = 0) = 1/2 .$$

Study the recurrence properties of the chain in terms of $\alpha$.

**4.32** Establish (i) and (ii) of Theorem 4.4.4.

(a) Use

$$K^n(x, A) \geq K^r(x, \alpha) K^s(\alpha, \alpha) K^t(\alpha, A)$$

for $r + s + t = n$ and $r$ and $s$ such that

$$K^r(x, \alpha) > 0 \qquad \text{and} \qquad K^s(\alpha, A) > 0$$

to derive from the Chapman–Kolmogorov equations that $\mathbb{E}_x[\eta_A] = \infty$ when $\mathbb{E}_\alpha[\eta_\alpha] = \infty$.

(b) To show (ii):

    (a) Establish that transience is equivalent to $P_\alpha(\tau_\alpha < \infty) < 1$.

    (b) Deduce that $\mathbb{E}_x[\eta_\alpha] < \infty$ by using a generating function as in the proof of Proposition 4.4.7.

    (c) Show that the covering of $\mathcal{X}$ is made of the

$$\overline{\alpha}_j = \{y; \sum_{n=1}^{j} K^n(y, \alpha) > j^{-1}\}.$$

**4.33** Show that if two states $x$ and $y$ are connected (that is, if there exist $m$ and $k$ such that $P_{xy}^m > 0$ and $P_{yx}^k > 0$), $x$ is transient if and only if $y$ is transient.

**4.34** Referring to Definition 4.4.8, show that if $P(\eta_A = \infty) \neq 0$ then $\mathbb{E}_x[\eta_A] = \infty$, but that $P(\eta_A = \infty) = 0$ does not imply $\mathbb{E}_x[\eta_A] < \infty$.

**4.35** In connection with Example 4.5.8, show that the chain is null recurrent when $f'(1) = 1$.

**4.36** Establish the equality (4.4.1).

**4.37** Consider the simple Markov chain $(X_n)$, where each $X_i$ takes on the values $-1$ and $1$ with $P(X_{i+1} = 1 | X_i = -1) = 1$, $P(X_{i+1} = -1 | X_i = 1) = 1$, and $P(X_0 = 1) = 1/2$.

(a) Show that this is a stationary Markov chain.

(b) Show that $\text{cov}(X_0, X_k)$ does not go to zero.

(c) The Markov chain is not strictly positive. Verify this by exhibiting a set that has positive unconditional probability but zero conditional probability.

(*Note:* The phenomena seen here is similar to what Seidenfeld and Wasserman 1993 call a *dilation*.)

**4.38** In the setup of Example 4.2.4, find the stationary distribution associated with the proposed transition when $\pi_i = \pi_j$ and in general.

**4.39** Show the decomposition of the "first entrance–last exit" equation (4.6.1).

**4.40** If $(a_n)$ is a sequence of real numbers converging to $a$, and if $b_n = (a_1 + \cdots + a_n)/n$, then show that

$$\lim_n b_n = a \ .$$

(*Note:* The sum $(1/n) \sum_{i=1}^n a_i$ is called a *Cesàro average*; see Billingsley 1995, Section A30.)

**4.41** Consider a sequence $(a_n)$ of positive numbers which is converging to $a^*$ and a convergent series with running term $b_n$. Show that the convolution

$$\sum_{j=1}^{n-1} a_j b_{n-j}$$

converges to

$$a^* \sum_{j=1}^{\infty} b_j .$$

(*Hint:* Use the Dominated Convergence Theorem.)

**4.42** Verify (4.6.4), namely that $\|\mu\|_{TV} = (1/2)\sup_{|h| \leq 1} \left| \int h(x)\mu(dx) \right|$.

**4.43** Show that if $(X_n)$ and $(X'_n)$ are coupled at time $N_0$ and if $X_0 \sim \pi$, then $X'_n \sim \pi$ for $n > N_0$ for any initial distribution of $X'_0$.

**4.44** Using the notation of §4.6.1, set

$$u(n) = \sum_{j=0}^{\infty} p^{j\star}(n)$$

with $p^{j\star}$ the distribution of the sum $S_1 + \cdots + S_j$, $p^{0\star}$ the Dirac mass at $0$, and

$$Z(n) = \mathbb{I}_{\exists j; S_j = n}.$$

(a) Show that $P_q(Z(n) = 1) = q \star u(n)$.

(b) Show that

$$|q \star u(n) - p \star u(n)| \leq 2\,P(T_{pq} > n).$$

(This bound is often called *Orey's inequality*, from Orey 1971.)

(c) Show that if $m_p$ is finite,

$$e(n) = \frac{\sum_{n+1}^{\infty} p(j)}{m_p}$$

is the invariant distribution of the renewal process in the sense that $P_e(Z(n) = 1) = 1/m_p$ for every $n$.

(d) Deduce from Lemma 4.6.3 that

$$\lim_n \left| q \star u(n) - \frac{1}{m_p} \right| = 0$$

when the mean renewal time is finite.

**4.45** Consider the so-called "forward recurrence time" process $V_n^+$, which is a Markov chain on $\mathbb{N}_+$ with transition probabilities

$$\begin{aligned} P(1, j) &= p(j), & j \geq 1, \\ P(j, j-1) &= 1, & j > 1, \end{aligned}$$

where $p$ is an arbitrary probability distribution on $\mathbb{N}_+$. (See Problem 4.14.)

(a) Show that $(V_n^+)$ is recurrent.

(b) Show that
$$P(V_n^+ = j) = p(j + n - 1).$$

(c) Deduce that the invariant measure satisfies
$$\pi(j) = \sum_{n \geq j} p(n)$$

and show it is finite if and only if
$$m_p = \sum_n np(n) < \infty.$$

**4.46** (Continuation of Problem 4.45) Consider two independent forward recurrence time processes $(V_n^+)$ and $(W_n^+)$ with the same generating probability distribution $p$.

(a) Give the transition probabilities of the joint process $V_n^* = (V_n^+, W_n^+)$.

(b) Show that $(V_n^*)$ is irreducible when $p$ is aperiodic. (*Hint:* Consider $r$ and $s$ such that g.c.d.$(r, s) = 1$ with $p(r) > 0$, $p(q) > 0$, and show that if $nr - ms = 1$ and $i \geq j$, then
$$P_{(i,j)}(V_{j+(i-j)nr}^* = (1, 1)) > 0 .)$$

(c) Show that $\pi^* = \pi \times \pi$, with $\pi$ defined in Problem 4.45 is invariant and, therefore, that $(V_n^*)$ is positive Harris recurrent when $m_p < \infty$.

**4.47** (Continuation of Problem 4.46) Consider $V_n^*$ defined in Problem 4.46 associated with $(S_n, S_n')$ and define $\tau_{1,1} = \min\{n; V_n^* = (1, 1)\}$.

(a) Show that $T_{pq} = \tau_{1,1} + 1$.

(b) Use (c) in Problem 4.46 to show Lemma 4.6.3.

**4.48** In the notation of §4.6.1, if $a$, $b$, and $p$ are distributions such that $m_a$, $m_b$, and $m_p$ are finite, show that
$$\sum |a \star u(n) - b \star u(n)| < \infty.$$

**4.49** Show that if two Markov chains $(S_k)$ and $(S_k')$ are coupled as in §4.6.1, the distribution of $S_{T_{pq}}$ is not $p$, although $S_k' \sim p$ for every $k$.

**4.50** (Kemeny and Snell 1960) Establish (directly) the Law of Large Numbers for a finite irreducible state-space chain $(X_n)$ and for $h(x_n) = \mathbb{I}_j(x_n)$, if $j$ is a possible state of the chain; that is,
$$\frac{1}{N} \sum_{n=1}^{N} \mathbb{I}_j(x_n) \longrightarrow \pi_j$$

where $\pi = (\pi_1, \ldots, \pi_j, \ldots)$ is the stationary distribution.

**4.51** (Kemeny and Snell 1960) Let $\mathbb{P}$ be a regular transition matrix (see Problem 4.11) with limiting (stationary) matrix $A$; that is, each column of $A$ is equal to the stationary distribution.

(a) Show that the so-called *fundamental matrix*
$$\mathbb{Z} = (I - (\mathbb{P} - A))^{-1},$$

where $I$ denotes the identity matrix, exists.

(b) Show that

$$\mathbb{Z} = I + \sum_{n=1}^{\infty} (\mathbb{P}^n - A).$$

(c) Show that $\mathbb{Z}$ satisfies

$$\pi\mathbb{Z} = \pi \qquad \text{and} \qquad \mathbb{P}\mathbb{Z} = \mathbb{Z}\mathbb{P} ,$$

where $\pi$ denotes a row of $A$ (this is the stationary distribution).

**4.52** (Continuation of Problem 4.51) Let $N_j(n)$ be the number of times the chain is in state $j$ in the first $n$ instants.

(a) Show that for every initial distribution $\mu$,

$$\lim_{n \to \infty} \mathbb{E}_\mu[N_j(n)] - n\pi_j = \mu(\mathbb{Z} - A).$$

(*Note:* This convergence shows the strong stability of a recurrent chain since each term in the difference goes to infinity.)

(b) Show that for every pair of initial distributions, $(\mu, \nu)$,

$$\lim_{n \to \infty} \mathbb{E}_\mu[N_j(n)] - \mathbb{E}_\nu[N_j(n)] = (\mu - \nu)\mathbb{Z}.$$

(c) Deduce that for every pair of states, $(u, v)$,

$$\lim_{n \to \infty} \mathbb{E}_u[N_j(n)] - \mathbb{E}_v[N_j(n)] = z_{uj} - z_{vj},$$

which is called the *divergence* $\mathrm{div}_j(u, v)$.

**4.53** (Continuation of Problem 4.51) Let $f_j$ denote the number of steps before entering state $j$.

(a) Show that for every state $i$, $\mathbb{E}_i[f_j]$ is finite.

(b) Show that the matrix $M$ with entries $m_{ij} = \mathbb{E}_i[f_j]$ satisfies

$$M = \mathbb{P}(M - M_d) + E,$$

where $M_d$ is the diagonal matrix with same diagonal as $M$ and $E$ is the matrix made of 1's.

(c) Deduce that $m_{ii} = 1/\pi_i$.

(d) Show that $\pi M$ is the vector of the $z_{ii}/\pi_i$'s.

(e) Show that for every pair of initial distributions, $(\mu, \nu)$,

$$\mathbb{E}_\mu[f_i] - \mathbb{E}_\nu[f_i] = (\mu - \nu)(I - \mathbb{Z})D,$$

where $D$ is the diagonal matrix $\mathrm{diag}(1/\pi_i)$.

**4.54** (Continuation of Problem 4.51) If $h$ is a function taking values on a finite state-space $\{1, \ldots, r\}$, with $h(i) = h_i$, and if $(X_n)$ is an irreducible Markov chain, show that

$$\lim_{n \to \infty} \frac{1}{n} \mathrm{var} \left( \sum_{t=1}^{n} h(x_t) \right) = \sum_{i,j} h_i c_{ij} h_j,$$

where $c_i j = \pi_i z_{ij} + \pi_j z_{ji} - \pi_i \delta_{ij} - \pi_i \pi_j$ and $\delta_{ij}$ is Kronecker's 0–1 function.

**4.55** (Continuation of Problem 4.53) For the two-state transition matrix

$$\mathbb{P} = \begin{pmatrix} 1 - \alpha & \alpha \\ \beta & 1 - \beta \end{pmatrix},$$

show that the stationary distribution is

$$\pi = \left( \frac{\beta}{\alpha + \beta}, \frac{\alpha}{\alpha + \beta} \right),$$

that the mean first passage matrix is

$$M = \begin{pmatrix} \dfrac{\alpha + \beta}{\beta} & \dfrac{1}{\alpha} \\ \dfrac{1}{\beta} & \dfrac{\alpha + \beta}{\alpha} \end{pmatrix},$$

and that the limiting variance for the number of times in state $j$ is

$$\frac{\alpha\beta(2 - \alpha - \beta)}{(\alpha + \beta)^3},$$

for $j = 1, 2$.

**4.56** Show that (4.6.4) is compatible with the definition of the total variation norm. Establish the relation with the alternative definition

$$\|\mu\|_{TV} = \sup_A \mu(A) - \inf_A \mu(A).$$

**4.57** Show that a finite state-space chain is always geometrically ergodic.

**4.58** (Kemeny and Snell 1960) Given a finite state-space Markov chain, with transition matrix $\mathbb{P}$, define a second transition matrix by

$$p_{ij}(n) = \frac{P_\mu(X_{n-1} = j)P(X_n = i | X_{n-1} = j)}{P_\mu(X_n = j)}.$$

(a) Show that $p_{ij}(n)$ does not depend on $n$ if the chain is stationary (i.e. if $\mu = \pi$).

(b) Explain why, in this case, the chain with transition matrix $\tilde{\mathbb{P}}$ made of the probabilities

$$\tilde{p}_{ij} = \frac{\pi_j p_{ji}}{\pi_i}$$

is called the *reverse* Markov chain.

(c) Show that the limiting variance $C$ is the same for both chains.

**4.59** (Continuation of Problem 4.58) A Markov chain is *reversible* if $\tilde{\mathbb{P}} = \mathbb{P}$. Show that every two-state ergodic chain is reversible and that an ergodic chain with symmetric transition matrix is reversible. Examine whether the matrix

$$\mathbb{P} = \begin{pmatrix} 0 & 0 & 1 & 0 & 0 \\ 0.5 & 0 & 0.5 & 0 & 0 \\ 0 & 0.5 & 0 & 0.5 & 0 \\ 0 & 0 & 0.5 & 0 & 0.5 \\ 0 & 0 & 1 & 0 & 0 \end{pmatrix}$$

is reversible. (*Hint:* Show that $\pi = (0.1, 0.2, 0.4, 0.2, 0.1)$.)

**4.60** (Continuation of Problem 4.59) Show that an ergodic random walk on a finite state-space is reversible.

**4.61** (Kemeny and Snell 1960) A Markov chain $(X_n)$ is *lumpable* with respect to a nontrivial partition of the state-space, $(A_1, \ldots, A_k)$, if, for every initial distribution $\mu$, the process

$$Z_n = \sum_{i=1}^{k} i \mathbb{I}_{A_i}(X_n)$$

is a Markov chain with transition probabilities independent of $\mu$.

(a) Show that a necessary and sufficient condition for lumpability is that

$$p_{uA_j} = \sum_{v \in A_j} p_{uv}$$

is constant (in $n$) on $A_i$ for every $i$.

(b) Examine whether

$$\mathbb{P} = \begin{pmatrix} 1 & 0 & 0 & 0 & 0 \\ 0 & 1 & 0 & 0 & 0 \\ 0.5 & 0 & 0 & 0.5 & 0 \\ 0 & 0 & 0.5 & 0 & 0.5 \\ 0 & 0.5 & 0 & 0.5 & 0 \end{pmatrix}$$

is lumpable for $A_1 = \{1, 2\}$, $A_2 = \{3, 4\}$, and $A_3 = \{5\}$.

**4.62** Consider the random walk on $\mathbb{R}^+$, $X_{n+1} = (X_n + \epsilon_n)^+$, with $\mathbb{E}[\epsilon_n] = \beta$.

(a) Establish Lemma 4.10.1. (*Hint:* Consider an alternative $V$ to $V^*$ and show by recurrence that

$$V(x) \geq \int_C K(x, y) V(y) dy + \int_{C^c} K(x, y) V(y) dy$$
$$\geq \cdots \geq V^*(x) .)$$

(b) Establish Theorem 4.10.3 by assuming that there exists $x^*$ such that $P_{x^*}(\tau_C < \infty) < 1$, choosing $M$ such that $M \geq V(x^*)/[1 - P_{x^*}(\tau_C < \infty)]$ and establishing that $V(x^*) \geq M[1 - P_{x^*}(\tau_C < \infty)]$.

**4.63** Show that

(a) a time-homogeneous Markov chain $(X_n)$ is stationary if the initial distribution is the invariant distribution.

(b) the invariant distribution of a stationary Markov chain is also the marginal distribution of any $X_n$.

**4.64** Show that if an irreducible Markov chain has a $\sigma$-finite invariant measure, this measure is unique up to a multiplicative factor. (*Hint:* Use Theorem 4.7.4.)

**4.65** (Kemeny and Snell 1960) Show that for an aperiodic irreducible Markov chain with finite state-space and with transition matrix $\mathbb{P}$, there always exists a stationary probability distribution which satisfies

$$\pi = \pi \mathbb{P}.$$

(a) Show that if $\beta < 0$, the random walk is recurrent. (*Hint:* Use the drift function $V(x) = x$ as in Theorem 4.10.2.)

(b) Show that if $\beta = 0$ and $\mathrm{var}(\epsilon_n) < \infty$, $(X_n)$ is recurrent. (*Hint:* Use $V(x) = \log(1 + x)$ for $x > R$ and $V(x) = 0$, otherwise, for an adequate bound $R$.)

(c) Show that if $\beta > 0$, the random walk is transient.

**4.66** Show that if there exist a finite potential function $V$ and a small set $C$ such that $V$ is bounded on $C$ and satisfies (4.10.3), the corresponding chain is Harris positive.

**4.67** Show that the random walk on $\mathbb{Z}$ is transient when $\mathbb{E}[W_n] \neq 0$. (*Hint:* Use $V(x) = 1 - \varrho^x$ for $x > 0$ and $0$ otherwise when $\mathbb{E}[W_n] > 0$.)

**4.68** Show that the chains defined by the kernels (4.10.9) and (4.10.11) are either both recurrent or both transient.

**4.69** We saw in §4.6.2 that a stationary Markov chain is *geometrically ergodic* if there is a non negative real-valued function $M$ and a constant $r < 1$ such that for any $A \in \mathcal{X}$,

$$|P(X_n \in A | X_0 \in B) - P(X_n \in A)| \leq M(x) r^n.$$

Prove that the following central limit theorem (due to Chan and Geyer 1994) can be considered a corollary to Theorem 4.10.13 (see Note 4.10.4):

**Corollary 4.9.1** *Suppose that the stationary Markov chain* $X_0, X_1, X_2, \ldots$ *is geometrically ergodic with* $M^* = \int |M(x)| f(x) dx < \infty$ *and satisfies the moment conditions of Theorem 4.10.13. Then,*

$$\sigma^2 = \lim_{n \to \infty} n \, \mathrm{var} \bar{X}_n < \infty$$

*and if* $\sigma^2 > 0$, $\sqrt{n} \bar{X}_n / \sigma$ *tends in law to* $\mathcal{N}(0, \sigma^2)$.

(*Hint:* Integrate (with respect to $f$) both sides of the definition of geometric ergodicity to conclude that the chain has exponentially fast $\alpha$-mixing, and apply Theorem 4.10.13.)

**4.70** Suppose that $X_0, X_1, \ldots, X_n$ have a common mean $\xi$ and variance $\sigma^2$ and that $\mathrm{cov}(X_i, X_j) = \rho_{j-i}$. For estimating $\xi$, show that:

(a) $\bar{X}$ may not be consistent if $\rho_{j-i} = \rho \neq 0$ for all $i \neq j$. (*Hint:* Note that $\mathrm{var}(\bar{X}) > 0$ for all sufficiently large $n$ requires $\rho \geq 0$ and determine the distribution of $\bar{X}$ in the multivariate normal case.)

(b) $\bar{X}$ is consistent if $|\rho_{j-i}| \leq M \gamma^{j-i}$ with $|\gamma| < 1$.

**4.71** For the situation of Example 4.8.1:

(a) Prove that the sequence $(X_n)$ is stationary provided $\sigma^2 = 1/(1 - \beta^2)$.

(b) Show that $\mathbb{E}(X_k | x_0) = \beta^k x_0$. (*Hint:* Consider $\mathbb{E}[(X_k - \beta X_{k-1}) | x_0]$.)

(c) Show that $\mathrm{cov}(X_0, X_k) = \beta^k / (1 - \beta^2)$.

**4.72** Show that in Note 4.10.3, if for some $k$ the function $f_k(x | x_0)$ is continuous in $x$ for each $x_0$, or if it is discrete, then Condition (i) will imply Condition (ii).

**4.73** Under the conditions of Theorem 4.8.2, it follows that $\mathbb{E}[\mathbb{E}(X_k | X_0)]^2 \to 0$. There are some other interesting properties of this sequence.

(a) Show that

$$\mathrm{var}[\mathbb{E}(X_k | X_0)] = \mathbb{E}[\mathbb{E}(X_k | X_0)]^2,$$
$$\mathrm{var}[\mathbb{E}(X_k | X_0)] \geq \mathrm{var}[\mathbb{E}(X_{k+1} | X_0)] .$$

(*Hint:* Write $f_{k+1}(y | x) = \int f_k(y | x') f(x' | x) dx'$ and use Fubini and Jensen.)

(b) Show that

$$\mathbb{E}[\mathrm{var}(X_k|X_0)] \leq \mathbb{E}[\mathrm{var}(X_{k+1}|X_0)]$$

and that

$$\lim_{k\to\infty} \mathbb{E}[\mathrm{var}(X_k|X_0)] = \sigma^2 .$$

## 4.10 Notes

### 4.10.1 Drift Conditions

Besides atoms and small sets, Meyn and Tweedie (1993) rely on another tool to check or establish various stability results, namely *drift criteria* which can be traced back to Lyapunov. Given a function $V$ on $\mathcal{X}$, the *drift of $V$* is defined by

$$\Delta V(x) = \int V(y) \, P(x, dy) - V(x) .$$

(Functions $V$ appearing in this setting are often referred to as *potentials*; see Norris 1997.) This notion is also used in the following chapters to verify the convergence properties of some MCMC algorithms (see, e.g., Theorem 6.3.6 or Mengersen and Tweedie 1996).

The following lemma is instrumental in deriving drift conditions for the transience or the recurrence of a chain $(X_n)$.

**Lemma 4.10.1** *If $C \in \mathcal{B}(\mathcal{X})$, the smallest positive function which satisfies the conditions*

(4.10.1)        $\Delta V(x) \leq 0 \quad \text{if} \quad x \notin C, \quad V(x) \geq 1 \quad \text{if} \quad x \in C$

*is given by*

$$V^*(x) = P_x(\sigma_C < \infty) ,$$

*when $\sigma_C$ denotes*

$$\sigma_C = \inf\{n \geq 0; x_n \in C\} .$$

Note that, if $x \notin C$, $\sigma_C = \tau_C$, while $\sigma_C = 0$ on C. We then have the following necessary and sufficient condition.

**Theorem 4.10.2** *The $\psi$-irreducible chain $(X_n)$ is transient if and only if there exist a bounded positive function $V$ and a real number $r \geq 0$ such that for every $x$ for which $V(x) > r$, we have*

(4.10.2)                                $\Delta V(x) > 0 .$

*Proof.* If $C = \{x; V(x) \leq r\}$ and $M$ is a bound on $V$, the conditions (4.10.1) are satisfied by

$$\tilde{V}(x) = \begin{cases} (M - V(x))/(M - r) & \text{if } x \in C^c \\ 1 & \text{if } x \in C. \end{cases}$$

Since $\tilde{V}(x) < 1$ for $x \in C^c$, $V^*(x) = P_x(\tau_C < \infty) < 1$ on $C^c$, and this implies the transience of $C$, therefore the transience of $(X_n)$. The converse can be deduced from a (partial) converse to Proposition 4.4.7 (see Meyn and Tweedie 1993, p. 190).                                                                    □□

Condition (4.10.2) describes an average increase of $V(x_n)$ once a certain level has been attained, and therefore does not allow a sure return to 0 of $V$. The condition is thus incompatible with the stability associated with recurrence. On the other hand, if there exists a potential function $V$ "attracted" to 0, the chain is recurrent.

**Theorem 4.10.3** *Consider* $(X_n)$ *a* $\psi$-*irreducible Markov chain. If there exist a small set* $C$ *and a function* $V$ *such that*

$$C_V(n) = \{x; V(x) \leq n\}$$

*is a small set for every* $n$, *the chain is recurrent if*

$$\Delta V(x) \leq 0 \text{ on } C^c.$$

The fact that $C_V(n)$ is small means that the function $V$ is not bounded outside small sets. The attraction of the chain toward smaller values of $V$ on the sets where $V$ is large is thus a guarantee of stability for the chain. The proof of the above result is, again, quite involved, based on the fact that $P_x(\tau_C < \infty) = 1$ (see Meyn and Tweedie 1993, p. 191).

**Example 4.10.4 (Continuation of Example 4.5.5)** If the distribution of $W_n$ has a finite support and zero expectation, $(X_n)$ *is recurrent.* When considering $V(x) = |x|$ and $r$ such that $\gamma_x = 0$ for $|x| > r$, we get

$$\Delta V(x) = \sum_{n=-r}^{r} \gamma_n(|x+n| - |x|) \,,$$

which is equal to

$$\sum_{n=-r}^{r} \gamma_n n \text{ if } x \geq r \qquad \text{and} \qquad -\sum_{n=-r}^{r} \gamma_n n \text{ if } x \leq -r \,.$$

Therefore, $\Delta V(x) = 0$ for $x \notin \{-r+1, \ldots, r-1\}$, which is a small set. Conversely, if $W_n$ has a nonzero mean, $X_n$ is transient.                    ‖

For Harris recurrent chains, positivity can also be related to a drift condition and to a "regularity" condition on visits to small sets.

**Theorem 4.10.5** *If* $(X_n)$ *is Harris recurrent with invariant measure* $\pi$, *there is equivalence between*

*(a)* $\pi$ *is finite;*

*(b)* *there exist a small set* $C$ *and a positive number* $M_C$ *such that*

$$\sup_{x \in C} \mathbb{E}_x[\tau_C] \leq M_C \,;$$

*(c)* *there exist a small set* $C$, *a function* $V$ *taking values in* $\mathbb{R} \cup \{\infty\}$, *and a positive real number* $b$ *such that*

(4.10.3)                    $\Delta V(x) \leq -1 + b\mathbb{I}_C(x) \,.$

See Meyn and Tweedie (1993, Chapter 11) for a proof and discussion of these equivalences. (If there exists $V$ finite and bounded on $C$ which satisfies (4.10.3), the chain $(X_n)$ is necessarily Harris positive.)

The notion of a *Kendall atom* introduced in §4.6.2 can also be extended to non-atomic chains by defining *Kendall sets* as sets $A$ such that

$$(4.10.4) \qquad\qquad \sup_{x \in A} \mathbb{E}_x \left[ \sum_{k=0}^{\tau_A - 1} \kappa^k \right] < \infty .$$

with $\kappa > 1$. The existence of a Kendall set guarantees a *geometric drift* condition. If $C$ is a Kendall set and if

$$V(x) = \mathbb{E}_x[\kappa^{\sigma_C}] ,$$

the function $V$ satisfies

$$(4.10.5) \qquad\qquad \Delta V(x) \le -\beta V(x) + b \, \mathbb{I}_C(x)$$

with $\beta > 0$ and $0 < b < \infty$. This condition also guarantees geometric convergence for $(X_n)$ in the following way:

**Theorem 4.10.6** *For a $\psi$-irreducible and aperiodic chain $(X_n)$ and a small Kendall set $C$, there exist $R < \infty$ and $r > 1$, $\kappa > 1$ such that*

$$(4.10.6) \qquad \sum_{n=1}^{\infty} r^n \| K^n(x, \cdot) - \pi(\cdot) \| \le R \, \mathbb{E}_x \left[ \sum_{k=0}^{\tau_C} \kappa^k \right] < \infty$$

*for almost every $x \in \mathcal{X}$.*

The three conditions (4.10.4), (4.10.5) and (4.10.6) are, in fact, equivalent for $\psi$-irreducible aperiodic chains if $A$ is a small set in (4.10.4) and if $V$ is bounded from below by 1 in (4.10.5) (see Meyn and Tweedie 1993, pp. 354–355). The drift condition (4.10.5) is certainly the simplest to check in practice, even though the potential function $V$ must be derived.

**Example 4.10.7 (Continuation of Example 4.3.8)** The condition $|\theta| < 1$ is necessary for the chain $X_n = \theta x_{n-1} + \varepsilon_n$ to be recurrent. Assume $\varepsilon_n$ has a strictly positive density on $\mathbb{R}$. Define $V(x) = |x| + 1$. Then,

$$\begin{aligned}
\mathbb{E}_x[V(X_1)] &= 1 + \mathbb{E}[|\theta X + \varepsilon_1|] \\
&\le 1 + |\theta| \, |x| + \mathbb{E}[|\varepsilon_1|] \\
&= |\theta| \, V(x) + \mathbb{E}[|\varepsilon_1|] + 1 - |\theta|
\end{aligned}$$

and

$$\begin{aligned}
\Delta V(x) &\le -(1 - |\theta|) \, V(x) + \mathbb{E}[|\varepsilon_1|] + 1 - |\theta| \\
&= -(1 - |\theta|) \, \gamma V(x) + \mathbb{E}[|\varepsilon_1|] + 1 - |\theta| - (1 - \gamma)(1 - |\theta|) \, V(x) \\
&\le -\beta V(x) + b \, \mathbb{I}_C(x)
\end{aligned}$$

for $\beta = (1 - |\theta|) \, \gamma$, $b = \mathbb{E}[|\varepsilon_1|] + 1 - |\theta|$, and $C$ equal to

$$C = \{x; \; V(x) < (\mathbb{E}[|\varepsilon_1|] + 1 - |\theta|)/(1 - \gamma)(1 - |\theta|)\} ,$$

if $|\theta| < 1$ and $\mathbb{E}[|\varepsilon_1|] < +\infty$. These conditions thus imply geometric ergodicity for $AR(1)$ models. ∥

Meyn and Tweedie (1994) propose, in addition, *explicit* evaluations of convergence rates $r$ as well as explicit bounds $R$ in connection with drift conditions (4.10.5), but the geometric convergence is evaluated under a norm induced by the very function $V$ satisfying (4.10.5), which makes the result somewhat artificial.

There also is an equivalent form of *uniform ergodicity* involving drift, namely that $(X_n)$ is aperiodic and there exist a small set $C$, a bounded potential function $V \geq 1$, and constants $0 < b < \infty$ and $\beta > 0$ such that

$$(4.10.7) \qquad \Delta V(x) \leq -\beta V(x) + b\, \mathbb{I}_C(x)\,, \qquad x \in \mathcal{X}\,.$$

In a practical case (see, e.g., Example 8.2.4), this alternative to the conditions of Theorem 4.6.13 is often the most natural approach.

As mentioned after Proposition 4.7.5, there exist alternative versions of the Central Limit Theorem based on drift conditions. Assume that there exist a function $f \geq 1$, a finite potential function $V$, and a small set $C$ such that

$$(4.10.8) \qquad \Delta V(x) \leq -f(x) + b\, \mathbb{I}_C(x), \qquad x \in \mathcal{X},$$

and that $\mathbb{E}^\pi[V^2] < \infty$. This is exactly condition (4.10.7) above, with $f = V$, which implies that (4.10.8) holds for an uniformly ergodic chain.

**Theorem 4.10.8** *If the ergodic chain $(X_n)$ with invariant distribution $\pi$ satisfies conditions (4.10.8), for every function $g$ such that $|g| \leq f$, then*

$$\gamma_g^2 = \lim_{n \to \infty} n \mathbb{E}_\pi[S_n^2(\bar{g})]$$

$$= \mathbb{E}_\pi[\bar{g}^2(x_0)] + 2 \sum_{k=1}^{\infty} \mathbb{E}_\pi[\bar{g}(x_0)\bar{g}(x_k)]$$

*is non negative and finite. If $\gamma_g > 0$, the Central Limit Theorem holds for $S_n(\bar{g})$. If $\gamma_g = 0$, $\sqrt{n}S_n(\bar{g})$ almost surely goes to 0.*

This theorem is definitely relevant for convergence control of Markov chain Monte Carlo algorithms since, when $\gamma_g^2 > 0$, it is possible to assess the convergence of the ergodic averages $S_n(g)$ to the quantity of interest $\mathbb{E}^\pi[g]$. Theorem 4.10.8 also suggests how to implement this monitoring through renewal theory, as discussed in detail in Chapter 8.

### 4.10.2 Eaton's Admissibility Condition

Eaton (1992) exhibits interesting connections, similar to Brown (1971), between the admissibility of an estimator and the recurrence of an associated Markov chain. The problem considered by Eaton (1992) is to determine whether, for a *bounded* function $g(\theta)$, a generalized Bayes estimator associated with a prior measure $\pi$ is admissible under quadratic loss. Assuming that the posterior distribution $\pi(\theta|x)$ is well defined, he introduces the transition kernel

$$(4.10.9) \qquad K(\theta, \eta) = \int_\mathcal{X} \pi(\theta|x) f(x|\eta)\, dx,$$

which is associated with a Markov chain $(\theta^{(n)})$ generated as follows: The transition from $\theta^{(n)}$ to $\theta^{(n+1)}$ is done by generating first $x \sim f(x|\theta^{(n)})$ and then $\theta^{(n+1)} \sim \pi(\theta|x)$. (Most interestingly, this is also a kernel used by Markov Chain Monte Carlo methods, as shown in Chapter 7.) Note that the prior measure $\pi$ is an invariant measure for the chain $(\theta^{(n)})$. For every measurable set $C$ such that $\pi(C) < +\infty$, consider

$$V(C) = \left\{ h \in \mathcal{L}^2(\pi); h(\theta) \geq 0 \text{ and } h(\theta) \geq 1 \text{ when } \theta \in C \right\}$$

and

$$\Delta(h) = \int \int \{h(\theta) - h(\eta)\}^2 \, K(\theta, \eta) \pi(\eta) \, d\theta \, d\eta.$$

The following result then characterizes admissibility for *all bounded functions* in terms of $\Delta$ and $V(C)$ (that is, independently of the estimated functions $g$).

**Theorem 4.10.9** *If for every $C$ such that $\pi(C) < +\infty$,*

(4.10.10)
$$\inf_{h \in V(C)} \Delta(h) = 0,$$

*then the Bayes estimator $\mathbb{E}^{\pi}[g(\theta)|x]$ is admissible under quadratic loss for every bounded function $g$.*

This result is obviously quite general but only mildly helpful in the sense that the practical verification of (4.10.10) for every set $C$ can be overwhelming. Note also that (4.10.10) always holds when $\pi$ is a proper prior distribution since $h \equiv 1$ belongs to $\mathcal{L}^2(\pi)$ and $\Delta(1) = 0$ in this case. The extension then considers approximations of 1 by functions in $V(C)$. Eaton (1992) exhibits a connection with the Markov chain $(\theta^{(n)})$, which gives a condition equivalent to Theorem 4.10.9. First, for a given set $C$, a stopping rule $\tau_C$ is defined as the first integer $n > 0$ such that $(\theta^{(n)})$ belongs to $C$ (and $+\infty$ otherwise), as in Definition 4.2.9.

**Theorem 4.10.10** *For every set $C$ such that $\pi(C) < +\infty$,*

$$\inf_{h \in V(C)} \Delta(h) = \int_C \left\{ 1 - P(\tau_C < +\infty | \theta^{(0)} = \eta) \right\} \pi(\eta) \, d\eta.$$

*Therefore, the generalized Bayes estimators of bounded functions of $\theta$ are admissible if and only if the associated Markov chain $(\theta^{(n)})$ is recurrent.*

Again, we refer to Eaton (1992) for extensions, examples, and comments on this result. Note, however, that the verification of the recurrence of the Markov chain $(\theta^{(n)})$ is much easier than the determination of the lower bound of $\Delta(h)$. Hobert and Robert (1999) consider the potential of using the dual chain based on the kernel

(4.10.11)
$$K'(x, y) = \int_{\Theta} f(y|\theta) \pi(\theta|x) d\theta$$

(see Problem 4.68) and derive admissibility results for various distributions of interest.

### 4.10.3 Alternative Convergence Conditions

Athreya *et al.* (1996) present a careful development of the basic limit theorems for Markov chains, with conditions stated that are somewhat more accessible in Markov chain Monte Carlo uses, rather than formal probabilistic properties.

Consider a time-homogeneous Markov chain $(X_n)$ where $f$ is the invariant density and $f_k(\cdot|\cdot)$ is the conditional density of $X_k$ given $X_0$. So, in particular, $f_1(\cdot|\cdot)$ is the transition kernel. For a basic limit theorem such as Theorem 4.6.5, there are two conditions that are required on the transition kernel, both of which have to do with the ability of the Markov chain to visit all sets $A$. Assume that the transition kernel satisfies:

(i) For any set $A$ for which $\int_A f(x)d\mu(x) > 0$ we have

$$\int_A f_1(x|x_0)d\mu(x) > 0 \text{ for all } x_0.$$

(ii) For some set $A$ for which $\int_A f(x)d\mu(x) > 0$, and some $k$, we have

$$\inf_{x,y \in A} f_k(y|x) > 0.$$

A transition kernel that satisfies Condition (i) is called *strictly positive*, which means that, with probability 1, in one step the chain can visit any set having positive measure. Note that in many situations Condition (i) will imply Condition (ii). This will happen, for example, if the function $f_1(x|x_0)$ is continuous in $x$ for each $x_0$, or if it is discrete (see Problem 4.72). To reconcile these conditions with the probabilistic ones, we note that they imply that the Markov chain is positive recurrent and aperiodic.

The limit theorem of Athreya *et al.* (1996) can be stated as follows:

**Theorem 4.10.11** *Suppose that the Markov chain* $(X_n)$ *has invariant density* $f(\cdot)$ *and transition kernel* $f_1(\cdot|\cdot)$ *that satisfies Conditions (i) and (ii). Then*

$$(4.10.12) \qquad \lim_{k \to \infty} \sup_A \left| \int_A f_k(x|x_0)dx - \int_A f(x)dx \right| = 0$$

*for $f$ almost all $x_0$.*

### 4.10.4 Mixing Conditions and Central Limit Theorems

In §4.7.2, we established a Central Limit Theorem using regeneration, which allowed us to use a typical independence argument. Other conditions, known as *mixing conditions*, can also result in a Central Limit Theorem. These mixing conditions guarantee that the dependence in the Markov chain decreases fast enough, and variables that are far enough apart are close to being independent. Unfortunately, these conditions are usually quite difficult to verify. Consider the property of $\alpha$-*mixing* (Billingsley 1995, Section 27).

**Definition 4.10.12** A sequence $X_0, X_1, X_2, \ldots$ is $\alpha$-*mixing* if

$$(4.10.13) \quad \alpha_n = \sup_{A,B} |P(X_n \in A, X_0 \in B) - P(X_n \in A)P(X_0 \in B)|$$

goes to 0 when $n$ goes to infinity.

So, we see that an $\alpha$-mixing sequence will tend to "look independent" if the variables are far enough apart. As a result, we would expect that Theorem 4.7.4 is a consequence of $\alpha$-mixing. This is, in fact, the case, as every positive recurrent aperiodic Markov chain is $\alpha$-mixing (Rosenblatt 1971, Section VII.3), and if the Markov chain is stationary and $\alpha$-mixing, the covariances go to zero (Billingsley 1995, Section 27).

However, for a Central Limit Theorem, we need even more. Not only must the Markov chain be $\alpha$-mixing, but we need the coefficient $\alpha_n$ to go to 0 fast enough; that is, we need the dependence to go away fast enough. One version of a Markov chain Central Limit Theorem is the following (Billingsley 1995, Section 27).

**Theorem 4.10.13** *Suppose that the Markov chain $(X_n)$ is stationary and $\alpha$-mixing with $\alpha_n = O(n^{-5})$ and that $\mathbb{E}[X_n] = 0$ and $\mathbb{E}[X_n^{12}] < \infty$. Then,*

$$\sigma^2 = \lim_{n \to \infty} \; n \, \mathrm{var} \bar{X}_n < \infty$$

*and if $\sigma^2 > 0$, $\sqrt{n} \bar{X}_n$ tends in law to $\mathcal{N}(0, \sigma^2)$.*

This theorem is not very useful because the condition on the mixing coefficient is very hard to verify. (Billingsley 1995 notes that the conditions are stronger than needed, but are imposed to avoid technical difficulties in the proof.) Others have worked hard to get the condition in a more accessible form and have exploited the relationship between mixing and ergodicity. Informally, if (4.10.13) goes to 0, dividing through by $P(X_0 \in B)$ we expect that

(4.10.14) $$|P(X_n \in A | X_0 \in B) - P(X_n \in A)| \to 0,$$

which looks quite similar to the assumption that the Markov chain is ergodic. (This corresponds, in fact, to a stronger type of mixing called $\beta$-mixing. See Bradley 1986). We actually need something stronger (see Problem 6.3), like uniform ergodicity, where there are constants $M$ and $r < 1$ such $|P(X_n \in A | X_0 \in B) - P(X_n \in A)| \leq Mr^n$. Tierney (1994) presents the following Central Limit Theorem.

**Theorem 4.10.14** *Let $(X_n)$ be a stationary uniformly ergodic Markov chain. For any function $h(\cdot)$ satisfying $\mathrm{var}\, h(X_i) = \sigma^2 < \infty$, there exists a real number $\tau_h$ such that*

$$\sqrt{n} \, \frac{\sum_{i=1}^{n} h(X_i)/n - \mathbb{E}[h(X)]}{\tau_h} \xrightarrow{\mathcal{L}} \mathcal{N}(0, 1).$$

Many other versions of the Central Limit Theorem exist. For example, Chan and Geyer (1994) relax the assumptions on Theorem 4.10.14 (see Problem 4.69). Robert (1994b) surveys other mixing conditions and their connections with the Central Limit Theorem.

CHAPTER 5

# Monte Carlo Optimization

"Remember, boy," Sam Nakai would sometimes tell Chee, "when you're tired of walking up a long hill you think about how easy it's going to be walking down."

—Tony Hillerman, *A Thief of Time*

## 5.1 Introduction

Similar to the problem of integration, differences between the numerical approach and the simulation approach to the problem

$$(5.1.1) \qquad\qquad \max_{\theta \in \Theta}\ h(\theta)$$

lie in the treatment of the function[1] $h$. (Note that (5.1.1) also covers minimization problems by considering $-h$.) In approaching an optimization problem using deterministic numerical methods, the analytical properties of the target function (convexity, boundedness, smoothness) are often paramount. For the simulation approach, we are more concerned with $h$ from a probabilistic (rather than analytical) point of view. Obviously, this dichotomy is somewhat artificial, as there exist simulation approaches where the probabilistic interpretation of $h$ is not used. Nonetheless, the use of the analytical properties of $h$ plays a lesser role in the simulation approach.

Numerical methods enjoy a longer history than simulation methods (see, for instance, Kennedy and Gentle 1980 or Thisted 1988), but simulation methods have gained in appeal due to the relaxation of constraints on both the regularity of the domain $\Theta$ and on the function $h$. Of course, there may exist an alternative numerical approach which provides an exact solution to (5.1.1), a property rarely achieved by a stochastic algorithm, but simulation has the advantage of bypassing the preliminary steps of devising an algorithm and studying whether some regularity conditions on $h$ hold. This is particularly true when the function $h$ is very costly to compute.

---

[1] Although we use $\theta$ as the running parameter and $h$ typically corresponds to a (possibly penalized) transform of the likelihood function, this setup applies to inferential problems other than likelihood or posterior maximization. As noted in the Introduction to Chapter 3, problems concerned with complex loss functions or confidence regions also require optimization procedures.

**Example 5.1.1 Signal processing.** Ó Ruanaidh and Fitzgerald (1996) study signal processing data, of which a simple model is ($i = 1, \ldots, N$)

$$x_i = \alpha_1 \cos(\omega t_i) + \alpha_2 \sin(\omega t_i) + \epsilon_i, \qquad \epsilon_i \sim \mathcal{N}(0, \sigma^2),$$

with unknown parameters $\alpha = (\alpha_1, \alpha_2), \omega$, and $\sigma$ and observation times $t_1, \ldots, t_N$. The likelihood function is then of the form

$$\sigma^{-N} \exp\left(-\frac{(\mathbf{x} - G\alpha^t)^t(\mathbf{x} - G\alpha^t)}{2\sigma^2}\right),$$

with $\mathbf{x} = (x_1, \ldots, x_N)$ and

$$G = \begin{pmatrix} \cos(\omega t_1) & \sin(\omega t_1) \\ \vdots & \vdots \\ \cos(\omega t_N) & \sin(\omega t_N) \end{pmatrix}.$$

The prior $\pi(\alpha, \omega, \sigma) = \sigma^{-1}$ then leads to the marginal distribution

$$(5.1.2) \quad \pi(\omega|\mathbf{x}) \propto \left(\mathbf{x}^t\mathbf{x} - \mathbf{x}^t G(G^t G)^{-1}G^t\mathbf{x}\right)^{(2-N)/2} (\det G^t G)^{-1/2},$$

which, although explicit in $\omega$, is not particularly simple to compute. This setup is also illustrative of functions with many modes, as shown by Ó Ruanaidh and Fitzgerald (1996).                                                    ‖

Following Geyer (1996), we want to distinguish between two approaches to Monte Carlo optimization. The first is an *exploratory* approach, in which the goal is to optimize the function $h$ by describing its entire range. The actual properties of the function play a lesser role here, with the Monte Carlo aspect more closely tied to the exploration of the entire space $\Theta$, even though, for instance, the slope of $h$ can be used to speed up the exploration. (Such a technique can be useful in describing functions with multiple modes, for example.) The second approach is based on a probabilistic *approximation* of the objective function $h$ and is somewhat of a preliminary step to the actual optimization. Here, the Monte Carlo aspect exploits the probabilistic properties of the function $h$ to come up with an acceptable approximation and is less concerned with exploring $\Theta$. We will see that this approach can be tied to *missing data methods*, such as the EM algorithm. We note also that Geyer (1996) only considers the second approach to be "Monte Carlo optimization." Obviously, even though we are considering these two different approaches separately, they might be combined in a given problem. In fact, methods like the EM algorithm (§5.3.3) or the Robbins–Monro algorithm (§5.5.3) take advantage of the Monte Carlo approximation to enhance their particular optimization technique.

## 5.2 Stochastic Exploration

### 5.2.1 A Basic Solution

There are a number of cases where the exploration method is particularly well suited. First, if $\Theta$ is bounded, which may sometimes be achieved by a reparameterization, a first approach to the resolution of (5.1.1) is to simulate from a uniform distribution on $\Theta$, $u_1, \ldots, u_m \sim \mathcal{U}_\Theta$, and to use the approximation $h_m^* = \max(h(u_1), \ldots, h(u_m))$. This method converges (as $m$ goes to $\infty$), but it may be very slow since it does not take into account any specific feature of $h$. Distributions other than the uniform, which can possibly be related to $h$, may then do better. In particular, in setups where the likelihood function is extremely costly to compute, the number of evaluations of the function $h$ is best kept to a minimum.

This leads to a second, and often more fruitful, direction, which relates $h$ to a probability distribution. For instance, if $h$ is positive and if

$$\int_\Theta h(\theta)\, d\theta < +\infty\,,$$

the resolution of (5.1.1) amounts to finding the *modes* of the density $h$. More generally, if these conditions are not satisfied, then we may be able to transform the function $h(\theta)$ into another function $H(\theta)$ that satisfies the following:

(i) The function $H$ is non-negative and satisfies $\int H < \infty$.

(ii) The solutions to (5.1.1) are those which maximize $H(\theta)$ on $\Theta$.

For example, we can take

$$H(\theta) = \exp(h(\theta)/T) \quad \text{or} \quad H(\theta) = \exp\{h(\theta)/T\}/(1 + \exp\{h(\theta)/T\})$$

and choose $T$ to accelerate convergence or to avoid local maxima (as in simulated annealing; see §5.2.3). When the problem is expressed in statistical terms, it becomes natural to then generate a sample $(\theta_1, \ldots, \theta_m)$ from $h$ (or $H$) and to apply a standard mode estimation method (or to simply compare the $h(\theta_i)$'s). (In some cases, it may be more useful to decompose $h(\theta)$ into $h(\theta) = h_1(\theta)h_2(\theta)$ and to simulate from $h_1$.)

**Example 5.2.1 Minimization of a complex function.** Consider minimizing the (artificially constructed) function in $\mathbb{R}^2$

$$h(x, y) = (x\sin(20y) + y\sin(20x))^2 \cosh(\sin(10x)x)$$
$$+ (x\cos(10y) - y\sin(10x))^2 \cosh(\cos(20y)y)\,,$$

whose global minimum is 0, attained at $(x, y) = (0, 0)$. Since this function has many local minima, as shown by Figure 5.2.1, it does not satisfy the conditions under which standard minimization methods are guaranteed to provide the global minimum. On the other hand, the distribution on $\mathbb{R}^2$ with density proportional to $\exp(-h(x, y))$ can be simulated, even though this is not a standard distribution, by using, for instance, Markov chain Monte Carlo techniques (introduced in Chapters 6 and 7), and a convergent approximation of the minimum of $h(x, y)$ can be derived from the minimum of the resulting $h(x_i, y_i)$'s. An alternative is to simulate from the density proportional to

$$h_1(x, y) = \exp\{-(x\sin(20y) + y\sin(20x))^2 - (x\cos(10y) - y\sin(10x))^2\},$$

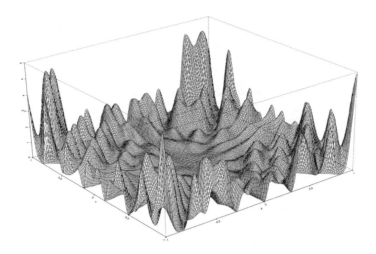

Figure 5.2.1.  *Grid representation of the function* $h(x, y)$ *of Example 5.2.1 on* $[-1, 1]^2$.

which eliminates the computation of both cosh and sinh in the simulation step.                                                                                    ‖

   Exploration may be particularly difficult when the space $\Theta$ is not convex (or perhaps not even connected). In such cases, the simulation of a sample $(\theta_1, \ldots, \theta_m)$ may be much faster than a numerical method applied to (5.1.1). The appeal of simulation is even clearer in the case when $h$ can be represented as

$$h(\theta) = \int H(x, \theta) dx .$$

In particular, if $H(x, \theta)$ is a density and if it is possible to simulate from this density, the solution of (5.1.1) is the mode of the marginal distribution of $\theta$. (Although this setting may appear contrived or even artificial, we will see in §5.3.1 that it includes the case of missing data models.)
   We now look at several methods to find maxima that can be classified as *exploratory methods*.

### 5.2.2 Gradient Methods

As mentioned in §1.4, the *gradient method* is a deterministic numerical approach to the problem (5.1.1). It produces a sequence $(\theta_j)$ that converges to the exact solution of (5.1.1), $\theta^*$, when the domain $\Theta \subset \mathbb{R}^d$ and the function $(-h)$ are both convex. The sequence $(\theta_j)$ is constructed in a recursive manner through

$$(5.2.1) \qquad\qquad \theta_{j+1} = \theta_j + \alpha_j \nabla h(\theta_j) , \qquad \alpha_j > 0 ,$$

where $\nabla h$ is the gradient of $h$. For various choices of the sequence $(\alpha_j)$ (see Thisted 1988), the algorithm converges to the (unique) maximum.

In more general setups (that is, when the function or the space is less regular), equation (5.2.1) can be modified by stochastic perturbations to again achieve convergence, as described in detail in Rubinstein (1981) or Duflo (1996, pp. 61–63). One of these stochastic modifications is to choose a second sequence $(\beta_j)$ to define the chain $(\theta_j)$ by

$$(5.2.2) \qquad \theta_{j+1} = \theta_j + \frac{\alpha_j}{2\beta_j} \, \Delta h(\theta_j, \beta_j \zeta_j) \, \zeta_j \ .$$

The variables $\zeta_j$ are uniformly distributed on the unit sphere $||\zeta|| = 1$ and $\Delta h(x, y) = h(x + y) - h(x - y)$ approximates $2||y||\nabla h(x)$. In contrast to the deterministic approach, this method does not necessarily proceed along the steepest slope in $\theta_j$, but this property is sometimes a *plus* in the sense that it may avoid being trapped in local maxima or in saddlepoints of $h$.

The convergence of $(\theta_j)$ to the solution $\theta^*$ again depends on the choice of $(\alpha_j)$ and $(\beta_j)$. We note in passing that $(\theta_j)$ can be seen as a *nonhomogeneous Markov chain* (see Definition 4.2.3) which almost surely converges to a given value. The study of these chains is quite complicated given their ever-changing transition kernel (see Winkler 1995 for some results in this direction). However, sufficiently strong conditions such as the decrease of $\alpha_j$ toward 0 and of $\alpha_j/\beta_j$ to a nonzero constant are enough to guarantee the convergence of the sequence $(\theta_j)$.

**Example 5.2.2 (Continuation of Example 5.2.1)** We can apply the iterative construction (5.2.2) to the multimodal function $h(x, y)$ with different sequences of $\alpha_j$'s and $\beta_j$'s. Figure 5.2.2 and Table 5.2.1 illustrate that, depending on the starting value, the algorithm converges to different local minima of the function $h$. Although there are occurrences when the sequence $h(\theta_j)$ increases and avoids some local minima, the solutions are quite distinct for the three different sequences, both in location and values.

As shown by Table 5.2.1, the number of iterations needed to achieve stability of $\theta_T$ also varies with the choice of $(\alpha_j, \beta_j)$. Note that Case 1 results in a very poor evaluation of the minimum, as the fast decrease of $(\alpha_j)$ is associated with big jumps in the first iterations. Case 2 converges to the closest local minima, and Case 3 illustrates a general feature of the stochastic gradient method, namely that slower decrease rates of the sequence $(\alpha_j)$ tend to achieve better minima. The final convergence along a valley of $h$ after some initial big jumps is also noteworthy.               ‖

This approach is still quite close to numerical methods in that it requires a precise knowledge on the function $h$, which may not necessarily be available.

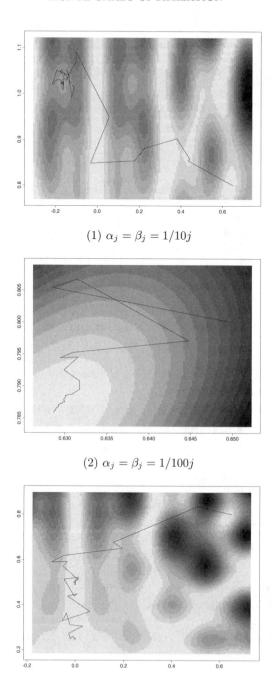

(1) $\alpha_j = \beta_j = 1/10j$

(2) $\alpha_j = \beta_j = 1/100j$

(3) $\alpha_j = 1/10\log(1+j), \ \beta_j = 1/j$

Figure 5.2.2. *Stochastic gradient paths for three different choices of the sequences* $(\alpha_j)$ *and* $(\beta_j)$ *and starting point* $(0.65, 0.8)$ *for the same sequence* $(\zeta_j)$ *in (5.2.2). The gray levels are such that darker shades mean higher elevations. The function* $h$ *to be minimized is defined in Example 5.2.1.*

| $\alpha_j$ | $\beta_j$ | $\theta_T$ | $h(\theta_T)$ | $\min_t h(\theta_t)$ | Iteration $T$ |
|---|---|---|---|---|---|
| $1/10j$ | $1/10j$ | $(-0.166, 1.02)$ | $1.287$ | $0.115$ | $50$ |
| $1/100j$ | $1/100j$ | $(0.629, 0.786)$ | $0.00013$ | $0.00013$ | $93$ |
| $1/10\log(1+j)$ | $1/j$ | $(0.0004, 0.245)$ | $4.24 \times 10^{-6}$ | $2.163 \times 10^{-7}$ | $58$ |

Table 5.2.1. *Results of three stochastic gradient runs for the minimization of the function h in Example 5.2.1 with different values of $(\alpha_j, \beta_j)$ and starting point $(0.65, 0.8)$. The iteration T is obtained by the stopping rule $\|\theta_T - \theta_{T-1}\| < 10^{-5}$.*

### 5.2.3 Simulated Annealing

The simulated annealing algorithm[2] was introduced by Metropolis *et al.* (1953) to minimize a criterion function on a finite set with very large size[3], but it also applies to optimization on a continuous set and to simulation (see Kirkpatrick *et al.* 1983, Ackley *et al.* 1985, Aldous 1987, 1990, and Neal 1993, 1995).

The fundamental idea of simulated annealing methods is that a change of scale, called *temperature*, allows for faster moves on the surface of the function $h$ to maximize, whose negative is called *energy*. Therefore, rescaling partially avoids the trapping attraction of local maxima. Given a temperature parameter $T > 0$, a sample $\theta_1^T, \theta_2^T, \ldots$ is generated from the distribution

$$\pi(\theta) \propto \exp(h(\theta)/T)$$

and can be used as in §5.2.1 to come up with an approximate maximum of $h$. As $T$ decreases toward 0, the values simulated from this distribution become concentrated in a narrower and narrower neighborhood of the local maxima of $h$ (see Theorem 5.2.7, Problem 5.4, and Winkler 1995).

The fact that this approach has a moderating effect on the attraction of the local maxima of $h$ becomes more apparent when we consider the simulation method proposed by Metropolis *et al.* (1953). Starting from $\theta_0$, $\zeta$ is generated from a uniform distribution on a neighborhood $\mathcal{V}(\theta_0)$ of $\theta_0$ or, more generally, from a distribution with density $g(|\zeta - \theta_0|)$, and the new value of $\theta$ is generated as follows:

$$\theta_1 = \begin{cases} \zeta & \text{with probability } \rho = \exp(\Delta h/T) \wedge 1 \\ \theta_0 & \text{with probability } 1 - \rho, \end{cases}$$

where $\Delta h = h(\zeta) - h(\theta_0)$. Therefore, if $h(\zeta) \geq h(\theta_0)$, $\zeta$ is accepted with

---

[2] This name is borrowed from Metallurgy: A metal manufactured by a slow decrease of temperature (*annealing*) is stronger than a metal manufactured by a fast decrease of temperature. There is also input from Physics, as the function to be minimized is called *energy* and the variance factor $T$, which controls convergence, is called *temperature*. We will try to keep these idiosyncrasies to a minimal level, but they are quite common in the literature.

[3] This paper is also the originator of the *Markov chain Monte Carlo methods* developed in the following chapters.

probability 1; that is, $\theta_0$ is always changed into $\zeta$. On the other hand, if $h(\zeta) < h(\theta_0)$, $\zeta$ may still be accepted with probability $\rho \neq 0$ and $\theta_0$ is then changed into $\zeta$. This property allows the algorithm to escape the attraction of $\theta_0$ if $\theta_0$ is a local maximum of $h$, with a probability which depends on the choice of the scale $T$, compared with the range of the density $g$. (This method is in fact the *Metropolis algorithm*, which simulates the density proportional to $\exp\{h(\theta)/T\}$, as the limiting distribution of the chain $\theta_0, \theta_1, \ldots$, as described and justified in Chapter 6.)

In its most usual implementation, the simulated annealing algorithm modifies the temperature $T$ at each iteration; it is then of the form

### Algorithm A.20 –Simulated Annealing–

1. Simulate $\zeta$ from an instrumental distribution
   with density $g(|\zeta - \theta_i|)$;

2. Accept $\theta_{i+1} = \zeta$ with probability $\rho_i = \exp\{\Delta h_i / T_i\} \wedge 1$;     [A.20]
   take $\theta_{i+1} = \theta_i$ otherwise.

3. Update $T_i$ to $T_{i+1}$.

The Markov chain $(\theta_i)$ thus created is no longer homogeneous. However, there still exist convergence results in the case of finite spaces, as Theorem 5.2.4, which was proposed by Hàjek (1988). (See Problem 5.2 and Winkler 1995, for extensions.) Consider the following notions, which are used to impose restrictions on the decrease rate of the temperature:

**Definition 5.2.3** Given a finite state-space $\mathcal{E}$ and a function $h$ to be maximized:

(i) a state $e_j \in \mathcal{E}$ can be *reached at altitude $\underline{h}$* from state $e_i \in \mathcal{E}$ if there exists a sequence of states $e_1, \ldots, e_n$ linking $e_i$ and $e_j$, such that $h(e_k) \geq \underline{h}$ for $k = 1, \ldots, n$;

(ii) the *height of a maximum* $e_i$ is the largest eigenvalue $d_i$ such that there exists a state $e_j$ such that $h(e_j) > h(e_i)$ which can be reached at altitude $h(e_i) + d_i$ from $e_i$.

Thus, $h(e_i) + d_i$ is the altitude of the highest pass linking $e_i$ and $e_j$ through an optimal sequence. (In particular, $h(e_i) + d_i$ can be larger than the altitude of the closest pass relating $e_i$ and $e_j$.) By convention, we take $d_i = -\infty$ if $e_i$ is a global maximum. If $\mathcal{O}$ denotes the set of local maxima of $E$ and $\underline{\mathcal{O}}$ is the subset of $\mathcal{O}$ of global maxima, Hàjek (1988) establishes the following result:

**Theorem 5.2.4** *Consider a system in which it is possible to link two arbitrary states by a finite sequence of states. If, for every $\underline{h} > 0$ and every pair $(e_i, e_j)$, $e_i$ can be reached at altitude $\underline{h}$ from $e_j$ if and only if $e_j$ can be reached at altitude $\underline{h}$ from $e_i$, and if $(T_i)$ decreases toward $0$, the sequence $(\theta_i)$ defined by the algorithm [A.20] satisfies*

$$\lim_{i \to \infty} P(\theta_i \in \underline{\mathcal{O}}) = 1$$

*if and only if*

$$\sum_{i=1}^{\infty} \exp(-D/T_i) = +\infty ,$$

*with* $D = \min\{d_i : e_i \in \mathcal{O} - \underline{\mathcal{O}}\}$.

This theorem therefore gives a necessary and sufficient condition, on the rate of decrease of the temperature, so that the simulated annealing algorithm converges to the set of global maxima. This remains a relatively formal result since $D$ is, in practice, unknown. For example, if $T_i = \Gamma / \log i$, there is convergence to a global maximum if and only if $\Gamma \geq D$. Numerous papers and books have considered the practical determination of the sequence $(T_n)$ (see Geman and Geman 1984, Mitra *et al.* 1986, Van Laarhoven and Aarts 1987, Aarts and Kors 1989, Winkler 1995, and references therein). Instead of the above logarithmic rate, a geometric rate, $T_i = \alpha^i T_0$ $(0 < \alpha < 1)$, is also often adopted in practice, with the constant $\alpha$ calibrated at the beginning of the algorithm so that the acceptance rate is high enough in the Metropolis algorithm (see §6.4).

The fact that approximate methods are necessary for optimization problems in finite state-spaces may sound rather artificial and unnecessary, but the spaces involved in some modeling can be huge. For instance, a black-and-white TV image with $256 \times 256$ pixels corresponds to a state-space with cardinality $2^{256 \times 256} \simeq 10^{20,000}$. Similarly, the analysis of DNA sequences may involve 600 thousand bases (A, C, G, or T), which corresponds to state-spaces of size $4^{600,000}$ (see Churchill 1989, 1995).

**Example 5.2.5 Ising model.** The *Ising model* can be applied in electromagnetism (Cipra 1987) and in image processing (Geman and Geman 1984). It models two-dimensional tables $s$, of size $D \times D$, where each term of $s$ takes the value $+1$ or $-1$. The distribution of the entire table is related to the (so-called energy) function

$$(5.2.3) \qquad h(s) = -J \sum_{(i,j) \in \mathcal{N}} s_i s_j - H \sum_i s_i ,$$

where $i$ denotes the index of a term of the table and $\mathcal{N}$ is an equivalence neighborhood relation, for instance, when $i$ and $j$ are neighbors either vertically or horizontally. (The scale factors $J$ and $H$ are supposedly known.) The model (5.2.3) is a particular case of models used in Spatial Statistics (Cressie 1993) to describe multidimensional correlated structures.

Note that the conditional representation of (5.2.3) is equivalent to a *logit* model on $\tilde{s}_i = (s_i + 1)/2$,

$$(5.2.4) \qquad P(\tilde{S}_i = 1 | s_j, j \neq i) = \frac{e^g}{1 + e^g} ,$$

with $g = g(s_j) = 2(H + J \sum_j s_j)$, the sum being taken on the neighbors of $i$. For known parameters $H$ and $J$, the inferential question may be to obtain the most likely configuration of the system; that is, the minimum

| Case | $T_i$ | $\theta_T$ | $h(\theta_T)$ | $\min_t h(\theta_t)$ | Accept. rate |
|------|-------|------------|---------------|----------------------|--------------|
| 1 | $1/10i$ | $(-1.94, -0.480)$ | 0.198 | $4.02\,10^{-7}$ | 0.9998 |
| 2 | $1/\log(1+i)$ | $(-1.99, -0.133)$ | 3.408 | $3.823 \times 10^{-7}$ | 0.96 |
| 3 | $100/\log(1+i)$ | $(-0.575, 0.430)$ | 0.0017 | $4.708 \times 10^{-9}$ | 0.6888 |
| 4 | $1/10\log(1+i)$ | $(0.121, -0.150)$ | 0.0359 | $2.382 \times 10^{-7}$ | 0.71 |

Table 5.2.2. *Results of simulated annealing runs for different values of $T_i$ and starting point* $(0.5, 0.4)$.

of $h(s)$. The implementation of the Metropolis *et al.* (1953) approach in this setup, starting from an initial value $s^{(0)}$, is to modify the sites of the table $s$ one at a time using the conditional distributions (5.2.4), with probability $\exp(-\Delta h/T)$, ending up with a modified table $s^{(1)}$, and to iterate this method by decreasing the temperature $T$ at each step. The reader can consult Swendson and Wang (1987) and Swendson *et al.* (1992) for their derivation of efficient simulation algorithms in these models and accelerating methods for the Gibbs sampler (see Problem 6.50).    ‖

Duflo (1996, pp. 264–271) also proposed an extension of these simulated annealing methods to the general (continuous) case. Andrieu and Doucet (1998) give a detailed proof of convergence of the simulated annealing algorithm, as well as sufficient conditions on the cooling schedule, in the setup of hidden Markov models (see §9.5.1). Their proof, which is beyond our scope, is based on the uniform ergodicity of the transition kernel and developments of Haario and Sacksman (1991).

**Example 5.2.6 (Continuation of Example 5.2.1)** We can apply the algorithm [A.20] to find a local minimum of the function $h$ of Example 5.2.1, or equivalently a maximum of the function $\exp(-h(x, y)/T_i)$. We choose a uniform distribution on $[-0.1, 0.1]$ for $g$, and different rates of decrease of the temperature sequence $(T_i)$. As illustrated by Figure 5.2.3 and Table 5.2.2, the results change with the rate of decrease of the temperature $T_i$. Case 3 leads to a very interesting exploration of the valleys of $h$ on both sides of the central zone. Since the theory (see Duflo 1996) states that rates of the form $\Gamma/\log(i+1)$ are satisfactory for $\Gamma$ large enough, this shows that $\Gamma = 100$ should be acceptable. Note also the behavior of the acceptance rate in Table 5.2.2 for Step 2 in algorithm [A.20]. This is indicative of a rule we will discuss further in Chapter 6 with Metropolis–Hastings algorithms, namely that superior performances are not always associated with higher acceptance rates.    ‖

### 5.2.4 Prior Feedback

Another approach to the maximization problem (5.1.1) is based on the result of Hwang (1980) of convergence (in $T$) of the so-called *Gibbs measure*

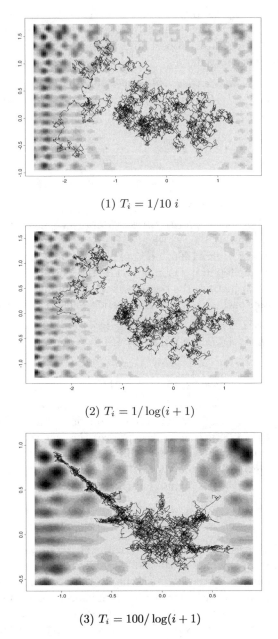

(1) $T_i = 1/10\ i$

(2) $T_i = 1/\log(i+1)$

(3) $T_i = 100/\log(i+1)$

Figure 5.2.3. *Simulated annealing sequence of* 5000 *points for three different choices of the temperature* $T_i$ *in* [A.20] *and starting point* $(0.5, 0.4)$, *aimed at minimizing the function h of Example 5.2.1.*

$\exp(h(\theta)/T)$ (see §5.5.3) to the uniform distribution on the set of global maxima of $h$. This approach, called *recursive integration* or *prior feedback* in Robert (1993) (see also Robert and Soubiran 1993), is based on the following convergence result.

**Theorem 5.2.7** *Consider $h$ a real-valued function defined on a closed and bounded set, $\Theta$, of $\mathbb{R}^p$. If there exists a unique solution $\theta^*$ satisfying*

$$\theta^* = \arg\max_{\theta \in \Theta} \ h(\theta) \ ,$$

*then*

$$\lim_{\lambda \to \infty} \frac{\int_\Theta \theta \, e^{\lambda h(\theta)} \, d\theta}{\int_\Theta e^{\lambda h(\theta)} \, d\theta} = \theta^* \ ,$$

*provided $h$ is continuous at $\theta^*$.*

A proof based on *the Laplace approximation* (see §3.5) of both integrals can be found in Pincus (1968) (see also Robert 1993 for the case of exponential families and Duflo 1996, pp. 244–245, for a sketch of a proof). A related result can be found in D'Epifanio (1989, 1996). A direct corollary to Theorem 5.2.7 then justifies the recursive integration method which results in a Bayesian approach to maximizing the log-likelihood, $\ell(\theta|x)$.

**Corollary 5.2.8** *Let $\pi$ be a positive density on $\Theta$. If there exists a unique maximum likelihood estimator $\theta^*$, it satisfies*

$$\lim_{\lambda \to \infty} \frac{\int \theta e^{\lambda \ell(\theta|x)} \, \pi(\theta) d\theta}{\int e^{\lambda \ell(\theta|x)} \, \pi(\theta) d\theta} = \theta^* \ .$$

This result uses the same technique as in Theorem 5.2.7, namely the Laplace approximation of the numerator and denominator integrals (see also Tierney *et al.* 1989). It mainly expresses the fact that the maximum likelihood estimator can be written as a limit of Bayes estimators associated with an *arbitrary* distribution $\pi$ and with *virtual* observations corresponding to the $\lambda$th power of the likelihood, $\exp\{\lambda\ell(\theta|x)\}$. When $\lambda \in \mathbb{N}$,

$$\delta_\lambda^\pi(x) = \frac{\int \theta e^{\lambda \ell(\theta|x)} \, \pi(\theta) d\theta}{\int e^{\lambda \ell(\theta|x)} \, \pi(\theta) d\theta}$$

is simply the Bayes estimator associated with the prior distribution $\pi$ and a corresponding sample which consists of $\lambda$ replications of the initial sample $x$. The intuition behind these results is that as the size of the sample goes to infinity, the influence of the prior distribution vanishes and the distribution associated with $\exp(\lambda\ell(\theta|x))\pi(\theta)$ gets more and more concentrated around the global maxima of $\ell(\theta|x)$ when $\lambda$ increases (see, e.g., Schervish 1995).

From a practical point of view, the recursive integration method can be implemented as follows:

**Algorithm A.21 –Recursive Integration–**
For integers $\lambda_1 < \lambda_2 < \cdots$
         Compute the Bayes estimates $\delta_\lambda^\pi(x)$
until they stabilize.

| $\lambda$ | 5 | 10 | 100 | 1000 | 5000 | $10^4$ |
|---|---|---|---|---|---|---|
| $\delta_\lambda^\pi$ | 2.02 | 2.04 | 1.89 | 1.98 | 1.94 | 2.00 |

Table 5.2.3. *Sequence of Bayes estimators of $\delta_\lambda^\pi$ for the estimation of $\alpha$ when $X \sim \mathcal{G}(\alpha, 1)$ and $x = 1.5$.*

Obviously, it is only interesting to maximize the likelihood by this method when more standard methods like the ones above are difficult or impossible to implement and the computation of Bayes estimators is straightforward. (Chapters 6 and 7 show that this second condition is actually very mild.) It is, indeed, necessary to compute the Bayes estimators $\delta_\lambda^\pi(x)$ corresponding to a sequence of $\lambda$'s until they stabilize. Note that when iterative algorithms are used to compute $\delta_\lambda^\pi(x)$, the previous solution (in $\lambda$) of $\delta_\lambda^\pi(x)$ can serve as the new initial value for the computation of $\delta_\lambda^\pi(x)$ for a larger value of $\lambda$. This feature increases the analogy with simulated annealing. The differences with simulated annealing are:

(i) for a fixed temperature $(1/\lambda)$, the algorithm converges to a fixed value, $\delta_\lambda^\pi$;

(ii) a continuous decrease of $1/\lambda$ is statistically meaningless;

(iii) the speed of convergence of $\lambda$ to $+\infty$ does not formally matter[4] for the convergence of $\delta_\lambda^\pi(x)$ to $\theta^*$;

(iv) the statistical motivation of this method is obviously stronger, in particular because of the meaning of the parameter $\lambda$;

(v) the only analytical constraint on $\ell(\theta|x)$ is the existence of a global maximum, $\theta^*$ (see Robert and Titterington 1998 for extensions).

**Example 5.2.9 Gamma shape estimation.** Consider the estimation of the shape parameter, $\alpha$, of a $\mathcal{G}(\alpha, \beta)$ distribution with $\beta$ known. Without loss of generality, take $\beta = 1$. For a constant (improper) prior distribution on $\alpha$, the posterior distribution satisfies

$$\pi_\lambda(\alpha|x) \propto x^{\lambda(\alpha-1)} e^{-\lambda x} \Gamma(\alpha)^{-\lambda} .$$

For a fixed $\lambda$, the computation of $\mathbb{E}[\alpha|x, \lambda]$ can be obtained by simulation with the Metropolis–Hastings algorithm (see Chapter 6 for details) based on the instrumental distribution $\mathcal{E}xp(1/\alpha^{(n-1)})$, where $\alpha^{(n-1)}$ denotes the previous value of the associated Markov chain. Table 5.2.3 presents the evolution of $\delta_\lambda^\pi(x) = \mathbb{E}[\alpha|x, \lambda]$ against $\lambda$, for $x = 1.5$.

An analytical verification (using a numerical package like `Mathematica`) shows that the maximum of $x^\alpha/\Gamma(\alpha)$ is, indeed, close to 2.0 for $x = 1.5$. ‖

The appeal of recursive integration is also clear in the case of constrained parameter estimation.

---

[4] However, we note that if $\lambda$ increases too quickly, the performance is affected in that there may be convergence to a local mode (see Robert and Titterington 1998 for an illustration).

| ACT | 1–12 | 13–15 | 16–18 | 19–21 | 22–24 | 25–27 | 28–30 | 31–33 | 34–36 |
|------|------|-------|-------|-------|-------|-------|-------|-------|-------|
| 91- 99 | 1.57 (4) | 2.11 (5) | 2.73 (18) | 2.96 (39) | 2.97 (126) | 3.13 (219) | 3.41 (232) | 3.45 (47) | 3.51 (4) |
| 81- 90 | 1.80 (6) | 1.94 (15) | 2.52 (30) | 2.68 (65) | 2.69 (117) | 2.82 (143) | 2.75 (70) | 2.74 (8) | – (0) |
| 71- 80 | 1.88 (10) | 2.32 (13) | 2.32 (51) | 2.53 (83) | 2.58 (115) | 2.55 (107) | 2.72 (24) | 2.76 (4) | – (0) |
| 61- 70 | 2.11 (6) | 2.23 (32) | 2.29 (59) | 2.29 (84) | 2.50 (75) | 2.42 (44) | 2.41 (19) | – (0) | – (0) |
| 51- 60 | 1.60 (11) | 2.06 (16) | 2.12 (49) | 2.11 (63) | 2.31 (57) | 2.10 (40) | 1.58 (4) | 2.13 (1) | – (0) |
| 41- 50 | 1.75 (6) | 1.98 (12) | 2.05 (31) | 2.16 (42) | 2.35 (34) | 2.48 (21) | 1.36 (4) | – (0) | – (0) |
| 31- 40 | 1.92 (7) | 1.84 (6) | 2.15 (5) | 1.95 (27) | 2.02 (13) | 2.10 (13) | 1.49 (2) | – (0) | – (0) |
| 21- 30 | 1.62 (1) | 2.26 (2) | 1.91 (5) | 1.86 (14) | 1.88 (11) | 3.78 (1) | 1.40 (2) | – (0) | – (0) |
| 00- 20 | 1.38 (1) | 1.57 (2) | 2.49 (5) | 2.01 (7) | 2.07 (7) | – (0) | 0.75 (1) | – (0) | – (0) |

Table 5.2.4. *Average grades of first year students at the University of Iowa given their rank at the end of high school (HSR) and at the ACT exam. Numbers in parentheses indicate the number of students in each category. (Source: Robertson et al. 1988).*

| ACT | 1 − 12 | 13 − 15 | 16 − 18 | 19 − 21 | 22 − 24 | 25 − 27 | 28 − 39 | 31 − 32 | 34 − 36 |
|------|--------|---------|---------|---------|---------|---------|---------|---------|---------|
| 91 − 99 | 1.87 | 2.18 | 2.73 | 2.96 | 2.97 | 3.13 | 3.41 | 3.45 | 3.51 |
| 81 − 89 | 1.87 | 2.17 | 2.52 | 2.68 | 2.69 | 2.79 | 2.79 | 2.80 | — |
| 71 − 79 | 1.86 | 2.17 | 2.32 | 2.53 | 2.56 | 2.57 | 2.72 | 2.76 | — |
| 61 − 69 | 1.86 | 2.17 | 2.29 | 2.29 | 2.46 | 2.46 | 2.47 | — | — |
| 51 − 59 | 1.74 | 2.06 | 2.12 | 2.13 | 2.24 | 2.24 | 2.24 | 2.27 | — |
| 41 − 49 | 1.74 | 1.98 | 2.05 | 2.13 | 2.24 | 2.24 | 2.24 | — | — |
| 31 − 39 | 1.74 | 1.94 | 1.99 | 1.99 | 2.02 | 2.06 | 2.06 | — | — |
| 21 − 29 | 1.62 | 1.93 | 1.97 | 1.97 | 1.98 | 2.05 | 2.06 | — | — |
| 00 − 20 | 1.38 | 1.57 | 1.97 | 1.97 | 1.97 | — | 1.97 | — | — |

Table 5.2.5. *Maximum likelihood estimates of the mean grades under lexicographical constraint. (Source: Robert and Hwang 1996).*

**Example 5.2.10 Isotonic regression.** Consider a table of normal observations $X_{i,j} \sim \mathcal{N}(\theta_{i,j}, 1)$ with means that satisfy

$$\theta_{i-1,j} \vee \theta_{i,j-1} \le \theta_{i,j} \le \theta_{i+1,j} \wedge \theta_{i,j+1}.$$

Dykstra and Robertson (1982) have developed an efficient *deterministic* algorithm which maximizes the likelihood under these restrictions (see Problems 1.22 and 1.23.) However, a direct application of recursive integration also provides the maximum of likelihood estimator of $\theta = (\theta_{ij})$, requiring neither an extensive theoretical study nor high programming skills.

Table 5.2.4 presents the data of Robertson *et al.* (1988), which relates the notes at the end of first year with two entrance exams at the University of Iowa. Although these values are bounded, it is possible to use a normal model if the function to minimize is the least squares criterion (1.2.3), as already pointed out in Example 1.2.2. Table 5.2.5 provides the solution obtained by recursive integration in Robert and Hwang (1996); it coincides with the result of Robertson *et al.* (1988). ‖

## 5.3 Stochastic Approximation

We next turn to methods that work more directly with the objective function rather than being concerned with fast explorations of the space. Informally speaking, these methods are somewhat preliminary to the true optimization step, in the sense that they utilize approximations of the objective function $h$. We note that these approximations have a different purpose than those we have previously encountered (for example, Laplace and saddlepoint approximations in §3.5 and §3.6). In particular, the methods described here may sometimes result in an additional level of error by looking at the maximum of an approximation to $h$.

Since most of these approximation methods only work in so-called *missing data models*, we start this section with a brief introduction to these models (see Chapter 9 for more details). We return to the assumption that the objective function $h$ satisfies $h(x) = \mathbb{E}[H(x, Z)]$ and (as promised) to show that this assumption arises in many realistic setups. Moreover, note that artificial extensions (or *demarginalization*), which use this representation, are only computational devices and do not invalidate the overall inference.

### 5.3.1 Missing Data Models and Demarginalization

In the previous chapters, we have already met structures where some missing (or latent) element greatly complicates the observed model. Examples include the obvious censored data models (Example 1.1.1), mixture models (Example 1.1.2), where we do not observe the indicator of the component generating the observation, or logistic regression (Example 1.3.3), where the observation $Y_i$ can be interpreted as an indicator that a continuous variable with logistic distribution is less than $X_i^t \beta$.

Missing data models are best thought of as models where the likelihood can be expressed as

$$(5.3.1) \qquad g(x|\theta) = \int_{\mathcal{Z}} f(x, z|\theta) \, dz$$

or, more generally, where the function $h(x)$ to be optimized can be expressed as the expectation

$$(5.3.2) \qquad h(x) = \mathbb{E}[H(x, Z)] \, .$$

This assumption is relevant, and useful, in the setup of censoring models:

**Example 5.3.1 Censored data likelihood.** Suppose that we observe $Y_1$, ..., $Y_n$, iid, from $f(y - \theta)$ and we have ordered the observations so that $\mathbf{y} = (y_1, \cdots, y_m)$ are uncensored and $(y_{m+1}, \dots, y_n)$ are censored (and equal to $a$). The likelihood function is

$$(5.3.3) \qquad L(\theta|\mathbf{y}) = \prod_{i=1}^{m} [1 - F(a - \theta)]^{n-m} f(y_i - \theta) \, ,$$

where $F$ is the cdf associated with $f$. If we had observed the last $n - m$ values, say $\mathbf{z} = (z_{m+1}, \dots, z_n)$, with $z_i > a$ $(i = m + 1, \dots, n)$, we could have constructed the (complete data) likelihood

$$L^c(\theta|\mathbf{y},\mathbf{z}) = \prod_{i=1}^m f(y_i - \theta) \prod_{i=m+1}^n f(z_i - \theta),$$

with which it often is easier to work. Note that

$$L(\theta|\mathbf{y}) = \mathbb{E}[L^c(\theta|\mathbf{y},\mathbf{Z})] = \int_{\mathcal{Z}} L^c(\theta|\mathbf{y},\mathbf{z}) f(\mathbf{z}|\mathbf{y},\theta)\,d\mathbf{z},$$

where $f(\mathbf{z}|\mathbf{y},\theta)$ is the density of the missing data conditional on the observed data.                                                                    ‖

When (5.3.1) holds, the $z$ vector merely serves to simplify calculations, and the way $z$ is selected to satisfy (5.3.2) should not affect the value of the estimator. This is a *missing data model*, and we refer to the function $L^c(\theta|x,z)) = f(x,z|\theta)$ as the "complete-model" or "complete-data" likelihood, which corresponds to the observation of the *complete data* $(x,z)$. This complete model is often within the exponential family framework, making it much easier to work with (see Problem 5.16).

More generally, we refer to the representation (5.3.1) as *demarginalization*, a setting where a function (or a density) of interest can be expressed as an integral of a more manageable quantity. We will meet such setups again in Chapter 7. They cover models such as missing data models (censoring, grouping, mixing, etc.), latent variable models (tobit, probit, arch, stochastic volatility, etc.) and also artificial embedding, where the variable $Z$ in (5.3.2) has no meaning for the inferential or optimization problem, as illustrated by *slice sampling* (§7.1.2). Chapter 9 also contains more illustrations of where demarginalizations are useful.

### 5.3.2 Monte Carlo Approximation

In cases where $h(x)$ can be written as $\mathbb{E}[H(x,Z)]$ but is not directly computable, it can be approximated by the empirical (Monte Carlo) average

$$(5.3.4) \qquad \hat{h}(x) = \frac{1}{m} \sum_{i=1}^m H(x, z_i)\,,$$

where the $Z_i$'s are generated from the conditional distribution $f(z|x)$. This approximation yields a convergent estimator of $h(x)$ for every value of $x$, but its use in optimization setups is not recommended for at least two related reasons:

(i) Since the function $h(x)$ is to be optimized and as most optimization algorithms are iterative, presumably $h(x)$ needs to be evaluated at many points. This will involve the generation of many samples of $Z_i$'s of size $m$.

(ii) Since the sample changes with every value of $x$, the resulting sequence of evaluations of $h$ will usually not be smooth enough to preserve the convergence properties of the optimization technique based on a fixed $h$.

(For instance, we may lose the monotonicity or convexity characteristics of the true function.[5])

These difficulties prompted Geyer (1996) to suggest, instead, an *importance sampling* approach to this problem and use a single sample of $Z_i$'s simulated from a distribution $g(z)$ (which may or may not correspond to a specific value of $x$; that is, $g(z) = f(z|x_0)$). A better estimate for optimization purposes is thus

$$(5.3.5) \qquad \hat{h}_m(x) = \frac{1}{m} \sum_{i=1}^{m} \frac{f(z_i|x)}{g(z_i)} H(x, z_i) \, ,$$

where the $Z_i$'s are simulated from $g(z)$. Since this evaluation of $h$ does not depend on $x$, points (i) and (ii) above are answered.

The problem then shifts from (5.1.1) to

$$(5.3.6) \qquad \max_{x} \hat{h}_m(x) \, ,$$

which leads to a convergent solution in most cases and also allows for the use of regular optimization techniques, since the function $\hat{h}_m$ does not vary with each iteration. However, three remaining drawbacks of this approach are as follows:

(i) As $\hat{h}_m$ is expressed as a sum, it most often enjoys fewer analytical properties than the original function $h$. For example, the smoothing effect of the integral on the integrand $H(x, z)$ is no longer enjoyed by $\hat{h}_m$, and properties such as unimodality or concavity may vanish. Regular optimization methods will thus have to expend more effort to determine the solution to (5.3.6) than for other approximations of $h(x)$.

(ii) The choice of the importance function $g$ can be very influential in obtaining a good approximation of the *function* $h(x)$, in the sense that the estimator (5.3.5) does not converge uniformly well over all values of $x$. In particular, it may be slow to converge when $g(z)$ is of the form $f(z|x_0)$ and $x_0$ is outside a neighborhood of the true maximum $x^*$.

(iii) The number of points $z_i$ used in the approximation (5.3.4) should vary with $x$ to achieve the same precision in the approximation of $h(x)$, but this is usually impossible to assess in advance.

In the case $g(z) = f(z|x_0)$, Geyer's (1996) solution to (ii) is to use a recursive process in which $x_0$ is updated by the solution of the last optimization at each step. The Monte Carlo maximization algorithm then looks like the following:

**Algorithm A.22 –Monte Carlo Maximization–**
At step $i$

---

[5] In other words, if $x^*$ is a solution to $\hat{h}'(x) = 0$, it most often depends on the sample of $Z_i$'s. Therefore, iterative optimization algorithms which involve simulation of the $Z_i$'s at each step may fail to converge.

1. Generate $z_1, \ldots, z_m \sim f(z|x_i)$
   and compute $\hat{h}_{g_i}$ with $g_i = f(\cdot|x_i)$.
2. Find $x^* = \arg\max \hat{h}_{g_i}(x)$.
3. Update $x_i$ to $x_{i+1} = x^*$.                                    [A.22]
   Repeat until $x_i = x_{i+1}$.

**Example 5.3.2 Maximum likelihood estimation for exponential families.** Geyer and Thompson (1992) take advantage of this technique to derive maximum likelihood estimators in exponential families; that is, for functions

$$h(x|\theta) = c(\theta)e^{\theta x - \nu(x)} = c(\theta)\tilde{h}(x|\theta) .$$

The difficulty encountered in these problems is that, outside of the most standard examples, the normalizing constant $c(\theta)$ may be unknown or difficult to compute. (Recall the case of the beta distribution of Example 1.2.4.) This can complicate computation of the maximum likelihood estimator or likelihood regions.

Direct computation of $c(\theta)$ can be avoided by using the identity

$$(5.3.7) \qquad \log \frac{h(x|\theta)}{h(x|\eta)} = \log \frac{\tilde{h}(x|\theta)}{\tilde{h}(x|\eta)} - \log \mathbb{E}_\eta \left[ \frac{\tilde{h}(X|\theta)}{\tilde{h}(X|\eta)} \right] ,$$

and Geyer and Thompson (1992, 1995) propose to optimize the approximation

$$\log \frac{\tilde{h}(x|\theta)}{\tilde{h}(x|\eta)} - \log \frac{1}{m} \sum_{i=1}^m \frac{\tilde{h}(x_i|\theta)}{\tilde{h}(x_i|\eta)} ,$$

where the $X_i$'s are generated from $h(x|\eta)$ (see also Geyer 1993, 1994).

This representation also extends to setups where the likelihood $L(\theta|x)$ is known up to a multiplicative constant (that is, where $L(\theta|x) \propto h(\theta|x)$). ∥

In some missing-data models, as those detailed in Billio *et al.* (1998) for econometric models, the likelihood function $L(\theta|x)$ can be written as the marginal of $f(x, z|\theta)$ (as in (5.3.3)) and the likelihood ratio is thus available in the form

$$\frac{L(\theta|x)}{L(\eta|x)} = \mathbb{E}_\eta \left[ \frac{f(x, Z|\theta)}{f(x, Z|\eta)} \bigg| x \right] ,$$

where $Z \sim f(x, z|\eta)$, since

$$\mathbb{E}_\eta \left[ \frac{f(x, Z|\theta)}{f(x, Z|\eta)} \bigg| x \right] = \int_\mathcal{Z} \frac{f(x,z|\theta)}{f(x,z|\eta)} \frac{f(x,z|\eta)}{f(x|\eta)} dz$$

$$= \frac{\displaystyle\int_\mathcal{Z} f(x, z|\theta)dz}{\displaystyle\int_\mathcal{Z} f(x, z|\eta)dz} .$$

Complex likelihoods can thus be maximized through approximations like

$$(5.3.8) \qquad \hat{h}_\eta(\theta) = \frac{1}{m} \sum_{i=1}^m \frac{f(x, z_i|\theta)}{f(x, z_i|\eta)} ,$$

where the $z_i$'s are generated from $f(z|x, \eta)$.

Besides the drawback mentioned above, that the approximation (5.3.8) is a sum and can be difficult to maximize, an additional difficulty associated with this method is that the estimator (5.3.8) is an importance sampling estimator and, therefore its variance is not necessarily finite. In many cases, there exists a range of values of $\theta$, dependent on $\eta$, where (5.3.8) has an infinite variance. This peculiarity implies that the subset

$$\Theta_\eta = \{\theta;\ \text{var}(\hat{h}_\eta(\theta)) < \infty\}$$

must be determined and that the maximization of $\hat{h}_\eta(\theta)$ must be done under the constraint that $\theta \in \Theta_\eta$. Although the complete-data likelihood is usually more manageable, the construction of $\Theta_\eta$ is not always feasible.

**Example 5.3.3 ARCH Models.** A Gaussian ARCH (*Auto Regressive Conditionally Heteroscedastic*) model is defined, for $t = 2, \ldots, T$, by

$$(5.3.9) \qquad \begin{cases} Z_t = (\alpha + \beta Z_{t-1}^2)^{1/2}\varepsilon_t^*, \\ X_t = aZ_t + \varepsilon_t, \end{cases}$$

where $a \in \mathbb{R}^p$, $\varepsilon_t^* \overset{iid}{\sim} \mathcal{N}(0, 1)$, and $\varepsilon_t \overset{iid}{\sim} \mathcal{N}_p(0, \Sigma)$ independently. Let $\theta = (a, \beta, \sigma)$ denote the parameter of the model and assume, in addition, that $Z_1 \sim \mathcal{N}(0, 1)$. The sample is denoted by $x^T = (x_1, \ldots, x_T)$. (For details on the theory and use of ARCH models, see the review papers by Bollerslev *et al.* 1992 and Kim *et al.* 1998, or the books by Enders 1994 and Gouriéroux 1996.)

For simplicity and identifiability reasons, we consider the following special case

$$(5.3.10) \qquad \begin{cases} Z_t = (1 + \beta Z_{t-1}^2)^{1/2}\varepsilon_t^*, & \varepsilon_t^* \sim \mathcal{N}(0, 1), \\ X_t = aZ_t + \varepsilon_t, & \varepsilon_t \sim \mathcal{N}_p(0, \sigma^2 I_p). \end{cases}$$

Of course, the difficulty in estimation under this model comes from the fact that only the $x_t$'s are observed. The approximation of the likelihood ratio (5.3.8) is then based on the simulation of the missing data $Z^T = (Z_1, \ldots, Z_T)$ from

$$\tilde{f}(z^T|x^T, \theta) \propto f(z^T, x^T|\theta)$$

$$(5.3.11) \propto \sigma^{-2T} \exp\left\{-\sum_{t=1}^{T} ||x_t - az_t||^2/2\sigma^2\right\} e^{-z_1^2/2} \prod_{t=2}^{T} \frac{e^{-z_t^2/2(1+\beta z_{t-1}^2)}}{(1 + \beta z_{t-1}^2)^{1/2}},$$

whose implementation using a Metropolis–Hastings algorithm (see Chapter 6) is detailed in Example 7.3.5. Given a sample $(z_1^T, \ldots, z_m^T)$ simulated from (5.3.11), with $\theta = \eta$, and an observed sample $x^T$, the approximation (5.3.8) is given by

$$(5.3.12) \qquad \frac{1}{m}\sum_{i=1}^{m} \frac{f(z_i^T, x^T|\theta)}{f(z_i^T, x^T|\eta)},$$

where $f(z_i^T, x^T|\theta)$ is defined in (5.3.11).

We consider a simulated sample of size $T = 100$ with $p = 2$, $a = (-0.2, 0.6)$, $\beta = 0.8$, and $\sigma^2 = 0.2$. The above approximation method is

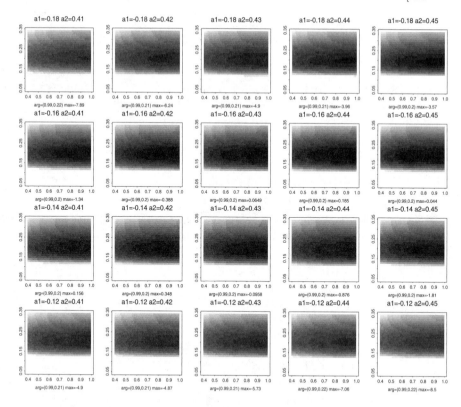

Figure 5.3.1. *Estimated slices of the log-likelihood ratio for the factor ARCH model (5.3.10) for a simulated sample of size $T = 100$ and $50,000$ iterations of the completion step, with a batch size of $50$. See Table 5.3.1 for the parameter values. (Source: Billio et al. 1998.)*

quite sensitive to the value of $\eta$ and a good choice is the noninformative Bayes estimate associated with the prior $\pi(a, \beta, \sigma) = 1/\sigma$, which can be obtained by a Metropolis–Hastings algorithm (see Example 7.3.5). Table 5.3.1 gives the result of the maximization of (5.3.12) for $m = 50,000$.

As shown in Figure 5.3.1, which provides a slice representation of the estimated log-likelihood ratio, the appeal of this Monte Carlo approximation is that it goes beyond the derivation of an approximate maximum likelihood estimator and provides information on the likelihood function itself. (The maximum likelihood estimator $\hat{\beta}$ is far from the true value $\beta$, but the estimated log-likelihood ratio at $(\hat{a}, \hat{\beta}, \hat{\sigma})$ is 0.348, which indicates that the likelihood is rather flat in this region.)          ‖

| | True $\theta$ | Starting value $\eta$ | Approximate maximum likelihood estimate $\hat{\theta}$ |
|---|---|---|---|
| $a_1$ | -0.2 | -0.153 | -0.14 |
| $a_2$ | 0.6 | 0.43 | 0.42 |
| $\beta$ | 0.8 | 0.86 | 0.99 |
| $\sigma^2$ | 0.2 | 0.19 | 0.2 |

Table 5.3.1. *Estimation result for the factor ARCH model (5.3.10) with a simulated sample of size $T = 100$ and a Bayes estimate as starting value. (Source: Billio et al. 1998.)*

### 5.3.3 The EM Algorithm

The EM (*Expectation–Maximization*) algorithm was originally introduced by Dempster *et al.* (1977) to overcome the difficulties in maximizing likelihoods by taking advantage of the representation (5.3.1) and solving a sequence of easier maximization problems whose limit is the answer to the original problem. It thus fits naturally in this demarginalization section, even though it is not a stochastic algorithm in its original version. Monte Carlo versions are examined in §5.3.4 and in Note 5.5.1. Moreover, the EM algorithm relates to MCMC algorithms in the sense that it can be seen as a forerunner of the Gibbs sampler in its Data Augmentation version (§7.1.2), replacing simulation by maximization.

Suppose that we observe $X_1, \ldots, X_n$, iid from $g(x|\theta)$ and want to compute $\hat{\theta} = \arg\max L(\theta|\mathbf{x}) = \prod_{i=1}^{n} g(x_i|\theta)$. We augment the data with $\mathbf{z}$, where $\mathbf{X}, \mathbf{Z} \sim f(\mathbf{x}, \mathbf{z}|\theta)$ and note the identity (which is a basic identity for the EM algorithm)

$$(5.3.13) \qquad k(\mathbf{z}|\theta, \mathbf{x}) = \frac{f(\mathbf{x}, \mathbf{z}|\theta)}{g(\mathbf{x}|\theta)},$$

where $k(\mathbf{z}|\theta, \mathbf{x})$ is the conditional distribution of the missing data $\mathbf{Z}$ given the observed data $\mathbf{x}$. The identity (5.3.13) leads to the following relationship between the complete-data likelihood $L^c(\theta|\mathbf{x}, \mathbf{z})$ and the observed data likelihood $L(\theta|\mathbf{x})$. For any value $\theta_0$,

$$(5.3.14) \quad \log L(\theta|\mathbf{x}) = \mathbb{E}_{\theta_0}[\log L^c(\theta|\mathbf{x}, \mathbf{z})|\theta_0, \mathbf{x}] - \mathbb{E}_{\theta_0}[\log k(\mathbf{z}|\theta, \mathbf{x})|\theta_0, \mathbf{x}],$$

where the expectation is with respect to $k(\mathbf{z}|\theta_0, \mathbf{x})$. We now see the EM algorithm as a demarginalization model. However, the strength of the EM algorithm is that it can go further. In particular, to maximize $\log L(\theta|\mathbf{x})$, we only have to deal with the first term on the right side of (5.3.14), as the other term can be ignored.

Common EM notation is to denote the expected log-likelihood by

$$(5.3.15) \qquad Q(\theta|\theta_0, \mathbf{x}) = \mathbb{E}_{\theta_0}[\log L^c(\theta|\mathbf{x}, \mathbf{z})|\theta_0, \mathbf{x}].$$

We then maximize $Q(\theta|\theta_0, \mathbf{x})$, and if $\hat{\theta}_{(1)}$ is the value of $\theta$ maximizing $Q(\theta|\theta_0, \mathbf{x})$, the process can then be repeated with $\theta_0$ replaced by the updated value $\hat{\theta}_{(1)}$. In this manner, a sequence of estimators $\hat{\theta}_{(j)}$, $j = 1, 2, \ldots$, is obtained iteratively where $\hat{\theta}_{(j)}$ is defined as the value of $\theta$ maximizing $Q(\theta | \hat{\theta}_{(j-1)}, \mathbf{x})$; that is,

$$(5.3.16) \qquad Q(\hat{\theta}_{(j)} | \hat{\theta}_{(j-1)}, \mathbf{x}) = \max_{\theta} Q(\theta | \hat{\theta}_{(j-1)}, \mathbf{x}).$$

The iteration described above contains both an expectation step and a maximization step, giving the algorithm its name. At the $j$th step of the iteration, we calculate the expectation (5.3.15), with $\theta_0$ replaced by $\hat{\theta}_{(j-1)}$ (*the E-step*), and then maximize it (*the M-step*).

**Algorithm A.23 –The EM Algorithm–**

1. Compute
$$Q(\theta | \hat{\theta}_{(m)}, \mathbf{x}) = \mathbb{E}_{\hat{\theta}_{(m)}} [\log L^c(\theta | \mathbf{x}, \mathbf{z})],$$

   where the expectation is with respect to $k(\mathbf{z} | \hat{\theta}_m, \mathbf{x})$ (*the E-step*)

   .

2. Maximize $Q(\theta | \hat{\theta}_{(m)}, \mathbf{x})$ in $\theta$ and take (*the M-step*)          [A.23]

$$\theta_{(m+1)} = \arg\max_{\theta} Q(\theta | \hat{\theta}_{(m)}, \mathbf{x}).$$

The iterations are conducted until a fixed point of $Q$ is obtained.

The theoretical core of the EM Algorithm is based on the fact that by maximizing $Q(\theta | \hat{\theta}_{(m)}, \mathbf{x})$ at each step, the likelihood on the left side of (5.3.14) is increased at each step. The following theorem was established by Dempster *et al.* (1977).

**Theorem 5.3.4** *The sequence* $(\hat{\theta}_{(j)})$ *defined by (5.3.16) satisfies*

$$L(\hat{\theta}_{(j+1)} | \mathbf{x}) \geq L(\hat{\theta}_{(j)} | \mathbf{x}),$$

*with equality holding if and only if* $Q(\hat{\theta}_{(j+1)} | \hat{\theta}_{(j)}, \mathbf{x}) = Q(\hat{\theta}_{(j)} | \hat{\theta}_{(j)}, \mathbf{x})$.

*Proof.* On successive iterations, it follows from the definition of $\hat{\theta}_{(j+1)}$ that

$$Q(\hat{\theta}_{(j+1)} | \hat{\theta}_{(j)}, \mathbf{x}) \geq Q(\hat{\theta}_{(j)} | \hat{\theta}_{(j)}, \mathbf{x}).$$

Thus, if we can show that

$$(5.3.17) \ \mathbb{E}_{\hat{\theta}_{(j)}} [\log k(\mathbf{Z} | \hat{\theta}_{(j+1)}, \mathbf{x}) | \hat{\theta}_{(j)}, \mathbf{x}] \leq \mathbb{E}_{\hat{\theta}_{(j)}} [\log k(\mathbf{V} | \hat{\theta}_{(j)}, \mathbf{x}) | \hat{\theta}_{(j)}, \mathbf{x}],$$

it will follow from (5.3.14) that the value of the likelihood is increased at each iteration.

Since the difference of the logarithms is the logarithm of the ratio, (5.3.17) can be written as

$$\mathbb{E}_{\hat{\theta}_{(j)}} \left[ \log \left( \frac{k(\mathbf{Z} | \hat{\theta}_{(j+1)}, \mathbf{x})}{k(\mathbf{Z} | \hat{\theta}_j, \mathbf{x})} \right) | \hat{\theta}_{(j)}, \mathbf{x} \right]$$

$$(5.3.18) \qquad \leq \log \mathbb{E}_{\hat{\theta}_{(j)}} \left[ \frac{k(\mathbf{Z} | \hat{\theta}_{(j+1)}, \mathbf{x})}{k(\mathbf{Z} | \hat{\theta}_j, \mathbf{x})} | \hat{\theta}_{(j)}, \mathbf{x} \right] = 0,$$

where the inequality follows from Jensen's inequality (see Problem 5.17). The theorem is therefore established. □□

Although Theorem 5.3.4 guarantees that the likelihood will increase at each iteration, we still may not be able to conclude that the sequence $(\hat{\theta}_{(j)})$ converges to a maximum likelihood estimator. To ensure convergence we require further conditions on the mapping $\hat{\theta}_{(j)} \to \hat{\theta}_{(j+1)}$. These conditions are investigated by Boyles (1983) and Wu (1983). The following theorem is, perhaps, the most easily applicable condition to guarantee convergence to a *stationary point*, a zero of the first derivative that may be a local maximum or saddlepoint.

**Theorem 5.3.5** *If the expected complete-data likelihood $Q(\theta|\theta_0, \mathbf{x})$ is continuous in both $\theta$ and $\theta_0$, then every limit point of an EM sequence $(\hat{\theta}_{(j)})$ is a stationary point of $L(\theta|\mathbf{x})$, and $L(\hat{\theta}_{(j)}|\mathbf{x})$ converges monotonically to $L(\hat{\theta}|\mathbf{x})$ for some stationary point $\hat{\theta}$.*

Note that convergence is only guaranteed to a stationary point. Techniques such as running the EM algorithm a number of times with different, random starting points, or algorithms such as *simulated annealing* (see, for example, Finch *et al.* 1989) attempt to give some assurance that the global maximum is found. Wu (1983) states another theorem that guarantees convergence to a local maximum, but its assumptions are difficult to check. It is usually better, in practice, to use empirical methods (graphical or multiple starting values) to check that a maximum has been reached.

**Example 5.3.6 (Continuation of Example 5.3.1)** If $f(x - \theta)$ is the $\mathcal{N}(\theta, 1)$ density, the likelihood function (5.3.3) is

$$L(\theta|\mathbf{x}) = \frac{1}{(2\pi)^{m/2}} \exp\left\{-\frac{1}{2}\sum_{i=1}^{m}(x_i - \theta)^2\right\}[1 - \Phi(a - \theta)]^{n-m}$$

and the complete-data log-likelihood is

$$\log L^c(\theta|\mathbf{x}, \mathbf{z}) \propto -\frac{1}{2}\sum_{i=1}^{m}(x_i - \theta)^2 - \frac{1}{2}\sum_{i=m+1}^{n}(z_i - \theta)^2,$$

where the $z_i$'s are observations from the truncated normal distribution

$$k(z|\theta, \mathbf{x}) = \frac{\exp\{-\frac{1}{2}(z - \theta)^2\}}{\sqrt{2\pi}[1 - \Phi(a - \theta)]} = \frac{\varphi(z - \theta)}{1 - \Phi(a - \theta)}, \qquad a < z.$$

At the $j$th step in the EM sequence, we have

$$Q(\theta|\hat{\theta}_{(j)}, \mathbf{x}) \propto -\frac{1}{2}\sum_{i=1}^{m}(x_i - \theta)^2 - \frac{1}{2}\sum_{i=m+1}^{n}\int_a^{\infty}(z_i - \theta)^2 k(z|\hat{\theta}_{(j)}, \mathbf{x})\,dz_i,$$

and differentiating with respect to $\theta$ yields

$$m(\bar{x} - \theta) + (n - m)\left(\mathbb{E}[Z|\hat{\theta}_{(j)}] - \theta\right) = 0 ;$$

that is

$$\hat{\theta}_{(j+1)} = \frac{m\bar{x} + (n-m)\mathbb{E}[Z|\hat{\theta}_{(j)}]}{n} ,$$

where

$$\mathbb{E}[Z|\hat{\theta}_{(j)}] = \int_a^\infty zk(z|\hat{\theta}_{(j)}, \mathbf{x}) \, dz = \hat{\theta}_{(j)} + \frac{\varphi(a - \hat{\theta}_{(j)})}{1 - \Phi(a - \hat{\theta}_{(j)})}.$$

Thus, the EM sequence is defined by

$$\hat{\theta}_{(j+1)} = \frac{m}{n}\bar{x} + \frac{n-m}{n}\left[\hat{\theta}_{(j)} + \frac{\varphi(a - \hat{\theta}_{(j)})}{1 - \Phi(a - \hat{\theta}_{(j)})}\right],$$

which converges to the MLE $\hat{\theta}$.                                   ‖

The books by Little and Rubin (1987) and Tanner (1996) provide good overviews of the EM literature. Other references include Louis (1982), Little and Rubin 1983, Laird *et al.* (1987), Meng and Rubin (1991), Quian and Titterington (1992) Smith and Roberts (1993), Liu and Rubin (1994), MacLachlan and Krishnan (1997), and Meng and van Dyk (1997).

### 5.3.4 Monte Carlo EM

A difficulty with the implementation of the EM algorithm is that each "E-step" requires the computation of the expected log likelihood $Q(\theta|\theta_0, \mathbf{x})$. Wei and Tanner (1990a, b) propose to use a Monte Carlo approach (MCEM) to overcome this difficulty, by simulating $Z_1, \ldots, Z_m$ from the conditional distribution $k(\mathbf{z}|\mathbf{x}, \theta)$ and then maximizing the approximate complete data log-likelihood

(5.3.19)                    $$\hat{Q}(\theta|\theta_0, \mathbf{x}) = \frac{1}{m}\sum_{i=1}^m \log L^c(\theta|\mathbf{x}, \mathbf{z}) .$$

When $m$ goes to infinity, this quantity, indeed, converges to $Q(\theta|\theta_0, \mathbf{x})$ and the limiting form of the *Monte Carlo EM* algorithm is, thus, the regular EM algorithm. The authors suggest that $m$ should be increased along with the iterations. Although the maximization of a sum like (5.3.19) is, in general, rather involved, exponential family settings often allow for closed-form solutions.

**Example 5.3.7 Genetic linkage.** A classic (perhaps overused) example of the EM Algorithm is the genetics problem (see Rao 1973, Dempster *et al.* 1977, or Tanner 1996), where observations $(x_1, x_2, x_3, x_4)$ are gathered from the multinomial distribution

$$\mathcal{M}\left(n; \frac{1}{2} + \frac{\theta}{4}, \frac{1}{4}(1-\theta), \frac{1}{4}(1-\theta), \frac{\theta}{4}\right).$$

Estimation is easier if the $x_1$ cell is split into two cells, so we create the augmented model

$$(z_1, z_2, x_2, x_3, x_4) \sim \mathcal{M}\left(n; \frac{1}{2}, \frac{\theta}{4}, \frac{1}{4}(1-\theta), \frac{1}{4}(1-\theta), \frac{\theta}{4}\right),$$

with $x_1 = z_1 + z_2$. The complete-data likelihood function is then simply $\theta^{z_2+x_4}(1-\theta)^{x_2+x_3}$, as opposed to the observed likelihood function $(2 + \theta)^{x_1}\theta^{x_4}(1-\theta)^{x_2+x_3}$. The expected complete log-likelihood function is

$$\mathbb{E}_{\theta_0}[(Z_2 + x_4)\log\theta + (x_2 + x_3)\log(1-\theta)]$$

$$= \left(\frac{\theta_0}{2+\theta_0}x_1 + x_4\right)\log\theta + (x_2 + x_3)\log(1-\theta),$$

which can easily be maximized in $\theta$, leading to

$$\hat{\theta}_1 = \frac{\dfrac{\theta_0 x_1}{2+\theta_0} + x_4}{\dfrac{\theta_0 x_1}{2+\theta_0} + x_2 + x_3 + x_4}.$$

If we instead use the Monte Carlo EM algorithm, $\theta_0 x_1/(2+\theta_0)$ is replaced with the average

$$\bar{z}_m = \frac{1}{m}\sum_{i=1}^{m} z_i,$$

where the $z_i$'s are simulated from a binomial distribution $\mathcal{B}(x_1, \theta_0/(2+\theta_0))$. The maximum in $\theta$ is then

$$\theta_1 = \frac{\bar{z}_m + x_4}{\bar{z}_m + x_2 + x_3 + x_4}. \qquad \|$$

This example is merely an illustration of the Monte Carlo EM algorithm since EM also applies. The next example, however, details a situation in which the expectation is quite complicated and the Monte Carlo EM algorithm works quite nicely.

**Example 5.3.8 Capture–Recapture models revisited.** A generalization of a capture–recapture model (see Example 2.3.11) is to assume that an animal $i$, $i = 1, 2, \ldots, n$ can be captured at time $j$, $j = 1, 2, \ldots, t$, in one of $m$ locations, where the location is a multinomial random variable

$$H \sim \mathcal{M}_m(\theta_1, \ldots, \theta_m).$$

Of course, the animal may not be captured (it may not be seen, or it may have died). As we track each animal through time, we can model this process with two random variables. The random variable $H$ can take values in the set $\{1, 2, \ldots, m\}$ with probabilities $\{\theta_1, \ldots, \theta_m\}$. Given $H = k$, the random variable is $X \sim \mathcal{B}(p_k)$, where $p_k$ is the probability of capturing the animal in location $k$. (See Dupuis 1995b for details.)

As an example, for $t = 6$, a typical realization for an animal might be

$$\mathbf{h} = (4, 1, -, 8, 3, -), \quad \mathbf{x} = (1, 1, 0, 1, 1, 0),$$

where $\mathbf{h}$ denotes the sequence of observed $h_j$'s and of non-captures. Thus, we have a missing data problem. If we had observed all of $\mathbf{h}$, the maximum likelihood estimation would be trivial, as the MLEs would simply be the cell means. For animal $i$, we define the random variables $X_{ijk} = 1$ if animal $i$ is captured at time $j$ in location $k$, and 0 otherwise, $Y_{ijk} = \mathbb{I}(H_{ij} = k)\mathbb{I}(X_{ijk} = 1)$ (the observed data), and $Z_{ijk} = \mathbb{I}(H_{ij} = k)\mathbb{I}(X_{ijk} = 0)$ (the missing data). The likelihood function is

$$
\begin{aligned}
L(\theta_1, &\ldots, \theta_m, p_1, \ldots, p_m | \mathbf{y}, \mathbf{x}) \\
&= \sum_{\mathbf{z}} L(\theta_1, \ldots, \theta_m, p_1, \ldots, p_m | \mathbf{y}, \mathbf{x}, \mathbf{z}) \\
&= \sum_{\mathbf{z}} \prod_{k=1}^{m} p_k^{\sum_{i=1}^{n} \sum_{j=1}^{t} x_{ijk}} \\
&\quad \times (1 - p_k)^{nt - \sum_{i=1}^{n} \sum_{j=1}^{t} x_{ijk}} \theta_k^{\sum_{i=1}^{n} \sum_{j=1}^{t} (y_{ijk} + z_{ijk})},
\end{aligned}
$$

where the sum over $\mathbf{z}$ represents the expectation over all the states that could have been visited. This can be a complicated expectation, but the likelihood can be calculated by first using an EM strategy and working with the complete data likelihood $L(\theta_1, \ldots, \theta_k, p_1, \ldots, p_k | \mathbf{y}, \mathbf{x}, \mathbf{z})$, then using MCEM for the calculation of the expectation. Note that calculation of the MLEs of $p_1, \ldots, p_k$ is straightforward, and for $\theta_1, \ldots, \theta_k$, we use

1. (M-step) Take $\hat{\theta}_k = \frac{1}{nt} \sum_{i=1}^{n} \sum_{j=1}^{t} y_{ijk} + \hat{z}_{ijk}$.

2. (Monte Carlo E-step) If $x_{ijk} = 0$, for $\ell = 1, \ldots, L$, generate

$$
\hat{z}_{ijk\ell} \sim \mathcal{M}_k(\hat{\theta}_1, \ldots, \hat{\theta}_m)
$$

   and calculate

$$
\hat{z}_{ijk} = \sum_{\ell} \hat{z}_{ijk\ell} / L.
$$

Scherrer (1997) examines the performance of more general versions of this algorithm and shows, in particular, that they outperform the conditional likelihood approach of Brownie *et al.* (1993). ‖

   Note that the MCEM approach does not enjoy the EM monotonicity any longer and may even face some smoothness difficulties when the sample used in (5.3.19) is different for each new value $\theta_j$. In more involved likelihoods, Markov chain Monte Carlo methods can also be used to generate the missing data sample, usually creating additional dependence structures between the successive values produced by the algorithm.

## 5.4 Problems

**5.1** (Ó Ruanaidh and Fitzgerald 1996) Consider the setup of Example 5.1.1.

   (a) Show that (5.1.2) holds.

(b) Discuss the validity of the approximation

$$\mathbf{x}^t \mathbf{x} - \mathbf{x}^t G (G^t G)^{-1} G^t \mathbf{x} \simeq \sum_{i=1}^{N} x_i^2 - \frac{2}{N} \left( \sum_{i=1}^{N} x_i \cos(\omega t_i) + \sum_{i=1}^{N} x_i \sin(\omega t_i) \right)^2$$

$$= \sum_{i=1}^{N} x_i^2 - 2 S_N.$$

(c) Show that $\pi(\omega|\mathbf{x})$ can be approximated by

$$\pi(\omega|\mathbf{x}) \propto \left[ 1 - \frac{2 S_N}{\sum_{i=1}^{N} x_i^2} \right]^{(2-N)/2}.$$

**5.2** (Winkler 1995) Establish the following theorem: *For a sequence of Markov kernels $K_n$ on a finite state-space, with respective invariant distributions $\mu_n$, assume that*

$$\sum_n \| \mu_n - \mu_{n+1} \| < \infty$$

*and*

$$\lim_{n \to \infty} \max_{x,y} \| P_1(x, \cdot) \circ \cdots \circ P_n - P_1(y, \cdot) \circ \cdots \circ P_n \| = 0$$

*for every $i \geq 1$. Then, the limit $\mu_\infty$ of $\mu_n$ exists and, for every initial distribution $\nu$,*

$$\nu P_1 \cdots P_n \to \mu_\infty.$$

(*Hint*: Use the Cauchy criterion to show that $\mu_\infty$ exists.)

**5.3** (Continuation of Problem 5.2) Consider $\mathbb{P}_n$, the transition matrix

$$\mathbb{P}_n = \begin{pmatrix} 1 - a_n & a_n \\ a_n & 1 - a_n \end{pmatrix}.$$

(a) Show that the invariant distribution is $\mu = (1/2, 1/2)$.

(b) Show that if

$$\prod_{n \geq 1} (1 - 2 a_n) \geq 3/4$$

and $\nu = (1, 0)$, $\nu \mathbb{P}_1 \cdots \mathbb{P}_n$ does not converge to $\mu$.

**5.4** (Winkler 1995) Consider a finite space $\mathcal{X}$ and a function $H$ defined on $\mathcal{X}$.

(a) Show that

$$\lim_{\lambda \to \infty} \frac{\exp\{\lambda H(x)\}}{\sum_{\mathcal{X}} \exp\{\lambda H(x)\}} = \begin{cases} \frac{1}{M} & \text{if } x \in \mathcal{O} \\ 0 & \text{otherwise,} \end{cases}$$

where $\mathcal{O}$ is the set of global maxima of $H$ and $M$ is the cardinality of $\mathcal{O}$. (*Hint*: Introduce $\omega = \max_{\mathcal{X}} H(x)$.)

(b) Show that for $x \in \mathcal{O}$, $\pi_\lambda(x) \propto \exp\{\lambda H(x)\}$ increases, and for $x \notin \mathcal{O}$, $\pi_\lambda(x)$ decreases eventually.

**5.5** (Ripley 1987) Consider the linear system $Ax = b$, with $x, b \in \mathbb{R}^p$.

(a) Show that the equation $Ax = b$ can be rewritten as $x = Hx + b$.

(b) Show that the sequence $x_{t+1} = Hx_t + b$ converges to the solution $x^*$ of $Ax = b$ as $t$ goes to infinity if $\lambda_{\max}(H) < 1$, where $\lambda_{\max}$ denotes the largest absolute eigenvalue of $H$.

(c) Consider a Markov chain $X_t$ with transition matrix $\mathbb{P} = (p_{ij})$ such that $p_{ij} > 0$ if $h_{ij} \neq 0$. If

$$V_k = \left( \prod_{i=1}^{k} \frac{h_{x_i x_{i+1}}}{p_{x_i x_{i+1}}} \right) b_{x_{k+1}},$$

show that $\mathbb{E}[V_k | X_0 = i] = (H^{k+1}b)_i$.

(d) Derive a simulation approach to the solution of $Ax = b$ and discuss its merits.

**5.6** *(Continuation of Problem 5.5)* Given a matrix $A$ with real eigenvalues, find a technique similar to the above Markov approach to derive the largest eigenvalue of $A$. (*Hint:* An eigenvalue $\lambda$ with eigenvector $x$ satisfies $Ax = \lambda x$. See Ripley 1987, pp. 188–189.)

**5.7** Given a simple Ising model with energy function

$$h(s) = \beta \sum_{(u,v) \in \mathcal{N}} s_u s_v,$$

where $\mathcal{N}$ is the neighborhood relation that $u$ and $v$ are neighbors either horizontally or vertically, apply the algorithm [A.20] to the cases $\beta = 0.4$ and $\beta = 4.5$.

**5.8** Using the Laplace approximation of (3.5.4), outline the proofs of Theorem 5.2.7 and Corollary 5.2.8. (You may disregard the approximation error in your calculations.)

**5.9** For the Student's $t$ distribution $\mathcal{T}_\nu$:

(a) Show that $\mathcal{T}_\nu$ can be expressed as the mixture of a normal and of a chi squared distribution. (*Hint:* See §2.2.)

(b) When $X \sim \mathcal{T}_\nu$, given a function $h(x)$ derive a representation of $h(x) = \mathbb{E}[H(x, Z)|x]$, where $Z \sim \mathcal{G}a((\alpha - 1)/2, \alpha/2)$.

**5.10** Let $X \sim \mathcal{T}_\nu$ and consider the function

$$h(x) = \frac{\exp(-x^2/2)}{[1 + (x - \mu)^2]^\nu} .$$

(a) Show that $h(x)$ can be expressed as the conditional expectation $\mathbb{E}[H(x, Z)|x]$, when $Z \sim \mathcal{G}a(\nu, 1)$.

(b) Apply the direct Monte Carlo method of §5.3.2 to maximize (5.3.4) and determine whether or not the resulting sequence converges to the true maximum of $h$.

(c) Compare the implementation of (b) with an approach based on (5.3.5) for (i) $g = \mathcal{E}xp(\lambda)$ and (ii) $g = f(z|\mu)$, the conditional distribution of $Z$ given $X = \mu$. For each choice, examine if the approximation (5.3.5) has a finite variance.

(d) Run [A.22] to see if the recursive scheme of Geyer (1996) improves the convergence speed.

**5.11** Consider the sample $\mathbf{x} = (0.12, 0.17, 0.32, 0.56, 0.98, 1.03, 1.10, 1.18, 1.23,$
$1.67, 1.68, 2.33)$, generated from an exponential mixture

$$p\,\mathcal{E}xp(\lambda) + (1-p)\,\mathcal{E}xp(\mu).$$

(a) Show that the likelihood $h(p, \lambda, \mu)$ can be expressed as $\mathbb{E}[H(x, Z)]$, where
$z = (z_1, \ldots, z_{12})$ corresponds to the vector of allocations of the observations
$x_i$ to the first and second components of the mixture; that is, for $i = 1, \ldots, 12$,

$$P(z_i = 1) = 1 - P(z_i = 2) = \frac{p\lambda \exp(-\lambda x_i)}{p\lambda \exp(-\lambda x_i) + (1-p)\mu \exp(-\mu x_i)}.$$

(b) For $g(z)$ a Bernoulli distribution, compare the performance of the approximations (5.3.5) and (5.3.8) in obtaining the maximum of $h(p, \lambda, \mu)$. In
particular, determine which values of $\theta = (p, \lambda, \mu)$ in (5.3.8) lead to a finite
variance for a given value $\eta = (p_0, \lambda_0, \mu_0)$.

(c) Compare the performances of [A.22] with those of the EM algorithm in this
setup.

**5.12** Establish (5.3.7).

**5.13** Let $h(x) = \mathbb{E}[H(x, Z)]$, with $H$ a twice-differentiable function:

(a) Show that the first and second derivatives of $h$ can be approximated by
extensions of (5.3.5).

(b) Study the performance of a gradient algorithm (see §5.2.2) based on these
approximations in the case of the function $h$ of Problem 5.9.

**5.14** Consider $h(\alpha)$, the likelihood of the beta $\mathcal{B}(\alpha, \alpha)$ distribution associated
with the observation $x = 0.345$.

(a) Express the normalizing constant, $c(\alpha)$, in $h(\alpha)$ and show that it cannot
be easily computed when $\alpha$ is not an integer.

(b) Examine the approximation of the ratio $c(\alpha)/c(\alpha_0)$, for $\alpha_0 = 1/2$ by the
method of Geyer and Thompson (1992) (see Example 5.3.2).

(c) Compare this approach with the alternatives of Chen and Shao (1997),
detailed in Problems 3.15 and 3.16.

**5.15** Consider the function

$$h(\theta) = \frac{||\theta||^2(p + ||\theta||^2)(2p - 2 + ||\theta||^2)}{(1 + ||\theta||^2)(p + 1 + ||\theta||^2)(p + 3 + ||\theta||^2)},$$

when $\theta \in \mathbb{R}^p$ and $p = 10$.

(a) Show that the function $h(\theta)$ has a unique maximum.

(b) Show that $h(\theta)$ can be expressed as $\mathbb{E}[H(\theta, Z)]$, where $z = (z_1, z_2, z_3)$ and
$Z_i \sim \mathcal{E}xp(1/2)$ ($i = 1, 2, 3$). Deduce that $f(z|x)$ does not depend on $x$ in
(5.3.5).

(c) When $g(z) = \exp(-\alpha\{z_1 + z_2 + z_3\})$, show that the variance of (5.3.5) is
infinite for some values of $t = ||\theta||^2$ when $\alpha > 1/2$. Identify $A_2$, the set of
values of $t$ for which the variance of (5.3.5) is infinite when $\alpha = 2$.

(d) Study the behavior of the estimate (5.3.5) when $t$ goes from $A_2$ to its complement $A_2^c$ to see if the infinite variance can be detected in the evaluation
of $h(t)$.

**5.16** In the exponential family, EM computations are somewhat simplified. Show that if the complete data density $f$ is of the form

$$f(y, z|\theta) = h(y, z) \exp\{\eta(\theta)T(y, z) - B(\theta)\},$$

then we can write

$$Q(\theta|\theta^*, \mathbf{y}) = \mathbb{E}_{\theta^*}[\log h(\mathbf{y}, \mathbf{Z})] + \sum \eta_i(\theta)\mathbb{E}_{\theta^*}[T_i|\mathbf{y}] - B(\theta).$$

Deduce that calculating the complete-data MLE only involves the simpler expectation $\mathbb{E}_{\theta^*}[T_i|\mathbf{y}]$.

**5.17** For density functions $f$ and $g$, we define the *entropy distance* between $f$ and $g$, with respect to $f$ (also known as *Kullback–Leibler information of g at f* or *Kullback–Leibler distance between g and f*) as

$$\mathbb{E}_f[\log(f(X)/g(X))] = \int \log\left[\frac{f(x)}{g(x)}\right] f(x)\, dx.$$

(a) Use Jensen's inequality to show that $\mathbb{E}_f[\log(f(X)/g(X))] \geq 0$ and, hence, that the entropy distance is always non-negative, and equals zero if $f = g$.

(b) The inequality in part (a) implies that $\mathbb{E}_f \log[g(X)] \leq \mathbb{E}_f \log[f(X)]$. Show that this yields (5.3.18).

(*Note:* Entropy distance was explored by Kullback 1968; for an exposition of its properties, see, for example, Brown 1986. Entropy distance has, more recently, found many uses in Bayesian analysis see, for example, Berger 1985, Bernardo and Smith 1994, or Robert 1996c.)

**5.18** In Example 5.3.6, show directly that the EM sequence converges to the MLE.

**5.19** Consider the following 12 observations from $\mathcal{N}_2(0, \Sigma)$, with $\sigma_1^2, \sigma_2^2$, and $\rho$ unknown:

| $x_1$ | 1 | 1 | $-1$ | $-1$ | 2 | 2 | $-2$ | $-2$ | $-$ | $-$ | $-$ | $-$ |
|-------|---|---|------|------|---|---|------|------|-----|-----|-----|-----|
| $x_2$ | 1 | $-1$ | 1 | $-1$ | $-$ | $-$ | $-$ | $-$ | 2 | 2 | $-2$ | $-2$ |

where "$-$" represents a missing value.

(a) Show that the likelihood function has global maxima at $\rho = \pm 1/2$, $\sigma_1^2 = \sigma_2^2 = 8/3$, and a saddlepoint at $\rho = 0$, $\sigma_1^2 = \sigma_2^2 = 5/2$.

(b) Show that if an EM sequence starts with $\rho = 0$, then it remains at $\rho = 0$ for all subsequent iterations.

(c) Show that if an EM sequence starts with $\rho$ bounded away from zero, it will converge to a maximum.

(d) Take into account roundoff errors; that is, the fact that $[x_i]$ is observed instead of $x_i$.

(*Note:* This problem is due to Murray 1977 and is discussed by Wu 1983.)

**5.20** (Ó Ruanaidh and Fitzgerald 1996) Consider an $AR(p)$ model

$$X_t = \sum_{j=1}^{p} \theta_j X_{t-j} + \epsilon_t,$$

with $\epsilon_t \sim \mathcal{N}(0, \sigma^2)$, observed for $t = p+1, \ldots, m$. The future values $X_{m+1}, \ldots, X_n$ are considered to be missing data. The initial values $x_1, \ldots, x_p$ are taken to be zero.

(a) Give the expression of the observed and complete-data likelihoods.

(b) Give the conditional maximum likelihood estimators of $\theta$, $\sigma$ and $\mathbf{z} = (X_{m+1}, \ldots, X_n)$; that is, the maximum likelihood estimators when the two other parameters are fixed.

(c) Detail the E- and M-steps of the EM algorithm in this setup, when applied to the future values $\mathbf{z}$ and when $\sigma$ is fixed.

(d) Compare with the Kalman filter (see Problem 1.43) prediction.

**5.21** Suppose that the random variable $X$ has a mixture distribution; that is, the $X_i$ are independently distributed as

$$X_i \sim \theta g(x) + (1 - \theta)h(x), \quad i = 1, \ldots, n,$$

where $g(\cdot)$ and $h(\cdot)$ are known. An EM algorithm can be used to find the ML estimator of $\theta$. Introduce $Z_1, \ldots, Z_n$, where $Z_i$ indicates from which distribution $X_i$ has been drawn, so

$$X_i|Z_i = 1 \sim g(x),$$
$$X_i|Z_i = 0 \sim h(x).$$

(a) Show that the complete data likelihood can be written

$$L^c(\theta|\mathbf{x}, \mathbf{z}) = \prod_{i=1}^n \left[ z_i g(x_i) + (1 - z_i)h(x_i) \right] \theta^{z_i} (1 - \theta)^{1-z_i}.$$

(b) Show that $\mathbb{E}[Z_i|\theta, x_i] = \theta g(x_i)/[\theta g(x_i) + (1-\theta)h(x_i)]$ and, hence, that the EM sequence is given by

$$\hat{\theta}_{(j+1)} = \frac{1}{n} \sum_{i=1}^n \frac{\hat{\theta}_{(j)} g(x_i)}{\hat{\theta}_{(j)} g(x_i) + (1 - \hat{\theta}_{(j)})h(x_i)}.$$

(c) Show that $\hat{\theta}_{(j)}$ converges to $\hat{\theta}$, a maximum likelihood estimator of $\theta$.

**5.22** We observe independent Bernoulli variables $X_1, \ldots, X_n$, which depend on unobservable variables $Z_i$ distributed independently as $\mathcal{N}(\zeta, \sigma^2)$, where

$$X_i = \begin{cases} 0 & \text{if } Z_i \le u \\ 1 & \text{if } Z_i > u. \end{cases}$$

Assuming that $u$ is known, we are interested in obtaining MLEs of $\zeta$ and $\sigma^2$.

(a) Show that the likelihood function is

$$p^S(1 - p)^S,$$

where $S = \sum x_i$ and

$$p = P(Z_i > u) = \Phi\left(\frac{\zeta - u}{\sigma}\right).$$

(b) If we consider $z_1, \ldots, z_n$ to be the complete data, show that the complete data likelihood is

$$\prod_{i=1}^n \frac{1}{\sqrt{2\pi}\sigma} \exp\left\{ -\frac{1}{2\sigma^2}(z_i - \zeta)^2 \right\}$$

and the expected complete-data log-likelihood is

$$-\frac{n}{2}\log(2\pi\sigma^2) - \frac{1}{2\sigma^2}\sum_{i=1}^{n}\left(\mathbb{E}[Z_i^2|x_i] - 2\zeta\mathbb{E}[Z_i|x_i] + \zeta^2\right).$$

(c) Show that the EM sequence is given by

$$\hat{\zeta}_{(j+1)} = \frac{1}{n}\sum_{i=1}^{n}t_i(\hat{\zeta}_{(j)}, \hat{\sigma}_{(j)}^2)$$

$$\hat{\sigma}_{(j+1)}^2 = \frac{1}{n}\left[\sum_{i=1}^{n}v_i(\hat{\zeta}_{(j)}, \hat{\sigma}_{(j)}^2) - \frac{1}{n}\left(\sum_{i=1}^{n}t_i(\hat{\zeta}_{(j)}, \hat{\sigma}_{(j)}^2)\right)^2\right],$$

where

$$t_i(\zeta, \sigma^2) = \mathbb{E}[Z_i|x_i, \zeta, \sigma^2] \quad \text{and} \quad v_i(\zeta, \sigma^2) = \mathbb{E}[Z_i^2|x_i, \zeta, \sigma^2].$$

(d) Show that

$$\mathbb{E}[Z_i|x_i, \zeta, \sigma^2] = \zeta + \sigma H_i\left(\frac{u-\zeta}{\sigma}\right),$$

$$\mathbb{E}[Z_i^2|x_i, \zeta, \sigma^2] = \zeta^2 + \sigma^2 + \sigma(u+\zeta)H_i\left(\frac{u-\zeta}{\sigma}\right),$$

where

$$H_i(t) = \begin{cases} \frac{\varphi(t)}{1-\Phi(t)} & \text{if } X_i = 1 \\ -\frac{\varphi(t)}{\Phi(t)} & \text{if } X_i = 0. \end{cases}$$

(e) Show that $\hat{\zeta}_{(j)}$ converges to $\hat{\zeta}$ and that $\hat{\sigma}_{(j)}^2$ converges to $\hat{\sigma}^2$, the MLEs of $\zeta$ and $\sigma^2$, respectively.

**5.23** Recall the censored Weibull model of Problems 1.10–1.12. By extending the likelihood technique of Example 5.3.1, use the EM algorithm to fit the Weibull model, accounting for the censoring. Use the data of Problem 1.12 and fit the three cases outlined there.

**5.24** An alternate implementation of the Monte Carlo EM might be, for $Z_1, \ldots, Z_m \sim k(\mathbf{z}|\mathbf{x}, \theta)$, to iteratively maximize

$$\log \hat{L}(\theta|\mathbf{x}) = \frac{1}{m}\sum_{i=1}^{m}\{\log L^c(\theta|\mathbf{x}, \mathbf{z_i}) - \log k(\mathbf{z_i}|\theta, x)\}$$

(which might more accurately be called Monte Carlo maximum likelihood).

(a) Show that $\hat{L}(\theta|\mathbf{x}) \to L(\theta|\mathbf{x})$ as $m \to \infty$.
(b) Show how to use $\hat{L}(\theta|\mathbf{x})$ to obtain the MLE in Example 5.3.6.
(c) Show how to use $\hat{L}(\theta|\mathbf{x})$ to obtain the MLE in Example 5.3.8 (this is difficult).

**5.25** For the situation of Example 5.3.7, data $(x_1, x_2, x_3, x_4) = (125, 18, 20, 34)$ are collected.

(a) Use the EM algorithm to find the MLE of $\theta$.
(b) Use the Monte Carlo EM algorithm to find the MLE of $\theta$. Compare your results to those of part (a).

**5.26** For the situation of Example 5.3.8,

(a) Verify the formula for the likelihood function.

(b) Show that the complete-data MLEs are given by $\hat{\theta}_k = \frac{1}{nt} \sum_{i=1}^{n} \sum_{j=1}^{t} y_{ijk} + \hat{z}_{ijk}$.

**5.27** For the model of Example 5.3.8, Table 5.4.1 contains data on the movement between 5 zones of 18 tree swallows with $m = t = 5$, where a 0 denotes that the bird was not captured.

| | Time | | | | | | Time | | | | | | Time | | | | |
|---|---|---|---|---|---|---|---|---|---|---|---|---|---|---|---|---|---|
| | 1 | 2 | 3 | 4 | 5 | | 1 | 2 | 3 | 4 | 5 | | 1 | 2 | 3 | 4 | 5 |
| a | 2 | 2 | 0 | 0 | 0 | g | 1 | 1 | 1 | 5 | 0 | m | 1 | 1 | 1 | 1 | 1 |
| b | 2 | 2 | 0 | 0 | 0 | h | 4 | 2 | 0 | | 0 | n | 2 | 2 | 1 | 0 | 0 |
| c | 4 | 1 | 1 | 2 | 0 | i | 5 | 5 | 5 | 5 | 0 | o | 4 | 2 | 2 | 0 | 0 |
| d | 4 | 2 | 0 | 0 | 0 | j | 2 | 2 | 0 | 0 | 0 | p | 1 | 1 | 1 | 1 | 0 |
| e | 1 | 1 | 0 | 0 | 0 | k | 2 | 5 | 0 | 0 | 0 | q | 1 | 0 | 0 | 4 | 0 |
| f | 1 | 1 | 0 | 0 | 0 | l | 1 | 1 | 0 | 0 | 0 | s | 2 | 2 | 0 | 0 | 0 |

Table 5.4.1. *Movement histories of 18 tree swallows over 5 time periods (Source: Scherrer 1997).*

(a) Using the MCEM algorithm of Example 5.3.8, calculate the MLEs for $\theta_1, \ldots, \theta_5$ and $p_1, \ldots, p_5$.

(b) Assume now that state 5 represents the death of the animal. Rewrite the MCEM algorithm to reflect this, and recalculate the MLEs. Compare them to the answer in part (a).

**5.28** (For baseball fans only) It is typical for baseball announcers to report biased information, intended to overstate a player's ability. If we consider a sequence of at-bats as Bernoulli trials, we are likely to hear the report of a maximum (the player is 8-out-of-his-last-17) rather than an ordinary average. Assuming that $X_1, X_2, \ldots, X_n$ are the Bernoulli random variables representing a players sequence of at-bats (1=hit, 0=no hit), a biased report is the observance of $k^*$, $m^*$, and $r^*$ where

$$r^* = \frac{k^*}{m^*} \geq \max_{m^* \leq i < n} \frac{X_n + X_{n-1} + \cdots + X_{n-i}}{i+1}.$$

If we assume that $\mathbb{E}[X_i] = \theta$, then $\theta$ is the player's true batting ability and the parameter of interest. Estimation of $\theta$ is difficult using only $k^*$, $m^*$, and $r^*$, but it can be accomplished with an EM algorithm. With observed data $(k^*, m^*, r^*)$, let $\mathbf{z} = (z_1, \ldots, z_{n-m^*-1})$ be the augmented data. (This is a sequence of 0's and 1's that are commensurate with the observed data. Note that $X_{n-m^*}$ is certain to equal 0.)

(a) Show that the EM sequence is given by

$$\hat{\theta}_{j+1} = \frac{k^* + \mathbb{E}[S_Z|\hat{\theta}_j]}{n},$$

where $\mathbb{E}(S_Z|\hat{\theta}_j)$ is the expected number of successes in the missing data, assuming that $\hat{\theta}_j$ is the true value of $\theta$.

(b) Give an algorithm for computing the sequence $(\hat{\theta}_j)$. Use a Monte Carlo approximation to evaluate the expectation.

| At-Bat | $k^*$ | $m^*$ | $\hat{\theta}$ | EM MLE |
|--------|-------|-------|----------------|--------|
| 339 | 12 | 39 | 0.298 | 0.240 |
| 340 | 47 | 155 | 0.297 | 0.273 |
| 341 | 13 | 41 | 0.299 | 0.251 |
| 342 | 13 | 42 | 0.298 | 0.245 |
| 343 | 14 | 43 | 0.300 | 0.260 |
| 344 | 14 | 44 | 0.299 | 0.254 |
| 345 | 4 | 11 | 0.301 | 0.241 |
| 346 | 5 | 11 | 0.303 | 0.321 |

Table 5.4.2. *A portion of the 1992 batting record of major league baseball player Dave Winfield.*

(c) For the data given in Table 5.4.2, implement the Monte Carlo EM algorithm and calculate the EM estimates.

(*Note:* The "true batting average" $\hat{\theta}$ cannot be computed from the given data and is only included for comparison. The selected data EM MLEs are usually biased downward, but also show a large amount of variability. See Casella and Berger 1994 for details.)

**5.29** The following dataset gives independent observations of $Z = (X, Y) \sim \mathcal{N}_2(0, \Sigma)$ with missing data $(-)$.

| $x$ | 1.17 | $-0.98$ | 0.18 | 0.57 | 0.21 | $-$ | $-$ | $-$ |
|-----|------|---------|------|------|------|-----|-----|-----|
| $y$ | 0.34 | $-1.24$ | $-0.13$ | $-$ | $-$ | $-0.12$ | $-0.83$ | 1.64 |

(a) Show that the observed likelihood is

$$\prod_{i=1}^{3} \left\{ |\Sigma|^{-1/2}\, e^{-z_i^t \Sigma^{-1} z_i/2} \right\} \sigma_1^{-2}\, e^{-(x_4^2+x_5^2)/2\sigma_1^2} \sigma_2^{-3}\, e^{-(y_6^2+y_7^2+y_8^2)/2\sigma_2^2}.$$

(b) Examine the consequence of the choice of $\pi(\Sigma) \propto |\Sigma|^{-1}$ on the posterior distribution of $\Sigma$.

(c) Show that the missing data can be simulated from

$$X_i^* \sim \mathcal{N}\left( \rho \frac{\sigma_1}{\sigma_2} y_i,\, \sigma_1^2(1-\rho^2) \right) \qquad (i = 6, 7, 8),$$

$$Y_i^* \sim \mathcal{N}\left( \rho \frac{\sigma_2}{\sigma_1} x_i,\, \sigma_2^2(1-\rho^2) \right) \qquad (i = 4, 5),$$

to derive a stochastic EM algorithm.

(d) Derive an efficient simulation method to obtain the MLE of the covariance matrix $\Sigma$.

**5.30** The EM algorithm can also be implemented in a Bayesian hierarchical model to find a posterior mode. Suppose that we have the hierarchical model

$$X|\theta \sim f(x|\theta) \,,$$
$$\theta|\lambda \sim \pi(\theta|\lambda) \,,$$
$$\lambda \sim \gamma(\lambda) \,,$$

where interest would be in estimating quantities from $\pi(\theta|x)$. Since

$$\pi(\theta|x) = \int \pi(\theta, \lambda|x) d\lambda,$$

where $\pi(\theta, \lambda|x) = \pi(\theta|\lambda, x)\pi(\lambda|x)$, the EM algorithm is a candidate method for finding the mode of $\pi(\theta|x)$, where $\lambda$ would be used as the augmented data.

(a) Define $k(\lambda|\theta, x) = \pi(\theta, \lambda|x)/\pi(\theta|x)$ and show that

$$\log \pi(\theta|x) = \int \log \pi(\theta, \lambda|x)k(\lambda|\theta^*, x)d\lambda - \int \log k(\lambda|\theta, x)k(\lambda|\theta^*, x)d\lambda.$$

(b) If the sequence $(\hat{\theta}_{(j)})$ satisfies

$$\max_\theta \int \log \pi(\theta, \lambda|x)k(\lambda|\theta_{(j)}, x)d\lambda = \int \log \pi(\theta_{(j+1)}, \lambda|x)k(\lambda|\theta_{(j)}, x)d\lambda,$$

show that $\log \pi(\theta_{(j+1)}|x) \geq \log \pi(\theta_{(j)}|x)$. Under what conditions will the sequence $(\hat{\theta}_{(j)})$ converge to the mode of $\pi(\theta|x)$?

(c) For the hierarchy

$$X|\theta \sim \mathcal{N}(\theta, 1) \,,$$
$$\theta|\lambda \sim \mathcal{N}(\lambda, 1) \,,$$

with $\pi(\lambda) = 1$, show how to use the EM algorithm to calculate the posterior mode of $\pi(\theta|x)$.

## 5.5 Notes

### 5.5.1 Variations on EM

Besides a possible difficult computation in the E-step (see §5.3.4), problems with the EM algorithm can occur in the case of multimodal likelihoods. The increase of the likelihood function at each step of the algorithm ensures its convergence to the maximum likelihood estimator in the case of unimodal likelihoods but implies a dependence on initial conditions for multimodal likelihoods. Several proposals can be found in the literature to overcome this problem, one of which we now describe.

Broniatowski *et al.* (1984) and Celeux and Diebolt (1985, 1992) have tried to overcome the dependence of EM methods on the starting value by replacing step 1 in [A.23] with a *simulation* step, the missing data $z$ being generated conditionally on the observation $x$ and on the current value of the parameter $\theta_m$. The maximization in step 2 is then done on the (simulated) complete-data log-likelihood, $\tilde{H}(x, z_m|\theta)$. The appeal of this approach is that it allows for a more systematic exploration of the likelihood surface by partially avoiding the fatal attraction of the closest mode. Unfortunately, the theoretical convergence results for these methods are limited: The Markov chain $(\theta_m)$ produced by this variant of EM called SEM (for *Stochastic EM*) is often ergodic, but the relation of the stationary distribution with the maxima of the observed likelihood is rarely known (see Diebolt and Ip 1996). Moreover, the authors mainly study the behavior of the "ergodic" average,

$$\frac{1}{M} \sum_{m=1}^{M} \theta_m,$$

instead of the "global" mode,

$$\hat{\theta}_{(M)} = \arg \max_{1 \leq m \leq M} \ell(\theta_m | x),$$

which is more natural in this setup. Celeux and Diebolt (1990) have, however, solved the convergence problem of SEM by devising a hybrid version called SAEM (for *Simulated Annealing EM*), where the amount of randomness in the simulations decreases with the iterations, ending up with an EM algorithm. This version actually relates to the simulated annealing methods, described in §5.2.3. Celeux *et al.* (1996) also propose a hybrid version, where SEM produces $\hat{\theta}_{(M)}$, the starting point of EM, when the later applies. See Lavielle and Moulines (1997) for an approach similar to Celeux and Diebolt (1990), where the authors obtain convergence conditions which are equivalent to those of EM.

Meng and Rubin (1991, 1992), Liu and Rubin (1994), and Meng and van Dyk (1997) have also developed versions of EM called ECM which take advantage of Gibbs sampling advances by maximizing the complete-data likelihood along successive given directions (that is, through conditional likelihoods).

### 5.5.2 Neural Networks

Neural networks provide another type of missing data model where simulation methods are almost always necessary. These models are frequently used in classification and pattern recognition, as well as in robotics and computer vision (see Cheng and Titterington 1994 for a review on these models). Barring the biological vocabulary and the idealistic connection with actual neurons, the theory of neural networks covers

(i) modeling nonlinear relations between explanatory and dependent (explained) variables,

(ii) estimation of the parameters of these models based on a (training) sample

Although the neural network literature usually avoids probabilistic modeling, these models can be analyzed and estimated from a statistical point of view (see Neal 1996 or Ripley 1994, 1996). They can also be seen as a particular type of nonparametric estimation problem, where a major issue is then identifiability.

A simple classical example of a neural network is the *multilayer* model (also called the *backpropagation model*) which relates explanatory variables $x = (x_1, \ldots, x_n)$ with dependent variables $y = (y_1, \ldots, y_n)$ through a hidden "layer", $h = (h_1, \ldots, h_p)$, where $(k = 1, \ldots, p, \ell = 1, \ldots, n)$

$$h_k = f\left(\alpha_{k0} + \sum_j \alpha_{kj} x_j\right),$$

$$\mathbb{E}[Y_\ell | h] = g\left(\beta_{\ell 0} + \sum_{k=1}^{p} \beta_{\ell k} h_k\right),$$

and $\mathrm{var}(Y_\ell) = \sigma^2$. The functions $f$ and $g$ are known (or arbitrarily chosen) from categories such as *threshold*, $f(t) = \mathbb{I}_{t>0}$, *hyperbolic*, $f(t) = \tanh(t)$, or *sigmoid*, $f(t) = 1/(1 + e^{-t})$.

As an example, consider the problem of character recognition, where hand-written manuscripts are automatically deciphered. The $x$'s may correspond to geometric characteristics of a digitized character, or to pixel gray levels, and the $y$'s are the 26 letters of the alphabet, plus side symbols. (See Le Cun *et al.* 1989 for an actual modeling based on a sample of 7291, $16 \times 16$ pixel images, for 9760 parameters.) The likelihood of the multilayer model then includes the parameters $\alpha = (\alpha_{kj})$ and $\beta = (\beta_{lk})$ in a non-linear structure. Assuming normality, for observations $(y_t, x_t)$, $t = 1, 2, \ldots, T$, the log-likelihood can be written

$$\ell(\alpha, \beta | x, y) = -\sum_{t=1}^{T} \sum_{l=1}^{n} (y_{tl} - \mathbb{E}[y_{tl} | x_{tl}])^2 / 2\sigma^2 \ .$$

A similar objective function can be derived using a least squares criterion. The maximization of $\ell(\alpha, \beta | x, y)$ involves the detection and the elimination of numerous local modes.

### 5.5.3 The Robbins–Monro procedure

The Robbins–Monro algorithm (1951) is a technique of stochastic approximation to solve for $x$ in equations of the form

(5.5.1)                          $$h(x) = \beta \ ,$$

when $h(x)$ can be written as in (5.3.2). It was also extended by Kiefer and Wolfowitz (1952) to the more general setup of (5.1.1). In the case of (5.5.1), the Robbins–Monro method proceeds by generating a Markov chain of the form

(5.5.2)                  $$X_{j+1} = X_j + \gamma_j(\beta - H(Z_j, X_j)) \ ,$$

where $Z_j$ is *simulated* from the conditional distribution defining (5.3.2). The following result then describes sufficient conditions on the $\gamma_j$'s for the algorithm to be convergent (see Bouleau and Lépingle 1994 for a proof).

**Theorem 5.5.1** *If $(\gamma_n)$ is a sequence of positive numbers such that*

$$\sum_{n=1}^{\infty} \gamma_n = +\infty \quad and \quad \sum_{n=1}^{\infty} \gamma_n^2 < +\infty,$$

*if the $x_j$'s are simulated from $H$ conditionally on $\theta_j$ such that*

$$\mathbb{E}[H(X_j, \theta_j) | \theta_j] = h(\theta_j)$$

*and $|x_j| \leq B$ for a fixed bound $B$, and if there exists $\theta^* \in \Theta$ such that*

$$\inf_{\delta \leq |\theta - \theta^*| \leq 1/\delta} (\theta - \theta^*) \cdot (h(\theta) - \beta) > 0$$

*for every $0 < \delta < 1$, the sequence $(\theta_j)$ converges to $\theta^*$ almost surely.*

The solution of the maximization problem (5.1.1) can be expressed in terms of the solution of the equation $\nabla h(\theta) = 0$ if the problem is sufficiently regular;

that is, if the maximum is not achieved on the boundary of the domain $\Theta$. Note then the similarity between (5.2.1) and (5.5.2). Since its proposal, this method has seen numerous variations. Besides Bouleau and Lépingle (1994), see Wasan (1969), Kersting (1987), Winkler (1995), or Duflo (1996) for more detailed references.

When the function $h$ has several local maxima, the Robbins–Monro procedure converges to one of these maxima. In the particular case $H(x, \theta) = h(\theta) + x/\sqrt{a}$ ($a > 0$), Pflug (1994) examines the relation

$$(5.5.3) \qquad \theta_{j+1} = \theta_j + a\, h(\theta_j) + \sqrt{a/2}\, x_j\ ,$$

when the $X_j$ are iid with $\mathbb{E}[X_j] = 0$ and $\mathrm{var}(X_j) = T$. The relevance of this particular case is that, under the conditions

(i) $\exists k > 0,\ k_1 > 0$ such that $\theta \cdot h(\theta) \leq -k_1|\theta|$ for $|\theta| > k$ ;
(ii) $|h(\theta)| \leq k_2|h(\theta_0)|$ for $K < |\theta| < |\theta_0|$ ;
(iii) $|h'(\theta)| \leq k_3$

the stationary measure $\nu_a$ associated with the Markov chain (5.5.3) weakly converges to the distribution with density

$$c(T)\, \exp\left(E(\theta)/T\right)\ ,$$

when $a$ goes to 0, $T$ remaining fixed and $E$ being the primitive of $h$ (see Duflo 1996). The hypotheses (i)–(iii) ensure, in particular, that $\exp\{E(\theta)/T\}$ is integrable on $\mathbb{R}$. Therefore, the limiting distribution of $\theta_j$ (when $a$ goes to 0) is the so-called *Gibbs measure* with *energy function* $E$, which is a pivotal quantity for the *simulated annealing* method introduced in §5.2.3. In particular, when $T$ goes to 0, the Gibbs measure converges to the uniform distribution on the set of (global) maxima of $E$ (see Hwang 1980). This convergence is interesting more because of the connections it exhibits with the notions of Gibbs measure and of simulated annealing rather than for its practical consequences. The assumptions (i)–(iii) are rather restrictive and difficult to check for implicit $h$'s, and the representation (5.5.3) is rather specialized! (Note, however, that the completion of $h(\theta)$ in $H(x, \theta)$ is free, since the conditions (i)-(iii) only relate to $h$.)

# CHAPTER 6

# The Metropolis–Hastings Algorithm

"What's changed, except what needed changing?" And there was something in that, Cadfael reflected. What was changed was the replacement of falsity by truth...

—Ellis Peter, *The Confession of Brother Haluin*

## 6.1 Monte Carlo Methods Based on Markov Chains

It was shown in Chapter 3 that it is not necessary to use a sample from the distribution $f$ to approximate the integral

$$\int h(x)f(x)dx \ ,$$

since *importance sampling* techniques can be used. This chapter develops a somewhat different strategy and shows that it is possible to obtain a sample $X_1, \ldots, X_n$ approximately distributed from $f$ without directly simulating from $f$. The basic principle underlying the methods described in this chapter and the following chapters is *to use an ergodic Markov chain with stationary distribution $f$*.

For an arbitrary starting value $x^{(0)}$, a chain $(X^{(t)})$ is generated using a transition kernel with stationary distribution $f$, which ensures the convergence in distribution of $(X^{(t)})$ to a random variable from $f$. (Given that the chain is ergodic, the starting value $x^{(0)}$ is, in principle, unimportant.) Thus, for a "large enough" $T_0$, $X^{(T_0)}$ can be considered as distributed from $f$ and the methods studied in this chapter produce a *dependent* sample $X^{(T_0)}, X^{(T_0+1)}, \ldots$, which is generated from $f$.

**Definition 6.1.1** A *Markov chain Monte Carlo (MCMC) method* for the simulation of a distribution $f$ is any method producing an ergodic Markov chain $(X^{(t)})$ whose stationary distribution is $f$.

In comparison with the techniques developed in Chapter 3, the above strategy may sound suboptimal, as it relies on asymptotic convergence properties. For example, the number of iterations required to obtain a good approximation of $f$ is *a priori* important. The appeal to Markov chains is nonetheless justified from at least two points of view: First, in Chapter 5, we have seen that some stochastic optimization algorithms (for example, the Robbins–Monro procedure or the SEM algorithm) naturally produce

Markov chain structures. These can be generalized and, if possible, optimized. Second, the appeal to Markov chains allows for greater generality than the methods presented in Chapter 2, where we described only one general method of simulation, the ARS algorithm (§2.3.3) (which is moreover restricted to log-concave densities). In fact, some generic algorithms, like the Metropolis–Hastings algorithms, use simulations from virtually any density $g$ to actually generate from a given density $f$, while allowing for the dependence of $g$ on the previous simulation. Moreover, even when an Accept–Reject algorithm is available, it is sometimes more efficient to use the pair $(f, g)$ through a Markov chain, as detailed in §6.3.1.

It must be stressed that the (re)discovery of Markov chain Monte Carlo methods by statisticians in the 1990s has produced considerable progress in simulation-based inference and, in particular, in Bayesian inference, since it has allowed the analysis of models that were too complex to be satisfactorily processed by previous schemes.[1]

This chapter covers a very general MCMC method, namely the *Metropolis–Hastings algorithm*, whereas Chapter 7 specializes in *the Gibbs sampler* which, although a particular case of Metropolis–Hastings algorithm (see Theorem 7.1.17), has fundamentally different methodological and historical motivations.

Despite its formal aspect, Definition 6.1.1 implies that the use of a chain $(X^{(t)})$ resulting from a Markov chain Monte Carlo algorithm with stationary distribution $f$ is similar to the use of an iid sample from $f$ in the sense that the ergodic theorem (Theorem 4.7.4) guarantees the convergence of the empirical average

$$(6.1.1) \qquad \frac{1}{T} \sum_{t=1}^{T} h(X^{(t)})$$

to the quantity $\mathbb{E}_f[h(X)]$. A sequence $(X^{(t)})$ produced by a Markov chain Monte Carlo algorithm can thus be employed just as an iid sample. If there is no particular requirement of independence but rather the purpose of the simulation study is to examine the properties of the distribution $f$, there is no need for the generation of $n$ independent chains $(X_i^{(t)})$ $(i = 1, \ldots, n)$, where only the "terminal" values $X_i^{(T_0)}$ are kept.[2] In other words, a single Markov chain is enough to ensure a proper approximation through estimates like (6.1.1) of $\mathbb{E}_f[h(X)]$ for the functions $h$ of interest (and sometimes even of the density $f$, as detailed in Chapter 7). Obviously, handling this sequence is somewhat more arduous than in the iid case because of the dependence structure. (Some approaches to the convergence assessment of (6.1.1) are given in §6.4 and in Chapter 8.)

---

[1] For example, Chapter 9 considers the case of latent variable models, which prohibit both analytic processing and numerical approximation in either classical (maximum likelihood) or Bayesian setups (see also Example 1.1.2).

[2] For one thing, the determination of the "proper" length $T_0$ is still under debate, as we will see in Chapter 8. For another, this approach can result in the considerable waste of $n(T_0 - 1)$ simulations out of $nT_0$.

Given the principle stated in Definition 6.1.1, one can propose a infinite number of practical implementations based, for instance, on methods used in statistical physics. The Metropolis–Hastings algorithms described in this chapter have the advantage of imposing minimal requirements on the target density $f$ and allowing for a wide choice of possible implementations. In contrast, the Gibbs sampler described in Chapter 7 is more restrictive, in the sense that it requires some knowledge on the target density to derive some conditional densities.

## 6.2 The Metropolis–Hastings algorithm

Before illustrating the universality of Metropolis–Hastings algorithms and demonstrating their straightforward implementation, we first address the (important) issue of theoretical validity. Since the results presented below are valid for all types of Metropolis–Hastings algorithms, we do not include examples in this section, but rather wait for §6.3, which presents a collection of specific algorithms.

### 6.2.1 Definition

The Metropolis–Hastings algorithm starts with the objective (target) density $f$. A conditional density $q(y|x)$, defined with respect to the dominating measure for the model (see §6.5.1 for a nontrivial example), is then chosen. The Metropolis–Hastings algorithm can be implemented in practice when $q(\cdot|x)$ is easy to simulate from and is either explicitly available (up to a multiplicative constant *independent of x*) or *symmetric*; that is, such that $q(x|y) = q(y|x)$. (A more general requirement is that the ratio $f(y)/q(y|x)$ is known up to a constant *independent* of $x$.)

The Metropolis–Hastings algorithm associated with the objective (target) density $f$ and the conditional density $q$ produces a Markov chain $(X^{(t)})$ through the following transition:

**Algorithm A.24 –Metropolis–Hastings–**
Given $x^{(t)}$,
1. Generate $Y_t \sim q(y|x^{(t)})$.
2. Take

$$X^{(t+1)} = \begin{cases} Y_t & \text{with probability} \quad \rho(x^{(t)}, Y_t), \\ x^{(t)} & \text{with probability} \quad 1 - \rho(x^{(t)}, Y_t), \end{cases}$$

where                                                                                    [A.24]

$$\rho(x, y) = \min \left\{ \frac{f(y)}{f(x)} \, \frac{q(x|y)}{q(y|x)}, 1 \right\} .$$

The distribution $q$ is called the *instrumental* (or *proposal*) *distribution*.

This algorithm always accepts values $y_t$ such that the ratio $f(y_t)/q(y_t|x^{(t)})$ is increased, compared with the previous value $f(x^{(t)})/(q(x^{(t)}|y_t))$. It is only in the symmetric case that the acceptance is driven by the likelihood ratio $f(y_t)/f(x^{(t)})$. An important feature of the algorithm [A.24] is that it may accept values $y_t$ such that the ratio is decreased, similar to stochastic optimization methods (see §5.2.2). Like the Accept–Reject method, the Metropolis–Hastings algorithm only depends on the ratios

$$f(y_t)/f(x^{(t)}) \quad \text{and} \quad q(x^{(t)}|y_t)/q(y_t|x^{(t)})$$

and is, therefore, independent of normalizing constants, assuming that $q(\cdot|y)$ is known up to a constant that is *independent* of $y$ (which is not always the case).

Obviously, the probability $\rho(x^{(t)}, y_t)$ is only defined when $f(x^{(t)}) > 0$. However, if the chain starts with a value $x^{(0)}$ such that $f(x^{(0)}) > 0$, it follows that $f(x^{(t)}) > 0$ for every $t \in \mathbb{N}$ since the values of $y_t$ such that $f(y_t) = 0$ lead to $\rho(x^{(t)}, y_t) = 0$ and are, therefore, rejected by the algorithm. We will make the *convention* that the ratio $\rho(x, y)$ is equal to 0 when both $f(x)$ and $f(y)$ are null, in order to avoid theoretical difficulties.

There are similarities between [A.24] and the Accept–Reject methods of §2.3, and it is possible to use the algorithm [A.24] as an alternative to an Accept–Reject algorithm for a given pair $(f, g)$. These approaches are compared in §6.3.1.

A sample produced by [A.24] obviously differs from an iid sample. For one thing, such a sample may involve repeated occurrences of the same value, since rejection of $Y_t$ leads to repetition of $X^{(t)}$ at time $t + 1$ (an impossible occurrence in continuous iid settings). Thus, in calculating a mean such as (6.1.1), the $Y_t$'s generated by the algorithm [A.24] can be associated with weights of the form $m_t/T$ $(m_t = 0, 1, \ldots)$, where $m_t$ counts the number of times the subsequent values have been rejected. (This makes the comparison with importance sampling somewhat more relevant, as discussed in §6.4.)

It is obviously necessary to impose minimal regularity conditions on both $f$ and the conditional distribution $q$ for $f$ to be the limiting distribution of the chain $(X^{(t)})$ produced by [A.24]. For instance, it is easier if $\mathcal{E}$, the support of $f$, is *connected*. (This assumption is often omitted in the literature, but an unconnected support $\mathcal{E}$ can invalidate the Metropolis–Hastings algorithm. For such supports, it is necessary to proceed on one connected component at a time and show that the different connected components of $\mathcal{E}$ are linked by the kernel of [A.24].) If the support of $\mathcal{E}$ is truncated by $q$ (that is, if there exists $A \subset \mathcal{E}$ such that

$$\int_A f(x)dx > 0 \quad \text{and} \quad \int_A q(y|x)dy = 0 , \quad \forall x \in \mathcal{E} \Bigg) ,$$

the algorithm [A.24] does not have $f$ as a limiting distribution since, for $x^{(0)} \notin A$, the chain $(X^{(t)})$ never visits $A$. Thus, a minimal necessary condition is that

$$\bigcup_{x \in \text{supp } f} \text{supp } q(\cdot|x) \supset \text{supp } f .$$

To see that $f$ is the stationary distribution of the Metropolis chain, we first examine the Metropolis kernel more closely and find that it satisfies the following property.

**Definition 6.2.1** A Markov chain with transition kernel $K$ satisfies the *detailed balance condition* if there exists a function $f$ satisfying

$$K(y,x)f(y) = K(x,y)f(x)$$

for every $(x,y)$.

**Theorem 6.2.2** *Suppose that a Markov chain with transition function $K$ satisfies the detailed balance condition with $f$ a probability density function. Then:*

*(i) The density $f$ is the invariant density of the chain.*

*(ii) The chain is reversible.*

*Proof.* Part (i) follows by noting that, by the detailed balance condition, for any measurable set $B$,

$$\int_y K(y, B)f(y)dy = \int_y \int_B K(y,x)f(y)dxdy$$
$$= \int_y \int_B K(x,y)f(x)dxdy = \int_B f(x)dx,$$

since $\int K(x,y)dy = 1$. With the existence of the kernel $K$ and invariant density $f$, it is clear that detailed balance and reversibility are the same property. □□

We now have the following result for the Metropolis–Hastings chain.

**Theorem 6.2.3** *For every conditional distribution $q$, whose support includes $\mathcal{E}$, $f$ is a stationary distribution of the chain $(X^{(t)})$ produced by [A.24].*

*Proof.* The transition kernel associated with [A.24] is

$$(6.2.1) \qquad K(x,y) = \rho(x,y)q(y|x) + (1 - r(x))\delta_x(y) \ ,$$

where $r(x) = \int \rho(x,y)q(y|x)dy$ and $\delta_x$ denotes the Dirac mass in $x$. It is straightforward to verify that

$$\rho(x,y)q(y|x)f(x) = \rho(y,x)q(x|y)f(y)$$

and

$$(1 - r(x))\delta_x(y)f(x) = (1 - r(y))\delta_y(x)f(y) \ ,$$

which together establish detailed balance for the Metropolis–Hastings chain. The result now follows from Theorem 6.2.2. □□

The stationarity of $f$ is therefore established for almost any conditional distribution $q$, a fact which indicates the universality of Metropolis–Hastings algorithms.

### 6.2.2 Convergence Properties

To show that the Markov chain of [A.24] indeed converges and that statements such as (6.1.1) hold, we apply the theory developed in Chapter 4.

Since the Metropolis–Hastings Markov chain has, by construction, an invariant probability distribution $f$, if it is also an aperiodic Harris chain (see Definition 4.4.8), then the ergodic theorem (Theorem 4.7.4) would apply to establish a result like (6.1.1).

A sufficient condition for the Metropolis–Hastings Markov chain to be *aperiodic* is that the algorithm [A.24] allows events such as $\{X^{(t+1)} = X^{(t)}\}$; that is, that the probability of such events is not zero, and thus

$$(6.2.2) \qquad P\left[f(X^{(t)})\, q(Y_t|X^{(t)}) \leq f(Y_t)\, q(X^{(t)}|Y_t)\right] < 1.$$

Interestingly, this condition implies that $q$ is not the transition kernel of a reversible Markov chain with stationary distribution $f$.[3] (Note that $q$ is not the transition kernel of the Metropolis–Hastings chain, given by (6.2.1), which *is* reversible.)

The fact that [A.24] only works when (6.2.2) is satisfied is not overly troublesome, since it merely states that it is useless to further perturb a Markov chain with transition kernel $q$ if the latter already converges to the distribution $f$. It is then sufficient to directly study the chain associated with $q$.

The property of *irreducibility* of the Metropolis–Hastings chain $(X^{(t)})$ follows from sufficient conditions such as positivity of the conditional density $q$; that is,

$$(6.2.3) \qquad q(y|x) > 0 \text{ for every } (x, y) \in \mathcal{E} \times \mathcal{E},$$

since it then follows that every set of $\mathcal{E}$ with positive Lebesgue measure can be reached in a single step. As the density $f$ is the invariant measure for the chain, the chain is *positive* (see Definition 4.5.1) and Proposition 4.5.2 implies that the chain is recurrent. We can also establish the following stronger result for the Metropolis–Hastings chain.

**Lemma 6.2.4** *If the Metropolis–Hastings chain $(X^{(t)})$ is $f$-irreducible, it is Harris recurrent.*

*Proof.* This result can be established by using the fact that a characteristic of Harris recurrence is that the only bounded harmonic functions are constant (see Proposition 4.7.2).

If $h$ is a harmonic function, it satisfies

$$h(x_0) = \mathbb{E}[h(X^{(1)})|x_0] = \mathbb{E}[h(X^{(t)})|x_0] .$$

Because the Metropolis-Hastings chain is positive recurrent and aperiodic, we can use Theorem 4.10.11 and, as in the discussion surrounding (4.6.5), can conclude that $h$ is $f$-almost everywhere constant and equal to

---

[3] For instance, (6.2.2) is not satisfied by the successive steps of the Gibbs sampler (see Theorem 7.1.17).

$\mathbb{E}_f[h(X)]$. Since

$$\mathbb{E}[h(X^{(1)})|x_0] = \int \rho(x_0, x_1)\, q(x_1|x_0)\, h(x_1) dx_1 + (1 - r(x_0))\, h(x_0)\, ,$$

it follows that

$$\mathbb{E}_f[h(X)]\, r(x_0) + (1 - r(x_0))\, h(x_0) = h(x_0)\, ;$$

that is, $(h(x_0) - \mathbb{E}_f[h(X)])\, r(x_0) = 0$ for every $x_0 \in \mathcal{E}$. Since $r(x_0) > 0$ for every $x_0 \in \mathcal{E}$, by virtue of the $f$-irreducibility, $h$ is necessarily constant and the chain is Harris recurrent. ◻◻

We therefore have the following convergence result for Metropolis–Hastings Markov chains.

**Theorem 6.2.5** *Suppose that the Metropolis–Hastings Markov chain $(X^{(t)})$ is $f$-irreducible.*

*(i) If $h \in L^1(f)$, then*

$$\lim_{T \to \infty} \frac{1}{T} \sum_{t=1}^{T} h(X^{(t)}) = \int h(x) f(x) dx \qquad a.e.\ f.$$

*(ii) If, in addition, $(X^{(t)})$ is aperiodic, then*

$$\lim_{n \to \infty} \left\| \int K^n(x, \cdot)\mu(dx) - f \right\|_{TV} = 0$$

*for every initial distribution $\mu$, where $K^n(x, \cdot)$ denotes the kernel for $n$ transitions, as in (4.2.4).*

*Proof.* If $(X^{(t)})$ is $f$-irreducible, it is Harris recurrent by Lemma 6.2.4 , and part (i) then follows from Theorem 4.7.4 (the Ergodic Theorem). Part (ii) is an immediate consequence of Theorem 4.6.5. ◻◻

As the $f$-irreducibility of the Metropolis–Hastings chain follows from the above-mentioned positivity property of the conditional density $q$, we have the following immediate corollary, whose proof is left as an exercise.

**Corollary 6.2.6** *The conclusions of Theorem 6.2.5 hold if the Metropolis–Hastings Markov chain $(X^{(t)})$ has conditional density $q(x|y)$ that satisfies (6.2.2) and (6.2.3).*

Although condition (6.2.3) may seem restrictive, it is often satisfied in practice. (Note that, typically, conditions for irreducibility involve the transition kernel of the chain, as in Theorem 4.3.3 or Note 4.10.3.)

We close this section with a result due to Roberts and Tweedie (1996), which gives a somewhat less restrictive condition for irreducibility and aperiodicity.

**Lemma 6.2.7** *Assume $f$ is bounded and positive on every compact set of its support $\mathcal{E}$. If there exist positive numbers $\varepsilon$ and $\delta$ such that*

$$(6.2.4) \qquad\qquad q(y|x) > \varepsilon \quad if \quad |x - y| < \delta\, ,$$

*then the Metropolis–Hastings Markov chain chain* $(X^{(t)})$ *is* $f$*-irreducible and aperiodic. Moreover, every nonempty compact set is a small set.*

The rationale behind this result is the following. If the conditional distribution $q(y|x)$ allows for moves in a neighborhood of $x^{(t)}$ with diameter bounded from below and if $f$ is such that $\rho(x^{(t)}, y)$ is positive in this neighborhood, then any subset of $\mathcal{E}$ can be visited in $k$ steps for $k$ large enough. (This property obviously relies on the assumption that $\mathcal{E}$ is connected.)

*Proof.* Consider $x^{(0)}$ an arbitrary starting point and $A \subset \mathcal{E}$ an arbitrary measurable set. The connectedness of $\mathcal{E}$ implies that there exist $m \in \mathbb{N}$ and a sequence $x^{(i)} \in \mathcal{E}$ $(1 \le i \le m)$ such that $x^{(m)} \in A$ and $|x^{(i+1)} - x^{(i)}| < \delta$. It is therefore possible to link $x^{(0)}$ and $A$ through a sequence of balls with radius $\delta$. The assumptions on $f$ imply that the acceptance probability of a point $x^{(i)}$ of the $i$th ball starting from the $(i-1)$st ball is positive and, therefore, $P^m_{x^{(0)}}(A) = P(X^{(m)} \in A | X^{(0)} = x^{(0)}) > 0$. By Theorem 4.3.3, the $f$-irreducibility of $(X^{(t)})$ is established.

For an arbitrary value $x^{(0)} \in \mathcal{E}$ and for every $y \in B(x^{(0)}, \delta/2)$ (the ball with center $x^{(0)}$ and radius $\delta/2$) we have

$$P_y(A) \ge \int_A \rho(y, z)\, q(z|y)\, dz$$
$$= \int_{A \cap D_y} \frac{f(z)}{f(y)}\, q(y|z)\, dz + \int_{A \cap D_y^c} q(z|y)\, dz\, ,$$

where $D_y = \{z; f(z)q(y|z) \le f(y)q(z|y)\}$. It therefore follows that

$$P_y(A) \ge \int_{A \cap D_y \cap B} \frac{f(z)}{f(y)}\, q(y|z)dz + \int_{A \cap D_y^c \cap B} q(z|y)dz$$
$$\ge \frac{\inf_B f(x)}{\sup_B f(x)} \int_{A \cap D_y \cap B} q(y|z)dz + \int_{A \cap D_y^c \cap B} q(z|y)dz$$
$$\ge \varepsilon\, \frac{\inf_B f(x)}{\sup_B f(x)} \lambda(A \cap B)\, ,$$

where $\lambda$ denotes the Lebesgue measure on $\mathcal{E}$. The balls $B(x^{(0)}, \delta/2)$ are small sets associated with uniform distributions on $B(x^{(0)}, \delta/2)$. This simultaneously implies the aperiodicity of $(X^{(t)})$ and the fact that every compact set is small. $\square\square$

**Corollary 6.2.8** *The conclusions of Theorem 6.2.5 hold if the Metropolis–Hastings Markov chain* $(X^{(t)})$ *has invariant probability density* $f$ *and conditional density* $q(x|y)$ *that satisfy the assumptions of Lemma 6.2.7.*

## 6.3 A Collection of Metropolis–Hastings Algorithms

One of the most fascinating aspects of the algorithm [A.24] is its universality; that is, the fact that an arbitrary conditional distribution $q$ with support $\mathcal{E}$ can lead to the simulation of an arbitrary distribution $f$ on $\mathcal{E}$.

On the other hand, this universality may be only a formality if the instrumental distribution $q$ only rarely simulates points in the main portion of $\mathcal{E}$; that is to say, in the region where most of the mass of the density $f$ is located. This problem of a good choice of $q$ for a given $f$ is detailed in §6.4.

Since we have provided no examples so far, we now proceed to describe several specific approaches used in the literature, with some probabilistic properties and corresponding examples. Note that a complete classification of the Metropolis–Hastings algorithms is impossible, given the versatility of the method and the possibility of creating even more hybrid methods (see, for instance, Roberts and Tweedie 1995 and Stramer and Tweedie 1997b).

### 6.3.1 The Independent Case

This method appears as a straightforward generalization of the Accept–Reject method in the sense that the instrumental distribution $q$ is independent of $X^{(t)}$ and is denoted $g$ by analogy. The algorithm [A.24] will then produce the following transition from $x^{(t)}$ to $X^{(t+1)}$.

**Algorithm A.25 –Independent Metropolis–Hastings–**
Given $x^{(t)}$

1. Generate $Y_t \sim g(y)$.

2. Take                                                                           [A.25]

$$X^{(t+1)} = \begin{cases} Y_t & \text{with probability} \quad \min\left\{ \dfrac{f(Y_t)\, g(x^{(t)})}{f(x^{(t)})\, g(Y_t)}, 1 \right\} \\ x^{(t)} & \text{otherwise.} \end{cases}$$

Although the $Y_t$'s are generated independently, the resulting sample is not iid, if only because the probability of acceptance of $Y_t$ depends on $X^{(t)}$.

The convergence properties of the chain $(X^{(t)})$ follow from properties of the density $g$ in the sense that $(X^{(t)})$ is irreducible and aperiodic (thus, ergodic according to Corollary 6.2.6) if and only if $g$ is almost everywhere positive on the support of $f$. Stronger properties of convergence like geometric and uniform ergodicity are also clearly described by the following result of Mengersen and Tweedie (1996).

**Theorem 6.3.1** *The algorithm [A.25] produces a uniformly ergodic chain if there exists a constant $M$ such that*

(6.3.1)                    $$f(x) \leq M g(x), \quad x \in \text{supp } f.$$

*In this case,*

(6.3.2)                    $$\|K^n(x, \cdot) - f\|_{TV} \leq 2 \left( 1 - \frac{1}{M} \right)^n,$$

*where $\| \cdot \|_{TV}$ denotes the total variation norm introduced in Definition 4.6.1. On the other hand, if for every $M$, there exists a set of positive measure where (6.3.1) does not hold, $(X^{(t)})$ is not even geometrically ergodic.*

*Proof.* If (6.3.1) is satisfied, the transition kernel satisfies

$$K(x, x') \geq g(x') \, \min \left\{ \frac{f(x')g(x)}{f(x)g(x')} , 1 \right\}$$

$$= \min \left\{ f(x') \frac{g(x)}{f(x)} , g(x') \right\} \geq \frac{1}{M} \, f(x') \, .$$

The set $\mathcal{E}$ is therefore small and the chain is uniformly ergodic (Theorem 4.6.13).

To establish the bound on $\|K^n(x, \cdot) - f\|_{TV}$, first write

$$\|K(x, \cdot) - f\|_{TV} = 2 \sup_A \left| \int_A (K(x, y) - f(y))dy \right|$$

$$= 2 \int_{\{y; f(y) \geq K(x,y)\}} (f(y) - K(x, y))dy$$

(6.3.3)
$$\leq 2 \left(1 - \frac{1}{M}\right) \int_{\{y; f(y) \geq K(x,y)\}} f(y)dy$$

$$\leq 2 \left(1 - \frac{1}{M}\right) .$$

We now continue with a recursion argument to establish (6.3.2). We can write

$$\int_A (K^2(x, y) - f(y))dy = \int_{\mathcal{E}} \left[ \int_A (K(u, y) - f(y))dy \right]$$

(6.3.4)
$$\times (K(x, u) - f(u))du,$$

and an argument like that in (6.3.3) leads to

(6.3.5)
$$\|K^2(x, \cdot) - f\|_{TV} \leq 2 \left(1 - \frac{1}{M}\right)^2 .$$

We next write a general recursion relation

$$\int_A (K^{n+1}(x, y) - f(y))dy$$

(6.3.6)
$$= \int_{\mathcal{E}} \left[ \int_A (K^n(u, y) - f(y))dy \right] (K(x, u) - f(u))du,$$

and proof of (6.3.2) is established by induction (Problem 6.8).

If (6.3.1) does not hold, then the sets

$$D_n = \left\{ x; \frac{f(x)}{g(x)} \geq n \right\} .$$

satisfy $P_f(D_n) > 0$ for every $n$. If $x \in D_n$, then

$$P(x, \{x\}) = 1 - \mathbb{E}_g \left[ \min \left\{ \frac{f(Y)g(x)}{g(Y)f(x)} , 1 \right\} \right]$$

$$= 1 - P_g \left( \frac{f(Y)}{g(Y)} \geq \frac{f(x)}{g(x)} \right) - \mathbb{E}_g \left[ \frac{f(Y)g(x)}{f(x)g(Y)} \, \mathbb{I}_{\frac{f(Y)}{g(Y)} < \frac{f(x)}{g(x)}} \right]$$

$$\geq 1 - P_g\left(\frac{f(Y)}{g(Y)} \geq n\right) - \frac{g(x)}{f(x)} \geq 1 - \frac{2}{n},$$

since Markov inequality implies that

$$P_g\left(\frac{f(Y)}{g(Y)} \geq n\right) \leq \frac{1}{n}\mathbb{E}_g\left[\frac{f(Y)}{g(Y)}\right] = \frac{1}{n}.$$

Consider a small set $C$ such that $D_n \cap C^c$ is not empty for $n$ large enough and $x_0 \in D_n \cap C^c$. The return time to $C$, $\tau_C$, satisfies

$$P_{x_0}(\tau_C > k) \geq \left(1 - \frac{2}{n}\right)^k;$$

therefore, the radius of convergence of the series (in $\kappa$) $\mathbb{E}_{x_0}[\kappa^{\tau_C}]$ is smaller than $n/(n-2)$ for every $n$, and this implies that $(X^{(t)})$ cannot be geometrically ergodic, according to Theorem 4.10.6. □□

This particular class of Metropolis–Hastings algorithms naturally suggests a comparison with Accept–Reject methods since every pair $(f, g)$ satisfying (6.3.1) can also induce an Accept–Reject algorithm. Note first that the expected acceptance probability for the variable simulated according to $g$ is larger in the case of the algorithm [A.25].

**Lemma 6.3.2** *If (6.3.1) holds, the expected acceptance probability associated with the algorithm [A.25] is at least $\frac{1}{M}$ when the chain is stationary.*

*Proof.* The expected acceptance probability is

$$\mathbb{E}\left[\min\left\{\frac{f(Y_t)g(X^{(t)})}{f(X^{(t)})g(Y_t)}, 1\right\}\right] = \int \mathbb{I}_{\frac{f(y)g(x)}{g(y)f(x)}>1}\, f(x)g(y)\, dxdy$$

$$+ \int \frac{f(y)g(x)}{g(y)f(x)}\, \mathbb{I}_{\frac{f(y)g(x)}{g(y)f(x)}\leq 1}\, f(x)g(y)\, dxdy$$

$$= 2\int \mathbb{I}_{\frac{f(y)g(x)}{g(y)f(x)}\geq 1}\, f(x)g(y)\, dxdy$$

$$\geq 2\int \mathbb{I}_{\frac{f(y)}{g(y)}\geq\frac{f(x)}{g(x)}}\, f(x)\frac{f(y)}{M}\, dxdy$$

$$= \frac{2}{M}\, P\left(\frac{f(X_1)}{g(X_1)} \geq \frac{f(X_2)}{g(X_2)}\right) = \frac{1}{M},$$

since $X_1$ and $X_2$ are both generated from $f$ in this probability. □□

Thus, the independent Metropolis–Hastings algorithm [A.25] is more efficient than the Accept–Reject algorithm [A.5] in its handling of the sample produced by $g$, since, on the average, it accepts more proposed values. A more advanced comparison between these approaches is about as difficult as the comparison between Accept–Reject and importance sampling proposed in §3.3.3, namely that the size of one of the two samples is random and this complicates the computation of the variance of the resulting estimator. In addition, the correlation between the $X_i$'s resulting from [A.25] prohibits a closed-form expression of the joint distribution. We, therefore, study the consequence of the correlation on the variance of both estimators through an example. (See also Liu 1996 and Problem 6.33 for a comparison based on the eigenvalues of the transition operators in the discrete case, which also shows the advantage of the Metropolis–Hastings algorithm.)

**Example 6.3.3 Generating gamma variables.** Using the algorithm of Example 2.3.4 (see also Example 3.3.7), an Accept–Reject method can be derived to generate random variables from the $\mathcal{G}a(\alpha, \beta)$ distribution using a Gamma $\mathcal{G}a([\alpha], b)$ candidate (where $[a]$ denotes the integer part of $a$). When $\beta = 1$, the optimal choice of $b$ is $b = [\alpha]/\alpha$. The algorithms to compare are then

**Algorithm A.26 –Gamma Accept–Reject–**

1. Generate $Y \sim \mathcal{G}a([\alpha], [\alpha]/\alpha)$.
2. Accept $X = Y$ with probability [A.26]

$$\left( \frac{ey \exp(-y/\alpha)}{\alpha} \right)^{\alpha - [\alpha]}$$

and

**Algorithm A.27 –Gamma Metropolis–Hastings–**

1. Generate $Y_t \sim \mathcal{G}a([\alpha], [\alpha]/\alpha)$.
2. Take [A.27]

$$X^{(t+1)} = \begin{cases} Y_t & \text{with probability } \varrho_t \\ x^{(t)} & \text{otherwise,} \end{cases}$$

where

$$\varrho_t = \min \left[ \left( \frac{Y_t}{x^{(t)}} \exp \left\{ \frac{x^{(t)} - Y_t}{\alpha} \right\} \right)^{\alpha - [\alpha]}, 1 \right].$$

Note that (6.3.1) does apply in this particular case with $\exp(x/\alpha)/x > e/\alpha$.

A first comparison is based on a sample $(y_1, \ldots, y_n)$, of fixed size $n$, generated from $\mathcal{G}a([\alpha], [\alpha]/\alpha)$ with $x^{(0)}$ generated from $\mathcal{G}a(\alpha, 1)$. The number $t$ of values accepted by [A.26] is then random. Figure 6.3.1 describes the convergence of the estimators of $\mathbb{E}_f[X^2]$ associated with both algorithms for the same sequence of $y_i$'s and exhibits strong agreement between the approaches, with the estimator based on [A.27] being closer to the exact value 8.33 in this case.

On the other hand, the number $t$ of values accepted by [A.26] can be fixed and [A.27] can then use the resulting sample of random size $n$, $y_1, \ldots, y_n$. Figure 6.3.2 reproduces the comparison in this second case and exhibits a behavior rather similar to Figure 6.3.1, with another close agreement between estimators and, the scale being different, a smaller variance (which is due to the larger size of the effective sample).

Note, however, that both comparisons are biased. In the first case, the sample of $X^{(i)}$ produced by [A.26] does not have the distribution $f$ and, in

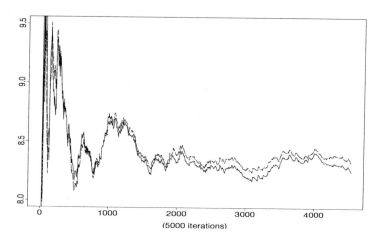

Figure 6.3.1. *Convergence of Accept–Reject (solid line) and Metropolis–Hastings (dotted line) estimators to* $\mathbb{E}_f[X^2] = 8.33$, *for* $\alpha = 2.43$ *based on the same sequence* $y_1, \dots, y_{5000}$ *simulated from* $\mathcal{G}a(2, 2/2.43)$. *The number of acceptances in* [A.26] *is then random. The final values of the estimators are 8.25 for* [A.26] *and 8.32 for* [A.27].

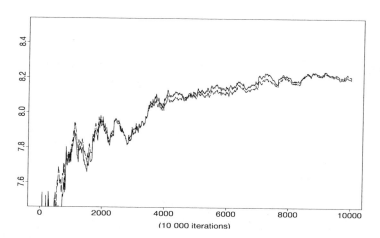

Figure 6.3.2. *Convergence to* $\mathbb{E}_f[X^2] = 8.33$ *of Accept–Reject (full line) and Metropolis–Hastings (dots) estimators for* 10,000 *acceptances in* [A.26], *the same sequence of* $y_i$ *'s simulated from* $\mathcal{G}a(2, 2/2.43)$ *being used in* [A.26] *and* [A.27]. *The final values of the estimators are 8.20 for* [A.26] *and 8.21 for* [A.27].

the second case, the sample of $Y_i$'s in [A.27] is not iid. In both cases, this is due to the use of a stopping rule which modifies the distribution of the samples.    ‖

**Example 6.3.4 Saddlepoint tail area approximation.** In Example 3.6.3, we saw an approximation to noncentral chi squared tail areas based

| Interval | Renormalized | Exact | Monte Carlo |
|----------|--------------|-------|-------------|
| $(36.225, \infty)$ | 0.0996 | 0.1 | 0.0992 |
| $(40.542, \infty)$ | 0.0497 | 0.05 | 0.0497 |
| $(49.333, \infty)$ | 0.0099 | 0.01 | 0.0098 |

Table 6.3.1. *Monte Carlo saddlepoint approximation of a noncentral chi squared integral for $p = 6$ and $\lambda = 9$, based on $10,000$ simulated random variables.*

on the regular and renormalized saddlepoint approximations. Such an approximation requires numerical integration, both to calculate the constant and to evaluate the tail area.

An alternative is to produce a sample $Z_1, \ldots, Z_m$, from the saddlepoint distribution, and then approximate the tail area using

$$P(\bar{X} > a) = \int_{\hat{\tau}(a)}^{1/2} \left(\frac{n}{2\pi}\right)^{1/2} [K_X''(t)]^{1/2} \exp\left\{n\left[K_X(t) - tK_X'(t)\right]\right\} dt$$

(6.3.7)
$$\approx \frac{1}{m} \sum_{i=1}^{m} \mathbb{I}[Z_i > \hat{\tau}(a)],$$

where $K_X(\tau)$ is the cumulant generating function of $X$ and $\hat{\tau}(x)$ is the solution of the saddlepoint equation $K'(\hat{\tau}(x)) = x$ (see §3.6.1).

Note that we are simulating from the transformed density. It is interesting (and useful) that we can easily derive an instrumental density to use in a Metropolis–Hastings algorithm. Using a Taylor series approximation, we find that

(6.3.8)
$$\exp\left\{n\left[K_X(t) - tK_X'(t)\right]\right\} \approx \exp\left\{-nK_X''(0)\frac{t^2}{2}\right\},$$

so a first choice for an instrumental density is the $\mathcal{N}(0, 1/nK_X''(0))$ distribution (see Problem 6.20 for details). Booth *et al.* (1999) use a Student's $t$ approximation instead.

We can now simulate the noncentral chi squared tail areas using a normal instrumental density with $K_X''(t) = 2[p(1 - 2t) + 4\lambda]/(1 - 2t)^3$. The results are presented in Table 6.3.1, where we see that the approximations are quite good. Note that the same set of simulated random variables can be used for all the tail area probability calculations. Moreover, by using the Metropolis–Hastings algorithm, we have avoided calculating the normalizing constant for the saddlepoint approximation.                                        ‖

As an aside, note that the usual classification of "Hastings" for the algorithm [A.25] is somewhat inappropriate, since Hastings (1970) considers the algorithm [A.24] in general, using random walks (§6.3.2) rather than independent distributions in his examples. It is also interesting to recall that Hastings (1970) proposes a theoretical justification of these methods for finite state-space Markov chains based on the finite representation of real numbers in a computer. However, a complete justification of this physical discretization needs to take into account the effect of the approximation in the entire analysis. In particular, it needs to be verified that the computer choice of discrete approximation to the continuous distribution has no effect on the resulting stationary distribution or irreducibility of the chain. Since Hastings (1970) does not go into such detail, but keeps to the simulation level, we prefer to study the theoretical properties of these algorithms by bypassing the finite representation of numbers in a computer and by assuming flawless pseudo-random generators, namely algorithms producing variables which are uniformly distributed on $[0, 1]$. See Roberts *et al.* (1995) for a theoretical study of some effects of the computer discretization.

### 6.3.2 Random Walks

A second natural approach for the practical construction of a Metropolis–Hastings algorithm is to take into account the value previously simulated to generate the following value. This idea is used in algorithms such as the simulated annealing algorithm [A.20] and the stochastic gradient method given in (5.2.2).

Since the candidate $g$ in algorithm [A.24] is allowed to depend on the current state $X^{(t)}$, a first choice to consider is to simulate $Y_t$ according to

$$Y_t = X^{(t)} + \varepsilon_t,$$

where $\varepsilon_t$ is a random perturbation with distribution $g$, independent of $X^{(t)}$. In terms of the algorithm [A.24], $q(y|x)$ is now of the form $g(y - x)$. The Markov chain associated with $q$ is a *random walk* (see Example 4.5.5) on $\mathcal{E}$.

The convergence results of §6.2.2 naturally apply in this particular case. Following Lemma 6.2.7, if $g$ is positive in a neighborhood of 0, the chain $(X^{(t)})$ is $f$-irreducible and aperiodic, therefore ergodic. The most common distributions in this setup are the uniform distributions on spheres centered at the origin or standard distributions like the normal and the Student's $t$ distributions. (Note that these distributions usually need to be scaled. We discuss this problem in §6.4.) At this point, we note that the choice of a *symmetric function* $g$ (that is, such that $g(-t) = g(t)$), leads to the following original expression of [A.24], as proposed by Metropolis *et al.* (1953).

### Algorithm A.28 –Random walk Metropolis–Hastings–
Given $x^{(t)}$,

1. Generate $Y_t \sim g(y - x^{(t)})$.

2. Take                                                            [A.28]

$$X^{(t+1)} = \begin{cases} Y_t & \text{with probability } \min\left\{1, \dfrac{f(Y_t)}{f(x^{(t)})}\right\} \\ x^{(t)} & \text{otherwise.} \end{cases}$$

| $\delta$ | 0.1 | 0.5 | 1.0 |
|---|---|---|---|
| Mean | 0.399 | −0.111 | 0.10 |
| Variance | 0.698 | 1.11 | 1.06 |

Table 6.3.2. *Estimators of the mean and the variance of a normal distribution* $\mathcal{N}(0,1)$ *based on a sample obtained by a Metropolis–Hastings algorithm using a random walk on* $[-\delta, \delta]$ *(15,000 simulations).*

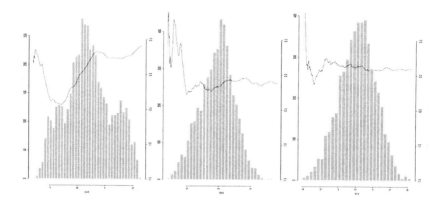

Figure 6.3.3. *Histograms of three samples produced by the algorithm* [A.28] *for a random walk on* $[-\delta, \delta]$ *with (a)* $\delta = 0.1$, *(b)* $\delta = 0.5$, *and (c)* $\delta = 1.0$, *with the convergence of the means (6.1.1), superimposed with scales on the right of the graphs (15,000 simulations).*

**Example 6.3.5 A random walk normal generator.** Hastings (1970) considers the generation of the normal distribution $\mathcal{N}(0,1)$ based on the uniform distribution on $[-\delta, \delta]$. The probability of acceptance is then $\rho(x^{(t)}, y_t) = \exp\{(x^{(t)^2} - y_t^2)/2\} \wedge 1$. Figure 6.3.3 describes three samples of 15,000 points produced by this method for $\delta = 0.1, 0.5$, and 1. The corresponding estimates of the mean and variance are provided in Table 6.3.2. Figure 6.3.3 clearly shows the different speeds of convergence of the averages associated with these three values of $\delta$, with an increasing regularity (in $\delta$) of the corresponding histograms and a faster exploration of the support of $f$. ‖

Despite its simplicity and natural features, the random walk Metropolis–Hastings algorithm does not enjoy uniform ergodicity properties. Mengersen and Tweedie (1996) have shown that in the case where supp $f = \mathbb{R}$, this algorithm cannot produce a uniformly ergodic Markov chain on $\mathbb{R}$ (Problem 6.13).

Although uniform ergodicity cannot be obtained with random walk Metropolis–Hastings algorithms, it is possible to derive necessary and sufficient conditions for geometric ergodicity. Mengersen and Tweedie (1996) have proposed a condition based on the *log-concavity of* $f$ in the tails; that is, if there exist $\alpha > 0$ and $x_1$ such that

(6.3.9)                    $\log f(x) - \log f(y) \geq \alpha |y - x|$

for $y < x < -x_1$ or $x_1 < x < y$.

**Theorem 6.3.6** *Consider a symmetric density $f$ which is log-concave with associated constant $\alpha$ in (6.3.9) for $|x|$ large enough. If the density $g$ is positive and symmetric, the chain $(X^{(t)})$ of $[A.28]$ is geometrically ergodic. If $f$ is not symmetric, a sufficient condition for geometric ergodicity is that $g(t)$ be bounded by $b \exp\{-\alpha|t|\}$ for a sufficiently large constant $b$.*

The proof of this result is based on the use of the drift function $V(x) = \exp\{\alpha|x|/2\}$ (see Note 4.10.1) and the verification of a geometric drift condition of the form

(6.3.10)              $\Delta V(x) \leq -\lambda V(x) + b \mathbb{I}_{[-x^*, x^*]}(x)$ ,

for a suitable bound $x^*$. Mengersen and Tweedie (1996) have shown, in addition, that this condition on $g$ is also necessary in the sense that if $(X^{(t)})$ is geometrically ergodic, there exists $s > 0$ such that

(6.3.11)                  $\int e^{s|x|} f(x)dx < \infty$ .

**Example 6.3.7 A comparison of tail effects.** In order to assess the practical effect of this theorem, Mengersen and Tweedie (1996) considered two random-walk Metropolis–Hastings algorithms based on a $\mathcal{N}(0,1)$ instrumental distribution for the generation of (a) a $\mathcal{N}(0,1)$ distribution and (b) a distribution with density $\psi(x) \propto (1 + |x|)^{-3}$. Applying Theorem 6.3.6 (see Problem 6.15), it can be shown that the first chain associated is geometrically ergodic, whereas the second chain is not. Figures 6.3.4(a) and 6.3.4(b) represent the average behavior of the sums

$$\frac{1}{T} \sum_{t=1}^{T} X^{(t)}$$

over 500 chains initialized at $x^{(0)} = 0$. The 5% and 95% quantiles of these chains show a larger variability of the chain associated with the distribution $\psi$, both in terms of width of the confidence region and in precision of the resulting estimators.                                                            ‖

We next look at a discrete example where Algorithm $[A.28]$ generates a geometrically ergodic chain.

**Example 6.3.8 Random walk geometric generation.** Consider generating a geometric[4] distribution, $\mathcal{G}eo(\theta)$ using $[A.28]$ with $(Y^{(t)})$ having

---

[4] The material used in the current example refers to the drift condition introduced in Note 4.10.1.

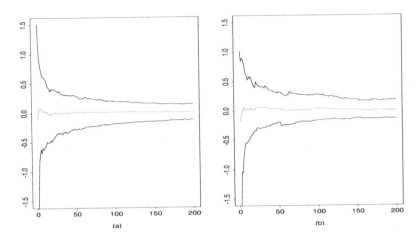

Figure 6.3.4. *90% confidence envelopes of the means produced by the random walk Metropolis–Hastings algorithm [A.24] based on a instrumental distribution $\mathcal{N}(0,1)$ for the generation of (a) a normal distribution $\mathcal{N}(0,1)$ and (b) a distribution with density $\psi$. These envelopes are derived from 500 parallel independent chains and with identical uniform samples on both distributions.*

transition probabilities $q(i,j) = P(Y^{(t+1)} = i|Y^{(t)} = j)$ given by

$$
q(j|i) = \begin{cases} 1/2 & j = i-1, i+1 \text{ and } i = 1,2,3,\dots \\ 1/2 & j = 0,1 \text{ and } i = 0 \\ 0 & \text{otherwise;} \end{cases}
$$

that is, $q$ is the transition kernel of a symmetric random walk on the non-negative integers *with reflecting boundary at* 0.

Now, $X \sim \mathcal{G}eo(\theta)$ implies $P(X = x) = (1-\theta)^x \theta$ for $x = 0,1,2,\dots$. The transition matrix has a band diagonal structure and is given by

$$
T = \begin{pmatrix} \frac{1+\theta}{2} & \frac{1-\theta}{2} & 0 & 0 & \cdots \\ \frac{1}{2} & \frac{\theta}{2} & \frac{1-\theta}{2} & 0 & \cdots \\ 0 & \frac{1}{2} & \frac{\theta}{2} & \frac{1-\theta}{2} & \cdots \\ & \ddots & \ddots & \ddots & \cdots \end{pmatrix} .
$$

Consider the potential function $V(i) = \beta^i$ where $\beta > 1$, and recall that $\Delta V(y^{(0)}) = \mathbb{E}[V(Y^{(1)})|y^{(0)}] - V(y^{(0)})$. For $i > 0$, we have

$$
\mathbb{E}[V(Y^{(1)})|Y^{(0)} = i] = \frac{1}{2}\beta^{i-1} + \frac{\theta}{2}\beta^i + \frac{1-\theta}{2}\beta^{i+1}
$$

$$
= V(i)\left( \tfrac{1}{2\beta} + \tfrac{\theta}{2} + \tfrac{1-\theta}{2}\beta \right) .
$$

Thus, $\Delta V(i) = V(i)(1/(2\beta) + \theta/2 - 1 + \beta(1-\theta)/2) = V(i)g(\theta,\beta)$. For a fixed value of $\theta$, $g(\theta,\beta)$ is minimized by $\beta = 1/\sqrt{1-\theta}$. In this case, $\Delta V(i) = (\sqrt{1-\theta} + \theta/2 - 1)V(i)$ and $\lambda = \sqrt{1-\theta} + \theta/2 - 1$ is the geometric rate of convergence. The closer $\theta$ is to 1, the faster the convergence.     ‖

Tierney (1994) proposed a modification of the previous algorithm with a proposal density of the form $g(y - a - b(x - a))$; that is,

$$y_t = a + b(x^{(t)} - a) + z_t , \qquad z_t \sim g .$$

This autoregressive representation can be seen as intermediary between the independent version ($b = 0$) and the random walk version ($b = 1$) of the Metropolis–Hastings algorithm. Moreover, when $b < 0$, $X^{(t)}$ and $X^{(t+1)}$ are negatively correlated, and this may allow for faster excursions on the surface of $f$ if the symmetry point $a$ is well chosen. Hastings (1970) also considers an alternative to the uniform distribution on $[x^{(t)} - \delta, x^{(t)} + \delta]$ (see Example 6.3.5) with the uniform distribution on $[-x^{(t)} - \delta, -x^{(t)} + \delta]$: The convergence of the empirical average to 0 is then faster in this case, but the choice of 0 as center of symmetry is obviously crucial and requires some a priori information on the distribution $f$. In a general setting, $a$ and $b$ can be calibrated during the first iterations. (See also Problem 6.17.) (See also Chen and Schmeiser 1993, 1998 for the alternative "hit-and-run" algorithm, which proceeds by generating a random direction in the space and moves the current value by a random distance along this direction.)

### 6.3.3 ARMS: A General Metropolis–Hastings Algorithm

The ARS algorithm, which provides a general Accept–Reject method for log-concave densities (see §2.3.3), can be generalized to the ARMS method (which stands for *Adaptive Rejection Metropolis Sampling*) following the approach developed by Gilks *et al.* (1995). This generalization applies to the simulation of arbitrary densities, instead of being restricted to log-concave densities as the ARS algorithm, by simply adapting the ARS algorithm for densities $f$ that are not log-concave. The algorithm progressively fits a function $g$, which plays the role of a pseudo-envelope of the density $f$. In general, this function $g$ does not provide an upper bound on $f$, but the introduction of a Metropolis–Hastings step in the algorithm justifies the procedure.

Using the notation from §2.3.3, take $h(x) = \log f_1(x)$ with $f_1$ proportional to the density $f$. For a sample $S_n = \{x_i, 0 \leq i \leq n + 1\}$, the equations of the lines between $(x_i, h(x_i))$ and $(x_{i+1}, h(x_{i+1}))$ are denoted by $y = L_{i,i+1}(x)$. Consider

$$\tilde{h}_n(x) = \max\{L_{i,i+1}(x), \min[L_{i-1,i}(x), L_{i+1,i+2}(x)]\} ,$$

for $x_i \leq x < x_{i+1}$, with

$$\tilde{h}_n(x) = L_{0,1}(x) \qquad\qquad \text{if} \quad x < x_0,$$
$$\tilde{h}_n(x) = \max[L_{0,1}(x), L_{1,2}(x)] \qquad \text{if} \quad x_0 \leq x < x_1,$$
$$\tilde{h}_n(x) = \max[L_{n,n+1}(x), L_{n-1,n}(x)] \quad \text{if} \quad x_n \leq x < x_{n+1},$$
$$\text{and} \qquad \tilde{h}_n(x) = L_{n,n+1}(x) \qquad\qquad \text{if} \quad x \geq x_{n+1}.$$

The resulting proposal distribution is $g_n(x) \propto \exp\{\tilde{h}_n(x)\}$. The ARMS algorithm is based on $g_n$ and it can be decomposed into two parts, a first step which is a standard Accept–Reject step for the simulation from the instrumental distribution

$$\psi_n(x) \propto \min\left[f_1(x), \exp\{\tilde{h}_n(x)\}\right],$$

based on $g_n$, and a second part, which is the acceptance of the simulated value by a Metropolis–Hastings procedure:

### Algorithm A.29 –ARMS Metropolis–Hastings–

1. Simulate $Y$ from $g_n(y)$ and $U \sim \mathcal{U}_{[0,1]}$
   until

   $$U \leq f_1(Y)/\exp\{\tilde{h}_n(Y)\}.$$

2. Generate $V \sim \mathcal{U}_{[0,1]}$ and take                    [A.29]

   $$(6.3.12) \qquad X^{(t+1)} = \begin{cases} Y & \text{if} \quad V < \dfrac{f_1(Y)\,\psi_n(x^{(t)})}{f_1(x^{(t)})\,\psi_n(Y)} \wedge 1 \\ x^{(t)} & \text{otherwise.} \end{cases}$$

The Accept–Reject step indeed produces a variable distributed from $\psi_n(x)$ and this justifies the expression of the acceptance probability in the Metropolis–Hastings step. Note that [A.29] is a particular case of the approximate Accept–Reject algorithms considered by Tierney (1994) (see Problem 6.6). The probability (6.3.12) can also be written

$$\begin{cases} \min\left[1, \dfrac{f_1(Y)\exp\{\tilde{h}_n(x^{(t)})\}}{f_1(x^{(t)})\exp\{\tilde{h}_n(Y)\}}\right] & \text{if } f_1(Y) > \exp\{\tilde{h}_n(Y)\}, \\[2ex] \min\left[1, \dfrac{\exp\{\tilde{h}_n(x^{(t)})\}}{f_1(x^{(t)})}\right] & \text{otherwise,} \end{cases}$$

which implies a sure acceptance of $Y$ when $f_1(x^{(t)}) < \exp\{\tilde{h}_n(x^{(t)})\}$; that is, when the bound is correct.

Each simulation of $Y \sim g_n$ in Step 1 of [A.29] provides, in addition, an update of $S_n$ in $S_{n+1} = S_n \cup \{y\}$, and therefore of $g_n$, when $Y$ is rejected. As in the case of the ARS algorithm, the initial $S_n$ set must be chosen so that $g_n$ is truly a probability density. If the support of $f$ is not bounded from below, $L_{0,1}$ must be increasing and, similarly, if the support of $f$ is not bounded from above, $L_{n,n+1}$ must be decreasing. Note also that the simulation of $g_n$ detailed in §2.3.3 is valid in this setting.

Since the algorithm [A.29] appears to be a particular case of independent Metropolis–Hastings algorithm, the convergence and ergodicity results obtained in §6.3.1 should apply for [A.29]. This is not the case, however, because of the lack of time homogeneity of the chain (see Definition 4.2.3) produced by [A.29]. The transition kernel, based on $g_n$, can change at each step with a positive probability. Since the study of nonhomogeneous chains is quite delicate, the algorithm [A.29] can only be justified by reverting to the homogeneous case; that is, by fixing the function $g_n$ and the set $S_n$ after a warm-up period of length $n_0$. The constant $n_0$ need not to be fixed in advance as this warm-up period can conclude when the approximation of $f_1$ by $g_n$ is satisfactory, for instance when the rejection rate in Step 1 of $[A.29]$ is sufficiently small. The algorithm $[A.29]$ must then start with an initializing (or calibrating) step which adapts the parameters at hand (in this case, $g_n$) to the function $f_1$. This adaptive structure is generalized in §6.4.

The ARMS algorithm is useful when a precise analytical study of the density $f$ is impossible, as, for instance, in the setup of generalized linear models (see §9.7.3). In fact, $f$ (or $f_1$) only needs to be computed in a few points to initialize the algorithm, which thus does not require the search for "good" density $g$ which approximates $f$. This feature should be contrasted to the cases of the independent Metropolis–Hastings algorithm and of sufficiently fast random walks as in the case of $[A.28]$.

**Example 6.3.9 Poisson logistic model.** For the generalized linear model in Example 2.3.12, consider a logit dependence between explanatory and dependent (observations) variables,

$$Y_i | x_i \sim \mathcal{P}\left(\frac{\exp(bx_i)}{1 + \exp(bx_i)}\right), \qquad i = 1, \ldots, n,$$

which implies the restriction $\lambda_i < 1$ on the parameters of the Poisson distribution, $Y_i \sim \mathcal{P}(\lambda_i)$. When $b$ has the prior distribution $\mathcal{N}(0, \tau^2)$, the posterior distribution is

$$\pi(b|\mathbf{x}, \mathbf{y}) \propto \frac{\exp\{\sum_i y_i(bx_i)\}}{\prod_i(1 + \exp(bx_i))} \exp\left\{-\sum_i \frac{e^{bx_i}}{1 + e^{bx_i}}\right\} e^{-b^2/2\tau^2}.$$

Next, $b$ can be simulated according to the conditional distribution $\pi(b|\mathbf{x}, \mathbf{y})$. This conditional is not easy to simulate from, but one can use the ARMS Metropolis–Hastings algorithm of $[A.29]$. ‖

## 6.4 Optimization and Control

The previous sections have established the theoretical validity of the Metropolis–Hastings algorithms by showing that under suitable (and not very restrictive) conditions on the transition kernel, the chain produced by $[A.24]$ is ergodic and, therefore, that the mean (6.1.1) converges to the expectation $\mathbb{E}_f[h(X)]$. In §6.3 however, we showed that the most common algorithms only rarely enjoy strong ergodicity properties (geometric or uniform ergodicity). In particular, there are simple examples (see Problem 6.2) that show how slow convergence can be.

This section addresses the problem of choosing the transition kernel $q(y|x)$ and illustrates a general acceleration method for Metropolis–Hastings algorithms, which extends the conditioning techniques presented in §3.7.3.

### 6.4.1 Optimizing the Acceptance Rate

When considering only the classes of algorithms described in §6.3, the most common alternatives are to use the following:

(a) a fully automated algorithm like ARMS ([A.29]);

(b) an instrumental density $g$ which approximates $f$, such that $f/g$ is bounded for uniform ergodicity to apply to the algorithm [A.25];

(c) a random walk as in [A.28].

In case (a), the automated feature of [A.29] reduces "parameterization" to the choice of initial values, which are theoretically of limited influence on the efficiency of the algorithm. In both of the other cases, the choice of $g$ is much more critical, as it determines the performances of the resulting Metropolis–Hastings algorithm. As we will see below, the few pieces of advice available on the choice of $g$ are, in fact, contrary! Depending on the type of Metropolis–Hastings algorithm selected, one would want high acceptance rates in case (b) and low acceptance rates in case (c).

Consider, first, the independent Metropolis–Hastings algorithm introduced in §6.3.1. Its similarity with the Accept–Reject algorithm suggests a choice of $g$ that maximizes the average *acceptance rate*

$$\rho = \mathbb{E}\left[\min\left\{\frac{f(Y)\,g(X)}{f(X)\,g(Y)}, 1\right\}\right]$$
$$= 2P\left(\frac{f(Y)}{g(Y)} \geq \frac{f(X)}{g(X)}\right), \qquad X \sim f,\; Y \sim g,$$

as seen in Lemma 6.3.2. In fact, the optimization associated with the choice of $g$ is related to the speed of convergence of $\frac{1}{T}\sum_{t=1}^{T} h(X^{(t)})$ to $\mathbb{E}_f[h(X)]$ and, therefore, to the ability of the algorithm [A.25] to quickly explore any complexity of $f$ (see, for example, Theorem 6.3.1).

If this optimization is to be generic (that is, independent of $h$), $g$ should reproduce the density $f$ as faithfully as possible, which implies the maximization of $\rho$. For example, a density $g$ that is either much less or much more concentrated, compared with $f$, produces a ratio

$$\frac{f(y)\,g(x)}{f(x)\,g(y)} \wedge 1$$

having huge variations and, therefore, leads to a low acceptance rate.

The acceptance rate $\rho$ is typically impossible to compute, and one solution is to use the minorization result $\rho \geq 1/M$ of Lemma 6.3.2 to minimize $M$ as in the case of the Accept–Reject algorithm.

Alternatively, we can consider a more *empirical* approach that consists of choosing a parameterized instrumental distribution $g(\cdot|\theta)$ and adjusting the corresponding parameters $\theta$ based on the evaluated acceptance rate, now $\rho(\theta)$; that is, first choose an initial value for the parameters, $\theta_0$, and estimate the corresponding acceptance rate, $\hat{\rho}(\theta_0)$, based on $m$ iterations of [A.25], then modify $\theta_0$ to obtain an increase in $\rho$.

In the simplest cases, $\theta_0$ will reduce to a scale parameter which is increased or decreased depending on the behavior of $\hat{\rho}(\theta)$. In multidimensional settings, $\theta_0$ can also include a position parameter or a matrix acting as a scale parameter, which makes optimizing $\rho(\theta)$ a more complex task. Note that $\hat{\rho}(\theta)$ can be obtained by simply counting acceptances or through

$$\frac{2}{m} \sum_{i=1}^{m} \mathbb{I}_{\{f(y_i)g(x_i) > f(x_i)g(y_i)\}} \,,$$

where $x_1, \ldots, x_m$ is a sample from $f$, obtained, for instance, from a first MCMC algorithm, and $y_1, \ldots, y_m$ is an iid sample from $g$. Therefore, if $\theta$ is composed of location and scale parameters, a sample $((x_1, y_1), \ldots, (x_m, y_m))$ corresponding to a value $\theta_0$ can be used repeatedly to evaluate different values of $\theta$ by a deterministic modification of $y_i$, which facilitates the maximization of $\rho(\theta)$.

**Example 6.4.1 Inverse Gaussian distribution.** The *inverse Gaussian distribution* has the density

$$(6.4.1) \quad f(z|\theta_1, \theta_2) \propto z^{-3/2} \, \exp\left\{-\theta_1 z - \frac{\theta_2}{z} + 2\sqrt{\theta_1 \theta_2} + \log \sqrt{2\theta_2}\right\}$$

on $\mathbb{R}_+$ $(\theta_1 > 0, \theta_2 > 0)$. Denoting $\psi(\theta_1, \theta_2) = 2\sqrt{\theta_1 \theta_2} + \log \sqrt{2\theta_2}$, it follows from a classical result on exponential families (see Brown 1986, Chapter 2, Robert 1994a, Section 3.2.2, or Problem 1.48) that

$$\mathbb{E}[(Z, 1/Z)] = \nabla \psi(\theta_1, \theta_2)$$
$$= \left(\sqrt{\frac{\theta_2}{\theta_1}}, \sqrt{\frac{\theta_1}{\theta_2}} + \frac{1}{2\theta_2}\right).$$

A possible choice for the simulation of (6.4.1) is the Gamma distribution $\mathcal{G}a(\alpha, \beta)$ in algorithm [A.25], taking $\alpha = \beta\sqrt{\theta_2/\theta_1}$ so that the means of both distributions coincide. Since

$$\frac{f(x)}{g(x)} \propto x^{-\alpha-1/2} \, \exp\left\{(\beta - \theta_1)x - \frac{\theta_2}{x}\right\} \,,$$

the ratio $f/g$ is bounded for $\beta < \theta_1$. The value of $x$ which maximizes the ratio is the solution of

$$(\beta - \theta_1)x^2 - \left(\alpha + \frac{1}{2}\right)x + \theta_2 = 0 \,;$$

that is,

$$x_\beta^* = \frac{(\alpha + 1/2) - \sqrt{(\alpha + 1/2)^2 + 4\theta_2(\theta_1 - \beta)}}{2(\beta - \theta_1)} \,.$$

The analytical optimization (in $\beta$) of

$$M(\beta) = (x_\beta^*)^{-\alpha-1/2} \, \exp\left\{(\beta - \theta_1)x_\beta^* - \frac{\theta_2}{x_\beta^*}\right\}$$

| $\beta$ | 0.2 | 0.5 | 0.8 | 0.9 | 1 | 1.1 | 1.2 | 1.5 |
|---------|------|------|------|------|------|------|------|------|
| $\hat{\rho}(\beta)$ | 0.22 | 0.41 | 0.54 | 0.56 | 0.60 | 0.63 | 0.64 | 0.71 |
| $\mathbb{E}[Z]$ | 1.137 | 1.158 | 1.164 | 1.154 | 1.133 | 1.148 | 1.181 | 1.148 |
| $\mathbb{E}[1/Z]$ | 1.116 | 1.108 | 1.116 | 1.115 | 1.120 | 1.126 | 1.095 | 1.115 |

Table 6.4.1. *Estimation of the means of $Z$ and of $1/Z$ for the inverse Gaussian distribution $\mathcal{IN}(\theta_1, \theta_2)$ by the Metropolis–Hastings algorithm [A.25] and evaluation of the acceptance rate for the instrumental distribution $\mathcal{G}a(\sqrt{\theta_2/\theta_1}\ \beta, \beta)$ ($\theta_1 = 1.5, \theta_2 = 2$, and $m = 5000$).*

is not possible, although, in this specific case the curve $M(\beta)$ can be plotted for given values of $\theta_1$ and $\theta_2$ and the optimal value $\beta^\star$ can be approximated numerically. Typically, the influence of the choice of $\beta$ must be assessed empirically; that is, by approximating the acceptance rate $\rho$ via the method described above.

Note that a new sample $(y_1, \ldots, y_m)$ must be simulated for every new value of $\beta$. Whereas $y \sim \mathcal{G}a(\alpha, \beta)$ is equivalent to $\beta y \sim \mathcal{G}a(\alpha, 1)$, the factor $\alpha$ depends on $\beta$ and it is not possible to use the same sample for several values of $\beta$. Table 6.4.1 provides an evaluation of the rate $\rho$ as a function of $\beta$ and gives estimates of the means of $Z$ and $1/Z$ for $\theta_1 = 1.5$ and $\theta_2 = 2$. The constraint on the ratio $f/g$ then imposes $\beta < 1.5$. The corresponding theoretical values are respectively 1.155 and 1.116, and the optimal value of $\beta$ is $\beta^\star = 1.5$.    ||

The *random walk* version of the Metropolis–Hastings algorithm, introduced in §6.3.2, requires a different approach to acceptance rates, given the dependence of the instrumental distribution on the current state of the chain. In fact, a high acceptance rate does not necessarily indicate that the algorithm is moving correctly since it may indicate that the random walk is moving too slowly on the surface of $f$. If $x^{(t)}$ and $y_t$ are close, in the sense that $f(x^{(t)})$ and $f(y_t)$ are approximately equal, the algorithm [A.28] leads to the acceptance of $y$ with probability

$$\min\left(\frac{f(y_t)}{f(x^{(t)})}, 1\right) \simeq 1 \ .$$

A higher acceptance rate may therefore correspond to a slower convergence as the moves on the support of $f$ are more limited. In the particular case of multimodal densities whose modes are separated by zones of extremely small probability, the negative effect of limited moves on the surface of $f$ clearly shows. While the acceptance rate is quite high for a distribution $g$ with small variance, the probability of jumping from one mode to another may be arbitrarily small. This phenomenon occurs, for instance, in the case of mixtures of distributions (see Chapter 9) and in overparameterized models (see, e.g., Tanner and Wong 1987 and Besag *et al.* 1995). In contrast, if the average acceptance rate is low, the successive values of $f(y_t)$ tend to be small compared with $f(x^{(t)})$, which means that the random walk moves quickly on the surface of $f$ since it often reaches the "borders" of the support of $f$ (or, at least, that the random walk explores regions with low probability under $f$).

The above analysis seems to require an advanced knowledge of the density of interest, since an instrumental distribution $g$ with too narrow a range will slow down the convergence rate of the algorithm. On the other hand, a distribution $g$ with a wide range results in a waste of simulations of points outside the range of $f$ without improving the probability of visiting all of the modes of $f$. It is unfortunate that an automated parameterization of $g$ cannot guarantee uniformly optimal performances for the algorithm [A.28], and that the rules for choosing the rate presented in Note 6.7.4 are only heuristic.

### 6.4.2 Conditioning and Accelerations

Similar[5] to the Accept–Reject method, the Metropolis–Hastings algorithm does not take advantage of the total set of random variables that are generated. Lemma 6.3.2 shows that the "rate of waste" of these variables $y_t$ is lower than for the Accept–Reject method, but it still seems inefficient to ignore the rejected $y_t$'s. As the rejection mechanism relies on an independent uniform random variable, it is reasonable to expect that the rejected variables bring, although indirectly, some relevant information on the distribution $f$. As in the conditioning method introduced in §3.7.3, the *Rao–Blackwellization* technique applies in the case of the Metropolis–Hastings algorithm. (Other approaches to Metropolis–Hastings acceleration can be found in Green and Han 1992, Gelfand and Sahu 1994, or McKeague and Wefelmeyer 1996.)

First, note that a sample produced by the Metropolis–Hastings algorithm, $x^{(1)}, \ldots, x^{(T)}$, is based on two samples, $y_1, \ldots, y_T$ and $u_1, \ldots, u_T$, with $y_t \sim q(y|x^{(t-1)})$ and $u_t \sim \mathcal{U}_{[0,1]}$. The mean (6.1.1) can then be written

$$\delta^{MH} = \frac{1}{T} \sum_{t=1}^{T} h(x^{(t)}) = \frac{1}{T} \sum_{t=1}^{T} \sum_{i=1}^{t} \mathbb{I}_{x^{(t)}}$$

$$= \frac{1}{T} \sum_{t=1}^{T} h(y_t) \sum_{i=t}^{T} \mathbb{I}_{x^{(i)}=y_t}$$

and the conditional expectation

$$\delta^{RB} = \frac{1}{T} \sum_{t=1}^{T} h(y_t) \, \mathbb{E}\left[ \sum_{i=t}^{T} \mathbb{I}_{X^{(i)}=y_t} \Big| y_1, \ldots, y_T \right]$$

---

[5] This section presents material related to nonparametric Rao–Blackwellization, as in §3.7.3, and may be skipped on a first reading.

$$= \frac{1}{T} \sum_{t=1}^{T} h(y_t) \left( \sum_{i=t}^{T} P(X^{(i)} = y_t | y_1, \ldots, y_T) \right)$$

dominates the empirical mean, $\delta^{MH}$, under quadratic loss. This is a consequence of the Rao–Blackwell Theorem (see Lehmann and Casella 1998, Section 1.7), resulting from the fact that $\delta^{RB}$ integrates out the variation due to the uniform sample.

The practical interest of this alternative to $\delta^{MH}$ is that the probabilities $P(X^{(i)} = y_t | y_1, \ldots, y_T)$ can be explicitly computed. Casella and Robert (1996) have established the two following results, which provide the weights for $h(y_t)$ in $\delta^{RB}$ both for the independent Metropolis–Hastings algorithm and the general Metropolis–Hastings algorithm. In both cases, the computational complexity of these weights is of order $T^2$, which is a manageable order of magnitude.

Consider first the case of the independent Metropolis–Hastings algorithm associated with the instrumental distribution $g$. For simplicity's sake, assume that $X^{(0)}$ is simulated according to the distribution of interest, $f$, so that the chain is stationary, and the mean (6.1.1) can be written

$$\delta^{MH} = \frac{1}{T+1} \sum_{t=0}^{T} h(x^{(t)}) \, ,$$

with $x^{(0)} = y_0$. If we denote

$$w_i = \frac{f(y_i)}{g(y_i)}, \qquad \rho_{ij} = \frac{w_j}{w_i} \wedge 1 \qquad (0 \le i < j),$$

$$\zeta_{ii} = 1, \qquad \zeta_{ij} = \prod_{t=i+1}^{j} (1 - \rho_{it}) \qquad (i \; < \; j),$$

we have the following theorem, whose proof is left to Problem 6.25.

**Theorem 6.4.2** *The estimator* $\delta^{RB}$ *can be written*

$$\delta^{RB} = \frac{1}{T+1} \sum_{i=0}^{T} \varphi_i \, h(y_i),$$

*where*

$$\varphi_i = \tau_i \sum_{j=i}^{T} \zeta_{ij} \, ,$$

*and the conditional probability* $\tau_i = P(X^{(i)} = y_i | y_0, y_1, \ldots, y_T)$, $i = 0, \ldots, T$, *is given by* $\tau_0 = 1$ *and* $\tau_i = \sum_{j=0}^{i-1} \tau_j \, \zeta_{j(i-1)} \rho_{ji}$ *for* $i > 0$.

The computation of $\zeta_{ij}$ for a fixed $i$ requires $(T - i)$ multiplications since $\zeta_{i(j+1)} = \zeta_{ij}(1 - \rho_{i(j+1)})$; therefore, the computation of all the $\zeta_{ij}$'s require $T(T + 1)/2$ multiplications. The derivations of $\tau_i$ and $\varphi_i$ are of the same order of complexity.

| $n$   | 10    | 25    | 50    | 100   |
|-------|-------|-------|-------|-------|
| $h_1$ | 50.11 | 49.39 | 48.27 | 46.68 |
| $h_2$ | 42.20 | 44.75 | 45.44 | 44.57 |

Table 6.4.2. *Decrease (in percentage) of squared error risk associated with $\delta^{RB}$ for the evaluation of $\mathbb{E}[h_i(X)]$, evaluated over 7500 simulations for different sample sizes $n$. (Source: Casella and Robert 1996).*

**Example 6.4.3 Rao–Blackwellization improvement for a $\mathcal{T}_3$ simulation.** Suppose the target distribution is $\mathcal{T}_3$ and the instrumental distribution is Cauchy, $\mathcal{C}(0,1)$. The ratio $f/g$ is bounded, which ensures a geometric rate of convergence for the associated Metropolis–Hastings algorithm. Table 6.4.2 illustrates the improvement brought by $\delta^{RB}$ for some functions of interest $h_1(x) = x$ and $h_2(x) = \mathbb{I}_{(1.96,+\infty)}(x)$, whose (exact) expectations $\mathbb{E}[h_i(X)]$ ($i = 1, 2$) are 0 and 0.07, respectively. Over the different sample sizes selected for the experiment, the improvement in mean square error brought by $\delta^{RB}$ is of the order 50%. ‖

We next consider the general case, with an arbitrary instrumental distribution $q(y|x)$. The dependence between $Y_i$ and the set of previous variables $Y_j$ ($j < i$) (since $X^{(i-1)}$ can be equal to $Y_0$, $Y_1, \ldots$, or $Y_{i-1}$) complicates the expression of the joint distribution of $Y_i$ and $U_i$, which cannot be obtained in closed form for arbitrary $n$. In fact, although $(X^{(t)})$ is a Markov chain, $(Y_t)$ is not.

Let us denote

$$\rho_{ij} = \frac{f(y_j)/q(y_j|y_i)}{f(y_i)/q(y_i|y_j)} \wedge 1 \qquad (j > i),$$

$$\overline{\rho}_{ij} = \rho_{ij} q(y_{j+1}|y_j), \quad \underline{\rho}_{ij} = (1 - \rho_{ij}) q(y_{j+1}|y_i) \qquad (i < j < T),$$

$$\zeta_{jj} = 1, \quad \zeta_{jt} = \prod_{l=j+1}^{t} \underline{\rho}_{jl} \qquad (i < j < T),$$

$$\tau_0 = 1, \quad \tau_j = \sum_{t=0}^{j-1} \tau_t \zeta_{t(j-1)} \overline{\rho}_{tj}, \quad \tau_T = \sum_{t=0}^{T-1} \tau_t \zeta_{t(T-1)} \rho_{tT} \qquad (i < T),$$

$$\omega_T^i = 1, \quad \omega_i^j = \overline{\rho}_{ji} \omega_{i+1}^i + \underline{\rho}_{tj} \omega_{i+1}^j \qquad (0 \le j < i < T).$$

Casella and Robert (1996) derive the following expression for the weights of $h(y_i)$ in $\delta^{RB}$. We again leave the proof as a problem (Problem 6.26).

**Theorem 6.4.4** *The estimator $\delta^{RB}$ satisfies*

$$\delta^{RB} = \frac{\sum_{i=0}^{T} \varphi_i\, h(y_i)}{\sum_{i=0}^{T-1} \tau_i\, \zeta_{i(T-1)}},$$

| $n$ | 10 | 25 | 50 | 100 |
|---|---|---|---|---|
| | $h_1$ | 10.7 | 8.8 | 7.7 | 7.7 |
| $\sigma = 0.4$ | | (1.52) | (0.98) | (0.63) | (0.3) |
| | $h_2$ | 23.6 | 25.2 | 25.8 | 25.0 |
| | | (0.02) | (0.01) | (0.006) | (0.003) |
| | $h_1$ | 0.18 | 0.15 | 0.11 | 0.07 |
| $\sigma = 3.0$ | | (2.28) | (1.77) | (1.31) | (0.87) |
| | $h_2$ | 0.99 | 0.94 | 0.71 | 1.19 |
| | | (0.03) | (0.02) | (0.014) | (0.008) |

Table 6.4.3. *Improvement brought by $\delta^{RB}$ (in %) and quadratic risk of the empirical average (in parentheses) for different sample sizes and* 50,000 *simulations of the random walk based on $\mathcal{C}(0, \sigma^2)$ (Source: Casella and Robert 1996).*

with $(i < T)$

$$\varphi_i = \tau_i \left[ \sum_{j=i}^{T-1} \zeta_{ij} \omega_{j+1}^i + \zeta_{i(T-1)}(1 - \rho_{iT}) \right]$$

and $\varphi_T = \tau_T$.

Although these estimators are more complex than in the independent case, the complexity of the weights is again of order $T^2$ since the computations of $\bar{\rho}_{ij}$, $\rho_{ij}$, $\zeta_{ij}$, $\tau_i$, and $\omega_j^i$ involve $T(T + 1)/2$ multiplications. Casella and Robert (1996) give algorithmic advice toward easier and faster implementation.

**Example 6.4.5 (Continuation of Example 6.4.3)** Consider now the simulation of a $t$ distribution $\mathcal{T}_3$ based on a random walk with perturbations distributed as $\mathcal{C}(0, \sigma^2)$. The choice of $\sigma$ determines the acceptance rate for the Metropolis–Hastings algorithm: When $\sigma = 0.4$, it is about 0.33, and when $\sigma = 3.0$, it increases to 0.75.

As explained in §6.4.1, the choice $\sigma = 0.4$ is undoubtedly preferable in terms of efficiency of the algorithm. Table 6.4.3 confirms this argument, since the quadratic risk of the estimators (6.1.1) is larger for $\sigma = 3.0$. The gains brought by $\delta^{RB}$ are smaller, compared with the independent case. They amount to approximately 8% and 25% for $\sigma = 0.4$ and 0.1% and 1% for $\sigma = 3$. Casella and Robert (1996) consider an additional comparison with an importance sampling estimator based on the same sample $y_1, \ldots, y_n$. ‖

## 6.5 Further Topics

We now[6] describe two types of Metropolis–Hastings algorithms with some distinctive features.

### 6.5.1 Reversible Jumps

The algorithms of §6.3 provide simulation methods for distributions in fixed-dimensional spaces, but they cannot handle settings where the dimension of the space is itself one of the parameters to be simulated. Such settings naturally occur in model choice problems, where different models with variable dimensions are compared. Aside from the usual case of variable selection in regression setups, important examples are the estimation of the number of components in a mixture (see §9.3), the order of an $ARMA(p,q)$ sequence, or of the number of changepoints in a piecewise stationary sequence (see §9.5.2).

Consider, therefore, a distribution with density

$$(6.5.1) \qquad f(k, \theta^{(k)}) = f(\theta^{(k)}|k)p(k) ,$$

where $k \in \mathbb{N}$ and $\theta^{(k)} \in \Xi_k$, so the dimension of $\theta^{(k)}$ changes with $k$. The function $p(k)$ is a density with respect to the counting measure on $\mathbb{N}$ and $f(\theta^{(k)}|k)$ is typically a density with respect to Lebesgue measure on $\Xi_k$. The density (6.5.1) is then a density with respect to Lebesgue measure on the sum of spaces

$$\Xi = \bigoplus_{k=1}^{\infty} \Xi_k .$$

The problem in implementing a Markov chain Monte Carlo algorithm is the need to be able to move from a submodel $\mathcal{H}_k = \{k\} \times \Xi_k$ to another submodel $\mathcal{H}_{k'}$, where $k \neq k'$, with particular difficulty being encountered when $k < k'$. If we let $x = (k, \theta^{(k)})$, a solution proposed by Green (1995) is based on a *reversible* transition kernel $K$, that is, a kernel satisfying

$$\int_A \int_B K(x, dy)\pi(x)dx = \int_B \int_A K(y, dx)\pi(y)dy$$

for some invariant density $\pi$. If $q_m$ is a transition measure to the submodel $m$, and $\rho_m$ the probability of accepting the jump, the kernel can be decomposed as

$$K(x, B) = \sum_{m=1}^{\infty} \int_B \rho_m(x, y')q_m(x, dy') + \omega(x)\mathbb{I}_B(x),$$

where

$$\omega(x) = 1 - \sum_m q_m(x, \mathcal{H}), \qquad \mathcal{H} = \bigcup_k \mathcal{H}_k,$$

---

[6] This section may be skipped at first reading, as it contains more advanced material, which will not be used in the remainder of the book.

is the probability of no jump between dimensions. Typically, the jumps are limited to moves between models with dimensions similar to the dimension of $\mathcal{H}_k$, possibly including $\mathcal{H}_k$. The definition of $\rho_m$ (and the verification of the reversibility assumption) relies on the following constraint: *The joint measure $\pi(dx)q_m(x, dy)$ must be absolutely continuous with respect to a symmetric measure $\xi_m(dx, dy)$ on $\mathcal{H}^2$.* If $g_m(x, y)$ denotes the density of $q_m(x, , dy)\pi(dx)$ and $\rho_m$ is written in the usual form

$$\rho_m(x, y) = \min\left\{1, \frac{g_m(y, x)}{g_m(x, y)}\right\} ,$$

then reversibility is ensured by the symmetry of the measure $\xi_m$,

$$\int_A \int_B \rho_m(x, y)q_m(x, dy)\pi(dx) = \int_A \int_B \rho_m(x, y)g_m(x, y)\xi_m(dx, dy)$$
$$= \int_A \int_B \rho_m(y, x)g_m(y, x)\xi_m(dy, dx)$$
$$= \int_A \int_B \rho_m(y, x)q_m(y, dx)\pi(dy) ,$$

as $\rho_m(x, y)g_m(x, y) = \rho_m(y, x)g_m(y, x)$ by construction.

The main difficulty of this approach lies in the determination of the measure $\xi_m$, given the symmetry constraint. If the jumps can be decomposed into moves between two models only, $\mathcal{H}_{k_1}$ and $\mathcal{H}_{k_2}$, the (clever!) idea of Green (1995) is to supplement both spaces $\mathcal{H}_{k_1}$ and $\mathcal{H}_{k_2}$ and to create a bijection between both completions. For instance, if $\dim(\mathcal{H}_{k_1}) > \dim(\mathcal{H}_{k_2})$ and if the move from $\mathcal{H}_{k_1}$ to $\mathcal{H}_{k_2}$ can be represented by a deterministic transformation of $\theta^{(k_1)}$ (that is, $\theta^{(k_2)} = T(\theta^{(k_1)})$), Green (1995) imposes a *dimension matching* condition which is that the opposite move from $\mathcal{H}_{k_2}$ to $\mathcal{H}_{k_1}$ is concentrated on the curve

$$\{\theta^{(k_1)}; \theta^{(k_2)} = T(\theta^{(k_1)})\} .$$

In the general case, if $\theta^{(k_1)}$ is completed to $(\theta^{(k_1)}, u_1)$ and $\theta^{(k_2)}$ to $(\theta^{(k_2)}, u_2)$ so that the mapping between $(\theta^{(k_1)}, u_1)$ and $(\theta^{(k_2)}, u_2)$ is a bijection, with, say,

(6.5.2)        $$(\theta^{(k_2)}, u_2) = T(\theta^{(k_1)}, u_1),$$

the probability of acceptance for the move from $\mathcal{H}_{k_1}$ to $\mathcal{H}_{k_2}$ is then

$$\min\left(\frac{af(k_2, \theta^{(k_2)})\pi_{12}g_2(u_2)}{f(k_1, \theta^{(k_1)})\pi_{21}g_1(u_1)} \left|\frac{\partial T(\theta^{(k_1)}, u_1)}{\partial(\theta^{(k_1)}, u_1)}\right|, 1\right) ,$$

involving the Jacobian of the transform (6.5.2), the probabilities $\pi_{ij}$ of choosing $\mathcal{H}_{k_j}$ while in $\mathcal{H}_{k_i}$, and $g_i$, the density of $u_i$.

**Example 6.5.1 A linear Jacobian.** To illustrate the procedure, Green (1995) considers the small example of switching between the parameters $(1, \theta)$ and $(2, \theta_1, \theta_2)$ using the following moves:

(i) To go from $(2, \theta_1, \theta_2)$ to $(1, \theta)$, set $\theta = (\theta_1 + \theta_2)/2$.

(ii) To go from $(1, \theta)$ to $(2, \theta_1, \theta_2)$, generate a random variable $u$ and set $\theta_1 = \theta - u$ and $\theta_2 = \theta + u$.

These moves represent one-to-one transformations of variables in $\mathbb{R}^2$; that is,

$$\left( \frac{\theta_1 + \theta_2}{2}, \theta_2 \right) = T_1(\theta_1, \theta_2), \qquad (\theta - u, \theta + u) = T_2(\theta, u),$$

with corresponding Jacobians

$$\frac{\partial T_1(\theta1, \theta_2)}{\partial(\theta_1, \theta_2)} = \frac{1}{2}, \qquad \frac{\partial T_2(\theta, u)}{\partial(\theta, u)} = 2. \qquad\qquad \parallel$$

The construction of the dimension matching transform can be difficult; one could almost say this is a drawback with the reversible jump method. (Nothing is so easy as to write down the wrong Jacobian!) In particular, the transform $T$ must be differentiable. Moreover, the total freedom left by the reversible jump principle about the choice of the jumps, which are often referred to as *split* and *merge* moves in embedded models, creates a potential for inefficiency and requires tuning steps which may be quite demanding. (See Green 1995, Richardson and Green 1997, or Denison *et al.* 1998 for illustrations.)

As pointed out by Green (1995), the density $f$ does not need to be normalized, but the different component densities $f(\theta^{(l)}|k)$ must be known up to the same constant.

We close this section with two examples of the reversible jump algorithm.

**Example 6.5.2 Linear versus quadratic regression.** Instead of choosing a particular regression model, we can use the reversible jump algorithm to do model averaging, avoiding choosing between models and, we hope, obtaining a reasonable fit.

Suppose that the two candidate models are the simple linear and simple quadratic regression models; that is,

$$y_i = \beta_0 + \beta_1 x_i + \varepsilon_i \qquad \text{or} \qquad y_i = \beta_0 + \beta_1 x_i + \beta_2 x_i^2 + \varepsilon_i .$$

If we represent either regression by $\mathbf{y} = \mathbf{X}\boldsymbol{\beta} + \varepsilon$, where $\varepsilon \sim \mathcal{N}(0, \sigma^2 I)$, the least squares estimate $\hat{\boldsymbol{\beta}} = (\mathbf{X}'\mathbf{X})^{-1}\mathbf{X}'\mathbf{y}$ has distribution

$$\hat{\boldsymbol{\beta}} \sim \mathcal{N}(\boldsymbol{\beta}, \sigma^2(\mathbf{X}'\mathbf{X})^{-1}) .$$

Using normal prior distributions will result in normal posterior distributions, and the reversible jump algorithm will then be jumping between a two-dimensional and three-dimensional normal.

To jump between these models, it is sensible to first transform to orthogonal coordinates, as a jump that is made by simple adding or deleting a coefficient will then not affect the fit of the other coefficients. We thus find an orthogonal matrix $\mathbf{P}$ and diagonal matrix $D_\lambda$ satisfying $\mathbf{P}'\mathbf{X}'\mathbf{X}\mathbf{P} = D_\lambda$. The elements of $D_\lambda$, $\lambda_i$, are the eigenvalues of $\mathbf{X}'\mathbf{X}$ and the columns of $\mathbf{P}$ are the eigenvectors. We then write $\mathbf{X}^* = \mathbf{X}\mathbf{P}$ and $\boldsymbol{\alpha} = \mathbf{P}'\boldsymbol{\beta}$, and we work with the model $\mathbf{y} = \mathbf{X}^*\boldsymbol{\alpha} + \varepsilon$.

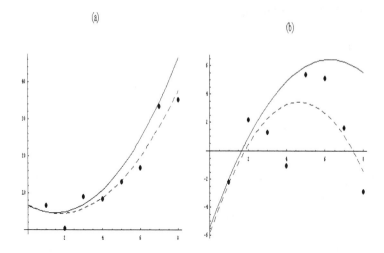

Figure 6.5.1. *Comparison of quadratic least squares (dashed line) and reversible jump (solid line) fits. In (a), the underlying model is quadratic, and in (b), the underlying model is linear.*

If each $\alpha_i$ has a normal prior distribution, $\alpha_i \sim \mathcal{N}(0, \tau^2)$, its posterior density, denoted by $f_i$, is $\mathcal{N}(b_i, b_i \sigma^2)$, where $b_i = \frac{\lambda_i \tau^2}{\lambda_i \tau^2 + \sigma^2}$. The possible moves are as follows:

(i) linear $\rightarrow$ linear: $(\alpha_0, \alpha_1) \rightarrow (\alpha_0', \alpha_1')$, where $(\alpha_0', \alpha_1') \sim f_0 f_1$,

(ii) linear $\rightarrow$ quadratic: $(\alpha_0, \alpha_1) \rightarrow (\alpha_0, \alpha_1, \alpha_2')$, where $\alpha_2' \sim f_2$,

(iii) quadratic $\rightarrow$ quadratic: $(\alpha_0, \alpha_1, \alpha_2) \rightarrow (\alpha_0', \alpha_1', \alpha_2')$, where $(\alpha_0', \alpha_1', \alpha_2') \sim f_0 f_1 f_2$,

(iv) quadratic $\rightarrow$ linear: $(\alpha_0, \alpha_1, \alpha_2) \rightarrow (\alpha_0', \alpha_1')$, where $(\alpha_0', \alpha_1') = (\alpha_0, \alpha_1)$.

The algorithm was implemented on simulated data with move probabilities $\pi_{ij}$ all taken to be $1/4$ and a prior probability of $1/2$ on each regression model. The resulting fits are given in Figure 6.5.1. It is interesting to note that when the model is quadratic, the reversible jump fit is close to that of quadratic least squares, but it deviates from quadratic least squares when the underlying model is linear. See Problem 6.28 for more details.          ‖

**Example 6.5.3 Piecewise constant densities.** Consider a density $\ell$ on $[0, 1]$ of the form

$$\ell(x) = \sum_{i=1}^{k} \omega_i \mathbb{I}_{[a_i, a_{i+1})}(x) ,$$

with

$$a_1 = 0, \quad a_{k+1} = 1, \quad \text{and} \quad \sum_{i=1}^{k} \omega_i (a_{i+1} - a_i) = 1 .$$

This means that the phenomenon under study has a constant behavior on intervals with different probabilities on these intervals. Assuming all parameters unknown (including $k$), define $p_i = \omega_i(a_{i+1} - a_i)$, so these are probabilities. Let the prior distribution be

$$\pi(k, p^{(k)}, a^{(k)}) = \frac{\lambda^k e^{-\lambda}}{k!} \frac{\Gamma(k/2)p_1^{-1/2} \cdots p_k^{-1/2}}{\Gamma(1/2)^k}(k-1)! \mathbb{I}_{a_2 \leq \ldots \leq a_k},$$

where $p^{(k)} = (p_1, \ldots, p_k)$ and $a^{(k)} = (a_2, \ldots, a_k)$, which implies a Poisson distribution on $k$, $\mathcal{P}(\lambda)$, a uniform distribution on $\{a_2, \ldots, a_k\}$, and a Dirichlet distribution $D_k(1/2, \ldots, 1/2)$ on the weights $p_i$ of the components $\mathcal{U}_{[a_i, a_{i+1}]}$ of $\ell$. (Note that the density integrates to 1 over $\Xi$.)

For a sample $x_1, \ldots, x_n$, the posterior distribution is

$$\pi(k, p^{(k)}, a^{(k)} | x_1, \ldots, x_n) \propto \frac{\lambda^k}{k} \frac{\Gamma(k/2)}{\Gamma(1/2)^k} p_1^{n_1 - 1/2} \cdots p_k^{n_k - 1/2}$$
$$\times \prod_{i=1}^k (a_{i+1} - a_i)^{-n_i} \mathbb{I}_{a_2 \leq \ldots \leq a_k},$$

where $n_j$ is the number of observations between $a_j$ and $a_{j+1}$. Let us choose only jumps between neighboring models; that is, models with one more or one less component in the partition of $[0, 1]$. We then represent the jump from model $\mathcal{H}_k$ to model $\mathcal{H}_{k-1}$ as a random choice of $i$ ($1 \leq i \leq k - 1$), and the aggregation of the $i$th and $(i+1)$st components as

$$a_2^{(k-1)} = a_2^{(k)}, \ldots, a_i^{(k-1)} = a_i^{(k)}, \; a_{i+1}^{(k-1)} = a_{i+2}^{(k)}, \ldots, a_{k-1}^{(k-1)} = a_k^{(k)}$$

and

$$p_1^{(k-1)} = p_1^{(k)}, \ldots, p_i^{(k-1)} = (p_i^{(k)} + p_{i+1}^{(k)}), \ldots, p_{k-1}^{(k-1)} = p_k^{(k)}.$$

For reasons of symmetry, the opposite (upward) jump implies choosing a component $i$ at random and breaking it by the procedure

1. Generate $u_1, u_2 \sim \mathcal{U}_{[0,1]}$;

2. Take $a_i^{(k)} = a_i^{(k-1)}$, $a_{i+1}^{(k)} = u_1 a_i^{(k-1)} + (1 - u_1)a_{i+1}^{(k-1)}$ and $a_{i+2}^{(k)} = a_{i+1}^{(k-1)}$;

3. Take $p_i^{(k)} = u_2 p_i^{(k-1)}$, and $p_{i+1}^{(k)} = (1 - u_2)p_i^{(k-1)}$.

The other quantities remain identical up to a possible index shift. The weight corresponding to the jump from model $\mathcal{H}_k$ to model $\mathcal{H}_{k+1}$ is then

$$\rho_m = \min\left\{1, \frac{\pi(k+1, p^{(k+1)}, a^{(k+1)})}{\pi(k, p^{(k)}, a^{(k)})} \left| \frac{\partial(p^{(k+1)}, a^{(k+1)})}{\partial(p^{(k)}, a^{(k)}, u_1, u_2)} \right| \right\}.$$

As an illustration, take $k = 3$ and consider the jump from model $\mathcal{H}_3$ to model $\mathcal{H}_4$. The transformation is given by

$$
\begin{pmatrix} p_1^{(3)} \\ p_2^{(3)} \\ p_3^{(3)} \\ a_2^{(3)} \\ a_3^{(3)} \\ u_1 \\ u_2 \end{pmatrix} \rightarrow \begin{pmatrix} p_1 \\ u_2 p_2 \\ (1-u_2)p_2 \\ p_3 \\ a_2 \\ u_1 a_2 + (1-u_1)a_3 \\ a_3 \end{pmatrix} = \begin{pmatrix} p_1^{(4)} \\ p_2^{(4)} \\ p_3^{(4)} \\ p_4^{(4)} \\ a_2^{(4)} \\ a_3^{(4)} \\ a_4^{(4)} \end{pmatrix}
$$

with Jacobian

$$
\left| \frac{\partial(p^{(4)}, a^{(4)})}{\partial(p^{(3)}, a^{(3)}, u_1, u_2)} \right| = \begin{vmatrix} 1 & 0 & 0 & 0 & 0 & 0 & 0 \\ 0 & u_2 & 1-u_2 & 0 & 0 & 0 & 0 \\ 0 & 0 & 0 & 1 & 0 & 0 & 0 \\ 0 & 0 & 0 & 0 & 1 & u_1 & 0 \\ 0 & 0 & 0 & 0 & 0 & 1-u_1 & 1 \\ 0 & 0 & 0 & 0 & 0 & a_2^{(3)} - a_3^{(3)} & 0 \\ 0 & p_2^{(3)} & p_2^{(3)} & 0 & 0 & 0 & 0 \end{vmatrix}
$$

$$
= \left| p_2^{(3)}(a_2^{(3)} - a_3^{(3)}) \right|.
$$

(See Problem 6.30 for extensions.)                                    ‖

   In conclusion, note that the complexity of the method is aggravated by the reversibility constraints imposed on the jumps, although these are not necessary for the convergence of Metropolis–Hastings algorithms. In particular, the symmetry between moves can be abandoned.[7] (Reversibility is actually subject to much debate in the literature—some authors defending its use for the Central Limit Theorem, others implying that nonreversible chains can be constructed to have faster convergence; see Holmes *et al.* 1997 and Stramer and Tweedie 1997a. See also Carlin and Chib 1995 for an alternative to reversible jump in model choice.)

### 6.5.2 Langevin Algorithms

Alternatives to the random walk Metropolis–Hastings algorithm can be derived from *diffusion theory* as proposed in Grenander and Miller (1994) and Phillips and Smith (1996). The basic idea of this approach is to seek a *diffusion equation* (or a *stochastic differential equation*) which produces a *diffusion* (or continuous-time process) with stationary distribution $f$ and

---

[7] There is also a difficulty of a more philosophical nature with the uses (or misuses) of reversible jump techniques, in that they allow for unmonitored or "black box" types of inference, where variable dimension models pile up on top of one another and result in conclusions where (i) the role of the prior modeling may be predominant, (ii) the variability of the inference is underestimated, and (iii) the related MCMC convergence issues are difficult to evaluate.

then discretize the process to implement the method. More specifically, the *Langevin diffusion* $L_t$ is defined by the stochastic differential equation

$$(6.5.3) \qquad\qquad dL_t = dB_t + \frac{1}{2}\nabla \log f(L_t)dt,$$

where $B_t$ is the standard *Brownian motion*; that is, a random function such that $B_0 = 0$, $B_t \sim \mathcal{N}(0, \omega^2 t)$, $B_t - B_{t'} \sim \mathcal{N}(0, \omega^2|t - t'|)$ and $B_t - B_{t'}$ is independent of $B_{t'}$ $(t > t')$. (This process is the limit of a simple random walk when the magnitude of the steps, $\Delta$, and the time between steps, $\tau$, both approach zero in such a way that

$$\lim \Delta/\sqrt{\tau} = \omega.$$

See, for example, Ethier and Kurz 1986, Resnick 1994, or Norris 1998.) As stressed by Roberts and Rosenthal (1998), the Langevin diffusion (6.5.3) is the only non-explosive diffusion which is reversible with respect to $f$.

The actual implementation of the diffusion algorithm involves a discretization step (as in §3.9.2) where (6.5.3) is replaced with the random-walk-like transition

$$(6.5.4) \qquad\qquad x^{(t+1)} = x^{(t)} + \frac{\sigma^2}{2}\nabla \log f(x^{(t)}) + \sigma\varepsilon_t,$$

where $\varepsilon_t \sim \mathcal{N}_p(0, I_p)$ and $\sigma^2$ corresponds to the discretization size (see Problem 6.38). Although this naive discretization aims at reproducing the convergence of the random walk to the Brownian motion, the behavior of the Markov chain (6.5.4) may be very different from that of the diffusion process (6.5.3). As shown by Roberts and Tweedie (1995), the chain (6.5.4) may well be transient! Indeed, a sufficient condition for this transience is that the limits

$$(6.5.5) \qquad\qquad \lim_{x \to \pm\infty} \sigma^2\nabla \log f(x)|x|^{-1}$$

exist and are larger than 1 and smaller than $-1$ at $-\infty$ and $+\infty$, respectively, since the moves are then necessarily one-sided for large values of $|x^{(t)}|$. Note also the strong similarity between (6.5.4) and the stochastic gradient equation inspired by (5.2.2),

$$(6.5.6) \qquad\qquad x^{(t+1)} = x^{(t)} + \frac{\sigma}{2}\nabla \log f(x^{(t)}) + \sigma\varepsilon_t .$$

As suggested by Besag (1994), a way to correct this negative behavior is to treat (6.5.4) as a regular Metropolis–Hastings instrumental distribution; that is, to accept the new value $Y_t$ with probability

$$\frac{f(Y_t)}{f(x^{(t)})} \cdot \frac{\exp\left\{-\left\|Y_t - x^{(t)} - \frac{\sigma^2}{2}\nabla \log f(x^{(t)})\right\|^2 \Big/ 2\sigma^2\right\}}{\exp\left\{-\left\|x^{(t)} - Y_t - \frac{\sigma^2}{2}\nabla \log f(Y_t)\right\|^2 \Big/ 2\sigma^2\right\}} \wedge 1 .$$

The corresponding Metropolis–Hastings algorithm will not necessarily outperform the regular random walk Metropolis–Hastings algorithm, since

Roberts and Tweedie (1995) show that the resulting chain is not geometrically ergodic when $\nabla \log f(x)$ goes to 0 at infinity, similar to the random walk, but the (basic) ergodicity of this chain is ensured.

Roberts and Rosenthal (1998) give further results about the choice of the scaling factor $\sigma$, which should lead to an acceptance rate of 0.574 to achieve optimal convergence rates in the special case where the components of $x$ are uncorrelated[8] under $f$. Note also that the proposal distribution (6.5.4) is rather natural from a Laplace approximation point of view since it corresponds to a second-order approximation of $f$. Indeed, by a standard Taylor expansion,

$$\log f(x^{(t+1)}) = \log f(x^{(t)}) + (x^{(t+1)} - x^{(t)})' \nabla \log f(x^{(t)})$$
$$+ \frac{1}{2}(x^{(t+1)} - x^{(t)})' \left[ \nabla' \nabla \log f(x^{(t)}) \right] (x^{(t+1)} - x^{(t)}),$$

the random-walk-type approximation to $f(x^{(t+1)})$ is

$$f(x^{(t+1)}) \propto \exp \left\{ (x^{(t+1)} - x^{(t)})' \nabla \log f(x^{(t)}) \right.$$
$$\left. - \frac{1}{2}(x^{(t+1)} - x^{(t)})' H(x^{(t)})(x^{(t+1)} - x^{(t)}) \right\}$$
$$\propto \exp \left\{ -\frac{1}{2}(x^{(t+1)})' H(x^{(t)}) x^{(t+1)} \right.$$
$$\left. + (x^{(t+1)})' \left[ \nabla \log f(x^{(t)}) + H(x^{(t)}) x^{(t)} \right] \right\}$$
$$\propto \exp \left\{ -\frac{1}{2}[x^{(t+1)} - x^{(t)} - (H(x^{(t)}))^{-1} \nabla \log f(x^{(t)})]' H(x^{(t)}) \right.$$
$$\left. \times [x^{(t+1)} - x^{(t)} - (H(x^{(t)}))^{-1} \nabla \log f(x^{(t)})] \right\},$$

where $H(x^{(t)}) = -\nabla' \nabla \log f(x^{(t)})$ is the Hessian matrix. If we simplify this approximation by replacing $H(x^{(t)})$ with $\sigma^{-2} I_p$, the Taylor approximation then leads to the random walk with a *drift* term

(6.5.7)          $$x^{(t+1)} = x^{(t)} + \sigma^2 \nabla \log f(x^{(t)}) + \sigma \varepsilon_t .$$

From an exploratory point of view, the addition of the gradient of $\log f$ is relevant, since it should improve the moves toward the modes of $f$, whereas only requiring a minimum knowledge of this density function (in particular, constants are not needed). Note also that, in difficult settings, exact gradients can be replaced by numerical derivatives.

Stramer and Tweedie (1997) start from the lack of uniform minimal performances of (6.5.4) to build up modifications (see Problem 6.43) which avoid some pathologies of the basic Langevin Metropolis–Hastings algorithm. They obtain general geometric and uniformly ergodic convergence results.

---

[8] The authors also show that this corresponds to a variance of order $p^{-1/3}$, whereas the optimal variance for the Metropolis–Hastings algorithm is of order $p^{-1}$ (see Roberts *et al.* 1997).

**Example 6.5.4 (Continuation of Example 6.3.5)** If we take $f$ to be the normal $\mathcal{N}(0,1)$ density, the Langevin transition is based on

$$Y_t \sim \mathcal{N}\left(x^{(t)}\left(1 - \frac{\sigma^2}{2}\right), \sigma^2\right),$$

with acceptance probability $\min(1, \exp\{(x^2 - y^2)\sigma^2/8\})$, while the regular random walk is

$$Y_t \sim \mathcal{N}(x^{(t)}, \tau^2),$$

with acceptance probability $\min(1, \exp(x^2 - y^2)/2)$. In addition, we can consider the stochastic gradient alternative,

$$(6.5.8) \qquad Y_t \sim \mathcal{N}\left(x^{(t)}\left(1 - \frac{\omega}{2}\right), \omega^2\right),$$

with acceptance probability $\min(1, \exp\{(x^2 - y^2)(5 - 4/\omega)/8\})$ (see Problem 6.42).

The calibration of the three variance coefficients leads to $\sigma = 2.01$, $\tau = 1.99$, and $\omega = 2.37$, for an average acceptance rate of $1/2$. Figure 6.5.2 provides a comparison of the three Metropolis–Hastings algorithms obtained by a Monte Carlo experiment run over 500 replications of the corresponding Markov chains. It gives the averages and the 90% (equal tail) ranges of the estimates of $\mathbb{E}[X^4]$ and shows that, in this case, the Langevin algorithm does not bring a significant improvement over the regular random walk. On the other hand, the stochastic-gradient-type algorithm leads to a much smaller variance, since the ratio of the ranges is about 1.6. Note that since $\sigma^2$ and $\omega$ are larger than 2, both Langevin and stochastic gradient algorithms lead to antithetic moves, since their proposed values are centered on the side of the origin opposite $x^{(t)}$. ‖

**Example 6.5.5 Nonidentifiable normal model.** To illustrate the performance of the Langevin diffusion method in a multimodal setting, consider the nonidentifiable model

$$Y \sim \mathcal{N}\left(\frac{\theta_1 + \theta_2}{2}, 1\right)$$

associated with the exchangeable prior $\theta_1, \theta_2 \sim \mathcal{N}(0,1)$. The posterior distribution

$$(6.5.9) \quad \pi(\theta_1, \theta_2 | y) \propto \exp\left(-\frac{1}{2}\left\{\theta_1^2 \frac{5}{4} + \theta_2^2 \frac{5}{4} + \frac{\theta_1 \theta_2}{2} - (\theta_1 + \theta_2)y\right\}\right)$$

is then well defined (and can be simulated directly, see Problem 6.40), with a bimodal structure due to the nonidentifiability. The Langevin transition is then based on the proposal

$$\mathcal{N}_2\left(\frac{\sigma^2}{2}\left(\frac{2y - 2\theta_2^{(t)} - 5\theta_1^{(t)}}{8}, \frac{4y - 2\theta_1^{(t)} - 5\theta_2^{(t)}}{8}\right) + (\theta_1^{(t)}, \theta_2^{(t)}), \sigma^2 I_2\right).$$

A preliminary calibration then leads to $\sigma = 1.46$ for an acceptance rate of $1/2$. As seen on Figure 6.5.3, the 90% range is larger for the output

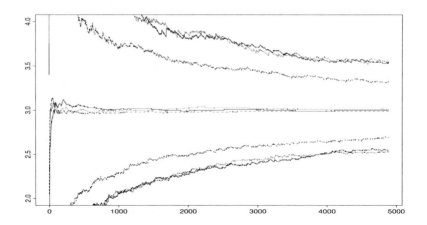

Figure 6.5.2. *Convergence of the empirical averages for the Langevin (full), random walk (dots), and stochastic gradient (dashes) Metropolis–Hastings algorithms for the estimation of* $\mathbb{E}[X^4]$ *and corresponding 90% equal tail confidence intervals. The final empirical confidence intervals are* $[2.69, 3.32]$, $[2.53, 3.55]$, *and* $[2.54, 3.54]$ *for stochastic gradient, random walk, and Langevin algorithms, respectively (based on 500 replications).*

of the Langevin algorithm than for an iid sample in the approximation of $\mathbb{E}[\theta_1|y]$, but the ratio of these ranges is only approximately 1.3, which shows a moderate loss of efficiency in using the Langevin approximation.$\|$

## 6.6 Problems

**6.1** For the transition kernel,

$$X^{(t+1)}|x^{(t)} \sim \mathcal{N}(\rho x^{(t)}, \tau^2)$$

gives sufficient conditions on $\rho$ and $\tau$ for the stationary distribution $\pi$ to exist. Show that, in this case, $\pi$ is a normal distribution and that (6.2.2) occurs.

**6.2** (Doukhan *et al.* 1994) The algorithm presented in this problem is used in Chapter 8 as a benchmark for slow convergence.

(a) Prove the following result:

**Lemma 6.6.1** *Consider a probability density $g$ on $[0, 1]$ and a function $0 < \rho < 1$ such that*

$$\int_0^1 \frac{g(x)}{1 - \rho(x)} \, dx < \infty .$$

*The Markov chain with transition kernel*

$$K(x, x') = \rho(x) \, \delta_x(x') + (1 - \rho(x)) \, g(x') ,$$

*where $\delta_x$ is the Dirac mass at $x$, has stationary distribution*

$$f(x) \propto g(x)/(1 - \rho(x)).$$

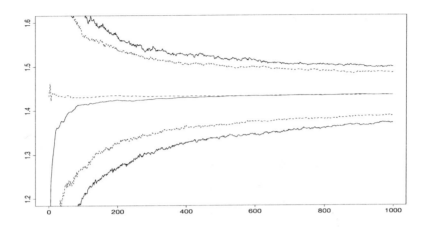

Figure 6.5.3. *Convergence of the empirical averages for the Langevin Metropolis–Hastings (full) and iid simulation (dots) of $(\theta_1, \theta_2)$ for the estimation of $\mathbb{E}[\theta_1]$ and corresponding 90% equal tail confidence intervals when $y = 4.3$. The final intervals are $[1.373, 1.499]$ and $[1.388, 1.486]$ for the Langevin and exact algorithms, respectively.*

(b) Show that an algorithm for generating the Markov chain associated with Lemma 6.6.1 is given by

**Algorithm A.30 –Repeat or Simulate–**

1. Take $X^{(t+1)} = x^{(t)}$ with probability $\rho(x^{(t)})$
2. Else, generate $X^{(t+1)} \sim g(y)$.                    [A.30]

(c) Highlight the similarity with the Accept–Reject algorithm and discuss in which sense they are complementary.

**6.3** (Continuation of Problem 6.2) Implement the algorithm of Problem 6.2 when $g$ is the density of the $\mathcal{B}e(\alpha + 1, 1)$ distribution and $\rho(x) = 1 - x$. Give the expression of the stationary distribution $f$. Study the acceptance rate as $\alpha$ varies around 1. (*Note:* Doukhan *et al.* 1994 use this example to derive $\beta$-mixing chains which do not satisfy the Central Limit Theorem.)

**6.4** (Continuation of Problem 6.2) Compare the algorithm [A.30] with the corresponding Metropolis–Hastings algorithm; that is, the algorithm [A.25] associated with the same pair $(f, g)$. (*Hint:* Take into account the fact that [A.30] only simulates the $y_t$'s which are not discarded and compare the computing times when a recycling version as in §6.4.2 is implemented.)

**6.5** Determine the distribution of $Y_t$ given $y_{t-1}, \ldots$ in [A.25].

**6.6** (Tierney 1994) Consider a version of [A.25] based on a "bound" $M$ on $f/g$ that is not a uniform bound; that is, $f(x)/g(x) > M$ for some $x$.

(a) If an Accept–Reject algorithm uses the density $g$ with acceptance probability $f(y)/Mg(y)$, show that the resulting variables are generated from

$$\tilde{f}(x) \propto \min\{f(x), Mg(x)\} ,$$

instead of $f$.

(b) Show that this error can be corrected, for instance by using the Metropolis–Hastings algorithm:

   1. Generate $Y_t \sim \tilde{f}$.
   2. Accept with probability

$$P(X^{(t+1)} = y_t | x^{(t)}, y_t) = \begin{cases} \min\left\{1, \dfrac{f(y_t)g(x^{(t)})}{g(y_t)f(x^{(t)})}\right\} & \text{if } \dfrac{f(y_t)}{g(y_t)} > M \\ \min\left\{1, \dfrac{Mg(x^{(t)})}{f(x^{(t)})}\right\} & \text{otherwise.} \end{cases}$$

to produce a sample from $f$.

**6.7** The inequality (6.3.2) can also be established using *Orey's inequality*. For two transitions $P$ and $Q$,

$$\|P^n - Q^n\|_{TV} \le 2P(X_n \ne Y_n), \qquad X_n \sim P^n, \quad Y_n \sim Q^n.$$

Deduce that when $P$ is associated with the stationary distribution $f$ and when $X_n$ is generated by [A.25], under the condition (6.3.1),

$$\|P^n - f\|_{TV} \le \left(1 - \frac{1}{M}\right)^n.$$

*Hint:* Use a coupling argument based on

$$X^n = \begin{cases} Y^n & \text{with probability } 1/M \\ Z^n \sim \dfrac{g(z) - f(z)/M}{1 - 1/M} & \text{otherwise.} \end{cases}$$

**6.8** Complete the proof of Theorem 6.3.1:

(a) Verify (6.3.4) and prove (6.3.5). (*Hint:* By (6.3.3), the inner integral is immediately bounded by $1 - \frac{1}{M}$. Then repeat the argument for the outer integral.)

(b) Verify (6.3.6) and prove (6.3.2).

**6.9** In the setup of Hastings (1970) uniform-normal example (see Example 6.3.5):

(a) Study the convergence rate (represented by the 90% interquantile range) and the acceptance rate when $\delta$ increases.

(b) Determine the value of $\delta$ which minimizes the variance of the empirical average. (*Hint:* Use a simulation experiment.)

**6.10** Show that, for an arbitrary Metropolis–Hastings algorithm, every compact set is a small set when $f$ and $q$ are positive and continuous everywhere.

**6.11** (Mengersen and Tweedie 1996) With respect to Theorem 6.3.6, define

$$A_x = \{y; \ f(x) \le f(y)\} \quad \text{and} \quad B_x = \{y; \ f(x) \ge f(y)\}.$$

(a) If $f$ is symmetric, show that $A_x = \{|y| < |x|\}$ for $|x|$ larger than a value $x_0$.

(b) Define $x_1$ as the value after which $f$ is log-concave and $x^* = x_0 \vee x_1$. For $V(x) = \exp s|x|$ and $s < \alpha$, show that

$$\frac{\mathbb{E}[V(X_1)|x_0 = x]}{V(x)} \le 1 + \int_0^x \left[e^{s(y-x)} - 1\right] g(x - y) dy$$

$$+ \int_x^{2x} e^{-\alpha(y-x)} \left[ e^{s(y-x)} - 1 \right] g(x-y) dy$$

$$+ 2 \int_x^{\infty} g(y) dy .$$

(c) Show that

$$\int_0^x \left( e^{-sy} - 1 + e^{-(\alpha-s)y} - e^{-\alpha y} \right) g(y) dy$$

$$= - \int_0^x \left[ 1 - e^{-sy} \right] \left[ 1 - e^{-(\alpha-s)y} \right] g(y) dy$$

and deduce that (6.3.10) holds for $x > x^*$ and $x^*$ large enough.

(d) For $x < x^*$, show that

$$\frac{\mathbb{E}[V(X_1)|x_0 = x]}{V(x)} \leq 1 + 2 \int_{x^*}^{\infty} g(y) dy + 2 e^{sx^*} \int_0^{x^*} g(z) dz$$

and thus establish the theorem.

**6.12** Examine whether the following distributions are log-concave in the tails: Normal, log-normal, Gamma, Student's $t$, Pareto, Weibull.

**6.13** The following theorem is due to Mengersen and Tweedie (1996).

**Theorem 6.6.2** *If the support of $f$ is not compact and if $g$ is symmetric, the chain $(X^{(t)})$ produced by [A.28] is not uniformly ergodic.*

Assume that the chain satisfies Doeblin's condition (Theorem 4.6.13).

(a) Take $x_0$ and $A_0 = ]-\infty, x_0]$ such that $\nu(A_0) > \varepsilon$ and consider the *unilateral* version of the random walk, with kernel

$$K^-(x, A) = \frac{1}{2} \, \mathbb{I}_A(x) + \int_{A \cap ]-\infty, x]} g(x-y) \, dy ;$$

that is, the random walk which only goes to the left. Show that for $y > x_0$,

$$P^m(y, A_0) \leq P_y(\tau \leq m) \leq P_y(\tau^- \leq m),$$

where $\tau$ and $\tau^-$ are the return times to $A_0$ for the chain $(X^{(t)})$ and for $K^-$, respectively,

(b) For $y$ sufficiently large to satisfy

$$(K^-)^m(y, A_0) = (K^-)^m(0, ]-\infty, x_0 - y]) < \frac{\delta}{m} ,$$

show that

$$P_y(\tau^- \leq m) \leq \sum_{j=1}^{m} (K^-)^j(y, A_0) \leq m(K^-)^m(y, A_0) ,$$

contradicting Doeblin's condition and proving the theorem.

(c) Formulate a version of Theorem 6.6.2 for higher-dimensional chains.

**6.14** Mengersen and Tweedie (1996) also establish the following theorem:

**Theorem 6.6.3** *If g is continuous and satisfies*

$$\int |x| \, g(x)dx < \infty \, ,$$

*the chain* $(X^{(t)})$ *of* $[A.28]$ *is geometrically ergodic if and only if*

$$\overline{\varphi} = \lim_{x \to \infty} \frac{d}{dx} \log f(x) < 0 \, .$$

(a) To establish sufficiency, show that for $x < y$ large enough, we have

$$\log f(y) - \log f(x) = \int_x^y \frac{d}{dt} \log f(t)dt \le \frac{\overline{\varphi}}{2}(y - x).$$

Deduce that this inequality ensures the log-concavity of $f$ and, therefore, the application of Theorem 6.3.6.

(b) For necessity, suppose $\overline{\varphi} = 0$. For every $\delta > 0$, show that you can choose $x$ large enough so that $\log f(x + z) - \log f(x) \ge -\delta z$, $z > 0$ and, therefore, $f(x + z) \exp(\delta z) \ge f(x)$. By integrating out the $z$'s, show that

$$\int_x^\infty f(y) \, e^{\delta y} \, dy = \infty \, ,$$

contradicting condition (6.3.11).

**6.15** For the situation of Example 6.3.7, show that

$$\lim_{x \to \infty} \frac{d}{dx} \log \varphi(x) = -\infty \quad \text{and} \quad \lim_{x \to \infty} \frac{d}{dx} \log \psi(x) = 0 \, ,$$

showing that the chain associated with $\varphi$ is geometrically ergodic and the chain associated with $\psi$ is not.

**6.16** Verify that the transition matrix associated with the geometric random walk in Example 6.3.8 is correct and that $\beta = \theta^{-1/2}$ minimizes $\lambda_\beta$.

**6.17** Check whether a negative coefficient $b$ in the random walk $Y_t = a + b(X^{(t)} - a) + Z_t$ induces a negative correlation between the $X^{(t)}$'s. Extend to the case where the random walk has an ARCH structure,

$$Y_t = a + b(X^{(t)} - a) + \exp(c + d(X^{(t)} - a)^2)Z_t.$$

**6.18** Implement the Metropolis–Hastings algorithm when $f$ is the normal $\mathcal{N}(0,1)$ density and $q(\cdot|x)$ is the uniform $\mathcal{U}[-x - \delta, -x + \delta]$ density. Check for negative correlation between the $X^{(t)}$'s when $\delta$ varies.

**6.19** In a setting of your choice, study the evolution of the number of updates of $\tilde{h}_n$ in the ARMS algorithm $[A.29]$ when $n$ increases.

**6.20** Referring to the situation of Example 6.3.4:

(a) Use the Taylor approximations $K_X(t) \approx K_X(0) + K'_X(0)t + K''_X(0)t^2/2$ and $K'_X(t) \approx K'_X(0) + K''_X(0)t$ to establish (6.3.8).

(b) Write out the Metropolis–Hastings algorithm that will produce random variables from the saddlepoint distribution.

(c) Apply the Metropolis saddlepoint approximation to the noncentral chi squared distribution and reproduce the tail probabilities in Table 6.3.1.

**6.21** Given a Cauchy $\mathcal{C}(0, \sigma)$ instrumental distribution:

(a) Experimentally select $\sigma$ to maximize (i) the acceptance rate when simulating a $\mathcal{N}(0,1)$ distribution and (ii) the squared error when estimating the mean (equal to 0).

(b) Same as (a), but when the instrumental distribution is $\mathcal{C}(x^{(t)}, \sigma)$.

**6.22** Show that the Rao–Blackwellized estimator $\delta^{RB}$ does not depend on the normalizing factors in $f$ and $g$.

**6.23** Reproduce the experiment of Example 6.4.3 in the case of a Student $\mathcal{T}_7$ distribution.

**6.24** In the setup of the Metropolis–Hastings algorithm [A.24], the $Y_t$'s are generated from the distributions $q(y|x^{(t)})$. Assume that $Y_1 = X^{(1)} \sim f$.

(a) Show that the estimator

$$\delta_0 = \frac{1}{T} \sum_{t=1}^{T} \frac{f(y_t)}{q(y_t|x^{(t)})} h(y_t)$$

is an unbiased estimator of $\mathbb{E}_f[h(X)]$.

(b) Derive, from the developments of §6.4.2, that the Rao–Blackwellized version of $\delta_0$ is

$$\delta_1 = \frac{1}{n} \left\{ h(x_1) + \frac{f(y_2)}{q(y_2|x_1)} h(y_2) \right.$$
$$\left. + \frac{\sum_{i=3}^{T} \sum_{j=1}^{i-1} f(y_i) \delta_j \zeta_{j(i-2)} (1 - \rho_{j(i-1)}) \omega_i^j}{\sum_{i=1}^{T-1} \delta_i \zeta_{i(T-1)}} h(y_i) \right\} .$$

(c) Compare $\delta_1$ with the Rao–Blackwellized estimator of Theorem 6.4.4 in the case of a $\mathcal{T}_3$ distribution for the estimation of $h(x) = \mathbb{I}_{x>2}$.

**6.25** Prove Theorem 6.4.2 as follows:

(a) Use the properties of the Markov chain to show that the conditional probability $\tau_i$ can be written as

$$\tau_i = \sum_{j=0}^{i-1} P(X_i = y_i | X_{i-1} = y_j)\, P(X_{i-1} = y_j) = \sum_{j=0}^{i-1} \rho_{ji}\, P(X_{i-1} = y_j).$$

(b) Show that

$$P(X_{i-1} = y_j) = P(X_j = y_j, X_{j+1} = y_j, \dots, X_{i-1} = y_j)$$
$$= (1 - \rho_{j(j+1)}) \cdots (1 - \rho_{j(i-1)})$$

and, hence, establish the expression for the weight $\varphi_i$.

**6.26** Prove Theorem 6.4.4 as follows:

(a) As in the independent case, the first step is to compute $P(X_j = y_i | y_0, y_1, \dots, y_n)$. The event $\{X_j = y_i\}$ can be written as the set of all the $i$-tuples $(u_1, \dots, u_i)$ leading to $\{X_i = y_i\}$, of all the $(j-i)$-tuples $(u_{i+1}, \dots, u_j)$ corresponding to the rejection of $(y_{i+1}, \dots, y_j)$ and of all the $(n-j)$-tuples $u_{j+1}, \dots, u_n$ following after $X_j = y_i$.

Define $B_0^1 = \{u_1 > \rho_{01}\}$ and $B_1^1 = \{u_1 < \rho_{01}\}$, and let $B_k^t(u_1, \ldots, u_t)$ denote the event $\{X_t = y_k\}$. Establish the relation

$$B_k^t(u_1, \ldots, u_t) = \bigcup_{m=0}^{k-1} \left[ B_m^{k-1}(u_1, \ldots, u_{k-1}) \right.$$
$$\left. \cap \{u_k < \rho_{mk}, u_{k+1} > \rho_{k(t+1)}, \ldots, u_t > \rho_{kt}\} \right],$$

and show that

$$\{x_j = y_i\} = \bigcup_{k=0}^{i-1} \left[ B_k^{i-1}(u_1, \ldots, u_{i-1}) \right.$$
$$\left. \cap \{u_i < \rho_{ki}, u_{i+1} > \rho_{i(i+1)}, \ldots, u_j > \rho_{ij}\} \right].$$

(b) Let $p(u_1, \ldots, u_T, y_1, \ldots, y_T) = p(\mathbf{u}, \mathbf{y})$ denote the joint density of the $U_i'$s and the $Y_i'$s. Show that $\tau_i = \int_A p(\mathbf{u}, \mathbf{y}) du_1 \cdots du_i$, where $A = \bigcup_{k=0}^{i-1} B_k^{i-1}(u_1, \ldots, u_{i-1}) \cap \{u_i < \rho_{ki}\}$.

(c) Show that $\omega_{j+1}^i = \int_{\{x_j = y_i\}} p(\mathbf{u}, \mathbf{y}) du_{j+1} \cdots du_T$ and, using part (b), establish the identity

$$\tau_i \prod_{t=i+1}^{j} (1 - \rho_{it}) q(y_{t+1}|y_i) \omega_{j+1}^i = \tau_i \prod_{t=i+1}^{j} \bar{\rho}_{it} \omega_{j+1}^i = \tau_i \zeta_{ij} \omega_{j+1}^i.$$

(d) Verify the relations

$$\tau_i = \sum_{t=0}^{i-1} \tau_i \zeta_{t(i-1)j} \bar{\rho}_{ti} \quad \text{and} \quad \omega_{j+1}^i = \omega_{j+2}^{j+1} \bar{\rho}_{i(j+1)} + \omega_{j+2}^i \bar{\rho}_{i(j+1)},$$

which provide a recursion relation on the $\omega_j^i$'s depending on acceptance or rejection of $y_{j+1}$. The case $j = T$ must be dealt with separately, since there is no generation of $y_{T+1}$ based on $q(y|x_T)$. Show that $\omega_T^i$ is equal to $\rho_{iT} + (1 - \rho_{iT}) = 1$

(e) The probability $P(X_j = y_i)$ can be deduced from part (d) by computing the marginal distribution of $(Y_1, \ldots, Y_T)$. Show that $1 = \sum_{i=0}^{T} P(X_T = y_i)$ $= \sum_{i=0}^{T-1} \tau_i \zeta_{i(T-1)}$, and, hence, the normalizing constant for part (d) is $\left( \sum_{i=0}^{T-1} \tau_i \zeta_{i(T-1)} \right)^{-1}$, which leads to the expression for $\varphi$.

**6.27** In the situation of Example 6.3.9, if a second explanatory variable, $z$, is potentially influential in

$$\mathbb{E}[Y|x, z] = \frac{\exp(a + bx + cz)}{1 + \exp(a + bx + cz)},$$

the two competing models are $\mathcal{H}_1 = \{1\} \times \mathbb{R}^2$ and $\mathcal{H}_2 = \{2\} \times \mathbb{R}^3$. If the jump from $\mathcal{H}_2$ to $\mathcal{H}_1$ is a (deterministic) transform of $(1, a, b, c)$ to $(2, a, b)$, the jump from $\mathcal{H}_1$ to $\mathcal{H}_2$ must necessarily preserve the values of $a$ and $b$ and only modify $c$. Write out the details of the reversible jump algorithm.

**6.28** For the situation of Example 6.5.2, show that the four possible moves are made with the following probabilities:

(i) $(\alpha_0, \alpha_1) \to (\alpha_0', \alpha_1')$, with probability $\pi_{11} \min \left( 1, \frac{f_0(\alpha_0')f_1(\alpha_1')}{f_0(\alpha_0)f_1(\alpha_1)} \right)$,

(ii) $(\alpha_0, \alpha_1) \rightarrow (\alpha_0, \alpha_1, \alpha_2')$, with probability $\pi_{12} \min\left(1, \frac{f_2(\alpha_2')}{f_2(\alpha_2)}\right)$,

(iii) $(\alpha_0, \alpha_1, \alpha_2) \rightarrow (\alpha_0', \alpha_1', \alpha_2')$ with probability $\pi_{21} \min\left(1, \frac{f_0(\alpha_0')f_1(\alpha_1')f_2(\alpha_2')}{f_0(\alpha_0)f_1(\alpha_1)f_2(\alpha_2)}\right)$,

(iv) $(\alpha_0, \alpha_1, \alpha_2) \rightarrow (\alpha_0', \alpha_1')$ with probability $\pi_{22} \min\left(1, \frac{f_0(\alpha_0)f_1(\alpha_1)}{f_0(\alpha_0)f_1(\alpha_1)f_2(\alpha_2)}\right)$.

**6.29** Similar to the situation of Example 6.5.2, the data in Table 6.29 are a candidate for model averaging between a linear and quadratic regression. This "braking data" (Tukey 1977) is the distance needed to stop ($y$) when an automobile travels at a particular speed ($x$) and brakes.

| x | 4 | 7 | 8 | 9 | 10 | 11 | 12 |
|---|---|---|---|---|---|---|---|
| y | 2,10 | 4,22 | 16 | 10 | 18,26 | 17,28 | 14,20 |
|   |   |   |   |   | 34 |   | 24,28 |

| x | 13 | 14 | 15 | 16 | 17 | 18 | 19 |
|---|---|---|---|---|---|---|---|
| y | 26,34 | 26,36 | 20,26 | 32,40 | 32,40 | 42,56 | 36,46 |
|   | 34,46 | 60,80 | 54 |   | 50 | 76,84 | 68 |

| x | 20 | 22 | 23 | 24 | 25 |
|---|---|---|---|---|---|
| y | 32,48 | 66 | 54 | 70,92 | 85 |
|   | 52,56,64 |   |   | 93,120 |   |

Table 6.6.1. *Braking distances.*

(a) Fit the data using a reversible jump algorithm that jumps between a linear and quadratic model. Use normal priors with relatively large variances.

(b) Assess the robustness of the fit to specifications of the $\pi_{ij}$. More precisely, choose values of the $\pi_{ij}$ so that the algorithm spends 25%, 50%, and 75% of the time in the quadratic space. How much does the overall model average change?

**6.30** Consider the model introduced in Example 6.5.3.

(a) Show that the weight of the jump from $\mathcal{H}_{k+1}$ to $\mathcal{H}_k$ involves the Jacobian

$$\left| \frac{\partial(p^{(k)}, a^{(k)}, u_1, u_2)}{\partial(p^{(k+1)}, a^{(k+1)})} \right|.$$

(b) Compute the Jacobian for the jump from $\mathcal{H}_4$ to $\mathcal{H}_3$. (The determinant can be calculated by row-reducing the matrix to triangular form, or using the fact that $\begin{vmatrix} A & B \\ C & D \end{vmatrix} = |A||D - CA^{-1}B|$.)

**6.31** For the model of Example 6.5.3:

(a) Derive general expressions for the weight of the jump from $\mathcal{H}_k$ to $\mathcal{H}_{k+1}$ and from $\mathcal{H}_{k+1}$ to $\mathcal{H}_k$.

(b) Select a value of the Poisson parameter $\lambda$ and implement a reversible jump algorithm to fit a density estimate to the data of Table 6.6.2. (Previous investigation[9] and subject matter considerations indicate that there may be between three and six modes. Use this information to choose an appropriate value of $\lambda$.)

(c) Investigate the robustness of your density estimate to variations in $\lambda$.

| | | | | | |
|---|---|---|---|---|---|
| 9172 | 9350 | 9483 | 9558 | 9775 | 10227 |
| 10406 | 16084 | 16170 | 18419 | 18552 | 18600 |
| 18927 | 19052 | 19070 | 19330 | 19343 | 19349 |
| 19440 | 19473 | 19529 | 19541 | 19547 | 19663 |
| 19846 | 19856 | 19863 | 19914 | 19918 | 19973 |
| 19989 | 20166 | 20175 | 20179 | 20196 | 20215 |
| 20221 | 20415 | 20629 | 20795 | 20821 | 20846 |
| 20875 | 20986 | 21137 | 21492 | 21701 | 21814 |
| 21921 | 21960 | 22185 | 22209 | 22242 | 22249 |
| 22314 | 22374 | 22495 | 22746 | 22747 | 22888 |
| 22914 | 23206 | 23241 | 23263 | 23484 | 23538 |
| 23542 | 23666 | 23706 | 23711 | 24129 | 24285 |
| 24289 | 24366 | 24717 | 24990 | 25633 | 26960 |
| 26995 | 32065 | 32789 | 34279 | | |

Table 6.6.2.   *Velocity (km/second) of galaxies in the Corona Borealis Region. (Source: Roeder 1992.)*

**6.32** Consider a sequence $X_1, \ldots, X_n$ such that

$$X_i \sim \begin{cases} \mathcal{N}(\mu_1, \tau^2) & \text{if } i \le i_0 \\ \mathcal{N}(\mu_2, \tau^2) & \text{if } i > i_0. \end{cases}$$

(a) When $i_0$ is uniformly distributed on $\{1, \ldots, n-1\}$, determine whether the improper prior $\pi(\mu_1, \mu_2, \tau) = 1/\tau$ leads to a proper prior.

(b) Propose a reversible jump Monte Carlo algorithm for this setup.

(c) Use an importance sampling argument to recycle the sample produced by the algorithm in part (b) when the prior is

$$\pi(\mu_1, \mu_2, \tau) = \exp\left(-(\mu_1 - \mu_2)^2/\tau^2\right)/\tau^2.$$

(d) Show that an alternative implementation is to use $(n-1)$ latent variables $z_i$ which are the indicators of changepoints.

(e) Extend the analysis to the case where the number of changepoints is unknown.

**6.33** (Liu 1996) Consider a finite state-space $\mathcal{X} = \{1, \ldots, m\}$ and a Metropolis–Hastings algorithm on $\mathcal{X}$ associated with the stationary distribution $\pi = (\pi_1, \ldots, \pi_m)$ and the proposal distribution $p = (p_1, \ldots, p_m)$.

(a) For $\omega_i = \pi_i/p_i$ and $\lambda_k = \sum_{i=k}^{m}(p_i - \pi_i/\omega_k)$, express the transition matrix of the Metropolis–Hastings algorithm as $K = G + \mathbf{e}p^T$, where $\mathbf{e} = (1, \ldots, 1)^T$. Show that $G$ is upper triangular with diagonal elements the $\lambda_k$'s.

---

[9] The data in Table 6.6.2 have been analyzed by a number of people. See Roeder (1992), Chib (1995), Richardson and Green (1997), and Robert and Mengersen (1999).

(b) Deduce that the eigenvalues of $G$ and $K$ are the $\lambda_k$'s.

(c) Show that for $||p|| = \sum |p_i|$,

$$||K^n(x, \cdot) - \pi||^2 \leq \frac{\lambda_1^{2n}}{\pi(x)},$$

following a result by Diaconis and Hanlon (1992).

**6.34** (Ó Ruanaidh and Fitzgerald 1996) Given the model

$$y_i = A_1 e^{\lambda_1 t_i} + A_2 e^{\lambda_2 t_i} + c + \epsilon_i,$$

where $\epsilon_i \sim \mathcal{N}(0, \sigma^2)$ and the $t_i$'s are known observation times, study the estimation of $(A_1, A_2, \lambda_1, \lambda_2, \sigma)$ by recursive integration (see §5.2.4) with particular attention to the Metropolis–Hastings implementation.

**6.35** (Neal 1996) Consider a two-layer neural network (or *perceptron*), introduced in Note 5.5.2, where

$$\begin{cases} h_k = \tanh\left(\alpha_{k0} + \sum_j \alpha_{kj} x_j\right), & k = 1, \ldots, p, \\ \mathbb{E}[Y_\ell | h] = \beta_{\ell 0} + \sum_k v_{\ell k} h_k, & \ell = 1, \ldots, n. \end{cases}$$

Propose a reversible jump Monte Carlo method when the number of hidden units $p$ is unknown, based on observations $(x_j, y_j)$ $(j = 1, \ldots, n)$.

**6.36** (Roberts 1998) Take $f$ to be the density of the $\mathcal{E}xp(1)$ distribution and $g_1$ and $g_2$ the densities of the $\mathcal{E}xp(0.1)$ and $\mathcal{E}xp(5)$ distributions, respectively. The aim of this problem is to compare the performances of the independent Metropolis–Hastings algorithms based on the pairs $(f, g_1)$ and $(f, g_2)$.

(a) Compare the convergences of the empirical averages for both pairs, based on 500 replications of the Markov chains.

(b) Show that the pair $(f, g_1)$ leads to a geometrically ergodic Markov chain and $(f, g_2)$ does not.

**6.37** For the Markov chain of Example 6.7.2, define

$$\xi^{(t)} = \mathbb{I}_{[0,13]}(\theta^{(t)}) + 2\mathbb{I}_{[13,\infty)}(\theta^{(t)}).$$

(a) Show that $(\xi^{(t)})$ is not a Markov chain.

(b) Construct an estimator of the pseudo-transition matrix of $(\xi^{(t)})$.

The next six problems (6.38–6.43) deal with Langevin diffusions, as introduced in §6.5.2.

**6.38** Show that the naive discretization of (6.5.3) as $dt = \sigma^2$, $dL_t = X^{(t+\sigma^2)} - X^{(t)}$, and $dB_t = B_{t+\sigma^2} - B_t$ does lead to the representation (6.5.4).

**6.39** Consider $f$ to be the density of $\mathcal{N}(0, 1)$. Show that when $\sigma = 2$ in (6.5.4), the limiting distribution of the chain is $\mathcal{N}(0, 2)$.

**6.40** Show that (6.5.9) can be directly simulated as

$$\theta_2 \sim \mathcal{N}\left(\frac{y}{3}, \frac{5}{6}\right), \qquad \theta_1 | \theta_2 \sim \mathcal{N}\left(\frac{2y - \theta_2}{5}, \frac{4}{5}\right).$$

**6.41** Show that when (6.5.5) exists and is larger (smaller) than 1 ($-1$) at $-\infty$ ($\infty$), the random walk (6.5.4) is transient.

**6.42** In the setup of Example 6.5.4, examine whether the Markov chain associated with (6.5.8) is recurrent. Show that when the transition

$$y_t \sim \mathcal{N}\left(x^{(t)}(1-\omega), \omega^2\right)$$

is used instead, the Markov chain is transient.

**6.43** (Stramer and Tweedie 1997a) Show that the following stochastic differential equation still produces $f$ as the stationary distribution of the associated process:

$$dL_t = \sigma(L_t)\nabla \log f(L_t) + b(L_t)dt,$$

when

$$b(x) = \frac{1}{2}\nabla \log f(x)\sigma^2(x) + \sigma(x)\nabla\sigma(x).$$

Give a discretized version of this differential equation to derive a Metropolis–Hastings algorithm and apply to the case $\sigma(x) = \exp(\omega|x|)$.

**6.44** Show that the transition associated with the acceptance probability (6.7.1) also leads to $f$ as invariant distribution, for every symmetric function $s$. (*Hint:* Use the reversibility equation.)

**6.45** Show that the Metropolis–Hastings algorithm is, indeed, a special case of the transition associated with the acceptance probability (6.7.1) by providing the corresponding $s(x, y)$.

**6.46** (Peskun 1973) Let $\mathbb{P}_1$ and $\mathbb{P}_2$ be regular, reversible stochastic matrices with the same stationary distribution $\pi$ on $\{1, \dots, m\}$. Show that if $\mathbb{P}_1 \leq \mathbb{P}_2$ (meaning that the off-diagonal elements are smaller in the first case) for every function $h$,

$$\lim_{N\to\infty} \text{var}\left[\sum h(X_1^{(t)})\right]/N \geq \lim_{N\to\infty} \text{var}\left[\sum h(X_2^{(t)})\right]/N,$$

where $(X_i^{(t)})$ is a Markov chain with transition matrix $\mathbb{P}_i$ $(i = 1, 2)$. (*Hint:* Use Kemeny and Snell's (1960) result on the asymptotic variance in Problem 4.54.)

**6.47** (Continuation of Problem 6.46) Deduce from Problem 6.46 that for a given instrumental matrix $Q$ in a Metropolis–Hastings algorithm, the choice

$$p_{ij}^* = q_{ij}\left(\frac{\pi_j q_{ji}}{\pi_i q_{ij}} \wedge 1\right)$$

is optimal among the transitions such that

$$p_{ij} = \frac{q_{ij}s_{ij}}{1 + \frac{\pi_i q_{ij}}{\pi_j q_{ji}}} = q_{ij}\alpha_{ij},$$

where $s_{ij} = s_{ji}$ and $0 \leq \alpha_{ij} \leq 1$. (*Hint:* Give the corresponding $\alpha_{ij}$ for the Metropolis–Hastings algorithm and show that it is maximal for $i \neq j$. Tierney 1995, Mira and Geyer 1998, and Tierney and Mira 1999 propose extensions to the continuous case.)

**6.48** Show that $f$ is the stationary density associated with the acceptance probability (6.7.1).

**6.49** In the setting of Example 6.7.2, implement the simulated annealing algorithm to find the maximum of the likelihood $L(\theta|x_1, x_2, x_3)$. Compare with the performances based on $\log L(\theta|x_1, x_2, x_3)$.

**6.50** (Winkler 1995) A *Potts model* is defined on a set $S$ of "sites" and a finite set $G$ of "colors" by its energy

$$H(x) = -\sum_{(s,t)} \alpha_{st} \mathbb{I}_{x_s = x_t}, \qquad x \in G^S,$$

where $\alpha_{st} = \alpha_{ts}$, the corresponding distribution being $\pi(x) \propto \exp(H(x))$. An additional structure is introduced as follows: "Bonds" $b$ are associated with each pair $(s,t)$ such that $\alpha_{st} > 0$. These bonds are either active ($b = 1$) or inactive ($b = 0$).

(a) Defining the joint distribution

$$\mu(x,b) \propto \prod_{b_{st}=0} q_{st} \prod_{b_{st}=1} (1 - q_{st}) \mathbb{I}_{x_s = x_t},$$

with $q_{st} = \exp(\alpha_{st})$, show that the marginal of $\mu$ in $x$ is $\pi$. Show that the marginal of $\mu$ in $b$ is

$$\mu(b) \propto |G|^{c(b)} \prod_{b_{st}=0} q_{st} \prod_{b_{st}=1} (1 - q_{st}),$$

where $c(b)$ denotes the number of *clusters* (the number of sites connected by active bonds).

(b) Show that the *Swendson–Wang* (1987) algorithm

1. Take $b_{st} = 0$ if $x_s \neq x_t$ and, for $x_s = x_t$,

$$b_{st} = \begin{cases} 1 & \text{with probability } 1 - q_{st} \\ 0 & \text{otherwise.} \end{cases}$$

2. For every cluster, choose a color at random on $G$.

leads to simulations from $\pi$ (*Note:* This algorithm is acknowledged as accelerating convergence in image processing.)

## 6.7 Notes

### 6.7.1 A Bit of Background

The original Metropolis algorithm was introduced by Metropolis *et al.* (1953) in a setup of optimization on a discrete state-space, and it was later generalized by Hastings (1970) and Peskun (1973, 1981) to statistical simulation. Despite several other papers that highlighted its usefulness in specific settings (see, for example, Geman and Geman 1984, Tanner and Wong 1987, Besag 1989), the starting point for an intensive use of Markov chain Monte Carlo methods by the statistical community can be traced to the presentation of the *Gibbs sampler* by Gelfand and Smith (1990), as explained in Casella and George (1992) or Chib and Greenberg (1995).

The gap of more than 30 years between Metropolis *et al.* (1953) and Gelfand and Smith (1990) can be partially attributed to the lack of appropriate computing power, as most of the examples now processed by Markov chain Monte Carlo algorithms could not have been treated previously.

As shown by Hastings (1970), Metropolis–Hastings algorithms are a special case of a more general class of algorithms whose transition is associated with the acceptance probability

(6.7.1)
$$\varrho(x,y) = \frac{s(x,y)}{1 + \dfrac{f(x)q(y|x)}{f(y)q(x|y)}} \; ,$$

where $s$ is an arbitrary positive symmetric function such that $\varrho(x,y) \le 1$ (see also Winkler 1995). The particular case $s(x,y) = 1$ is also known as the *Boltzman algorithm* and is used in simulation for particle physics, although Peskun (1973) has shown that, in the discrete case, the performance of this algorithm is always suboptimal compared to the Metropolis–Hastings algorithm (see Problem 6.47).

### 6.7.2 Geometric Convergence of Metropolis–Hastings Algorithms

The sufficient condition (6.2.4) of Lemma 6.2.7 for the irreducibility of the Metropolis–Hastings Markov chain is particularly well adapted to *random walks*, with transition densities $q(y|x) = g(y - x)$. It is, indeed, enough that $g$ is positive in a neighborhood of 0 to ensure the ergodicity of [A.24] (see §6.3.2 for a detailed study of these methods). On the other hand, convergence results stronger than the simple ergodic convergence of (6.1.1) or than the (total variation) convergence of $\|P^n_{x^{(0)}} - f\|_{TV}$ are difficult to derive without introducing additional conditions on $f$ and $q$. For instance, it is impossible to establish *geometric convergence* without a restriction to the discrete case or without considering particular transition densities, since Roberts and Tweedie (1996) have come up with chains which are not geometrically ergodic. Defining, for every measure $\nu$ and every $\nu$-measurable function $h$, the *essential supremum*

$$\mathrm{ess}_\nu \sup \; h(x) = \inf\{w; \nu(h(x) > w) = 0\} \; ,$$

they established the following result, where $\bar\rho$ stands for the average acceptance probability.

**Theorem 6.7.1** *If the marginal probability of acceptance satisfies*

$$\mathrm{ess}_f \sup \; (1 - \bar\rho(x)) = 1,$$

*the algorithm* [A.24] *is not geometrically ergodic.*

Therefore, if $\bar\rho$ is not bounded from below on a set of measure 1, a geometric speed of convergence cannot be guaranteed for [A.24]. This result is important, as it characterizes Metropolis–Hastings algorithms which are weakly convergent (see the extreme case of Example 8.2.8); however, it cannot be used to establish nongeometric ergodicity, since the function $\bar\rho$ is almost always intractable.

When $\mathcal{E}$ is a small set (see Chapter 4), Roberts and Polson (1994) note that the chain $(X^{(t)})$ is uniformly ergodic. However, this is rarely the case when the state-space is uncountable. It is, in fact, equivalent to *Doeblin's condition*, as stated by Theorem 4.6.13. Chapter 7 exhibits examples of continuous Gibbs samplers for which uniform ergodicity holds (see Examples 7.1.23 and 7.1.6), but §6.3.2 has shown that in the particular case of random walks, uniform ergodicity almost never holds, even though this type of move is a natural choice for the instrumental distribution $q$.

### 6.7.3 A Reinterpretation of Simulated Annealing

Consider a function $E$ defined on a finite set $\mathcal{E}$ with such a large cardinality that a minimization of $E$ based on the comparison of the values of $E(\mathcal{E})$ is not feasible. The simulated annealing technique (see §5.2.3) is based on a conditional density $q$ on $\mathcal{E}$ such that $q(i|j) = q(j|i)$ for every $(i,j) \in \mathcal{E}^2$. For a given value $T > 0$, it produces a Markov chain $(X^{(t)})$ on $\mathcal{E}$ by the following transition:

1. Generate $\zeta_t$ according to $q(\zeta|x^{(t)})$.
2. Take

$$X^{(t+1)} = \begin{cases} \zeta_t & \text{with probability} \quad \exp\left(\{E(X^{(t)}) - E(\zeta_t)\}/T\right) \wedge 1 \\ X^{(t)} & \text{otherwise.} \end{cases}$$

As noted in §5.2.3, the simulated value $\zeta_t$ is automatically accepted when $E(\zeta_t) \le E(X^{(t)})$. The fact that the simulated annealing algorithm may accept a value $\zeta_t$ with $E(\zeta_t)$ larger than $E(X^{(t)})$ is a very positive feature of the method, since it allows for escapes from the attraction zone of local minima of $E$ when $T$ is large enough. The simulated annealing algorithm is actually a Metropolis–Hastings algorithm with stationary distribution $f(x) \propto \exp(-E(x)/T)$, provided that the matrix of the $q(i|j)$'s generates an irreducible chain. Note, however, that the theory of time-homogeneous Markov chains presented in Chapter 4 does not cover the extension to the case when $T$ varies with $t$ and converges to 0 "slowly enough" (typically in $\log t$).

### 6.7.4 Reference Acceptance Rates

Roberts *et al.* (1997) recommend the use of instrumental distributions with *acceptance rate close to 1/4 for models of high dimension and equal to 1/2 for the models of dimension 1 or 2*. This heuristic rule is based on the asymptotic behavior of an *efficiency criterion* equal to the ratio of the variance of an estimator based on an iid sample and the variance of the estimator (3.1.1); that is,

$$\left[1 + 2 \sum_{k>0} \text{cov}\left(X^{(t)}, X^{(t+k)}\right)\right]^{-1}$$

in the case $h(x) = x$. When $f$ is the density of the $\mathcal{N}(0, 1)$ distribution and $g$ is the density of a Gaussian random walk with variance $\sigma$, Roberts *et al.* (1997) have shown that the optimal choice of $\sigma$ is 2.4, with an asymmetry in the efficiency in favor of large values of $\sigma$. The corresponding acceptance rate is

$$\rho = \frac{2}{\pi} \arctan\left(\frac{2}{\sigma}\right),$$

equal to 0.44 for $\sigma = 2.4$. A second result by Roberts *et al.* (1997), based on an approximation of $X^{(t)}$ by a Langevin diffusion process (see §6.5.2) when the dimension of the problem goes to infinity, is that the acceptance probability converges to 0.234 (approximately 1/4). An equivalent version of this empirical rule is to *take the scale factor in $g$ equal to $2.38/\sqrt{d} \, \Sigma$*, where $d$ is the dimension of the model and $\Sigma$ is the asymptotic variance of $X^{(t)}$. This is obviously far from being an absolute optimality result since this choice is based on the particular case of the normal distribution, which is not representative (to say the least) of the distributions usually involved in the Markov chain Monte Carlo algorithms. In addition, $\Sigma$ is never known in practice.

| $\sigma$ | 0.1 | 0.2 | 0.5 | 1.0 | 5.0 | 8.0 | 10.0 | 12.0 |
|---|---|---|---|---|---|---|---|---|
| $\rho_\sigma$ | 0.991 | 0.985 | 0.969 | 0.951 | 0.893 | 0.890 | 0.891 | 0.895 |
| $t_\sigma$ | 106.9 | 67.1 | 32.6 | 20.4 | 9.37 | 9.08 | 9.15 | 9.54 |
| $h_1$ | 41.41 | 44.24 | 44.63 | 43.76 | 42.59 | 42.12 | 42.92 | 42.94 |
| $h_2$ | 0.035 | 0.038 | 0.035 | 0.036 | 0.036 | 0.036 | 0.035 | 0.036 |
| $h_3$ | 0.230 | 0.228 | 0.227 | 0.226 | 0.228 | 0.229 | 0.228 | 0.230 |

Table 6.7.1. *Performances of the algorithm* [A.28] *associated with* (6.7.2) *for* $x_1 = -8$, $x_1 = 8$, *and* $x_3 = 17$ *and the random walk* $\mathcal{C}(0, \sigma^2)$. *These performances are evaluated via the acceptance probability* $\rho_\sigma$, *the interjump time,* $t_\sigma$, *and the empirical variance associated with the approximation of the expectations* $\mathbb{E}^\pi[h_i(\theta)]$ (20,000 *simulations*).

The implementation of this heuristic rule follows the principle of an *algorithmic calibration*, first proposed by Müller (1991); that is, of successive modifications of scale factors followed by estimations of $\Sigma$ and of the acceptance rate until this rate is close to 1/4 and the estimation of $\Sigma$ remains stable. Note, again, that the convergence results of Metropolis–Hastings algorithms only apply for these adaptive versions when the different hyperparameters of the instrumental distribution are fixed: As long as these parameters are modified according to the simulation results, the resulting Markov chain is heterogeneous (or is not a Markov chain anymore if the parameters depend on the previous realizations). This cautionary notice signifies that, in practice, the use of a (true) Metropolis–Hastings algorithm must be preceded by a calibration step which determines an acceptable range for the simulation hyperparameters.

**Example 6.7.2 Cauchy posterior distribution.** Consider $X_1, X_2$, and $X_3$ iid $\mathcal{C}(\theta, 1)$ and $\pi(\theta) = \exp(-\theta^2/100)$. The posterior distribution on $\theta$, $\pi(\theta|x_1, x_2, x_3)$, is proportional to

$$(6.7.2) \qquad e^{-\theta^2/100}[(1 + (\theta - x_1)^2)(1 + (\theta - x_2)^2)(1 + (\theta - x_3)^2)]^{-1} ,$$

which is trimodal when $x_1, x_2$, and $x_3$ are sufficiently spaced out, as suggested by Figure 1.2.1 for $x_1 = 0$, $x_2 = 5$, and $x_3 = 9$. This distribution is therefore adequate to test the performances of the Metropolis–Hastings algorithm in a unidimensional setting. Given the dispersion of the distribution (6.7.2), we use a random walk based on a Cauchy distribution $\mathcal{C}(0, \sigma^2)$. (Chapter 7 proposes an alternative approach via the Gibbs sampler for the simulation of (6.7.2).)

Besides the probability of acceptance, $\rho_\sigma$, a parameter of interest for the comparison of algorithms is the *interjump* time, $t_\sigma$; that is, the average number of iterations that it takes the chain to move to a different mode. (For the above values, the three regions considered are $(-\infty, 0]$, $(0, 13]$, and $(13, +\infty)$.) Table 6.7.1 provides, in addition, an evaluation of [A.28], through the values of the standard deviation of the random walk $\sigma$, in terms of the variances of some estimators of $\mathbb{E}^\pi[h_i(\theta)]$ for the functions

$$h_1(\theta) = \theta, \quad h_2(\theta) = \left(\theta - \frac{17}{3}\right)^2 \quad \text{and} \quad h_3(\theta) = \mathbb{I}_{[4,8]}(\theta) .$$

The means of these estimators for the different values of $\sigma$ are quite similar and equal to 8.96, 0.063, and 0.35, respectively. A noteworthy feature of this

example is that the probability of acceptance never goes under 0.88 and, therefore the goal of Roberts *et al.* (1997) cannot be attained with this choice of
instrumental distribution, whatever $\sigma$ is. This phenomenon is quite common
in practice.                                                                    ‖

CHAPTER 7

# The Gibbs Sampler

In this place he was found by Gibbs, who had been sent for him in some haste. He got to his feet with promptitude, for he knew no small matter would have brought Gibbs in such a place at all.

—G.K. Chesterton, *The Innocence of Father Brown*

## 7.1 General Principles

The previous chapter developed simulation techniques that could be called "generic," since they require only a limited amount of information about the distribution to be simulated. For example, the generic algorithm ARMS (§6.3.3) aims at reproducing the density $f$ of this distribution in an automatic manner. However, Metropolis–Hastings algorithms can achieve higher levels of efficiency when they take into account the specifics of the target distribution $f$, in particular through the calibration of the acceptance rate (see §6.4.1). Moving even further in this direction, the properties and performance of the Gibbs sampling method presented in this chapter are very closely tied to the distribution $f$. This is because the choice of instrumental distribution is essentially reduced to a choice between a *finite* number of possibilities.

### 7.1.1 Definition

Suppose that for some $p > 1$, the random variable $\mathbf{X} \in \mathcal{X}$ can be written as $\mathbf{X} = (X_1, \ldots, X_p)$, where the $X_i$'s are either uni- or multidimensional. Moreover, suppose that we can simulate from the corresponding univariate conditional densities $f_1, \ldots, f_p$, that is, we can simulate

$$X_i | x_1, x_2, \ldots, x_{i-1}, x_{i+1}, \ldots, x_p \sim f_i(x_i | x_1, x_2, \ldots, x_{i-1}, x_{i+1}, \ldots, x_p)$$

for $i = 1, 2, \ldots, p$. The associated *Gibbs sampling* algorithm (or *Gibbs sampler*) is given by the following transition from $X^{(t)}$ to $X^{(t+1)}$:

**Algorithm A.31 –The Gibbs Sampler–**
Given $\mathbf{x}^{(t)} = (x_1^{(t)}, \ldots, x_p^{(t)})$, generate
1. $X_1^{(t+1)} \sim f_1(x_1 | x_2^{(t)}, \ldots, x_p^{(t)})$;

2. $X_2^{(t+1)} \sim f_2(x_2|x_1^{(t+1)}, x_3^{(t)}, \ldots, x_p^{(t)}),$                    [A.31]

$$\vdots$$

p. $X_p^{(t+1)} \sim f_p(x_p|x_1^{(t+1)}, \ldots, x_{p-1}^{(t+1)}).$

The densities $f_1, \ldots, f_p$ are called the *full conditionals*, and a particular feature of the Gibbs sampler is that these are the only densities used for simulation. Thus, even in a high-dimensional problem, *all of the simulations may be univariate*, which is usually an advantage.

**Example 7.1.1 Bivariate Gibbs sampler.** Let the random variables $X$ and $Y$ have joint density $f(x, y)$, and generate a sequence of observations according to the following:
Set $X_0 = x_0$, and for $t = 1, 2, \ldots$, generate

$$\begin{aligned} Y_t &\sim f_{Y|X}(\cdot|x_{t-1}), \\ X_t &\sim f_{X|Y}(\cdot|y_t), \end{aligned}$$

(7.1.1)

where $f_{Y|X}$ and $f_{X|Y}$ are the conditional distributions. The sequence $(X_t, Y_t)$, is a Markov chain, as is each sequence $(X_t)$ and $(Y_t)$ individually. For example, the chain $(X_t)$ has transition density

$$K(x, x^*) = \int f_{Y|X}(y|x) f_{X|Y}(x^*|y) dy,$$

with invariant density $f_X(\cdot)$. (Note the similarity to Eaton's transition (4.10.9).)

For the special case of the bivariate normal density,

$$(7.1.2) \qquad (X, Y) \sim \mathcal{N}_2\left(0, \begin{pmatrix} 1 & \rho \\ \rho & 1 \end{pmatrix}\right),$$

the Gibbs sampler is
Given $y_t$, generate

$$\begin{aligned} (7.1.3) \qquad X_{t+1} \mid y_t &\sim \mathcal{N}(\rho y_t, \ 1 - \rho^2), \\ Y_{t+1} \mid x_{t+1} &\sim \mathcal{N}(\rho x_{t+1}, \ 1 - \rho^2). \end{aligned}$$

The Gibbs sampler is obviously not necessary in this particular case, as iid copies of $(X, Y)$ can be easily generated using the Box–Muller algorithm (see Example 2.2.2). ‖

**Example 7.1.2 Autoexponential model.** The *autoexponential model* of Besag (1974) has been found useful in some aspects of spatial modeling. When $y \in \mathbb{R}_+^3$, the corresponding density is

$$f(y_1, y_2, y_3) \propto \exp\{-(y_1 + y_2 + y_3 + \theta_{12}y_1y_2 + \theta_{23}y_2y_3 + \theta_{31}y_3y_1)\},$$

with known $\theta_{ij} > 0$. The full conditional densities are exponential. For example,

$$Y_3|y_1, y_2 \sim \mathcal{E}xp\left(1 + \theta_{23}y_2 + \theta_{31}y_1\right).$$

They are, thus, very easy to simulate from. In contrast, the other conditionals and the marginal distributions have forms such as

$$f(y_2|y_1) \propto \frac{\exp\{-(y_1 + y_2 + \theta_{12}y_1y_2)\}}{1 + \theta_{23}y_2 + \theta_{31}y_1} ,$$

$$f(y_1) \quad \propto e^{-y_1} \int_0^{+\infty} \frac{\exp\{-y_2 - \theta_{12}y_1y_2\}}{1 + \theta_{23}y_2 + \theta_{31}y_1} \, dy_2 ,$$

which cannot be simulated easily.                                              ‖

**Example 7.1.3 Ising model.** For the *Ising model* of Example 5.2.5, where

$$f(s) \propto \exp\left\{-H \sum_i s_i - J \sum_{(i,j) \in \mathcal{N}} s_i s_j\right\}, \qquad s_i \in \{-1, 1\},$$

and where $\mathcal{N}$ denotes the neighborhood relation for the network, the full conditionals are given by

$$f(s_i|s_{j \neq i}) = \frac{\exp\{-Hs_i - Js_i \sum_{j:(i,j) \in \mathcal{N}} s_j\}}{\exp\{-H - J \sum_j s_{j:(i,j) \in \mathcal{N}}\} + \exp\{H + J \sum_j s_{j:(i,j) \in \mathcal{N}}\}}$$

$$(7.1.4) \qquad = \frac{\exp\{-(H + J \sum_{j:(i,j) \in \mathcal{N}} s_j)(s_i + 1)\}}{1 + \exp\{-2(H + J \sum_{j:(i,j) \in \mathcal{N}} s_j)\}} .$$

It is therefore particularly easy to implement the Gibbs sampler [A.31] for these conditional distributions by successively updating each node $i$ of the network.                                                                          ‖

Although the Gibbs sampler is, formally, a special case of the Metropolis–Hastings algorithm (or rather a combination of Metropolis–Hastings algorithms applied to different components; see Theorem 7.1.17), the Gibbs sampling algorithm has a number of distinct features:

(i) The acceptance rate of the Gibbs sampler is uniformly equal to 1. Therefore, every simulated value is accepted and the suggestions of §6.4.1 on the optimal acceptance rates do not apply in this setting. This also means that convergence assessment for this algorithm should be treated differently than for Metropolis–Hastings techniques.

(ii) The use of the Gibbs sampler implies limitations on the choice of instrumental distributions and requires a prior knowledge of some analytical or probabilistic properties of $f$.

(iii) The Gibbs sampler is, by construction, multidimensional. Even though some components of the simulated vector may be artificial for the problem of interest, or unnecessary for the required inference, the construction is still at least two dimensional.

(iv) The Gibbs sampler does not apply to problems where the number of parameters varies, as in §6.5.1, because of the obvious lack of irreducibility of the resulting chain.

### 7.1.2 Completion

Following the mixture method discussed in §2.2 (see Example 2.2.6), the Gibbs sampling algorithm can be generalized by a "demarginalization" or completion construction.

**Definition 7.1.4** Given a probability density $f$, a density $g$ that satisfies

$$\int_Z g(x, z)\, dz = f(x)$$

is called a *completion* of $f$.

The density $g$ is chosen so that the full conditionals of $g$ are easy to simulate from and the Gibbs algorithm $[A.31]$ is implemented on $g$ instead of $f$. For $p > 1$, write $y = (x, z)$ and denote the conditional densities of $g(y) = g(y_1, \ldots, y_p)$ by

$$Y_1 | y_2, \ldots, y_p \sim g_1(y_1 | y_2, \ldots, y_p),$$
$$Y_2 | y_1, y_3, \ldots, y_p \sim g_2(y_2 | y_1, y_3, \ldots, y_p),$$
$$\vdots$$
$$Y_p | y_1, \ldots, y_{p-1} \sim g_p(y_p | y_1, \ldots, y_{p-1}).$$

The move from $Y^{(t)}$ to $Y^{(t+1)}$ is then defined as follows:

**Algorithm A.32 –Completion Gibbs Sampler–**
Given $(y_1^{(t)}, \ldots, y_p^{(t)})$, simulate

1. $Y_1^{(t+1)} \sim g_1(y_1 | y_2^{(t)}, \ldots, y_p^{(t)})$,
2. $Y_2^{(t+1)} \sim g_2(y_2 | y_1^{(t+1)}, y_3^{(t)}, \ldots, y_p^{(t)})$,        $[A.32]$
   $\vdots$
p. $Y_p^{(t+1)} \sim g_p(y_p | y_1^{(t+1)}, \ldots, y_{p-1}^{(t+1)})$.

**Example 7.1.5 Truncated normal distribution.** As in Example 2.3.5, consider a truncated normal distribution,

$$f(x) \propto e^{-(x-\mu)^2/2\sigma^2}\, \mathbb{I}_{x \geq \mu}\,.$$

As mentioned previously, the naive simulation of a normal $\mathcal{N}(\mu, \sigma^2)$ rv until the outcome is above $\mu$ is suboptimal for large values of the truncation point $\mu$. However, devising and optimizing the Accept–Reject algorithm of Example 2.3.5 can be costly if the algorithm is only to be used a few times. An alternative is to use completion and base the simulation on

$$g(x, z) \propto \mathbb{I}_{x \geq \mu}\, \mathbb{I}_{0 \leq z \leq \exp\{-(x-\mu)^2/2\sigma^2\}}\,.$$

(This is a special case of *slice sampling*, see $[A.33]$.) The corresponding implementation of $[A.32]$ is then

Simulate

1. $X^{(t)}|z^{(t-1)} \sim \mathcal{U}([\underline{\mu}, \mu + \sqrt{-2\sigma^2 \log(z^{(t-1)})}])$,

2. $Z^{(t)}|x^{(t)} \sim \mathcal{U}([0, \exp\{-(x^{(t)} - \mu)^2/2\sigma^2\}])$.

Note that the initial value of $z$ must be chosen so that

$$\mu + \sqrt{-2\sigma^2 \log(z^{(0)})} > \underline{\mu}.$$

$\parallel$

The completion of a density $f$ into a density $g$ such that $f$ is the marginal density of $g$ corresponds to one of the first (historical) appearances of the Gibbs sampling in Statistics, namely the introduction of *data augmentation* by Tanner and Wong (1987) (see Note 7.6.1).

In cases when such completions seem necessary (for instance, when every conditional distribution associated with $f$ is not explicit, or when $f$ is uni-dimensional), there exist an infinite number of densities for which $f$ is a marginal density. We will not discuss this choice in terms of optimality, because, first, there are few practical results on this topic (as it is similar to finding an optimal density $g$ in the Metropolis–Hastings algorithm) and, second, because there exists, in general, a *natural* completion of $f$ in $g$. *Missing data models*, treated in Chapter 9, provide a series of examples of natural completion (see also §5.3.1.)

Note the similarity of this approach with the EM algorithm for maximizing a missing data likelihood (§5.3.3), and even more with recent versions of EM such as ECM and MCEM (see Meng and Rubin 1991, 1992, Liu and Rubin 1994, and Problem 7.40).

In principle, the Gibbs sampler does not require that the completion of $f$ into $g$ and of $x$ in $y = (x, z)$ should be related to the problem of interest. Indeed, there are settings where the vector $z$ has no meaning from a statistical point of view and is only a useful device.

**Example 7.1.6 Cauchy–normal posterior distribution.** As shown in Chapter 2 (§2.2), Student's $t$ distribution can be generated as a mixture of a normal distribution by a chi squared distribution. This decomposition is useful when the expression $[1 + (\theta - \theta_0)^2]^{-\nu}$ appears in a more complex distribution. Consider, for instance, the density

$$f(\theta|\theta_0) \propto \frac{e^{-\theta^2/2}}{[1 + (\theta - \theta_0)^2]^\nu}.$$

This is the posterior distribution resulting from the model

$$X|\theta \sim \mathcal{N}(\theta, 1) \quad \text{and} \quad \theta \sim \mathcal{C}(\theta_0, 1)$$

(see Example 8.3.2). A similar function arises in the estimation of the parameter of interest in a linear calibration model (see Example 1.3.6).

The density $f(\theta|\theta_0)$ can be written as the marginal density

$$f(\theta|\theta_0) \propto \int_0^\infty e^{-\theta^2/2} \, e^{-[1+(\theta-\theta_0)^2]\,\eta/2} \, \eta^{\nu-1} \, d\eta$$

and can, therefore, be completed as

$$g(\theta, \eta) \propto e^{-\theta^2/2} \, e^{-[1+(\theta-\theta_0)^2] \, \eta/2} \, \eta^{\nu-1},$$

which leads to the conditional densities

$$g_1(\eta|\theta) = \left(\frac{1+(\theta-\theta_0)^2}{2}\right)^{\nu} \frac{\eta^{\nu-1}}{\Gamma(\nu)} \exp\left\{-[1+(\theta-\theta_0)^2] \, \eta/2\right\},$$

$$g_2(\theta|\eta) = \sqrt{\frac{1+\eta}{2\pi}} \exp\left\{-\left(\theta - \frac{\eta\theta_0}{1+\eta}\right)^2 \frac{1+\eta}{2}\right\},$$

that is, to a gamma and a normal distribution on $\eta$ and $\theta$, respectively:

$$\eta|\theta \sim \mathcal{G}a\left(\nu, \frac{1+(\theta-\theta_0)^2}{2}\right), \qquad \theta|\eta \sim \mathcal{N}\left(\frac{\theta_0\eta}{1+\eta}, \frac{1}{1+\eta}\right).$$

Note that the parameter $\eta$ is completely meaningless for the problem at hand but serves to facilitate computations. (See Problem 7.7 for further properties.) ‖

We next look at a very general version of the Gibbs sampler, called the *slice sampler* (Wakefield *et al.* 1991, Besag and Green 1993, Damien and Walker 1996, Higdon 1996, Damien *et al.* 1999, and Tierney and Mira 1999) It applies to most distributions and is based on the simulation of specific uniform random variables (see also Problem 7.9). If $f(\theta)$ can be written as a product

$$\prod_{i=1}^{k} f_i(\theta),$$

where the $f_i$'s are positive functions, *not necessarily densities*, $f$ can be completed (or *demarginalized*) into

$$\prod_{i=1}^{k} \mathbb{I}_{0 \leq \omega_i \leq f_i(\theta)},$$

leading to the following Gibbs algorithm:

**Algorithm A.33 –Slice Sampler–**
Simulate

1. $\omega_1^{(t+1)} \sim \mathcal{U}_{[0, f_1(\theta^{(t)})]};$

   $\vdots$                                                                      [A.33]

k. $\omega_k^{(t+1)} \sim \mathcal{U}_{[0, f_k(\theta^{(t)})]};$

k+1. $\theta^{(t+1)} \sim \mathcal{U}_{A^{(t+1)}}$, with

$$A^{(t+1)} = \{y; \, f_i(y) \geq \omega_i^{(t+1)}, \, i = 1, \ldots, k\}.$$

Roberts and Rosenthal (1998) study the slice sampler [A.33] and show that it usually enjoys good theoretical properties. In particular, they prove

| $n$ | 0.0 | .67 | .84 | $q_\alpha$ 1.28 | 1.64 | 2.33 | 2.58 | 3.09 | 3.72 |
|------|------|------|------|------|------|------|------|------|------|
| $10^2$ | .5050 | .7822 | .8614 | .9406 | .9901 | 1 | 1 | 1 | 1 |
| $10^3$ | .4885 | .7522 | .7982 | .8951 | .9570 | .9930 | .9960 | .9990 | 1 |
| $10^4$ | .4939 | .7522 | .8015 | .8980 | .9499 | .9887 | .9943 | .9983 | .9999 |
| $10^5$ | .4993 | .7497 | .7987 | .8996 | .9504 | .9899 | .9949 | .999 | .9999 |
| $10^6$ | .4993 | .7496 | .7995 | .8996 | .9499 | .9899 | .9949 | .999 | .9999 |
| $10^7$ | .4997 | .7499 | .7998 | .8999 | .9499 | .99 | .995 | .999 | .9999 |
| true | .5 | .7500 | .7995 | .8997 | .9495 | .9901 | .9951 | .999 | .9999 |

Table 7.1.1. *Convergence of the estimated quantiles, based on a Markov chain simulated by the slice sampler [A.33].*

that the slice sampler allows small sets and that the associated chain is geometrically ergodic when the density $f$ is bounded. Tierney and Mira (1999) have also established uniform ergodicity when $k = 1$.

In practice, there may be problems. As $k$ increases, the determination of the set $A^{(t+1)}$ may get increasingly complex. Also, the Gibbs sampler is forced to work in a highly restricted (and nonlinear) space when $k$ is large, and this may create convergence slowdowns. (See Neal 1997 for partial solutions to these two problems.) In particular, latent variable models like stochastic volatility (see §9.5.3) can be quite greedy in their production of $\omega_i$'s and lead to deadlocks in the Gibbs sampler, based on our personal experience. Nonetheless, slice sampling is a readily available, in fact, almost automatic method to simulate from new densities and its availability should not be neglected.

**Example 7.1.7 Normal simulation.** For the standard normal density, $f(x) \propto \exp(-x^2/2)$, a slice sampler is associated with the two conditional distributions

$$(7.1.5) \quad \omega | x \sim \mathcal{U}_{[0,\exp(-x^2/2)]} , \qquad X | \omega \sim \mathcal{U}_{[-\sqrt{-2\log(\omega)},\sqrt{-2\log(\omega)}]} ,$$

which are quite easy to simulate. Figure 7.1.1 compares slice sampling with the Box–Muller algorithm. The two methods appear to give very similar results. Table 7.1.1 shows that the convergence of the empirical cdf takes place at the same speed as for iid sampling (see Example 3.2.3).     ‖

As mentioned above, the completion may be driven by latent variables in the model.

**Example 7.1.8 Censored data models.** As noted in §5.3.1, censored data models (Example 5.3.1) can be associated with a missing data structure. Consider $y^* = (y_1^*, y_2^*) = (y \wedge r, \mathbb{I}_{y<r})$, with $y \sim f(y|\theta)$ and $r \sim h(r)$. The observation $y$ is censored by the random variable $r$, which often is taken to be a constant. If we observe $(y_1^*, \ldots, y_n^*)$, the density of $y_i^* = (y_{1i}^*, y_{2i}^*)$ is

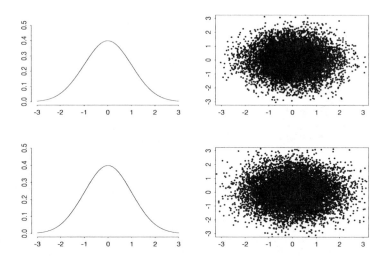

Figure 7.1.1. *Comparison of two samples of* 10,000 *points through the clouds of the pairs* $(x_t, x_{t+1})$ *(right) and of the density estimates (left), by the Box–Muller (top) and the slice sampler (7.1.5) (bottom) algorithms.*

then

$$(7.1.6) \quad \int_{y_{1i}^*}^{+\infty} f(y|\theta) \, dy \, h(y_{1i}^*) \mathbb{I}_{y_{2i}^*=0} + \int_{y_{1i}^*}^{+\infty} h(r) \, dr \, f(y_{1i}^*|\theta) \mathbb{I}_{y_{2i}^*=1} \,.$$

In the cases where (7.1.6) cannot be explicitly integrated, the likelihood and posterior distribution associated with this model may be too complex to be used. If $\theta$ has the prior distribution $\pi$, the posterior distribution satisfies

$$\pi(\theta|y_1^*, \ldots, y_n^*) \propto \pi(\theta) \prod_{\{i:y_{2i}^*=0\}} \left\{ h(y_{1i}^*) \int_{y_{1i}^*}^{+\infty} f(y|\theta) dy \right\}$$

$$\times \prod_{\{i:y_{2i}^*=1\}} \left\{ f(y_{1i}^*|\theta) \int_{y_{1i}^*}^{+\infty} h(r) dr \right\}.$$

If the *uncensored* model leads to explicit computations for the posterior distribution of $\theta$, a logical completion is to reconstruct the original data, $y_1, \ldots, y_n$, conditionally on the observed $y_i^*$'s $(i = 1, \ldots, n)$. Therefore, if $\delta_y$ represents the Dirac mass at $y$, a set of full conditionals that can be used to implement a Gibbs sampler are

$$\pi(\theta|y_1, \ldots, y_n) \propto \pi(\theta) \prod_{i=1}^{n} f(y_i|\theta),$$

$$f^*(y|\theta, y_i^*) \propto \delta_{y_{1i}^*}(y) \mathbb{I}_{y_{2i}^*=1} + f(y|\theta) \, \mathbb{I}_{y>y_{1i}^*} \mathbb{I}_{y_{2i}^*=0} \,.$$

Note that this Gibbs sampler does not involve the censoring distribution and that the distribution of $(\theta, Y_1, \ldots, , Y_n|y_1^*, \ldots, y_n^*)$ can be rewritten in closed form only if (7.1.6) can.                                    ‖

### 7.1.3 Convergence Properties

In this section we investigate the convergence properties of the Gibbs sampler. The first thing we show is that most Gibbs samplers satisfy the minimal conditions necessary for ergodicity. We have mentioned that Gibbs sampling is a special case of the Metropolis–Hastings algorithm (this notion is formalized in Theorem 7.1.17). Unfortunately, however, this fact is not much help to us in the current endeavor.

We will work with the general Gibbs sampler [A.32] and show that the Markov chain $(Y^{(t)})$ converges to the distribution $g$ and the subchain $(X^{(t)})$ converges to the distribution $f$. It is important to note that although $(Y^{(t)})$ is, by construction, a Markov chain, the subchain $(X^{(t)})$ is, typically, not a Markov chain, except in the particular case of Data Augmentation (see §7.2).

**Theorem 7.1.9** *For the Gibbs sampler of [A.32], if $(Y^{(t)})$ is ergodic, then the distribution $g$ is a stationary distribution for the chain $(Y^{(t)})$ and $f$ is the limiting distribution of the subchain $(X^{(t)})$ .*

*Proof.* The kernel of the chain $(Y^{(t)})$ is the product

$$(7.1.7) \quad K(\mathbf{y}, \mathbf{y}') = g_1(y_1'|y_2, \ldots, y_p)$$
$$\times g_2(y_2'|y_1', y_3, \ldots, y_p) \cdots g_p(y_p'|y_1', \ldots, y_{p-1}').$$

For the vector $\mathbf{y} = (y_1, y_2, \ldots, y_p)$, let $g^i(y_1, \ldots, y_{i-1}, y_{i+1}, \ldots, y_p)$ denote the marginal density of the vector $\mathbf{y}$ with $y_i$ integrated out. If $\mathbf{Y} \sim g$ and $A$ is measurable under the dominating measure, then

$$P(\mathbf{Y}' \in A) = \int \mathbb{I}_A(\mathbf{y}') \, K(\mathbf{y}, \mathbf{y}') \, g(\mathbf{y}) \, d\mathbf{y}'d\mathbf{y}$$

$$= \int \mathbb{I}_A(\mathbf{y}') \, [g_1(y_1'|y_2, \ldots, y_p) \cdots g_p(y_p'|y_1', \ldots, y_{p-1}')]$$
$$\times [g_1(y_1|y_2, \ldots, y_p)g^1(y_2, \ldots, y_p)] \, dy_1 \cdots dy_p dy_1' \ldots dy_p'$$

$$= \int \mathbb{I}_A(\mathbf{y}') \, g_2(y_2'|y_1', \ldots, y_p) \cdots g_p(y_p'|y_1', \ldots, y_{p-1}')$$
$$\times g(y_1', y_2, \ldots, y_p)dy_2 \cdots dy_p dy_1' \cdots dy_p'$$

where we have integrated out $y_1$ and combined $g_1(y_1'|y_2, \ldots, y_p)g^1(y_2, \ldots, y_p)$ $= g(y_1', y_2, \ldots, y_p)$. Next, write $g(y_1', y_2, \ldots, y_p) = g(y_2|y_1', y_3, \ldots, y_p)g^2(y_1', y_3, \ldots, y_p)$, and integrate out $y_2$ to obtain

$$P(\mathbf{Y}' \in A) = \int \mathbb{I}_A(\mathbf{y}') \, g_3(y_3'|y_1', y_2', \ldots, y_p) \cdots g_p(y_p'|y_1', \ldots, y_{p-1}')$$
$$\times g(y_1', y_2', y_3, \ldots, y_p) \, dy_3 \cdots dy_p dy_1' \cdots dy_p' \, .$$

By continuing in this fashion and successively integrating out the $y_i$'s, the above probability is

$$P(\mathbf{Y}' \in A) = \int_A g(y'_1, \ldots, y'_p) \, d\mathbf{y}',$$

showing that $g$ is the stationary distribution. Therefore, Theorem 4.6.4 implies that $Y^{(t)}$ is asymptotically distributed according to $g$ and $X^{(t)}$ is asymptotically distributed according to the marginal $f$, by integration.

□□

A shorter proof can be based on the stationarity of $g$ after *each* step in the $p$ steps of [A.32] (see Theorem 7.1.17).

We now consider the irreducibility of the Markov chain produced by the Gibbs sampler. The following example shows that a Gibbs sampler need not be irreducible:

**Example 7.1.10 Nonconnected support.** Let $\mathcal{E}$ and $\mathcal{E}'$ denote the disks of $\mathbb{R}^2$ with radius 1 and respective centers $(1, 1)$ and $(-1, -1)$. Consider the distribution with density

$$f(x_1, x_2) = \frac{1}{2\pi} \left\{ \mathbb{I}_{\mathcal{E}}(x_1, x_2) + \mathbb{I}_{\mathcal{E}'}(x_1, x_2) \right\} .$$

The natural decomposition of $f$ into

$$f_1(x_1|x_2) = \frac{1}{2\sqrt{1 - (1 - x_2)^2}} \, \mathbb{I}_{\left[-1 - \sqrt{1 - (1 - x_2)^2}, \, -1 + \sqrt{1 - (1 - x_2)^2}\right]}(x_1)$$

if $x_2 < 0$ and

$$f_1(x_1|x_2) = \frac{1}{2\sqrt{1 - (1 - x_2)^2}} \, \mathbb{I}_{\left[1 - \sqrt{1 - (1 - x_2)^2}, \, 1 + \sqrt{1 - (1 - x_2)^2}\right]}(x_1)$$

if $x_2 > 0$, and into

$$f_2(x_2|x_1) = \frac{1}{2\sqrt{1 - (1 - x_1)^2}} \, \mathbb{I}_{\left[-1 - \sqrt{1 - (1 - x_1)^2}, \, -1 + \sqrt{1 - (1 - x_1)^2}\right]}(x_2)$$

if $x_1 < 0$ and

$$f_2(x_2|x_1) = \frac{1}{2\sqrt{1 - (1 - x_1)^2}} \, \mathbb{I}_{\left[1 - \sqrt{1 - (1 - x_1)^2}, \, 1 + \sqrt{1 - (1 - x_1)^2}\right]}(x_2)$$

if $x_1 > 0$ cannot produce an irreducible chain through [A.32], since the resulting chain remains concentrated on the (positive or negative) quadrant on which it is initialized. (Note that a change of coordinates such as $z_1 = x_1 + x_2$ and $z_2 = x_1 - x_2$ is sufficient to remove this difficulty.) ‖

A sufficient condition for irreducibility of the Gibbs Markov chain is the following condition, introduced by Besag (1974) (see §7.1.5).

**Definition 7.1.11** Let $(Y_1, Y_2, \ldots, Y_p) \sim g(y_1, \ldots, y_p)$, where $g^{(i)}$ denotes the marginal distribution of $Y_i$. If $g^{(i)}(y_i) > 0$ for every $i = 1, \ldots, p$, implies that $g(y_1, \ldots, y_p) > 0$, then $g$ satisfies the *positivity condition*.

Thus, the support of $g$ is the Cartesian product of the supports of the $g^{(i)}$'s. Moreover, it follows that the conditional distributions will not reduce the range of possible values of $Y_i$ when compared with $g$. In this case, two arbitrary Borel subsets of the support can be joined in a single iteration of [A.32]. We therefore have the following theorem, whose proof is left to Problem 7.14.

**Theorem 7.1.12** *For the Gibbs sampler of* [A.32], *if the density $g$ satisfies the positivity condition, it is irreducible.*

Unfortunately, the condition of Theorem 7.1.12 is often difficult to verify, and Tierney (1994) gives a more manageable condition which we will use instead. We state this in the following lemma.

**Lemma 7.1.13** *If the transition kernel associated with* [A.32] *is absolutely continuous with respect to the dominating measure, the resulting chain is Harris recurrent.*

The proof of Lemma 7.1.13 is similar to that of Lemma 6.2.4, where Harris recurrence was shown to follow from irreducibility. Here, the condition on the Gibbs transition kernel yields an irreducible chain, and Harris recurrence follows. Once the irreducibility of the chain associated with [A.32] is established, more advanced convergence properties can be established (see Note 7.6.3).

This condition of absolute continuity on the kernel (7.1.7) is satisfied by most decompositions. However, if one of the $i$ steps $(1 \leq i \leq p)$ is replaced by a simulation from an Metropolis–Hastings algorithm, as in the hybrid algorithms described in §7.3, absolute continuity is lost and it will be necessary to either study the recursion properties of the chain or to introduce an additional simulation step to guarantee Harris recurrence.

From the property of Harris recurrence, we can now establish a result similar to Theorem 6.2.5.

**Theorem 7.1.14** *Suppose that the transition kernel of the Gibbs sampling Markov chain $(Y^{(t)})$ is absolutely continuous with respect to the dominating measure.*

*(i) If $h_1, h_2 \in L^1(g)$ with $\int h_2(y) dg(y) \neq 0$, then*

$$\lim_{n \to \infty} \frac{\sum_{t=1}^{T} h_1(Y^{(t)})}{\sum_{t=1}^{T} h_2(Y^{(t)})} = \frac{\int h_1(y) dg(y)}{\int h_2(y) dg(y)} \quad a.e. \; g.$$

*(ii) If, in addition, $(Y^{(t)})$ is aperiodic, then*

$$\lim_{n \to \infty} \left\| \int K^n(y, \cdot) \mu(dx) - g \right\|_{TV} = 0$$

*for every initial distribution $\mu$.*

We also state a more general condition, which follows from Lemma 6.2.7 by allowing for moves in a minimum neighborhood.

**Lemma 7.1.15** *Let* $\mathbf{y} = (y_1, \ldots, y_p)$ *and* $\mathbf{y}' = (y_1', \ldots, y_p')$ *and suppose there exists* $\delta > 0$ *for which* $\mathbf{y}, \mathbf{y}' \in \mathrm{supp}(g)$, $|\mathbf{y} - \mathbf{y}'| < \delta$ *and*

$$g_i(y_i|y_1, \ldots, y_{i-1}, y_{i+1}', \ldots, y_p') > 0, \quad i = 1, \ldots, p.$$

*If there exists* $\delta' < \delta$ *for which almost every pair* $(\mathbf{y}, \mathbf{y}') \in \mathrm{supp}(g)$ *can be connected by a finite sequence of balls with radius* $\delta'$ *having the (stationary) measure of the intersection of two consecutive balls positive, then the chain produced by* [A.32] *is irreducible and aperiodic.*

The laborious formulation of this condition is needed to accommodate settings where the support of $g$ is not connected, as in Example 7.1.10, since a necessary irreducibility condition is that two connected components of $\mathrm{supp}(g)$ can be linked by the kernel of [A.32]. The proof of Lemma 7.1.15 follows from arguments similar to those in the proof of Lemma 6.2.7.

**Corollary 7.1.16** *The conclusions of Theorem 7.1.14 hold if the Gibbs sampling Markov chain* $(Y^{(t)})$ *has conditional densities* $g_i(y_i|y_1, \ldots, y_{i-1}, y_{i+1}', \ldots, y_p')$ *that satisfy the assumptions of Lemma 7.1.15.*

### 7.1.4 Gibbs Sampling and Metropolis–Hastings

We now examine the exact relationship between the Gibbs sampling and Metropolis–Hastings algorithms by considering [A.32] as the composition of $p$ Markovian kernels. (As mentioned earlier, this representation is not sufficient to establish irreducibility since each of these separate Metropolis–Hastings algorithms does not produce an irreducible chain, given that it only modifies one component of $y$. In fact, these kernels are *never irreducible*, since they are constrained to subspaces of lower dimensions.)

**Theorem 7.1.17** *The Gibbs sampling method of* [A.32] *is equivalent to the composition of* $p$ *Metropolis–Hastings algorithms, with acceptance probabilities uniformly equal to* 1.

*Proof.* If we write [A.32] as the composition of $p$ "elementary" algorithms which correspond to the $p$ simulation steps from the conditional distribution, it is sufficient to show that each of these algorithms has an acceptance probability equal to 1. For $1 \leq i \leq p$, the instrumental distribution in step i. of [A.32] is given by

$$q_i(y'|y) = \delta_{(y_1, \ldots, y_{i-1}, y_{i+1}, y_p)}(y_1', \ldots, y_{i-1}', y_{i+1}', y_p') \\ \times g_i(y_i'|y_1, \ldots, y_{i-1}, y_{i+1}, y_p)$$

and the ratio defining the probability $\rho(y, y')$ is therefore

$$\frac{g(y') \, q_i(y|y')}{g(y) \, q_i(y'|y)} = \frac{g(y') \, g_i(y_i|y_1, \ldots, y_{i-1}, y_{i+1}, y_p)}{g(y) \, g_i(y_i'|y_1, \ldots, y_{i-1}, y_{i+1}, y_p)}$$

$$= \frac{g_i(y_i'|y_1, \ldots, y_{i-1}, y_{i+1}, y_p) \, g_i(y_i|y_1, \ldots, y_{i-1}, y_{i+1}, y_p)}{g_i(y_i|y_1, \ldots, y_{i-1}, y_{i+1}, y_p) \, g_i(y_i'|y_1, \ldots, y_{i-1}, y_{i+1}, y_p)}$$

$$= 1. \qquad \qquad \square\square$$

Note that Gibbs sampling is not the only MCMC algorithm to enjoy this property. As noted in §6.1, every kernel $q$ associated with a reversible Markov chain with invariant distribution $g$ has an acceptance probability uniformly equal to 1 (see also Barone and Frigessi 1989 and Liu and Sabbati 1999). However, the (global) acceptance probability for the vector $(y_1, \ldots, y_p)$ is usually different from 1 and a direct processing of [A.32] as a particular Metropolis–Hastings algorithm leads to a positive probability of rejection.

**Example 7.1.18 (Continuation of Example 7.1.2)** Consider the two-dimensional autoexponential model

$$g(y_1, y_2) \propto \exp\{-y_1 - y_2 - \theta_{12} y_1 y_2\} .$$

The kernel associated with [A.32] is then composed of the conditional densities

$$K(y, y') = g_1(y_1'|y_2) \, g_2(y_2'|y_1'),$$

with

$$g_1(y_1|y_2) = (1 + \theta_{12} y_2) \, \exp\left\{-(1 + \theta_{12} y_2) \, y_1\right\},$$
$$g_2(y_2|y_1) = (1 + \theta_{12} y_1) \, \exp\left\{-(1 + \theta_{12} y_1) \, y_2\right\}.$$

The ratio

$$\frac{g(y_1', y_2')}{g(y_1, y_2)} \, \frac{K((y_1', y_2'), (y_1, y_2))}{K((y_1, y_2), (y_1', y_2'))} = \frac{(1 + \theta_{12} y_2')(1 + \theta_{12} y_1)}{(1 + \theta_{12} y_2)(1 + \theta_{12} y_1')} \, e^{\theta_{12}(y_2' y_1 - y_1' y_2)}$$

is thus different from 1 for almost every vector $(y_1, y_2, y_1', y_2')$.      ‖

In this global analysis of [A.32], it is possible to reject some of the values produced by the sequence of Steps 1., . . ., **p.** of this algorithm. This version of the Metropolis–Hastings algorithm could then be compared with the original Gibbs sampler. No full-scale study has been yet undertaken in this direction, except for the modification introduced by Liu (1995,1998) and presented in §7.3. The fact that the first approach allows for rejections seems beneficial when considering the results of Gelman *et al.* (1996), but the Gibbs sampling algorithm cannot be evaluated in this way.

### 7.1.5 The Hammersley–Clifford Theorem

A most surprising feature of the Gibbs sampler is that the conditional distributions $g_i(y_i|y_{j \neq i})$ contain sufficient information to produce a sample from $g$. By comparison with optimization problems, this approach is like maximizing an objective function successively in every direction of a given basis. It is well known that this method does not necessarily lead to the global maximum, but may end up in a saddlepoint.

It is, therefore, somewhat remarkable that the full conditional distributions $g_i$ perfectly summarize the joint density $g(y)$, although the set of marginal distributions $g^{(i)}$ obviously fails to do so. Before we give a full treatment of the underlying phenomenon, we first consider the particular case $p = 2$, where [A.32] is implemented from two conditional distributions $g_1(y_1|y_2)$ and $g_2(y_2|y_1)$. The following result then shows that the joint density can be directly derived from the conditional densities.

**Theorem 7.1.19** *The joint distribution associated with the conditional densities $g_1$ and $g_2$ has the density*

$$g(y_1, y_2) = \frac{g_2(y_2|y_1)}{\int g_2(v|y_1)/g_1(y_1|v)\, dv}.$$

*Proof.* Recall that $g^{(1)}$ and $g^{(2)}$ denote the marginal densities of $Y_1$ and $Y_2$, respectively. We can write

$$g(y_1, y_2) = g_1(y_1|y_2)\, g^{(2)}(y_2) = g_2(y_2|y_1)\, g^{(1)}(y_1),$$

from which it follows that

$$g^{(2)}(y_2) = \frac{g_2(y_2|y_1)}{g_1(y_1|y_2)}\, g^{(1)}(y_1)$$

for every $y_1 \in \mathcal{Y}$. Since $g^{(2)}(y_2)$ is a density, $\int g^{(2)}(y_2)dy_2 = 1$ and, hence,

$$\int \frac{g_2(y_2|y_1)}{g_1(y_1|y_2)}\, dy_2 = \frac{1}{g^{(1)}(y_1)}$$

and the result follows.                                                   □□

This derivation of $g(y_1, y_2)$ obviously requires the existence and computation of the integral

$$\int \frac{g_2(v|y_1)}{g_1(y_1|v)}\, dv.$$

However, this result clearly demonstrates the fundamental feature that $g_1(y_1|y_2)$ and $g_2(y_2|y_1)$ are sufficiently informative to recover the joint density. Note, also, that this theorem and the remaining development here makes the implicit assumption that the joint density $g(y_1, y_2)$ exists. (See §7.4 for a discussion of what happens when this assumption is not satisfied.)

The general case of integer $p > 2$ follows from similar manipulations of the conditional distributions and leads to what is known as the *Hammersley–Clifford Theorem*[1] (Hammersley and Clifford 1970; see also Besag 1974 and Gelman and Speed 1993). To extend Theorem 7.1.19 to an arbitrary $p$, we need the condition of *positivity* (see Definition 7.1.11).

**Theorem 7.1.20 Hammersley–Clifford.** *Under the positivity condition, the joint distribution $g$ satisfies*

$$g(y_1, \ldots, y_p) \propto \prod_{j=1}^{p} \frac{g_{\ell_j}(y_{\ell_j}|y_{\ell_1}, \ldots, y_{\ell_{j-1}}, y'_{\ell_{j+1}}, \ldots, y'_{\ell_p})}{g_{\ell_j}(y'_{\ell_j}|y_{\ell_1}, \ldots, y_{\ell_{j-1}}, y'_{\ell_{j+1}}, \ldots, y'_{\ell_p})}$$

*for every permutation $\ell$ on $\{1, 2, \ldots, p\}$ and every $y' \in \mathcal{Y}$.*

---

[1] Clifford and Hammersley never published their result. Hammersley (1974) justifies their decision by citing the impossibility of extending this result to the nonpositive case, as shown by Moussouris (1974) through a counterexample.

*Proof.* For a given $y' \in \mathcal{Y}$,

$$
\begin{aligned}
g(y_1, \ldots, y_p) &= g_p(y_p | y_1, \ldots, y_{p-1}) g^p(y_1, \ldots, y_{p-1}) \\
&= \frac{g_p(y_p | y_1, \ldots, y_{p-1})}{g_p(y'_p | y_1 \ldots, y_{p-1})} \, g(y_1, \ldots, y_{p-1}, y'_p) \\
&= \frac{g_p(y_p | y_1, \ldots, y_{p-1})}{g_p(y'_p | y_1, \ldots, y_{p-1})} \, \frac{g_{p-1}(y_{p-1} | y_1, \ldots, y_{p-2}, y'_p)}{g_{p-1}(y'_{p-1} | y_1, \ldots, y_{p-2}, y'_p)} \\
&\quad \times g(y_1, \ldots, y'_{p-1}, y'_p) \, .
\end{aligned}
$$

A recursion argument then shows that

$$
g(y_1, \ldots, y_p) = \prod_{j=1}^{p} \frac{g_j(y_j | y_1, \ldots, y_{j-1}, y'_{j+1}, \ldots, y'_p)}{g_j(y'_j | y_1, \ldots, y_{j-1}, y'_{j+1}, \ldots, y'_p)} \, g(y'_1, \ldots, y'_p).
$$

The proof is identical for an arbitrary permutation $\ell$.  □□

The extension of Theorem 7.1.20 to the non-positive case is more delicate and requires additional assumptions, as shown by Example 7.1.10. Besag (1994) proposes a formal generalization which is not always relevant in the setup of Gibbs sampling algorithms. Hobert *et al.* (1997) modify Besag's (1994) condition to preserve the convergence properties of these algorithms and show, moreover, that the connectedness of the support of $g$ is essential for [A.32] to converge under every regular parameterization of the model. (See Problems 7.6 and 7.8.)

This result is also interesting from the general point of view of MCMC algorithms, since it shows that the density $g$ is known up to a multiplicative constant when the conditional densities $g_i(y_i | y_{j \neq i})$ are available. It is therefore possible to compare the Gibbs sampler with alternatives like Accept–Reject or Metropolis–Hastings algorithms. This also implies that the Gibbs sampler is never *the single available method* to simulate from $g$ and that it is always possible to include Metropolis–Hastings steps in a Gibbs sampling algorithm, following an hybrid strategy developed in §7.3.

### 7.1.6 Hierarchical Structures

To conclude this section, we investigate a structure for which Gibbs sampling is particularly well adapted, that of *hierarchical models*. These are structures in which the distribution $f$ can be decomposed as ($I \geq 1$)

$$
f(x) = \int f_1(x | z_1) f_2(z_1 | z_2) \cdots f_I(z_I | z_{I+1}) f_{I+1}(z_{I+1}) dz_1 \cdots dz_{I+1},
$$

either for structural or computational reasons. Such models naturally appear in the Bayesian analysis of complex models, where the diversity of the prior information or the variability of the observations may require the introduction of several levels of prior distributions (see Wakefield *et al.* 1994, Robert 1994a, Chapter 8, Bennett *et al.* 1996, Spiegelhalter *et al.* 1996, and Draper 1998). This is, for instance, the case of Example 7.1.23, where the first level represents the exchangeability of the parameters $\lambda_i$, and the second level represents the prior information. In the particular case where the prior information is sparse, hierarchical modeling is also useful since diffuse (or noninformative) distributions can be introduced at various levels of the hierarchy (see Problems 7.15 and 7.16).

The following two examples illustrate situations where hierarchical models are particularly useful (see also Problems 7.44–7.55).

**Example 7.1.21 Hierarchical models in animal epidemiology.** Research in animal epidemiology sometimes uses data from groups of animals, such as litters or herds. Such data may not follow some of the usual assumptions of independence, etc., and, as a result, variances of parameter estimates tend to be larger (this phenomenon is often referred to as "overdispersion"). Schukken *et al.* (1991) obtained counts of the number of cases of clinical mastitis[2] in 127 dairy cattle herds over a 1 year period.

If we assume that, in each herd, the occurrence of mastitis is a Bernoulli random variable, and if we let $X_i$, $i = 1, \ldots, m$, denote the number of cases in herd $i$, it is then reasonable to model $X_i \sim \mathcal{P}(\lambda_i)$, where $\lambda_i$ is the underlying rate of infection in herd $i$. However, there is lack of independence here (mastitis is infectious), which might manifest itself as overdispersion. To account for this, Schukken *et al.* (1991) put a gamma prior distribution on the Poisson parameter. A complete hierarchical specification is

$$X_i \sim \mathcal{P}(\lambda_i),$$
$$\lambda_i \sim \mathcal{G}a(\alpha, \beta_i),$$
$$\beta_i \sim \mathcal{G}a(a, b),$$

where $\alpha$, $a$, and $b$ are specified. The posterior density of $\lambda_i$, $\pi(\lambda_i|\mathbf{x}, \alpha)$, can now be obtained from the Gibbs sampler

$$\lambda_i \sim \pi(\lambda_i|\mathbf{x}, \alpha, \beta_i) = \mathcal{G}a(x_i + \alpha, 1 + \beta_i),$$
$$\beta_i \sim \pi(\beta_i|\mathbf{x}, \alpha, a, b, \lambda_i) = \mathcal{G}a(\alpha + a, \lambda_i + b).$$

See Eberly (1997) for more details.                    ∥

**Example 7.1.22 Hierarchical models in medicine.** Part of the concern of the study of *Pharmacokinetics* is the modeling of the relationship between the dosage of a drug and the resulting concentration in the blood. (More generally, Pharmacokinetics studies the different interactions of a drug and the body.) Gilks *et al.* (1993) introduce an approach for estimating pharmacokinetic parameters that uses the traditional mixed-effects model and nonlinear structure, but which is also robust to the outliers common to clinical trials. For a given dose $d_i$ administered at time 0 to patient $i$, the measured log concentration in the blood at time $t_{ij}$, $X_{ij}$, is assumed

---

[2] Mastitis is an inflammation usually caused by infection.

to follow a Student's $t$ distribution

$$\frac{X_{ij} - \log g_{ij}(\lambda_i)}{\sigma\sqrt{n/(n-2)}} \sim \mathcal{T}_n,$$

where $\lambda_i = (\log C_i, \log V_i)'$ is a vector of parameters for the $i$th individual, $\sigma^2$ is the measurement error variance, and $g_{ij}$ is given by

$$g_{ij}(\lambda_i) = \frac{d_i}{V_i} \exp\left(-\frac{C_i\, t_{ij}}{V_i}\right).$$

(The parameter $C_i$ represents *clearance* and $V_i$ represents *volume* for patient $i$.) Gilks *et al.* (1993) then complete the hierarchy by specifying an noninformative prior on $\sigma$, $\pi(\sigma) = 1/\sigma$ and

$$\lambda_i \sim \mathcal{N}(\boldsymbol{\theta}, \Sigma),$$

$$\boldsymbol{\theta} \sim \mathcal{N}(\tau_1, T_1), \quad \text{and} \quad \Sigma^{-1} \sim \mathcal{W}_2(\tau_2, T_2),$$

where the values of $\tau_1, T_1, \tau_2,$ and $T_2$ are specified. Conjugate structures can then be exhibited for most parameters by using the Dickey's decomposition of the Student's $t$ distribution (see Example 3.7.5); that is, by associating to each $x_{ij}$ an (artificial) variable $\omega_{ij}$ such that

$$X_{ij}|\omega_{ij} \sim \mathcal{N}\left(\log g_{ij}(\lambda_i), \omega_{ij}\sigma^2\frac{n}{n-2}\right).$$

Using this completion, the full conditional distributions on the $C_i$'s and $\theta$ are normal, while the full conditional distributions on $\sigma^2$ and $\Sigma$ are inverse gamma and inverse Wishart, respectively. The case of the $V_i$'s is more difficult to handle since the full conditional density is proportional to

$$(7.1.8)\exp\left(\frac{-1}{2}\left\{\sum_j \left(\log V_i + \frac{C_i t_{ij}}{V_i} - \mu_i\right)^2 /\xi + (\log V_i - \gamma_i)^2/\zeta\right\}\right),$$

where the hyperparameters $\mu_i, \xi, \gamma_i,$ and $\zeta$ depend on the other parameters and $x_{ij}$. Gilks *et al.* (1993) suggest using an Accept–Reject algorithm to simulate from (7.1.8), by removing the $C_i t_{ij}/V_i$ terms. Another possibility is to use a Metropolis–Hastings step, as described in §7.3.3.  ‖

For some cases of hierarchical models, it is possible to show that the associated Gibbs chains are *uniformly ergodic*. Typically, this can only be accomplished on a case-by-case study, as in the following example. (Note 7.6.3 studies the weaker property of *geometric convergence* of [A.32] in some detail, which requires conditions on the kernel that can be difficult to assess in practice.)

**Example 7.1.23 Nuclear pump failures.** Gaver and O'Muircheartaigh (1987) introduced a model that is frequently used (or even overused) in the Gibbs sampling literature to illustrate various properties (see, for instance, Gelfand and Smith 1990, Tanner 1996, or Guihenneuc–Jouyaux and Robert

| Pump | 1 | 2 | 3 | 4 | 5 | 6 | 7 | 8 | 9 | 10 |
|---|---|---|---|---|---|---|---|---|---|---|
| Failures | 5 | 1 | 5 | 14 | 3 | 19 | 1 | 1 | 4 | 22 |
| Time | 94.32 | 15.72 | 62.88 | 125.76 | 5.24 | 31.44 | 1.05 | 1.05 | 2.10 | 10.48 |

Table 7.1.2. *Numbers of failures and times of observation of 10 pumps in a nuclear plant. (Source: Gaver and O'Muircheartaigh 1987).*

1998). This model describes multiple failures of pumps in a nuclear plant, with the data given in Table 7.1.2.

The modeling is based on the assumption that the failures of the $i$th pump follow a Poisson process with parameter $\lambda_i$ $(1 \leq i \leq 10)$. For an observed time $t_i$, the number of failures $p_i$ is thus a Poisson $\mathcal{P}(\lambda_i t_i)$ random variable. The associated prior distributions are $(1 \leq i \leq 10)$

$$\lambda_i \overset{iid}{\sim} \mathcal{G}a(\alpha, \beta), \qquad \beta \sim \mathcal{G}a(\gamma, \delta),$$

with $\alpha = 1.8$, $\gamma = 0.01$, and $\delta = 1$ (see Gaver and O'Muircheartaigh 1987 for a motivation of these numerical values). The joint distribution is thus

$$\pi(\lambda_1, \ldots, \lambda_{10}, \beta | t_1, \ldots, t_{10}, p_1, \ldots, p_{10})$$

$$\propto \prod_{i=1}^{10} \left\{ (\lambda_i t_i)^{p_i} \, e^{-\lambda_i t_i} \, \lambda_i^{\alpha-1} e^{-\beta \lambda_i} \right\} \beta^{10\alpha} \beta^{\gamma-1} e^{-\delta\beta}$$

$$\propto \prod_{i=1}^{10} \left\{ \lambda_i^{p_i+\alpha-1} \, e^{-(t_i+\beta)\lambda_i} \right\} \beta^{10\alpha+\gamma-1} e^{-\delta\beta}$$

and a natural decomposition[3] of $\pi$ in conditional distributions is

$$\lambda_i | \beta, t_i, p_i \sim \mathcal{G}a(p_i + \alpha, t_i + \beta) \qquad (1 \leq i \leq 10),$$

$$\beta | \lambda_1, \ldots, \lambda_{10} \sim \mathcal{G}a\left(\gamma + 10\alpha, \delta + \sum_{i=1}^{10} \lambda_i\right).$$

The transition kernel on $\beta$ associated with [A.32] and this decomposition satisfies (see Problem 7.5)

$$K(\beta, \beta') = \int \frac{(\beta')^{\gamma+10\alpha-1}}{\Gamma(10\alpha + \gamma)} \left(\delta + \sum_{i=1}^{10} \lambda_i\right)^{\gamma+10\alpha} \exp\left\{-\beta'\left(\delta + \sum_{i=1}^{10} \lambda_i\right)\right\}$$

$$\times \prod_{i=1}^{10} \frac{(t_i + \beta)^{p_i+\alpha}}{\Gamma(p_i + \alpha)} \lambda_i^{p_i+\alpha-1} \exp\{-(t_i + \beta)\lambda_i\} \, d\lambda_1 \ldots d\lambda_{10}$$

$$(7.1.9) \quad \geq \frac{\delta^{\gamma+10\alpha}(\beta')^{\gamma+10\alpha-1}}{\Gamma(10\alpha + \gamma)} e^{-\delta\beta'} \prod_{i=1}^{10} \left(\frac{t_i}{t_i + \beta'}\right)^{p_i+\alpha}.$$

This minorization by a positive quantity which does not depend on $\beta$ im-

---

[3] This decomposition reflects the hierarchical structure of the model.

plies that the entire space ($\mathbb{R}_+$) is a small set for the transition kernel; thus, by Theorem 4.6.13, the chain $(\beta^t)$ is uniformly ergodic (see Definition 4.6.12).

We will show in §7.2.4 that uniform ergodicity directly extends to the dual chain $\lambda^t = (\lambda_1^t, \dots, \lambda_{10}^t)$.    ‖

Note that Examples 7.1.21 and 7.1.23 share the property that they generate two Markov chains, $(X^{(t)})$ and $(Z^{(t)})$, such that $X^{t+1}$ is generated conditionally on $z^t$ and $Z^{t+1}$ is generated conditionally on $x^{t+1}$. This corresponds to the technique of *Data Augmentation*, described in §7.2. This particular case of Gibbs sampler simplifies the study of the probabilistic properties of the algorithm since each chain can be examined separately by integrating out the other chain. This interesting feature disappears for $p \geq 3$, because the subchains $(Y_1^{(t)}), \dots, (Y_p^{(t)})$ are not Markov chains, although the vector $(\mathbf{Y}^{(t)})$ is. Therefore, there is no transition kernel associated with $(Y_i^{(t)})$, and the study of uniform ergodicity can only cover a grouped vector of $(p-1)$ of the $p$ components of $(\mathbf{Y}^{(t)})$ since the original kernel cannot, in general, be bounded uniformly from below. If we denote $\mathbf{z}_1 = (y_2, \dots, y_p)$, the transition from $\mathbf{z}_1$ to $\mathbf{z}_1'$ has the following kernel:

$$K_1(\mathbf{z}_1, \mathbf{z}_1') = \int_{\mathcal{Y}_1} g_1(y_1'|\mathbf{z}_1) g_2(y_2'|y_1', y_3, \dots, y_p) \cdots$$

(7.1.10)
$$g_p(y_p'|y_1', y_2', \dots, y_{p-1}') \, dy_1' \ .$$

While some setups result in a uniform bound on $K_1(\mathbf{z}_1, \mathbf{z}_1')$, it is often impossible to achieve a uniform minorization of $K$. For example, it is impossible to bound the transition kernel of the autoexponential model (Example 7.1.2) from below (see Problem 7.3).

## 7.2 The Two-Stage Gibbs Sampler

A recurrent problem in exploring the probabilistic properties of the Gibbs sampler is that this algorithm does not lend itself to componentwise study, but rather must be treated in a *global* fashion. As mentioned in §7.1, a componentwise study of [A.32] is rarely interesting since the corresponding Metropolis–Hastings algorithms are not irreducible and the associated sequences $(Y_i^{(t)})$ are not Markov chains. In the particular case $p = 2$, however, this decomposition into (two) steps enables us to more thoroughly evaluate the properties of the Gibbs sampler.

### 7.2.1 Dual Probability Structures

The *Data Augmentation method* corresponds to the particular case of [A.32], where $f$ is completed in $g$ and $x$ in $y = (y_1, y_2)$ such that both conditional distributions $g_1(y_1|y_2)$ and $g_2(y_2|y_1)$ are available. (Again, note that both $y_1$ and $y_2$ can be either scalars or vectors.)

**Algorithm A.34 –Data Augmentation–**
Given $y^{(t)}$,

1. Simulate $Y_1^{(t+1)} \sim g_1(y_1|y_2^{(t)})$ ;                    [A.34]
2. Simulate $Y_2^{(t+1)} \sim g_2(y_2|y_1^{(t+1)})$ .

This method was introduced independently (that is, unrelated to Gibbs sampling) by Tanner and Wong (1987), and is, perhaps, more closely related to the EM algorithm of Dempster *et al.* (1977) and methods of stochastic restoration (see Note 5.5.1), in the setup of *missing data structures*, when the distribution to be simulated can be written as the marginal distribution

$$\pi(\theta|x) \propto \int f(x,z|\theta)\,\pi(\theta)\,dz.$$

The range of application of this method is not just restricted to missing data problems, as shown by Examples 7.1.6 and 7.1.23. (Recall that *strong irreducibility* is introduced in Definition 4.3.1.)

**Lemma 7.2.1** *Each of the sequences $(Y_1^{(t)})$ and $(Y_2^{(t)})$ produced by [A.34] is a Markov chain with corresponding stationary distributions*

$$g^{(2)}(y_1) = \int g(y_1,y_2)\,dy_2 \quad and \quad g^{(1)}(y_2) = \int g(y_1,y_2)\,dy_1 \; .$$

*If the positivity constraint on $g$ holds, both chains are strongly irreducible.*

*Proof.* Note that $(Y_1^{(t)})$ is associated with the transition kernel

$$K_1(y_1,y_1') = \int g_1(y_1'|y_2)\,g_2(y_2|y_1)\,dy_2 \; .$$

Therefore, $Y_1^{(t)}$ is simulated conditionally only on $y_1^{(t-1)}$ and is independent from $Y_1^{(t-2)}, Y_1^{(t-3)}, \ldots$ conditionally on $y_1^{(t-1)}$. The chain $(Y_1^{(t)})$ is, thus, a Markov chain.

Under the positivity constraint, if $g$ is positive, $g_1(y_1'|y_2)$ is positive on the (projected) support of $g$ and every Borel set of $\mathcal{Y}_1$ can be visited in a single iteration of [A.34], establishing the strong irreducibility.

The previous development also applies to $(Y_2^{(t)})$.                    □□

This elementary reasoning shows, in addition, that if only the chain $(Y_1^{(t)})$ is of interest and if the condition $g_1(y_1'|y_2) > 0$ holds for every couple $(Y_1', Y_2)$, irreducibility is satisfied. As shown in §7.2.4, the "dual" chain $(Y_2^{(t)})$ can be used to establish some probabilistic properties of $(Y_1^{(t)})$.

**Example 7.2.2 Grouped counting data.** For 360 consecutive units, consider recording the number of passages of individuals, per unit time, past some sensor. This can be, for instance, the number of cars observed at a crossroad or the number of leucocytes in a region of a blood sample. Hypothetical results are given in Table 7.2.1.

This table, therefore, involves a grouping of the observations with four passages and more. If we assume that every observation is a Poisson $\mathcal{P}(\lambda)$

| Number of passages | 0 | 1 | 2 | 3 | 4 or more |
|---|---|---|---|---|---|
| Number of observations | 139 | 128 | 55 | 25 | 13 |

Table 7.2.1. *Frequencies of passage for 360 consecutive observations.*

random variable, the likelihood of the model corresponding to Table 7.2.1 is

$$\ell(\lambda|x_1,\ldots,x_5) \propto e^{-347\lambda} \lambda^{128+55\times2+25\times3} \left(1 - e^{-\lambda} \sum_{i=0}^{3} \frac{\lambda^i}{i!}\right)^{13},$$

for $x_1 = 139,\ldots,x_5 = 13$. An analytical maximization of this likelihood could therefore be delicate, even though the simple form of this likelihood allows for an easy derivation of the numerical value of the maximum likelihood estimate, namely $\hat{\lambda} = 1.0224$.

For $\pi(\lambda) = 1/\lambda$ and $(y_1,\ldots,y_{13})$, vector of the 13 units larger than 4, it is possible to complete the posterior distribution $\pi(\lambda|x_1,\ldots,x_5)$ in $\pi(\lambda, y_1,\ldots,y_{13}|x_1,\ldots,x_5)$ and to propose the associated Gibbs sampling algorithm,

**Algorithm A.35 –Poisson–Gamma Gibbs Sampler–**
Given $\lambda^{(t-1)}$,

1. Simulate $Y_i^{(t)} \sim \mathcal{P}(\lambda^{(t-1)}) \, \mathbb{I}_{y \geq 4} \ (i = 1,\ldots,13)$

2. Simulate                                          [A.35]

$$\lambda^{(t)} \sim \mathcal{G}a\left(313 + \sum_{i=1}^{13} y_i^{(t)}, \ 360\right).$$

Figure 7.2.1 describes the convergence of the so-called Rao–Blackwellized estimator (see §7.2.3 for motivation)

$$\delta_{rb} = \frac{1}{T} \sum_{t=1}^{T} \mathbb{E}\left[\lambda|x_1,\ldots,x_5,y_1^{(t)},\ldots,y_{13}^{(t)}\right]$$

$$= \frac{1}{360T} \sum_{t=1}^{T} \left(313 + \sum_{i=1}^{13} y_i^{(t)}\right),$$

based on the finite state-space chain $(y^{(t)})$, to $\delta^\pi = 1.0224$, the Bayes estimator of $\lambda$, along with an histogram of the $\lambda^{(t)}$'s simulated from [A.35]. The convergence to the Bayes estimator is particularly fast in this case. ‖

**Example 7.2.3 Grouped multinomial data.** Tanner and Wong (1987) consider the multinomial model

$$X \sim \mathcal{M}_5\left(n; a_1\mu + b_1, a_2\mu + b_2, a_3\eta + b_3, a_4\eta + b_4, c(1 - \mu - \eta)\right),$$

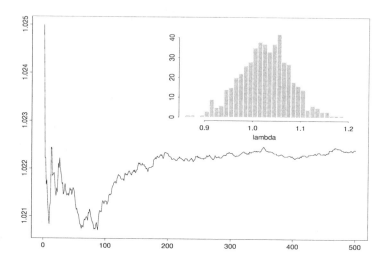

Figure 7.2.1. *Evolution of the estimator $\delta_{rb}$ against the number of iterations of* [A.35] *and (insert) histogram of the sample of* $\lambda^{(t)}$*'s for* 500 *iterations.*

with

$$0 \leq a_1 + a_2 = a_3 + a_4 = 1 - \sum_{i=1}^{4} b_i = c \leq 1,$$

where the $a_i, b_i \geq 0$ are known, based on genetic considerations. This model is equivalent to a sampling from

$$Y \sim \mathcal{M}_9\left(n; a_1\mu, b_1, a_2\mu, b_2, a_3\eta, b_3, a_4\eta, b_4, c(1 - \mu - \eta)\right),$$

where some observations are aggregated,

$$X_1 = Y_1 + Y_2, \ X_2 = Y_3 + Y_4, \ X_3 = Y_5 + Y_6, \ X_4 = Y_7 + Y_8, \ X_5 = Y_9.$$

A natural prior distribution on $(\mu, \eta)$ is the Dirichlet prior $\mathcal{D}(\alpha_1, \alpha_2, \alpha_3)$,

$$\pi(\mu, \eta) \propto \mu^{\alpha_1 - 1} \eta^{\alpha_2 - 1} (1 - \eta - \mu)^{\alpha_3 - 1},$$

where $\alpha_1 = \alpha_2 = \alpha_3 = 1/2$ corresponds to the noninformative case. The posterior distributions $\pi(\mu|x)$ and $\pi(\eta|x)$ are not readily available, as shown in Problem 7.25. If we define $Z = (Z_1, Z_2, Z_3, Z_4) = (Y_1, Y_3, Y_5, Y_7)$, the completed posterior distribution can be defined as

$$\begin{aligned}
\pi(\eta, \mu | x, z) &= \pi(\eta, \mu | y) \\
&\propto \mu^{\alpha_1 - 1} \eta^{\alpha_2 - 1} (1 - \eta - \mu)^{\alpha_3 - 1} \mu^{z_1} \mu^{z_2} \eta^{z_3} \eta^{z_4} (1 - \eta - \mu)^{x_5} \\
&= \mu^{z_1 + z_2 + \alpha_1 - 1} \eta^{z_3 + z_4 + \alpha_2 - 1} (1 - \eta - \mu)^{x_5 + \alpha_3 - 1}.
\end{aligned}$$

Thus,

$$(\mu, \eta, 1 - \mu - \eta) | x, z \sim \mathcal{D}(z_1 + z_2 + \alpha_1, z_3 + z_4 + \alpha_2, x_5 + \alpha_3).$$

Moreover,

$$Z_i|x, \mu, \eta \sim \mathcal{B}\left(x_i, \frac{a_i\mu}{a_i\mu + b_i}\right) \qquad (i = 1, 2),$$

$$Z_i|x, \mu, \eta \sim \mathcal{B}\left(x_i, \frac{a_i\eta}{a_i\eta + b_i}\right) \qquad (i = 3, 4).$$

Therefore, if $\theta = (\mu, \eta)$, this completion provides manageable conditional distributions $g_1(\theta|y)$ and $g_2(z|x, \theta)$. ‖

**Example 7.2.4 Capture–recapture uniform model.** In the setup of *capture–recapture models* seen in Examples 2.3.11 and 5.3.8, the simplest case corresponds to the setting where the size $N$ of the entire population is unknown and each individual has a probability $p$ of being captured in every capture experiment, whatever its past history. For two successive captures, a sufficient statistic is the triplet $(n_{11}, n_{10}, n_{01})$, where $n_{11}$ is the number of individuals which have been captured twice, $n_{10}$ the number of individuals captured only in the first experiment, and $n_{01}$ the number of individuals captured only in the second experiment. Writing $n = n_{10} + n_{01} + 2n_{11}$, the total number of captures, the likelihood of this model is given by

$$\ell(N, p|n_{11}, n_{10}, n_{01}) \propto \begin{pmatrix} N \\ n_{11} \quad n_{10} \quad n_{01} \end{pmatrix} p^n(1 - p)^{N-n}.$$

Castledine (1981) calls this a *uniform* likelihood (see also Wolter 1986 or George and Robert 1992).

It is easy to see that the likelihood function factors through $n$ and $n' = n_{10} + n_{11} + n_{01}$, the total number of different individuals captured in the experiments. If $\pi(p, N)$ corresponds to a Poisson distribution $\mathcal{P}(\lambda)$ on $N$ and a uniform distribution $\mathcal{U}_{[0,1]}$ on $p$,

$$\pi(p, N|n, n') \propto \frac{N!}{(N - n')!} p^n(1 - p)^{N-n} \frac{e^{-\lambda}\lambda^N}{N!}$$

implies that

$$(N - n')|p, n, n' \sim \mathcal{P}(\lambda),$$
$$p|N, n, n' \sim \mathcal{B}e(n + 1, N - n + 1);$$

and therefore that Data Augmentation is available in this setup (even though direct computation is possible, see Problem 7.26). ‖

**Example 7.2.5 Capture–Recapture models revisited again.** The estimation strategy that led to Monte Carlo EM estimates in Example 5.3.8 can be modified to yield a Gibbs sampler. Recall that in this extension of the uniform model, animal $i$, $i = 1, 2, \ldots, n$ may be captured at time $j$, $j = 1, 2, \ldots, t$, in one of $m$ locations, where the location is a multinomial random variable $H_{ij} \sim \mathcal{M}_m(\theta_1, \ldots, \theta_m)$. Given $H_{ij} = k$ ($k = 1, 2, \ldots, m$), the animal is captured with probability $p_k$, represented by the random variable $X \sim \mathcal{B}(p_k)$. Under the conjugate priors

$$(\theta_1, \ldots, \theta_m) \sim \mathcal{D}(\lambda_1, \ldots, \lambda_m) \quad \text{and} \quad p_k \sim \mathcal{B}e(\alpha, \beta),$$

the Gibbs sampler is straightforward to implement, as all of the full conditionals are conjugate. See Problem 7.27 for details.                    ‖

### 7.2.2 Reversible and Interleaving Chains

The two-stage Gibbs sampler or, equivalently, Data Augmentation, was shown by Liu *et al.* (1994) to have a very strong structural property. The two chains, the *chain of interest*, $(X^{(t)})$, and the *instrumental* (or *dual*) *chain*, $(Y^{(t)})$, satisfy a duality property they call *interleaving*. This property is characteristic of Data Augmentation setups, and the only Gibbs sampling chains satisfying this property are those associated with a Data Augmentation-type kernel.

**Definition 7.2.6** Two Markov chains $(X^{(t)})$ and $(Y^{(t)})$ are said to be *conjugate to each other with the interleaving property* (or *interleaved*) if

(i) $X^{(t)}$ and $X^{(t+1)}$ are independent conditionally on $Y^{(t)}$;

(ii) $Y^{(t-1)}$ and $Y^{(t)}$ are independent conditionally on $X^{(t)}$;

(iii) $(X^{(t)}, Y^{(t-1)})$ and $(X^{(t)}, Y^{(t)})$ are identically distributed under stationarity.

In most cases where we have interleaved chains, there is more interest in one of the chains. That the property of interleaving is always satisfied by the method of Data Augmentation is immediate, as shown below. Note that the (global) chain $(X^{(t)}, Y^{(t)})$ is not necessarily (time-)reversible.

**Lemma 7.2.7** *Each of the chains* $(X^{(t)})$ *and* $(Y^{(t)})$ *generated by a Data Augmentation algorithm is reversible, and the chain* $(X^{(t)}, Y^{(t)})$ *satisfies the interleaving property.*

*Proof.* We first establish reversibility for each chain, a property that is independent of interleaving: It follows from the detailed balance condition (Theorem 6.2.2). Consider $(X_0, Y_0)$ distributed from the stationary distribution $g$, with respective marginal distributions $g^{(2)}(x)$ and $g^{(1)}(y)$. Then, if

$$K_1(x_0, x_1) = \int g_2(y_0|x_0) g_1(x_1|y_0) dy_0$$

denotes the transition kernel of $(X^{(t)})$,

$$g^{(2)}(x_0) K_1(x_0, x_1) = g^{(2)}(x_0) \int g_2(y_0|x_0) g_1(x_1|y_0) dy_0$$

$$= \int g(x_0, y_0) g_1(x_1|y_0) dy_0$$

$$[g^{(2)}(x_0) = \int g(x_0, y_0) dy_0]$$

$$= \int g(x_0, y_0) \frac{g_2(y_0|x_1) g^{(2)}(x_1)}{g^{(1)}(y_0)} dy_0$$

$$\left[ g_1(x_1|y_0) = \frac{g_2(y_0|x_1)g^{(2)}(x_1)}{g^{(1)}(y_0)} \right]$$

$$= g^{(2)}(x_1) K_1(x_1, x_0),$$

where the last equality follows by integration. Thus, the reversibility of the $(X^{(t)})$ chain is established, and a similar argument applies for the reversibility of the $(Y^{(t)})$ chain; that is, if $K_2$ denotes the associated kernel, the detailed balance condition

$$g^{(1)}(y_0) K_2(y_0, y_1) = g^{(1)}(y_1) K_2(y_1, y_0)$$

is satisfied.

Turning to the interleaving property of Definition 7.2.6, the construction of each chain establishes (i) and (ii) directly. To see that property (iii) is satisfied, we note that the joint cdf of $X_0$ and $Y_0$ is

$$P(X_0 \le x, Y_0 \le y) = \int_{-\infty}^{x} \int_{-\infty}^{y} g(v, u)dudv$$

and the joint cdf of $X_1$ and $Y_0$ is

$$P(X_1 \le x, Y_0 \le y) = \int_{-\infty}^{x} \int_{-\infty}^{y} g_1(v|u)g^{(1)}(u)dudv.$$

Since

$$g_1(v|u)g^{(1)}(u) = \int g_1(v|u)g(v', u)dv'$$

$$= \int g_1(v|u)g_1(v'|u)g^{(1)}(u)dv'$$

$$= g(v, u),$$

the result follows, and $(X^{(t)})$ and $(Y^{(t)})$ are interleaved chains.  □□

To ensure that the entire chain $(X^{(t)}, Y^{(t)})$ is reversible, an additional step is necessary in $[A.34]$.

**Algorithm A.36 –Reversible Data Augmentation–**
   Given $y_2^{(t)}$,

1. Simulate $W \sim g_1(w|y_2^{(t)})$;
2. Simulate $Y_2^{(t+1)} \sim g_2(y_2|w)$;                    [A.36]
3. Simulate $Y_1^{(t+1)} \sim g_1(y_1|y_2^{(t+1)})$.

In this case,

$$(Y_1, Y_2, Y_1', Y_2') \sim g(y_1, y_2) \int g_1(w|y_2) \, g_2(y_2'|w) \, dw \, g_1(y_1'|y_2')$$

$$= g(y_1', y_2') \int \frac{g_2(y_2'|w)}{g^1(y_2')} \, g(w, y_2) \, dw \, g_1(y_1|y_2)$$

$$= g(y_1', y_2') \int g_1(w|y_2') \, g_2(y_2|w) \, dw \, g_1(y_1|y_2)$$

is distributed as $(Y_1', Y_2', Y_1, Y_2)$.

The Gibbs sampler of [A.32] has also been called the Gibbs sampler with *systematic scan* (or systematic sweep), as the path of iteration is to proceed systematically in one direction. Such a sampler results in a non-reversible Markov chain, but we can construct a reversible Gibbs sampler with *symmetric scan*. The following algorithm guarantees the reversibility of the chain $(Y^{(t)})$, as easily seen through a generalization of the previous proof.

**Algorithm A.37 –Reversible Gibbs Sampler–**

   Given $(y_2^{(t)}, \ldots, y_p^{(t)})$, generate

1. $Y_1^* \sim g_1(y_1|y_2^{(t)}, \ldots, y_p^{(t)})$

2. $Y_2^* \sim g_2(y_2|y_1^*, y_3^{(t)}, \ldots, y_p^{(t)})$

$$\vdots$$

p-1. $Y_{p-1}^* \sim g_{p-1}(y_{p-1}|y_1^*, \ldots, y_{p-2}^*, y_p^{(t)})$

p. $Y_p^{(t+1)} \sim g_p(y_p|y_1^*, \ldots, y_{p-1}^*)$                                     [A.37]

p+1. $Y_{p-1}^{(t+1)} \sim g_{p-1}(y_{p-1}|y_1^*, \ldots, y_{p-2}^*, y_p^{(t+1)})$

$$\vdots$$

2p-1. $Y_1^{(t+1)} \sim g_1(y_1|y_2^{(t+1)}, \ldots, y_p^{(t+1)})$

An alternative to [A.37] has been proposed by Liu *et al.* (1995) and is called Gibbs sampling with *random scan*, as the simulation of the components of $y$ is done in a random order following each transition. For the setup of [A.32], the modification [A.38] produces a reversible chain with stationary distribution $g$, and every simulated value is used.

**Algorithm A.38 –Random Sweep Gibbs Sampler–**

1. Generate a permutation $\sigma \in \mathcal{G}_p$;

2. Simulate $Y_{\sigma_1}^{(t+1)} \sim g_{\sigma_1}(y_{\sigma_1}|y_j^{(t)}, j \neq \sigma_1)$;                              [A.38]

$$\vdots$$

p+1. Simulate $Y_{\sigma_p}^{(t+1)} \sim g_{\sigma_p}(y_{\sigma_p}|y_j^{(t+1)}, j \neq \sigma_p)$.

This algorithm improves upon [A.37], which only uses one simulation out of two. (Recall that, following Theorem 4.7.7, the reversibility of $(Y^{(t)})$ allows application of the Central Limit Theorem.)

### 7.2.3 Monotone Covariance and Rao–Blackwellization

As shown in §3.7.3 and §6.4.2, conditioning on a subset of the simulated variables may produce considerable improvement upon the standard empirical estimator in terms of variance, by a simple "recycling" of the rejected variables (see also §3.3.3). Gibbs sampling does not permit this kind of recycling since every simulated value is accepted (Theorem 7.1.17). Nonetheless, Gelfand and Smith (1990) propose a type of conditioning christened *Rao–Blackwellization* in connection with the famous Rao–Blackwell Theorem (see Lehmann and Casella 1998, Section 1.7) and defined as *parametric* Rao–Blackwellization by Casella and Robert (1996) to differentiate from the form studied in §3.7.3 and §6.4.2.

*Rao–Blackwellization* is based on the identity

$$g_1(y_1) = \int g_1(y_1|y_2,\ldots,y_p)\, g^1(y_2,\ldots,y_p)\, dy_2\cdots dy_p,$$

and it replaces

$$\delta_0 = \frac{1}{T}\sum_{t=1}^{T} h\left(y_1^{(t)}\right)$$

with

$$\delta_{rb} = \frac{1}{T}\sum_{t=1}^{T} \mathbb{E}\left[h(Y_1)|y_2^{(t)},\ldots,y_p^{(t)}\right].$$

Both estimators converge to $\mathbb{E}[h(Y_1)]$ and, under the stationary distribution, they are unbiased. A simple application of the identity $\mathrm{var}(U) = \mathrm{var}(\mathbb{E}[U|V]) + \mathbb{E}[\mathrm{var}(U|V)]$ implies that

$$(7.2.1)\qquad \mathrm{var}\left(\mathbb{E}\left[h(Y_1)|Y_2^{(t)},\ldots,Y_p^{(t)}\right]\right) \leq \mathrm{var}(h(Y_1)).$$

This led Gelfand and Smith (1990) to suggest the use of $\delta_{rb}$ instead of $\delta_0$. However, inequality (7.2.1) is insufficient to conclude on the domination of $\delta_{rb}$ when compared with $\delta_0$, as it fails to take into account the correlation between the $Y^{(t)}$'s. The domination of $\delta_0$ by $\delta_{rb}$ can therefore be established only in a few cases; Liu *et al.* (1994) show in particular that it holds for Data Augmentation. (See also Geyer 1995 for necessary conditions.)

We establish the domination result in Theorem 7.2.10. We first need some preliminary results, beginning with a representation lemma yielding the interesting result that covariances are positive in an interleaved chain.

**Lemma 7.2.8** *If* $h \in \mathcal{L}_2(g_2)$ *and if* $(X^{(t)})$ *is interleaved with* $(Y^{(t)})$, *then*

$$\mathrm{cov}\left(h(Y^{(1)}), h(Y^{(2)})\right) = \mathrm{var}(\mathbb{E}[h(Y)|X]).$$

*Proof.* Assuming that $\mathbb{E}_{g_2}[h(\theta)] = 0$,

$$\mathrm{cov}\left(h(Y^{(1)}), h(Y^{(2)})\right) = \mathbb{E}\left[h(Y^{(1)})h(Y^{(2)})\right]$$
$$= \mathbb{E}\left\{\mathbb{E}\left[h(Y^{(1)})|X^{(2)}\right]\mathbb{E}\left[h(Y^{(2)})|X^{(2)}\right]\right\}$$
$$= \mathbb{E}\left[\mathbb{E}\left[h(Y^{(1)})|X^{(2)}\right]^2\right] = \mathrm{var}(\mathbb{E}[h(Y)|X]),$$

where the second equality follows from iterating the expectation and using the conditional independence of the interleaving property. The last equality uses reversibility (that is, condition (iii)) of the interleaved chains. □□

**Proposition 7.2.9** *If* $(Y^{(t)})$ *is a Markov chain with the interleaving property, the covariances*

$$\text{cov}(h(Y^{(1)}), h(Y^{(t)}))$$

*are positive and decreasing in t for every* $h \in \mathcal{L}_2(g_2)$.

*Proof.* Lemma 7.2.8 implies, by induction, that

$$\text{cov}\left(h(Y^{(1)}), h(Y^{(t)})\right) = \mathbb{E}\left[\mathbb{E}[h(Y)|X^{(2)}]\,\mathbb{E}[h(Y)|X^{(t)}]\right]$$

$$(7.2.2) \hspace{2cm} = \text{var}(\mathbb{E}[\cdots \mathbb{E}[\mathbb{E}[h(Y)|X]|Y]\cdots]),$$

where the last term involves $(t-1)$ conditional expectations, alternatively in $Y$ and in $X$. The decrease in $t$ directly follows from the inequality on conditional expectations, by virtue of the representation (7.2.2) and the inequality (7.2.1). □□

The result on the improvement brought by Rao–Blackwellization then easily follows from Proposition 7.2.9.

**Theorem 7.2.10** *If* $(X^{(t)})$ *and* $(Y^{(t)})$ *are two interleaved Markov chains, with stationary distributions* $g_1$ *and* $g_2$, *respectively, the estimator* $\delta_{rb}$ *dominates the estimator* $\delta_0$ *for every function* $h \in \mathcal{L}_2(g_i)$ $(i = 1, 2)$.

*Proof.* Assuming, again, $\mathbb{E}_{g_2}[h(X)] = 0$, and introducing the estimators

$$(7.2.3) \hspace{1cm} \delta_0 = \frac{1}{T}\sum_{t=1}^{T} h(X^{(t)}), \hspace{1cm} \delta_{rb} = \frac{1}{T}\sum_{t=1}^{T} \mathbb{E}[h(X)|Y^{(t)}],$$

it follows that

$$\text{var}(\delta_0) = \frac{1}{T^2}\sum_{t,t'} \text{cov}\left(h(X^{(t)}), h(X^{(t')})\right)$$

$$(7.2.4) \hspace{1.5cm} = \frac{1}{T^2}\sum_{t,t'} \text{var}(\mathbb{E}[\cdots \mathbb{E}[h(X)|Y]\cdots])$$

and

$$\text{var}(\delta_{rb}) = \frac{1}{T^2}\sum_{t,t'} \text{cov}\left(\mathbb{E}[h(X)|Y^{(t)}], \mathbb{E}[h(X)|Y^{(t')}]\right)$$

$$(7.2.5) \hspace{1.5cm} = \frac{1}{T^2}\sum_{t,t'} \text{var}(\mathbb{E}[\cdots \mathbb{E}[\mathbb{E}[h(X)|Y]|X]\cdots]),$$

according to the proof of Proposition 7.2.9, with $|t - t'|$ conditional expectations in the general term of (7.2.4) and $|t - t'| + 1$ in the general term of (7.2.5). It is then sufficient to compare $\text{var}(\delta_0)$ with $\text{var}(\delta_{rb})$ term by term to conclude that $\text{var}(\delta_0) \geq \text{var}(\delta_{rb})$. □□

One might question whether Rao–Blackwellization will always result in an appreciable variance reduction, even as the sample size (or the number of Monte Carlo iterations) increases. This point was addressed by Levine (1996), who formulated this problem in terms of the *asymptotic relative efficiency* (ARE) of $\delta_0$ with respect to its Rao–Blackwellized version $\delta_{rb}$, given in (7.2.3), where the pairs $(X^{(t)}, Y^{(t)})$ are generated from a bivariate Gibbs sampler. The ARE is a ratio of the variances of the limiting distributions for the two estimators, which are given by

$$(7.2.6) \qquad \sigma_{\delta_0}^2 = \text{var}(h(X^{(0)})) + 2 \sum_{k=1}^{\infty} \text{cov}(h(X^{(0)}), h(X^{(k)}))$$

and

$$\sigma_{\delta_{rb}}^2 = \text{var}(\mathbb{E}[h(X)|Y])$$

$$(7.2.7) \qquad\qquad + 2 \sum_{k=1}^{\infty} \text{cov}(\mathbb{E}[h(X^{(0)})|Y^{(0)}], \mathbb{E}[h(X^{(k)})|Y^{(k)}]).$$

Levine (1996) established that the ratio $\sigma_{\delta_0}^2 / \sigma_{\delta_1}^2 \geq 1$, with equality if and only if $\text{var}(h(X)) = \text{cov}(\mathbb{E}[h(X)|Y]) = 0$.

**Example 7.2.11 Improvement from Rao–Blackwellization.** Consider the case

$$(X, Y)' \sim \mathcal{N}\left( \begin{pmatrix} 0 \\ 0 \end{pmatrix}, \begin{pmatrix} 1 & \rho \\ \rho & 1 \end{pmatrix} \right),$$

where $-1 < \rho < 1$ and interest lies in estimating $\mu = \mathbb{E}(X)$. The Gibbs sampler alternately draws random variables from

$$X \mid y \sim \mathcal{N}(\rho y, \, 1 - \rho^2),$$
$$Y \mid x \sim \mathcal{N}(\rho x, \, 1 - \rho^2),$$

and it can be shown (Problem 7.17) that $\text{cov}(X^{(0)}, X^{(k)}) = \rho^{2k}$, for all $k$, and

$$\sigma_{\delta_0}^2 / \sigma_{\delta_1}^2 = \frac{1}{\rho^2} > 1.$$

So, if $\rho$ is small the amount of improvement, which is independent of the number of iterations, can be substantial.    ‖

Liu *et al.* (1995) are able to extend Proposition 7.2.9 to the case of the *random Gibbs sampler*, where every step only updates a single component of $y$. In the setup of [A.32], define a multinomial distribution $\sigma = (\sigma_1, \ldots, \sigma_p)$.

**Algorithm A.39 –Random Gibbs Sampler–**
    Given $y^{(t)}$,

1. Select a component $\nu \sim \sigma$.                                     [A.39]

2. Generate $Y_\nu^{(t+1)} \sim g_\nu(y_\nu | y_j^{(t)}, j \neq \nu)$ and take

$$y_j^{(t+1)} = y_j^{(t)} \qquad \text{for} \qquad j \neq \nu.$$

Note that, although [A.39] only generates one component of $y$ at each iteration, the resulting chain is strongly irreducible (Definition 4.3.1) because of the random choice of $\nu$. It also satisfies the following property:

**Proposition 7.2.12** *The chain* $(Y^{(t)})$ *generated by* [A.39] *has the property that for every function* $h \in \mathcal{L}_2(g)$, *the covariance* $\mathrm{cov}(h(Y^{(0)}), h(Y^{(t)}))$ *is positive and decreasing in* $t$.

*Proof.* Assume again $\mathbb{E}_g[h(Y)] = 0$; then

$$\mathbb{E}[h(Y^{(0)})h(Y^{(t)})] = \sum_{v=1}^{p} \sigma_v \ \mathbb{E}[\mathbb{E}[h(Y^{(0)})h(Y^{(1)})|\nu = v, (Y_j^{(0)}, j \neq v)]]$$

$$= \sum_{v=1}^{p} \sigma_v \ \mathbb{E}[\mathbb{E}[h(Y^{(0)})|(Y_j^{(0)}, j \neq v)]^2]$$

$$= \mathbb{E}[\mathbb{E}[h(Y)|\nu, (Y_j, j \neq \nu)]^2],$$

due to the reversibility of the chain and the independence between $Y^{(0)}$ and $Y^{(1)}$, conditionally on $\nu$ and $(y_j^{(0)}, j \neq v)$. A simple recursion implies

$$\mathbb{E}[h(Y^{(0)})h(Y^{(t)})] = \mathrm{var}(\mathbb{E}[\cdots \mathbb{E}[\mathbb{E}[h(Y)|\nu, (Y_j, j \neq \nu)]|Y]\cdots]),$$

where the second term involves $t$ conditional expectations, successively in $(\nu, (y_j, j \neq \nu))$ and in $Y$.                    □□

This proof suggests choosing a distribution $\sigma$ that more heavily weights components with small $\mathbb{E}[h(Y)|(y_j, j \neq v)]^2$, so the chain will typically visit components where $g_v(y_v|y_j, j \neq v)$ is not too variable. However, the opposite choice seems more logical when considering the *speed* of convergence to the stationary distribution, since components with high variability should be visited more often in order to accelerate the exploration of the support of $g$. This dichotomy between acceleration of the convergence to the stationary distribution and reduction of the empirical variance is typical of MCMC methods.

Another substantial benefit of Rao–Blackwellization is an elegant method for the approximation of densities of different components of $y$. Since

$$\frac{1}{T} \sum_{t=1}^{T} g_i(y_i|y_j^{(t)}, j \neq i)$$

is unbiased and converges to the marginal density $g_i(y_i)$ if these conditional densities are available in closed form, it is unnecessary (and inefficient) to use nonparametric density estimation methods such as *kernel methods* (see Fan and Gijbels 1996 or Wand and Jones 1995). This property can also be used in extensions of the Riemann sum method (see §3.4 and §8.3.2) to setups where the density $f$ needs to be approximated (see Problem 7.22).

Another consequence of Proposition 7.2.9 is a justification of the technique of *batch sampling* proposed in some MCMC algorithms (Geyer 1992, Raftery and Lewis 1992a, Diebolt and Robert 1994). Batch sampling involves subsampling the sequence $(Y^{(t)})_t$ produced by a Gibbs sampling method into $(Y^{(ks)})_s$ $(k > 1)$, in order to decrease the dependence between the points of the (sub-)sample. However, Lemma 8.1.1 describes a negative impact of subsampling on the variance of the corresponding estimators.

### 7.2.4 The Duality Principle

Rao–Blackwellization exhibits an interesting difference between statistical perspectives and simulation practice, in the sense that the approximations used in the estimator do not (directly) involve the chain of interest. We will see in this section that this phenomenon is more fundamental than a mere improvement of the variance of some estimators, as it provides a general technique to establish convergence properties for the chain of interest $(X^{(t)})$ based on the instrumental chain $(Y^{(t)})$, even when the latter is unrelated with the inferential problem. Diebolt and Robert (1993, 1994) have called this use of the dual chain the *Duality Principle* when they used it in the setup of mixtures of distributions. While it has been introduced in a Data Augmentation setup, this principle extends to other Gibbs sampling methods since it is sufficient that the chain of interest $(X^{(t)})$ be generated conditionally on another chain $(Y^{(t)})$, which supplies the probabilistic properties.

**Theorem 7.2.13** *Consider a Markov chain $(X^{(t)})$ and a sequence $(Y^{(t)})$ of random variables generated from the conditional distributions*

$$X^{(t)}|y^{(t)} \sim \pi(x|y^{(t)}) , \qquad Y^{(t+1)}|x^{(t)}, y^{(t)} \sim f(y|x^{(t)}, y^{(t)}) .$$

*If the chain $(Y^{(t)})$ is ergodic (resp. geometrically or uniformly ergodic) and if $\pi_{y^{(0)}}^t$ denotes the distribution of $(X^{(t)})$ associated with the initial value $y^{(0)}$, the norm $\|\pi_{y^{(0)}}^t - \pi\|_{TV}$ goes to 0 when $t$ goes to infinity (resp. goes to 0 at a geometric or uniformly bounded rate).*

*Proof.* The transition kernel is

$$K(y, y') = \int_{\mathcal{X}} \pi(x|y) \, f(x'|x, y) \, dx.$$

If $f_{y^{(0)}}^t(y)$ denotes the marginal density of $Y^{(t)}$, then the marginal of $X^{(t)}$ is

$$\pi_{y^{(0)}}^t(x) = \int_{\mathcal{Y}} \pi(x|y) \, f_{y^{(0)}}^t(y) \, dy$$

and convergence in the $(X^{(t)})$ chain can be tied to convergence in the $(Y^{(t)})$ chain by the following. The total variation in the $(X^{(t)})$ chain is

$$\|\pi_{y^{(0)}}^t - \pi\|_{TV} = \frac{1}{2} \int_{\mathcal{Y}} \left| \pi_{y^{(0)}}^t(x) - \pi(x) \right| \, dx$$

$$= \frac{1}{2} \int_{\mathcal{X}} \left| \int_{\mathcal{Y}} \pi(x|y) \left( f_{y^{(0)}}^t(y) - f(y) \right) \, dy \right| \, dx$$

$$\leq \frac{1}{2} \int_{\mathcal{X} \times \mathcal{Y}} \left| f_{y^{(0)}}^t(y) - f(y) \right| \, \pi(x|y) \, dx dy$$

$$= \|f^t_{y^{(0)}} - f\|_{TV},$$

so the convergence properties of $\|f^t_{y^{(0)}} - f\|_{TV}$ can be transfered to $\|\pi^t_{y^{(0)}} - \pi\|_{TV}$. Both sequences have in the same speed of convergence since

$$\|f^{t+1}_{y^{(0)}} - f\|_{TV} \leq \|\pi^t_{y^{(0)}} - \pi\|_{TV} \leq \|f^t_{y^{(0)}} - f\|_{TV}. \qquad \square\square$$

The duality principle is even easier to state, and stronger, when $(Y^{(t)})$ is a finite state-space Markov chain. (See Section 9.3 for more illustrations.)

**Theorem 7.2.14** *If $(Y^{(t)})$ is a finite state-space Markov chain, with state-space $\mathcal{Z}$, such that*

$$P(Y^{(t+1)} = k|y^{(t)}, x) > 0, \qquad \forall\, k \in \mathcal{Y}, \forall x \in \mathcal{X},$$

*the sequence $(X^{(t)})$ derived from $(Y^{(t)})$ by the transition $\pi(x|y^{(t)})$ is uniformly ergodic.*

Notice that this convergence result does not impose any constraint on the transition $\pi(x|y)$, which, for instance, is not necessarily everywhere positive.

*Proof.* The result follows from Theorem 7.2.13. First, write

$$p_{ij} = \int P(Y^{(t+1)} = j|x, y^{(t)} = i)\, \pi(x|y^{(t)} = i)\, dx, \qquad i, j \in \mathcal{Y}.$$

If we define the lower bound $\rho = \min_{i,j \in \mathcal{Y}} p_{ij} > 0$, then

$$P(Y^{(t)} = i|y^{(0)} = i) \geq \rho \sum_{k \in \mathcal{Y}} P(Y^{(t-1)} = k|y^{(0)} = i) = \rho\,,$$

and hence

$$\sum_{t=1}^{\infty} P(Y^{(t)} = i|y^{(0)} = i) = \infty$$

for every state $i$ of $\mathcal{Y}$. Thus, the chain is positive recurrent with limiting distribution

$$q_i = \lim_{t \to \infty} P(Y^{(t)} = i|y^{(0)} = i).$$

Uniform ergodicity follows from the finiteness of $\mathcal{Y}$ (see, e.g., Billingsley 1968). $\qquad \square\square$

**Example 7.2.15 (Continuation of Example 7.2.3)** The vector $(z_1, \ldots, z_4)$ takes its values in a finite space of size $(x_1+1) \times (x_2+1) \times (x_3+1) \times (x_4+1)$ and the transition is strictly positive. Corollary 7.2.16 therefore implies that the chain $(\eta^{(t)}, \mu^{(t)})$ is uniformly ergodic. $\qquad \|$

The statement of Theorems 7.2.13 and 7.2.14 is complicated by the fact that $(X^{(t)})$ is not necessarily a Markov chain and, therefore, the notions of ergodicity and of geometric ergodicity do not apply to this sequence.

However, there still exists a limiting distribution $\pi$ for $(X^{(t)})$, obtained by transformation of the limiting distribution of $(Y^{(t)})$ by the transition $\pi(x|y)$. The particular case of Data Augmentation allows for a less complicated formulation.

**Corollary 7.2.16** *For two interleaved Markov chains, $(X^{(t)})$ and $(Y^{(t)})$, if $(X^{(t)})$ is ergodic (resp. geometrically ergodic), $(Y^{(t)})$ is ergodic (resp. geometrically ergodic).*

Finally, we turn to the question of rate of convergence, and find that the duality principle still applies, and the rates transfer between chains.

**Proposition 7.2.17** *If $(Y^{(t)})$ is geometrically convergent with compact state-space and with convergence rate $\rho$, there exists $C_h$ such that*

$$(7.2.8) \qquad \left\| \mathbb{E}[h(X^{(t)})|y^{(0)}] - \mathbb{E}^{\pi}[h(X)] \right\| < C_h \, \rho^t \, ,$$

*for every function $h \in \mathcal{L}_1(\pi(\cdot|x))$ uniformly in $y^{(0)} \in \mathcal{Y}$.*

*Proof.* For $h(x) = (h_1(x), \ldots, h_d(x))$,

$$\left\| \mathbb{E}[h(X^{(t)})|y^{(0)}] - \mathbb{E}^{\pi}[h(X)] \right\|^2$$

$$= \sum_{i=1}^{d} \left( \mathbb{E}[h_i(X^{(t)})|y^{(0)}] - \mathbb{E}^{\pi}[h_i(X)] \right)^2$$

$$= \sum_{i=1}^{d} \left( \int h_i(x)\{\pi^t(x|y^{(0)}) - \pi(x)\} \, dx \right)^2$$

$$= \sum_{i=1}^{d} \left( \int h_i(x) \, \pi(x|y) \int \{f^t(y|y^{(0)}) - f(y)\} \, dy dx \right)^2$$

$$\leq \sum_{i=1}^{d} \sup_{y \in \mathcal{Y}} \mathbb{E}[|h_i(X)||y]^2 \, 4 \, \|f_{y^{(0)}}^t - f\|_{TV}^2$$

$$\leq 4d \max_{i} \sup_{y \in \mathcal{Y}} \mathbb{E}[|h_i(X)| \, |y]^2 \, \|f_{y^{(0)}}^t - f\|_{TV}^2 \, . \qquad \Box\Box$$

Unfortunately, this result has rather limited consequences for the study of the sequence $(X^{(t)})$ since (7.2.8) is a *average* property of $(X^{(t)})$, while MCMC algorithms such as [A.34] only produce *a single* realization from $\pi_{z^{(0)}}^t$. We will see in §8.2.3 a more practical implication of the Duality Principle since Theorem 7.2.13 may allow a control of convergence by renewal.

## 7.3 Hybrid Gibbs Samplers

### 7.3.1 Comparison with Metropolis–Hastings Algorithms

A comparison between a Gibbs sampling method and an arbitrary Metropolis–Hastings algorithm would seem *a priori* to favor Gibbs sampling since it derives its conditional distributions from the true distribution $f$, whereas a Metropolis–Hastings kernel is, at best, based on a approximation of this distribution $f$. In particular, we have noted several times that Gibbs sampling methods are, by construction, more straightforward than Metropolis–Hastings methods, since they cannot have a "bad" choice of the instrumental distribution and, hence, they avoid useless simulations (*rejections*). Although these algorithms can be formally compared, seeking a ranking of these two main types of MCMC algorithms is not only illusory but also somewhat pointless.

However, we do stress in this section that the availability and apparent objectivity of Gibbs sampling methods are not necessarily compelling arguments. If we consider the Gibbs sampler of Theorem 7.1.17, the Metropolis–Hastings algorithms which underly this method are not valid on an individual basis since they do not produce irreducible Markov chains $(Y^{(t)})$. Therefore, only a combination of a sufficient number of Metropolis–Hastings algorithms can ensure the validity of the Gibbs sampler. This composite structure is also a weakness of the method, since a decomposition of the joint distribution $f$ given a particular system of coordinates does not necessarily agree with the form of $f$. Example 7.1.10 illustrates this incompatibility in a pathological case: A wrong choice of the coordinates traps the corresponding Gibbs sampling in one of the two connected components of the support of $f$. Hills and Smith (1992, 1993) also propose examples where an incorrect parameterization of the model significantly increases the convergence time for the Gibbs sampler. We will see in Chapter 9 how parameterization influences the performances of the Gibbs sampler in the particular case of mixtures of distributions.

To draw an analogy, let us recall that when a function $v(y_1, \ldots, y_p)$ is maximized one component at a time, the resulting solution is not always satisfactory since it may correspond to a saddlepoint or to a local maximum of $v$. Similarly, the simulation of a single component at each iteration of [A.32] restricts the possible excursions of the chain $(Y^{(t)})$ and this implies that Gibbs sampling methods are generally *slow* to converge, since they are slow to explore the surface of $f$.

This intrinsic defect of the Gibbs sampler leads to phenomena akin to convergence to local maxima in optimization algorithms, which are expressed by strong attractions to the closest local modes and, in consequence, to difficulties in exploring the entire range of the support of $f$.

**Example 7.3.1 Two-dimensional mixture.** Consider a two-dimensional mixture of normal distributions,

$$(7.3.1) \qquad p_1 \mathcal{N}_2(\mu_1, \Sigma_1) + p_2 \mathcal{N}_2(\mu_2, \Sigma_2) + p_3 \mathcal{N}_2(\mu_3, \Sigma_3),$$

given in Figure 7.3.1 as a gray-level image. Both unidimensional conditionals are also mixtures of normal distributions and lead to a straightforward Gibbs sampler. The first 100 steps of the associated Gibbs sampler are represented on Figure 7.3.1; they show mostly slow moves along the two first components of the mixture and a single attempt to reach the third compo-

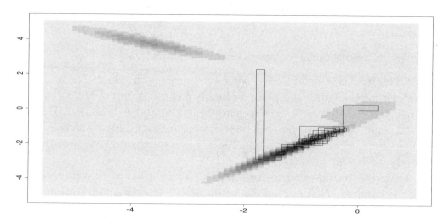

Figure 7.3.1. *Successive (full conditional) moves of the Gibbs chain on the surface of the stationary distribution (7.3.1), represented by gray levels (darker shades mean higher elevations).*

nent, which is too far in the tail of the conditional. Note that the numerical values chosen for this illustration are such that the third component has a 31% probability mass in (7.3.1). (Each step of the Gibbs sampler is given in the graph, which explains for the succession of horizontal and vertical moves.)                                                                                    ‖

### 7.3.2 Mixtures and Cycles

The drawbacks of Metropolis–Hastings algorithms are different from those of the Gibbs sampler, as they are more often related to a bad agreement between $f$ and the instrumental distribution. Moreover, the freedom brought by Metropolis–Hastings methods sometimes allows for remedies to these drawbacks through the modification of some scale (parameters or hyperparameters are particularly useful).

Compared to the Gibbs sampler, a failing of Metropolis–Hastings algorithms is to miss the finer details of the distribution $f$, if the simulation is "too coarse." However, following Tierney (1994), a way to take advantage of both algorithms is to implement a *hybrid* approach which uses both Gibbs sampling and Metropolis–Hastings algorithms.

**Definition 7.3.2** An *hybrid MCMC algorithm* is a Markov chain Monte Carlo method which simultaneously utilizes both Gibbs sampling steps and Metropolis–Hastings steps. If $K_1, \ldots, K_n$ are kernels which correspond to these different steps and if $(\alpha_1, \ldots, \alpha_n)$ is a probability distribution, a *mixture* of $K_1, K_2, \ldots, K_n$ is an algorithm associated with the kernel

$$\tilde{K} = \alpha_1 K_1 + \cdots + \alpha_n K_n$$

and a *cycle* of $K_1, K_2, \ldots, K_n$ is the algorithm with kernel

$$K^* = K_1 \circ \cdots \circ K_n,$$

where "$\circ$" denotes the composition of functions.

Of course, this definition is somewhat ambiguous since Theorem 7.1.17 states that the Gibbs sampling is already a composition of Metropolis–Hastings kernels; that is, a cycle according to the above definition. Definition 7.3.2 must, therefore, be understood as processing heterogeneous MCMC algorithms, where the chain under study is a subchain $(Y^{(kt)})_t$ of the chain produced by the algorithm.

From our initial perspective concerning the speed of convergence of the Gibbs sampler, Definition 7.3.2 leads us to consider modifications of the initial algorithm where, every $m$ iterations, [A.32] is replaced by a Metropolis–Hastings step with larger dispersion or, alternatively, at each iteration, this Metropolis–Hastings step is selected with probability $1/m$. These modifications are particularly helpful to escape "trapping effects" related to local modes of $f$.

Hybrid procedures are valid from a theoretical point of view, when the *heterogeneity* of the chains generated by cycles is removed by considering only the subchains $(Y^{(mt)})_t$ (although the whole chain should be exploited). A composition of kernels associated with an identical stationary distribution $f$ leads to a kernel with stationary distribution $f$. The irreducibility and aperiodicity of $\tilde{K}$ directly follows from the irreducibility and the aperiodicity of one of the $i$ kernels $K_i$. In the case of a *cycle*, we already saw that the irreducibility of $K^*$ for the Gibbs sampler does not require the irreducibility of its component kernels and a specific study of the algorithm at hand may sometimes be necessary. Tierney (1994) also proposes sufficient conditions for uniform ergodicity.

**Proposition 7.3.3** *If $K_1$ and $K_2$ are two kernels with the same stationary distribution $f$ and if $K_1$ produces a uniformly ergodic Markov chain, the mixture kernel*

$$\tilde{K} = \alpha K_1 + (1 - \alpha)K_2 \qquad (0 < \alpha < 1)$$

*is also uniformly ergodic. Moreover, if $\mathcal{X}$ is a small set for $K_1$ with $m = 1$, the kernel cycles $K_1 \circ K_2$ and $K_2 \circ K_1$ are uniformly ergodic.*

*Proof.* If $K_1$ produces a uniformly ergodic Markov chain, there exists $m \in \mathbb{N}$, $\varepsilon_m > 0$, and a probability measure $\nu_m$ such that $K_1$ satisfies

$$K_1^m(x, A) \geq \varepsilon_m \nu_m(A), \qquad \forall x \in \mathcal{X}, \quad \forall A \in \mathcal{B}(\mathcal{X}).$$

Therefore, we have the minorization condition

$$(\alpha K_1 + (1 - \alpha)K_2)^m(x, A) \geq \alpha^m K_1^m(x, A) \geq \alpha^m \varepsilon_m \nu_m(A),$$

which, from Theorem 4.6.13, establishes the uniform ergodicity of the mixture.

If $\mathcal{X}$ is a small set for $K_1$ with $m = 1$, we have the minorizations

$$(K_1 \circ K_2)(x, A) = \int_A \int_{\mathcal{X}} K_2(x, dy)\, K_1(y, dz)$$

$$\geq \varepsilon_1 \nu_1(A) \int_{\mathcal{X}} K_2(x, dy) = \varepsilon_1 \nu_1(A)$$

and

$$(K_2 \circ K_1)(x, A) = \int_A \int_{\mathcal{X}} K_1(x, dy)\, K_2(y, dz)$$

$$\geq \varepsilon_1 \int_A \int_{\mathcal{X}} \nu_1(dy) K_2(x, dy) = \varepsilon_1 (K_2 \circ \nu_1)(A).$$

From Theorem 4.6.13, both cycles are therefore uniformly ergodic. $\quad\Box\Box$

These results are not only formal since it is possible (see Theorem 6.3.1) to produce a uniformly ergodic kernel from an independent Metropolis–Hastings algorithm with instrumental distribution $g$ such that $f/g$ is bounded. Hybrid MCMC algorithms can, therefore, be used to impose uniform ergodicity in an almost automatic way.

The following example, due to Nobile (1998), shows rather clearly how the introduction of a Metropolis–Hastings step in the algorithm speeds up the exploration of the support of the stationary distribution.

**Example 7.3.4 Probit model.** A (dichotomous) probit model is defined by random variables $D_i$ $(1 \leq i \leq n)$ such that $(1 \leq j \leq 2)$

(7.3.2) $$P(D_i = 1) = 1 - P(D_i = 0) = P(Z_i \geq 0)$$

with $Z_i \sim \mathcal{N}(R_i\beta, \sigma^2)$, $\beta \in \mathbb{R}$, $R_i$ being a covariate. (Note that the $z_i$'s are *not* observed. This is a special case of a *latent variable model*, which is treated in more detail in Chapter 9, §9.2.1.) For the prior distribution

$$\sigma^{-2} \sim \mathcal{G}a(1.5, 1.5), \qquad \beta|\sigma \sim \mathcal{N}(0, 10^2),$$

a Gibbs sampler can be constructed as in [A.47] and Figure 7.3.2 plots the $20,000$ first iterations of the Gibbs chain $(\beta^{(t)}, \sigma^{(t)})$ against some contours of the true posterior distribution. The exploration is thus very poor, since the chain does not even reach the region of highest posterior density. (A reason for this behavior is that the likelihood is quite uninformative about $(\beta, \sigma)$, providing only a lower bound on $\beta/\sigma$, as explained in Nobile 1998. See also Problem 7.19.)

A hybrid alternative, proposed by Nobile (1998), is to insert a Metropolis–Hastings step after each Gibbs cycle. The proposal distribution merely rescales the current value of the Markov chain $y^{(t)}$ by a random scale factor $c$, drawn from an exponential $\mathcal{E}xp(1)$ distribution (which is similar to the "hit-and-run" method of Chen and Schmeiser 1993). The rescaled value $cy^{(t)}$ is then accepted or rejected by a regular Metropolis–Hastings scheme to become the current value of the chain. Figure 7.3.3 shows the improvement brought by this hybrid scheme, since the MCMC sample now covers

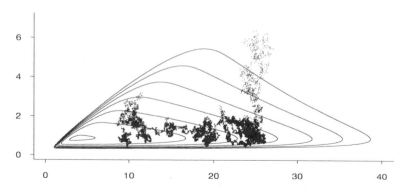

Figure 7.3.2. *Sample of* $(\beta^{(t)}, \sigma^{(t)})$ *obtained by the Gibbs sampler plotted with some contours of the posterior distribution of* $(\beta, \sigma)$ *for* 20, 000 *observations from (7.3.2), when the chain is started at* $(\beta, \sigma) = (25, 5)$.

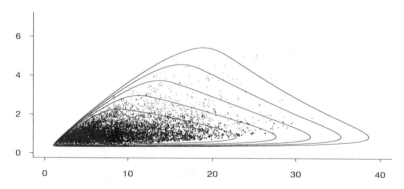

Figure 7.3.3. *Sample of* $(\beta^{(t)}, \sigma^{(t)})$ *obtained by an hybrid MCMC algorithm plotted with the posterior distribution for the same* 20, 000 *observations as in Figure 7.3.2 and the same starting point.*

most of the support of the posterior distribution for the same number of iterations as in Figure 7.3.2.                                                   ‖

### 7.3.3 Metropolizing the Gibbs Sampler

Hybrid MCMC algorithms are often useful at an elementary level of the simulation process; that is, when some components of [A.32] cannot be easily simulated. Rather than looking for a customized algorithm such as Accept–Reject in each of these cases or for alternatives to Gibbs sampling, there is a compromise suggested by Müller (1991, 1993) (sometimes called "Metropolis-within-Gibbs"). In any step i. of [A.32] with a difficult simulation from $g_i(y_i|y_j, j \neq i)$, substitute a simulation from an instrumental distribution $q_i$. In the setup of [A.32], Müller's modification is

**Algorithm A.40 –Hybrid MCMC–**
For $i = 1, \ldots, p$, given $(y_1^{(t+1)}, \ldots, y_{i-1}^{(t+1)}, y_i^{(t)}, \ldots, y_p^{(t)})$:

1. Simulate
$$\tilde{y}_i \sim q_i(y_i|y_1^{(t+1)}, \ldots, y_i^{(t)}, y_{i+1}^{(t)}, \ldots, y_p^{(t)})$$

2. Take $\hspace{8cm}$ [A.40]

$$y_i^{(t+1)} = \begin{cases} y_i^{(t)} & \text{with probability } 1 - \rho, \\ \tilde{y}_i & \text{with probability } \rho, \end{cases}$$

where

$$\rho = 1 \wedge \left\{ \frac{\left( \dfrac{g_i(\tilde{y}_i|y_1^{(t+1)}, \ldots, y_{i-1}^{(t+1)}, y_{i+1}^{(t)}, \ldots, y_p^{(t)})}{q_i(\tilde{y}_i|y_1^{(t+1)}, \ldots, y_{i-1}^{(t+1)}, y_i^{(t)}, y_{i+1}^{(t)}, \ldots, y_p^{(t)})} \right)}{\left( \dfrac{g_i(y_i^{(t)}|y_1^{(t+1)}, \ldots, y_{i-1}^{(t+1)}, y_{i+1}^{(t)}, \ldots, y_p^{(t)})}{q_i(y_i^{(t)}|y_1^{(t+1)}, \ldots, y_{i-1}^{(t+1)}, \tilde{y}_i, y_{i+1}^{(t)}, \ldots, y_p^{(t)})} \right)} \right\}.$$

An important point about this substitution is that the above Metropolis–Hastings step is only used *once* in an iteration from [A.32]. The modified step thus produces a *single* simulation $\tilde{y}_i$ instead of trying to approximate $g_i(y_i|y_j, j \neq i)$ more accurately by producing $T$ simulations from $q_i$. The reasons for this choice are twofold: First, the resulting hybrid algorithm is valid since $g$ is its stationary distribution (See §7.3.2 or Problem 7.20). Second, Gibbs sampling also leads to an approximation of $g$. To provide a more "precise" approximation of $g_i(y_i|y_j, j \neq i)$ in [A.40] does not necessarily lead to a better approximation of $g$ and the replacement of $g_i$ by $q_i$ may even be beneficial for the speed of excursion of the chain on the surface of $g$. (See also Chen and Schmeiser 1998.)

When several Metropolis–Hastings steps appear in a Gibbs algorithm, either because of some complex conditional distributions (see Problems 7.15 and 7.16) or because of convergence contingencies, a method proposed by Müller (1993) is to run a single acceptance step after the $p$ conditional simulations. This approach is more time-consuming in terms of simulation (since it may result in the rejection of the $p$ simulated components), but the resulting algorithm can be written as a simple Metropolis–Hastings method instead of a (hybrid) combination of Metropolis–Hastings algorithms. Moreover, it produces an approximation $q(y)$ of the distribution $g(y)$ rather than *local* approximations of the conditional distributions of the subvectors $y_i$.

**Example 7.3.5 ARCH models.** For model (5.3.10) of Example 5.3.3, the missing data $Z^T = (Z_1, \ldots, Z_T)$ has a conditional density given by

$$\tilde{f}(z^T|x^T, \theta) \propto f(x^T, z^T|\theta) = e^{-z_1^2/2} \prod_{t=2}^{T} \frac{e^{-z_t^2/2(1+\beta z_{t-1}^2)}}{(1 + \beta z_{t-1}^2)^{1/2}}$$

$$(7.3.3) \qquad \times (2\pi)^T \sigma^{-2T} \exp \left\{ - \sum_{t=1}^{T} ||x_t - az_t||^2 / 2\sigma^2 \right\},$$

where $\theta = (a, \beta, \sigma)$ and $x^T = (x_1, \ldots, x_T)$ is the observed sample.

Since the distribution (7.3.3) cannot be simulated, we consider the full conditional distributions of the latent variables $Z_t^*$,

$$\tilde{f}(z_t | x_t, z_{t-1}, z_{t+1}\theta) \propto \exp \left\{ -||x_t - az_t||^2 / 2\sigma^2 \right.$$
$$\left. -z_t^2 / 2(1 + \beta z_{t-1}^2) - z_{t+1}^2 / 2(1 + \beta z_t^2) \right\},$$

which are still difficult to simulate, because of the $z_{t+1}^2 / 2(1 + \beta z_t^2)$ term. A possible proposal for a Metropolis–Hastings step is to omit this term by using the instrumental distribution

$$\tilde{f}(z_t | x_t, z_{t-1}, \theta) \propto \exp \left\{ -||x_t - az_t||^2 / 2\sigma^2 - z_t^2 / 2(1 + \beta z_{t-1}^2) \right\};$$

that is, to simulate from

$$\mathcal{N} \left( \frac{(1 + \beta z_{t-1}^2) a' x_t}{(1 + \beta z_{t-1}^2)||a||^2 + \sigma^2}, \frac{(1 + \beta z_{t-1}^2)\sigma^2}{(1 + \beta z_{t-1}^2)||a||^2 + \sigma^2} \right).$$

The corresponding Metropolis–Hastings acceptance probability is $\rho_t = \min(\omega_t, 1)$, with

$$\omega_t = \sqrt{\frac{1 + \bar{\beta} z_t^2(s\text{-}1)}{1 + \bar{\beta} z_t^2(s)}} \exp \left\{ -\frac{1}{2} \frac{z_{t+1}^2(s\text{-}1)}{1 + \bar{\beta} z_t^2(s)} + \frac{1}{2} \frac{z_{t+1}^2(s\text{-}1)}{1 + \bar{\beta} z_t^2(s\text{-}1)} \right\},$$

where $z_t(s-1)$ and $z_t(s)$ denote the previous value of $z_t$ and the simulated value, respectively. (See Shephard and Pitt 1997 for alternatives.)  ‖

We conclude this section with a surprising result of Liu (1995), who shows that, in a discrete state setting, Gibbs sampling can be improved by Metropolis–Hastings steps. The improvement is expressed in terms of a reduction of the variance of the empirical mean of the $h(y^{(t)})$'s for the Metropolis–Hastings approach. This modification, called *Metropolization* by Liu (1995), is based on the following result, established by Peskun (1973), where $T_1 \ll T_2$ means that the non-diagonal elements of $T_2$ are larger that those of $T_1$ (see Problem 6.46 for a proof).

**Lemma 7.3.6** *Consider two reversible Markov chains on a countable state-space, with transition matrices $T_1$ and $T_2$ such that $T_1 \ll T_2$. The chain associated with $T_2$ dominates the chain associated with $T_1$ in terms of variances.*

Given a conditional distribution $g_i(y_i | y_j, j \neq i)$ on a discrete space, the modification proposed by Liu (1995) is to use an additional Metropolis–Hastings step.

**Algorithm A.41 –Metropolization of the Gibbs Sampler–**
   Given $y^{(t)}$,

1. Simulate $z_i \neq y_i^{(t)}$ with probability

$$\frac{g_i(z_i|y_j^{(t)}, j \neq i)}{1 - g_i(y_i^{(t)}|y_j^{(t)}, j \neq i)} \; ;$$

2. Accept $y_i^{(t+1)} = z_i$ with probability                              [A.41]

$$\frac{1 - g_i(y_i^{(t)}|y_j^{(t)}, j \neq i)}{1 - g_i(z_i|y_j^{(t)}, j \neq i)} \wedge 1.$$

The probability of moving from $y_i^{(t)}$ to a different value is then necessarily higher in [A.41] than in the original Gibbs sampling algorithm and Lemma 7.3.6 implies the following domination result:

**Theorem 7.3.7** *The modification* [A.41] *of the Gibbs sampler is more efficient in terms of variances.*

**Example 7.3.8 (Continuation of Example 7.2.3)** For the aggregated multinomial model of Tanner and Wong (1987), the completed variables $z_i$ $(i = 1, \ldots, 4)$ take values on $\{0, 1, \ldots, x_i\}$ and can, therefore, be simulated from [A.41]; that is, from a binomial distribution

$$\mathcal{B}\left(x_i, \frac{a_i\mu}{a_i\mu + b_i}\right) \quad \text{for } i = 1, 2 \quad \text{and} \quad \mathcal{B}\left(x_i, \frac{a_i\eta}{a_i\eta + b_i}\right) \quad \text{for } i = 3, 4,$$

until the simulated value $\tilde{z}_i$ is different from $z_i^{(t)}$. This new value $\tilde{z}_i$ is accepted with probability $(i = 1, 2)$

$$\frac{1 - \binom{z_i^{(t)}}{x_i}\left(\frac{a_i\mu}{a_i\mu + b_i}\right)^{z_i^{(t)}}\left(\frac{b_i}{a_i\mu + b_i}\right)^{x_i - z_i^{(t)}}}{1 - \binom{\tilde{z}_i}{x_i}\left(\frac{a_i\mu}{a_i\mu + b_i}\right)^{\tilde{z}_i}\left(\frac{b_i}{a_i\mu + b_i}\right)^{x_i - \tilde{z}_i}}$$

and $(i = 3, 4)$

$$\frac{1 - \binom{z_i^{(t)}}{x_i}\dfrac{(a_i\eta)^{z_i^{(t)}} b_i^{x_i - z_i^{(t)}}}{(a_i\eta + b_i)^{x_i}}}{1 - \binom{\tilde{z}_i}{x_i}\dfrac{(a_i\mu)^{\tilde{z}_i} b_i^{x_i - \tilde{z}_i}}{(a_i\eta + b_i)^{x_i}}} \; .$$

Figure 7.3.4 describes the convergence of estimations of $\mu$ and $\eta$ under the two simulation schemes for the following data:

$$(a_1, a_2, a_3, a_4) = (0.06, 0.14, 0.11, 0.09),$$
$$(b_1, b_2, b_3, b_4) = (0.17, 0.24, 0.19, 0.20),$$
$$(x_1, x_2, x_3, x_4, x_5) = (9, 15, 12, 7, 8) .$$

As shown by Table 7.3.1, the difference on the variation range of the approximations of both $\mathbb{E}[\mu|x]$ and $\mathbb{E}[\eta|x]$ is quite minor, in particular for the

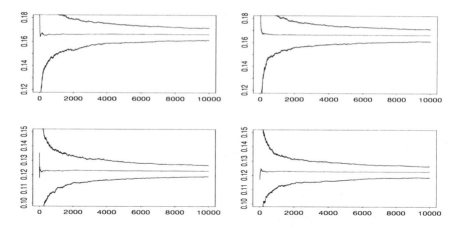

Figure 7.3.4. *Comparison of the Gibbs sampling (left) and of the modification of Liu (1995) (right), for the estimation of* $\mathbb{E}[\mu|x]$ *(top) and of* $\mathbb{E}[\eta|x]$ *(bottom). The 90% confidence regions on both methods have been obtained on 500 parallel chains.*

| $n$ | 10 | 100 | 1000 | 10,000 |
|---|---|---|---|---|
| $\mu$ | 0.302 | 0.0984 | 0.0341 | 0.0102 |
|  | 0.288 | 0.0998 | 0.0312 | 0.0104 |
| $\eta$ | 0.234 | 0.0803 | 0.0274 | 0.00787 |
|  | 0.234 | 0.0803 | 0.0274 | 0.00778 |

Table 7.3.1. *90% interquantile ranges for the original Gibbs sampling (top) and the modification of Liu (1995) (bottom), for the estimation of* $\mathbb{E}[\mu|x]$ *and* $\mathbb{E}[\eta|x]$.

estimation of $\eta$, and are not always larger for the original Gibbs sampler. ‖

### 7.3.4 Reparameterization

Convergence of both Gibbs sampling and Metropolis–Hastings algorithms may suffer from a poor choice of parameterization (see the extreme case of Example 7.1.10). As a result of this, following the seminal paper of Hills and Smith (1992), the MCMC literature has considered changes in the parameterization of a model as a way to speed up convergence in a Gibbs sampler. (See Gilks and Roberts 1996, for a more detailed review.) It seems, however, that most efforts have concentrated on the improvement of specific models, resulting in a lack of a general methodology for the choice of a "proper" parameterization. Nevertheless, the overall advice is to try to make the components "as independent as possible."

As noted in Example 7.3.1, convergence performances of the Gibbs sampler may be greatly affected by the choice of the coordinates. For instance, if the distribution $g$ is a $\mathcal{N}_2(0, \Sigma)$ distribution with a covariance matrix $\Sigma$ such that its eigenvalues satisfy $\lambda_{\min}(\Sigma) \ll \lambda_{\max}(\Sigma)$ and its eigenvectors correspond to the first and second diagonals of $\mathbb{R}^2$, the Gibbs sampler based on the conditionals $g(x_1|x_2)$ and $g(x_2|x_1)$ is very slow to explore the whole range of the support of $g$. However, if $y_1 = x_1 + x_2$ and $y_2 = x_1 - x_2$ is the selected parameterization (which corresponds to the coordinates in the eigenvector basis) the Gibbs sampler will move quite rapidly over the support of $g$. (Hobert $et$ $al.$ 1997 have shown that the influence of the parameterization on convergence performances may be so drastic that the chain can be irreducible for some parameterizations and not irreducible for others.)

Similarly, the geometry of the selected instrumental distribution in a Metropolis–Hastings algorithm must somehow correspond to the geometry of the support of $g$ for good acceptance rates to be obtained. In particular, as pointed out by Hills and Smith (1992), if a normal second-order approximation to the density $g$ is used, the structure of the Hessian matrix will matter in a Gibbs implementation, whereas the simplification brought by a Laplace approximation must be weighted against the influence of the parameterization. For instance, the value of the approximation based on the Taylor expansion

$$h(\theta) = h(\hat{\theta}) + \frac{1}{2}(\theta - \hat{\theta})^t (\nabla \nabla^t h)|_{\theta = \hat{\theta}}(\theta - \hat{\theta}) \,,$$

with $h(\theta) = \log g(\theta)$, depends on the choice of the parameterization.

Gelfand $et$ $al.$ (1994, 1995, 1996) have studied the effects of parameterization on different linear models.

**Example 7.3.9 Random effects model.** Consider the simple random effects model

$$Y_{ij} = \mu + \alpha_i + \varepsilon_{ij} \,, \qquad i = 1, \ldots, I, \quad j = 1, \ldots, J,$$

where $\alpha_i \sim \mathcal{N}(0, \sigma_\alpha^2)$ and $\varepsilon_{ij} \sim \mathcal{N}(0, \sigma_y^2)$. For a flat prior on $\mu$, the Gibbs sampler implemented for the $(\mu, \alpha_1, \ldots, \alpha_I)$ parameterization exhibits high correlation and consequent slow convergence if $\sigma_y^2/(IJ\sigma_\alpha^2)$ is large (see Problem 7.35). On the other hand, if the model is rewritten as the hierarchy

$$Y_{ij} \sim \mathcal{N}(\eta_i, \sigma_y^2), \qquad \eta_i \sim \mathcal{N}(\mu, \sigma_\alpha^2),$$

the correlations between the $\eta_i$'s and between $\mu$ and the $\eta_i$'s are lower (see Problem 7.36).

Another approach is suggested by Vines and Gilks (1994) and Vines $et$ $al.$ (1995), who eliminate the unidentifiability feature by so-called $sweeping;$ that is, by writing the model as

$$Y_{ij} = \nu + \varphi_i + \varepsilon_{ij},$$

with $(1 \leq i \leq I, 1 \leq j \leq J)$

$$\sum_i \varphi_i = 0, \qquad \varphi_i \sim \mathcal{N}\left(0, \sigma_\alpha^2 \left(1 - (1/I)\right)\right) \,,$$

and $\text{cov}(\varphi_i, \varphi_j) = -\sigma_\alpha^2/I$. This choice leads to even better correlation structures since $\nu$ is independent of the $\varphi_i$'s while $\text{corr}(\varphi_i, \varphi_j) = -1/I$, *a posteriori*. ‖

## 7.4 Improper Priors

This section discusses a particular danger resulting from careless use of Metropolis–Hastings algorithms, in particular the Gibbs sampler. We know that the Gibbs sampler is based on conditional distributions derived from $f(x_1, \ldots, x_q)$ or $g(y_1, \ldots, y_p)$. What is particularly insidious is that these conditional distributions may be well defined and may be simulated from, but may not correspond to any joint distribution $g$; that is, the function $g$ given by Lemma 7.1.20 is not integrable. The same problem may occur when using a proportionality relation as $\pi(\theta|x) \propto \pi(\theta)f(x|\theta)$ to derive a Metropolis–Hastings algorithm for $\pi(\theta)f(x|\theta)$.

   This problem is not a *defect* of the Gibbs sampler, nor even a simulation[4] problem, but rather a problem of carelessly using the Gibbs sampler in a situation for which the underlying assumptions are violated. It is nonetheless important to warn the user of MCMC algorithms against this danger, because it corresponds to a situation often encountered in Bayesian noninformative (or *"default"*) modelings.

   The construction of the Gibbs sampler directly from the conditional distributions is a strong incentive to bypass checking for the propriety of $g$, especially in complex setups. In particular, the function $g$ may well be undefined in a "generalized" Bayesian approach where the prior distribution is "improper," in the sense that it is a $\sigma$-finite measure instead of being a regular probability measure (see Berger 1985, Section 3.3, or Robert 1994a, Section 1.5).

**Example 7.4.1 Conditional exponential distributions.** The following model was used by Casella and George (1992) to point out the difficulty of assessing the impropriety of a posterior distribution through the conditional distributions. The pair of conditional densities

$$X_1|x_2 \sim \mathcal{E}xp(x_2) , \qquad\qquad X_2|x_1 \sim \mathcal{E}xp(x_1)$$

are well defined and are functionally compatible in the sense that the ratio

$$\frac{f(x_1|x_2)}{f(x_2|x_1)} = \frac{x_1 \exp(-x_1\, x_2)}{x_2 \exp(-x_2\, x_1)} = \frac{x_1}{x_2}$$

separates into a function of $x_1$ and a function of $x_2$. By virtue of Bayes theorem, this is a necessary condition for the existence of a joint density $f(x_1, x_2)$ corresponding to these conditional densities. However, recall that Theorem 7.1.19 requires that

$$\int \frac{f(x_1|x_2)}{f(x_2|x_1)} dx_1 < \infty,$$

---

[4] The "distribution" $g$ does not exist in this case.

but the function $x^{-1}$ cannot be integrated on $\mathbb{R}_+$. Moreover, if the joint density, $f(x_1, x_2)$, existed, it would have to satisfy $f(x_1|x_2) = f(x_1, x_2)/\int f(x_1, x_2)dx_1$. The only function $f(x_1, x_2)$ satisfying this equation is

$$f(x_1, x_2) \propto \exp(-x_1 x_2).$$

Thus, these conditional distributions do not correspond to a joint probability distribution (see Problem 7.37).                                    ‖

**Example 7.4.2 Normal improper posterior.** Consider $X \sim \mathcal{N}(\theta, \sigma^2)$. A Bayesian approach when there is little prior information on the parameters $\theta$ and $\sigma$ is Jeffreys' (1961). This approach is based on the derivation of the prior distribution of $(\theta, \sigma)$ from the Fisher information of the model as $I(\theta, \sigma)^{1/2}$, the square root being justified by invariance reasons (as well as information theory considerations; see Robert 1994a, Section 3.4, or Lehmann and Casella 1998). Therefore, in this particular case, $\pi(\theta, \sigma) = \sigma^{-2}$ and the posterior distribution follows by a (formal) application of Bayes' Theorem,

$$\pi(\theta, \sigma|x) = \frac{f(x|\theta, \sigma)\, \pi(\theta, \sigma)}{\int f(x|\zeta)\, \pi(\zeta)\, d\zeta}.$$

It is straightforward to check that

$$\int f(x|\zeta)\, \pi(\zeta)\, d\zeta = \int \frac{1}{\sqrt{2\pi}\sigma} e^{-(x-\theta)^2/(2\sigma^2)} \frac{d\theta\, d\sigma}{\sigma^2} = \infty,$$

and, hence, $\pi(\theta, \sigma|x)$ is not defined.

However, this impropriety may never be detected as it is common Bayesian practice to consider only the proportionality relation between $\pi(\theta, \sigma|x)$ and $f(x|\theta, \sigma)\, \pi(\theta, \sigma)$. Similarly, a derivation of a Gibbs sampler for this model is also based on the proportionality relation since the conditional distributions $\pi(\theta|\sigma, x)$ and $\pi(\sigma|\theta, x)$ can be obtained directly as

$$\pi(\theta|\sigma, x) \propto e^{-(\theta-x)^2/(2\sigma^2)}, \qquad \pi(\sigma|\theta, x) \propto \sigma^{-3} e^{-(\theta-x)^2/(2\sigma^2)},$$

which lead to the full conditional distributions $\mathcal{N}(x, \sigma^2)$ and $\mathcal{E}xp((\theta-x)^2/2)$ for $\theta$ and $\sigma^{-2}$, respectively.

This practice of omitting the computation of normalizing constants is justified as long as the joint distribution does exist and is particularly useful because these constants often cannot be obtained in closed form. However, the use of improper prior distributions implies that the derivation of conditional distributions from the proportionality relation is not always justified. In the present case, the Gibbs sampler associated with these two conditional distributions is

1. Simulate $\theta^{(t+1)} \sim \mathcal{N}(x, \sigma^{(t)2})$ ;

2. Simulate $\varepsilon \sim \mathcal{E}xp\left((\theta^{(t+1)} - x)^2/2\right)$ and take
   $\sigma^{(t+1)} = 1/\sqrt{\varepsilon}.$

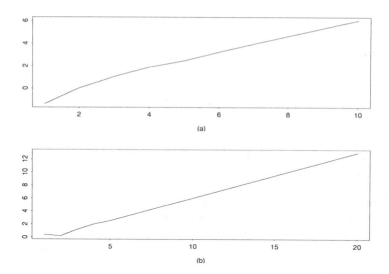

Figure 7.4.1.  *First iterations of the chains (a)* $\log(|\theta^{(t)}|)$ *and (b)* $\log(\sigma^{(t)})$ *for a divergent Gibbs sampling algorithm.*

For an observation $x = 0$, the behavior of the chains $(\theta^{(t)})$ and $(\sigma^{(t)})$ produced by this algorithm is illustrated in Figure 7.4.1, which exhibits an extremely fast divergence[5] to infinity (note that both graphs are plotted in log scales).                                                                ‖

The setup of improper posterior distributions and of the resulting behavior of the Gibbs sampler can be evaluated from a theoretical point of view, as in Hobert and Casella (1996, 1998). For example, the associated Markov chain cannot be positive recurrent since the $\sigma$-finite measure associated with the conditional distributions given by Lemma 7.1.20 is invariant. However, the major task in such settings is to come up with indicators to flag that something is wrong with the stationary measure, so that the experimenter can go back to his/her prior distribution and check for propriety.

Given the results of Example 7.4.2, it may appear that a simple graphical monitoring is enough to exhibit deviant behavior of the Gibbs sampler. However, this is not the case in general and there are many examples, some of which are published (see Casella 1996), where the output of the Gibbs sampler seemingly does not differ from a positive recurrent Markov chain. Often, this takes place when the divergence of the posterior density occurs "at 0," that is, at a specific point whose immediate neighborhood is rarely visited by the chain. Hobert and Casella (1996) illustrate this

---

[5] The impropriety of this posterior distribution is hardly surprising from a statistical point of view since this modeling means trying to estimate both parameters of the distribution $\mathcal{N}(\theta, \sigma^2)$ from a single observation without prior information.

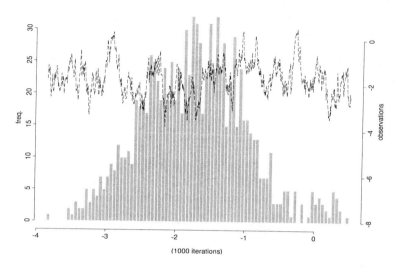

Figure 7.4.2. *Sequence $(\beta^{(t)})$ and corresponding histogram for the random effects model. The measurement scale for the $\beta^{(t)}$'s is on the right and the scale of the histogram is on the left.*

seemingly acceptable behavior on an example initially treated in Gelfand *et al.* (1990).

**Example 7.4.3 Improper random effects posterior.** Consider a random effects model,

$$Y_{ij} = \beta + U_i + \varepsilon_{ij}, \qquad i = 1, \ldots, I, \; j = 1, \ldots, J,$$

where $U_i \sim \mathcal{N}(0, \sigma^2)$ and $\varepsilon_{ij} \sim \mathcal{N}(0, \tau^2)$. The Jeffreys (improper) prior for the parameters $\beta$, $\sigma$, and $\tau$ is

$$\pi(\beta, \sigma^2, \tau^2) = \frac{1}{\sigma^2 \tau^2} .$$

However, the conditional distributions

$$U_i | y, \beta, \sigma^2, \tau^2 \sim \mathcal{N}\left( \frac{J(\bar{y}_i - \beta)}{J + \tau^2 \sigma^{-2}}, (J\tau^{-2} + \sigma^{-2})^{-1} \right),$$

$$\beta | u, y, \sigma^2, \tau^2 \sim \mathcal{N}(\bar{y} - \bar{u}, \tau^2 / JI),$$

$$\sigma^2 | u, \beta, y, \tau^2 \sim \mathcal{IG}\left( I/2, (1/2) \sum_i u_i^2 \right),$$

$$\tau^2 | u, \beta, y, \sigma^2 \sim \mathcal{IG}\left( IJ/2, (1/2) \sum_{i,j} (y_{ij} - u_i - \beta)^2 \right),$$

are well defined and a Gibbs sampling can be easily implemented in this setting. Figure 7.4.2 provides the sequence of the $\beta^{(t)}$ produced by [A.32] and the corresponding histogram for 1000 iterations. The trend of the sequence and the histogram do not indicate that the corresponding "joint distribution" does not exist (Problem 7.38). ‖

Under some regularity conditions on the transition kernel, Hobert and Casella (1996) have shown that if there exist a positive function $b$, $\varepsilon > 0$ and a compact set $C$ such that $b(x) < \varepsilon$ for $x \in C^c$, the chain $(y^{(t)})$ satisfies

$$(7.4.1) \qquad \liminf_{t \to +\infty} \frac{1}{t} \sum_{s=1}^{t} b(y^{(s)}) = 0.$$

A drawback of the condition (7.4.1) is the derivation of the function $b$, whereas the monitoring of a lim inf is delicate in a simulation experiment. Other references on the analysis of improper Gibbs samplers can be found in Besag *et al.* (1995) and in the subsequent discussion (see, in particular, Roberts and Sahu 1997), as well as in Natarajan and McCulloch (1998).

## 7.5 Problems

**7.1** For the Gibbs sampler (7.1.1):

(a) Show that the sequence $(X_i, Y_i)$ is a Markov chain, as is each sequence $(X_i)$ and $(Y_i)$.

(b) Show that $f_X(\cdot)$ and $f_Y(\cdot)$ are respectively the invariant densities of the $X$ and $Y$ sequences of (7.1.1).

**7.2** Propose a direct Accept–Reject algorithm for the conditional distribution of $Y_2$ given $y_1$ in Example 7.1.2. Compare with the Gibbs sampling implementation in a Monte Carlo experiment. (*Hint:* Run parallel chains and compute 90% confidence intervals for both methods.)

**7.3** For the autoexponential model of Example 7.1.2:

(a) If $Z_1 = (X_2, X_3)$, show that the transition kernel of $(Z_1^t)$ is

$$K_1(z_1, z_1') = \int_0^\infty (1 + \theta_{13}x_3 + \theta_{12}x_2) \, e^{-(1+\theta_{13}x_3+\theta_{12}x_2)y}(1 + \theta_{12}y + \theta_{23}x_3)$$

$$\times \, e^{-(1+\theta_{12}y+\theta_{23}x_3)x_2'}(1 + \theta_{23}x_2' + \theta_{13}y) \, e^{-(1+\theta_{23}x_2'+\theta_{13}y)x_3'} \, dy$$

$$= \frac{(1 + \theta_{13}x_3 + \theta_{12}x_2) \, e^{-(1+\theta_{23}x_3)x_2'} \, e^{-(1+\theta_{23}x_2')x_3'}}{1 + \theta_{13}x_3 + \theta_{12}x_2 + \theta_{12}x_2' + \theta_{13}x_3'}$$

$$\times \, \big\{ (1 + \theta_{23}x_3)(1 + \theta_{23}x_2') + [\theta_{12}(1 + \theta_{23}x_2')$$

$$+ \, \theta_{13}(1 + \theta_{23}x_3)]\psi(z_1, z_1') + 2\theta_{12}\theta_{13} \, \psi^2(z_1, z_1') \big\},$$

where $\psi(z_1, z_1') = (1 + \theta_{13}x_3 + \theta_{12}x_2 + \theta_{12}x_2' + \theta_{13}x_3')^{-1}$.

(b) For large $x_3$, show that $K_1(z_1, z_1')$ is equivalent to

$$\frac{\theta_{13}\theta_{23}(1 + \theta_{23}x_2')x_3^2}{\theta_{13}x_3} \, \exp\{-(1 + \theta_{23}x_3)x_2' - (1 + \theta_{23}x_2')x_3'\}$$

and, hence, converges to 0 when $x_3$ goes to infinity. It is therefore impossible to bound $K_1(z_1, z_1')$ uniformly from below in $z_1$.

(c) Show that the fact that $K_1(z_1, z_1')$ is not bounded from below implies that the original chain is not uniformly ergodic. (*Hint:* See (7.1.10) and Theorem 4.6.13.)

**7.4** Devise and implement a simulation algorithm for the logistic distribution (7.1.4).

**7.5** In the setup of Example 7.1.23:

(a) Evaluate the numerical value of the boundary constant $\epsilon$ derived from (7.1.9), given the data in Table 7.1.2.

(b) Establish the lower bound in (7.1.9). (*Hint:* Replace $\sum \lambda_i$ by 0 in the integrand [but not in the exponent] and do the integration.)

**7.6** (Besag 1994) Consider a distribution $f$ on a finite state-space $\mathcal{X}$ of dimension $k$ and conditional distributions $f_1, \ldots, f_k$ such that for every $(x, y) \in \mathcal{X}^2$, there exist an integer $m$ and a sequence $x_0 = x, \ldots, x_m = y$ where $x_i$ and $x_{i+1}$ only differ in a single component and $f(x_i) > 0$.

(a) Show that this condition extends Hammersley–Clifford Theorem (Theorem 7.1.20) by deriving $f(y)$ as

$$f(y) = f(x) \prod_{i=0}^{m-1} \frac{f(x_{i+1})}{f(x_i)}.$$

(b) Deduce that irreducibility and ergodicity holds for the associated Gibbs sampler.

(c) Show that the same condition on a continuous state-space is not sufficient to ensure ergodicity of the Gibbs sampler. (*Hint:* See Example 7.1.10.)

**7.7** A bound similar to the one in Example 7.1.23, established in Problem 7.5, can be obtained for the kernel of Example 7.1.6; that is, show that

$$K(\theta, \theta') = \int_0^\infty \sqrt{\frac{1+\eta}{2\pi}} \exp\left\{ -\left(\theta' - \frac{\theta_0 \eta}{1+\eta}\right)^2 \frac{1+\eta}{2} \right\} \left( \frac{1 + (\theta - \theta_0)^2}{2} \right)^\nu$$

$$\times \frac{\eta^{\nu-1}}{\Gamma(\nu)} \exp\left\{ \frac{-\eta}{2}(1 + (\theta - \theta_0)^2) \right\} d\eta$$

$$\geq \left[ 1 + (\theta' - \theta_0)^2 \right]^{-\nu} \frac{e^{-(\theta')^2/2}}{\sqrt{2\pi}}.$$

(*Hint:* Establish that

$$\exp\left\{ -\left(\theta' - \frac{\theta_0 \eta}{1+\eta}\right)^2 \frac{1+\eta}{2} \right\} \geq \exp\left\{ -\frac{1}{2}(\theta')^2 - \frac{\eta}{2}(\theta'^2 - 2\theta'\theta_0 + \theta_0^2) \right\}$$

and integrate.)

**7.8** (Hobert *et al.* 1997) Consider a distribution $f$ on $\mathbb{R}^k$ to be simulated by Gibbs sampling. A one-to-one transform $\Psi$ on $\mathbb{R}^k$ is called a *parameterization*. The convergence of the Gibbs sampler can be jeopardized by a bad choice of parameterization.

(a) Considering that a Gibbs sampler can be formally associated with the full conditional distributions for every choice of a parameterization, show that there always exist parameterizations such that the Gibbs sampler fails to converge.

(b) The fact that the convergence depends on the parameterization vanishes if
    (i) the support of $f$ is arcwise connected, that for every $(x, y) \in (\text{supp}(f))^2$,
    there exists a continuous function $\varphi$ on $[0, 1]$ with $\varphi(0) = x$, $\varphi(1) = y$,
    and $\varphi([0, 1]) \subset \text{supp}(f)$ and (ii) the parameterizations are restricted to be
    continuous functions.

(c) Show that condition (i) in (b) is necessary for the above property to hold.
    (*Hint:* See Example 7.1.10.)

(d) Show that a rotation of $\pi/4$ of the coordinate axes eliminates the irre-
    ducibility problem in Example 7.1.10.

**7.9** Compare the usual demarginalization of the Student's $t$ distribution discussed in Example 7.1.6 with an alternative using a slice sampler. Produce a
table like Table 7.1.1. (*Hint:* Use two uniform dummy variables.)

**7.10** Reproduce the comparison of Example 7.1.7 in the case of (a) the gamma
distribution and (b) the Poisson distribution.

**7.11** For the factor ARCH model of (5.3.9):

(a) Propose a noninformative prior distribution on the parameter $\theta = (\alpha, \beta, a, \Sigma)$
    that leads to a proper posterior distribution.

(b) Propose a completion step for the latent variables based on $f(y_t^* | y_t, y_{t-1}^*, \theta)$.

(*Note:* See Diebold and Nerlove 1989, Gouriéroux *et al.* 1993, Kim *et al.* 1998,
and Billio *et al.* 1998 for different estimation approaches to this model.)

**7.12** In the setup of Example 5.2.10, show that the Gibbs sampling simulation
of the ordered normal means $\theta_{ij}$ can either be done in $I \times J$ conditional steps
or in only two conditional steps. Conduct an experiment to compare both
approaches.

**7.13** The clinical mastitis data described in Example 7.1.21 is the number of
cases observed in 127 herds of dairy cattle (the herds are adjusted for size).
The data are given in Table 7.5.1.

| 0 | 0 | 0 | 0 | 0 | 0 | 0 | 1 | 1 | 1 | 1 | 1 | 1 | 1 | 1 |
|---|---|---|---|---|---|---|---|---|---|---|---|---|---|---|
| 1 | 1 | 1 | 1 | 2 | 2 | 2 | 2 | 2 | 2 | 2 | 2 | 3 | 3 | 3 |
| 3 | 3 | 3 | 3 | 3 | 3 | 4 | 4 | 4 | 4 | 4 | 4 | 4 | 4 | 5 |
| 5 | 5 | 5 | 5 | 5 | 5 | 5 | 6 | 6 | 6 | 6 | 6 | 6 | 6 | 6 |
| 6 | 7 | 7 | 7 | 7 | 7 | 7 | 8 | 8 | 8 | 8 | 8 | 9 | 9 | 9 |
| 10 | 10 | 10 | 10 | 11 | 11 | 11 | 11 | 11 | 11 | 11 | 12 | 12 | 12 | 12 |
| 13 | 13 | 13 | 13 | 13 | 14 | 14 | 15 | 16 | 16 | 16 | 16 | 17 | 17 | 17 |
| 18 | 18 | 18 | 19 | 19 | 19 | 19 | 20 | 20 | 21 | 21 | 22 | 22 | 22 | 22 |
| 23 | 25 | 25 | 25 | 25 | 25 | 25 | | | | | | | | |

Table 7.5.1. *Occurrences of clinical mastitis in 127 herds of dairy cattle. (Source: Schukken et al. 1991.)*

For these data, implement the Gibbs sampler described in Example 7.1.21,
and plot the posterior density $\pi(\lambda_1 | \mathbf{x}, \alpha)$. (Use the values $a = 0.1$ and 1, and
$\alpha = 1.25$.)

**7.14**(a) For the Gibbs sampler of $[A.32]$, if the Markov chain $(Y^{(t)})$ satisfies the positivity condition (see Definition 7.1.11), show that the conditional distributions $g(y_i|y_1, y_2, \ldots, y_{i-1}, y_{i+1}, \ldots, y_p)$ will not reduce the range of possible values of $Y_i$ when compared with $g$.

(b) Prove Theorem 7.1.12.

**7.15** A model studied by Besag *et al.* (1995) describes a mortality analysis related to cancer for some cohorts of individuals. The data are made of groups of ages $i$ $(1 \leq i \leq I)$ and of periods of observation $j$ $(1 \leq j \leq J)$, with observations $X_{ij} \sim \mathcal{B}(n_{ij}, p_{ij})$. The modeling of the mortality probabilities $p_{ij}$ is made through a logistic representation,

$$(7.5.1) \qquad \text{logit}(p_{ij}) = \mu + \theta_i + \varphi_j + \psi_{I-i+j} + \varepsilon_{ij},$$

and the extra-binomial variations of the observations (explained by latent variables) justify the introduction of a random effect $\varepsilon_{ij} \sim \mathcal{N}(0, \delta^2)$. (This representation is also proposed by Albert 1988 to account for deviance from a strict logit model.)

(a) Give the likelihood associated with this model.

The second level of the model represents the age, period, and cohort effects by a random walk on the *second differences* $(1 < i < I, \ 1 < j < J, \ 1 < k < K = I + J - 1)$

$$\Delta^2 \theta_{i+1} = \Delta\theta_{i+1} - \Delta\theta_i$$
$$= (\theta_{i+1} - \theta_i) - (\theta_i - \theta_{i-1}) \sim \mathcal{N}(0, \lambda^2),$$
$$\Delta^2 \varphi_{j+1} \sim \mathcal{N}(0, \nu^2), \qquad \Delta^2 \psi_{k+1} \sim \mathcal{N}(0, \zeta^2),$$

in order to remove a tendency effect on these parameters. The prior on the initial and terminal values $(i = 1, I, j = 1, J, k = 1, K)$ is uniform.

(b) Show that the conditional (on $\lambda, \nu, \zeta$) posterior distribution is proper.

The third level of the prior distribution includes the distributions of the hyperparameters $\delta^{-2}$, $\lambda^{-2}$, $\nu^{-2}$, and $\zeta^{-2}$ which are gamma distributions with corresponding parameters $(a, b)$, $(c, d)$, $(u, f)$, and $(g, h)$, the prior on $\mu$ being uniform.[6]

(c) Examine whether the joint posterior distribution is well defined.

(d) Determine the nonidentifiable directions of the parameter space and propose identifying constraints on the parameters $\theta_i$, $\varphi_i$ and $\psi_k$.

(e) Show that the Bayes estimates of the $p_{ij}$'s can still be obtained by marginalization.

(f) Show that the joint posterior distribution is

$$\prod_{i,j} \frac{e^{\text{logit}(p_{ij})y_{ij}}}{\left(1 + e^{\text{logit}(p_{ij})}\right)^{n_{ij}}} \frac{1}{\lambda^{I-2}} \exp\left\{ -\frac{1}{2\lambda^2} \sum_{i=2}^{I-1} (\theta_{i+1} - 2\theta_i + \theta_{i-1})^2 \right\}$$

$$\times \frac{1}{\nu^{J-2}} \frac{1}{\zeta^{K-2}} \frac{1}{\delta^{IJ}} \exp\left\{ -\frac{1}{2\nu^2} \sum_{j=2}^{J-1} (\varphi_{j+1} - 2\varphi_j + \varphi_{j-1})^2 \right\}$$

---

[6] This modeling is quite complex, in particular because of the nonidentifiability of the parameters of the age, period, and cohort effects, but it gives a more satisfactory representation of the dataset, contrary to simpler modelings (see Besag *et al.* 1995).

$$\times \exp\left\{-\frac{1}{2\zeta^2}\sum_{k=2}^{K-1}(\psi_{k+1} - 2\psi_k + \psi_{k-1})^2\right\}\exp\left\{-\frac{1}{2\delta^2}\sum_{i,j}\varepsilon_{ij}^2\right\}$$

$$\times \delta^{-2a-1}e^{-b/\delta^2}\lambda^{-2c-1}e^{-d/\lambda^2}\nu^{-2u-1}e^{-f/\nu^2}\zeta^{-2g-1}e^{-h/\zeta^2}.$$

(g) Define $k = I - i + j$, $\omega_{ij} = \text{logit}(p_{ij})$ and $M_k = k \wedge J - (k+1-I) \vee 1$. Verify that the full conditional distributions are as follows:

$$\mu \sim \mathcal{N}\left(\frac{1}{IJ}\sum_{i,j}(\omega_{ij} - \theta_i - \varphi_i - \psi_k), \frac{\delta^2}{IJ}\right),$$

$$\theta_i \sim \mathcal{N}\left(\frac{\delta^2\lambda^2}{J\lambda^2 + 6\delta^2}\left\{\lambda^{-2}[-\theta_{i-2} + 4(\theta_{i-1} + \theta_{i+1}) - \theta_{i+2}]\right.\right.$$
$$\left.\left. +\delta^{-2}\sum_j[\omega_{ij} - \mu - \varphi_j - \psi_k]\right\}, \frac{\delta^2\lambda^2}{J\lambda^2 + 6\delta^2}\right),$$

$$\varphi_j \sim \mathcal{N}\left(\frac{\delta^2\nu^2}{I\nu^2 + 6\delta^2}\left\{\nu^{-2}[-\varphi_{j-2} + 4(\varphi_{j-1} + \varphi_{j+1}) - \varphi_{j+2}]\right.\right.$$
$$\left.\left. +\delta^{-2}\sum_i[\omega_{ij} - \mu - \theta_i - \psi_k]\right\}, \frac{\delta^2\nu^2}{I\nu^2 + 6\delta^2}\right),$$

$$\psi_k \sim \mathcal{N}\left(\frac{\delta^2\zeta^2}{M_k\zeta^2 + 6\delta^2}\left\{\zeta^{-2}[-\psi_{k-2} + 4(\psi_{k-1} + \psi_{k+1}) - \psi_{k+2}]\right.\right.$$
$$\left.\left. +\delta^{-2}\sum_{(i,j);i-j=I-k}[\omega_{ij} - \mu - \theta_i - \varphi_j]\right\}, \frac{\delta^2\zeta^2}{M_k\zeta^2 + 6\delta^2}\right),$$

$$\omega_{ij} \sim \pi(\omega_{ij}) \propto \frac{e^{\omega_{ij}y_{ij}}}{(1 + e^{\omega_{ij}})^{n_{ij}}}\frac{1}{\delta^2}e^{-(\omega_{ij} - \mu - \theta_i - \varphi_j - \psi_{ij})^2/(2\delta^2)},$$

$$\delta^2 \sim \mathcal{IG}\left(a + \frac{IJ}{2} - 1, b + \frac{1}{2}\sum_{i,j}(\omega_{ij} - \mu - \theta_i - \varphi_j - \psi_k)^2\right),$$

$$\lambda^2 \sim \mathcal{IG}\left(c + \frac{I}{2} - 1, d + \frac{1}{2}\sum_{i=2}^{I-1}(\Delta^2\theta_{i+1})^2\right),$$

$$\nu^2 \sim \mathcal{IG}\left(u + \frac{J}{2} - 1, f + \frac{1}{2}\sum_{j=2}^{J-1}(\Delta^2\varphi_{j+1})^2\right),$$

$$\zeta^2 \sim \mathcal{IG}\left(g + \frac{K}{2} - 1, h + \frac{1}{2}\sum_{k=2}^{K-2}(\Delta^2\psi_{k+1})^2\right).$$

(h) Give the modifications of the hyperparameters of these distributions in the limiting cases.

(i) Verify that the conditional distributions of part (g) can be implemented in a Gibbs sampler to obtain the distribution of $\omega_{ij}$.

**7.16** (Continuation of Problem 7.15) The conditional posterior distribution of

$w_{ij}$ $(1 \leq i \leq I, 1 \leq j \leq J)$ in (g) is not directly available since

$$(7.5.2) \quad \pi(w_{ij}) \propto \frac{e^{w_{ij} y_{ij}}}{(1 + e^{w_{ij}})^{n_{ij}}} \exp\left\{-(w_{ij} - \mu - \theta_i - \varphi_j - \psi_{ij})^2/2\delta^2\right\}.$$

(a) Propose and calibrate an Accept–Reject algorithm for the simulation of (7.5.2).

(b) Show that the following Metropolis–Hastings step can be used instead:

  (a) Simulate $\varpi \sim \mathcal{N}(w_{ij}^{(t)}, \delta^2 + \tau^2)$, $u \sim \mathcal{U}_{[0,1]}$.
  (b) Take

$$w_{ij}^{(t+1)} = \begin{cases} \varpi & \text{if } u < \dfrac{e^{(\varpi - w_{ij}^{(t)})y_{ij}}}{\left(\dfrac{1 + e^{\varpi}}{1 + e^{w_{ij}^{(t)}}}\right)^{n_{ij}}} \varrho, \\[2em] w_{ij}^{(t)} & \text{otherwise}, \end{cases}$$

  with

$$\varrho = \frac{e^{-(\varpi - \mu - \theta_i - \varphi_j - \psi_{ij})^2/(2\delta^2)}}{e^{-(w_{ij}^{(t)} - \mu - \theta_i - \varphi_j - \psi_{ij})^2/(2\delta^2)}}.$$

(*Note:* The variance $\tau^2$ can be calibrated during a warmup stage, as suggested in Note 6.7.4.)

(c) Show that a slice sampler [A.33] can implemented for the simulation of (7.5.2). (*Hint:* It is possible to use three uniform auxiliary variables.)

(d) Compare the three approaches via a Monte Carlo experiment.

**7.17** For the situation of Example 7.2.11:

(a) Verify the variance representations (7.2.6) and (7.2.7).

(b) For the bivariate normal sampler, show that $\text{cov}(X_1, X_k) = \rho^{2k}$, for all $k$, and $\sigma^2_{\delta_0}/\sigma^2_{\delta_1} = 1/\rho^2$.

**7.18** The monotone decrease of the correlation seen in §7.2.3 does not hold uniformly for all Gibbs samplers as shown by the example of Liu *et al.* (1994): For the bivariate normal Gibbs sampler of Example 7.2.11, and $h(x, y) = x - y$, show that

$$\text{cov}[h(X_1, Y_1), h(X_2, Y_2)] = -\rho(1 - \rho)^2 < 0.$$

**7.19** Show that the model (7.3.2) is not identifiable in $(\beta, \sigma)$. (*Hint:* Show that it only depends on the ratio $\beta/\sigma$.)

**7.20** For the hybrid algorithm of [A.40], show that the resulting chain is a Markov chain and verify its stationary distribution. Is the chain reversible?

**7.21** For the algorithm of Example 6.3.9, an obvious candidate to simulate $\pi(a|\mathbf{x}, \mathbf{y}, b)$ and $\pi(b|\mathbf{x}, \mathbf{y}, a)$ is the ARS algorithm of §2.3.3.

(a) Show that

$$\log \pi(a|\mathbf{x}, \mathbf{y}, b) = \sum_i y_i a - \sum_i y_i \log\left(1 + e^{a + bx_i}\right) - \sum_i \frac{e^{a + bx_i}}{1 + e^{a + bx_i}} - \frac{a^2}{2\sigma^2},$$

and, thus,

$$\frac{\partial}{\partial a} \log \pi(a|\mathbf{x}, \mathbf{y}, b) = \sum_i y_i - \sum_i y_i \frac{e^{a + bx_i}}{1 + e^{a + bx_i}} - \sum_i \frac{e^{a + bx_i}}{(1 + e^{a + bx_i})^2} - \frac{a}{\sigma^2}$$

and

$$\frac{\partial^2}{\partial a^2} \log \pi(a|\mathbf{x}, \mathbf{y}, b) = -\sum_i \frac{y_i e^{a+bx_i}}{(1 + e^{a+bx_i})} - \sum_i \frac{e^{a+bx_i} - e^{2(a+bx_i)}}{(1 + e^{a+bx_i})^3} - \sigma^{-2}$$

$$= -\sum_i \frac{e^{a+bx_i}}{(1 + e^{a+bx_i})^3} \left\{ (1 + y_i) - (1 - y_i)e^{a+bx_i} \right\} - \sigma^{-2}.$$

Argue that this last expression is not always negative, so the ARS algorithm cannot be applied.

(b) Show that

$$\frac{\partial^2}{\partial b^2} \log \pi(b|\mathbf{x}, \mathbf{y}, a)$$

$$= -\sum_i \frac{e^{a+bx_i}}{(1 + e^{a+bx_i})^3} \left\{ (1 + y_i) - (1 - y_i)e^{a+bx_i} \right\} x_i^2 - \tau^{-2}$$

and deduce there is also no log-concavity in the $b$ direction.

(c) Even though distributions are not log-concave, they can be simulated with the ARMS algorithm [A.29]. Give details on how to do this.

**7.22** Show that the Riemann sum method of §3.4 can be used in conjunction with Rao–Blackwellization to cover multidimensional settings. (*Note:* See Philippe and Robert 1998b for details.)

**7.23** The *tobit* model is used in econometrics (see Tobin 1958, and Gouriéroux and Monfort 1996). It is based on a transform of a normal variable $y_i^* \sim \mathcal{N}(x_i^t \beta, \sigma^2)$ $(i = 1, \dots, n)$ by truncation,

$$y_i = \max(y_i^*, 0).$$

Show that the following algorithm provides a valid approximation of the posterior distribution of $(\beta, \sigma^2)$:

1. Simulate $y_i^* \sim \mathcal{N}_-(x_i^t \beta, \sigma^2, 0)$ if $y_i = 0$.
2. Simulate $(\beta, \sigma) \sim \pi(\beta, \sigma|y^*, x)$ with               $[A_{31}]$

$$\pi(\beta, \sigma|y^*, x) \propto \sigma^{-n} \exp \left\{ -\sum_i (y_i^* - x_i^t \beta)^2 / 2\sigma^2 \right\} \pi(\beta, \sigma).$$

**7.24** The situation of Example 7.2.2 also lends itself to the EM algorithm, similar to the Gibbs treatment (Example 7.2.3) and EM treatment (Example 5.3.7) of the grouped multinomial data problem. For the data of Table 7.2.1:

(a) Use the EM algorithm to calculate the MLE of $\lambda$.

(b) Compare your answer in part (a) to that from the Gibbs sampler of Algorithm [A.35]

(c) Establish that the Rao–Blackwellized estimator is correct.

**7.25** In the setup of Example 7.2.3, the (uncompleted) posterior distribution is available as

$$\pi(\eta, \mu|x) \propto (a_1\mu + b_1)^{x_1} (a_2\mu + b_2)^{x_2} (a_3\eta + b_3)^{x_3} (a_4\eta + b_4)^{x_4}$$
$$\times (1 - \mu - \eta)^{x_5 + \alpha_3 - 1} \mu^{\alpha_1 - 1} \eta^{\alpha_2 - 1}.$$

(a) Show that the marginal distributions $\pi(\mu|x)$ and $\pi(\eta|x)$ can be explicitly computed as polynomials when the $\alpha_i$'s are integers.

(b) Give the marginal posterior distribution of $\xi = \mu/(1 - \eta - \mu)$ (*Note:* See Robert 1995b for a solution.)

(c) Evaluate the Gibbs sampler proposed in Example 7.2.3 by comparing approximate moments of $\mu$, $\eta$, and $\xi$ with their exact counterpart, derived from the explicit marginal.

**7.26** In the setup of Example 7.2.4, the posterior distribution of $N$ can be evaluated by recursion.

(a) Show that
$$\pi(N) \propto \frac{(N - n_0)! \lambda^N}{N!(N - n_t)!} \ .$$

(b) Using the ratio $\pi(N)/\pi(N - 1)$, derive a recursion relation to compute $\mathbb{E}^\pi[N|n_0, n_t]$.

(c) In the case $n_0 = 112$, $n_t = 79$, and $\lambda = 500$, compare the computation time of the above device with the computation time of the Gibbs sampler. (See George and Robert 1992 for details.)

**7.27** For the situation of Example 7.2.5 (see also Example 5.3.8), define $y_{ijk} = \mathbb{I}(h_{ij} = k)\mathbb{I}(x_{ijk} = 1)$.

(a) Show that the full conditional posterior distributions are given by
$$\{\theta_1, \ldots, \theta_m\} \sim \mathcal{D}(\lambda_1 + \Sigma_{i=1}^n \Sigma_{j=1}^t y_{ij1}, \ldots, \lambda_m + \Sigma_{i=1}^n \Sigma_{j=1}^t y_{ijm})$$
and
$$p_k \sim \mathcal{B}e(\alpha + \Sigma_{i=1}^n \Sigma_{j=1}^t x_{ijk}, \beta + n - \Sigma_{i=1}^n \Sigma_{j=1}^t x_{ijk}).$$

(b) For the data of Problem 5.27, estimate the $\theta_i$'s and the $p_i$'s using the Gibbs sampler starting from the prior parameter values $\alpha = \beta = 5$ and $\lambda_i = 2$.

(Dupuis 1995b and Scherrer 1997 discuss how to choose the prior parameter values to reflect the anticipated movement of the animals.)

**7.28** (Smith and Gelfand 1992) For $i = 1, 2, 3$, consider $Y_i = X_{1i} + X_{2i}$, with
$$X_{1i} \sim \mathcal{B}(n_{1i}, \theta_1), \qquad X_{2i} \sim \mathcal{B}(n_{2i}, \theta_2).$$

(1) Give the likelihood $L(\theta_1, \theta_2)$ for $n_{1i} = 5, 6, 4$, $n_{2i} = 5, 4, 6$, and $y_i = 7, 5, 6$.

(2) For a uniform prior on $(\theta_1, \theta_2)$, derive the Gibbs sampler based on the natural parameterization.

(3) Examine whether an alternative parameterization or a Metropolis–Hastings algorithm may speed up convergence.

**7.29** Although the Gibbs sampler has a well-understood covariance structure (see Problem 7.30), it is still difficult to do exact calculations in general. However, for the special case of the bivariate normal density,
$$(X, Y)' \sim \mathcal{N}\left( \begin{pmatrix} 0 \\ 0 \end{pmatrix}, \begin{pmatrix} 1 & \rho \\ \rho & 1 \end{pmatrix} \right),$$

with Gibbs sampler
$$X \mid y \sim \mathcal{N}(\rho y, \ 1 - \rho^2),$$
$$Y \mid x \sim \mathcal{N}(\rho x, \ 1 - \rho^2),$$

some explicit calculations can be done.

(a) Show that for the $X$ chain, the transition kernel is

$$K(x, x^*) = \frac{1}{2\pi(1 - \rho^2)} \int e^{-\frac{1}{2(1-\rho^2)}(x-\rho y)^2} e^{-\frac{1}{2(1-\rho^2)}(y-\rho x^*)^2} dy.$$

(b) Show that $X \sim \mathcal{N}(0, 1)$ is the invariant distribution of the $X$ chain.

(c) Show that $X|x^* \sim \mathcal{N}(\rho^2 x^*, 1 - \rho^4)$. (*Hint:* Complete the square in the exponent of part (a).)

(d) Show that we can write $X_k = \rho^2 X_{k-1} + U_k$, $k = 1, 2, \ldots$, where the $U_k$ are iid $\mathcal{N}(0, 1 - \rho^4)$ and that $\text{cov}(X_0, X_k) = \rho^{2k}$, for all $k$. Deduce that the covariances go to zero.

**7.30** Let $(X_n)$ be a reversible stationary Markov chain (see Definition 4.7.6) with $\mathbb{E}[X_i] = 0$.

(a) Show that the distribution of $X_k|X_1$ is the same as the distribution of $X_1|X_k$.

(b) Show that the covariance between alternate random variables is positive. More precisely, show that $\text{cov}(X_0, X_{2\nu}) = \mathbb{E}[\mathbb{E}(X_0|X_\nu)^2] > 0$.

(c) Show that the covariance of alternate random variables is decreasing; that is, show that $\text{cov}(X_0, X_{2\nu}) \geq \text{cov}(X_0, X_{2(\nu+1)})$. (*Hint:* Use the fact that

$$\mathbb{E}[\mathbb{E}(X_0|X_\nu)^2] = \text{var}[\mathbb{E}(X_0|X_\nu)] \geq \text{var}[\mathbb{E}\{\mathbb{E}(X_0|X_\nu)|X_{\nu+1}\}],$$

and show that this latter quantity is $\text{cov}(X_0, X_{2(\nu+1)})$.)

**7.31** Show that a Gibbs sampling kernel with more than two full conditional steps cannot produce interleaving chains.

**7.32** Show that the inequality (7.6.3) implies

$$\|g_t - g\|_{TV} \leq \frac{1}{2} c_0 \alpha^t,$$

and thus that it truly gives an equivalent definition of geometric ergodicity.

**7.33** Establish that the distance (7.6.2) is correct and explain the difference when $x_0 = \mu$.

**7.34** In the setup of Example 7.1.3:

(a) Consider a grid in $\mathbb{R}^2$ with the simple nearest-neighbor relation $\mathcal{N}$ (that is, $(i, j) \in \mathcal{N}$ if and only if $\min(|i_1 - j_1|, |i_2 - j_2|) = 1$). Show that a Gibbs sampling algorithm with only two conditional steps can be implemented in this case.

(b) Implement the Metropolization scheme of Liu (1995) discussed in §7.3.3 and compare the results with those of part (a).

**7.35** (Gelfand *et al.* 1995) In the setup of Example 7.3.9, show that the original parameterization leads to the following correlations:

$$\rho_{\mu,\alpha_i} = \left(1 + \frac{I\sigma_y^2}{J\sigma_\alpha^2}\right)^{-1/2}, \qquad \rho_{\alpha_i,\alpha_j} = \left(1 + \frac{I\sigma_y^2}{J\sigma_\alpha^2}\right)^{-1},$$

for $i \neq j$.

**7.36** (Continuation of Problem 7.35) For the hierarchical parameterization, show that the correlations are

$$\rho_{\mu,\eta_i} = \left(1 + \frac{IJ\sigma_\alpha^2}{\sigma_y^2}\right)^{-1/2}, \qquad \rho_{\eta_i,\eta_j} = \left(1 + \frac{IJ\sigma_\alpha^2}{\sigma_y^2}\right)^{-1},$$

for $i \neq j$.

**7.37** In Example 7.4.1, we noted that a pair of conditional densities is *functionally compatible* if the ratio $f_1(x|y)/f_2(y|x) = h_1(x)/h_2(y)$, for some functions $h_1$ and $h_2$. This is a necessary condition for a joint density to exist, but not a sufficient condition. If such a joint density does exist, the pair $f_1$ and $f_2$ would be *compatible*.

(a) Formulate the definitions of compatible and functionally compatible for a set of densities $f_1, \ldots, f_m$.

(b) Show that if $f_1, \ldots, f_m$ are the full conditionals from a hierarchical model, they are functionally compatible.

(c) Prove the following theorem, due to Hobert and Casella (1998), which shows there cannot be any stationary probability distribution for the chain to converge to unless the densities are compatible:

**Theorem 7.5.1** *Let $f_1, \ldots, f_m$ be a set of functionally compatible conditional densities on which a Gibbs sampler is based. The resulting Markov chain is positive recurrent if and only if $f_1, \ldots, f_m$ are compatible.*

(Compatibility of a set of densities was investigated by Besag 1974, Arnold and Press 1989, and Gelman and Speed 1993.)

**7.38** For the situation of Example 7.4.3, by integrating out the (unobservable) random effects $u_i$, show that the full "posterior distribution" of $(\beta, \sigma^2, \tau^2)$ is

$$\pi(\beta, \sigma^2, \tau^2|y) \propto \sigma^{-2-I} \tau^{-2-IJ} \exp\left\{-\frac{1}{2\tau^2} \sum_{i,j} (y_{ij} - \bar{y}_i)^2\right\}$$

$$\times \exp\left\{-\frac{J\sum_i(\bar{y}_i - \beta)^2}{2(\tau^2 + J\sigma^2)}\right\} (J\tau^{-2} + \sigma^{-2})^{-I/2}.$$

Next, integrate out $\beta$ to get the marginal posterior density

$$\pi(\sigma^2, \tau^2|y) \propto \frac{\sigma^{-2-I} \tau^{-2-IJ}}{(J\tau^{-2} + \sigma^{-2})^{I/2}} (\tau^2 + J\sigma^2)^{1/2}$$

$$\times \exp\left\{-\frac{1}{2\tau^2} \sum_{i,j} (y_{ij} - \bar{y}_i)^2 - \frac{J}{2(\tau^2 + J\sigma^2)} \sum_i (\bar{y}_i - \bar{y})^2\right\}.$$

Finally, show that the full posterior is not integrable since, for $\tau \neq 0$, $\pi(\sigma^2, \tau^2|y)$ behaves like $\sigma^{-2}$ in a neighborhood of 0.

**7.39** The *traveling salesman problem* is a classic in combinatoric and operation research, where a salesman has to find the shortest route to visit each of his $N$ customers.

(a) Show that the problem can be described by (a) a permutation $\sigma$ on $\{1, \ldots, N\}$ and (b) a distance $d(i,j)$ on $\{1, \ldots, N\}$.

(b) Deduce that the traveling salesman problem is equivalent to minimization of the function

$$H(\sigma) = \sum_i d(i, \sigma(i)).$$

(c) Propose a Metropolis–Hastings algorithm to solve the problem with a simulated annealing scheme (§5.2.3).

**7.40** There is a connection between the EM algorithm and Gibbs sampling, in that both have their basis in Markov chain theory. One way of seeing this is to show that the incomplete data likelihood is a solution to the integral equation of successive substitution sampling and that Gibbs sampling can then be used to calculate the likelihood function. If $L(\theta|\mathbf{y})$ is the incomplete data likelihood and $L(\theta|\mathbf{y}, \mathbf{z})$ is the complete data likelihood, define

$$L^*(\theta|\mathbf{y}) = \frac{L(\theta|\mathbf{y})}{\int L(\theta|\mathbf{y})d\theta}, \quad L^*(\theta|\mathbf{y}, \mathbf{z}) = \frac{L(\theta|\mathbf{y}, \mathbf{z})}{\int L(\theta|\mathbf{y}, \mathbf{z})d\theta},$$

assuming both integrals to be finite.

(a) Show that $L^*(\theta|\mathbf{y})$ is the solution to

$$L^*(\theta|\mathbf{y}) = \int \left[ \int L^*(\theta|\mathbf{y}, \mathbf{z})k(\mathbf{z}|\theta', \mathbf{y})d\mathbf{z} \right] L^*(\theta'|\mathbf{y})d\theta',$$

where $k(\mathbf{z}|\theta, \mathbf{y}) = L(\theta|\mathbf{y}, \mathbf{z})/L(\theta|\mathbf{y})$.

(b) Show that the sequence $\theta_{(j)}$ from the Gibbs iteration

$$\theta_{(j)} \sim L^*(\theta|\mathbf{y}, \mathbf{z}_{(j-1)}),$$
$$\mathbf{z}_{(j)} \sim k(\mathbf{z}|\theta_{(j)}, \mathbf{y}),$$

converges to a random variable with density $L^*(\theta|\mathbf{y})$ as $j$ goes $\infty$. How can this be used to compute the likelihood function $L(\theta|\mathbf{y})$?

(*Note:* Based on the same functions $L(\theta|\mathbf{y}, \mathbf{z})$ and $k(\mathbf{z}|\theta, \mathbf{y})$ the EM algorithm will get the ML estimator from $L(\theta|\mathbf{y})$, whereas the Gibbs sampler will get us the entire function. This likelihood implementation of the Gibbs sampler was used by Casella and Berger 1994 and is also described by Smith and Roberts 1993. A version of the EM algorithm, where the Markov chain connection is quite apparent, was given by Baum and Petrie 1966 and Baum *et al.* 1971.)

**7.41** (Gelfand and Dey 1994) Consider a density function $f(x|\theta)$ and a prior distribution $\pi(\theta)$ such that the marginal $m(x) = \int_\Theta f(x|\theta)\pi(\theta)d\theta$ is finite a.e. The marginal density is of use in the comparison of models since it appears in the Bayes factor (see §1.3).

(a) Give a Laplace approximation of $m$ and derive the corresponding approximation of the Bayes factor. (See Tierney *et al.* 1989 for details.)

(b) Give the general shape of an importance sampling approximation of $m$.

(c) Detail this approximation when the importance function is the posterior distribution and when the normalizing constant is unknown.

(d) Show that for a proper density $\tau$,

$$m(x)^{-1} = \int_\Theta \frac{\tau(\theta)}{f(x|\theta)\pi(\theta)}\pi(\theta|x)d\theta,$$

and deduce that when the $\theta_i^*$'s are generated from the posterior,

$$\hat{m}(x) = \left\{ \frac{1}{T} \sum_{t=1}^{T} \frac{\tau(\theta_i^*)}{f(x|\theta_i^*)\pi(\theta_i^*)} \right\}^{-1}$$

is another importance sampling estimator of $m$.

**7.42** (Chib 1995) Consider the approximation of the marginal density $m(x) = f(x|\theta)\pi(\theta)/\pi(\theta|x)$, where $\theta = (\theta_1, \ldots, \theta_B)$ and the full conditionals $\pi(\theta_r|x, \theta_s, s \neq r)$ are available.

(a) In the case $B = 2$, show that an appropriate estimate of the marginal loglikelihood is

$$\hat{\ell}(x) = \log\, f(x|\theta_1^*) + \log\, \pi_1(\theta_1^*) - \log\, \left\{ \frac{1}{T} \sum_{t=1}^{T} \pi(\theta_1^*|x, \theta_2^{(t)}) \right\},$$

where $\theta_1^*$ is an arbitrary point, assuming that $f(x|\theta) = f(x|\theta_1)$ and that $\pi_1(\theta_1)$ is available.

(b) In the general case, rewrite the posterior density as

$$\pi(\theta|x) = \pi_1(\theta_1|x)\pi_2(\theta_2|x, \theta_1) \cdots \pi_B(\theta_B|x, \theta_1, \ldots, \theta_{B-1}).$$

Show that an estimate of $\pi_r(\theta_r^*|x, \theta_1^*, \ldots, \theta_{r-1}^*)$ is

$$\hat{\pi}_r(\theta_r^*|x, \theta_s^*, s < r) = \frac{1}{T} \sum_{t=1}^{T} \pi(\theta_r^*|x, \theta_1^*, \ldots, \theta_{r-1}^*, \theta_{r+1}^{(t)}, \ldots, \theta_B^{(t)}),$$

where the $\theta_\ell^{(t)}$ ($\ell > r$) are simulated from the full conditionals $\pi_\ell(\theta_\ell|\theta_1^*, \ldots, \theta_r^*, \theta_{r+1}, \ldots, \theta_B)$.

(c) Deduce that an estimate of the joint posterior density is $\prod_{r=1}^{B} \hat{\pi}_r(\theta_r^*|x, \theta_s^*, s < r)$ and that an estimate of the marginal loglikelihood is

$$\hat{\ell}(x) = \log\, f(x|\theta^*) + \log\, \pi(\theta^*) - \sum_{r=1}^{B} \log\, \hat{\pi}_r(\theta_r^*|x, \theta_s^*, s < r)$$

for an arbitrary value $\theta^*$.

(d) Discuss the computational cost of this method as a function of $B$.

(e) Extend the method to the approximation of the predictive density $f(y|x) = \int f(y|\theta)\pi(\theta|x)d\theta$.

**7.43** (Fishman 1996) Given a contingency table with cell sizes $N_{ij}$, row sums $N_i.$, and column sums $N_{.j}$, the chi squared statistics for independence is

$$\chi^2 = \sum_{(i,j)} \frac{(N_{ij} - N_i.N_{.j}/N)^2}{N_i.N_{.j}/N}.$$

Assuming fixed margins $N_i.$ and $N_{.j}$ ($i = 1, \ldots, I, j = 1, \ldots, J$), the goal is to simulate the distribution of $\chi^2$. Design a Gibbs sampling experiment to simulate a contingency table under fixed margins. (*Hint:* Show that the vector to be simulated is of dimension $(I - 1)(J - 1)$.) (Alternative methods are described by Aldous (1987) and Diaconis and Sturmfels (1998).)

**7.44** †[7](Gelfand *et al.* 1990) For a population of 30 rats, the weight $y_{ij}$ of rat $i$ is observed at age $x_j$ and is associated with the model

$$Y_{ij} \sim \mathcal{N}(\alpha_i + \beta_i(x_j - \bar{x}), \sigma_c^2).$$

(a) Give the Gibbs sampler associated with this model and the prior

$$\alpha_i \sim \mathcal{N}(\alpha_c, \sigma_\alpha^2), \qquad \beta_i \sim \mathcal{N}(\beta_c, \sigma_\beta^2),$$

with almost flat hyperpriors

$$\alpha_c, \beta_c \sim \mathcal{N}(0, 10^4), \quad \sigma_c^{-2}, \sigma_\alpha^{-2}, \sigma_\beta^{-2} \sim \mathcal{G}a(10^{-3}, 10^{-3}).$$

(b) Assume now that

$$Y_{ij} \sim \mathcal{N}(\beta_{1i} + \beta_{2i}x_j, \sigma_c^2) ,$$

with $\beta_i = (\beta_{1i}, \beta_{2i}) \sim \mathcal{N}_2(\mu_\beta, \Omega_\beta)$. Using a Wishart hyperprior $\mathcal{W}(2, R)$ on $\Omega_\beta$, with

$$R = \begin{pmatrix} 200 & 0 \\ 0 & 0.2 \end{pmatrix},$$

give the corresponding Gibbs sampler.

(c) Study whether the original assumption of independence between $\beta_{1i}$ and $\beta_{2i}$ holds.

**7.45** †(Spiegelhalter *et al.* 1996) Binomial observations

$$R_i \sim \mathcal{B}(n_i, p_i), \qquad i = 1, \dots, 12,$$

correspond to mortality rates for cardiac surgery on babies in hospitals. When the failure probability $p_i$ for hospital $i$ is modeled by a random effect structure

$$\text{logit}(p_i) \sim \mathcal{N}(\mu, \tau^2),$$

with almost flat hyperpriors

$$\mu \sim \mathcal{N}(0, 10^4), \quad \tau^{-2} \sim \mathcal{G}a(10^{-3}, 10^{-3}),$$

examine whether a Gibbs sampler can be implemented. (*Hint:* Consider the possible use of ARS as in §2.3.3.)

**7.46** †(Spiegelhalter *et al.* 1996) In an experiment about the batch variation in dyes, Gelfand *et al.* (1995) models five samples $y_{ij}$ from six batches as

$$Y_{ij} \sim \mathcal{N}(\mu_i, \sigma_w^2), \qquad \mu_i \sim \mathcal{N}(\theta, \sigma_b^2)$$

$(i = 1, \dots, 6, j = 1, \dots, 5)$.

(a) Give the Gibbs sampler corresponding to the almost flat hyperprior

$$\theta \sim \mathcal{N}(0, 10^{10}), \qquad \sigma_b^{-2}, \sigma_w^{-2} \sim \mathcal{G}a(10^{-3}, 10^{-3}) ,$$

and derive the posterior distribution of the ratio $\sigma_w^2/\sigma_b^2$.

(b) Using the alternative parameterization of Box and Tiao (1968),

$$Y_{ij} = \theta + \beta_i + \omega_{ij}, \qquad \beta_i \sim \mathcal{N}(0, \sigma_b^2), \quad \omega_{ij} \sim \mathcal{N}(0, \sigma_w^2) ,$$

compare the results of the associated Gibbs sampler with those of the previous sampler.

---

[7] Problems with this dagger symbol are studied in detail in the BUGS manual of Spiegelhalter *et al.* (1996). Corresponding datasets can also be obtained from this software (see Note 7.6.4).

**7.47** †(Carlin 1992) A meta-analysis of 22 evaluations of the effect of a drug on mortality after myocardial infarction leads to the observation of deaths $r_i^C$ and $r_i^T$ in both control and treatment groups, with distributions

$$r_i^C \sim \mathcal{B}(n_i^C, p_i^C), \qquad r_i^T \sim \mathcal{B}(n_i^T, p_i^T) \qquad (i = 1, \ldots, 22)$$

with

$$\text{logit}(p_i^C) = \mu_i, \qquad \text{logit}(p_i^T) = \mu_i + \delta_i.$$

If $\delta_i \sim \mathcal{N}(\theta, \sigma^2)$, with almost flat hyperpriors on the $\mu_i$'s, $\theta$, and $\sigma^2$, construct the Gibbs sampler and derive the posterior distribution of $\theta$.

(*Note:* Almost flat priors are used as a substitute for improper priors by BUGS; see Note 7.6.4.)

**7.48** †(Boch and Aitkin 1981) Data from the Law School Aptitude Test (LSAT) corresponds to multiple-choice test answers $(y_{j1}, \ldots, y_{j5})$ in $\{0, 1\}^5$. The $y_{jk}$'s are modeled as $\mathcal{B}(p_{jk})$, with $(j = 1, \ldots, 1000, k = 1, \ldots, 5)$

$$\text{logit}(p_{jk}) = \theta_j - \alpha_k, \qquad \theta_j \sim \mathcal{N}(0, \sigma^2).$$

(This is the *Rasch model.*) Using vague priors on the $\alpha_k$'s and $\sigma^2$, give the marginal distribution of the probability $P_i$ to answer $i \in \{0, 1\}^5$. (*Hint:* Show that $P_i$ is the posterior expectation of the probability $P_{i|\theta}$ conditional on ability level.)

**7.49** †(Dellaportas and Smith 1993) Observations $t_i$ on survival time are related to covariates $z_i$ by a Weibull model

$$T_i | z_i \sim \mathcal{W}e(r, \mu_i), \qquad \mu_i = \exp(\beta^t z_i),$$

with possible censoring. The prior distributions are

$$\beta_j \sim \mathcal{N}(0, 10^4), \qquad r \sim \mathcal{G}a(1, 10^{-4}).$$

(a) Construct the associated Gibbs sampler and derive the posterior expectation of the median, $\log(2 \exp(-\beta^t z_i))^{1/r}$.

(b) Compare with an alternative implementation using a slice sampler [A.33]

**7.50** †(Carlin and Gelfand 1991) The following nonlinear model relates the length $y_i$ of dugong $i$ (an aquatic herbivorous mammal living in Asian coastal regions) to its age $x_i$ $(i = 1, \ldots, 27)$,

$$Y_i \sim \mathcal{N}(\alpha - \beta \gamma^{x_i}, \sigma^2) \qquad \alpha, \beta > 1, \ \gamma \in [0, 1].$$

The parameters $\alpha$, $\beta$, and $\tau$ are associated with almost flat hyperpriors, and $\gamma \sim \mathcal{U}([0, 1])$.

(a) Show that the full conditional distribution on $\gamma$ is not log-concave.

(b) Spiegelhalter *et al.* (1996) propose a Gibbs sampler based on a discretization of $\gamma$. Derive this Gibbs sampler and study the sensitivity to the discretization step.

(c) Demarginalize the conditional posterior on $\gamma$

$$\exp\left(-\sum_{i=1}^{27} (y_i - \alpha + \beta \gamma e^{x_i})^2 / 2\sigma^2\right)$$

by using uniform variables $u_i$ and construct a Gibbs sampler for the simulation of $\alpha, \beta, \tau, \gamma, u_1, \ldots, u_{27}$. Compare with the previous implementation.

**7.51** †(Spiegelhalter *et al.* 1996) In the study of the effect of a drug on a heart disease, the number of contractions per minute for patient $i$ is recorded before treatment $(x_i)$ and after treatment $(y_i)$. The full model is $X_i \sim \mathcal{P}(\lambda_i)$ and for uncured patients, $Y_i \sim \mathcal{P}(\beta\lambda_i)$, whereas for cured patients, $y_i = 0$.

(a) Show that the conditional distribution of $Y_i$ given $t_i = x_i + y_i$ is $\mathcal{B}(t_i, \beta/(1+\beta))$ for the uncured patients.

(b) Express the distribution of the $Y_i$'s as a mixture model and derive the Gibbs sampler.

**7.52** †(Breslow and Clayton 1993) In the modeling of breast cancer cases $y_i$ according to age $x_i$ and year of birth $d_i$, an exchangeable solution is

$$Y_i \sim \mathcal{P}(\mu_i),$$
$$\log(\mu_i) = \log(d_i) + \alpha_{x_i} + \beta_{d_i},$$
$$\beta_k \sim \mathcal{N}(0, \sigma^2).$$

(a) Derive the Gibbs sampler associated with almost flat hyperpriors on the parameters $\alpha_j, \beta_k$, and $\sigma$.

(b) Breslow and Clayton (1993) consider a dependent alternative where for $k = 3, \ldots, 11$ we have

(7.5.3) $$\beta_k | \beta_1, \ldots, \beta_{k-1} \sim \mathcal{N}(2\beta_{k-1} - \beta_{k-2}, \sigma^2),$$

while $\beta_1, \beta_2 \sim \mathcal{N}(0, 10^5\sigma^2)$. Construct the associated Gibbs sampler and compare with the previous results.

(c) An alternative representation of (7.5.3) is

$$\beta_k | \beta_j, j \neq k \sim \mathcal{N}(\bar{\beta}_k, n_k\sigma^2).$$

Determine the value of $\bar{\beta}_k$ and compare the associated Gibbs sampler with the previous implementation.

**7.53** †(Dobson 1983) The effect of a pesticide is tested against its concentration $x_i$ on $n_i$ beetles, $R_i \sim \mathcal{B}(n_i, p_i)$ of which are killed. Three generalized linear models are in competition:

$$p_i = \frac{\exp(\alpha + \beta x_i)}{1 + \exp(\alpha + \beta x_i)},$$
$$p_i = \Phi(\exp(\alpha + \beta x_i)),$$
$$p_i = 1 - \exp(-\exp(\alpha + \beta x_i));$$

that is, the logit, probit, and log-log models, respectively. For each of these models, construct a Gibbs sampler and compute the expected posterior *deviance*; that is, the posterior expectation of

$$D = 2\left(\sum_{i=1}^{n} \hat{\ell}_i - \sum_{i=1}^{n} \ell_i\right),$$

where

$$\ell_i = r_i \log(p_i) + (n_i - p_i)\log(1 - p_i), \qquad \hat{\ell}_i = \max_{p_i} \ell_i.$$

**7.54** †(Spiegelhalter *et al.* 1996) Consider a standard Bayesian ANOVA model $(i = 1, \ldots, 4, j = 1, \ldots, 5)$

$$Y_{ij} \sim \mathcal{N}(\mu_{ij}, \sigma^2),$$

$$\mu_{ij} = \alpha_i + \beta_j,$$
$$\alpha_i \sim \mathcal{N}(0, \sigma_\alpha^2),$$
$$\beta_j \sim \mathcal{N}(0, \sigma_\beta^2),$$
$$\sigma^{-2} \sim \mathcal{G}a(a, b),$$

with $\sigma_\alpha^2 = \sigma_\beta^2 = 5$ and $a = 0$, $b = 1$. Gelfand *et al.* (1992) impose the constraints $\alpha_1 > \ldots > \alpha_4$ and $\beta_1 < \cdots < \beta_3 > \cdots > \beta_5$.

(a) Give the Gibbs sampler for this model. (*Hint:* Use the optimal truncated normal Accept–Reject algorithm of Example 2.3.5.)

(b) Change the parameterization of the model as ($i = 2, \ldots, 4, j = 1, 2, 4, 5$)

$$\alpha_i = \alpha_{i-1} + \epsilon_i, \qquad \epsilon_i > 0,$$
$$\beta_j = \beta_3 - \eta_j, \qquad \eta_j > 0,$$

and modify the prior distribution in

$$\alpha_1 \sim \mathcal{N}(0, \sigma_\alpha^2), \quad \beta_3 \sim \mathcal{N}(0, \sigma_\alpha^2), \quad \epsilon_i, \eta_j \sim \mathcal{G}a(0, 1).$$

Check whether the posterior distribution is well defined and compare the performances of the corresponding Gibbs sampler with the previous implementation.

**7.55** †(Spiegelhalter *et al.* 1996) Consider a simple normal regression model with a changepoint at time $k$,

$$Y_i \sim \mathcal{N}(\alpha + \beta_1(x_i - x_k), \sigma^2), \qquad i \leq k,$$
$$Y_i \sim \mathcal{N}(\alpha + \beta_2(x_i - x_k), \sigma^2), \qquad i > k.$$

Using a uniform prior on $k \in \{1, \ldots, n-1\}$ and almost flat priors on $\alpha, \beta_1$, and $\beta_2$, construct a Gibbs sampler and derive the posterior distribution of $k$.

**7.56** In the setup of Note 7.6.5:

(a) Evaluate the time requirements for the computation of the exact formulas for the $\mu_i - \mu_{i+1}$.

(b) Devise an experiment to test the maximal value of $x_{(n)}$ which can be processed on your computer.

(c) In cases when the exact value can be computed, study the convergence properties of the corresponding Gibbs sampler.

**7.57** In the setup of Note 7.6.5, we define the *canonical moments* of a distribution and show that they can be used as a representation of this distribution.

(a) Show that the two first moments $\mu_1$ and $\mu_2$ are related by the two following inequalities:

$$\mu_1^2 \leq \mu_2 \leq \mu_1 \,,$$

and that the sequence $(\mu_k)$ is monotonically decreasing to 0.

(b) Consider a $k$th-degree polynomial

$$P_k(x) = \sum_{i=0}^{k} a_i x^i.$$

Deduce from

$$\int_0^1 P_k^2(x) dG(x) \, dx \geq 0$$

that

(7.5.4) $$a^t C_k a \geq 0, \qquad \forall a \in \mathbb{R}^{k+1},$$

where

$$C_k = \begin{pmatrix} 1 & \mu_1 & \mu_2 & \cdots & \mu_k \\ \mu_1 & \mu_2 & \mu_3 & \cdots & \mu_{k+1} \\ \cdots & \cdots & \cdots & \cdots & \cdots \\ \mu_k & \mu_{k+1} & & \cdots & \mu_{2k} \end{pmatrix}$$

and $a^t = (a_0, a_1, \ldots, a_k)$.

(c) Show that for every distribution $g$, the moments $\mu_k$ satisfy

$$\begin{vmatrix} 1 & \mu_1 & \mu_2 & \cdots & \mu_k \\ \mu_1 & \mu_2 & \mu_3 & \cdots & \mu_{k+1} \\ \cdots & \cdots & \cdots & \cdots & \cdots \\ \mu_k & \mu_{k+1} & & \cdots & \mu_{2k} \end{vmatrix} \geq 0.$$

(*Hint:* Interpret this as a property of $C_k$.)

(d) Using inequalities similar to (7.5.4) for the polynomials $t(1-t)P_k^2(t)$, $tP_k^2(t)$, and $(1-t)P_k^2(t)$, derive the following inequalities on the moments of $G$:

$$\begin{vmatrix} \mu_1 - \mu_2 & \mu_2 - \mu_3 & \cdots & \mu_{k-1} - \mu_k \\ \mu_2 - \mu_3 & \mu_3 - \mu_4 & \cdots & \mu_k - \mu_{k+1} \\ \cdots & \cdots & \cdots & \cdots \\ \mu_{k-1} - \mu_k & & \cdots & \mu_{2k-1} - \mu_{2k} \end{vmatrix} \geq 0,$$

$$\begin{vmatrix} \mu_1 & \mu_2 & \cdots & \mu_k \\ \mu_2 & \mu_3 & \cdots & \mu_{k+1} \\ \cdots & \cdots & \cdots & \cdots \\ \mu_k & \mu_{k+1} & \cdots & \mu_{2k-1} \end{vmatrix} \geq 0,$$

$$\begin{vmatrix} 1 - \mu_1 & \mu_1 - \mu_2 & \cdots & \mu_{k-1} - \mu_k \\ \mu_1 - \mu_2 & \mu_2 - \mu_3 & \cdots & \mu_k - \mu_{k+1} \\ \cdots & \cdots & \cdots & \cdots \\ \mu_{k-1} - \mu_k & & \cdots & \mu_{2k-2} - \mu_{2k-1} \end{vmatrix} \geq 0.$$

(e) Show that the bounds in parts (c) and (d) induce a lower (resp. upper) bound $\underline{c}_{2k}$ (resp. $\bar{c}_{2k}$) on $\mu_{2k}$ and that part (c) [resp. (d)] induces a lower (resp. upper) bound $\underline{c}_{2k-1}$ (resp. $\bar{c}_{2k-1}$) on $\mu_{2k-1}$.

(f) Defining $p_k$ as

$$p_k = \frac{c_k - \underline{c}_k}{\bar{c}_k - \underline{c}_k},$$

show that the relation between $(p_1, \ldots, p_n)$ and $(\mu_1, \ldots, \mu_n)$ is one-to-one for every $n$ and that the $p_i$ are independent.

(g) Show that the inverse transform is given by the following recursive formulas. Define

$$q_i = 1 - p_i, \quad \zeta_1 = p_1, \quad \zeta_i = p_i q_{i-1} \qquad (i \geq 2).$$

then

$$\begin{cases} S_{1,k} = \zeta_1 + \cdots + \zeta_k & (k \geq 1), \\ S_{j,k} = \sum_{i=1}^{k-j+1} \zeta_i S_{j-1, i+j-1} & (j \geq 2), \\ c_n = S_{n,n}. \end{cases}$$

(See Dette and Studden 1997 for details and complements on canonical moments.)

## 7.6  Notes

*7.6.1  A Bit of Background*

Although somewhat removed from statistical inference in the classical sense and based on earlier techniques used in statistical Physics, the landmark paper by Geman and Geman (1984) brought Gibbs sampling into the arena of statistical application. This paper is also responsible for the name *Gibbs sampling*, because it implemented this method for the Bayesian study of *Gibbs random fields* which, in turn, derive their name from the physicist Josiah Willard Gibbs (1839–1903). This original implementation of the Gibbs sampler was applied to a discrete image processing problem and did not involve completion.

The work of Geman and Geman (1984) built on that of Metropolis *et al.* (1953) and Hastings (1970) and his student Peskun (1973), influenced Gelfand and Smith (1990) to write a paper that sparked new interest in Bayesian methods, statistical computing, algorithms, and stochastic processes through the use of computing algorithms such as the Gibbs sampler and the Metropolis–Hastings algorithm. It is interesting to see, in retrospect, that earlier papers had proposed similar solutions but did not find the same response from the statistical community. Among these, one may quote Besag (1974, 1986), Besag and Clifford (1989), Broniatowski *et al.* (1981), Quian and Titterington (1990), and Tanner and Wong (1987).

*7.6.2  Original Data Augmentation*

The original and more general version of Data Augmentation proposed by Tanner and Wong (1987) is based on a Rao–Blackwell approximation of the marginal densities.

**Algorithm A.42  –Monte Carlo Data Augmentation–**
Given $y_{21}^{(t)}, \ldots, y_{2V}^{(t)}$, $(V \geq 1)$

1. Simulate $Y_{11}^{(t+1)}, \ldots, Y_{1V}^{(t+1)} \overset{iid}{\sim} \dfrac{1}{V} \sum_{v=1}^{V} g_1\big(y_1 | y_{2v}^{(t)}\big).$ $\qquad\qquad$ [A.42]

2. Simulate $Y_{21}^{(t+1)}, \ldots, Y_{2V}^{(t+1)} \overset{iid}{\sim} \dfrac{1}{V} \sum_{v=1}^{V} g_2\big(y_2 | y_{1v}^{(t+1)}\big).$

A motivation for [A.42] is that the points of the subchain $(Y_{11}^{(t)})$ are more or less independent. Each value $(Y_{1j}^{(t+1)})$ is simulated according to $g_1(y_1 | y_{2v}^{(t)})$, with $v$ chosen at random from $\{1, \ldots, V\}$. The advantage of this approach over the parallel simulation via [A.34] of $V$ chains which are truly independent is not obvious since [A.42] does not guarantee independence, and, most importantly, it may slow down convergence to the stationary distribution since the $Y_{iv}^{(t)}$'s are chosen at random at stage $i$. Moreover, the subchains $(Y_{iv}^{(t)})$ are no longer Markov chains and a rigorous study must involve the entire vector $(Y_{i1}^{(t)}, \ldots, Y_{iV}^{(t)})$. These are some of the reasons why we only consider the case $V = 1$ (that is, the algorithm [A.34]).

*7.6.3  Geometric Convergence*

While[8] the geometric ergodicity conditions of Chapter 6 do not apply for

---

[8] This section presents some results on the properties of the Gibbs sampler algorithms from a *functional analysis* point of view. It may be skipped on a first reading since it will not be used in the book and remains at a rather theoretical level.

Gibbs sampling algorithms, some sufficient conditions can also be found in the literature, although these are not so easy to implement in practice. The approach presented below is based on results by Schervish and Carlin (1992), Roberts and Polson (1994), Liu *et al.* (1995) and Polson (1996), who work with a *functional representation* of the transition operator of the chain.

Schervish and Carlin (1992) define the measure $\mu$ as the measure with density $1/g$ with respect to the Lebesgue measure, where $g$ is the density of the stationary distribution of the chain. They then define a scalar product on $\mathcal{L}_2(\mu)$

$$\langle r, s \rangle = \int r(y)\, s(y)\, \mu(dy)$$

and define the operator $T$ on $\mathcal{L}_2(\mu)$ by

$$(Tr)(y) = \int r(y')\, K(y', y)\, g(y')\, \mu(dy'),$$

where $K(y, y')$ is the transition kernel (7.1.7) with stationary measure $g$. In this setting, $g$ is an *eigenvector associated with the eigenvalue* 1. The other eigenvalues of $T$ are characterized by the following result:

**Lemma 7.6.1** *The eigenvalues of $T$ are all within the unit disk of* $\mathbb{C}$.

*Proof.* Consider an eigenvector $r$ associated with the eigenvalue $c$, that is, such that $Tr = cr$. Since

$$\int K(y', y)\, g(y')\, dy' = g(y),$$

$r$ satisfies

$$|c| \int |r(y)|\, dy = \int |Tr(y)|\, dy$$

$$= \int \left| \int r(y')\, K(y', y)\, dy' \right| dy$$

$$\leq \int \int |r(y')|\, K(y', y)\, dy'\, dy$$

$$= \int |r(y')|\, dy'$$

and, therefore, $|c| \leq 1$.                                                    □□

The main requirement in Schervish and Carlin (1992) is the *Hilbert-Schmidt condition*

(7.6.1) $$\int \int K^2(y, y')\, dy\, dy' < \infty,$$

which ensures both the compactness of $T$ and the geometric convergence of the chain $(Y^{(t)})$. The *adjoint operator* associated with $T$, $T^*$, is defined by

$$\langle Tr, s \rangle = \langle r, T^* s \rangle$$

and satisfies

$$(T^* s)(y') = \int \frac{K(y', y)}{g(y)}\, g(y')\, s(y)\, dy,$$

since

$$\langle Tr, s \rangle = \int \int r(y') \, K(y, y') \, \frac{s(y)}{g(y)} \, dy \, dy'$$

$$= \int r(y') \int s(y) \, \frac{K(y', y)}{g(y)} \, dy \, \frac{g(y')}{g(y')} \, dy'$$

for every $(r, s)$. Note that the modification of $T$ into $T^*$ corresponds to the inversion of time for the chain $(Y^{(t)})$, by Bayes' Theorem, and therefore that $T^*$ is the Gibbs sampler for the reverse order of conditional distributions.

**Proposition 7.6.2** *Under the hypothesis (7.6.1), there exists $0 < C < 1$ such that if $y^{(0)} \sim g_0$ and $g_t = Tg_{t-1}$,*

$$\|g_t - g\| \leq \|g_0\| \, C^t.$$

*Proof.* Define $U = T^* \circ T$, which is the operator associated with the reversible Gibbs sampler [A.37]. By construction, it is self-adjoint and compact, and, therefore, enjoys an orthonormal basis of eigenfunctions. Moreover, the subspace corresponding to the eigenvalue 1 is spanned by the stationary distribution $g$. In fact, if $s$ is an eigenvector associated with the eigenvalue 1 which is orthogonal with $g$,

$$\langle s, g \rangle = \int g(y) s(y) d\mu(y) = \int s(y) dy = 0.$$

Consider $B$ (resp. $D$) as the set of $x$'s such that $s(x) > 0$ (resp. $s(x) < 0$) and $s(x) = s^+(x) - s^-(x)$. Then, since $Us = s$, for $x \in B$, $s(x) = s^+(x) < (Us^+)(x)$ and, for $x \in D$, $s^-(x) < (Us^-)(x)$. Therefore, on $B \cup D$, $U(|s|) > |sg|$, which implies that $B \cup D$ is of measure zero.

In addition, $-1$ cannot be an eigenvalue of $U$ (consider $\tilde{U} = U \circ U$). The other eigenvalues of $U$ are, therefore, all less than 1 in absolute value. If $V$ is the operator defined by $Vf = Uf - \langle f, g \rangle g$, its eigenvalues are also less than 1 in absolute value and

$$\|V\|_\infty = \sup_{\|f\|=1} \|Vf\| = |\rho| < 1.$$

Consider $W$ defined by $Wf = Tf - \langle f, g \rangle g$. Then $V = W^* \circ W$ and $\|W\|_\infty = \|V\|_\infty^{1/2}$. If $f$ is a density, $\langle g, f \rangle = 1$ and since $Wf = Tf - g$,

$$\|Tf - g\| = \|Wf\| \leq \|W\|_\infty \|f\| = |\rho|^{1/2} \|f\|,$$

which establishes the result for $C = |\rho|^{1/2}$.                                    □□

The corresponding Markov chain is also geometrically convergent in $\mathcal{L}_1(\mu)$.

**Corollary 7.6.3** *If (7.6.1) holds, there exists $0 < C < 1$ such that*

$$\|g_t - g\|_{TV} \leq C^t \|g_0\|.$$

*Proof.* For a Borel set $A$ of $\mathcal{Y}$, consider $v(y) = \mathbb{I}_A(y) g(y)$. Let $\tilde{g}$ denote the measure induced by $g$, which satisfies $\tilde{g}(A) = \langle g, v \rangle$, and, similarly, $\tilde{g}_t(A) = \langle g_t, v \rangle$. The Cauchy–Schwarz inequality then implies

$$|\tilde{g}_t(A) - \tilde{g}(A)| = |\langle g_t - g, v \rangle|$$
$$\leq \|g_t - g\| \, \|v\|$$
$$= \|g_t - g\| \, \tilde{g}(A) \leq C^t \|g_0\|.$$                                    □□

Similar results can obviously be obtained for other types of Gibbs samplers. In fact, as one can check in the previous proofs, the Hilbert–Schmidt condition is at the basis of the geometric convergence results since if (7.6.1) is satisfied by the kernel $K$, the self-adjoint operator $U = T^* \circ T$ is such that the sub-space associated with the eigenvalue 1 is spanned by $g$ and the other eigenvalues are less than 1 in absolute value. Then,

$$||T g_0 - g|| \leq \sup_{|\lambda_i| \neq 1} |\lambda_i| \, ||g_0||,$$

where $\lambda_1, \ldots, \lambda_n, \ldots$ are the eigenvalues of $T$. For instance, in the case of Data Augmentation, the kernel $K_1(y_1, y_1')$ satisfies the detailed balance condition and is therefore symmetric in $(y_1, y_1')$. Since the chain $(Y_1^{(t)})$ is reversible, the corresponding operator $T$ is self-adjoint and if

$$\int K_1^2(y_1, y_1') d\mu(y_1) d\mu(y_1') < +\infty,$$

geometric convergence holds for Data Augmentation.

Before describing the alternative approach of Liu $et\ al.$ (1994, 1995), we illustrate the previous results for a normal model treated by Schervish and Carlin (1992).

**Example 7.6.4 Normal geometric convergence.** Consider

$$\begin{pmatrix} X \\ Y \end{pmatrix} \sim \mathcal{N}_2 \left( \begin{pmatrix} \mu \\ \mu \end{pmatrix}, \begin{bmatrix} 2 & 1 \\ 1 & 1 \end{bmatrix} \right),$$

with the associated conditional distributions

$$X|y \sim \mathcal{N}(y, 1), \qquad Y|x \sim \mathcal{N}((x + \mu)/2, 1/2).$$

Since

$$K_2(y, y') = \varphi \left( \sqrt{4/3} \{ y' - (y + \mu)/2 \} \right),$$

(7.6.1) is

$$\int \int \exp(-(2y' - y - \mu)^2/3) \exp((y - \mu)^2/2) \exp((y' - \mu)^2/2) dy dy' < +\infty,$$

which is satisfied. Since the marginal distributions are known, they can be compared with the distributions at the $n$th iteration of the Gibbs sampler; that is, the distributions of $Y^{(t)}$ and of $X^{(t)}$,

$$\mathcal{N} \left( (1 - 2^{-t})\mu + 2^{-t} x^{(0)}, 1 - 2^{1-2t} \right),$$

$$\mathcal{N} \left( (1 - 2^{-t})\mu + 2^{-t} x^{(0)}, 2 - 2^{1-2t} \right).$$

Moreover, it is easily shown that if $g_2^t$ is the marginal density of $y^{(t)}$,

$$||g_2^t - g_2||^2 = \int_{-\infty}^{+\infty} \frac{(g_2^t(y) - g_2(y))^2}{g_2(y)} dy$$

(7.6.2)

$$= \frac{1}{\sqrt{1 - 2^{2-4t}}} e^{(x_0 - \mu)^2/(2 + 2^{2t})} - 1,$$

which is of order $2^{-2t}$ if $x_0 \neq \mu$ and of order $2^{-4t}$ if $x_0 = \mu$, as can be seen by a series expansion.                    $\|$

Liu *et al.* (1995) also obtain sufficient conditions for geometric convergence, while considering the transition from an "inverse" perspective; that is, by exploring the properties of the induced operator on conditional expectations rather than the densities of the transforms.[9] They consider instead the Hilbert space $\mathcal{L}_2(g)$ with the scalar product

$$\langle r, s \rangle_2 = \mathbb{E}^g[r(Y)s(Y)] = \int_{\mathcal{Y}} r(y)s(y)g(y)dy,$$

whose norm is denoted $|| \cdot ||_2$. The space $\mathcal{L}_2^0(g)$ is the subspace of $\mathcal{L}_2(g)$ of centered functions; that is, such that $\mathbb{E}^g[f(Y)] = 0$. The operators associated with the transition kernel $K(y, y')$ are $F$ and $B$:

$$(Ff)(y) = \int f(y')K(y, y')dy' = \mathbb{E}[f(Y^{(t+1)})|y^{(t)} = y],$$

$$(Bf)(y') = \int f(y')\frac{K(y', y)g(y')}{g(y)}dy' = \mathbb{E}[f(Y^{(t-1)})|y^{(t)} = y].$$

They thus both differ from the operator $T$ which operates in $\mathcal{L}_2(\mu)$. In this setup, the Markov chain with initial value $Y^{(0)} \sim g_0$ is *geometrically convergent* if there exist $0 < \alpha < 1$ and $c_0$ such that, for every $r \in \mathcal{L}_2(g)$,

$$\left| \mathbb{E}_{g_0}[r(Y^{(t)})] - \mathbb{E}_g[r(Y)] \right| = \left| \mathbb{E}_{g_0}[(F^t r)(Y)] - \mathbb{E}_g[r(Y)] \right| ,$$

(7.6.3)                                    $$\leq c_0 \alpha^t \sqrt{\text{var}_g r(Y)}.$$

(This definition truly coincides with the notion introduced in Chapter 4; see Problem 7.32). Moreover, this geometric convergence also implies the $\mathcal{L}_2(g)$ convergence of $g_t/g$ to 1; that is,

$$\left\| \frac{g_t}{g} - 1 \right\|_2 \leq c_0 \alpha^t.$$

Since

$$\left\| \frac{g_t}{g} - 1 \right\|_2^2 = \int (g_t(y) - g(y))^2 d\mu(y) = ||g_t - g||^2 ,$$

the correspondence between the approaches of Schervish and Carlin (1992) and of Liu *et al.* (1995) is complete (in terms of convergence properties). When studying the geometric convergence of $(Y^{(t)})$, Liu *et al.* (1995) introduce the condition

(7.6.4)                            $$\int \frac{g_0^2(y)}{g(y)} dy < +\infty,$$

which is equivalent to $||g_0|| < +\infty$ in $\mathcal{L}_2(\mu)$ (and is, therefore, implicitly assumed in Schervish and Carlin 1992).

**Proposition 7.6.5** *Under the condition (7.6.4), if the spectral radius of the operators $F$ and $B$ is strictly less than 1 on $\mathcal{L}_2^0(g)$, the chain $(Y^{(t)})$ is geometrically convergent.*

Note that $F$ and $B$ are *adjoint* operators in $\mathcal{L}_2(g)$. Moreover, by an elementary inequality, $||Ff||_2 \leq ||f||_2$ for every $f \in \mathcal{L}_2^0(g)$. The spectral radii of $F$ and of $B$ are therefore equal and less than or equal to 1 in $\mathcal{L}_2^0(g)$.

---

[9] See Duflo (1996) for a derivation of these results by the reversibilization technique of Fill (1991).

*Proof.* Denote by $\rho$ the spectral radius[10] of $F$,

$$\rho = \lim_{t \to +\infty} ||F^t||_\infty^{1/t} .$$

Therefore there exists $t_0$ such that $||F^t||_\infty < 1$ for $t \geq t_0$. Moreover, if $\tilde{g}(y) = g_0(y)/g(y)$,

$$\left| \left| \mathbb{E}[r(Y^{(t)})] - \mathbb{E}^g[r(Y)] \right| \right| = |\text{cov}_g(r(Y^{(t)}), \tilde{g}(Y^{(t)}))|$$
$$= |\text{cov}_g((F^t r)(Y), \tilde{g}(Y))|$$
$$\leq ||F^t||_\infty \, ||r||c_0,$$

with $c_0^2 = \text{var}_g(\tilde{g}(Y))$, which is finite if (7.6.4) is satisfied.                    □□

The previous condition on the spectral radius of $F$ implies that the chain $(Y^{(t)})$ must be Harris recurrent to be geometrically convergent. In the particular case of Data Augmentation, Liu *et al.* (1995) show that a sufficient condition for the spectral radius of $F$ to be less than 1 on $\mathcal{L}_2^0(g)$ is that

$$\int \int \frac{g(y_1, y_2)^2}{g_1(y_1)g_2(y_2)} dy_1 dy_2 < \infty,$$

which is equivalent to

$$\int \int g_1(y_1|y_2)g_2(y_2|y_1)dy_1 dy_2 < \infty.$$

This condition may appear to be less restrictive than (7.6.1), but they are equivalent. If the algorithm is initialized at a fixed value $y_1^{(0)}$, (7.6.3) can be written

$$\int \frac{K_1(y_1^{(0)}, y_1)}{g_1(y_1)} dy_1 < +\infty.$$

Similarly, for [A.31], Liu *et al.* (1995) introduce the condition

(7.6.5)                    $$\int \frac{K(y, y')^2}{g(y')} g(y)dy dy' < +\infty,$$

which is that the only functions $f$ satisfying

(7.6.6)    $\mathbb{E}[f(Y)|y_1, \ldots, y_{i-1}, y_{i+1}, \ldots, y_p] = f(Y) , \qquad i = 1, \ldots, p,$

are constant. This is actually a stronger form of Harris recurrence (see Lemma 4.7.3). Obviously, if (7.6.6) is satisfied by a nonconstant function, the operator $F$ has an eigenvector associated with the eigenvalue 1. This second condition on harmonic functions is naturally satisfied if $g$ has $\mathbb{R}^p$ as support, since a sufficient condition for (7.6.6) to be only satisfied by constant functions is the positivity of $(Y^{(t)})$, which is equivalent to (7.6.1).

Liu *et al.* (1994) establish that under (7.6.4) and (7.6.5) and the constraint corresponding to (7.6.6), the chain $(Y^{(t)})$ is geometrically ergodic. Therefore, their results are equivalent to those of Schervish and Carlin (1992). For the other simulation schemes (random sweeps, random symmetric, symmetric), these conditions can be adapted to the corresponding transition kernels.

---

[10] The difference $(1 - \rho)$ is called the *gap* and is instrumental in some theoretical convergence studies, see Polson (1996).

Other papers have developed the perspectives of Schervish and Carlin (1992) and of Liu *et al.* (1994, 1995). See, for instance, Polson (1996) for a review of finer convergence properties.

We still conclude with the warning that the practical consequences of this theoretical evaluation may be minimal. In fact, setups where eigenvalues of these operators are available often correspond to case where Gibbs sampling is not necessary (see, e.g., Example 7.6.4). Moreover, the Hilbert–Schmidt condition (7.6.1) is particularly difficult to check when $K(y, y')$ is not explicit. Finally, we also consider that the eigenvalues of the operators $T$ and $F$ and the convergence of norms $\|g_t - g\|_{TV}$ to 0 are only marginally relevant in the study of the convergence of the average

$$(7.6.7) \qquad\qquad \frac{1}{T} \sum_{t=1}^{T} h(y_t)$$

to $\mathbb{E}[h(Y)]$, even though there exists a theoretical connection between the asymptotic variance $\gamma_h^2$ of (7.6.7) and the spectrum $(\lambda_k)$ of $F$ through

$$\gamma_h^2 = \sum_{k \geq 2} w_k \frac{1 + \lambda_k}{1 - \lambda_k} \leq \frac{1 + \lambda_2}{1 - \lambda_2}$$

(see Geyer 1993 and Besag and Green 1993), where the weight $w_k$ depend on $h$ and $F$, and the $\lambda_k$ are the (decreasing) eigenvalues of $F$ (with $\lambda_1 = 1$).

### 7.6.4  The BUGS Software

The acronym BUGS stands for *Bayesian inference using Gibbs sampling*. This software has been developed by Spiegelhalter, *et al.* (1995a,b,c) at the MRC Biostatistics Unit in Cambridge, England. As shown by its name, it has been designed to take advantage of the possibilities of the Gibbs sampler in Bayesian analysis. BUGS includes a language which is C or S-Plus like and involves declarations about the model, the data, and the prior specifications, for single or multiple levels in the prior modeling. For instance, for the benchmark nuclear pump failures dataset of Example 7.1.23, the model and priors are defined by

```
for (i in 1:N) {
    theta[i] ~ dgamma(alpha,beta)
    lambda[i]~ theta[i] * t[i]
    x[i]     ~ dpois(lambda[i])
}
alpha ~ dexp(1.0)
beta  ~ dgamma(0.1,1.0)
```

(see Spiegelhalter *et al.* 1995b, p. 9). Most standard distributions are recognized by BUGS (21 are listed in Spiegelhalter *et al.* 1995a), which also allows for a large range of transforms. BUGS also recognizes a series of commands like compile, data, out, and stat. The output of BUGS is a table of the simulated values of the parameters after an open number of warmup iterations, the batch size being also open.

A major restriction of this software is the use of the conjugate priors or, at least, log-concave distributions for the Gibbs sampler to apply. However, more complex distributions can be handled by discretization of their support and assessment of the sensitivity to the discretization step. In addition, improper priors are not accepted and must be replaced by proper priors with small precision, like dnorm(0,0.0001), which represents a normal modeling with mean 0 and *precision* (inverse variance) 0.0001.

The BUGS manual (Spiegelhalter *et al.* 1995a) is quite informative and well written.[11] In addition, the authors have written an extended and most helpful example manual (Spiegelhalter *et al.* 1995b,c), which exhibits the ability of BUGS to deal with an amazing number of models, including meta-analysis, latent variable, survival analysis, nonparametric smoothing, model selection and geometric modeling, to name a few. (Some of these models are presented in Problems 7.44–7.55.) The BUGS software is also compatible with the convergence diagnosis software CODA presented in Note 8.6.4.

### 7.6.5 Nonparametric Mixtures

Consider $X_1, \ldots, X_n$ distributed from a mixture of geometric distributions,

$$X_1, \ldots, X_n \sim \int_0^1 \theta^x (1 - \theta) \, dG(\theta), \qquad X_i \in \mathbb{N},$$

where $G$ is an arbitrary distribution on $[0, 1]$. In this nonparametric setup, the likelihood can be expressed in terms of the moments

$$\mu_i = \int_0^1 \theta^i \, dG(\theta), \qquad i = 1, \ldots$$

since $G$ is then identified by the $\mu_i$'s. The likelihood can be written

$$L(\mu_1, \mu_2, \ldots | x_1, \ldots, x_n) = \prod_{i=1}^n \left( \mu_{x_i} - \mu_{1+x_i} \right).$$

A direct Bayesian modeling of the $\mu_i$'s is impossible because of the constraint between the moments, such as $\mu_i \geq \mu_1^i$ $(i \geq 2)$, which create dependencies between the different moments (Problem 7.57). The *canonical moments* technique (see Olver 1974 and Dette and Studden 1997) can overcome this difficulty by expressing the $\mu_i$'s as transforms of a sequence $(p_j)$ on $[0, 1]$ (see Problem 7.57). Since the $p_j$'s are not constrained, they can be modeled as uniform on $[0, 1]$. The connection between the $\mu_i$'s and the $p_j$'s is given by recursion equations. Let $q_i = 1 - p_i$, $\zeta_1 = p_1$, and $\zeta_i = p_i q_{i+1}$, $(i \geq 2)$, and define

$$\begin{aligned} S_{1,k} &= \zeta_1 + \cdots + \zeta_k, & (k \geq 1) \\ S_{j,k} &= \sum_{u=1}^{k-j+1} \zeta_u S_{j-1,u+j-1}, & (j \geq 2). \end{aligned}$$

It is then possible to show that $\mu_j - \mu_{j+1} = q_1 S_{j,j}$ (see Problem 7.57).

From a computational point of view, the definition of the $\mu_i$'s via recursion equations complicates the exact derivation of Bayes estimates, and they become too costly when max $x_i > 5$. This setting where numerical complexity prevents the analytical derivation of Bayes estimators can be solved via Gibbs sampling.

The complexity of the relations $S_{j,k}$ is due to the action of the sums in the recursion equations; for instance,

$$\mu_3 - \mu_4 = q_1 p_2 q_2 \{ p_1 q_2 (p_1 q_2 + p_2 q_3) + p_2 q_3 (p_1 q_2 + p_2 q_3 + p_3 q_4) \}.$$

---

[11] At the present time, that is, Winter 1998!, the BUGS software is available as a freeware on the Web site http://www.mrc-bsu.cam.ac.uk/bugs for a wide variety of platforms.

This complexity can be drastically reduced if, through a demarginalization (or completion) device, every sum $S_{j,k}$ in the recursion equation is replaced by one of its terms $\zeta_u S_{j-1,u+j-1}$ $(1 \leq u \leq k-j+1)$. In fact, $\mu_k - \mu_{k+1}$ is then a product of $p_i$'s and $q_j$'s, which leads to a beta distribution on the parameters $p_i$. To achieve such simplification, the expression

$$P(X_i = k) = \mu_k - \mu_{k+1} = \zeta_1 S_{k-1,k}$$
$$= \zeta_1 \{\zeta_1 S_{k-2,k-1} + \zeta_2 S_{k-2,k}\}$$

can be interpreted as a marginal distribution of the $X_i$ by introducing $Z_1^i \in \{0,1\}$ such that

$$P(X_i = k, Z_1^i = 0) = \zeta_1^2 S_{k-2,k-1},$$
$$P(X_i = k, Z_1^i = 1) = \zeta_1 \zeta_2 S_{k-2,k}.$$

Then, in a similar manner, introduce $Z_2^i \in \{0,1,2\}$ such that the density of $(X_i, Z_1^i, Z_2^i)$ (with respect to counting measure) is

$$f(x_i, z_1^i, z_2^i) = \zeta_1 \zeta_{z_1^i+1} \zeta_{z_2^i+1} S_{x_i-3, x_i-2+z_2^i}.$$

The replacement of the $S_{k,j}$'s thus requires the introduction of $(k-1)$ variables $Z_s^i$ for each observation $x_i = k$. Once the model is completed by the $Z_s^i$'s, the posterior distribution of $p_j$,

$$\prod_{i=1}^{n} (q_1 p_1 q_2 p_{z_1^i+1} q_{z_1^i+2} \cdots q_{z_{x_i-1}^i+2}),$$

is a product of beta distributions on the $p_j$'s, which are easily simulated.

Similarly, the distribution of $Z_s^i$ conditionally on $p_j$ and the other dummy variables $Z_r^i$ $(r \neq s)$ is given by

$$\pi(z_s^i|p, z_{s-1}^i = v, z_{s+1}^i = w) \propto p_w q_{w+1} \mathbb{I}_{z_s^i=w-1} + \cdots + p_{v+2} q_{v+3} \mathbb{I}_{z_s^i=v+1}.$$

The Gibbs sampler thus involves a large number of (additional) steps in this case, namely $1 + \sum_i (x_i - 1)$ simulations, since it imposes the "local" generation of the $z_s^i$'s. In fact, an arbitrary grouping of $z_s^i$ would make the simulation much more difficult, except for the case of a division of $z^i = (z_1^i, \ldots, z_{x_i-1}^i)$ into two subvectors corresponding to the odd and even indices, respectively.

Suppose that the parameter of interest is $\mathbf{p} = (p_1, \ldots, p_{K+1})$, where $K$ is the largest observation. (The distribution of $p_i$ for indices larger than $K+1$ is unchanged by the observations. See Robert 1994a.) Although $\mathbf{p}$ is generated conditionally on the complete data $(x_i, z^i)$ $(i = 1, \ldots, n)$, this form of Gibbs sampling is not a Data Augmentation scheme since the $z^i$'s are not simulated conditionally on $\theta$, but rather one component at a time, from the distribution

$$f(z_s^i|\mathbf{p}, z_{s-1}^i = v, z_{s+1}^i = w) \propto p_w q_{w+1} \mathbb{I}_{z_s^i=w-1} + \cdots + p_{v+2} q_{v+3} \mathbb{I}_{z_s^i=v+1}.$$

However, this complexity does not prevent the application of Theorem 7.2.14 since the sequence of interest is generated conditionally on the $z_s^i$'s. Geometric convergence thus applies.

### 7.6.6 Graphical Models

Graphical models use graphs to analyze statistical models. They have been developed mainly to represent conditional independence relations, primarily in the field of *expert systems* (Whittaker 1990; Spiegelhalter *et al.*, 1993). The

Bayesian approach to these models, as a way to incorporate model uncertainty, has been aided by the advent of MCMC techniques, as stressed by Madigan and York (1995) in an expository paper on which this note is based.

The construction of a graphical model is based on a collection of independence assumptions represented by a graph. We briefly recall here the essentials of graph theory and refer to Lauritzen (1996) for details. A *graph* is defined by a set of *vertices* or *nodes*, $\alpha \in \mathcal{V}$, which represents the random variables or factors under study, and by a set of *edges*, $(\alpha, \beta) \in \mathcal{V}^2$, which can be ordered (the graph is then said to be *directed*) or not (the graph is *undirected*). For a directed graph, $\alpha$ is a *parent* of $\beta$ if $(\alpha, \beta)$ is an edge (and $\beta$ is then a *child* of $\alpha$).[12] Graphs are also often assumed to be *acyclic*; that is, without directed paths linking a node $\alpha$ with itself. This leads to the notion of *directed acyclic graphs*, introduced by Kiiveri and Speed (1982), often represented by the acronym DAG.

For the construction of probabilistic models on graphs, an important concept is that of a *clique*. A *clique* $C$ is a maximal subset of nodes which are all joined by an edge (in the sense that there is no subset containing $C$ and satisfying this condition). An ordering of the cliques of an undirected graph $(C_1, \ldots, C_n)$ is *perfect* if the nodes of each clique $C_i$ contained in a previous clique are all members of one previous clique (these nodes are called the *separators*, $\alpha \in S_i$). In this case, the joint distribution of the random variable $V$ taking values in $\mathcal{V}$ is

$$p(V) = \prod_{v \in V} p(v | \mathcal{P}(v)) \,,$$

where $\mathcal{P}(v)$ denotes the parents of $v$. This can also be written as

(7.6.8)
$$p(V) = \frac{\displaystyle\prod_{i=1}^{n} p(C_i)}{\displaystyle\prod_{i=1}^{n} p(S_i)} \,,$$

and the model is then called *decomposable*; see Spiegelhalter and Lauritzen (1990), Dawid and Lauritzen (1993) or Lauritzen (1996). As stressed by Spiegelhalter *et al.* (1993), the representation (7.6.8) leads to a *principle of local computation*, which enables the building of a prior distribution, or the simulation from a conditional distribution on a single clique. (In other words, the distribution is *Markov with respect to the undirected graph*, as shown by Dawid and Lauritzen 1993.) The appeal of this property for a Gibbs implementation is then obvious.

When the densities or probabilities are parameterized, the parameters are denoted by $\theta_A$ for the marginal distribution of $V \in A$, $A \subset \mathcal{V}$. (In the case of discrete models, $\theta = \theta_V$ may coincide with $p$ itself; see Example 7.6.6.) The prior distribution $\pi(\theta)$ must then be compatible with the graph structure:

---

[12] Directed graphs can be turned into undirected graphs by adding edges between nodes which share a child and dropping the directions.

Dawid and Lauritzen (1993) show that a solution is of the form

$$(7.6.9) \qquad \pi(\theta) = \frac{\displaystyle\prod_{i=1}^{n} \pi_i(\theta_{C_i})}{\displaystyle\prod_{i=1}^{n} \tilde{\pi}_i(\theta_{S_i})},$$

thus reproducing the clique decomposition (7.6.8).

**Example 7.6.6 Discrete event graph.** Consider a decomposable graph such that the random variables corresponding to all the nodes of $\mathcal{V}$ are discrete. Let $w \in W$ be a possible value for the vector of these random variables and $\theta(w)$ be the associated probability. For the perfect clique decomposition $(C_1, \ldots, C_n)$, $\theta(w_i)$ denotes the marginal probability that the subvector $(v, v \in C_i)$ takes the value $w_i$ $(\in W_i)$ and, similarly, $\theta(w_i^s)$ is the probability that the subvector $(v, v \in S_i)$ takes the value $w_i^s$ when $(S_1, \ldots, S_n)$ is the associated sequence of separators. In this case,

$$\theta(w) = \frac{\displaystyle\prod_{i=1}^{n} \theta(w_i)}{\displaystyle\prod_{i=1}^{n} \theta(w_i^s)}.$$

As illustrated by Madigan and York (1995), a Dirichlet prior can be constructed on $\theta_W = (\theta(w), w \in W)$, which leads to genuine Dirichlet priors on the $\theta_{W_i} = (\theta(w_i), w_i \in W_i)$, under the constraint that the Dirichlet weights are identical over the intersection of two cliques. Dawid and Lauritzen (1993) demonstrate that this prior is unique, given the marginal priors on the cliques. ‖

**Example 7.6.7 Graphical Gaussian model.** Giudici and Green (1998) provide another illustration of prior specification in the case of a *graphical Gaussian model*, $\mathbf{X} \sim \mathcal{N}_p(0, \Sigma)$, where the precision matrix $K = \{k_{ij}\} = \Sigma^{-1}$ must comply with the conditional independence relations on the graph. For instance, if $\mathbf{X}_v$ and $\mathbf{X}_w$ are independent given the rest of the graph, then $k_{vw} = 0$. The likelihood can then be factored as

$$f(\mathbf{x}|\Sigma) = \frac{\displaystyle\prod_{i=1}^{n} f(\mathbf{x}_{C_i}|\Sigma^{C_i})}{\displaystyle\prod_{i=1}^{n} f(\mathbf{x}_{S_i}|\Sigma^{S_i})},$$

with the same clique and separator notations as above, where $f(\mathbf{x}_C|\Sigma^C)$ is the normal $\mathcal{N}_{p_C}(0, \Sigma^C)$ density, following (7.6.8). The prior on $\Sigma$ can be chosen as the conjugate inverse Wishart priors on the $\Sigma^{C_i}$'s, under some compatibility conditions. ‖

Madigan and York (1995) discuss an MCMC approach to model choice and model averaging in this setup, whereas Dellaportas and Foster (1996) and Giudici and Green (1998) implement reversible jump algorithms for determining

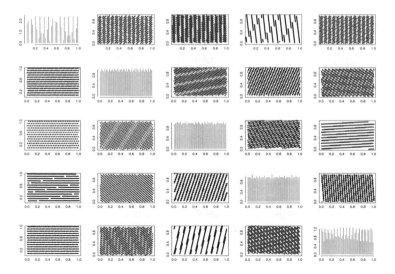

Figure 7.6.1.  *Cross-representation of the quasi-random sequence* $(y_k)$ *produced by (7.6.10) for* $p = s = 5$ *and* 1000 *iterations. The graphs on the diagonal are the histograms of the coordinates of* $(y_k)$, *while the off-diagonal graphs plot the spread of the pairs* $(\lfloor \gamma_i k \rfloor, \lfloor \gamma_j k \rfloor)$ $(i, j = 1, \ldots, 5)$.

the probable graph structures associated with a given dataset, the latter under a Gaussian assumption.

### 7.6.7 Quasi-Monte Carlo Methods for the Gibbs Sampler

Liao (1998) discusses the difficult extension of quasi-Monte Carlo methods for the Gibbs sampler. In fact, while quasi-Monte Carlo methods (see Note 2.5.1) allow for a faster coverage of the unit cube $[0, 1]^s$, the dependence resulting from this construction can have very negative consequences in settings where the successive values to be generated are dependent, a remark which prohibits the direct use of quasi-random sequences for the Gibbs sampler. For instance, Fang and Wang (1994) propose to select a prime integer $p$ and, for $q = p^{1/(s+1)}$, to define $\gamma_j = \lfloor q^j \rfloor$ as the fractional part of $q^j$. The quasi-random sequence $(y_k)$ is then produced by $(k = 1, 2, \ldots)$

$$(7.6.10) \qquad y_k = (\lfloor \gamma_1 k \rfloor, \lfloor \gamma_2 k \rfloor, \ldots, \lfloor \gamma_s k \rfloor),$$

whose coordinates are quite dependent. Figure 7.6.1 illustrates this dependence in the case $p = s = 5$ by exhibiting pairs $(i, j)$ where the coverage of $[0, 1]^2$ by the sequence $(\lfloor \gamma_i k \rfloor, \lfloor \gamma_j k \rfloor)$ is quite poor. (See, for instance, the cases $(i, j) = (1, 4)$ and $(i, j) = (5, 3)$.)

Liao's (1998) proposal to overcome this difficulty is to first generate a sequence $y_1, \ldots, y_n$ from (7.6.10) and then to randomly permute the indices $1, \ldots, n$. If the Gibbs sampler is associated with the conditional distributions $\pi_j(\theta_j | \theta_\ell, \ell \neq j)$ $(j = 1, \ldots, n)$, his quasi-Gibbs sampler then requires the selection of the dimension $s$, which must be large enough if some of the $\pi'_j$'s require Accept–Reject steps, but not too large because large values of $s$ require large $n$'s for a good coverage of $[0, 1]^s$. Liao (1998) shows a significant reduction in the variability of the Gibbs estimates on some examples but warns against possible biases due to a poor coverage of $[0,1]^s$ by $(y_1, \ldots, y_n)$ when $n$ is too small, a serious drawback which cannot always be detected on single runs of iterations. Note also that the random permutation modification requires fixed numbers of iterations, which should not be the case in a good usage of the Gibbs sampler (see Chapter 8), while it does not eliminate the dependence structure since Figure 7.6.1 remains exactly the same after the random permutation.

# Diagnosing Convergence

"Why does he insist that we must have a diagnosis? Some things are not meant to be known by man."

—Susanna Gregory, *An Unholy Alliance*

## 8.1 Stopping the Chain

The two previous chapters have presented the theoretical foundations of MCMC algorithms and showed that under fairly general conditions, the chains produced by these algorithms are ergodic, or even geometrically ergodic. While such developments are obviously necessary, they are nonetheless insufficient from the point of view of the implementation of MCMC methods. They do not directly result in methods of controlling *the* chain produced by an algorithm (in the sense of a *stopping rule* to guarantee that the number of iterations is sufficient). In other words, general convergence results do not tell us when to stop the MCMC algorithm and produce our estimates.

The goal of this chapter is to present, in varying amounts of detail, a catalog of the numerous monitoring methods (or *diagnostics*) proposed in the literature. (See the review papers of Cowles and Carlin 1996, Robert 1995b, Brooks 1998a, Brooks and Roberts 1999, and Mengersen *et al.* 1999.) Most of the techniques presented in this chapter have not yet withstood the test of time and are somewhat exploratory in nature. We are, therefore, in the situation of describing a sequence of noncomparable techniques with widely varying degrees of theoretical justification and usefulness.

### 8.1.1 Convergence Criteria

From a general point of view, there are three (increasingly stringent) types of convergence for which assessment is necessary:

(i) *Convergence to the Stationary Distribution*

This criterion considers convergence of the chain $\theta^{(t)}$ to the stationary distribution $f$ (or *stationarization*), which seems to be a minimal requirement for an algorithm that is supposed to approximate simulation from $f$! Unfortunately, it seems that this approach to convergence issues is not particularly fruitful. In fact, from a theoretical point of view, $f$

is only the *limiting* distribution of $\theta^{(t)}$. This means that stationarity is only achieved asymptotically.

For instance, the original implementation of the Gibbs sampler was based on the generation of $n$ independent initial values $\theta_i^{(0)}$ $(i = 1, \ldots, n)$, and the storage of the last simulation $\theta_i^{(T)}$ in each chain. (Strictly speaking, this approach thus requires a corresponding stopping rule for the correct determination of $T$; see, for instance, Tanner and Wong 1987. This method also results in a waste of resources, as most of the generated variables are discarded.)

If $\mu_0$ is the (initial) distribution of $\theta^{(0)}$, then the $\theta_i^{(T)}$'s are distributed from $f_{\mu_0}^T$. If we, instead, only consider a single realization (or *path*) of the chain $(\theta^{(t)})$, the question of *convergence* to the limiting distribution is not really relevant! Indeed, it is possible[1] to obtain the initial value $\theta^{(0)}$ from the distribution $f$, therefore to act as if the chain is already in its stationary regime from the start. (In cases where this assumption is unrealistic, there are methods like the exact simulation [or *perfect simulation*] of Propp and Wilson 1996, discussed in Note 8.6.5, where stationarity can be rigorously achieved from the start.)

This seeming dismissal of the first type of control may appear rather cavalier, but we do think that convergence to $f$ *per se* is not the major issue for most MCMC algorithms, in the sense that the chain truly produced by the algorithm often behaves like a chain initialized from $f$. The issues at stake are rather the speed of exploration of the support of $f$ and the degree of correlation between the $\theta^{(t)}$'s. This is not to say that stationarity should not be tested at all, as we will see in §8.2.2, since, notwithstanding the starting distribution, $\mu_0$, the chain may be slow to explore the different regions of the support of $f$, with lengthy stays in each of these regions (for example, the modes of the distribution $f$), and a stationarity test may be useful in detecting such difficulties.

(ii) *Convergence of Averages*

Here, we are concerned with convergence of the empirical average

(8.1.1)
$$\frac{1}{T} \sum_{t=1}^{T} h(\theta^{(t)})$$

to $\mathbb{E}_f[h(\theta)]$ for an arbitrary function $h$. This type of convergence is most relevant in the implementation of MCMC algorithms. Indeed, even when $\theta^{(0)} \sim f$, the exploration of the complexity of $f$ by the chain $(\theta^{(t)})$ can be more or less lengthy, depending on the transition chosen for the algorithm. The purpose of the convergence assessment is, therefore, to determine whether the chain has exhibited all the facets of $f$ (for instance,

---

[1] We consider a standard statistical setup where the support of $f$ is approximately known. This may not be the case for high-dimensional setups or complex structures where the algorithm is initialized at random.

all the modes). Brooks and Roberts (1999) relate this convergence to the *mixing* speed of the chain, in the informal sense of a strong (or weak) dependence on initial conditions and of a slow (or fast) exploration of the support of $f$ (see also Asmussen *et al.* 1992). A formal version of convergence monitoring in this setup is the convergence assessment of §8.1.1. While the ergodic theorem guarantees the convergence of this average from a theoretical point of view, the relevant issue at this stage is to determine a minimal value for $T$ which justifies the approximation of $\mathbb{E}_f[h(\theta)]$ by (8.1.1).

(iii) *Convergence to iid Sampling*

This convergence criterion measures how close a sample $(\theta_1^{(t)}, \ldots, \theta_n^{(t)})$ is to being iid. This type of convergence looks at *independence* requirements for the simulated values. Rather than approximating integrals such as $\mathbb{E}_f[h(\theta)]$, the goal is to produce variables $\theta_i$ which are (quasi-)independent. While the solution based on parallel chains mentioned above is not satisfactory, an alternative is to use *subsampling* (or *batch sampling*) to reduce correlation between the successive points of the Markov chain. This technique, which is customarily used in numerical simulation (see, for instance, Schmeiser 1989), subsamples the chain $(\theta^{(t)})$ with a batch size $k$, considering only the values $\eta^{(t)} = \theta^{(kt)}$. If the covariance $\text{cov}_f(\theta^{(0)}, \theta^{(t)})$ decreases monotonically with $t$ (see §7.2.3), the motivation for subsampling is obvious. In particular, if the chain $(\theta^{(t)})$ satisfies an interleaving property (see §7.2.2), subsampling is justified. However, checking for the monotone decrease of $\text{cov}_f(\theta^{(0)}, \theta^{(t)})$—which also justifies Rao–Blackwellization (see §7.2.3)—is not always possible and, in some settings, the covariance oscillates with $t$, which complicates the choice of $k$. Section §8.4.1 describes how Raftery and Lewis (1992a,b) estimate this batch size $k$.

We note that subsampling necessarily leads to losses in efficiency with regard to the second convergence goal. In fact, as shown by MacEachern and Berliner (1994), it is always preferable to use the entire sample for the approximation of $\mathbb{E}_f[h(\theta)]$. Nonetheless, for convergence assessment, subsampling may be beneficial (see, e.g., Robert *et al.* 1999).

**Lemma 8.1.1** *Consider $h \in \mathcal{L}^2(f)$. For every $k > 1$, if $(\theta^{(t)})$ is a Markov chain with stationary distribution $f$ and if*

$$\delta_1 = \frac{1}{Tk} \sum_{t=1}^{Tk} h(\theta^{(t)}) \quad and \quad \delta_k = \frac{1}{T} \sum_{\ell=1}^{T} h(\theta^{(k\ell)}),$$

*the variance of $\delta_1$ satisfies*

$$\text{var}(\delta_1) \leq \text{var}(\delta_k).$$

*Proof.* Define $\delta_k^1, \ldots, \delta_k^{k-1}$ as the shifted versions of $\delta_k = \delta_k^0$; that is,

$$\delta_k^i = \frac{1}{T} \sum_{t=1}^{T} h(\theta^{(tk-i)}), \qquad i = 0, 1, \ldots, k-1.$$

The estimator $\delta_1$ can then be written as $\delta_1 = \frac{1}{k} \sum_{i=0}^{k-1} \delta_k^i$, and hence

$$
\begin{aligned}
\mathrm{var}(\delta_1) &= \mathrm{var}\left( \frac{1}{k} \sum_{i=0}^{k-1} \delta_k^i \right) \\
&= \mathrm{var}(\delta_k^0)/k + \sum_{i \neq j} \mathrm{cov}(\delta_k^i, \delta_k^j)/k^2 \\
&\leq \mathrm{var}(\delta_k^0)/k + \sum_{i \neq j} \mathrm{var}(\delta_k^0)/k^2 \\
&= \mathrm{var}(\delta_k) \, ,
\end{aligned}
$$

where the inequality follows from the Cauchy–Schwarz inequality

$$
|\mathrm{cov}(\delta_k^i, \delta_k^j)| \leq \mathrm{var}(\delta_k^0). \qquad \square\square
$$

In the remainder of the chapter, we consider independence issues only in cases where they have bearing on the control of the chain, as in renewal theory (see §8.2.3).

### 8.1.2 Multiple Chains

Aside from distinguishing between convergence to stationarity (§8.2) and convergence of the average (§8.3), we also distinguish between the methods involving the simulation in parallel of $M$ independent chains $(\theta_m^{(t)})$ $(1 \leq m \leq M)$ and those based on a single "on-line" chain. The motivation of the former is intuitively sound. By simulating several chains, variability and dependence on the initial values are reduced and it should be easier to control convergence to the stationary distribution by comparing the estimation, using different chains, of quantities of interest. The dangers of a naive implementation of this principle should be obvious, namely that the slower chain governs convergence and that the choice of the initial distribution is extremely important in guaranteeing that the different chains are well dispersed.

Many multiple-chain convergence diagnostics are quite elaborate (Gelman and Rubin 1992, Liu et al. 1995, and Johnson 1996) and seem to propose convergence evaluations that are more robust than single-chain methods. Geyer (1992) points out that this robustness is illusory from several points of view. In fact, good performances of these parallel methods require a degree of a priori knowledge on the distribution $f$ in order to construct an initial distribution on $\theta_m^{(0)}$ which takes into account the specificities of $f$ (modes, shape of high density regions, etc.). For example, an initial distribution which is too concentrated around a local mode of $f$ does not contribute significantly more than a single chain to the exploration of the specificities of $f$. Moreover, slow algorithms, like Gibbs sampling used in highly nonlinear setups, usually favor single chains, in the sense that a unique chain with $MT$ observations and a slow rate of mixing is more likely to get closer to the stationary distribution than $M$ chains of size $T$, which will presumably stay in the neighborhood of the starting point with higher probability.

An additional practical drawback of parallel methods is that they require a modification of the original MCMC algorithm to deal with the processing of parallel outputs. (See Tierney 1994 and Raftery and Lewis 1996 for other criticisms.) On the other hand, single-chain methods suffer more severely from the defect that *"you've only seen where you've been,"* in the sense that the part of the support of $f$ which has not been visited by the chain at time $T$ is almost impossible to detect. Moreover, a single chain may present probabilistic pathologies which are possibly avoided by parallel chains.

## 8.1.3 Conclusions

We agree with many authors[2] that it is somewhat of an illusion to think we can control the flow of a Markov chain and assess its convergence behavior from a few realizations of this chain. There always are settings (transition kernels) which, for most realizations, will invalidate an arbitrary indicator (whatever its theoretical justification) and the randomness inherent to the nature of the problem prevents any categorical guarantee of performance. The heart of the difficulty is actually similar to Statistics, where the uncertainty due to the observations prohibits categorical conclusions and final statements. Far from being a failure acknowledgment, these remarks only aim at warning the reader about the *relative* value of the indicators developed below. As noted by Cowles and Carlin (1996), it is simply inconceivable, in the light of recent results, to envision *automated stopping rules*. Brooks and Roberts (1999) also stress that the prevalence of a given control method strongly depends on the model and on the inferential problem under study. It is, therefore, even more crucial to develop robust and general evaluation methods which extend and complement the present battery of stopping criteria. One goal is the development of "convergence diagnostic spreadsheets," in the sense of computer graphical outputs, which would graph several different features of the convergence properties of the chain under study (see Cowles and Carlin 1996, Best *et al.* 1995, Robert 1997, 1998, or Robert *et al.* 1999 for illustrations).

The criticisms presented in the wake of the techniques proposed below serve to highlight the incomplete aspect of each method. They do not aim at preventing their use but rather to warn against a selective interpretation of their results.

---

[2] To borrow from the injunction of Hastings (1970), *"even the simplest of numerical methods may yield spurious results if insufficient care is taken in their use... The setting is certainly no better for the Markov chain methods and they should be used with appropriate caution."*

## 8.2 Monitoring Convergence to the Stationary Distribution

### 8.2.1 Graphical Methods

A natural empirical approach to convergence control is to draw pictures of the output of simulated chains, in order to detect deviant or nonstationary behaviors. For instance, as in Gelfand and Smith (1990), a first plot is to draw the sequence of the $\theta^{(t)}$'s against $t$. However, this plot is only useful for strong nonstationarities of the chain.

**Example 8.2.1 Witch's hat distribution.** Consider the posterior distribution

$$\pi(\theta|y) \propto \left\{ (1 - \delta)\, \sigma^{-d} e^{-\|y-\theta\|^2/(2\sigma^2)} + \delta \right\} \mathbb{I}_C(\theta), \qquad y \in \mathbb{R}^d,$$

when $\theta$ is restricted to the unit cube $C = [0,1]^d$.

Figure 8.2.1 shows that this density has a mode which is very concentrated around $y$ for small $\delta$ and $\sigma$, hence the "witch's hat" christening. A naive implementation of the Gibbs sampler leads to simulation of the components of $(\theta^{(t)})$ according to the conditional distribution on $[0,1]$,

$$\pi(\theta_i|\theta_{j\neq i}, y) \propto \delta + (1 - \delta)\sigma^{-d} \exp\left\{ -\sum_{j\neq i} (y_j - \theta_j)^2/(2\sigma^2) \right\}$$
$$\times \exp\{-(y_i - \theta_i)^2/(2\sigma^2)\},$$

which can be expressed as

(8.2.1) $$\left( w_i \frac{\exp\{-(y_i - \theta_i)^2/(2\sigma^2)\}}{\sqrt{2\pi}\sigma \left[ \Phi\left(\frac{1-y_i}{\sigma}\right) - \Phi\left(\frac{-y_i}{\sigma}\right) \right]} + \delta \right) \mathbb{I}_{[0,1]}(\theta_i),$$

with

$$w_i = (1 - \delta)\, \sigma^{-d+1}\sqrt{2\pi} \exp\left\{ -\sum_{j\neq i} (y_j - \theta_j)^2/(2\sigma^2) \right\}$$
$$\times \left[ \Phi\left(\frac{1 - y_i}{\sigma}\right) - \Phi\left(\frac{-y_i}{\sigma}\right) \right].$$

The simulation of (8.2.1) is therefore straightforward:

**Algorithm A.43 Witch's Hat Distribution.**

1. Generate $U_i \sim \mathcal{U}_{[0,1]}$.
2. If $U_i < \delta/(w_i + \delta)$, generate $\theta_i \sim \mathcal{U}_{[0,1]}$;       [A.43]
   otherwise, generate

$$\theta_i \sim \mathcal{N}_-^+(y_i, \sigma^2, 0, 1),$$

where $\mathcal{N}_-^+(\mu, \sigma^2, \underline{\mu}, \overline{\mu})$ denotes the normal distribution $\mathcal{N}(\mu, \sigma^2)$ restricted to $[\underline{\mu}, \overline{\mu}]$ (see Example 2.3.5). Figure 8.2.2 describes the evolution of the subchain $(\theta_1^{(t)})$ for two different initial values. The waiting time before the

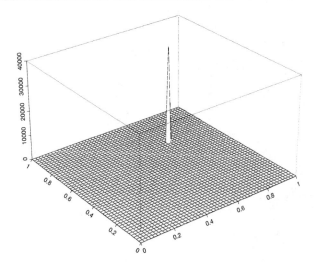

Figure 8.2.1. *Witch's hat distribution in* $\mathbb{R}^2$ *for* $y = (0.7, 0.7)$, $\delta = 0.01$, *and* $\sigma = 0.005$.

attraction by the mode at $(0.7, 0.7)$ is clearly exhibited by the difference of behavior of the sequence of $\theta_1^{(t)}$'s before $t = 25$ for the first case and $t = 50$ for the second case, and after. For most initial values, the sequence $(\theta_1^{(t)})$ has a behavior similar to that of the sequence with initial value 0.0217, namely a quick convergence to a neighborhood of the mode and an absence of visits to the remainder of the space $C$, although $\pi$ has a uniform weight of

$$\left( \frac{(1-\delta)(2\pi)^{d/2}}{\delta} + 1 \right)^{-1} = \left( \frac{0.992\pi}{0.01} + 1 \right)^{-1} = 0.002 ,$$

all over $C$. The attraction of the mode is thus such that it gives the impression of stationarity for the chain simulated by Gibbs sampling. The chain with initial value 0.9098, which achieves a momentary escape from the mode, is actually atypical. ‖

This example, while being quite artificial, has been proposed by Matthews (1993) as a calibration (*benchmark*) for MCMC algorithms since the choice of $\delta$, $\sigma$, and $d$ can lead to arbitrarily small probabilities of escaping the attraction of the mode. It will be used in the remainder of the chapter to evaluate the performances of different methods of convergence control. Note that some authors (see, e.g., Tanner 1996) use this benchmark example in an opposite spirit, namely as a failure to detect the relevant mode of the distribution (see Problem 8.1).

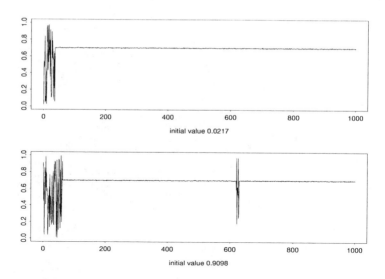

initial value 0.0217

initial value 0.9098

Figure 8.2.2. *Evolution of the chain* $(\theta_1^{(t)})$ *around the mode 0.7 of the witch's hat distribution for two initial values, 0.0217 (top) and 0.9098 (bottom).*

### 8.2.2 Nonparametric Tests of Stationarity

Standard nonparametric tests, such as Kolmogorov–Smirnov or Kuiper tests (Lehmann 1975), can also be applied in the stationarity assessment of a single output of the chain $\theta^{(t)}$. In fact, when the chain is stationary, $\theta^{(t_1)}$ and $\theta^{(t_2)}$ have the same marginal distribution for arbitrary times $t_1$ and $t_2$. Given an MCMC sample $\theta^{(1)}, \ldots, \theta^{(T)}$, it thus makes sense to compare the distributions of the two halves of this sample, $(\theta^{(1)}, \ldots, \theta^{(T/2)})$ and $(\theta^{(T/2+1)}, \ldots, \theta^{(T)})$. Since usual nonparametric tests are devised and calibrated in terms of iid samples, there needs to be a correction for the correlation between the $\theta^{(t)}$'s.

This correction can be achieved by the introduction of a batch size $G$ leading to the construction of two (quasi-) independent samples. For each of the two halves above, select subsamples $(\theta_1^{(G)}, \theta_1^{(2G)}, \ldots)$ and $(\theta_2^{(G)}, \theta_2^{(2G)}, \ldots)$. Then, for example, the *Kolmogorov-Smirnov statistic* is

$$(8.2.2) \qquad K = \frac{1}{M} \sup_\eta \left| \sum_{g=1}^{M} \mathbb{I}_{(0,\eta)}(\theta_1^{(gG)}) - \sum_{g=1}^{M} \mathbb{I}_{(0,\eta)}(\theta_2^{(gG)}) \right|$$

in the case of a one-dimensional chain. For multidimensional chains, (8.2.2) can be computed on either the function of interest or on each component of the vector $\theta^{(t)}$.

The statistic $K$ can be processed in several ways to derive a stopping rule. First, under the stationarity assumption as $M$ goes to infinity, the limiting distribution of $\sqrt{M}\, K$ has the following cdf:

$$(8.2.3) \qquad\qquad R(x) = 1 - \sum_{k=1}^{\infty}(-1)^{k-1}e^{-2k^2x^2} ,$$

which can be easily approximated by a finite sum (see Problem 8.2). The corresponding $p$-value can therefore be computed for each $T$ until it gets below a given level. An approximation of the 95% quantile, $q_\alpha = 1.36$ (for $M \geq 100$), simplifies this stage. Second, the above procedure can be improved by taking into account (i) the sequential nature of the test and (ii) the fact that $K$ may be computed as the infimum over all components of $\theta^{(t)}$ of the corresponding values of (8.2.2). An exact derivation of the level of the derived test is however extremely difficult given the correlation between the $\theta^{(t)}$'s and the influence of the subsampling mechanism. A third use of (8.2.2), which is more graphic, is to represent the sample of $\sqrt{M}\, K_T$'s against $T$ and to check visually for a stable distribution around small values.

Obviously, an assessment of stationarity based on a single chain is open to criticism: In cases of strong attraction from a local mode, the chain will most likely behave as if it was simulated from the restriction of $f$ to the neighborhood of this mode and thus lead to a convergence diagnosis (this is the *"you've only seen where you've been"* defect mentioned in the Introduction). However, in more intermediate cases, where the chain $(\theta^{(t)})$ stays for a while in the neighborhood of a mode before visiting another modal region, the subsamples $(\theta_1^{(t)})$ and $(\theta_2^{(t)})$ should exhibit different features until the chain explores every modal region.

**Example 8.2.2 Nuclear pump failures.** In the model of Gaver and O'Muircheartaigh (1987), described in Example 7.1.23, we consider the subchain $(\beta^{(t)})$ produced by the algorithm. Figure 8.2.3 gives the values of the Kolmogorov-Smirnov statistics $K$ for $T = 1000$ and $T = 10,000$ iterations, with $M = 100$ and 100 values of $10K_T$. Although both cases lead to similar proportions of about 95% values under the level 1.36, the first case clearly lacks the required homogeneity, since the statistic is almost monotone in $t$. This behavior may correspond to the local exploration of a modal region for $t \leq 400$ and to the move to another region of importance for $400 \leq t \leq 800$. ‖

### 8.2.3 Renewal Methods

Chapter 4 has presented the bases of *renewal theory* in §4.3.2 through the notion of *small set* (Definition 4.3.7). This theory is then used in §4.7 for a direct derivation of many limit theorems (Theorems 4.7.4 and 4.7.5). It is also possible to take advantage of this theory for the purpose(s) of convergence control, either through small sets as in Robert (1995b) or through the alternative of Mykland *et al.* (1995), which is based on a less restrictive representation of renewal called *regeneration*. (See also Gilks *et al.* 1998

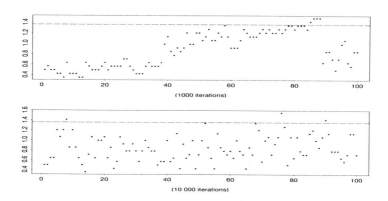

Figure 8.2.3. *Plot of the successive Kolmogorov–Smirnov statistics for $T = 1000$ and $T = 10,000$ iterations. The dotted line corresponds to the 95% level.*

for a related updating technique.) The use of small sets is mainly relevant in the control of convergence of averages, as presented in §8.3.3, and as a correct discretization technique; see §8.4.2.

Mykland *et al.* (1995) replaced small sets and the corresponding minorization condition with a generalization through functions $s$ such that

$$(8.2.4) \quad P(\theta^{(t+1)} \in B|\theta^{(t)}) \geq s(\theta^{(t)}) \, \nu(B), \qquad \theta^{(t)} \in \Theta, \, B \in \mathcal{B}(\Theta).$$

Small sets thus correspond to the particular case $s(x) = \varepsilon \mathbb{I}_C(x)$. If we define

$$r(\theta^{(t)}, \xi) = \frac{s(\theta^{(t)}) \, \nu(\xi)}{K(\theta^{(t)}, \xi)},$$

then each time $t$ is a renewal time with probability $r(\theta^{(t-1)}, \xi)$.

Mykland *et al.* (1995) propose a graphical convergence assessment based on the regeneration rate, which must remain about constant when an MCMC algorithm is in its stationary regime (by virtue of stationarity) and has correctly explored the support of $f$. They, thus, plot approximations of the regeneration rate,

$$\hat{r}_T = \frac{1}{T} \sum_{t=1}^{T} r(\theta^{(t)}, \theta^{(t+1)})$$

or

$$\tilde{r}_T = \frac{1}{T} \sum_{t=1}^{T} s(\theta^{(t)}),$$

(when the normalizing constant of $\nu$ is available against $T$). An additional recommendation of the authors is to smooth the graph of $\hat{r}_T$ by nonparametric techniques, but the influence of the smoothing parameter must be assessed to avoid overoptimistic conclusions (see §8.6.2 for a global criticism of functional nonparametric techniques).

Metropolis–Hastings algorithms provide a natural setting for the implementation of this regeneration method, since they allow for complete freedom in the choice of the transition kernel. Problems related to the Dirac mass in the kernel can be eliminated by considering only its absolutely continuous part,

$$\rho(\theta^{(t)}, \xi) \, q(\xi|\theta^{(t)}) = \min \left\{ \frac{f(\xi)}{f(\theta^{(t)})} \, q(\theta^{(t)}|\xi), q(\xi|\theta^{(t)}) \right\} ,$$

using the notation of [A.24]. The determination of $s$ and $\nu$ is facilitated in the case of a so-called *pseudo-reversible* transition; that is, when there exists a positive function $\tilde{f}$ with

(8.2.5)  $$\tilde{f}(\theta) \, q(\xi|\theta) = \tilde{f}(\xi) \, q(\theta|\xi) .$$

Equation (8.2.5) thus looks like a detailed balance condition for a reversible Markov chain, but $\tilde{f}$ is not necessarily a probability density. We denote $w(\theta) = f(\theta)/\tilde{f}(\theta)$.

**Lemma 8.2.3** *Suppose that $q$ satisfies (8.2.5) and for the transition induced by $q$, the $s_q$ and $\nu_q$ satisfy (8.2.4). Then, for every $c > 0$, the function*

$$s(\theta) = s_q(\theta) \, \min \left\{ \frac{c}{w(\theta)}, 1 \right\}$$

*and the density*

$$\nu(\theta) = \nu_q(\theta) \, \min \left\{ \frac{w(\theta)}{c}, 1 \right\}$$

*satisfies (8.2.4) for the Metropolis–Hastings algorithm associated with $q$.*

*Proof.* Since

$$P(\theta^{(t+1)} \in B|\theta^{(t)}) \geq \int_B \rho(\theta^{(t)}, \xi) \, q(\xi|\theta^{(t)}) \, d\xi,$$

we have

$$P(\theta^{(t+1)} \in B|\theta^{(t)}) \geq \int_B \min \left\{ \frac{w(\theta^{(t)})}{w(\xi)}, 1 \right\} q(\xi|\theta^{(t)}) \, d\xi$$

$$\geq \min \left\{ \frac{w(\theta^{(t)})}{c}, 1 \right\} \int_B \min \left\{ \frac{c}{w(\xi)}, 1 \right\} s_q(\theta^{(t)}) \, \nu_q(\xi) \, d\xi$$

$$= s(\theta^{(t)}) \int_B \nu(\xi) \, d\xi.$$

$\square\square$

In the particular case of an *independent* Metropolis–Hastings algorithm, $q(\xi|\theta) = g(\xi)$ and (8.2.5) applies for $\tilde{f} = g$. Therefore, $s_q \equiv 1$, $\nu_q = g$, and

$$\nu(\xi) \propto g(\xi) \, \min \left\{ \frac{f(\xi)}{cg(\xi)}, 1 \right\}$$

$$\propto \min \{ f(\xi), cg(\xi) \},$$

which behaves as a truncation of the instrumental distribution $g$ depending on the true density $f$. If $\xi \sim g$ is accepted, the probability of regeneration is then

$$(8.2.6) \quad r(\theta^{(t)}, \xi) = \begin{cases} \dfrac{c}{w(\theta^{(t)}) \wedge w(\xi)} & \text{if } w(\xi) \wedge w(\theta^{(t)}) > c \\[2ex] \dfrac{w(\theta^{(t)}) \vee w(\xi)}{c} & \text{if } w(\xi) \vee w(\theta^{(t)}) < c \\[2ex] 1 & \text{otherwise.} \end{cases}$$

Note that $c$ is a free parameter. Therefore, it can be selected (calibrated) to optimize the renewal probability (or the mixing rate).

For a *symmetric* Metropolis–Hastings algorithm, $q(\xi|\theta) = q(\theta|\xi)$ implies that $\tilde{f} \equiv 1$ satisfies (8.2.5). The parameters $s_q$ and $\nu_q$ are then determined by the choice of a set $D$ and of a value $\tilde{\theta}$ in the following way:

$$\nu_q(\xi) \propto q(\theta|\tilde{\theta}) \, \mathbb{I}_D(\xi), \qquad s_q(\theta) = \inf_{\xi \in D} \frac{q(\xi|\theta)}{q(\theta|\xi)} .$$

The setting is therefore less interesting than in the independent case since $D$ and $\tilde{\theta}$ have first to be found, as in a regular minorization condition. Note that the choice of $\tilde{\theta}$ can be based on an initial simulation of the algorithm, using the mode or the median of $w(\theta^{(t)})$.

**Example 8.2.4 (Continuation of Example 8.2.2)** In the model of Gaver and O'Muircheartaigh (1987), $D = \mathbb{R}_+$ is a small set for the chain $(\beta^{(t)})$. Indeed,

$$K(\beta, \beta') \geq \frac{\delta^{\gamma+10\alpha}(\beta')^{\gamma+10\alpha-1}}{\Gamma(10\alpha + \gamma)} \, e^{-\delta\beta'} \prod_{i=1}^{10} \left( \frac{t_i}{t_i + \beta'} \right)^{p_i + \alpha}$$

(see Example 7.1.23). The regeneration (or renewal) probability is then

$$r(\beta, \beta') = \frac{\delta^{\gamma+10\alpha}(\beta')^{\gamma+10\alpha-1}}{K(\beta, \beta') \, \Gamma(\gamma + 10\alpha)} \, e^{-\delta\beta'} \prod_{i=1}^{10} \left( \frac{t_i}{t_i + \beta'} \right)^{p_i + \alpha} ,$$

where $K(\beta, \beta')$ can be approximated by

$$\frac{1}{M} \sum_{m=1}^{M} \frac{\left( \delta + \sum_{i=1}^{10} \lambda_{im} \right)^{\gamma+10\alpha} (\beta')^{\gamma+10\alpha-1} e^{-\beta'\left( \delta + \sum_{i=1}^{10} \lambda_{im} \right)}}{\Gamma(\gamma + 10\alpha)} ,$$

with $\lambda_{im} \sim \mathcal{G}a(p_i + \alpha, t_i + \beta)$. Figure 8.2.4 provides the plot of $r(\beta^{(t)}, \beta^{(t+1)})$ against $t$, and the graph of the averages $\hat{r}_T$, based on only the four first observations. (The renewal probabilities associated with 10 observations are too small to be of practical use.) Mykland *et al.* (1995) thus replace $\mathbb{R}_+$ with the interval $D = [2.3, 3.1]$ to achieve renewal probabilities of reasonable magnitude. ‖

Although renewal methods involve a detailed study of the chain $(\theta^{(t)})$ and may require modifications of the algorithms, as in the hybrid algorithms

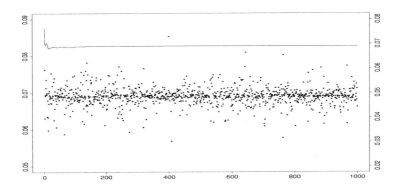

Figure 8.2.4.  *Plot of the probabilities* $r(\beta^{(t)}, \beta^{(t+1)})$ *for the pump failure data when* $\mathbb{R}_+$ *is chosen as the small set. Superimposed (and scaled to the right), is the average* $\hat{r}_T$.

of Mykland *et al.* (1995) (see also §8.3.3), their basic independence structure brings a certain amount of robustness in the monitoring of MCMC algorithms, more than the other approaches presented in this chapter. The specificity of the monitoring obviously prevents a completely automated implementation, but this drawback cannot detract from its attractive theoretical properties.

### 8.2.4 Distance Evaluations

Roberts (1992) considers convergence from a functional point of view, as in §7.6.3 (and only for Gibbs sampling). Using the norm induced by $f$ defined in §4.6.1, he proposes an unbiased estimator of the distance $\| f_t - f \|$, where $f_t$ denotes the marginal density of the *symmetrized* chain $\theta^{(t)}$. The symmetrized chain is obtained by adding to the steps 1., 2.,...,k. of the Gibbs sampler the additional steps k.,k-1.,...,1., as in the proof of Lemma 7.6.1. This device leads to a reversible chain, creating in addition a *dual* chain $(\tilde{\theta}^{(t)})$, which is obtained by the inversion of the steps of the Gibbs sampler: Starting with $\theta^{(t)}$, $\tilde{\theta}^{(t)}$ is generated conditionally on $\theta^{(t)}$ by steps 1.,2.,...,k., then $\theta^{(t+1)}$ is generated conditionally on $\tilde{\theta}^{(t)}$ by steps k.,k-1.,...,1.

   Using $m$ parallel chains $(\theta_\ell^{(t)})$ $(\ell = 1, \ldots, m)$ started with the same initial value $\theta^{(0)}$, Roberts (1992) shows that an unbiased estimator of $\| f_t - f \| + 1$ is

$$J_t = \frac{1}{m(m-1)} \sum_{1 \le \ell \neq s \le m} \frac{K_-(\tilde{\theta}_\ell^{(0)}, \theta_s^{(t)})}{f(\theta_s^{(t)})} \, ,$$

where $K_-$ denote the transition kernel for the steps k.,k-1.,...,1. of [A.31] (see Problem 8.6). Since the distribution $f$ is typically only known up to a multiplicative constant, the limiting value of $J_t$ is unknown. Thus, the

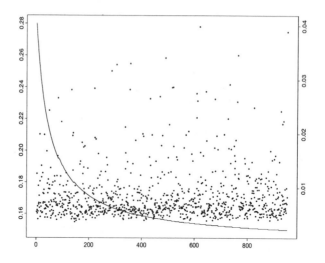

Figure 8.2.5. *Stationarity of the indicator* $J_t$ *for* $m = 50$ *parallel chains converging to the witch's hat distribution. The superimposed curve (scaled on the right) describes the convergence of the average over the 50 chains of estimators of the probability* $P((\theta_1, \theta_2) \in [0, 0.5]^2)$.

diagnostic based on this unbiased estimation of $\|f_t - f\| + 1$ is to evaluate the stabilization of $J_t$ graphically. This method requires both $K_-(\theta, \theta')$ and the normalizing constant of $K_-(\theta, \theta')$ to be known, as in the method of Ritter and Tanner (1992) (see Note 8.6.1).

An additional diagnostic can be derived from the convergence of

$$I_t = \frac{1}{m} \sum_{i=1}^{m} \frac{K_-\left(\tilde{\theta}_\ell^{(0)}, \theta_i^{(t)}\right)}{f\left(\theta_i^{(t)}\right)}$$

to the same limit as $J_t$, which is equal to 1 when the normalizing constant is available. Variations and generalizations of this method are considered by Roberts (1994), Brooks *et al.* (1997), and Brooks and Roberts (1999).

**Example 8.2.5 (Continuation of Example 8.2.1)** As shown by (8.2.1), the conditional densities $\pi(\theta_i | \theta_{j \neq i}, y)$ are known,

$$\theta_i | \theta_{j \neq i}, y \sim \frac{w_i}{w_i + \delta}\, \mathcal{N}_-^+(y_i, \sigma^2, 0, 1) + \frac{\delta}{w_i + \delta}\, \mathcal{U}_{[0,1]},$$

where $\mathcal{N}_-^+(\mu, \sigma^2, \underline{\mu}, \overline{\mu})$ denotes the normal distribution $\mathcal{N}(\mu, \sigma^2)$ truncated in $\underline{\mu}$ and $\overline{\mu}$. Therefore, $J_t$ can be easily computed and Figure 8.2.5 gives the sequence of $J_t$ for $m = 50$ parallel chains, the initial values being generated from $f$. The convergence of this criterion is therefore fast, with variations around the true value 0.16, which can be attributed to the small value of $m$.

The superimposed curve describes the convergence, to the true value

$5 \times 10^{-4}$, of the average of the quantities

$$(8.2.7) \qquad \frac{1}{t} \sum_{\ell=1}^{t} \mathbb{I}_{\theta_1^{(\ell)}<0.5} \mathbb{I}_{\theta_2^{(\ell)}<0.5}$$

over the 50 chains. Convergence of an empirical average for a given chain is therefore poorly evaluated by the method, whose primary goal is to assess convergence to the stationary distribution, rather than convergence of averages like (8.2.7).                                                                      ∥

From a theoretical point of view, this method of evaluation of the distance to the stationary distribution is quite satisfactory, but it does not exactly meet the convergence requirements of point (ii) of §8.1 for the following reasons:

(a) It requires parallel runs of several Gibbs sampler algorithms and thus results in a loss of efficiency in the execution (see §8.4).

(b) The convergence control is based on $f_t$, the marginal distribution of $(\theta^{(t)})$, which is typically of minor interest for Markov chain Monte Carlo purposes, when a single realization of $f_t$ is observed.

(c) The dependence on $m$ parallel chains implies that the slower chain governs the convergence of the group, hence another loss of efficiency.

(d) The computation (or the approximation) of the normalizing constant of $K_-$ can be time-consuming.

(e) A stabilizing of $J_t$ around a mean value does not imply that the chain has explored all the modes of $f$, even though the production of $m$ copies of the Gibbs sampler algorithm reduces this risk (see also §8.4 and Figure 9.5.4).

(f) As pointed out at the beginning of this chapter, convergence to the stationary distribution $f$ does not directly relate to the speed of convergence of empirical means.

Liu *et al.* (1992) also evaluate convergence to the stationary distribution through the difference between $f$, the limiting distribution, and $f_t$, the distribution of $\theta^{(t)}$. Their method is based on an expression of the variance of $f_t(\theta)/f(\theta)$ which is close to the $J_t$'s of Roberts (1992).

**Lemma 8.2.6** *Let $\theta_1^-, \theta_2^- \sim f_{(t-1)}$ be independent, and generate $\theta_1 \sim K(\theta_1^-, \theta_1)$, $\theta_2 \sim K(\theta_2^-, \theta_2)$. Then, the quantity*

$$U = \frac{f(\theta_1)}{f(\theta_2)} \frac{K(\theta_1^-, \theta_2)}{K(\theta_1^-, \theta_1)},$$

*satisfies*

$$\mathbb{E}_f[U] = \mathrm{var}_f \left( \frac{f_t(\theta)}{f(\theta)} \right) + 1.$$

*Proof.* The independence of the $\theta_i^-$'s implies that

$$\mathbb{E}_f \left[ \frac{f(\theta_1)}{f(\theta_2)} \frac{K(\theta_1^-, \theta_2)}{K(\theta_1^-, \theta_1)} \right] = \int \frac{f(\theta_1)}{f(\theta_2)} K(\theta_1^-, \theta_2) f_{t-1}(\theta_1^-) \, d\theta_1^- \, d\theta_1 d\theta_2$$

$$= \int \frac{f_t(\theta_2)^2}{f(\theta_2)} \, d\theta_2$$

$$= 1 + \int \left( \frac{f_t(\theta_2)}{f(\theta_2)} - 1 \right)^2 f(\theta_2) \, d\theta_2 \,,$$

with

$$\mathbb{E}_f \left[ \frac{f_t(\theta_2)}{f(\theta_2)} \right] = 1. \qquad \Box\Box$$

Given $M$ parallel chains, each iteration $t$ provides $M(M-1)$ values $U^{(i,j,t)}$ $(i \neq j)$. These can be processed either graphically or using the method of Gelman and Rubin (1992) (see §8.3.4) for $M/2$ independent values $U^{(i,j,t)}$ (that is, using disjoint couples $(i,j)$). Note that the computation of $U$ does not require the computation of the normalizing constant for the kernel $K$.

**Example 8.2.7 (Continuation of Example 8.2.4)** If the initial distribution on $\beta$ is the inverse Gamma distribution $\mathcal{IG}(0.1, 1)$, Gelman and Rubin's (1992) method (see §8.3.4) applies for both sequences $\beta^{(m,t)}$ and

$$U^{(i,j,t)} = \left[ \frac{\theta^{(i,t)}}{\theta^{(j,t)}} \right]^{10\alpha+\gamma} \exp \left\{ [\theta^{(j,t)} - \theta^{(i,t)}]/\beta^{(j,t-1)} \right\} ,$$

with

$$\theta^{(i,t)} = \delta + \sum_{k=1}^{10} \lambda_k^{(i,t)}.$$

Figure 8.2.6 represents the evolution of the coefficients $R_T$ for both quantities, showing a satisfactory convergence of the criterion after only 250 iterations.                                                                    ‖

**Example 8.2.8 Pathological beta generator.** In the setup of the algorithm [A.30], consider the particular case where $g$ is the density of the beta distribution $\mathcal{Be}(\alpha+1, 1)$ and $\rho(x) = 1-x$. Then, the stationary distribution is given by

$$f(x) \propto x^{\alpha+1-1}/\{1 - (1-x)\} = x^{\alpha-1} .$$

Lemma 6.6.1 thus implies that an algorithm for the generation of the distribution $\mathcal{Be}(\alpha, 1)$ $(\alpha > 0)$ is to simulate $Y \sim \mathcal{Be}(\alpha + 1, 1)$ and accept $y$ with probability $x^{(t)}$. (Note that a Metropolis–Hastings algorithm based on the same pair $(f, g)$ would lead to accept $y$ with probability $x^{(t)}/y$, which is larger than $x^{(t)}$ but requires a simulation of $y$ before deciding on its acceptance.)

This algorithm produces a Markov chain that does not satisfy the assumptions of the Central Limit Theorem. (Establishing this is beyond our scope. See, for example, Doukhan *et al.* 1994.)

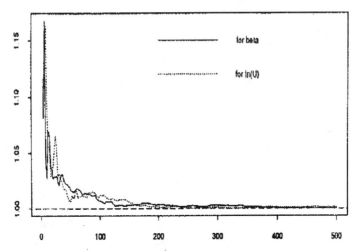

Figure 8.2.6.  *Evolutions of $R_T$ for $\beta^{(m,t)}$ (solid lines) and $\log(U^{(i,j,t)})$ (dotted lines) for the model of Gaver and O'Muircheartaigh (1987) (Source: Liu et al. 1992).*

For $h(x) = x^{1-\alpha}$, Figure 8.2.7 describes the particularly slow convergence of the empirical average of the $h(x^{(t)})$'s to $\alpha$ (see Problem 8.4), a convergence that is guaranteed by the ergodic theorem (Proposition 4.7.4). It is not achieved after $5 \times 10^6$ iterations. Figure 8.2.7 also provides a comparison with a realization of the Metropolis–Hastings algorithm based on the same pair. Both graphs correspond to identical generations of $y_t$ in [A.30], with a higher probability of acceptance for the Metropolis–Hastings algorithm. Note that the graph associated with the Metropolis–Hastings algorithm, while more regular that the graph for the algorithm [A.30], is also very slow to converge. The final value is still closer to the exact expectation $\alpha = 0.20$.

A simple explanation (due to Jim Berger) of the bad behavior of both algorithms is the considerable difference between the distributions $\mathcal{B}e(0.2, 1)$ and $\mathcal{B}e(1.2, 1)$. In fact, the ratio of the corresponding cdf's is $x^{0.2}/x^{1.2} = 1/x$ and, although the quantile at level 10% is $10^{-5}$ for $\mathcal{B}e(0.2, 1)$, the probability to reach the interval $[0, 10^{-5}]$ under the $\mathcal{B}e(1.2, 1)$ distribution is $10^{-6}$. Therefore, the number of simulations from $\mathcal{B}e(1.2, 1)$ necessary to obtain an adequate coverage of this part of the support of $\mathcal{B}e(0.2, 1)$ is enormous. Moreover, note that the probability of leaving the interval $[0, 10^{-5}]$ using [A.30] is less than $10^{-6}$, and it is of the same order of magnitude for the corresponding Metropolis–Hastings algorithm.  ||

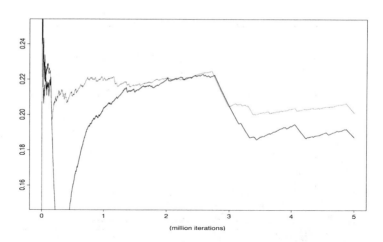

Figure 8.2.7. *Convergence of the empirical average of* $h(x^{(t)})$ *to* $\alpha = 0.2$ *for the algorithm* [A.30] *(solid lines) and the Metropolis–Hastings algorithm (dotted lines).*

## 8.3 Monitoring Convergence of Averages

### 8.3.1 Graphical Methods

Yu (1994) and Yu and Mykland (1998) propose a purely graphical evaluation of the convergence of (8.1.1) based on cumulative sums (CUSUM), graphing the partial differences

$$(8.3.1) \qquad D_T^i = \sum_{t=1}^{i} [h(\theta^{(t)}) - S_T], \qquad i = 1, \dots, T,$$

where

$$(8.3.2) \qquad S_T = \frac{1}{T} \sum_{t=1}^{T} h(\theta^{(t)}).$$

They derive a qualitative evaluation of the mixing speed of the chain and the correlation between the $\theta^{(t)}$'s: When the mixing of the chain is high, the graph of $D_T^i$ is highly irregular and concentrated around 0. Slowly mixing chains (that is, chains with a slow pace of exploration of the stationary distribution) produce regular graphs with long excursions away from 0. Figure 8.3.1 contains the graph which corresponds to the dataset of Example 8.2.8, exhibiting a slow convergence behavior already indicated in Figure 8.2.7.

Figure 8.3.2 analyzes the data of the witch's hat distribution given in Figure 8.2.2 for the initial value 0.0207 (see Example 8.2.1). The strongly perturbed aspect of the graph, close to the "ideal" shape of Yu and Mykland (1998), does not indicate that the chain has not yet left the attraction zone of the mode $(0.7, 0.7)$ and the same phenomenon occurs for a larger number of iterations. (See Brooks 1998c for a more quantitative approach

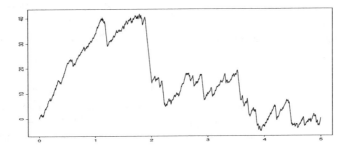

Figure 8.3.1. *Evolution of the $D_t^i$ criterion for the chain produced by [A.30] and for $\alpha = 0.2$ and $h(x) = x^{0.8}$ (millions of iterations).*

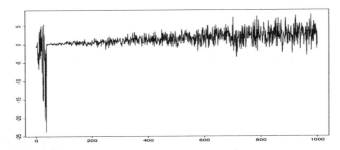

Figure 8.3.2. *Evolution of the CUSUM criterion $D_t^i$ for the chain produced by [A.43] and for $h(\theta) = \theta$.*

to CUSUM.)

This difficulty is common to most "on-line" methods; that is, to diagnoses based on a single chain. It is almost impossible to detect the existence of other modes of $f$ (or of other unexplored regions of the space with positive probability); this is the *"you've only seen where you've been"* defect. (Exceptions are the Riemann sum indicator (8.3.5) which "estimates" 1 and can detect probability losses in the region covered by the sample, and the related method developped in Brooks 1998b.)

### 8.3.2 Multiple Estimates

In most cases, the graph of the raw sequence $(\theta^{(t)})$ is unhelpful in the detection of stationarity or convergence. Indeed, it is only when the chain has explored different regions of the state-space during the observation time that lack of stationarity can be detected. (Gelman and Rubin 1992 also illustrate the use of the raw sequence graphs when using chains in parallel with quite distinct supports.)

Given some quantities of interest $\mathbb{E}_f[h(\theta)]$, a more helpful indicator is the behavior of the averages (8.3.2) in terms of $T$. A necessary condition for convergence is the stationarity of the sequence $(S_T)$, even though the stabilizing of this sequence may correspond to the influence of a single mode of the density $f$, as shown by Gelman and Rubin (1992).

Robert (1995b) proposes a more robust approach to graphical convergence monitoring. The idea is to use simultaneously several convergent estimators of $\mathbb{E}_f[h(\theta)]$ based on the same chain $(\theta^{(t)})$, until all estimations coincide (up to a given precision). The most common estimation techniques in this setup are, besides the empirical average $S_T$, the *conditional* (or *Rao–Blackwellized*) version of this average, either in its nonparametric version (§6.4.2) for Metropolis–Hastings algorithms, or in its parametric version (§7.2.3) for Gibbs sampling,

$$S_T^C = \frac{1}{T} \sum_{t=1}^T \mathbb{E}[h(\theta)|\eta^{(t)}] ,$$

where $(\eta^{(t)}, \theta^{(t)})$ is the Markov chain produced by the algorithm.

A second technique providing convergent estimators is *importance sampling* (§3.3). If the density $f$ is available up to a constant,[3] the importance sampling alternative is

$$S_T^P = \sum_{t=1}^T w_t \, h(\theta^{(t)}) ,$$

where $w_t \propto f(\theta^{(t)})/g_t(\theta^{(t)})$ and $g_t$ is the true density used for the simulation of $\theta^{(t)}$. In particular, in the case of the Gibbs sampling,

$$g_t(\theta^{(t)}) = f_1(\theta_1^{(t)}|\theta_2^{(t-1)}, \ldots, \theta_k^{(t-1)}) \cdots f_k(\theta_k^{(t)}|\theta_1^{(t)}, \ldots, \theta_{k-1}^{(t)}) .$$

If, on the other hand, the chain $(\theta^{(t)})$ is produced by a Metropolis–Hastings algorithm, the variables actually simulated, $\eta^{(t)} \sim q(\eta|\theta^{(t-1)})$, can be recycled through the estimator

$$S_T^{MP} = \sum_{t=1}^T w_t \, h(\eta^{(t)})$$

with $w_t \propto f(\eta^{(t)})/q(\eta^{(t)}|\theta^{(t-1)})$.

An interesting property of importance sampling is that it removes the correlation between the $\theta^{(t)}$'s (or the $\eta^{(t)}$'s).

**Lemma 8.3.1** *Consider* $(\theta^{(t)})$ *a Markov chain associated with a Metropolis–Hastings algorithm with transition kernel* $q$. *Then,*

$$\mathrm{var}\left( \sum_{t=1}^T h(\theta^{(t)}) \frac{f(\theta^{(t)})}{q(\theta^{(t)}|\theta^{(t-1)})} \right) = \sum_{t=1}^T \mathrm{var}\left( h(\theta^{(t)}) \frac{f(\theta^{(t)})}{q(\theta^{(t)}|\theta^{(t-1)})} \right)$$

*provided these quantities are well defined.*

---

[3] As it is, for instance, in all cases where the Gibbs sampler applies, as shown by the Hammersley–Clifford theorem (§7.1.5).

*Proof.* Assume, without loss of generality, that $\mathbb{E}_f[h(\theta)] = 0$. If we define $w_t = f(\theta^{(t)})/q(\theta^{(t)}|\theta^{(t-1)})$, the covariance between $w_t h(\theta^{(t)})$ and $w_{t+r} h(\theta^{(t+r)})$ satisfies

$$\mathbb{E}\left[w_t h(\theta^{(t)}) w_{t+r} h(\theta^{(t+r)})\right] = \mathbb{E}\left[w_t h(\theta^{(t)})\, \mathbb{E}\left[w_{t+r} h(\theta^{(t+r)})|\theta^{(t+r-1)}\right]\right]$$

$$= \mathbb{E}\left[w_t h(\theta^{(t)}) \int h(x) \frac{f(x)}{q(x|\theta^{(t+r-1)})}\, q(x|\theta^{(t+r-1)})\, dx\right]$$

$$= 0.$$

□□

This result does not formally apply to the estimators $S_T^P$ and $S_T^{MP}$ since the weights $w_t$ are known up to a multiplicative constant and are, therefore, normalized by the inverse of their sum. However, it can be assumed that the effect of this normalization on the correlation vanishes when $T$ is large.

The consequences of Lemma 8.3.1 are twofold. First, it indicates that, up to second order, $S_T^P$ and $S_T^{MP}$ behave as in the independent case, and thus allow for a more traditional convergence control on these quantities. Second, Lemma 8.3.1 implies that the variance of $S_T^P$ (or of $S_T^{MP}$), when it exists, decreases at speed $1/T$ in *stationarity settings*. Thus, nonstationarity can be detected if the decrease of the variations of $S_T^P$ does not fit in a confidence parabola of order $1/\sqrt{T}$. Note also that the density $f$ can sometimes be replaced (in $w_t$) by an approximation; in particular, in settings where Rao–Blackwellization applies,

$$(8.3.3) \qquad \hat{f}_T(\theta) = \frac{1}{T} \sum_{t=1}^{T} f_1(\theta|\eta^{(t)})$$

provides an *unbiased estimator* of $f$ (see Wei and Tanner 1990a and Tanner 1996). A parallel chain $(\eta^{(t)})$ should then be used to ensure the independence of $\hat{f}_T$ and of $h(\theta^{(t)})$.

The fourth estimator based on $(\theta^{(t)})$ is the *Riemann approximation* (§3.4); that is,

$$(8.3.4) \qquad S_T^R = \sum_{t=1}^{T-1} [\theta_{(t+1)} - \theta_{(t)}]\, h(\theta_{(t)})\, f(\theta_{(t)}),$$

which estimates $\mathbb{E}_f[h(\theta)]$, where $\theta_{(1)} \leq \cdots \leq \theta_{(T)}$ denotes the *ordered* chain $(\theta^{(t)})_{1 \leq t \leq T}$. This estimator is mainly studied in the iid case (see Proposition 3.4.2), but it can, nonetheless, be included as an alternative estimator in the present setup, since its performances tend to be superior to those of the previous estimators.

The main drawback of $S_T^R$ lies in the unidimensionnality requirement, the quality of multidimensional extensions of the Riemann approximation deceasing quickly with the dimension (see Yakowitz *et al.* 1978). When $h$ only involves one component of $\theta$, the marginal distribution of this component should therefore be used or replaced with an approximation like

(8.3.3), which increases the computation burden, and is not always available. For extensions and alternatives, see Robert (1995b), Philippe and Robert (1998b), and Problem 8.9.

Note that the Riemann approximation method also provides a simple convergence diagnosis since, when $h(\theta) = 1$, the quantity (8.3.4) converges to 1. When the density $f$ is completely known (that is, including the normalizing constant), it is straightforward to monitor the convergence of

$$(8.3.5) \qquad S_T^R = \sum_{t=1}^{T-1} [\theta_{(t+1)} - \theta_{(t)}]\, f(\theta_{(t)})$$

to 1. When the density is only known up to a multiplicative constant, this method can also provide an efficient estimate of the normalizing constant.

**Example 8.3.2 Cauchy posterior distribution.** For the posterior distribution of Example 6.4.1 (that is,

$$(8.3.6) \quad \pi(\theta|x_1, x_2, x_3) \propto e^{-\theta^2/2\sigma^2} \left[ (1 + (\theta - x_1)^2) \right.$$
$$\left. \times\ (1 + (\theta - x_2)^2)(1 + (\theta - x_3)^2) \right]^{-1}, \Big)$$

a Gibbs sampling algorithm which approximates this distribution can be derived by demarginalization, namely by introducing three artificial variables, $\eta_1, \eta_2,$ and $\eta_3,$ such that

$$\pi(\theta, \eta_1, \eta_2, \eta_3|x_1, x_2, x_3) \propto e^{-\theta^2/2\sigma^2}\, e^{-(1+(\theta-x_1)^2)\eta_1/2}$$
$$\times\ e^{-(1+(\theta-x_2)^2)\eta_2/2}\, e^{-(1+(\theta-x_3)^2)\eta_3/2}.$$

In fact, similar to the $t$ distribution (see Example 7.1.2), expression (8.3.6) appears as the marginal distribution of $\pi(\theta, \eta_1, \eta_2, \eta_3|x_1, x_2, x_3)$ and the conditional distributions of the Gibbs sampler ($i = 1, 2, 3$)

$$\eta_i|\theta, x_i \sim \mathcal{E}xp\left( \frac{1 + (\theta - x_i)^2}{2} \right),$$

$$\theta|x_1, x_2, x_3, \eta_1, \eta_2, \eta_3 \sim \mathcal{N}\left( \frac{\eta_1 x_1 + \eta_2 x_2 + \eta_3 x_3}{\eta_1 + \eta_2 + \eta_3 + \sigma^{-2}}, \frac{1}{\eta_1 + \eta_2 + \eta_3 + \sigma^{-2}} \right),$$

are easy enough to simulate. Figure 8.3.3 illustrates the efficiency of this algorithm by exhibiting the agreement between the histogram of the simulated $\theta^{(t)}$'s and the true posterior distribution (8.3.6).

For simplicity's sake, let us introduce

$$\mu(\eta_1, \eta_2, \eta_3) = \frac{\eta_1 x_1 + \eta_2 x_2 + \eta_3 x_3}{\eta_1 + \eta_2 + \eta_3 + \sigma^{-2}}$$

and

$$\tau^{-2}(\eta_1, \eta_2, \eta_3) = \eta_1 + \eta_2 + \eta_3 + \sigma^{-2}.$$

If the function of interest is $h(\theta) = \exp(-\theta/\sigma)$ (with known $\sigma$), the different approximations of $\mathbb{E}_\pi[h(\theta)]$ proposed above can be derived. The

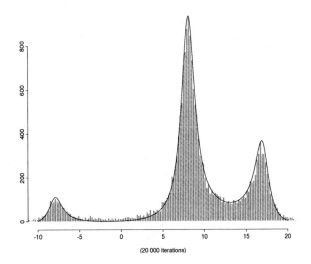

Figure 8.3.3. *Comparison of the density (8.3.6) and of the histogram from a sample of* 20,000 *points simulated by Gibbs sampling, for* $x_1 = -8$, $x_2 = 8$, $x_3 = 17$, *and* $\sigma = 50$.

conditional version of the empirical mean is

$$S_T^C = \frac{1}{T} \sum_{t=1}^T \exp\left\{ -\mu\left(\eta_1^{(t)}, \eta_2^{(t)}, \eta_3^{(t)}\right) + \tau^2\left(\eta_1^{(t)}, \eta_2^{(t)}, \eta_3^{(t)}\right)/2 \right\},$$

importance sampling is associated with the weights

$$w_t \propto \frac{\exp\left\{ -\dfrac{(\theta^{(t)})^2}{2\sigma^2} + \dfrac{\left(\theta^{(t)} - \mu\left(\eta_1^{(t)}, \eta_2^{(t)}, \eta_3^{(t)}\right)\right)^2}{2\tau^2\left(\eta_1^{(t)}, \eta_2^{(t)}, \eta_3^{(t)}\right)} \right\}}{\tau\left(\eta_1^{(t)}, \eta_2^{(t)}, \eta_3^{(t)}\right) \displaystyle\prod_{i=1}^3 [1 + (x_i - \theta^{(t)})^2]},$$

and the Riemann approximation is

$$S_T^R = \frac{\displaystyle\sum_{t=1}^{T-1} (\theta_{(t+1)} - \theta_{(t)})\, e^{-\theta_{(t)}/\sigma - \theta_{(t)}^2/(2\sigma^2)} \prod_{i=1}^3 [1 + (x_i - \theta_{(t)})^2]^{-1}}{\displaystyle\sum_{t=1}^{T-1} (\theta_{(t+1)} - \theta_{(t)})\, e^{-\theta_{(t)}^2/(2\sigma^2)} \prod_{i=1}^3 [1 + (x_i - \theta_{(t)})^2]^{-1}},$$

where $\theta_{(1)} \leq \cdots \leq \theta_{(T)}$ denotes the ordered sample of the $\theta^{(t)}$'s.

Figure 8.3.4 graphs the convergence of the four estimators versus $T$. As is typical, $S_T$ and $S_T^C$ are similar almost from the start (and, therefore, their difference cannot be used to diagnose [lack of] convergence). The strong

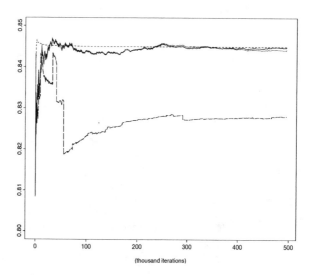

Figure 8.3.4.   *Convergence of four estimators $S_T$ (solid line), $S_T^C$ (dotted line), $S_T^R$ (mixed) and $S_T^P$ (long dashes) of the expectation under (8.3.6) of $h(\theta) = \exp(-\theta/\sigma)$, for $\sigma^2 = 50$ and $(x_1, x_2, x_3) = (-8, 8, 17)$. The graphs of $S_T$ and $S_T^C$ are identical. The final values are 0.845, 0.844, 0.828 and 0.845 for $S_T$, $S_T^C$, $S_T^P$ and $S_T^R$ respectively.*

stability of $S_T^R$, which is the central value of $S_T$ and $S_T^C$, indicates that convergence is achieved after a small number of iterations. On the other hand, the behavior of $S_T^P$ indicates that importance sampling does not perform satisfactorily in this case and is most likely associated with an infinite variance. (The ratio $f(\theta)/f_1(\theta|\eta)$ associated with the Gibbs sampling usually leads to an infinite variance (see §3.3.2), since $f$ is the marginal of $f_1(\theta|\eta)$ and has generally fatter tails than $f_1$. For example, this is the case when $f$ is a $t$ density and $f_1(\cdot|\eta)$ is a normal density with variance $\eta$.)     ‖

**Example 8.3.3 (Continuation of Example 8.2.8)**   The chain $(X^{(t)})$ produced by

$$X^{(t+1)} = \begin{cases} x^{(t)} & \text{with probability } 1 - x^{(t)} \\ Y^{(t)} \sim \mathcal{B}e(\alpha + 1, 1) & \text{otherwise,} \end{cases}$$

allows for a conditional version when $h(x) = x^{1-\alpha}$ since

$$\mathbb{E}\left[\left(X^{(t+1)}\right)^{1-\alpha} \Big| x^{(t)}\right] = \left(1 - x^{(t)}\right)\left(x^{(t)}\right)^{1-\alpha} + x^{(t)}\, \mathbb{E}[Y^{1-\alpha}]$$

$$= \left(1 - x^{(t)}\right)\left(x^{(t)}\right)^{1-\alpha} + x^{(t)}\,\frac{1+\alpha}{2}\,.$$

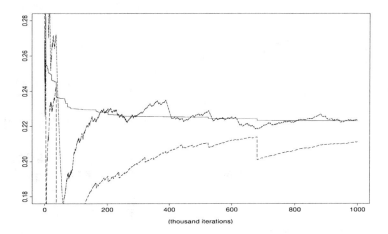

Figure 8.3.5. *Convergence of four estimators $S_T$ (solid line), $S_T^C$ (dotted line), $S_T^R$ (mixed dashes) and $S_T^P$ (long dashes) of $\mathbb{E}[(X^{(t)})^{0.8}]$ for the $\mathcal{B}e(0.2,1)$ distribution, after elimination of the first $200,000$ iterations. The graphs of $S_T$ and $S_T^C$ are identical. The final values are 0.224, 0.224, 0.211, and 0.223 for $S_T$, $S_T^C$, $S_T^R$, and $S_T^P$ respectively, to compare with a theoretical value of $0.2$.*

The importance sampling estimator based on the simulated $y^{(t)}$'s is

$$
S_T^P = \sum_{t=1}^{T} \frac{(y^{(t)})^{\alpha-1}}{(y^{(t)})^{\alpha}} (y^{(t)})^{1-\alpha} \Big/ \sum_{t=1}^{T} \frac{(y^{(t)})^{\alpha-1}}{(y^{(t)})^{\alpha}}
$$
$$
= \sum_{t=1}^{T} (y^{(t)})^{-\alpha} \Big/ \sum_{t=1}^{T} (y^{(t)})^{-1}
$$

and the Riemann approximation based on the same $y^{(t)}$'s is

$$
S_T^R = \sum_{t=1}^{T-1} (y_{(t+1)} - y_{(t)}) \Big/ \sum_{t=1}^{T-1} (y_{(t+1)} - y_{(t)})\, y_{(t)}^{\alpha-1}
$$
$$
= \frac{y_{(T)} - y_{(1)}}{\sum_{t=1}^{T} (y_{(t+1)} - y_{(t)})\, y_{(t)}^{\alpha-1}} .
$$

Figure 8.3.5 gives an evaluation of these different estimators, exhibiting the incredibly slow convergence of (8.1.1), since this experiment uses 1 million iterations. Note that the importance sampling estimator $S_T^P$ approaches the true value $\alpha = 0.2$ faster than $S_T$ and $S_T^R$ ($S_T^C$ being again indistinguishable from $S_T$), although the repeated jumps in the graph of $S_T^P$ do not indicate convergence. Besides, eliminating the $200,000$ first iterations from [A.30] does not stabilize convergence, which shows that the very slow mixing of the algorithm is responsible for this phenomenon rather than a lack of convergence to stationarity.     ‖

Both examples highlight the limitations of the method of multiple estimates:

(1) The method does not always apply (parametric Rao–Blackwellization and Riemann approximation are not often available).

(2) It is intrinsically conservative (since the speed of convergence is determined by the slower estimate).

(3) It cannot detect missing modes, except when the Riemann approximation can be implemented.

(4) It is often the case that Rao–Blackwellization cannot be distinguished from the standard average and that importance sampling leads to an infinite variance estimator.

### 8.3.3 Renewal Theory

As mentioned in §8.2.3, it is possible to implement small sets as a theoretical control, although their applicability is more restricted than for other methods. Recall that Theorem 4.3.9, which was established by Asmussen (1979), guarantees the existence of small sets for ergodic Markov chains. Assuming $(\theta^{(t)})$ is strongly aperiodic, there exist a set $C$, a positive number $\varepsilon < 1$, and a probability measure $\nu$ such that

$$(8.3.7) \qquad P(\theta^{(t+1)} \in B | \theta^{(t)}) \geq \varepsilon \nu(B) \, \mathrm{I}_C(\theta^{(t)}) .$$

Recall also that the *augmentation technique* of Athreya and Ney (1978) consists in a modification of the transition kernel $K$ into a kernel associated with the transition

$$\theta^{(t+1)} = \begin{cases} \xi \sim \nu & \text{with probability } \varepsilon \\ \xi \sim \dfrac{K(\theta^{(t)}, \xi) - \varepsilon \nu(\xi)}{1 - \varepsilon} & \text{with probability } (1 - \varepsilon), \end{cases}$$

when $\theta^{(t)} \in C$. Thus, every passage in $C$ results in an independent generation from $\nu$ with probability $\varepsilon$. Denote by $\tau_C(k)$ the time of the $(k+1)$st visit to $C$ associated with an independent generation from $\nu$; $(\tau_C(k))$ is then a sequence of renewal time. Therefore, the partial sums

$$S_k = \sum_{t=\tau_C(k)+1}^{\tau_C(k+1)} h(\theta^{(t)})$$

are iid in the stationary regime. Under conditions (4.7.1) in Proposition 4.7.5, it follows that the $S_k$'s satisfy a version of the Central Limit Theorem

$$\frac{1}{\sqrt{K}} \sum_{k=1}^{K} (S_k - \mu_C \, \mathbb{E}_f[h(\theta)]) \xrightarrow{\mathcal{L}} \mathcal{N}(0, \sigma_C^2) ,$$

with $K$ the number of renewals out of $T$ iterations and $\mu_C = \mathbb{E}_f[\tau_C(2) - \tau_C(1)]$. Since these variables are iid, it is even possible to estimate $\sigma_C^2$ by the usual estimator

$$\tilde{\sigma}_C^2 = \frac{1}{K} \sum_{k=1}^{K} \left( S_k - \frac{\lambda_k}{\tau_C(K)} \sum_{\ell=1}^{K} S_\ell \right)^2 ,$$

where $\lambda_k = \tau_C(k+1) - \tau_C(k)$.

These probabilistic results, although elementary, have nonetheless a major consequence on the monitoring of MCMC algorithms since they lead to an estimate of limit variance $\gamma_h^2$ of the Central Limit Theorem (Proposition 4.7.5), as well as a stopping rule for these algorithms. This result thus provides an additional criterion for convergence assessment, since, *for different small sets $C$, the ratios $K_C\tilde{\sigma}_C^2/T$ must converge to the same value.* Again, this criterion is not foolproof since it requires sufficiently dispersed small sets.

**Proposition 8.3.4** *When the Central Limit Theorem applies to the Markov chain under study, the variance $\gamma_h^2$ of the limiting distribution associated with the sum (8.1.1) can be estimated by*

$$K_C\tilde{\sigma}_C^2/T ,$$

*where $K_C$ is the number of renewal events before time $T$.*

*Proof.* The Central Limit Theorem result

$$\frac{1}{\sqrt{T}} \sum_{t=1}^{T} (h(\theta^{(t)}) - \mathbb{E}_f[h(\theta)]) \xrightarrow{\mathcal{L}} \mathcal{N}(0, \gamma_h^2)$$

can be written

$$\frac{1}{\sqrt{T}} \sum_{t=1}^{\tau_C(1)} (h(\theta^{(t)}) - \mathbb{E}_f[h(\theta)]) + \frac{1}{\sqrt{T}} \sum_{k=1}^{K_C} (S_k - \lambda_k \, \mathbb{E}_f[h(\theta)])$$

$$\text{(8.3.8)} \qquad + \frac{1}{\sqrt{T}} \sum_{t=\tau_C(K_C)+1}^{K} (h(\theta^{(t)}) - \mathbb{E}_f[h(\theta)]) \xrightarrow{\mathcal{L}} \mathcal{N}(0, \gamma_h^2).$$

Under the conditions

$$\mathbb{E}_f[\lambda_1^2] < \infty \qquad \text{and} \qquad \mathbb{E}_f\left[ \left( \sum_{t=1}^{\tau_C(1)} h(\theta^{(t)}) \right)^2 \right] < \infty ,$$

which are, in fact, the sufficient conditions (4.7.1) for the Central Limit Theorem (see Proposition 4.7.5), the first and the third terms of (8.3.8) almost surely converge to 0. Since

$$\frac{1}{\sqrt{K_C}} \sum_{k=1}^{K_C} (S_k - \lambda_k \, \mathbb{E}_f[h(\theta)]) \xrightarrow{\mathcal{L}} \mathcal{N}(0, \sigma_C^2)$$

and $T/K_C$ almost surely converges to $\mu_C$, the average number of excursions between two passages in $C$, it follows that

$$\frac{1}{\sqrt{T}} \sum_{k=1}^{K_C} (h(\theta^{(t)}) - \lambda_k \, \mathbb{E}_f[h(\theta)]) \xrightarrow{\mathcal{L}} \mathcal{N}\left(0, \frac{\sigma_C^2}{\mu_C}\right),$$

which concludes the proof.                                                            □□

The fundamental factor in this monitoring method is the triplet $(C, \varepsilon, \nu)$ which must be determined for every application and which (strongly) depends on the problem at hand. Moreover, if the renewal probability $\varepsilon\pi(C)$ is too small, this approach is useless as a practical control device. Latent variable and missing data setups often lead to this kind of difficulty as $\varepsilon$ decreases as a power of the number of observations. The same problem often occurs in high dimensional problems (see Gilks *et al.* 1998). Mykland *et al.* (1995) eliminate this difficulty by modifying the algorithm, as detailed in §8.3.3. If we stick to the original version of the algorithm, the practical construction of $(C, \varepsilon, \nu)$ seems to imply a detailed study of the chain $(\theta^{(t)})$ and of its transition kernel. Examples 7.1.2 and 7.1.23 in §7.1.3 have however shown that this study can be conducted for some Gibbs samplers and we see, below, how a more generic version can be constructed for an important class of problems.

**Example 8.3.5 (Continuation of Example 8.3.2)**  The form of the posterior distribution on $\theta$ suggests using a small set $C$ equal to $[r_1, r_2]$ with $x_2 \in [r_1, r_2]$, $x_1 < r_1$, and $x_3 > r_2$ (if $x_1 < x_2 < x_3$). This choice leads to the following bounds

$$\rho_{11} = r_1 - x_1 < |\theta - x_1| < \rho_{12} = r_1 - x_1,$$

$$0 < |\theta - x_2| < \rho_{22} = \max(r_2 - x_2, x_2 - r_1),$$

$$\rho_{31} = x_3 - r_2 < |\theta - x_3| < \rho_{32} = x_3 - r_1,$$

for $\theta \in C$, which give $\varepsilon$ and $\nu$. In fact,

$$\begin{aligned}
K(\theta, \theta') \geq & \int_{\mathbb{R}_+^3} \frac{1}{\sqrt{2\pi}} \, \tau(\eta_1, \eta_2, \eta_3)^{-1} \\
& \times \exp\left\{-(\theta' - \mu(\eta_1, \eta_2, \eta_3))^2 / (2\tau^2(\eta_1, \eta_2, \eta_3))\right\} \\
& \times \frac{1 + \rho_{11}^2}{2} \exp\{-(1 + \rho_{12}^2)\eta_1/2\}\frac{1}{2} \exp\{-(1 + \rho_{22}^2)\eta_2/2\} \\
& \times \frac{1 + \rho_{31}^2}{2} \exp\{-(1 + \rho_{32}^2)\eta_3/2\} \, d\eta_1 d\eta_2 d\eta_3 \\
= & \frac{1 + \rho_{11}^2}{1 + \rho_{12}^2} \frac{1}{1 + \rho_{22}^2} \frac{1 + \rho_{31}^2}{1 + \rho_{32}^2} \nu(\theta') = \varepsilon\nu(\theta'),
\end{aligned}$$

where $\nu$ is the marginal density of $\theta$ for

$$(\theta, \eta_1, \eta_2, \eta_3) \sim \mathcal{N}\left(\mu(\eta_1, \eta_2, \eta_3), \tau^2(\eta_1, \eta_2, \eta_3)\right)$$
$$\times \mathcal{E}xp\left(\frac{1 + \rho_{12}^2}{2}\right) \mathcal{E}xp\left(\frac{1 + \rho_{22}^2}{2}\right) \mathcal{E}xp\left(\frac{1 + \rho_{32}^2}{2}\right).$$

One can, thus, evaluate the frequency $1/\mu_C$ of visits to $C$ and calibrate $r_1$ and $r_2$ to obtain optimal renewal probabilities.     ‖

Another difficulty with this method is that the splitting technique requires the generation of variables distributed as

$$(8.3.9) \qquad \frac{K(\theta^{(t)}, \xi) - \varepsilon \nu(\xi)}{1 - \varepsilon}.$$

It is actually enough to know the ratio $K(\theta^{(t)}, \xi)/\varepsilon\nu(\xi)$ since, following from Lemma 2.4.1, the algorithm

**Algorithm A.44 –Negative Weight Mixture Simulation–**

1. Simulate $\xi \sim K(\theta^{(t)}, \xi)$.

2. Accept $\xi$ with probability $1 - \dfrac{\varepsilon\nu(\xi)}{K(\theta^{(t)}, \xi)}$,     [A.44]

   otherwise go to 1.

provides simulations from the distribution (8.3.9). Since the densities $K(\theta, \xi)$ and $\nu$ are in general unknown, it is necessary to use approximations, as in §8.2.4.

**Example 8.3.6 (Continuation of Example 8.3.2)** The integral forms of $K(\theta, \xi)$ and of $\nu(\theta)$ (that is, their representation as marginals of other distributions) allows for a Rao–Blackwellized approximation. Indeed,

$$\hat{K}(\xi) = \frac{1}{M} \sum_{i=1}^{M} \varphi(\xi | \eta_1^i, \eta_2^i, \eta_3^i)$$

and

$$\hat{\nu}(\xi) = \frac{1}{M} \sum_{i=1}^{M} \varphi(\xi | \tilde{\eta}_1^i, \tilde{\eta}_2^i, \tilde{\eta}_3^i) \,,$$

with

$$\varphi(\xi | \eta_1, \eta_2, \eta_3)$$
$$= \exp\left\{ -\left(\xi - \mu(\eta_1, \eta_2, \eta_3)\right)^2 / (2\tau^2(\eta_1, \eta_2, \eta_3)) \right\} \tau^{-1}(\eta_1, \eta_2, \eta_3)$$

and

$$\eta_j^i \sim \mathcal{E}xp\left(\frac{1 + (\xi - x_j)^2}{2}\right) \,, \quad \tilde{\eta}_j^i \sim \mathcal{E}xp\left(\frac{1 + \rho_{j2}^2}{2}\right) \quad (j = 1, 2, 3)$$

provide convergent (in $M$) approximations of $K(\theta, \xi)$ and $\nu(\xi)$, respectively, by the usual Law of Large Numbers and, therefore, suggest using the approximation

$$\frac{K(\theta, \xi)}{\nu(\xi)} \simeq \frac{\sum_{i=1}^{M} \varphi(\xi | \eta_1^i, \eta_2^i, \eta_3^i)}{\sum_{i=1}^{M} \varphi(\xi | \tilde{\eta}_1^i, \tilde{\eta}_2^i, \tilde{\eta}_3^i)}.$$

| $r$ | 0.1 | 0.21 | 0.32 | 0.43 | 0.54 | 0.65 | 076 | 0.87 | 0.98 | 1.09 |
|---|---|---|---|---|---|---|---|---|---|---|
| $\varepsilon$ | 0.92 | 0.83 | 0.73 | 0.63 | 0.53 | 0.45 | 0.38 | 0.31 | 0.26 | 0.22 |
| $\mu_C$ | 25.3 | 13.9 | 10.5 | 9.6 | 8.8 | 9.6 | 9.8 | 10.4 | 11.4 | 12.7 |
| $\gamma_h^2$ | 1135 | 1138 | 1162 | 1159 | 1162 | 1195 | 1199 | 1149 | 1109 | 1109 |

Table 8.3.1. *Estimation of the asymptotic variance $\gamma_h^2$ for $h(\theta) = \theta$ and renewal control for $C = [x_2 - r, x_2 + r]$ with $x_1 = -8$, $x_2 = 8$, $x_3 = 17$, and $\sigma^2 = 50$ (1 million simulations).*

The variables $\eta_j^i$ must be simulated at every iteration of the Gibbs sampler, whereas the variables $\tilde{\eta}_j^i$ can be simulated only once when starting the algorithm.

Table 8.3.1 gives the performances of the diagnostic for different values of $r$, when $C = [x_2 - r, x_2 + r]$. The decrease of the bound $\varepsilon$ as a function of $r$ is quite slow and indicates a high renewal rate. It is, thus, equal to 11% for $r = 0.54$. The stabilizing of the variance $\gamma_h^2$ occurs around of 1160 for $h(x) = x$, but the number of simulations before stabilizing is huge, since one million iterations do not provide convergence to the same quantity, except for values of $r$ between 0.32 and 0.54.                    ‖

The previous example shows how *conservative* this method can be and, thus, how many iterations it requires. In addition, the method does not guarantee convergence since the small sets can always omit important parts of the support of $f$.

A particular setup where the renewal method behaves quite satisfactorily is the case of Markov chains with *finite state-space*. In fact, the choice of a small set is then immediate: If $\Theta$ can be written as $\{i : i \in I\}$, $C = \{i_0\}$ is a small set, with $i_0$ the modal state under the stationary distribution $\tilde{\pi} = (\pi_i)_{i \in I}$. If $\mathbb{P} = (p_{ij})$ denotes the transition matrix of the chain,

$$C = \{i_0\}, \qquad \varepsilon = 1, \qquad \nu = (p_{i_0 i})_{i \in I},$$

and renewal occurs *at each visit* in $i_0$. Considering in parallel other likely states under $\tilde{\pi}$, we then immediately derive a convergence indicator.

**Example 8.3.7 A finite Markov chain.** Consider $\Theta = \{0, 1, 2, 3\}$ and

$$\mathbb{P} = \begin{pmatrix} 0.26 & 0.04 & 0.08 & 0.62 \\ 0.05 & 0.24 & 0.03 & 0.68 \\ 0.11 & 0.10 & 0.08 & 0.71 \\ 0.08 & 0.04 & 0.09 & 0.79 \end{pmatrix}.$$

The stationary distribution is then $\tilde{\pi} = (0.097, 0.056, 0.085, 0.762)$, with mean 2.51. Table 8.3.2 compares estimators of $\gamma_h^2$, with $h(\theta) = \theta$, for the four small sets $C = \{i_0\}$ and shows that stabilizing is achieved for $T = 500,000$.                    ‖

| $T/i_0$ | 0 | 1 | 2 | 3 |
|---|---|---|---|---|
| 5000 | 1.19 | 1.29 | 1.26 | 1.21 |
| 500,000 | 1.344 | 1.335 | 1.340 | 1.343 |

Table 8.3.2. *Estimators of $\gamma_h^2$ for $h(x) = x$, obtained by renewal in $i_0$.*

This academic example illustrates how easily the criterion can be used in finite environments. Finite state-spaces are, nonetheless, far from trivial since, as shown by the *Duality Principle* (§7.2.4), one can use the simplicity of some subchains to verify convergence for the Gibbs sampler. Therefore, renewal can be created for a (possibly artificial) subchain $(z^{(t)})$ and later applied to the parameter of interest if the Duality Principle applies. (See also Chauveau *et al.* 1998 for a related use of the Central Limit Theorem for convergence assessment.)

**Example 8.3.8 Grouped multinomial model.** The Gibbs sampler which corresponds to the grouped multinomial model of Example 7.2.3,

$$X \sim \mathcal{M}_5(n; a_1\mu + b_1, a_2\mu + b_2, a_3\eta + b_3, a_4\eta + b_4, c(1 - \mu - \eta)),$$

actually enjoys a data augmentation structure and satisfies the conditions leading to the Duality Principle. Moreover, the missing data $(z_1, \ldots, z_4)$ has finite support since

$$Z_i \sim \mathcal{B}\left(x_i, \frac{a_i\mu}{a_i\mu + b_i}\right) \quad (i = 1, 2), \qquad Z_i \sim \mathcal{B}\left(x_i, \frac{a_i\eta}{a_i\eta + b_i}\right) \quad (i = 3, 4).$$

Therefore, the support of $(Z^{(t)})$ is of size $(x_1 + 1) \times \cdots \times (x_4 + 1)$. A preliminary simulation indicates that $(0, 1, 0, 0)$ is the modal state for the chain $(Z^{(t)})$, with an average excursion time of 27.1. The second most frequent state is $(0, 2, 0, 0)$, with a corresponding average excursion time of 28.1. In order to evaluate the performances of renewal control, we also used the state $(1, 1, 0, 0)$, which appears, on average, every 49.1 iterations. Table 8.3.3 describes the asymptotic variances obtained for the functions

$$h_1(\mu, \eta) = \mu - \eta, \quad h_2(\mu, \eta) = \mathbb{I}_{\mu > \eta}, \quad h_3(\mu, \eta) = \frac{\mu}{1 - \mu - \eta}.$$

The estimator of $h_3$ is obtained by Rao–Blackwellization,

$$\frac{1}{T} \sum_{t=1}^{T} \frac{0.5 + z_1^{(t)} + z_2^{(t)}}{x_5 - 0.5},$$

which increases the stability of the estimator and reduces its variance.

The results thus obtained are in good agreement for the three states under study, even though they cannot rigorously establish convergence. ‖

| $i$ | $\mathbb{E}^{\pi}[h_i(\mu, \eta)|x]$ | $\hat{\gamma}_i^2(1)$ | $\hat{\gamma}_i^2(2)$ | $\hat{\gamma}_i^2(3)$ |
|-----|-----|-----|-----|-----|
| $h_1$ | $5 \cdot 10^{-4}$ | 0.758 | 0.720 | 0.789 |
| $h_2$ | 0.496 | 1.24 | 1.21 | 1.25 |
| $h_3$ | 0.739 | 1.45 | 1.41 | 1.67 |

Table 8.3.3. *Approximations by Gibbs sampling of posterior expectations and evaluation of variances by the renewal method for three different states (1 for $(0, 1, 0, 0)$, 2 for $(0, 2, 0, 0)$, and 3 for $(1, 1, 0, 0)$) in the grouped multinomial model (500,000 iterations).*

### 8.3.4 Within and Between Variances

The control strategy devised by Gelman and Rubin (1992) starts with the derivation of a distribution $\mu$ related with the modes of $f$, which are supposedly obtained by numerical methods. They suggest using a mixture of Student's $t$-distributions centered around the identified modes of $f$, the scale being derived from the second derivatives of $f$ at these modes. With the possible addition of an importance sampling step (see Problem 3.10), they then generate $M$ chains $(\theta_m^{(t)})$ $(1 \leq m \leq M)$. For *every quantity of interest* $\xi = h(\theta)$, the stopping rule is based on the difference between a weighted estimator of the variance and the variance of estimators from the different chains.

Denote

$$B_T = \frac{1}{M} \sum_{m=1}^{M} (\bar{\xi}_m - \bar{\xi})^2 \, ,$$

$$W_T = \frac{1}{M} \sum_{m=1}^{M} s_m^2 = \frac{1}{M} \sum_{m=1}^{M} \frac{1}{T} \sum_{t=1}^{T} (\xi_m^{(t)} - \bar{\xi}_m)^2 \, ,$$

with

$$\bar{\xi}_m = \frac{1}{T} \sum_{t=1}^{T} \xi_m^{(t)}, \qquad \bar{\xi} = \frac{1}{M} \sum_{m=1}^{M} \bar{\xi}_m \, ,$$

and $\xi_m^{(t)} = h(\theta_m^{(t)})$. The quantities $B_T$ and $W_T$ represent the *between-* and *within-chains* variances. A first estimator of the posterior variance of $\xi$ is

$$\hat{\sigma}_T^2 = \frac{T-1}{T} W_T + B_T \, .$$

Gelman and Rubin (1992) compare $\hat{\sigma}_T^2$ and $W_T$, which are asymptotically equivalent (Problem 8.13). Gelman (1996) notes that $\hat{\sigma}_T^2$ overestimates the variance of $\xi_m^{(t)}$ because of the large dispersion of the initial distribution, whereas $W_T$ underestimates this variance, as long as the different sequences $(\xi_m^{(t)})$ remain concentrated around their starting values.

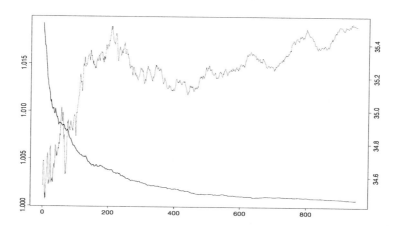

Figure 8.3.6.  *Evolutions of $R_T$ (solid lines and scale on the left) and of $W_T$ (dotted lines and scale on the right) for the posterior distribution (8.3.6) and $h(\theta) = \theta$ (M = 100).*

The recommended criterion of Gelman and Rubin (1992) is to monitor

$$R_T = \frac{\hat{\sigma}_T^2 + \frac{B_T}{M}}{W_T} \frac{\nu_T}{\nu_T - 2}$$

$$= \left( \frac{T-1}{T} + \frac{M+1}{M} \frac{B_T}{W_T} \right) \frac{\nu_T}{\nu_T - 2},$$

where $\nu_T = 2(\hat{\sigma}_T^2 + \frac{B_T}{M})^2/W_T$ and the approximate distribution of $R_T$ is derived from the approximation $T B_T/W_T \sim \mathcal{F}(M-1, 2W_T^2/\varpi_T)$ with

$$\varpi_T = \frac{1}{M^2} \left[ \sum_{m=1}^{M} s_m^4 - \frac{1}{M} \left( \sum_{m=1}^{M} s_m^2 \right)^2 \right].$$

(The approximation ignores the variability due to $\nu_T/(\nu_T - 2)$.) The stopping rule is based on either testing that the mean of $R_T$ is equal to 1 or on confidence intervals on $R_T$.

**Example 8.3.9 (Continuation of Example 8.3.2)**  Figure 8.3.6 describes the evolution of $R_T$ for $h(\theta) = \theta$ and $M = 100$ and 1000 iterations. As the scale of the graph of $R_T$ is quite compressed, one can conclude there is convergence after about 600 iterations. (The 50 first iterations have been eliminated.) On the contrary, the graph of $W_T$ superimposed in this figure does not exhibit the same stationarity after 1000 iterations. However, the study of the associated histogram (see Figure 8.3.3) shows that the distribution of $\theta^{(t)}$ is stationary after a few hundred iterations. In this particular case, the criterion is therefore conservative, which shows that the method is difficult to calibrate.    ‖

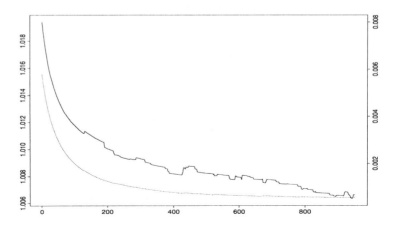

Figure 8.3.7. *Evolutions of $R_T$ (solid lines and scale on the left) and of $W_T$ (dotted lines and scale on the right) for the witch's hat distribution and $h(\theta) = \theta_1$ (M = 100).*

**Example 8.3.10 (Continuation of Example 8.2.1)** A preliminary study of the density $\pi(\theta|y)$ obviously signals the presence of a very concentrated mode around $y$, but, in a less artificial setting, this mode would not be detected. It therefore makes sense to propose the uniform distribution on $C = [0,1]^d$ as the initial distribution $\mu$. Figure 8.3.7 presents the evolution of the ratio $R_T$ and of the estimation $W_T$. The scale of the graph in $R_T$ is much more concentrated than in Example 8.3.9 and the stability of $R_T$ (and of $W_T$) leads us to conclude that the chain converges, although it has not left the neighborhood of the mode $(0.7, 0.7)$. ‖

This method has enjoyed wide usage, in particular because of its simplicity and of its connections with the standard tools of linear regression. However, we point out the following:

(a) Gelman and Rubin (1992) also suggest removing the first half of the simulated sample to reduce the dependence on the initial distribution $\mu$. By comparison with a single-chain method, the number of wasted simulations is thus (formally) multiplied by $M$.

(b) The accurate construction of the initial distribution $\mu$ can be quite delicate and time-consuming. Also, in some models, the number of modes is too great to allow for a complete identification and important modes may be missed.

(c) The method relies on normal approximations, whereas the MCMC algorithms are used in settings where these approximations are, at best, difficult to satisfy and, at worst, not valid. The use of Student's $t$ distributions by Gelman and Rubin (1992) does not modify this. More importantly, there is no embedded test for the validity of this approximation.

(d) The criterion is *unidimensional*; therefore, it gives a poor evaluation of the correlation between variables and the necessary slower convergence of the joint distribution. Moreover, the stopping rule must be modified for every function of interest, with very limited recycling of results obtained for other functions. (Brooks and Gelman 1998 studied multidimensional extensions of this criterion based on the same approximations.)

## 8.4 Simultaneous Monitoring

### 8.4.1 Binary Control

Raftery and Lewis (1992a,b) (see also Raftery and Banfield 1991) attempt to reduce the study of the convergence of the chain $(\theta^{(t)})$ to the study of the convergence of a *two-state* Markov chain, where an explicit analysis is possible. They then evaluate three parameters for the control of convergence, namely $k$, the minimum "batch" (subsampling) step, $t_0$, the number of "warm-up" iterations necessary to achieve stationarity (to eliminate the effect of the starting value) and $T$, total number of iterations "ensuring" convergence (giving a chosen precision on the empirical average). The control is understood to be at the level of the derived two-state Markov chain (rather than for the original chain of interest).

The binary structure at the basis of the diagnosis is derived from the chain $\theta^{(t)}$ by

$$Z^{(t)} = \mathbb{I}_{\theta^{(t)} \leq \underline{\theta}} \, ,$$

where $\underline{\theta}$ is an arbitrary value in the support of $f$. Unfortunately, the sequence $(Z^{(t)})$ *does not form a Markov chain*, even in the case where $\theta^{(t)}$ has a finite support (see Kemeny and Snell 1960 and Problem 4.61). Raftery and Lewis (1992a) determined the batch size $k$ by testing if $(Z^{(kt)})$ is a Markov chain against the alternative that $(Z^{(kt)})$ is a *second order* Markov chain (that is, that the vector $(Z^{(kt)}, Z^{(k(t+1))})$ is a Markov chain). This determination of $k$ therefore has limited theoretical foundations. In particular:

(a) The rejection of the alternative hypothesis (which only accounts for the order 2 dependency) does not support the null hypothesis.

(b) It is possible to use more rigorous nonparametric tests of the independence of the $Z^{(kt)}$'s (for example, *Spearman*'s test (see Lehmann 1975) and to use stopping rules derived from sequential statistics.

(c) The test depends on the choice of $\underline{\theta}$ and on the representativeness of the iterations leading to the first observations. If the chain $(\theta^{(t)})$ has not exhibited the complete characteristics of $f$, the test leads to an erroneous diagnosis.

(d) Using renewal arguments, it is possible to construct a valid Markov chain with two or more states, based on $(\theta^{(t)})$ (see §8.4.2).

(e) Examples studied by Raftery and Lewis (1992a) lead to very small values of $k$ which are often equal to 1 (see also Cowles and Carlin 1996 and Brooks and Roberts 1997, 1999). It therefore makes sense, when accounting for the efficiency loss detailed in Lemma 8.1.1, to suggest working with the complete sequence $Z^{(t)}$ and not trying to justify the Markov approximation.

If $(Z^{(t)})$ is treated as a homogeneous Markov chain, with (pseudo-) transition matrix

$$\mathbb{P} = \begin{pmatrix} 1 - \alpha & \alpha \\ \beta & 1 - \beta \end{pmatrix},$$

it converges to the stationary distribution

$$P(Z^{(\infty)} = 0) = \frac{\beta}{\alpha + \beta}, \qquad P(Z^{(\infty)} = 1) = \frac{\alpha}{\alpha + \beta}$$

(see Problem 4.55). It is therefore possible to determine the *warm-up* size by requiring that

$$(8.4.1) \qquad \left| P(Z^{(t_0)} = i | z^{(0)} = j) - P(Z^{(\infty)} = i) \right| < \varepsilon$$

for $i, j = 0, 1$. Raftery and Lewis (1992a) show that this condition is equivalent to

$$(8.4.2) \qquad t_0 \geq \log \left( \frac{(\alpha + \beta)\varepsilon}{\alpha \vee \beta} \right) \Big/ \log |1 - \alpha - \beta|$$

(see Problem 8.14). The sample size related with the (acceptable) convergence of

$$\delta_T = \frac{1}{T} \sum_{t=t_0}^{t_0+T} z^{(t)}$$

to $\dfrac{\alpha}{\alpha + \beta}$ can be determined by a normal approximation of $\delta_T$, with variance

$$\frac{1}{T} \frac{(2 - \alpha - \beta)\, \alpha\beta}{(\alpha + \beta)^3}.$$

If, for instance, we require

$$P \left( \left| \delta_T - \frac{\alpha}{\alpha + \beta} \right| < q \right) \geq \varepsilon',$$

the value of $T$ is

$$(8.4.3) \qquad T \geq \frac{\alpha\beta(2 - \alpha - \beta)}{q^2(\alpha + \beta)^3} \, \Phi^{-1} \left( \frac{\varepsilon' + 1}{2} \right)$$

(see Problem 8.16).

This analysis relies on knowledge of the parameters $(\alpha, \beta)$. These are unknown in most settings of interest and must be estimated from the simulation of a test sample. Based on the independent case, Raftery and Lewis

| $q(\times 10^{-2})$ | 0.25 | 0.5 | 1.0 | 1.5 | 2.0 |
|---|---|---|---|---|---|
| $T_{\min}$ | 14,932 | 3748 | 936 | 416 | 234 |

Table 8.4.1. *Minimum test sample sizes as a function of the precision q for the 2.5% quantile (Source: Raftery and Lewis 1992a).*

(1992a) suggest using a sample size which is at least

$$T_{\min} \geq \Phi^{-1} \left( \frac{\varepsilon' + 1}{2} \right)^2 \frac{\alpha\beta}{(\alpha + \beta)^2} \, q^{-1} \, .$$

Table 8.4.1 gives values of $T_{\min}$ for $\varepsilon' = 0.95$, $\beta/(\alpha + \beta) = 0.025$, and different values of $q$, implying that for a precision of 99.5%, the minimum size is about 15,000.

This method, called *binary control*, is quite popular, in particular because general programs are available (in Statlib) and also because its implementation is quite easy.[4] However, there are drawbacks to using it as a convergence indicator:

(a) The preliminary estimation of the coefficients $\alpha$ and $\beta$ requires a chain $(\theta^{(t)})$ which is already (almost) stationary and which has, we hope, sufficiently explored the characteristics of the distribution $f$. If $\alpha$ and $\beta$ are not correctly estimated, the (limited) validity of the method vanishes.

(b) The approach of Raftery and Lewis (1992a) is intrinsically *unidimensional* and, hence, does not assess the correlations between components. It can thus conclude there is convergence, based on the marginal distributions, whereas the joint distribution is not correctly estimated.

(c) Brooks and Roberts (1997) warn about a degradation of the performances of this method when $\underline{\theta}$ is located in the tails of the marginal distribution of $\theta^{(t)}$.

(d) Once $\alpha$ and $\beta$ are estimated, the stopping rules are completely independent of the model under study and of the selected MCMC algorithm, as shown by formulas (8.4.2) and (8.4.3). This generic feature is appealing for an automated implementation but it cannot guarantee global efficiency.

**Example 8.4.1 (Continuation of Example 8.3.8)**  For the multinomial model of Example 7.2.3, the parameters are

$$a = (0.1, 0.14, 0.7, 0.9) \,, \qquad b = (0.17, 0.24, 0.19, 0.20) \,,$$

and

$$(x_1, x_2, x_4, x_5) = (4, 15, 12, 7, 4).$$

Based on the completion proposed in Example 7.2.3, the corresponding Gibbs sampling algorithm can be subjected to binary control. Table 8.4.2

---

[4] In the examples to follow, the batch size $k$ is fixed at 1 and the method has been directly implemented in C, instead of calling the Statlib program.

| Chain | $\underline{\theta}$ | $\alpha$ | $\beta$ | $q_0$ | $t_0$ | $T$ |
|-------|------|------|------|------|----|--------|
| $\mu$ | 0.13 | 0.19 | 0.34 | 0.36 | 6 | 42,345 |
| $\eta$ | 0.35 | 0.38 | 0.18 | 0.68 | 6 | 36,689 |
| $\xi$ | 0.02 | 0.29 | 0.40 | 0.42 | 4 | 30,906 |

Table 8.4.2. *Parameters of the binary control method for three different param-eters of the grouped multinomial model, with control parameters $\varepsilon = q = 0.005$ and $\varepsilon' = 0.999$ (5000 preliminary simulations).*

gives the values of $\alpha$, $\beta$, $q_0 = \beta/(\alpha + \beta)$, $t_0$, and $T$ for binary variables based on $\mu$, $\eta$, and $\xi = \mu\eta$. The preliminary sample is of size 5000. Since the control parameters are particularly severe, the corresponding sample sizes are rather large, but the warm-up parameter $t_0$ is quite small. In this particular case, the role of the parameterization is negligible.          ‖

**Example 8.4.2 (Continuation of Example 8.3.3)** Algorithm [A.30] generates a $\mathcal{Be}(0.2, 1)$ distribution from a $\mathcal{Be}(1.2, 1)$ distribution. The prob-ability that $(X^{(t)})^{0.8}$ is less than 0.2, with $\mathbb{E}[(X^{(t)})^{0.8}] = 0.2$, can then be approximated by this algorithm and the two-state chain is directly derived from

$$Z^{(t)} = \mathbb{I}_{(X^{(t)})^{0.8} < 0.2} \cdot$$

Based on a preliminary sample of size $50,000$, the initial values of $\alpha$ and $\beta$ are $\alpha_0 = 0.0425$ and $\beta_0 = 0.0294$. These approximations lead to $t_0 = 55$ and $T = 430,594$ for $\varepsilon = q = 0.01$ and $\varepsilon' = 0.99$. If we run the algorithm for $t_0 + T$ iterations, the estimates of $\alpha$ and $\beta$ are then $\alpha^* = 0.0414$ and $\beta^* = 0.027$. Repeating this recursive evaluation of $(t_0, T)$ as a function of $(\hat{\alpha}, \hat{\beta})$ and the update of $(\hat{\alpha}, \hat{\beta})$ after $t_0 + T$ additional iterations, we obtain the results in Table 8.4.3. These exhibit a relative stability in the evaluation of $(\alpha, \beta)$ (except for the second iteration) and, thus, of the corresponding $(t_0, T)$'s.

In addition, Figure 8.4.1 represents the evolution of

$$\hat{\varrho}_t = \frac{1}{t} \sum_{\ell=t_0}^{t_0+t} \mathbb{I}_{(x^{(\ell)})^{0.8} \leq 0.2}$$

over $4,500,000$ iterations (that is, 10 times as many iterations as suggested by Table 8.4.3). An elementary computation shows that the probability $P(x^{0.8} \leq 0.2)$ is equal to $0.2^{0.2/0.8}$ (that is, 0.669). However, Figure 8.4.1 suggests the value 0.64 at the end of the $4,500,000$ iterations and this erroneous convergence is not indicated by the binary control method in this pathological setup.          ‖

| Round | $t_0$ | $T\ (\times 10^3)$ | $\hat{\alpha}$ | $\hat{\beta}$ |
|-------|-------|--------------------|----------------|----------------|
| 1     | 55    | 431                | 0.041          | 0.027          |
| 2     | 85    | 452                | 0.040          | 0.001          |
| 3     | 111   | 470                | 0.041          | 0.029          |
| 4     | 56    | 442                | 0.041          | 0.029          |
| 5     | 56    | 448                | 0.040          | 0.027          |
| 6     | 58    | 458                | 0.041          | 0.025          |
| 7     | 50    | 455                | 0.042          | 0.025          |
| 8     | 50    | 452                | 0.041          | 0.028          |

Table 8.4.3.  *Evolution of initializing and convergence times, and parameters* $(\alpha, \beta)$ *estimated after* $t_0 + T$ *iterations obtained from the previous round, in the* $\mathcal{Be}(\alpha, 1)/\mathcal{Be}(\alpha + 1, 1)$ *example.*

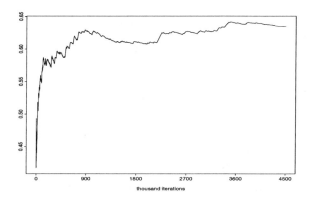

Figure 8.4.1. *Convergence of the mean* $\hat{\varrho}_t$ *for a chain generated from* [A.30].

### 8.4.2 Valid Discretization

A fundamental problem with the binary control technique of Raftery and Lewis (1992a) in §8.4.1 is that it relies on an approximation, namely that the discretized sequence $(z^{(t)})$ is a Markov chain. Guihenneuc–Jouyaux and Robert (1998) have shown that there exists a rigorous discretization of Markov chains which produces Markov chains. The idea at the basis of this discretization is to use several disjoint *small sets* $A_i$ $(i = 1, \ldots, k)$ for the chain $(\theta^{(t)})$, with corresponding parameters $(\epsilon_i, \nu_i)$, and to subsample only at renewal times $\tau_n$ $(n > 1)$. The $\tau_n$'s are defined as the successive instants when the Markov chain enters one of these small sets with splitting, that is, by

$$\tau_n = \inf\{\, t > \tau_{n-1};\ \exists\, 1 \leq i \leq k,\ \theta^{(t-1)} \in A_i \text{ and } \theta^{(t)} \sim \nu_i \,\}.$$

(Note that the $A_i$'s $(i = 1, \ldots, k)$ need not be a partition of the space.)

The discretized Markov chain is then derived from the finite-valued se-

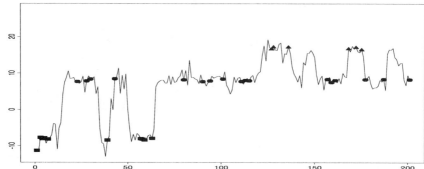

Figure 8.4.2. *Discretization of a continuous Markov chain, based on three small sets. The renewal events are represented by triangles for B (circles for C and squares for D, respectively). (Source: Guihenneuc–Jouyaux and Robert 1998.)*

quence

$$\eta^{(t)} = \sum_{i=1}^{k} i \mathbb{I}_{A_i}(\theta^{(t)})$$

as

$$(\xi^{(n)}) = (\eta^{(\tau_n)}) \, .$$

The resulting chain is then described by the sequence of small sets encountered by the original chain $(\theta^{(t)})$ at renewal times. It can be shown that the sequence $(\xi^{(n)})$ is a homogeneous Markov chain on the finite state-space $\{1, \ldots, k\}$ (Problem 8.19).

**Example 8.4.3 (Continuation of Example 8.3.2)** Figure 8.4.2 illustrates discretization for the subchain $(\theta^{(t)})$, with three small sets $C = [7.5, 8.5]$, derived in §8.3.5, and $B = [-8.5, -7.5]$, and $D = [17.5, 18.5]$, which can be constructed the same way. Although the chain visits the three sets quite often, renewal occurs with a much smaller frequency, as shown by the symbols.                                                                ‖

This result justifies control of Markov chains through their discretized counterparts. Guihenneuc–Jouyaux and Robert (1998) propose further uses of the discretized chain, including the evaluation of mixing rates. (See Cowles and Rosenthal 1998 for a different approach to this evaluation, based on the drift conditions of 4.10.1, following the theoretical developments of Rosenthal 1995.)

## 8.5 Problems

**8.1** For the witch's hat distribution of Example 8.2.1, find a set of parameters $(\delta, \sigma, y)$ for which the mode at $y$ takes many iterations to be detected. Apply the various convergence controls to this case.

**8.2** Establish that the cdf of the statistic (8.2.2) is (8.2.3).

**8.3** Reproduce the experiment of Example 8.2.8 in the cases $\alpha = 0.9$ and $\alpha = 0.99$.

**8.4** In the setup of Example 8.2.8:

(a) Show that $\mathbb{E}[X^{1-\alpha}] = \alpha$ when $X \sim \mathcal{B}e(\alpha, 1)$.

(b) Show that $P(X^{1-\alpha} < \varepsilon) = \varepsilon^{\alpha/1-\alpha}$.

(c) Show that the Riemann approximation $S_T^R$ of (8.3.4) has a very specific shape in this case, namely that it is equal to the normalizing constant of $f$.

**8.5** (Liu *et al.* 1992) Show that if $(\theta_1^{(t)})$ and $(\theta_2^{(t)})$ are independent Markov chains with transition kernel $K$ and stationary distribution $\pi$,

$$\omega_t = \frac{\pi(\theta_1^{(t)})K(\theta_1^{(t-1)}, \theta_2^{(t)})}{\pi(\theta_2^{(t)})K(\theta_2^{(t-1)}, \theta_1^{(t)})}$$

is an unbiased estimator of the $L_2$ distance between $\pi$ and

$$\pi^t(\theta) = \int K(\eta, \theta)\pi^{t-1}(\eta)d\eta\ ,$$

in the sense that $\mathbb{E}[\omega_t] = 1 + \text{var} \dfrac{\pi^t(\theta)}{\pi(\theta)}$.

**8.6** (Roberts 1994) Show that if $(\theta_1^{(t)})$ and $(\theta_2^{(t)})$ are independent Markov chains with transition kernel $K$ and stationary distribution $\pi$,

$$\chi^t = \frac{K(\theta_1^{(0)}, \theta_2^{(2t-1)})}{\pi(\theta_2^{(2t-1)})}$$

is such that

(8.5.1) $$\mathbb{E}[\chi^t] = \int \frac{\pi_1^t(\theta)\pi_2^t(\theta)}{\pi(\theta)} d\theta \overset{t\to\infty}{\longrightarrow} 1$$

and

$$\lim_{t\to\infty} \text{var}(\chi_t) = \int (K(\theta_1^{(0)}, \theta) - \pi(\theta))^2 \frac{1}{\pi(\theta)} d\theta,$$

where $\pi_i^t$ denotes the density of $\theta_i^{(t)}$. (*Hint*: Use the equilibrium relation

$$K(\theta_2^{(2t-2)}, \theta_2^{(2t-1)})\pi(\theta_2^{(2t-2)}) = K(\theta_2^{(2t-1)}, \theta_2^{(2t-2)})\pi(\theta_2^{(2t-1)})$$

to prove (8.5.1).)

**8.7** (Continuation of Problem 8.6) Define

$$I_t = \frac{1}{m(m-1)} \sum_{i\neq j} \chi_{ij}^t, \quad J_t = \frac{1}{m}\sum_{i=1}^m \chi_{ii}^t,$$

with

$$\chi_{ij}^t = \frac{K(\theta_i^{(0)}, \theta_j^{(2t-1)})}{\pi(\theta_j^{(2t-1)})}\ ,$$

based on $m$ parallel chains $(\theta_j^{(0)})$ $(j = 1, \ldots, m)$.

(a) Show that if $\theta_j^{(0)} \sim \pi$, $\mathbb{E}[I_t] = \mathbb{E}[J_t] = 1$ and that for every initial distribution on the $\theta_j^{(0)}$'s, $\mathbb{E}[I_t] \leq \mathbb{E}[J_t]$.

(b) Show that
$$\text{var}(I_t) \leq \text{var}(J_t)/(m-1) .$$

**8.8** (Dellaportas 1995) Show that
$$\mathbb{E}_g \left[ \min \left( 1, \frac{f(x)}{g(x)} \right) \right] = \int |f(x) - g(x)| dx .$$

Derive an estimator of the $L_1$ distance between the stationary distribution $\pi$ and the distribution at time $t$, $\pi^t$.

**8.9** (Robert 1995b) Give the marginal distribution of $\xi = h_3(\mu, \nu)$ in Example 8.3.8 by doing the following:

(a) The Jacobian of the transform of $(\mu, \nu)$ in $(\xi, \nu)$ is
$$\frac{1 - \eta}{(1 + \xi)^2} .$$

(b) Express the marginal density of $\xi$ as a $x_1 + x_2$ degree polynomial in $\xi/(1+\xi)$.

(c) Show that the weights $\varpi_j$ of this polynomial can be obtained, up to a multiplicative factor, in closed form.)

Deduce that a Riemann sum estimate of $\mathbb{E}[h_3(\mu, \nu)]$ is available in this case.

**8.10** For the model of Example 8.3.8, show that a small set is available in the $(\mu, \eta)$ space and derive the corresponding renewal probability.

**8.11** Given a mixture distribution $p\mathcal{N}(\mu, 1) + (1 - p)\mathcal{N}(\theta, 1)$, with conjugate priors
$$p \sim \mathcal{U}_{[0,1]} \ , \ \mu \sim \mathcal{N}(0, \tau^2) \ , \ \theta \sim \mathcal{N}(0, \tau^2),$$
show that $[\underline{p}, \bar{p}] \times [\underline{\mu}, \bar{\mu}] \times [\underline{\theta}, \bar{\theta}]$ is a small set and deduce that $\varepsilon$ in (8.3.7) decreases as a power of the sample size.

**8.12** For the estimator $S_T^P$ given in Example 8.3.3:

(a) Show that the variance of $S_T^P$ in Example 8.3.3 is infinite.

(b) Propose an estimator of the variance of $S_T^P$ (when it exists) and derive a convergence diagnostic.

(c) Check whether the importance sampling $S_T^P$ is (i) available and (ii) with finite variance for Examples 7.1.23 and 7.2.3 of Chapter 7.

**8.13** Referring to §8.3.4:

(a) Show that $\hat{\sigma}_T^2$ and $W_T$ have the same expectation.

(b) Show that $\hat{\sigma}_T^2$ and $W_T$ are both asymptotically normal. Find the limiting variances.

**8.14** (Raftery and Lewis 1992a) Establish the equivalence between (8.4.1) and (8.4.2) by showing first that it is equivalent to
$$|1 - \alpha - \beta|^{t_0} \leq \frac{(\alpha + \beta)\varepsilon}{\alpha \vee \beta} .$$

**8.15** For the transition matrix
$$\mathbb{P} = \begin{pmatrix} 1 - \alpha & \alpha \\ \beta & 1 - \beta \end{pmatrix} ,$$

show that the second eigenvalue is $\lambda = 1 - \alpha - \beta$. Deduce that

$$|P(x^{(t)} = i|x^{(0)} = j) - P(x^{(\infty)} = i)| < \varepsilon$$

for every $(i, j)$ if $|1-\alpha-\beta|^t \leq \varepsilon(\alpha+\beta)/\max(\alpha, \beta)$ (*Hint*: Use the representation $P(x^{(t)} = i|x^{(0)} = j) = e_j P^t e_i'$, with $e_0 = (1, 0)$ and $e_1 = (0, 1)$.)

**8.16** (Raftery and Lewis 1992a) Deduce (8.4.3) from the normal approximation of $\delta_T$. *Hint*: Show that

$$\Phi\left(\sqrt{T}\frac{(\alpha + \beta)^{3/2}q}{\sqrt{\alpha\beta(2 - \alpha - \beta)}}\right) \geq \frac{\varepsilon' + 1}{2} .$$

**8.17** (Raftery and Lewis 1996) Consider the logit model

$$\log\frac{\pi_i}{1 - \pi_i} = \eta + \delta_i ,$$

$$\delta_i \sim \mathcal{N}(0, \sigma^2) ,$$

$$\sigma^{-2} \sim \mathcal{G}a(0.5, 0.2) .$$

Study the convergence of the associated Gibbs sampler and the dependence on the starting value.

**8.18** Given a three-state Markov chain with transition matrix $\mathbb{P}$, determine sufficient conditions on $\mathbb{P}$ for a derived two-state chain to be a Markov chain.

**8.19** Show that the sequence $(\xi^{(n)})$ defined in §8.4.2 is a Markov chain. (*Hint*: Show that

$$P(\xi^{(n)} = i|\xi^{(n-1)} = j, \xi^{(n-2)} = \ell, \ldots)$$
$$= \mathbb{E}_{\theta^{(0)}}\left[\mathbb{I}_{A_i}\left(\theta^{(\tau_{n-1}-1+\Delta_n)}\right)\Big|\theta^{(\tau_{n-1}-1)} \in A_j, \theta^{(\tau_{n-2}-1)} \in A_\ell, \ldots\right] ,$$

and apply the strong Markov property.)

**8.20** (Cellier 1996) Consider a finite Markov chain with transition matrix

$$\mathbb{P} = \begin{pmatrix} 1/3 & 1/3 & 1/3 \\ 1 & 0 & 0 \\ 1 & 0 & 0 \end{pmatrix} .$$

(a) Show that the Markov chain is ergodic with invariant measure $(3/5, 1/5, 1/5)$.

(b) Show that if three chains $x_1^{(t)}$, $x_2^{(t)}$, and $x_3^{(t)}$ are *coupled*, that is, the three chains start from the three different states and

$$\tau = \inf\{t \geq 1; x_1^{(t)} = x_2^{(t)} = x_3^{(t)}\} ,$$

then the distribution of $x_i^{(\tau)}$ is not the invariant measure but the Dirac mass in the first state.

**8.21** (Tanner 1996) Show that, if $\theta^{(t)} \sim \pi^t$ and if the stationary distribution is the posterior density associated with $f(x|\theta)$ and $\pi(\theta)$, the weight

$$\omega_t = \frac{f(x|\theta^{(t)})\pi(\theta^{(t)})}{\pi^t(\theta^{(t)})}$$

converges to the marginal $m(x)$. Derive the *Gibbs stopper* of Note 8.6.1 by proposing an estimate of $\pi^t$.

**8.22** For the model of Problem 7.46, study the effect of the parameterization on the convergence diagnostics. (*Hint*: Use CODA.)

**8.23** (Continuation of Problem 8.22) Apply the convergence diagnostics of CODA to the models of Problems 7.52 and 7.54.

**8.24** Construct an algorithm which correctly simulates the witch's hat distribution. (*Hint:* Show that direct simulation is possible.)

**8.25** (Propp and Wilson 1996) The *backward coupling* algorithm (or *coupling from the past*) is based on the following algorithm ($k$ is the cardinality of the state-space):

1. At time $-t$ $(t \geq 1)$, simulate the $k$ transitions corresponding to the $k$ values of $x^{(-t)}$.
2. If the $k$ trajectories starting from the $k$ values of $x^{(-t)}$ coincide at time $t = 0$, stop the algorithm. Otherwise, decrease $-t$ to $-(t+1)$.

(a) Show that the resulting value $x^{(0)}$ is distributed from the stationary distribution of the chain.

(b) Show that the above result does not depend on the correlation between the simulations at time $-t$ and deduce a coupling strategy to accelerate convergence.

**8.26** (Continuation of Problem 8.25) The above algorithm can be accelerated if there exists some partial ordering of the state-space $\mathcal{X}$ of the chain $(x^{(t)})$.

(a) Show that if $\tilde{\mathcal{X}}$ denotes the set of *extremal points* of $\mathcal{X}$ (that is, for every $y \in \mathcal{X}$, there exist $x_1$, $x_2 \in \tilde{\mathcal{X}}$ such that, $\forall k \in \mathcal{X}$,

$$P_{x_1}(x^{(t)} \leq k) \leq P_y(x^{(t)} \leq k) \leq P_{x_2}(x^{(t)} \leq k)\Big),$$

the backward coupling algorithm only needs to consider the points of $\tilde{\mathcal{X}}$ at each step.

(b) Study the improvement brought by the above modification in the case of the transition matrix

$$\mathbb{P} = \begin{pmatrix} 0.34 & 0.22 & 0.44 \\ 0.28 & 0.26 & 0.46 \\ 0.17 & 0.31 & 0.52 \end{pmatrix}.$$

**8.27** For the CFTP algorithm [A.45], assume that the chain $(\theta^{(t)})$ is irreducible and strongly aperiodic, namely that $p_{uv} > 0$ for every $u$ and $v$.

(a) Show that the probability that $f_{-t}$ is constant satisfies

$$P(f_{-t} = f_{-t}(1)) = \sum_v \prod_u p_{uv} = \alpha > 0.$$

Use the fact that the $f_{-t}$'s are independent, and the Borel–Cantelli lemma to deduce that coalescence takes place with probability 1.

(b) Considering the resulting variable $F_{-\infty} = \theta^{(0)}$, show that

$$P(\theta^{(1)} = x) = \sum_y P(\theta^{(1)} = x|\theta^{(0)} = y)P(\theta^{(0)} = y)$$

$$= \sum_y p_{yx} P(x^{(0)} = y)$$

and deduce that $\theta^{(0)}$ is distributed from the stationary distribution of interest.

**8.28** In the setup of Example 7.1.23, recall that the chain $(\beta^{(t)})$ is uniformly ergodic, with lower bound

$$r(\beta') = \frac{\delta^{\gamma+10\alpha}(\beta')^{\gamma+10\alpha-1}}{\Gamma(10\alpha+\gamma)} e^{-\delta\beta'} \prod_{i=1}^{10} \left(\frac{t_i}{t_i+\beta'}\right)^{p_i+\alpha}.$$

Implement a simulation method to determine the value of $\rho$ and to propose an algorithm to simulate from this distribution. (*Hint:* Use the Riemann approximation method of §3.4 or one of the normalizing constant estimation techniques of Chen and Shao 1997; see Problem 3.15.)

**8.29** Consider a transition kernel $K$ which satisfies *Doeblin's condition*,

$$K(x,y) \geq r(x), \quad x,y \in \mathcal{X},$$

which is equivalent to uniform ergodicity (see, e.g., Example 1.2.4). As a perfect simulation technique, Murdoch and Green (1998) propose that the continuum of Markov chains coalesces into a single chain at each time with probability

$$\rho = \int r(x)dx.$$

(a) Using a coupling argument, show that if for all chains the *same uniform variable* is used to decide whether the generation from $r(x)/\rho$ occurs, then at time 0 the resulting random variable is distributed from the stationary distribution $f$.

(b) Deduce that the practical implementation of the method implies that a single chain is started at a random moment $-N$ in the past from the bounding probability $\rho^{-1}r(x)$ with $N \sim \mathcal{G}eo(\rho)$.

**8.30** (Continuation of Problem 8.29) When *Doeblin's condition* does not hold, show that $\mathcal{X}$ can be represented by a partition of $m$ sets $\mathcal{A}_i$ $(i = 1, \dots, m)$ such that

$$K(y|x) \geq r_i(x), \quad y \in \mathcal{X}, \quad x \in \mathcal{A}_i.$$

(*Hint:* See Theorem 4.3.9.)
Deduce an extension of the above perfect simulation based on $m$ chains started from the bounding distribution $\rho_i^{-1}r_i$ at a random time in the past and coupled until they coalesce. (*Hint:* Use the renewal decomposition

$$K(y|x) = \rho_i[\rho_i^{-1}r_i(x)] + (1-\rho_i)\left[\frac{K(y|x) - r_i(x)}{1-\rho_i}\right]$$

and a common sequence of uniform variables in the selection and the generation for all chains.)

**8.31** (Hobert *et al.* 1999) For a mixture of known distributions, $pf_1(x) + (1-p)f_2(x)$, show that a monotone CFTP can be implemented for the simulation of the posterior distribution on $p$, under the uniform prior. (Hobert *et al.* 1999 also consider an extension to the case of three components.)

## 8.6 Notes

### 8.6.1 *Other Stopping Rules*

(i) *The Gibbs Stopper*

Following Tanner and Wong (1987), who first approximated the distribution of $\theta^{(t)}$ to derive a convergence assessment, Ritter and Tanner (1992) proposed a control method based on the distribution of the weight $\omega_t$, which evaluates the difference between the limiting distribution $f$ and an approximation of $f$ at iteration $t$, $\hat{f}_t$,

$$w_t = \frac{f(\theta^{(t)})}{\hat{f}_t(\theta^{(t)})} \ .$$

Note that this evaluation aims more at controlling the convergence in distribution of $(\theta^{(t)})$ to $f$ (first type of control) than at measuring the mixing speed of the chain (second type of control).

In some settings, the approximation $\hat{f}_t$ can be computed by Rao–Blackwellization as in (8.3.3) but Ritter and Tanner (1992) developed a general approach (called the *Gibbs stopper*) and proposed using the weight $w_t$ through a stopping rule based on the evolution of the histogram of the $w_t$'s, until these weights are sufficiently concentrated near a constant which is the normalizing constant of $f$. Unfortunately, the example treated in Tanner (1996) does not indicate how this quantitative assessment is obtained, as it used typically unknown characteristics of the distribution $f$ for calibration. Cowles and Carlin (1996) and Brooks and Roberts (1999) contain further discussion, and Zellner and Min (1995) consider a similar approach.

(ii) *Coupling Techniques*

Diagnostics can also be based on the notion of *coupling*, briefly described in §4.6.1. This probabilistic concept can be useful not only in detecting convergence but also in constructing a stopping rule (by displaying any independence from initial conditions).

If $(\theta_1^{(t)})$ and $(\theta_2^{(t)})$ are two chains associated with the same transition kernel, and if there exists a recurrent atom $\alpha$, both chains $\theta_1^{(t)}$ and $\theta_2^{(t)}$ are identically distributed for

$$t \geq \tau = \inf \left\{ \ell; \theta_1^{(\ell)} \in \alpha \text{ and } \theta_2^{(\ell)} \in \alpha \right\} \ .$$

Once both chains have simultaneously visited the atom $\alpha$, they do not differ from a probabilistic point of view. In particular, if one of the two chains is stationary, *both* chains are stationary after $\tau$.

Johnson (1996) proposes to use coupling by simulating $M$ parallel chains $(\theta_m^{(t)})$ *from the same sample* of uniform variables. In the setup of Gibbs sampling, given by [A.31], the coupling convergence diagnostic can be written as

1. Choose $M$ initial values $\theta_m^{(0)}$ $(1 \leq m \leq M)$;
2. For $1 \leq i \leq p$, $1 \leq m \leq M$, generate $\theta_{i,m}^{(t)}$ by

$$\theta_{i,m}^{(t)} = F_i^{-1}(u_i^{(t)}|\theta_{1,m}^{(t)}, \ldots, \theta_{i-1,m}^{(t)}, \theta_{i+1,m}^{(t-1)}, \ldots, \theta_{p,m}^{(t-1)})$$

3. Stop the iterations when

$$\theta_1^{(t)} = \cdots = \theta_M^{(t)} \ ,$$

where $F_i$ denotes the cdf of the conditional distribution $P(\theta_i|\theta_1, \ldots, \theta_{i-1}, \theta_{i+1}, \ldots, \theta_p)$. In a continuous setup, Johnson (1996) suggests replacing the stopping rule **3**. by the weaker constraint $\max_{i,j} |\theta_i^{(t)} - \theta_j^{(t)}| < \epsilon$, for $1 \le i \ne j \le M$.

A first requirement of this method is thus that the conditional distributions must be simulated by the *inversion* technique (see Chapter 1, Lemma 2.1.2), a relatively rare occurrence. As noted by Johnson, this constraint directly excludes not only Metropolis–Hastings algorithms but also all setups involving Accept–Reject steps.

There are other problems. For example, although the validity of this stopping rule (that is, the fact that $P_f(\inf\{t; \theta_1^{(t)} = \cdots = \theta_M^{(t)}\} < \infty) = 1$) is indicated, it is not established in any generality. There are also problems with bias and dependence on initial conditions $\theta_m^{(0)}$, the latter of which invalidates conclusions about the stationarity of the chain.

Nonetheless, the idea of a method based on coupling should be explored further. For instance, the meeting of independent parallel chains in a small set and the link with the renewal methods of §8.2.3 should certainly be explored (see also §8.4.2). The connection with *perfect simulation* (see Note 8.6.5) also deserves further investigation.

### 8.6.2 Spectral Analysis

As already mentioned in Hastings (1970), the chain $(\theta^{(t)})$ or a transformed chain $(h(\theta^{(t)}))$ can be considered from a *time series* point of view (see Brockwell and Davis 1996 for an introduction). For instance, under an adequate parameterization, we can model $(\theta^{(t)})$ as an $ARMA(p,q)$ process, estimating the parameters $p$ and $q$, and then use partially empirical convergence control methods. Geweke (1992) proposed using the *spectral density* of $h(\theta^{(t)})$,

$$S_h(w) = \frac{1}{2\pi} \sum_{t=-\infty}^{t=\infty} \text{cov}\left(h(\theta^{(0)}), h(\theta^{(t)})\right) e^{itw},$$

where $i$ denotes the complex square root of 1 (that is,

$$e^{itw} = \cos(tw) + i\sin(tw)\bigr).$$

The spectral density is related to the asymptotic variance of (8.1.1) since the limiting variance $\gamma_h$ of Proposition 4.10.8 is given by

$$\gamma_h^2 = S_h^2(0).$$

Estimating $S_h$ by nonparametric methods like *the kernel method* (see Silverman 1986), Geweke (1992) takes the first $T_A$ observations and the last $T_B$ observations from a sequence of length $T$ to derive

$$\delta_A = \frac{1}{T_A} \sum_{t=1}^{T_A} h(\theta^{(t)}), \qquad \delta_B = \frac{1}{T_B} \sum_{t=T-T_B+1}^{T} h(\theta^{(t)}),$$

and the estimates $\sigma_A^2$ and $\sigma_B^2$ of $S_h(0)$ based on both subsamples, respectively. Asymptotically (in $T$), the difference

(8.6.1)
$$\frac{\sqrt{T}(\delta_A - \delta_B)}{\sqrt{\dfrac{\sigma_A^2}{\tau_A} + \dfrac{\sigma_B^2}{\tau_B}}}$$

is a standard normal variable (with $T_A = \tau_A T$, $T_B = \tau_B T$, and $\tau_A + \tau_B < 1$). We can therefore derive a convergence diagnostic from (8.6.1) and a determination of the size $t_0$ of the training sample. The values suggested by Geweke (1992) are $\tau_A = 0.1$ and $\tau_B = 0.5$.

A global criticism of this spectral approach also applies to all the methods using a nonparametric intermediary step to estimate a parameter of the model, namely that they necessarily induce losses in efficiency in the processing of the problem (since they are based on a less constrained representation of the model). Moreover, the calibration of nonparametric estimation methods (as the choice of the *window* in the kernel method) is always delicate since it is not standardized. We therefore refer to Geweke (1992) for a more detailed study of this method, which is used in some software (see Best *et al.* 1995, for instance). Other approaches based on spectral analysis are given in Heidelberger and Welsh (1988) and Schruben *et al.* (1983) to test the stationarity of the sequence by Kolmogorov–Smirnov tests (see Cowles and Carlin 1996 and Brooks and Roberts 1999 for a discussion). Note that Heidelberger and Welch (1983) test stationarity via a Kolmogorov–Smirnov test, based on

$$B_T(S) = \frac{S_{[Ts]} - [Ts]\bar\theta}{(T\hat\psi(0))^{1/2}} , \qquad 0 \le s \le 1,$$

where

$$S_t = \sum_{t=1}^{t} \theta^{(t)}, \qquad \bar\theta = \frac{1}{T}\sum_{t=1}^{T} \theta^{(t)} ,$$

and $\hat\psi(0)$ is an estimate of the spectral density. For large $T$, $B_T$ is approximately a Brownian bridge and can be tested as such. Their method thus provide the theoretical background to Yu and Mykland's (1998) CUSUM criterion (see §8.3.1).

### 8.6.3 Further Discretizations

Garren and Smith (1993) use the same discretization $Z^{(t)}$ as Raftery and Lewis (1992a,b). They show that under some conditions, there exist $\alpha$ and $1 > |\lambda_2| > |\lambda_3|$ such that the quantity

$$\begin{aligned} \varrho_t &= \mathbb{E}[Z^{(t)}] \\ &= \varrho + \alpha\lambda_2^t + O(|\lambda_3|^t), \end{aligned}$$

with $\varrho$ the limiting value of $\varrho_t$. (These conditions are related with the eigenvalues of the functional operator associated with the transition kernel of the original chain $(\theta^{(t)})$ and with Hilbert–Schmidt conditions; see §7.6.3.) In their study of the convergence of $\varrho_t$ to $\varrho$, Garren and Smith (1993) proposed using $m$ parallel chains $(\theta_\ell^{(t)})$, with the same initial value $\theta_\ell^{(0)}$ ($1 \le \ell \le m$). They then approximated $\varrho_t$ by

$$\hat\varrho_t = \frac{1}{m}\sum_{\ell=1}^{m} \mathbb{I}_{\theta_\ell^{(t)} < \underline\theta}$$

and derived some estimations of $\varrho, \alpha$ and $\lambda_2$ from

$$\min_{(\rho,\alpha,\lambda_2)} \sum_{t=n_0+1}^{T} (\hat\varrho_t - \varrho + \alpha\lambda_2^t)^2,$$

where $n_0$ and $T$ need to be calibrated. When $T$ is too high, the estimators of $\alpha$ and $\lambda_2$ are unstable and Garren and Smith (1993) suggested choosing $T$ such that $\hat{\alpha}$ and $\hat{\lambda}_2$ remain stable. When compared with the original binary control method, the approach of Garren and Smith (1993) does not require a preliminary evaluation of $(\alpha, \beta)$, but it is quite costly in simulations. Moreover, the expansion of $\varrho_t$ around $\varrho$ is only valid under conditions which cannot be verified in practice.

### 8.6.4 The CODA Software

While the methods presented in this chapter are at various stages of their development, some of the most common techniques have been aggregated in an S-Plus software package called CODA, developed by Best *et al.* (1995). While originally intended as an output processor for the BUGS software (see §7.6.4), this software can also be used to analyze the output of Gibbs sampling and Metropolis–Hastings algorithms. The techniques selected by Best *et al.* (1996) are mainly those described in Cowles and Carlin (1995) (that is, the convergence diagnostics of Gelman and Rubin (1992) (§8.3.4), Geweke (1992) (§8.6.2), Heidelberger and Welch (1983) (§8.6.2), Raftery and Lewis (1992a) (§8.4.1), plus plots of autocorrelation for each variable and of cross-correlations between variables). The MCMC output must, however, be presented in a very specific S-plus format to be processed by CODA, unless it is produced by BUGS.

### 8.6.5 Perfect Simulation

Although the following topic is in contrast with the theme of the previous chapters, in the sense that MCMC methods have been precisely introduced to overcome the difficulties of simulating directly from a given distribution, it is essential, in some settings, to start the Markov chain in its stationary regime, $\theta^{(0)} \sim f(x)$, because the bias caused by the initial value/distribution may be far from negligible. Moreover, if it becomes feasible to start from the stationary distribution, the convergence issues are reduced to the determination of an acceptable batch size $k$, so that $\theta^{(0)}, \theta^{(k)}, \theta^{(2k)}, \ldots$ are nearly independent, and to the accuracy of an ergodic average. In other cases, one needs to know, as put by Fill (1998a,b), *"how long is long enough?,"* in order to evaluate the necessary computing time or the mixing properties of the chain.

Following Propp and Wilson (1996), several authors have proposed devices to sample directly from the stationary distribution $f$ (that is, algorithms such that $\theta^{(0)} \sim f$), at varying computational costs[5] and for specific distributions and/or transitions. The denomination of *perfect simulation* for such techniques was coined by Kendall (1998), replacing the *exact sampling* terminology of Propp and Wilson (1996) with a more triumphant qualification! The main bulk of the work on perfect simulation deals, so far, with finite state spaces; this is due, for one thing, to the greater simplicity of these spaces and, for another, to statistical physics motivations related to the Ising model (see Example 7.1.3). The appeal of these methods for mainstream statistical problems is yet unclear from both points of view of (a) speeding up convergence and (b) controlling convergence, but Murdoch and Green (1998) have shown that some standard examples in continuous settings, like the nuclear pump failure model of Example 7.1.23 (see Problem 8.28), do allow for *perfect simulation*. Note

---

[5] In most cases, the computation time required to produce $\theta^{(0)}$ exceeds by orders of magnitude the computation time of a $\theta^{(t)}$ from the transition kernel.

also that in settings when the Duality Principle of §7.2 applies, the stationarity of the finite chain obviously transfers to the dual chain, even if the latter is continuous.

In a finite state-space $\mathcal{X}$ of size $k$, the method proposed by Propp and Wilson (1996) is called *coupling from the past* (CFTP). It runs, in parallel, $k$ chains corresponding to all possible starting points in $\mathcal{X}$ farther and farther back in time until all chains take the same value (or *coalesce*) at time 0 (or earlier). (See Problem 8.20 for an example where the corresponding forward scheme produces bias.) *Coupling* between the chains (see §4.6.1 and (ii) of §8.6.1) may improve convergence speed and also ensures that all chains are indistinguishable after time 0. If the transition matrix has cumulative probabilities $p_{uv}$, $u, v \in \mathcal{X}$, the CFTP algorithm runs as follows (see also Problem 8.25):

### Algorithm A.45 –Perfect Simulation–

1. Generate uniform variables $\omega_{-1}, \omega_{-2}, \ldots$.

2. Define local transition functions, called *maps*, $f_{-t}(u)$, on $\mathcal{X}$, as                    [A.45]

   (8.6.2)          $f_{-t}(u) = v$ iff $p_{uv} \le \omega_{-t} < p_{u(v+1)}$.

3. Define maps $F_{-1}, F_{-2}, \ldots$, on $\mathcal{X}$, as

   $$F_{-N}(u) = F_{-(N-1)}(f_{-N}(u)).$$

The functions $f_{-t}$ are thus realizations of the transitions at time $-t$, with $\theta^{(-t+1)} = f_{-t}(u)$ when $\theta^{(-t)} = u$. This formulation is equivalent to the generation of $k$ parallel chains for all possible starting points in $\mathcal{X}$. The chains are furthermore coupled (see §4.6.1) since they are based on the same uniform variable $\omega_{-t}$. The functions $F_{-N}$ are then realizations of the $N$-fold chain, that is, such that $F_{-N}(u) = \theta^{(0)}$ when $\theta^{(-N)} = u$. (An important feature of [A.45] is that, for the generation of the transition at time $-N$, the transitions at time $-N+1, -N+2, \ldots, -1$ are fixed once and for all.)

In the implementation of this algorithm, the excursion back in time stops at time $-N$ when $F_{-N}$ is a constant function; that is, when $x^{(0)}$ is the same for every starting value $x^{(-N)}$. (We are thus spared the task of running the algorithm until $-\infty$!) Note that by virtue of the separate definitions of the maps $f_{-t}$, for a given $\theta_{-N}$, the value of $\theta^{(0)}$ is also given: This not only saves simulation time but more importantly ensures the validity of the method (see Problem 8.27).

Propp and Wilson (1996) suggested monitoring the maps only at dates $-1$, $-2, -4, \ldots$, until coalescence occurs (at time 0), as this updating scheme has nearly optimal properties. An important remark of these authors (which has been emphasized in many of the following papers) is that convergence and running time are improved up when a *monotonicity* structure is available, namely when there exist a (stochastically) *larger* state $x_1$ and a (stochastically) *smaller* state $x_0$. In this case, it is sufficient to consider chains starting from the two states $x_0$ and $x_1$ until they coincide at time 0 since $F_{-N}(x_0) = F_{-N}(x_1)$ implies coalescence, given that all the intermediary paths are located between the two extreme cases (see Problem 8.26). Note also that the validity of the method extends to data augmentation settings (see §7.2.1), even where the second chain in the structure has a continuous support. The construction of the maps $F_{-N}$ is then more complex than in (8.6.2), since it implies generating the second chain $(Z^{(-t)})$ to go from $u$ to $v$ in (8.6.2), but, once this generation done, the transition from $x$ to $y$ at time $-t$ is well determined and the generation works as a black box.[6] (See Muri *et al.* 1998 for an illustration in the setup of hidden Markov chains.)

Despite the general convergence result that coalescence takes place over a finite number of backward iterations (see Problem 8.27), there are still problems with the CFTP algorithm, in that the running time is nonetheless unbounded and that aborting long runs does create bias. Fill (1998a) proposed an alternative algorithm which can be interrupted while preserving the central feature of CFTP. It is, however, restricted to monotonous and discrete settings.

Murdoch and Green's (1998) extension to the continuous case is based on renewal theory (see §8.2.3), either for uniformly ergodic chains or for Metropolis–Hastings algorithms with atoms, including, paradoxically, the independent case (see §8.2.3). It is also a CFTP technique, formally using a continuum of chains starting from all possible values in $\mathcal{X}$ at time $-1, -2, -4, \ldots$ until they coalesce into a single chain at time 0 (or earlier). The validity result of Propp and Wilson (1996) extends to this setting (see Problems 8.29 and 8.30). See Green and Murdoch (1998) for other extensions, including the promising *multigamma coupling*.

---

[6] Obviously, the generation of the $Z^{(-t)}$'s may weaken the coupling between the chains if it requires variable numbers of uniform variables, as the Accept–Reject method, but this does not invalidate the method in any way.

CHAPTER 9

# Implementation in Missing Data Models

The images so vague in his mind were now quite clear and distinct but he had to be certain: the solution must be presented like a concise clear legal document, everything in its right place, and, unfortunately, there were still gaps to be filled.

—P.C. Doherty, *Crown in Darkness*

## 9.1 Introduction

Missing data models (introduced in §5.3.1) are a natural application for simulation, since they use it to replace the missing data part so that one can proceed with a "classical" inference on the complete model. However, this idea was slow in being formalized; that is, in going beyond *ad hoc* solutions with no theoretical justification. It is only with the EM algorithm that Dempster *et al.* (1977) (see §5.3.3) described a rigorous and general formulation of statistical inference through completion of missing data (by expectation rather than simulation, though). The original algorithm could require a difficult analytic computation for the expectation (E) step and therefore cannot be used in all settings. As mentioned in §5.3.4 and §5.5.1, stochastic versions of EM (Broniatowski *et al.* 1983, Celeux and Diebolt 1985, 1993, Wei and Tanner 1990b, Qian and Titterington 1991, Lavielle and Moulines 1997) have come closer to simulation goals by replacing the E-step with a simulated completion of missing data (but without preserving the entire range of EM convergence properties).

This chapter attempts to illustrate the potential of Markov chain Monte Carlo algorithms in the analysis of missing data models, although it does not intend to provide an exhaustive treatment of these models. Rather, it must be understood as a sequence of examples on a common theme, the statistical processing of most of these models requiring MCMC methods. See Everitt (1984), Titterington *et al.* (1985), Little and Rubin (1987), MacLachlan and Krishnan (1997), or Tanner (1996) for deeper perspectives in this domain.

## 9.2 First Examples

### 9.2.1 Discrete Data Models

Numerous settings (surveys, medical experiments, epidemiological studies, design of experiment, quality control, etc.) produce a *grouping* of the original observations into less informative categories, often for reasons beyond the control of the experimenter. Heitjan and Rubin (1991) (see also Rubin 1987) call the resulting process *data coarsening* and study the effect of data aggregation on the inference. (In particular, they examine whether the grouping procedure has an effect on the likelihood and thus must be taken into account.)

**Example 9.2.1 Rounding effect.** Heitjan and Rubin (1991) model the approximation bias in a study on smoking habits, under the assumption that this bias increases with the daily consumption of cigarettes $[y_i]$. Assume that $Y_i \sim \mathcal{E}xp(\theta)$ are grouped in observations $x_i$ ($1 \le i \le n$) according to the procedure

$$G_i|y_i \sim \mathcal{B}(1, \Phi(\gamma_1 - \gamma_2 y_i)), \qquad X_i|g_i, y_i = \begin{cases} [y_i] & \text{if } g_i = 1 \\ 20[y_i/20] & \text{otherwise,} \end{cases}$$

where $\Phi$ is the cdf of the normal distribution $\mathcal{N}(0,1)$ and $[a]$ denotes the integer part of $a$. This means that as $y_i$ (which is unobserved) increases, the probability of rounding up the answer $x_i$ to the nearest full pack of cigarettes also increases, under the constraint $\gamma_2 > 0$. If the $g_i$'s are known, the completion of the model is straightforward. Otherwise, the conditional distribution

$$\begin{aligned}
\pi(y_i|x_i, \theta, \gamma_1, \gamma_2) &\propto e^{-\theta y_i} \left\{ \mathbb{I}_{[x_i, x_i+1]}(y_i)\Phi(\gamma_1 - \gamma_2 y_i) \right. \\
&\left. + \mathbb{I}_{[x_i, x_i+20]}(y_i)\Phi(-\gamma_1 + \gamma_2 y_i) \right\} \\
&= e^{-\theta y_i} \left\{ \mathbb{I}_{[x_i, x_i+1]}(y_i) + \mathbb{I}_{[x_i+1, x_i+20]}(y_i)\Phi(-\gamma_1 + \gamma_2 y_i) \right\}
\end{aligned}$$

is useful for the completion of the model through Gibbs sampling. This distribution can be simulated directly by an Accept–Reject algorithm (which requires a good approximation of $\Phi$—see Example 2.1.9) or by introducing an additional artificial variable $T_i$ such that

$$T_i|y_i \sim \mathcal{N}_+(\gamma_1 - \gamma_2 y_i, 1, 0),$$

$$Y_i|t_i \sim \theta e^{-\theta y_i} \left\{ \mathbb{I}_{[x_i, x_i+1]}(y_i) + \mathbb{I}_{[x_i+1, x_i+20]}(y_i) \frac{e^{-t_i^2/2}}{\sqrt{2\pi}} \right\},$$

where $\mathcal{N}_+(\mu, \sigma^2, 0)$ denotes the normal distribution $\mathcal{N}(\mu, \sigma^2)$ truncated to $\mathbb{R}_+$ (see Example 2.3.5). The two distributions above can then be completed by

$$(\theta, \gamma_1, \gamma_2)|\mathbf{y}, \mathbf{t} \sim \theta^n \exp\left\{ -\theta \sum_{i=1}^n y_i \right\} \mathbb{I}_{\gamma_1 > \max(\gamma_2 y_i + t_i)} \, \pi(\theta, \gamma_1, \gamma_2) \,,$$

|          | Diameter (inches) | |
|          | ≤ 4.0 | > 4.0 |
|----------|-------|-------|
| **Height** | | |
| **(feet)** > 4.75 | 32 | 11 |
| ≤ 4.75 | 86 | 35 |

Table 9.2.1. *Observation of two characteristics of the habitat of 164 lizards (Source: Schoener 1968).*

where $\mathbf{y} = (y_1, \ldots, y_n)$ and $\mathbf{t} = (t_1, \ldots, t_n)$, to provide a Gibbs sampling algorithm.                                                                                 ‖

When several variables are studied simultaneously in a sample, each corresponding to a grouping of individual data, the result is a *contingency table*. If the context is sufficiently informative to allow for a modeling of the individual data, the completion of the contingency table (by reconstruction of the individual data) may facilitate inference about the phenomenon under study

**Example 9.2.2 Lizard habitat.** Schoener (1968) studies the habitat of lizards, in particular the relationship between height and diameter of the branches on which they sleep. Table 9.2.1 provides the information available on these two parameters.

To test the independence between these factors, Fienberg (1977) proposes a classical solution based on a $\chi^2$ test. A possible alternative is to assume a parametric distribution on the individual observations $x_{ijk}$ of diameter and of height $(i, j = 1, 2, k = 1, \ldots, n_{ij})$. Defining $y_{ijk} = \log(x_{ijk})$, we could assume $y_{ijk}$ follows a bivariate normal distribution with mean $\theta = (\theta_1, \theta_2)$ and covariance matrix

$$\Sigma = \sigma^2 \begin{pmatrix} 1 & \rho \\ \rho & 1 \end{pmatrix} = \sigma^2 \Sigma_0.$$

The likelihood associated with Table 9.2.1 is then implicit, even in the case $\rho = 0$. However, the model can be completed by simulation of the individual values. If we undertake a Bayesian analysis with prior distribution $\pi(\theta, \sigma, \rho) = \frac{1}{\sigma} \mathbb{I}_{[-1,1]}(\rho)$, completion allows for the approximation of the posterior distribution through a Markov chain Monte Carlo algorithm. In fact, if $n_{11} = 32$, $n_{12} = 11$, $n_{21} = 86$, and $n_{22} = 35$, and if $\mathcal{N}_2^T(\theta, \Sigma; Q_{ij})$ represents the normal distribution restricted to one of the four quadrants $Q_{ij}$ induced by $(\log(4.75), \log(4))$, the steps of the Gibbs sampler are:

**Algorithm A.46 –Contingency Table Completion–**

1. Simulate $y_{ijk} \sim \mathcal{N}_2^T(\theta, \Sigma; Q_{ij})$     $(i, j = 1, 2, \ k = 1, \ldots, n_{ij});$

2. Simulate $\theta \sim \mathcal{N}_2(\bar{y}, \Sigma/164);$

3. Simulate $\sigma^2$ from the inverted Gamma distribution          [A.46]

$$\mathcal{IG}\left(164, \frac{1}{2}\sum_{i,j,k}(y_{ijk} - \theta)^t \, \Sigma_0^{-1}(y_{ijk} - \theta)\right) \, ;$$

4. Simulate $\rho$ according to

$$(9.2.1) \quad (1 - \rho^2)^{-164/2} \, \exp\left\{-\frac{1}{2}\sum_{i,j,k}(y_{ijk} - \theta)^t \, \Sigma^{-1}(y_{ijk} - \theta)\right\} \, ,$$

where $\bar{y} = \sum_{i,j,k} y_{ijk}/164$. We have seen the truncated normal distribution many times (see Example 2.3.5), but the distribution (9.2.1) requires a Metropolis–Hastings step based, for instance, on an inverse Wishart distribution. ∥

Another situation where grouped data appear in a natural fashion is that of *qualitative models*. As the logit model is treated in Problems 7.15 and 9.3, we now look at the probit model. The probit model is often considered as a threshold model in the following sense. We observe binary variables $y_i$ that take values in $\{0, 1\}$ and link them to a vector of covariates $x_i \in \mathbb{R}^p$ by the equation

(9.2.2)          $p_i = \Phi(x_i^t\beta) \, , \qquad \beta \in \mathbb{R}^p.$

where $p_i$ is the probability of success and $\Phi$ is the standard normal cdf.

The $Y_i$'s can be thought of as delimiting a threshold reached by latent (unobservable) continuous random variables $Y_i^*$, where we observe $y_i = 1$ if $y_i^* > 0$. Thus, $p_i = P(Y_i = 1) = P(Y_i^* > 0)$, and we have an automatic way to complete the model. Given the latent variables $y_i^*$,

$$y_i = \begin{cases} 1 & \text{if } y_i^* > 0 \\ 0 & \text{otherwise.} \end{cases}$$

Given a conjugate distribution $\mathcal{N}_p(\beta_0, \Sigma)$ on $\beta$, the algorithm which approximates the posterior distribution $\pi(\beta|y_1, \ldots, y_n, x_1, \ldots, x_n)$ is then

**Algorithm A.47 –Probit Posterior Distribution–**

1. Simulate

$$Y_i^* \sim \begin{cases} \mathcal{N}_+(x_i^t\beta, 1, 0) & \text{if } y_i = 1 \\ \mathcal{N}_-(x_i^t\beta, 1, 0) & \text{if } y_i = 0 \end{cases} \qquad (i = 1, \ldots, n) \, . \quad [A.47]$$

2. Simulate

$$\beta \sim \mathcal{N}_p\left((\Sigma^{-1} + XX^t)^{-1}(\Sigma^{-1}\beta_0 + \sum_i y_i^*x_i), (\Sigma^{-1} + XX^t)^{-1}\right),$$

where $\mathcal{N}_+(\mu, \sigma^2, \underline{u})$ and $\mathcal{N}_-(\mu, \sigma^2, \bar{u})$ denote the normal distribution truncated on the left in $\underline{u}$ and the normal distribution truncated on the right in $\bar{u}$, respectively (Example 2.3.5), and where $X$ is the matrix whose columns are the $x_i$'s. See Albert and Chib (1993b) for some details on the implementation of [A.47] and extension to hierarchical binary dat and extension to hierarchical binary data.

| | Men | | Women | |
|---|---|---|---|---|
| Age | Single | Married | Single | Maried |
| < 30 | 20.0 | 21.0 | 16.0 | 16.0 |
| | 24/1 | 5/11 | 11/1 | 2/2 |
| > 30 | 30.0 | 36.0 | 18.0 | – |
| | 15/5 | 2/8 | 8/4 | 0/4 |

Table 9.2.2. *Average incomes and numbers of responses/nonresponses to a survey on the income by age, sex, and marital status. (Source: Little and Rubin 1987.)*

### 9.2.2 Data Missing at Random

Incomplete observations arise in numerous settings. For instance, a survey with multiple questions may include nonresponses to some personal questions; a calibration experiment may lack observations for some values of the calibration parameters; a pharmaceutical experiment on the after-effects of a toxic product may skip some doses for a given patient; etc. The analysis of such structures is complicated by the fact that the missing data mechanism is not always explained. If these missing observations are entirely due to chance, it follows that the incompletely observed data only play a role through their marginal distribution. However, these distributions are not always explicit and a natural approach leading to a Gibbs sampler algorithm is to replace the missing data by simulation.

**Example 9.2.3 Nonignorable nonresponse.** Table 9.2.2 describes the (fictitious) results of a survey relating income to age, sex, and family status. The observations are grouped by average, and we assume an exponential shape for the individual data,

$$Y_{a,s,m,i}^* \sim \mathcal{E}xp(\mu_{a,s,m}) \quad \text{with} \quad \mu_{a,s,m} = \mu_0 + \alpha_a + \beta_s + \gamma_m \,,$$

where $1 \leq i \leq n_{a,s,m}$ and $\alpha_a$ $(a = 1, 2)$, $\beta_s$ $(s = 1, 2)$, and $\gamma_m$ $(m = 1, 2)$ correspond to the *age* (junior/senior), *sex* (female/male), and *family* (single/married) effects, respectively.

The model is unidentifiable, but that can be remedied with the constraints $\alpha_1 = \beta_1 = \gamma_1 = 0$. A more difficult and important problem appears when nonresponse depends on the income, say in the shape of a logit model,

$$p_{a,s,m,i} = \exp\{w_0 + w_1 y_{a,s,m,i}^*\} \Big/ 1 + \exp\{w_0 + w_1 y_{a,s,m,i}^*\},$$

where $p_{a,s,m,i}$ denotes the probability of nonresponse and $(w_0, w_1)$ are the logit parameters.

If the prior distribution on the parameter set is the Lebesgue measure on $\mathbb{R}^2$ for $(w_0, w_1)$ and on $\mathbb{R}_+$ for $\mu_0, \alpha_2, \beta_2$, and $\gamma_2$, a direct analysis of the likelihood is not feasible analytically. However, the likelihood of the complete model is much more tractable, being given by

$$\prod_{\substack{a=1,2\\s=1,2\\m=1,2}} \prod_{i=1}^{n_{a,s,m}} \frac{\exp\{z_{a,s,m,i}^*(w_0 + w_1 y_{a,s,m,i}^*)\}}{1 + \exp\{w_0 + w_1 y_{a,s,m,i}^*\}} (\mu_0 + \alpha_a + \beta_s + \gamma_m)^{r_{a,s,m}}$$

$$(9.2.3) \qquad \times \exp\left\{-r_{a,s,m}\overline{y}_{a,s,m}(\mu_0 + \alpha_a + \beta_s + \gamma_m)\right\},$$

where $z_{a,s,m,i}z_{a,s,m,i}^*$ is the indicator of a missing observation and $n_{a,s,m}$, $r_{a,s,m}$ and $\overline{y}_{a,s,m}$ denote the number of people in each category of the survey, the number of responses, and the average of these responses by category, respectively.

The completion of the data then proceeds by simulating the $y_{a,s,m,i}^*$'s from

$$\pi(y_{a,s,m,i}^*) \propto \exp(-y_{a,s,m,i}^* \mu_{a,s,m}) \frac{\exp\{z_{a,s,m,i}^*(w_0 + w_1 y_{a,s,m,i}^*)\}}{1 + \exp\{w_0 + w_1 y_{a,s,m,i}^*\}},$$

which requires a Metropolis–Hastings step. Note that the parameter step of the corresponding MCMC algorithm involves simulations from

$$\prod_{\substack{a=1,2\\s=1,2\\m=1,2}} (\mu_0 + \alpha_a + \beta_s + \gamma_m)^{r_{a,s,m}} \exp\left\{-r_{a,s,m}\overline{y}_{a,s,m}(\mu_0 + \alpha_a + \beta_s + \gamma_m)\right\}$$

for $\mu_0, \alpha_2, \beta_2$, and $\gamma_2$, possibly using a Gamma instrumental distribution (see Problem 9.4), and from

$$\prod_{\substack{a=1,2\\s=1,2\\m=1,2}} \prod_{i=1}^{n_{a,s,m}} \frac{\exp\{z_{a,s,m,i}^*(w_0 + w_1 y_{a,s,m,i}^*)\}}{1 + \exp\{w_0 + w_1 y_{a,s,m,i}^*\}}$$

for $(w_0, w_1)$, which corresponds to a logit model.          ‖

*Longitudinal* studies (that is, following of individuals through time), often result in missing data problems. This may result from censoring, where an individual leaves the domain of study for a (typically unknown) reason, often assumed to be independent of the observed phenomenon. Another cause is simply to have irregular observations on an individual.

Some longitudinal studies such as *capture–recapture models* (see Examples 7.2.4 and 7.2.5) are necessarily subject to irregularities. See Seber (1983, 1992), Wolter (1986), or Burnham *et al.* (1987) for introductions to these models.

**Example 9.2.4 Lizard captures.** A population of lizards *(Lacerta vivipara)* can move between three spatially connected zones and it is of interest to model this movement. At time $t = 1$, a (random) number of lizards are captured. The captured lizards are marked and released. This operation is repeated at times $2 \leq t \leq m$ by tagging the newly captured animals and by recording at each capture the position of the recaptured animals (zones

1, 2, or 3). Therefore, the model consists of a longitudinal study of capture and position of $m - 1$ cohorts of animals marked at times $1 \leq t \leq m - 1$.

In addition to the three spatial zones, we introduce the absorbing † *(dagger)* state, which corresponds to a permanent departure from the system, either by definitive migration or by death, but which is never observed. (Such a population is said to be *open.*)

The parameters of the model are the capture probabilities $p_t(r)$, indexed by the capture zone $r$ $(r = 1, 2, 3)$, and the movement probabilities $q_t(r, s)$ $(r, s = 1, 2, 3, †)$, which are the probabilities that the lizard is in $s$ at time $(t + 1)$ given that it is in $r$ at time $t$. For instance, the probability $q_t(†, †)$ is equal to 1, because of the absorbing nature of †. We also denote by $\varphi_t(r)$ the *survival* probability at time $t$ in zone $r$ (that is,

$$\varphi_t(r) = 1 - q_t(r, †) \, ,$$

and by $\psi_t(r, s)$ the effective probability of movement for the animals remaining in the system, that is

$$\psi_t(r, s) = q_t(r, s)/\varphi_t(r)) \, .$$

If we denote $\psi_t(r) = (\psi_t(r, 1), \psi_t(r, 2), \psi_t(r, 3))$, the prior distributions are

$$p_t(r) \sim \mathcal{B}e(a, b), \quad \varphi_t(r) \sim \mathcal{B}e(\alpha, \beta), \quad \psi_t(r) \sim \mathcal{D}_3(\gamma_1, \gamma_2, \gamma_3),$$

where the hyperparameters $a, b, \ldots, \gamma_3$ are known.

The probabilities of capture $p_t(r)$ depend on the zone of capture $r$ and the missing data structure of the model, which must be taken into account. If we let $y_{ti}^*$ be the position of animal $i$ at time $t$ and $x_{ti}$ its capture indicator, the observations can be written in the form $y_{ti} = x_{ti} y_{ti}^*$, where $y_{ti} = 0$ corresponds to a missing observation. The sequence of $y_{ti}^*$'s for a given $i$ then yields a *nonhomogeneous Markov chain*, with transition matrix $q_t = (q_t(r, s))$, and the $X_{ti}$'s are Bernoulli variables *conditionally on* $y_{ti}^*$,

$$x_{ti} | y_{ti}^* \sim \mathcal{B}(p_t(y_{ti}^*)) \, .$$

The Gibbs sampler associated with the completion of this model has the following steps:

**Algorithm A.48** –**Capture–Recapture Posterior Generation–**

1. Simulate $y_{ti}^*$ for $x_{ti} = 0$      $(2 \leq t \leq m)$.
2. Generate $(1 \leq t \leq m)$                                       [A.48]

$$p_t(r) \sim \mathcal{B}e(a + u_t(r), b + v_t(r)) \, ,$$

$$\varphi_t(r) \sim \mathcal{B}e(\alpha + w_t(r), \beta + w_t(r, †)) \, ,$$

$$\psi_t(r) \sim \mathcal{D}_3(\gamma_1 + w_t(r, 1), \gamma_2 + w_t(r, 2), \gamma_3 + w_t(r, 3))$$

where $u_t(r)$ denotes the number of captures in $r$ at time $t$, $v_t(r)$ the number of animals unobserved at time $t$ for which the simulated $y_{ti}$ is equal to $r$, $w_t(r, s)$ the number of passages (observed or simulated) from $r$ to $s$, $w_t(r, †)$ the number of (simulated) passages from $r$ to †, and

$$w_t(r) = w_t(r, 1) + w_t(r, 2) + w_t(r, 3) \, .$$

Step 1. must be decomposed into conditional substeps to account for the Markovian nature of the observations (see §9.5.1); this means that $y_{ti}^*$ must be simulated conditionally on $y_{(t-1)i}^*$ and $y_{(t+1)i}^*$ when $x_{ti} = 0$. If $t \neq m$, the missing data are simulated according to

$$P(y_{ti}^* = s | y_{(t-1)i}^* = r, y_{(t+1)i}^* = \ell, x_{ti} = 0) \propto q_{t-1}(r, s)(1 - p_t(s))q_t(s, \ell)$$

and

$$P(y_{mi}^* = s | y_{(m-1)i}^* = r, x_{ti} = 0) \propto q_{m-1}(r, s)(1 - p_m(s)) .$$

See Dupuis (1995a,b, 1999) for a detailed study of the algorithm [A.48], which is convergent "despite" the presence of an absorbing[1] state † in the chain $(Y_{ti}^*)$.                                                                    ‖

## 9.3 Finite Mixtures of Distributions

Although they may seem to apply only for some very particular sets of random phenomena, *mixtures of distributions*

$$(9.3.1) \qquad\qquad \tilde{f}(x) = \sum_{j=1}^{k} p_j \, f(x|\xi_j) ,$$

where $p_1 + \cdots + p_k = 1$, are of wide use in practical modeling. However, as already noticed in Examples 1.1.2 and 3.2.4, they can be challenging from an inferential point of view (that is, when estimating the parameters $p_j$ and $\xi_j$). Everitt (1984), Titterington et al. (1985), MacLachlan and Basford (1988), West (1992), Titterington (1995), and Robert (1996a) all provide different perspectives on mixtures of distributions, discuss their relevance for modeling purposes, and give illustrations of their use in various setups.

This section focuses on the MCMC resolution of the Bayesian estimation of the parameters in (9.3.1), as well as on the practical problems encountered in the implementation of this method. We also illustrate an improvement brought by a reparameterization of (9.3.1).

We assume, without a considerable loss of generality, that $f(\cdot|\xi)$ belongs to an exponential family

$$f(x|\xi) = h(x) \, \exp\{\xi \cdot x - \psi(\xi)\} ,$$

and we consider the associated conjugate prior on $\xi$ (see Robert 1994a, Section 3.2)

$$\pi(\xi|\alpha_0, \lambda) \propto \exp\{\lambda(\xi \cdot \alpha_0 - \psi(\xi))\} , \qquad \lambda > 0, \quad \alpha_0 \in \mathcal{X}.$$

For the mixture (9.3.1), it is therefore possible to associate with each component $f(\cdot|\xi_j)$ $(j = 1, \ldots, k)$ a conjugate prior $\pi(\xi_j|\alpha_j, \lambda_j)$. We also select

---

[1] It is quite important to distinguish between the observed Markov chain, which may have absorbing or transient states, and the simulated (MCMC) Markov chain, which must satisfy minimal properties such as irreducibility and ergodicity.

for $(p_1, \ldots, p_k)$ the standard Dirichlet conjugate prior; that is,

$$(p_1, \ldots, p_k) \sim \mathcal{D}_k(\gamma_1, \ldots, \gamma_k) \,.$$

Given a sample $(x_1, \ldots, x_n)$ from (9.3.1), the posterior distribution associated with this model is formally explicit (see Problem 9.6). However, it is virtually useless for large, or even moderate, values of $n$. In fact, the posterior distribution,

$$\pi(p, \xi | x_1, \ldots, x_n) \propto \prod_{i=1}^{n} \left\{ \sum_{j=1}^{k} p_j\, f(x_i | \xi_j) \right\} \pi(p, \xi),$$

can be expressed as a sum of $k^n$ terms which correspond to the different allocations of the observations $x_i$ to the components of (9.3.1). Although each term is conjugate, the number of terms involved in the posterior distribution makes the computation of the normalizing constant and of posterior expectations totally infeasible for large sample sizes (see Diebolt and Robert 1990a). The complexity of this model is such that there are virtually no other solutions than using the Gibbs sampler (see, for instance, Smith and Makov 1978 or Bernardo and Giròn 1986, 1988 for pre-Gibbs approximations).

A solution proposed by Lavine and West (1992), Verdinelli and Wasserman (1992), Diebolt and Robert (1990b, c, 1994), and Escobar and West (1995) is to take advantage of the missing data structure inherent to (9.3.1). In fact, one can associate with every observation $x_i$ an indicator variable $z_i \in \{1, \ldots, k\}$ that indicates which component of the mixture is associated with $x_i$. The demarginalization (or *completion*) of model (9.3.1) is then

$$Z_i \sim \mathcal{M}_k(1; p_1, \ldots, p_k), \qquad x_i | z_i \sim f(x | \xi_{z_i}) \,.$$

Thus, considering $x_i^* = (x_i, z_i)$ (instead of $x_i$) entirely eliminates the mixture structure since the likelihood of the completed model is

$$\ell(p, \xi | x_i^*, \ldots, x_i^*) \propto \prod_{i=1}^{n} p_{z_i}\, f(x_i | \xi_{z_i})$$

$$= \prod_{j=1}^{k} \prod_{i; z_i = j} p_j\, f(x_i | \xi_j) \,.$$

(This latent structure is also exploited in the original implementation of the EM algorithm; see §5.3.3.) The two steps of the Gibbs sampler are

**Algorithm A.49 –Mixture Posterior Simulation–**

1. Simulate $Z_i$ $(i = 1, \ldots, n)$ from

$$P(Z_i = j) \propto p_j\, f(x_i | \xi_j) \qquad (j = 1, \ldots, k)$$

and compute the statistics ⁣[A.49]

$$n_j = \sum_{i=1}^{n} \mathbb{I}_{z_i = j} \,, \qquad n_j \bar{x}_j = \sum_{i=1}^{n} \mathbb{I}_{z_i = j} x_i \,.$$

2. Generate $(j = 1, \ldots, k)$

$$\xi \sim \pi \left( \xi \,\middle|\, \frac{\lambda_j \alpha_j + n_j \overline{x}_j}{\lambda_j + n_j}, \lambda_j + n_j \right),$$

$$p \sim \mathcal{D}_k(\gamma_1 + n_1, \ldots, \gamma_k + n_k).$$

**Example 9.3.1 Normal mixtures.** In the case of a mixture of normal distributions,

$$(9.3.2) \qquad \tilde{f}(x) = \sum_{j=1}^{k} p_j \, \frac{e^{-(x - \mu_j)^2 / (2\tau_j^2)}}{\sqrt{2\pi}\, \tau_j},$$

the conjugate distribution on $(\mu_j, \tau_j)$ is

$$\mu_j | \tau_j \sim \mathcal{N}\left( \alpha_j, \tau_j^2 / \lambda_j \right), \qquad \tau_j^2 \sim \mathcal{IG}\left( \frac{\lambda_j + 3}{2}, \frac{\beta_j}{2} \right)$$

and the two steps of the Gibbs sampler are as follows:

**Algorithm A.50 –Normal Mixture Posterior Simulation–**

1. Simulate $(i = 1, \ldots, n)$

$$Z_i \sim P(Z_i = j) \propto p_j \, \exp\left\{ -(x_i - \mu_j)^2 / (2\tau_j^2) \right\} \tau_j^{-1}$$

and compute the statistics $(j = 1, \ldots, k)$ [A.50]

$$n_j = \sum_{i=1}^{n} \mathbb{I}_{z_i = j}, \quad n_j \overline{x}_j = \sum_{i=1}^{n} \mathbb{I}_{z_i = j} x_i, \quad s_j^2 = \sum_{i=1}^{n} \mathbb{I}_{z_i = j} (x_i - \overline{x}_j)^2.$$

2. Generate

$$\mu_j | \tau_j \sim \mathcal{N}\left( \frac{\lambda_j \alpha_j + n_j \overline{x}_j}{\lambda_j + n_j}, \frac{\tau_j^2}{\lambda_j + n_j} \right),$$

$$\tau_j^2 \sim \mathcal{IG}\left( \frac{\lambda_j + n_j + 3}{2}, \frac{\beta_j + s_j^2}{2} \right),$$

$$p \sim \mathcal{D}_k(\gamma_1 + n_1, \ldots, \gamma_k + n_k).$$

Robert and Soubiran (1993) use this algorithm to derive the maximum likelihood estimators by recursive integration (see §5.2.4), showing that the Bayes estimators converge to the local maximum, which is the closest to the initial Bayes estimator (see also Robert 1993). ‖

Good performance of the Gibbs sampler is guaranteed by the above setup since the *Duality Principle* of §7.2.4 applies. One can also deduce geometric convergence and a Central Limit Theorem. Moreover, Rao–Blackwellization is justified (see Problem 9.6).

The practical implementation of [A.49] might, however, face serious convergence difficulties, in particular because of the phenomenon of the "absorbing component" (Diebolt and Robert 1990b, Mengersen and Robert 1996, Robert 1996b). When only a small number of observations are allocated to a given component $j_0$, the following probabilities are quite small:

(1) The probability of allocating new observations to the component $j_0$.
(2) The probability of reallocating, to another component, observations already allocated to $j_0$.

Even though the chain $(z^{(t)}, \xi^{(t)})$ corresponding to [A.49] is irreducible, the practical setting is one of an almost-absorbing state, which is called a *trapping state* as it requires an enormous number of iterations of [A.49] to escape from this state. In the extreme case, the probability of escape is below the minimal precision of the computer and the trapping state is truly absorbing, due to computer "rounding errors."

This problem can be linked with a potential difficulty of this model, namely that it does not allow a noninformative (or improper) Bayesian approach, and therefore necessitates the elicitation of the hyperparameters $\gamma_j$, $\alpha_j$ and $\lambda_j$. Moreover, a vague choice of these parameters (taking for instance, $\gamma_j = 1/2$, $\alpha_j = \alpha_0$, and small $\lambda_j$'s) has often the effect of increasing the occurrence of trapping states (Chib 1995).

**Example 9.3.2 (Continuation of Example 9.3.1)** Consider the case where $\lambda_j \ll 1$, $\alpha_j = 0$ $(j = 1, \dots, k)$, and where a *single observation* $x_{i_0}$ is allocated to $j_0$. Using the algorithm [A.50], we get the approximation

$$\mu_{j_0} | \tau_{j_0} \sim \mathcal{N}\left(x_{i_0}, \tau_{j_0}^2\right), \qquad \tau_{j_0}^{-2} \sim \mathcal{IG}(2, \beta_j/2),$$

so $\tau_{j_0}^2 \approx \beta_j/4 \ll 1$ and $\mu_{j_0} \approx x_{i_0}$. Therefore,

$$P\left(Z_{i_0} = j_0\right) \propto p_{j_0} \tau_{j_0}^{-1} \gg p_{j_1} \, e^{-\left(x_{i_0} - \mu_{j_1}\right)^2 / 2\tau_{j_1}^2} \, \tau_{j_1}^{-1} \propto P\left(Z_{i_0} = j_1\right)$$

for $j_1 \neq j_0$. On the other hand, if $i_1 \neq i_0$, it follows that

$$P\left(Z_{i_1} = j_0\right) \propto p_{j_0} \, e^{-\left(x_{i_1} - \mu_{j_0}\right)^2 / 2\tau_{j_0}^2} \, \tau_{j_0}^{-1}$$

$$\ll p_{j_1} \, e^{-\left(x_{i_1} - \mu_{j_1}\right)^2 / 2\tau_{j_1}^2} \, \tau_{j_1}^{-1} \propto P\left(Z_{i_1} = j_1\right),$$

given the very rapid decrease of $\exp\{-t^2/2\tau_{j_0}^2\}$.                    ‖

An attempt at resolution of the paradox of trapping states may be to blame the Gibbs sampler, which moves too slowly on the likelihood surface, and to replace it by an Metropolis–Hastings algorithm with wider moves. This solution remains to be found, since the instrumental distribution must take into account the complex nature of the likelihood surface (Celeux and Diebolt 1993). The trapping phenomenon is also related to the lack of a maximum likelihood estimator in this setup, since the likelihood is not bounded. (See Problem 9.7 or Lehmann and Casella 1998.)

## 9.4 A Reparameterization of Mixtures

This section deals with a special feature of mixture estimation, whose relevance to MCMC algorithms shows that a change in the parameterization of a model can accelerate convergence.[2]

---

[2] This section can be omitted at first reading.

As a starting point, consider the paradox of trapping states, discussed in §9.3. We can reassess the relevance of the selected prior distribution, which may not endow the posterior distribution with sufficient stability. The conjugate distribution used in Example 9.3.1 ignores the specific nature of the mixture (9.3.1) since it only describes the completed model, and creates an independence between the components of the mixture. An alternative proposed in Mengersen and Robert (1996) is to relate the various components through a common reference, namely a scale, location, or scale *and* location parameter.

**Example 9.4.1 (Continuation of Example 9.3.1)** Normal distributions are location–scale distributions. Each component can thus be expressed in terms of divergence from a global location, $\mu$, and a global scale, $\tau$. Using somewhat abusive (but convenient) notation, we write

$$(9.4.1) \qquad \sum_{j=1}^{k} p_j \, \mathcal{N}(\mu + \tau\theta_j, \tau^2\sigma_j^2),$$

also requiring that $\theta_1 = 0$ and $\sigma_1 = 1$, which avoids overparameterization. Although the model (9.4.1) is an improvement over (9.3.1), it is still unable to ensure proper stability of the algorithm.[3] The solution proposed in Robert and Mengersen (1999) is rather to express each component as a perturbation of the previous component, which means to use the location and scale parameters of the previous component as a *local reference*. Starting from the normal distribution $\mathcal{N}(\mu, \tau^2)$, the two-component mixture

$$p\mathcal{N}(\mu, \tau^2) + (1 - p)\mathcal{N}(\mu + \tau\theta, \tau^2\sigma^2)$$

is thus modified into

$$p\mathcal{N}(\mu, \tau^2) + (1-p)q\mathcal{N}(\mu+\tau\theta, \tau^2\sigma^2) + (1-p)(1-q)\mathcal{N}(\mu+\tau\theta+\tau\sigma\varepsilon, \tau^2\sigma^2\omega^2)$$

for a three-component mixture.

The $k$ component version of the reparameterization of (9.3.1) is

$$p\mathcal{N}(\mu, \tau^2) + (1 - p) \cdots (1 - q_{k-2}) \, \mathcal{N}(\mu + \ldots + \tau \cdots \sigma_{k-2}\theta_{k-1}, \tau^2 \ldots \sigma_{k-1}^2)$$

$$(9.4.2) + \sum_{j=1}^{k-2} (1 - p) \cdots (1 - q_{j-1})q_j \, \mathcal{N}(\mu + \cdots + \tau \cdots \sigma_{j-1}\theta_j, \tau^2 \cdots \sigma_j^2) \, .$$

The mixture (9.3.1) being invariant under permutations of indices, we can impose an *identifiability constraint*, for instance,

$$(9.4.3) \qquad\qquad \sigma_1 \leq 1, \ldots, \sigma_{k-1} \leq 1 \, .$$

The prior distribution can then be modified into

$$(9.4.4) \quad \pi(\mu, \tau) = \tau^{-1}, \quad p, q_j \sim \mathcal{U}_{[0,1]}, \quad \sigma_j \sim \mathcal{U}_{[0,1]}, \quad \theta_j \sim \mathcal{N}(0, \zeta^2) \, .$$

---

[3] It reproduces the feature of the original parameterization, namely of independence between the parameters conditional on the global location–scale parameter.

The representation (9.4.2) allows the use of an improper distribution on $(\mu, \tau)$, and (9.4.3) justifies the use of a uniform distribution on the $\sigma_i$'s (Robert and Titterington 1998). The influence of the only hyperparameter, $\zeta$, on the resulting estimations of the parameters is, moreover, moderate, if present. (See Problems 9.10 and 9.12.)        ‖

**Example 9.4.2 Acidity level in lakes.** In a study of acid neutralization capacity in lakes, Crawford *et al.* (1992) used a Laplace approximation to the normal mixture model

$$(9.4.5) \qquad \tilde{f}(x) = \sum_{j=1}^{3} p_j \, \frac{e^{-(x-\mu_j)^2/2w_j^2}}{\sqrt{2\pi} \, w_j}.$$

This model can also be fit with the algorithm of Example 9.4.1, and Figures 9.4.1 and 9.4.2 illustrate the performances of this algorithm on a benchmark example introduced in Crawford *et al.* (1992). Figure 9.4.1 shows the (lack of) evolution of the estimated density based on the averaged parameters from the Gibbs sampler when the number of iterations $T$ increases. Figure 9.4.2 gives the corresponding convergence of these averaged parameters and shows that the stability is less obvious at this level. This phenomenon occurs quite often in mixture settings and is to be blamed on the weak identifiability of these models, for which quite distinct sets of parameters lead to very similar densities, as shown by Figure 9.4.1.        ‖

In the examples above, the reparameterization of mixtures and the corresponding correction of prior distributions result in a higher stability of Gibbs sampler algorithms and, in particular, with the disappearance of the phenomenon of trapping states

The identifiability constraint used in (9.4.3) has an advantage over equivalent constraints, such as $p_1 \geq p_2 \geq \cdots \geq p_k$ or $\mu_1 \leq \mu_2 \leq \cdots \leq \mu_k$. It not only automatically provides a compact support for the parameters but also allows the use of a uniform prior distribution. However, it sometimes slows down convergence of the Gibbs sampler.

Although the new parameterization helps the Gibbs sampler to distinguish between the homogeneous components of the sample (that is, to identify the observations with the same indicator variable), this reparameterization can encounter difficulties in moving from homogeneous subsamples. When the order of the component indices does not correspond to the constraint (9.4.3), the probability of simultaneously permuting all the observations of two ill-ordered but homogeneous subsamples is very low. One solution in the normal case is to identify the homogeneous components without imposing the constraint (9.4.3). The uniform prior distribution $\mathcal{U}_{[0,1]}$ on $\sigma_j$ must be replaced by $\sigma_j \sim \frac{1}{2}\mathcal{U}_{[0,1]} + \frac{1}{2}\mathcal{P}a(2,1)$, which is equivalent to assuming that either $\sigma_j$ or $\sigma_j^{-1}$ is distributed from a $\mathcal{U}_{[0,1]}$ distribution. The simulation steps for $\sigma_j$ are modified, but there exists a direct Accept–Reject algorithm to simulate from the conditional posterior

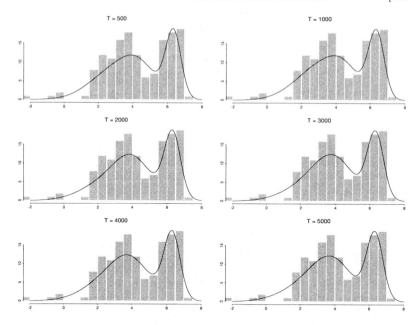

Figure 9.4.1. *Evolution of the estimation of the density (9.4.1) for three components and T iterations of algorithm (9.6.1). The estimations are based on 149 observations of acidity levels for lakes in the American Northeast, used in Crawford et al. (1992) and represented by the histograms.*

distribution of the $\sigma_j$'s. An alternative solution is to keep the identifiability constraints and to use a hybrid Markov chain Monte Carlo algorithm, where, in every $U$ iterations, a random permutation of the values of $z_i$ is generated via a Metropolis–Hastings scheme:

**Algorithm A.51 –Hybrid Allocation Mixture Estimation–**

0. Simulate $Z_i$ $(i = 1, \ldots, n)$.

1. Generate a random permutation $\psi$ on $\{1, 2, \ldots, k\}$ and derive

$$\tilde{z} = (\psi(z_i))_i \quad \text{and} \quad \tilde{\xi} = (\tilde{p}, \tilde{\mu}, \tilde{w}) = (p_{\psi(j)}, \mu_{\psi(j)}, w_{\psi(j)})_j. \quad [A.51]$$

2. Generate $\xi'$ conditionally on $(\tilde{\xi}, \tilde{z})$ by a standard Gibbs sampler iteration.

3. Accept $(\xi', \tilde{z})$ with probability

$$(9.4.6) \qquad \frac{\pi(\xi', \tilde{z})\pi_1(\psi^{-1}(z^-)|\xi')\,\pi_2(\xi|z^-, \psi(\xi'))}{\pi(\xi, z^-)\pi_1(\psi^{-1}(\tilde{z})|\xi)\,\pi_2(\xi'|\tilde{z}, \tilde{\xi})} \wedge 1 \,,$$

where $\pi(\xi, z)$ denotes the distribution of $(\xi, z)$,

$$\pi(\xi, z) \propto \prod_{i=1}^{n} \left( \sum_{j=1}^{k} p_j \frac{e^{-(\mu_j - x_i)^2/(2w_j^2)}}{w_j} \mathbb{I}_{z_i=j} \right) \pi(\xi) \,,$$

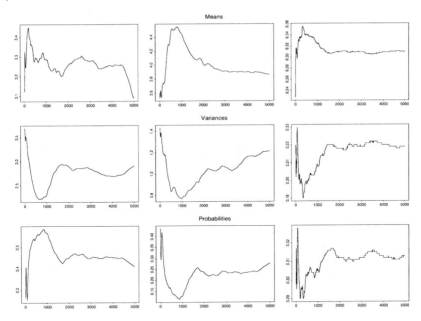

Figure 9.4.2. *Convergence of estimators of the parameters of the mixture (9.4.5) for the same iterations as in Figure 9.4.1.*

where $\pi_1(z|\xi)$ and $\pi_2(\xi'|z,\xi)$ are the conditional distributions used in the Gibbs sampler and $z^-$ is the previous value of the allocation vector. Note that, in (9.4.6), $\pi_1(\psi^{-1}(\tilde{z})|\xi) = \pi_1(\tilde{z}|\tilde{\xi})$.

This additional Metropolis–Hastings step may result in delicate computations in terms of manipulation of indices (see Problem 9.11), but it is of the same order of complexity as an iteration of the Gibbs sampler since the latter also requires the computation of the sums $\sum_{j=1}^{k} p_j \, f(x_i|\xi_j)$, $i = 1,\ldots,n$. The value of $U$ in [A.51] can be arbitrarily fixed (for instance, at 50) and later modified depending on the average acceptance rate corresponding to (9.4.6). A high acceptance rate means that the Gibbs sampler lacks a sufficient number of iterations to produce a stable allocation to homogeneous classes; a low acceptance rate suggests reduction of $U$ so as to more quickly explore the possible permutations.

## 9.5 Extensions

The models of §9.3 can be generalized in several ways. For instance, relaxing the independence assumption between observations can lead to hidden Markov chains (§9.5.1), whereas different constraints on the changes of components corresponds to changepoint models (§9.5.2) or to switching state models (§9.7.1). We do not deal here with the case of infinite mixtures which are related with *nonparametric* modeling nor with their resolution via Dirichlet priors (see Escobar 1994, Escobar and West 1995 or Müller, Erkanli and West 1996).

### 9.5.1 Hidden Markov Chains

A hidden Markov chain results from a type of demarginalization (see §5.3.1), where the augmented variable has a Markov chain structure. For example, the mixture (9.3.1), modeled in §9.3, can be generalized to a hidden Markov model, where we add a Markov dependence structure to $(z_t)$. For the completed model, $X_t^* = (X_t, Z_t)$ depends on $x_{t-1}^*$ through the conditional distribution

$$(9.5.1) \quad Z_t | z_{t-1} \sim P(Z_t = i | z_{t-1} = j) = p_{ji}, \qquad X_t | z_t \sim f(x|\xi_{z_t}).$$

The hidden structure $(Z_t)$ is therefore a Markov chain, with transition matrix $\mathbf{P} = (p_{ji})$, linked with the observations through the sequence of the $\xi_{z_t}$'s. Numerous phenomena can be modeled this way; see Note 9.7.1.

The Gibbs sampler algorithm associated with (9.5.1), using a conjugate prior on the $\xi_j$'s and on the rows of $\mathbf{P}$ (via Dirichlet distributions) is based on the same completion as [A.49]. Although step **2.** remains identical, the completion **1.** is more delicate because of the Markovian structure of the model. The solution used in Robert *et al.* (1993) is to successively generate every indicator $Z_t$, conditional on the parameters and the other indicators $Z_{t'}$ ($t' \neq t$):

**Algorithm A.52 –Hidden Markov Allocation–**
1. Generate[4] $(1 < t < n)$

$$Z_t | z_{j, j \neq t} \sim P(Z_t = u | z_{t-1}, z_{t+1}) \propto p_{z_{t-1} u} p_{u z_{t+1}} f(x_t | \xi_u),$$
$$Z_n | z_{n-1} \sim P(Z_n = u | z_{n-1}) \propto p_{z_{n-1} u} f(x_n | \xi_u). \qquad [A.52]$$

This extension does not fit the data augmentation structure (see §7.2) since $Z = (Z_t)$ is not generated conditionally on $\xi$. However, the *Duality Principle* (see §7.2.4) still applies, thus guaranteeing good convergence properties, at least theoretically. For extensions and generalizations, see Shephard (1994), Robert and Titterington (1998), and Robert *et al.* (1999).

**Example 9.5.1 Switching AR model.** A model studied in the literature (see Albert and Chib 1993a, McCulloch and Tsay 1994, Billio and Monfort 1995, Chib 1995, Billio *et al.* 1999) is the *switching AR* model, an autoregressive model where the mean can switch between two values $\mu_0$ and $\mu_1$. This structure can allow for modeling time series with several regimes or local nonstationarities and can be represented as

$$X_t | x_{t-1}, z_t, z_{t-1} \sim \mathcal{N}(\mu_{z_t} + \varphi(x_{t-1} - \mu_{z_{t-1}}), \sigma^2),$$
$$Z_t | z_{t-1} \sim \rho_{z_{t-1}} \mathbb{I}_{z_{t-1}}(z_t) + (1 - \rho_{z_{t-1}}) \mathbb{I}_{1 - z_{t-1}}(z_t),$$

---

[4] For identifiability reasons, we assume the first state arbitrarily fixed to be 1, when a constraint like (9.4.3) is not already present in the model.

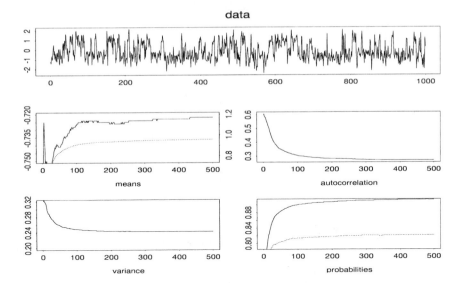

Figure 9.5.1. *Simulated sample of a switching AR model with parameters* $\mu_0 = -0.72$, $\mu_1 = 0.92$, $\varphi = 0.09$, $\sigma = 0.5$, $\rho_0 = 0.89$, $\rho_1 = 0.92$ *(top) and convergence of the approximations of the Bayes estimators of these parameters over 500 iterations (bottom). (The graph for the approximations of the means involves two scales corresponding to* $\mu_0$ *(left) and* $\mu_1$ *(right).)*

where $Z_t$ takes values in $\{0, 1\}$, with initial values $z_0 = 0$ and $x_0 = \mu_0$. The states $z_t$ are not observed and the algorithm completes the sample $x_1, \ldots,$ $x_T$ as in [A.52], namely by simulating every $z_t$ $(1 < t < T)$ from

$$P(Z_t | z_{t-1}, z_{t+1}, x_t, x_{t-1}, x_{t+1}) \propto$$
$$\exp\left\{-\frac{1}{2\sigma^2}\left[(x_t - \mu_{z_t} - \varphi(x_{t-1} - \mu_{z_{t-1}}))^2\right.\right.$$
$$\left.\left.+(x_{t+1} - \mu_{z_{t+1}} - \varphi(x_t - \mu_{z_t}))^2\right]\right\}$$
$$\times \left(\rho_{z_{t-1}}\mathbb{I}_{z_{t-1}}(z_t) + (1 - \rho_{z_{t-1}})\mathbb{I}_{1-z_{t-1}}(z_t)\right)$$
$$\times \left(\rho_{z_t}\mathbb{I}_{z_t}(z_{t+1}) + (1 - \rho_{z_t})\mathbb{I}_{1-z_t}(z_{t+1})\right),$$

with appropriate modifications for the limiting cases $Z_1$ and $Z_T$. Once the indicators $z_i$ are simulated, given a prior such as

$$(\mu_1 - \mu_0) \sim \mathcal{N}(0, \zeta^2), \quad \pi(\varphi, \sigma^2) = 1/\sigma, \quad \rho_0, \rho_1 \sim \mathcal{U}_{[0,1]},$$

the generation of the parameters is straightforward because of the conjugate structure (Problem 9.15).

Figure 9.5.1 illustrates the performance of the algorithm for a simulated sample of 1000 points, for the parameters

$$\mu_0 = -0.72, \quad \mu_1 = 0.92, \quad \varphi = 0.09, \quad \sigma = 0.5, \quad \rho_0 = 0.89, \quad \rho_1 = 0.92,$$

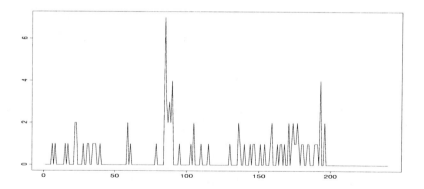

Figure 9.5.2. *Data of Leroux and Putterman (1992) on the number of moves of a lamb fetus during* 240 *successive* 5-*second periods.*

also used in Billio and Monfort (1995). It shows the fast convergence of the approximation to the Bayes estimators of these parameters, which happen to be quite close to the true values.    ‖

**Example 9.5.2 Hidden Markov Poisson model.** Robert and Titterington (1998) consider the estimation of a hidden Markov Poisson model; that is, of observed $x_t$'s depending on an unobserved Markov chain $(Z_t)$ such that $(i, j = 1, 2)$

$$X_t|z_t \sim \mathcal{P}(\lambda_{z_t}), \qquad P(Z_t = i|z_{t_1} = j) = p_{ji} .$$

They use a noninformative approach, with the prior

$$\pi(\lambda_1, \lambda_2, p_{11}, p_{22}) = \frac{1}{\lambda_1} \mathbb{I}_{\lambda_2 < \lambda_1},$$

and derive the maximum likelihood estimator using recursive integration (§5.2.4). They apply this technique to the dataset of Leroux and Putterman (1992), who analyze moves of a lamb fetus (see Figure 9.5.2). The modeling via a Poisson structure on both hidden states leads to the following estimates: $\hat{\lambda}_1 = 2.84$, $\hat{\lambda}_2 = 0.25$, $\hat{p}_{11} = .68$, and $\hat{p}_{22} = .985$. (See Problem 9.16.)    ‖

### 9.5.2 Changepoint models

Another extension of mixture models covers *changepoint models*, where the distribution of a temporal (or causal) phenomenon changes at one (or several) unknown time. A simple version of a changepoint model is based on a sample $(x_1, \ldots, x_n)$ associated with a latent index $\tau$ such that

$$\tau \sim \pi_0(\tau), \quad X_1, \ldots, X_\tau|\tau \overset{iid}{\sim} f(x|\xi_1) , \qquad X_{\tau+1}, \ldots, X_n|\tau \overset{iid}{\sim} f(x|\xi_2),$$

where the support of $\pi_0$ is $\{1, \ldots, n\}$. This model is less complex than a model like (9.3.1) since the completed observations are divided in only two subsamples $(x_1, \ldots, x_\tau)$ and $(x_{\tau+1}, \ldots, x_n)$.

The $X_i$'s are marginally distributed from a mixture with variable weights,

$$X_i \sim P(\tau > i) f(x|\xi_1) + P(\tau \leq i) f(x|\xi_2),$$

and the parameter of interest in these models is $\tau$. One can derive the marginal distribution of $\tau$ by integration, since the likelihood only involves $n$ terms. However, this is typically only possible when the prior distribution on $(\xi_1, \xi_2)$ is conjugate, with independent $\xi_1$ and $\xi_2$. For example, a *hierarchical* prior distribution like the one proposed by Carlin *et al.* (1992) prevents an explicit integration and requires the use of an MCMC algorithm. For instance, if

$$\xi_1, \xi_2 \overset{iid}{\sim} \pi_1(\xi|\alpha), \qquad \alpha \sim \pi_2(\alpha),$$

the corresponding steps of the Gibbs sampler are

**Algorithm A.53 –Hierarchical Changepoint Model–**

1. Generate $\tau$ from

$$\pi_0(\tau|x, \xi_1, \xi_2) \propto \prod_{j=1}^{\tau} f(x_j|\xi_1) \prod_{j=\tau+1}^{n} f(x_j|\xi_2) \, \pi_0(\tau).$$

2. Generate $\xi_i$ $(i = 1, 2)$ from                                                   [A.53]

$$\pi_1(\xi_i|x, \tau, \alpha) \propto \begin{cases} \displaystyle\prod_{j \leq \tau} f(x_j|\xi_1) \, \pi_1(\xi_1|\alpha) & \text{if } i = 1 \\ \displaystyle\prod_{j > \tau} f(x_j|\xi_2) \, \pi_1(\xi_2|\alpha) & \text{if } i = 2. \end{cases}$$

3. Generate $\alpha$ from

$$\pi_2(\alpha|\xi_1, \xi_2) \propto \pi_2(\alpha) \, \pi_1(\xi_1|\alpha) \, \pi_1(\xi_2|\alpha).$$

**Example 9.5.3 Longitudinal data.** The analysis of longitudinal counting data is sometimes associated with a changepoint representation (see Problem 9.19 and Table 9.6.2). The modeling is, for instance, based on a Poisson process

$$Y_i \sim \mathcal{P}(\theta t_i) \ (i \leq \tau), \qquad Y_i \sim \mathcal{P}(\lambda t_i) \qquad (i > \tau),$$

with

$$\theta \sim \mathcal{G}a(\alpha_1, \beta_1), \qquad \lambda \sim \mathcal{G}a(\alpha_2, \beta_2),$$

and

$$\beta_1 \sim \mathcal{G}a(\delta_1, \varepsilon_1), \qquad \beta_2 \sim \mathcal{G}a(\delta_2, \varepsilon_2),$$

where $\alpha_i, \epsilon_i$, and $\delta_i$ are assumed to be known $(i = 1, 2)$. The steps of [A.53] which correspond to these prior distributions are then

**Algorithm A.54 –Changepoint Poisson Model–**

**1. Generate** $(1 \leq k \leq n)$

$$\tau \sim P(\tau = k) \propto \exp \left\{ (\lambda - \theta) \sum_{i=1}^{k} t_i \right\} \exp \left\{ \log(\theta/\lambda) \sum_{i=1}^{k} y_i \right\}.$$

**2. Generate** [A.54]

$$\theta \sim \mathcal{G}a \left( \alpha_1 + \sum_{i=1}^{\tau} y_i, \beta_1 + \sum_{i=1}^{\tau} t_i \right),$$

$$\lambda \sim \mathcal{G}a \left( \alpha_2 + \sum_{i=\tau+1}^{n} y_i, \beta_2 + \sum_{i=\tau+1}^{n} t_i \right).$$

**3. Generate**

$$\beta_1 \sim \mathcal{G}a(\delta_1 + \alpha_1, \theta + \varepsilon_1), \quad \beta_2 \sim \mathcal{G}a(\delta_2 + \alpha_2, \lambda + \varepsilon_2).$$

In the particular case of the data in Table 9.6.2, which summarizes the number of mining accidents in England from 1851 to 1962, Raftery and Akman (1986) propose the hyperparameters $\alpha_1 = \alpha_2 = 0.5$, $\delta_1 = \delta_2 = 0$, and $\varepsilon_1 = \varepsilon_2 = 1$. The different analyses lead to estimates of $k$ located between 1889 and 1892. ‖

**Example 9.5.4 Changepoint regression.** Carlin *et al.* (1992) also study a *changepoint regression model* where

$$Y_i \sim \begin{cases} \mathcal{N}(\alpha_1 + \beta_1 x_i, \sigma_1^2) & \text{for } i = 1, \ldots, \tau \\ \mathcal{N}(\alpha_2 + \beta_2 x_i, \sigma_2^2) & \text{for } i = \tau + 1, \ldots, n, \end{cases}$$

$\tau$ being generated from a uniform distribution on $\{1, \ldots, n\}$. The prior distributions are conjugate,

$$\theta_j = (\alpha_j, \beta_j)^t \sim \mathcal{N}_2(\theta_0, \Sigma), \qquad \sigma_j^2 \sim \mathcal{IG}(\delta_0, \gamma_0) \qquad (j = 1, 2)$$

and

$$\Sigma^{-1} \sim \mathcal{W}_2 \left( \rho, \frac{1}{\rho} V^{-1} \right), \qquad \theta_0 \sim \mathcal{N}(\mu, C),$$

with $\delta_0, \gamma_0, \rho, \mu, V$, and $C$ known. The distribution $\mathcal{W}_p(\nu, A)$ is the Wishart distribution on the space of $p \times p$ symmetric matrices (Problem 2.32). For a given $\tau$, this model can be transformed into a standard normal model since the two parts of the sample factor through the sufficient statistics $(j = 1, 2)$

$$\hat{\theta}_j = (X_j^t X_j)^{-1} X_j^t y^j,$$

$$S_j^2 = (y^j - X_j(X_j^t X_j)^{-1} X_j^t y^j)(y^j - X_j(X_j^t X_j)^{-1} X_j^t y^j)^t$$

(see Robert 1994a, Section 4.3.3), where

$$y^1 = (y_1, \ldots, y_\tau)^t, \quad y^2 = (y_{\tau+1}, \ldots, y_n)^t,$$

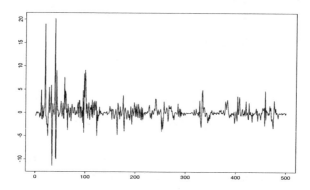

Figure 9.5.3. *Simulated sample of a stochastic volatility process (9.5.2) with $\sigma = 1$ and $\varrho = 0.9$. (Source: Mengersen et al. 1999.)*

$$X_1 = \begin{pmatrix} 1 & \cdots & 1 \\ x_1 & \cdots & x_\tau \end{pmatrix}^t, \quad X_2 = \begin{pmatrix} 1 & \cdots & 1 \\ x_{\tau+1} & \cdots & x_n \end{pmatrix}^t,$$
$$B_j = (\sigma_j^{-2} X_j^t X_j + \Sigma^{-1})^{-1}, \quad \xi_j = \sigma_j^{-2} X_j^t y^j + \Sigma^{-1} \theta_0,$$

which implicitly depend on $\tau$. (See Problem 9.17.)                          ‖

### 9.5.3 Stochastic Volatility

Stochastic volatility models are quite popular in financial applications, especially in describing series with sudden changes in the magnitude of variation of the observed values (see, e.g., Jacquier *et al.* 1994 or Kim *et al.* 1998). They use a latent linear process $(Y_t^*)$, called the *volatility*, to model the variance of the observables $Y_t$ in the following way: Let $Y_0^* \sim \mathcal{N}(0, \sigma^2)$ and, for $t = 1, \ldots, T$, define

(9.5.2)
$$\begin{cases} Y_t^* = \varrho Y_{t-1}^* + \sigma \epsilon_{t-1}^*, \\ Y_t = e^{Y_t^*/2} \epsilon_t, \end{cases}$$

where $\epsilon_t$ and $\epsilon_t^*$ are iid $\mathcal{N}(0,1)$ random variables. Figure 9.5.3 shows a typical stochastic volatility behavior for $\sigma = 1$ and $\varrho = 0.9$.

The observed likelihood $L(\varrho, \sigma | y_0, \ldots, y_T)$ is obtained by integrating the complete-data likelihood

$$L(\varrho, \sigma | y_0, \ldots, y_T) = \mathbb{E}[L^c(\varrho, \sigma | y_0, \ldots, y_T, Y_0^*, \ldots, Y_T^*) | y_0, \ldots, y_T],$$

where the expectation is with respect to the distribution of the latent variables $Y_t^*$. This complete likelihood is available in closed form as

$$L^c(\varrho, \sigma | y_0, \ldots, y_T, y_0^*, \ldots, y_T^*) \propto \exp - \sum_{t=0}^{T} \left\{ y_t^2 e^{-y_t^*} + y_t^* \right\} / 2$$

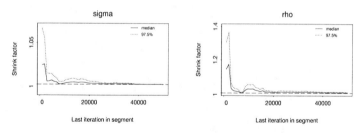

Figure 9.5.4. *Gelman and Rubin's (1992) shrink factors for a stochastic volatility model (9.5.2), with $\varrho = 0.9$ and $\sigma = 0.5$, obtained by* CODA. *(Source: Mengersen et al. 1999.)*

$$(9.5.3) \qquad (\sigma)^{-T+1} \exp - \left\{ (y_0^*)^2 + \sum_{t=1}^{T} (y_t^* - \varrho y_{t-1}^*)^2 \right\} /2(\sigma)^2 .$$

Bayesian inference on this model, as well as likelihood computation based on the approximation (5.3.8), require the simulation of the volatilities $Y_t^*$. This can be done by a Metropolis–Hastings step, based on an approximation of the full conditional

$$f(y_t^* | y_{t-1}^*, y_{t+1}^*, y^t, \varrho, \sigma) \propto \exp \left\{ -(y_t^* - \varrho y_{t-1}^*)^2 /2\sigma^2 \right.$$
$$(9.5.4) \qquad\qquad \left. -(y_{t+1}^* - \varrho y_t^*)^2 /2\sigma^2 - y_t^*/2 - y_t^2 e^{-y_t^*} /2 \right\} .$$

For instance, the expression $y_t^*/2 + y_t^2 e^{-y_t^*}/2$ in the exponential can be approximated by $(y_t^* - \log(y_t^2))^2/2$ (Problem 9.20.) The instrumental distribution is then a normal distribution

$$(9.5.5) \quad \mathcal{N} \left( \frac{[\varrho y_{t-1}^* + \varrho y_{t+1}^*]c^{-2} + \log(y_t^2)/2}{(1 + \varrho^2)c^{-2} + 1/2}, \frac{1}{(1 + \varrho^2)c^{-2} + 1/2} \right),$$

with the appropriate Metropolis–Hastings acceptance probability.[5] In a Bayesian framework, given a non-informative prior $\pi(\varrho, \sigma) = 1/\sigma$, the simulation of $(\varrho, \sigma)$ based on the complete likelihood (9.5.3) is straightforward, since it corresponds to an inverted Gamma distribution on $\sigma^2$ and a truncated normal distribution on $\varrho$.

Mengersen *et al.* (1999) study the performance of various convergence diagnostics in this setup. A first sample of size 500, simulated with $\varrho = 0.9$ and $\sigma = 0.5$, leads to unanimous agreement on convergence after a few thousand iterations.

Figures 9.5.4 and 9.5.5 illustrate this for the criteria of Gelman and Rubin (1992) (§8.3.4) and of Geweke (1992) (§8.6.2). Figure 9.5.6 represents

---

[5] Another solution, based on a Gamma approximation of the distribution of $\omega_t = \exp(-y_t^*)$, can be found in Jacquier *et al.* (1994). See also Kim *et al.* (1998) for a different normal approximation.

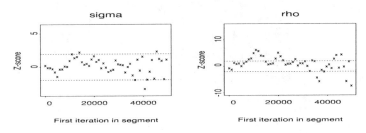

Figure 9.5.5. *Geweke's (1992) convergence indicators for a stochastic volatility model (9.5.2), with* $\varrho = 0.9$ *and* $\sigma = 0.5$, *obtained by* CODA. *(Source: Mengersen et al. 1999.)*

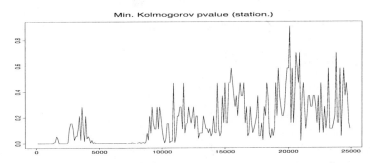

Figure 9.5.6. *Successive values of the minimum p-value evaluating the stationarity of the sample of* $(\varrho^{(t)}, \sigma^{(t)})$ *'s, by comparing both halves, in the case of the stochastic volatility model (9.5.2), with* $\varrho = 0.9$ *and* $\sigma = 0.5$. *(Source: Mengersen et al. 1999.)*

the evolution of the $p$-value of a nonparametric stationarity test (§8.2.2), with a gap between iteration $10,000$ and iteration $15,000$, which indicates that the chain is exploring another region in the support of the stationary distribution. (Note a similar feature in Figure 9.5.4 for the shrink factor on $\varrho$.)

Figure 9.5.7 reproduces the so-called *allocation map*, first introduced in Robert (1997), which represents the simulated volatilities against the iterations by gray levels. The vertical stripes on this image are pointing out a strong stability in the successive values of the $y_t^*$'s. (Since this is a simulated sample, the lower part of Figure 9.5.7 compares the average of the simulated $y_t^*$'s with the true values, showing satisfactory closeness.)

Using the diagnostics of §8.4.1, the method of Raftery and Lewis (1992a) concludes that no more than $20,000$ total iterations are needed, and $\varrho$ and $\sigma$ pass the test of Heidelberger and Welch (1983). The complexity of the kernel prohibits the use of distance evaluations, while the latent variable structure leads to very low renewal probabilities.

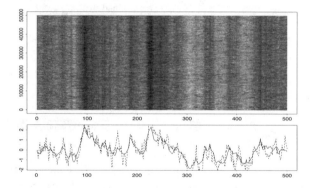

Figure 9.5.7. *Allocation map (top) and average vs. true allocation (bottom) for the missing data $(y_t^*)$ in the case of the stochastic volatility model (9.5.2). (The true allocations are represented by dashes.) (Source: Mengersen et al. 1999.)*

## 9.6 Problems

**9.1** In the setup of Example 9.2.2, check whether the posterior distribution simulated by [A.46] is well defined.

**9.2** Propose an alternative to [A.47] based on the slice sampler.

**9.3** In the setup of Example 9.2.3, assume instead that the missing data structure depends on the social status of the nonrespondent and that the probability of nonresponse is modeled through a logit model,

$$p_{a,s,m} = \exp\{w_0 + \varepsilon_a + \delta_s + \nu_m\}/(1 + \exp\{w_0 + \varepsilon_a + \delta_s + \nu_m\}) \,,$$

where $\varepsilon_1 = \delta_1 = \nu_1 = 0$ for identifiability reasons.

(a) Show that the likelihood of this model is

$$\prod_{\substack{a=1,2 \\ s=1,2 \\ m=1,2}} \frac{e^{r_{a,s,m}(w_0 + \varepsilon_a + \delta_s + \nu_m)}}{[1 + e^{w_0 + \varepsilon_a + \delta_s + \nu_m}]^{n_{a,s,m}}} (\mu_0 + \alpha_a + \beta_s + \gamma_m)^{r_{a,s,m}}$$

$$\times \exp\left\{-r_{a,s,m}\overline{y}_{a,s,m}(\mu_0 + \alpha_a + \beta_s + \gamma_m)\right\}$$

with the notation of Example 9.2.3.

(b) Justify the completion of the model by the logistic variables $z_{a,s,m,i}^*$ such that

$$P(z_{a,s,m,i}^* \leq w_0 + \varepsilon_a + \delta_s + \nu_m) = \frac{\exp(w_0 + \varepsilon_a + \delta_s + \nu_m)}{1 + \exp(w_0 + \varepsilon_a + \delta_s + \nu_m)} \,;$$

that is,

$$z_{a,s,m,i}^* \sim g(z) = \frac{e^{-z}}{(1 + e^{-z})^2} \,.$$

(c) Show that a possible algorithm to generate from the posterior distribution of $(w_0, \varepsilon_2, \delta_2, \nu_2)$ is then

1. For $a = 1, 2$, $s = 1, 2$ and $m = 1, 2$, **generate**

$$z_{a,s,m,i}^* \sim g(z) \, \mathbb{I}_{z < w_0 + \varepsilon_a + \delta_s + \nu_m} \qquad (1 \leq i \leq r_{a,s,m}),$$

$$z_{a,s,m,i}^* \sim g(z) \, \mathbb{I}_{z > w_0 + \varepsilon_a + \delta_s + \nu_m} \qquad (r_{a,s,m} < i \leq n_{a,s,m}).$$

2. Simulate

$$w_0' \sim \mathcal{U}_{[\underline{w}_0(\varepsilon_2,\delta_2,\nu_2),\ \overline{w}_0(\varepsilon_2,\delta_2,\nu_2)]},$$

$$\varepsilon_2' \sim \mathcal{U}_{[\underline{\varepsilon}(w_0',\delta_2,\nu_2),\ \overline{\varepsilon}(w_0',\delta_2,\nu_2)]},$$

$$\delta_2' \sim \mathcal{U}_{[\underline{\delta}(w_0',\varepsilon_2',\nu_2),\ \overline{\delta}(w_0',\varepsilon_2',\nu_2)]},$$

$$\nu_2' \sim \mathcal{U}_{[\underline{\nu}(w_0',\varepsilon_2',\delta_2'),\ \overline{\nu}(w_0',\varepsilon_2',\delta_2')]}.$$

where the lower and upper bounds on the quantities $w_0 + \varepsilon_a + \delta_s + \nu_m$, are derived from the constraints on the $z_{a,s,m,i}^*$'s.

**9.4** In the setup of Example 9.2.3:

(a) Establish the following majorization:

$$\pi(\alpha_2|\mu_0,\beta_2,\gamma_2,y) \propto (\mu_0 + \alpha_2)^{r_{211}} \cdots (\mu_0 + \alpha_2 + \beta_2 + \gamma_2)^{m_{222}}$$
$$\times \exp\{-(m_{211}\overline{y}_{111} + \cdots + m_{222}\overline{y}_{222})\alpha_2\}$$
$$\leq (\mu_0 + \alpha_2 + \gamma_2 + \beta_2)^{m_{211} + \cdots + m_{222}}$$
$$\times e^{-(m_{211}\overline{y}_{211} + \cdots + m_{222}\overline{y}_{222})\alpha_2}$$
$$= (\alpha_2 + \underline{\alpha})^t\ e^{-(\alpha_2 + \underline{\alpha})v}\ e^{\underline{\alpha}v}\ ,$$

with $t = m_{211} + \cdots + m_{222}$ and $v = m_{211}\overline{y}_{211} + \cdots + m_{222}\overline{y}_{222}$.

(b) Deduce that the inference on the parameters $\mu_0$, $\alpha_2$, $\beta_2$, and $\gamma_2$ can be done by Accept–Reject from a truncated Gamma distribution. (*Hint:* Verify that a Gamma $\mathcal{G}a(t+1, v)$ distribution truncated on the left in $\underline{\alpha}$ can be used as the instrumental distribution.)

(c) Describe the additional slice sampling step.

**9.5** A typical setup of *convolution* consists in the observation of

$$Y_t = \sum_{j=0}^{k} h_j Y_{t-j} + \varepsilon_t\ , \qquad t = 1,\ldots,n,$$

where $\varepsilon_t \sim \mathcal{N}(0,\sigma^2)$, the $x_i$'s are unknown and with finite support, $\mathcal{X} = \{s_1,\ldots,s_m\}$, and where the parameters of interest $h_j$ are unknown. The discrete variables $X_t$ are iid with distribution $\mathcal{M}_m(\theta_1,\ldots,\theta_m)$ on $\mathcal{X}$, the probabilities $\theta_i$ are distributed from a Dirichlet prior distribution, $\mathcal{D}_m(\gamma_1,\ldots,\gamma_m)$, and the coefficients $h_j$ are from a conjugate distribution $h = (h_0,\ldots,h_k) \sim \mathcal{N}_{k+1}(\mu,\Sigma)$.

(a) Show that the completed likelihood is

$$L(x,h,\sigma) \propto \sigma^{-n} \exp\left\{-\frac{1}{2\sigma^2} \sum_{t=1}^{n} \left(y_t - \sum_{j=0}^{k} h_j x_{t-j}\right)^2\right\}.$$

(b) Verify that the completion of $Y_t$ in $(Y_t, X_t)$ can be done as follows:

**Algorithm A.55 –Deconvolution Completion–**

1. Generate $(t = 1 - k,\ldots,n)$

$$X_t \sim P(x_t = s_\ell) \propto \frac{\theta_\ell s_\ell}{\sigma^2} \exp\left\{-s_\ell^2 \sum_{j=0}^{k} h_j^2/2\sigma^2\right.$$

$$+ \sum_{j=0}^{k} h_j \left( y_{t+j} - \sum_{\substack{i=0 \\ i \neq j}}^{k} h_i x_{t+j-i} \right) \Bigg) \Bigg\}$$

and derive                                                                                      [A.55]

$$X = \begin{pmatrix} x_1 & x_0 & \cdots & x_{1-k} \\ & & \cdots & \\ x_n & x_{n-1} & \cdots & x_{n-k} \end{pmatrix} , \qquad \hat{h} = (X^t X)^{-1} X^t y ,$$

where $y = (y_1, \ldots, y_n)$.

(c) Verify that the update of the parameters $(h, \sigma^2)$ is given by

**2. Generate**

$$h \sim \mathcal{N}_{k+1} \left( (\Sigma^{-1} + \sigma^{-2} X^t X)^{-1} (\Sigma^{-1} \mu + \sigma^{-2} X^t X \hat{h}), (\Sigma^{-1} + \sigma^{-2} X^t X)^{-1} \right),$$

$$\sigma^2 \sim \mathcal{IG} \left( \frac{n}{2}, \frac{1}{2} \sum_{t=1}^{n} \left( y_t - \sum_{j=0}^{k} h_j x_{t-j} \right)^2 \right) ,$$

while the step for hyperparameters $\theta_\ell$ $(\ell = 1, \ldots, m)$ corresponds to

**3. Generate**

$$(\theta_1, \ldots, \theta_m) \sim \mathcal{D} \left( \gamma_1 + \sum_{t=1-k}^{n} \mathbb{I}_{x_t = s_1}, \ldots, \gamma_m + \sum_{t=1-k}^{n} \mathbb{I}_{x_t = s_m} \right)$$

(*Note:* See Kong et al. 1994, Gassiat 1995, Liu and Chen 1995, and Gamboa and Gassiat 1997 for more detailed introductions.)

**9.6** (Diebolt and Robert 1994))

(a) Show that, for a mixture of distributions from exponential families, there exist conjugate priors.

(b) For the conjugate priors, show that the posterior expectation of the mean parameters of the components can be written in a closed form.

(c) Show that the convergence to stationarity is geometric for all the chains involved in the Gibbs sampler for the mixture model.

(d) Show that Rao–Blackwellization applies in the setup of normal mixture models and that it theoretically improves upon the naive average.

(*Hint:* Use the Duality Principle.)

**9.7** In the setup of Example 9.3.2, show that the likelihood function is not bounded and deduce that, formally, there is no maximum likelihood estimator. (*Hint:* Take $\mu_{j_0} = x_{i_0}$ and let $\tau_{j_0}$ go to 0.)

**9.8** A model consists in the partial observation of normal vectors $Z = (X, Y) \sim \mathcal{N}_2(0, \Sigma)$ according to a mechanism of random censoring. The corresponding data are given in Table 9.6.1.

(a) Show that inference can formally be based on the likelihood

$$\prod_{i=1}^{3} \left\{ |\Sigma|^{-1/2} e^{-z_i^t \Sigma^{-1} z_i / 2} \right\} \sigma_1^{-2} e^{-(x_4^2 + x_5^2)/2\sigma_1^2} \sigma_2^{-3} e^{-(y_6^2 + y_7^2 + y_8^2)/2\sigma_2^2}.$$

| $x$ | 1.17 | $-0.98$ | 0.18 | 0.57 | 0.21 | $-$ | $-$ | $-$ |
|---|---|---|---|---|---|---|---|---|
| $y$ | 0.34 | $-1.24$ | $-0.13$ | $-$ | $-$ | $-0.12$ | $-0.83$ | 1.64 |

Table 9.6.1. *Independent observations of* $Z = (X, Y) \sim \mathcal{N}_2(0, \Sigma)$ *with missing data (denoted* $-$*).*

(b) Show that the choice of the prior distribution $\pi(\Sigma) \propto |\Sigma|^{-1}$ leads to difficulties given that $\sigma_1$ and $\sigma_2$ are isolated in the likelihood.

(c) Show that the missing components can be simulated through the following algorithm:

**Algorithm A.56 –Normal Completion–**

1. Simulate

$$X_i^\star \sim \mathcal{N}\left(\rho\frac{\sigma_1}{\sigma_2}y_i, \sigma_1^2(1-\rho^2)\right) \qquad (i = 6, 7, 8),$$

$$Y_i^\star \sim \mathcal{N}\left(\rho\frac{\sigma_2}{\sigma_1}x_i, \sigma_2^2(1-\rho^2)\right) \qquad (I = 4, 5).$$

2. Generate

$$\Sigma^{-1} \sim \mathcal{W}_2(8, X^{-1}),$$

with $X = \sum_{i=1}^{8} z_i^\star z_i^{\star t}$, the dispersion matrix of the completed data.

to derive the posterior distribution of the quantity of interest, $\rho$.

(d) Propose a Metropolis–Hastings alternative based on a slice sampler.

**9.9** Consider the normal mixture model (9.4.2) under the prior (9.4.4).

(a) Give the prior distribution under the original parameterization.

(b) Show that the Gibbs sampler can be used to simulate the parameters of the original parameterization.

**9.10** The modification of [A.50] corresponding to the reparameterization discussed in Example 9.4.1 only involves a change in the generation of the parameters. In the case $k = 2$, show that it is given by

Simulate

$$p|\mu, \theta, \sigma, \tau \sim \mathcal{B}e(n_1 + 1, n_2 + 1);$$

$$\theta|\mu, \sigma, \tau, p \sim \mathcal{N}\left(\frac{(\bar{x}_2 - \mu)}{\tau}\left[1 + \frac{\sigma^2\zeta^{-2}}{n_2}\right]^{-1}, \frac{\sigma^2}{n_2 + \sigma^2\xi^{-2}}\right);$$

$$\mu|\theta, \sigma, \tau, , p \sim \mathcal{N}\left(\frac{n_1\bar{x}_1 + \sigma^{-2}n_2(\bar{x}_2 - \tau\theta)}{n_1 + n_2\sigma^{-2}}, \frac{\tau^2\sigma^2}{n_2\sigma^2 + n_2}\right);$$

$$(9.6.1) \quad \sigma^{-2}|\mu, \theta, \tau, p \sim \mathcal{G}a\left(\frac{n_2 - 1}{2}, \frac{s_2^2 + n_2(\bar{x}_2 - \mu - \tau\theta)^2}{2\tau^2}\right) \mathbb{I}_{\sigma^2 < 1};$$

$$\tau^{-2}|\mu, \theta, \sigma, p \sim \mathcal{G}a\left(\frac{n}{2}, \frac{1}{2}\left(s_1^2 + n_1(\bar{x}_1 - \mu)^2 + \frac{s_2^2}{\sigma^2} + \frac{n_2(\bar{x}_2 - \mu)^2}{n_2\zeta^2 + \sigma^2}\right)\right).$$

**9.11** Show that if the distribution of $\sigma^2$ is of the form

$$\sigma^{-\alpha} \exp(-\beta\sigma^2) \left\{ \mathbb{I}_{[0,1]}(\sigma) + \sigma^{-2} \mathbb{I}_{[0,1]^c}(\sigma) \right\},$$

an Accept–Reject algorithm based on the inverse Gamma distribution is available.

**9.12** (Roeder and Wasserman 1997) In the setup of normal mixtures of Example 9.3.1:

(a) Derive the posterior distribution associated with the prior $\pi(\mu, \tau)$, where the $\tau_j^2$'s are inverted Gamma $\mathcal{IG}(\nu, A)$ and $\pi(\mu_j | \mu_{j-1}, \tau)$ is a left-truncated normal distribution

$$\mathcal{N}\left(\mu_{j-1}, B(\tau_j^{-2} + \tau_{j-1}^{-2}))^{-1}\right),$$

except for $\pi(\mu_1) = 1/\mu_1$. Assume that the constant $B$ is known and $\pi(A) = 1/A$.

(b) Show that the posterior is always proper.

(c) Derive the posterior distribution using the noninformative prior of (9.4.4), and compare.

**9.13** (Gruet et al. 1999) As in §9.4, consider a reparameterization of a mixture of exponential distributions, $\sum_{j=1}^{k} p_j \, \mathcal{E}xp(\lambda_j)$.

(a) Use the identifiability constraint, $\lambda_1 \geq \lambda_2 \geq \cdots \geq \lambda_k$, to write the mixture with the "cascade" parameterization

$$p \, \mathcal{E}xp(\lambda) + \sum_{j=1}^{k-1} (1-p) \cdots q_j \, \mathcal{E}xp(\lambda\sigma_1 \cdots \sigma_j)$$

with $q_{k-1} = 1$, $\sigma_1 \leq 1, \ldots, \sigma_{k-1} \leq 1$.

(b) For the prior distribution $\pi(\lambda) = \frac{1}{\lambda}$, $p, q_j \sim \mathcal{U}_{[0,1]}$, $\sigma_j \sim \mathcal{U}_{[0,1]}$, show that the corresponding posterior distribution is always proper.

(c) For the prior of part (b), show that [A.49] leads to the following algorithm:

**Algorithm A.57 –Exponential Mixtures–**

2. Generate

$$\lambda \sim \mathcal{G}a\left(n, n_0\bar{x}_0 + \sigma_1 n_1\bar{x}_1 + \cdots + \sigma_1 \cdots \sigma_{k-1} n_{k-1}\bar{x}_{k-1}\right),$$

$$\sigma_1 \sim \mathcal{G}a\left(n_1 + \cdots + n_{k-1}, \lambda\{n_1\bar{x}_1 + \sigma_2 n_2\bar{x}_2 + \cdots + \sigma_2 \cdots \sigma_{k-1} n_{k-1}\bar{x}_{k-1}\}\right) \mathbb{I}_{\sigma_1 \leq 1},$$

$$\vdots$$

$$\sigma_{k-1} \sim \mathcal{G}a\left(n_{k-1}, \lambda n_{k-1}\bar{x}_{k-1}\right) \mathbb{I}_{\sigma_{k-1} \leq 1},$$

$$p \sim \mathcal{B}e(n_0 + 1, n - n_0 + 1),$$

$$\vdots$$

$$q_{k-2} \sim \mathcal{B}e(n_{k-2} + 1, n_{k-1} + 1),$$

where $n_0, n_1, \ldots, n_{k-1}$ denote the size of subsamples allocated to the components $\mathcal{E}xp(\lambda), \mathcal{E}xp(\lambda\sigma_1), \ldots, \mathcal{E}xp(\lambda\sigma_1 \ldots \sigma_{k-1})$ and $n_0\bar{x}_0, n_1\bar{x}_1, \ldots, n_{k-1}\bar{x}_{k-1}$ are the sums of the observations allocated to these components.

**9.14** Consider the following mixture of uniforms,[6]

$$p\mathcal{U}_{[\lambda,\lambda+1]} + (1-p)\mathcal{U}_{[\mu,\mu+1]},$$

and an ordered sample $x_1 \leq \cdots \leq x_{n_1} \leq \cdots \leq x_n$ such that $x_{n_1} + 1 < x_{n_1+1}$.

(a) Show that the chain $(\lambda^{(t)}, \mu^{(t)})$ associated with the Gibbs sampler corresponding to [A.49] is not irreducible.

(b) Show that the above problem disappears if $\mathcal{U}_{[\mu,\mu+1]}$ is replaced with $\mathcal{U}_{[\mu,\mu+0.5]}$ (for $n$ large enough).

**9.15** For the situation of Example 9.5.1, verify that the switching $AR$ parameters can be estimated with the following algorithm:

Generate

$$\mu_0 \sim \mathcal{N}\left([n_{00}(1-\varphi)(\overline{y}_{00} - \varphi\tilde{y}_{00}) + n_{01}\varphi(\mu_1 - \overline{y}_{01} + \varphi\tilde{y}_{01})\right.$$
$$+n_{10}(\varphi\mu_1 + \overline{y}_{10} - \varphi\tilde{y}_{10}) + \zeta^{-2}\sigma^2\mu_1\left][n_{00}(1-\varphi)^2 + n_{01}\varphi^2\right.$$
$$\left.+n_{10} + \zeta^2\sigma^2\right]^{-1}, \sigma^2[n_{00}(1-\varphi)^2 + n_{01}\varphi^2 + n_{10} + \zeta^2\sigma^2]^{-1}\right),$$

$$\mu_1 \sim \mathcal{N}\left([n_{11}(1-\varphi)(\overline{y}_{11} - \varphi\tilde{y}_{11}) + n_{10}\varphi(\mu_0 - \overline{y}_{10} + \varphi\tilde{y}_{10})\right.$$
$$+n_{01}(\varphi\mu_0 + \overline{y}_{01} - \varphi\tilde{y}_{01}) + \zeta^{-2}\sigma^2\mu_0\left][n_{11}(1-\varphi)^2 + n_{10}\varphi^2\right.$$
$$\left.+n_{01} + \zeta^2\sigma^2\right]^{-1}, \sigma^2[n_{11}(1-\varphi)^2 + n_{10}\varphi^2 + n_{01} + \zeta^2\sigma^2]^{-1}\right),$$

$$\varphi \sim \mathcal{N}\left(\frac{\sum_{t=1}^T(y_t - \mu_{z_t})(y_{t-1} - \mu_{z_{t-1}})}{\sum_{t=1}^T(y_{t-1} - \mu_{z_{t-1}})^2}, \frac{\sigma^2}{\sum_{t=1}^T(y_{t-1} - \mu_{z_{t-1}})^2}\right),$$

$$\sigma^2 \sim \mathcal{IG}\left(\frac{n+1}{2}, \frac{1}{2}\sum_{t=1}^T(y_t - \mu_{z_t} - \varphi(y_{t-1} - \mu_{z_{t-1}}))^2\right),$$

$$\rho_0 \sim \mathcal{B}e(n_{00}+1, n_{01}+1),$$
$$\rho_1 \sim \mathcal{B}e(n_{11}+1, n_{10}+1),$$

where $n_{ij}$ denotes the number of jumps from $i$ to $j$ and

$$n_{ij}\overline{y}_{ij} = \sum_{\substack{z_{t-1}=i \\ z_t=j}} y_t, \qquad n_{ij}\tilde{y}_{ij} = \sum_{\substack{z_{t-1}=i \\ z_t=j}} y_{t-1}.$$

**9.16** Consider a hidden Markov Poisson distribution as in Example 9.5.2, such that, for a state $j$, the associated prior distribution on $\lambda_j$ is the Gamma distribution $\mathcal{G}a(\alpha_j, \beta_j)$. Show that step 2. of the Gibbs sampler corresponding to [A.52] is then

**Algorithm A.58 –Poisson Hidden Markov Model–**
Generate $(j = 1, \ldots, k)$

$$\lambda_j \sim \mathcal{G}a(\alpha_j + n_j\overline{x}_j, \beta_j + n_j)$$

and

$$p_j \sim \mathcal{D}(\gamma_1 + n_{j1}, \ldots, \gamma_k + n_{jk}),$$

---

[6] This problem was suggested by Eric Moulines.

where

$$n_{ji} = \sum_{t=2}^{n} \mathbb{I}_{z_t=i}\mathbb{I}_{z_{t-1}=j}, \quad n_j = \sum_{t=1}^{n} \mathbb{I}_{z_t=j} \text{ and } n_j\overline{x}_j = \sum_{t=1}^{n} \mathbb{I}_{z_t=j}x_t.$$

**9.17** For the changepoint regression model of Example 9.5.4, verify that the steps of the associated Gibbs sampler are

### Algorithm A.59 –Changepoint Regression Model–

1. Simulate $(j = 1, 2)$

$$\theta_j \sim \mathcal{N}_2(B_j\xi_j, B_j) .$$

2. Simulate

$$\sigma_1^2 \sim \mathcal{IG}\left(\delta_0 + \frac{\tau}{2}, \gamma_0 + \frac{1}{2}(y^1 - X_1\theta_1)^t(y^1 - X_1\theta_1)\right) ,$$

$$\sigma_2^2 \sim \mathcal{IG}\left(\delta_0 + \frac{n-\tau}{2}, \gamma_0 + \frac{1}{2}(y^2 - X_2\theta_2)^t(y^2 - X_2\theta_2)\right) .$$

3. Simulate

$$\theta_0 \sim \mathcal{N}_2\left((2\Sigma^{-1} + C^{-1})^{-1}\{\Sigma^{-1}(\theta_1 + \theta_2) + C^{-1}\mu\}, \right.$$
$$\left. (2\Sigma^{-1} + C^{-1})^{-1}\right) .$$

4. Simulate                                                                     [A.59]

$$\Sigma^{-1} \sim \mathcal{W}_2\left(\rho + 2, \{(\theta_1 - \theta_0)(\theta_1 - \theta_0)^t + (\theta_2 - \theta_0)(\theta_2 - \theta_0)^t + \rho V\}^{-1}\right) .$$

5. Simulate

$$\tau \sim P(\tau = k) \propto \exp\left\{-\frac{(y^1 - X_1\theta_1)^t(y^1 - X_1\theta_1)}{2\sigma_1^2}\right.$$
$$\left. -\frac{(y^2 - X_2\theta_2)^t(y^2 - X_2\theta_2)}{2\sigma_2^2}\right\} \frac{1}{\sigma_1^k\sigma_2^{n-k}} .$$

**9.18** Show that a semi-Markov chain, as introduced in Note 9.7.1, where the duration in each state is a geometric random variable is, in fact, a Markov chain.

**9.19** Following Example 9.5.3, analyze the dataset on the number of mining accidents in England treated in Carlin *et al.* (1992) and given in Table 9.6.2. In particular, follow the modeling of Raftery and Akman (1986), who propose the hyperparameters $\alpha_1 = \alpha_2 = 0.5$, $\delta_1 = \delta_2 = 0$, and $\varepsilon_1 = \varepsilon_2 = 1$. (*Hint:* Check first that the posterior distribution is well defined.)

**9.20** For the stochastic volatility model (9.5.3):

(a) Examine the identifiability of the parameters.

(b) When $\pi(\varrho, \sigma) = 1/(\sigma)$, show that the posterior distribution is well defined.

(c) Verify that (9.5.4) is the conditional distribution of $Y_t^*$ given $y_{t-1}^*, y_{t+1}^*, t$.

(d) Referring to (9.5.4), compute the second-order Taylor expansion of $y_t^*/2 + y_t^2 e^{-y_t^*}/2$.

| Decade/Year | 0 | 1 | 2 | 3 | 4 | 5 | 6 | 7 | 8 | 9 |
|:---:|:---:|:---:|:---:|:---:|:---:|:---:|:---:|:---:|:---:|:---:|
| 1850 | − | 4 | 5 | 4 | 1 | 0 | 4 | 3 | 4 | 0 |
| 1860 | 6 | 3 | 3 | 4 | 0 | 2 | 6 | 3 | 3 | 5 |
| 1870 | 4 | 5 | 3 | 1 | 4 | 4 | 1 | 5 | 5 | 3 |
| 1880 | 4 | 2 | 5 | 2 | 2 | 3 | 4 | 2 | 1 | 3 |
| 1890 | 2 | 2 | 1 | 1 | 1 | 1 | 3 | 0 | 0 | 1 |
| 1900 | 0 | 1 | 1 | 0 | 0 | 3 | 1 | 0 | 3 | 2 |
| 1910 | 2 | 0 | 1 | 1 | 1 | 0 | 1 | 0 | 1 | 0 |
| 1920 | 0 | 0 | 2 | 1 | 0 | 0 | 0 | 1 | 1 | 0 |
| 1930 | 2 | 3 | 3 | 1 | 1 | 2 | 1 | 1 | 1 | 1 |
| 1940 | 2 | 4 | 2 | 0 | 0 | 0 | 1 | 4 | 0 | 0 |
| 1950 | 0 | 1 | 0 | 0 | 0 | 0 | 0 | 1 | 0 | 0 |
| 1960 | 1 | 0 | 1 | − | − | − | − | − | − | − |

Table 9.6.2. *Yearly number of mining accidents in England for the years* 1851 *to* 1962, *from Maguire et al.* (1952) *and Jarrett* (1979) *datasets.* (*Source: Carlin et al.* 1992.)

(e) Given the full conditional (9.5.4), propose a normal approximation based on the representation

$$\log(Y_t) = Y_t^*/2 + \log(\epsilon_t),$$

which takes into account the bias in mean and variance of $\log(\epsilon_t)$, and compare with the proposal (9.5.5). (*Note:* See Shephard and Pitt 1997 for another proposal.)

**9.21** (Zeger and Karim 1991) For the generalized linear model (GLM) of Note 9.7.3:

(a) Show that the Gibbs sampler is given by

**Algorithm A.60 –Generic Gibbs Sampler for GLMs–**

1. Simulate $\beta$ from

$$\pi(\beta|b) \propto \exp\left(\sum_{i,j} \{y_j\theta_{ij}(\beta,b_i) - \psi(\theta_{ij}(\beta,b_i))\}\right).$$

2. Simulate                                                                              [A.60]

$$\Sigma^{-1} \sim \mathcal{W}_q\left(I,\left(\sum_i b_i b_i'\right)^{-1}\right).$$

3. Simulate $b_i$ $(i = 1, \ldots, I)$ from

$$\pi_i(b_i|\beta,\Sigma) \propto \exp\left(\sum_{j=1}^{n_i} \{y_{ij}\theta_{ij}(\beta,b_i) - \psi(\theta_{ij}(\beta,b_i))\}\right)$$
$$\times \exp\left\{-b_i'\Sigma^{-1}b_i/2\right\},$$

where $\theta_{ij}(\beta,b_i)$ is the transform of $x_{ij}'\beta + z_{ij}'b_i$ such that $h(\psi'(\theta_{ij})) = x_{ij}'\beta + z_{ij}'b_i$.

(b) Show that Steps 1. and 3. can be replaced by Metropolis–Hastings steps, where $\beta$ and the $b_i$'s are simulated from instrumental distributions $\tilde{\pi}$ and $\tilde{\pi}_i$ and accepted with the probabilities

$$\frac{\pi(\beta^*)/\pi(\beta)}{\tilde{\pi}(\beta^*|\beta)/\tilde{\pi}(\beta|\beta^*)} \wedge 1 \quad \text{and} \quad \frac{\pi_i(b_i^*)/\pi_i(b_i)}{\tilde{\pi}_i(b_i^*|b_i)/\tilde{\pi}_i(b_i|b_i^*)} \wedge 1,$$

respectively, where $\beta^*$ and $b_i^*$ denote the simulated values.

**9.22** For the generalized linear model of Note 9.7.3, consider specializing (9.7.2) to the case of a Poisson distribution, $Y_{ij}|\lambda_{ij} \sim \mathcal{P}(\lambda_{ij})$, with $\log(\lambda_{ij}) = (\alpha + a_i)x_{ij} + \beta$, $\alpha, \beta \in \mathbb{R}$, and $a_i \sim \mathcal{N}(0, \sigma^2)$ being the random effect.

(a) Show that Step 1. of [A.60] can be decomposed into

    1. Simulate

$$\alpha \sim \pi_{11}(\alpha) \propto \exp\left\{\alpha \sum_{i,j} x_{ij}y_{ij} - e^\beta \sum_{i,j} e^{(\alpha+a_i)x_{ij}}\right\},$$

$$\beta \sim \pi_{12}(\beta) \propto \exp\left\{\beta \sum y_{ij} - e^\beta \sum_{i,j} e^{(\alpha+a_i)x_{ij}}\right\}$$

(b) Show that Step 3. corresponds to
    3. Simulate $(i = 1, \ldots, I)$

$$a_i \sim \pi_{3i}(a_i) \propto \exp\left\{a_i \sum_j x_{ij}y_{ij} - e^\beta \sum_j e^{(\alpha+a_i)x_{ij}} - a_i^2/(2\sigma^2)\right\}.$$

(c) Show that the distribution of $\beta$ in Step 1. has

$$\hat{\beta} = \log\left(\sum_{i,j} y_{ij} \Big/ \sum_{i,j} \exp\{(\alpha+a_i)x_{ij}\}\right)$$

as its mode, and a possible instrumental distribution is, therefore, $\mathcal{N}(\hat{\beta}, \tau_\beta^2)$.

(d) Although analytical maximization of the distribution of $\alpha$ in Step 1. is not possible, show that $\alpha$ can be simulated from a random walk with variance $\tau_\alpha^2$, where $\tau_\alpha^2$ is calibrated through the acceptance rate.

(e) Show that

$$\exp\left\{a_i \sum_j x_{ij}y_{ij} - a_i^2/2\sigma^2 - e^\beta \sum_j e^{(\alpha+a_i)x_{ij}}\right\}$$

$$\leq \exp\left\{a_i \sum_j x_{ij}y_{ij} - a_i^2/(2\sigma^2)\right\};$$

so the distribution of $a_i$ can be simulated using an Accept–Reject algorithm, and that the normal distribution $\mathcal{N}(\sigma^2 \sum_j x_{ij}y_{ij}, \sigma^2)$ can be used as the instrumental distribution.

**9.23** (McCulloch 1997) For the generalized linear mixed model of Note 9.7.3, take $n_i = 1$ so the link function is $h(\xi_i) = x_i'\beta + z_i'\mathbf{b}$, and further assume that $\mathbf{b} = (b_1, \ldots, b_I)$ where $\mathbf{b} \sim f_\mathbf{b}(\mathbf{b}|D)$. (Here, we assume $\varphi$ to be unknown.)

(a) Show that the usual (incomplete-data) likelihood is

$$L(\theta, \varphi, D|y) = \int \prod_{i=1}^{n} f(y_i|\theta_i) f_{\mathbf{b}}(\mathbf{b}|D) d\mathbf{b} \ .$$

(b) Denote the complete data by $\mathbf{w} = (\mathbf{y}, \mathbf{b})$, and show that

$$\log L_W = \sum_{i=1}^{n} \log f(y_i|\theta_i) + \log f_{\mathbf{b}}(b_i|D) \ .$$

(c) Show that the EM algorithm, given by

1. Choose starting values $\beta^{(0)}$, $\varphi^{(0)}$, and $D^{(0)}$.
2. Calculate (expectations evaluated under $\beta^{(m)}$, $\varphi^{(m)}$, and $D^{(m)}$)

   $\beta^{(m+1)}$ and $\varphi^{(m+1)}$, which maximize $\mathbb{E}[\log f(y_i|\theta_i, u, \beta, \varphi)|y]$.
3. $D^{(m+1)}$ maximizes $\mathbb{E}[f_{\mathbf{b}}(\mathbf{b}|D)|y]$.
4. Set $m$ to $m+1$.

converges to the MLE.

(d) The expectations in part (c) are difficult to evaluate and require the Monte Carlo EM algorithm (§5.3.4). This latter algorithm has a difficulty in the generation of $\mathbf{b}$. Show that the Metropolis–Hastings algorithm

1. Set $\mathbf{b}$ equal to the previous draw.
2. Generate $b_k^*$ from the $k$th component of the candidate distribution and accept $\mathbf{b}^* = (b_1, \ldots, b_{k-1}, b_k^*, b_{k+1}, \ldots, b_I)$ with probability

$$\min \left\{ \frac{f_{\mathbf{b}|\mathbf{y}}(\mathbf{b}^*|\mathbf{y}, \beta, \varphi, D) f_{\mathbf{b}}(\mathbf{b}|D)}{f_{\mathbf{b}|\mathbf{y}}(\mathbf{b}|\mathbf{y}, \beta, \varphi, D) f_{\mathbf{b}}(\mathbf{b}^*|D)}, 1 \right\} \ .$$

will produce random variables $\mathbf{b}$ with the desired distribution.

(e) Show that if the candidate distribution is, in fact, $f_{\mathbf{b}}(\mathbf{b}|D)$ (which is easy to compute if this is a normal distribution), then the acceptance probability becomes

$$\min \left\{ \frac{\prod_{i=1}^{I} f(y_i|\theta_i) f_{\mathbf{b}}(\mathbf{b}^*|D)}{\prod_{i=1}^{I} f(y_i|\theta_i) f_{\mathbf{b}}(\mathbf{b}|D)}, 1 \right\}$$

which is particularly easy to compute.

**9.24** (Billio, Monfort and Robert 1998) A *dynamic desequilibrium* model is defined as the observation of

$$Y_t = \min(Y_{1t}^*, Y_{2t}^*),$$

where the $Y_{it}^*$ are distributed from a parametric joint model, $f(y_{1t}^*, y_{2t}^*)$.

(a) Give the distribution of $(Y_{1t}^*, Y_{2t}^*)$ conditional on $Y_t$.

(b) Show that a possible completion of the model is to first draw the regime (1 versus 2) and then draw the missing component.

(c) Show that when $f(y_{1t}^*, y_{2t}^*)$ is Gaussian, the above steps can be implemented without approximation.

## 9.7 Notes

### 9.7.1 References on Hidden Markov Models

Hidden Markov models and their generalizations have enjoyed an extremely widespread use in applied problems. For example, in Econometrics, regression models and switching time series are particular cases of (9.5.1) (see Goldfeld and Quant 1973, Albert and Chib 1993a, McCulloch and Tsay 1994, Shephard 1994, Billio and Monfort 1995, Chib 1996). Hidden Markov chains also appear in character and speech processing (Juang and Rabiner 1991, Stuchlik *et al.* 1994), in medicine (Celeux and Clairambault 1992, Guihenneuc-Jouyaux *et al.* 1998), in genetics (Churchill 1989, 1995, Muri *et al.* 1998), in engineering (Cocozza-Thivent and Guédon 1990) and in neural networks (Juang 1984). Chib (1996) proposes an alternative approach to [A.52]. The missing data vector is simulated from the conditional distribution $g(z|\xi, x)$. (See also Archer and Titterington 1995 and Robert *et al.* 1998.)

Decoux (1997) deals with the extension to *hidden semi-Markov chains*, where the observations remain in a given state during a random number of epochs following a Poisson distribution and then move to another state (see Problem 9.18). Guihenneuc-Jouyaux and Richardson (1996) also propose a Markov chain Monte Carlo algorithm for the processing of a Markov process on a finite state-space which corresponds to successive degrees of seropositivity.

### 9.7.2 More on Changepoint Models

As mentioned in §9.5.2, changepoint models are an extension of hidden Markov models. Barry and Hartigan (1992, 1993) study an even more complex model where the number $\nu$ of changepoints is itself random (see also Green 1995). They introduce the notion of a *partition product*, assuming that the probability that $\nu$ is equal to $k$ and that the changepoints are in $1 < i_1, \ldots, i_k < n$, this is given by

$$\rho(k, i_1, \ldots, i_k) \propto c_{1i_1} c_{i_1 i_2} \ldots c_{i_k n},$$

where the *cohesions* $c_{ij}$ are known (or may depend on a small number of parameters). Assume, in addition, that the distributions of the parameters corresponding to $(i_1, \ldots, i_k)$, $\xi_1, \ldots, \xi_{k+1}$, depend on $\xi$, in the constrained form

$$\xi_{i_j} \sim f_{i_{j-1} i_j}(\xi)$$

for $x_i \sim f(x|\xi_{i_j})$ when $i_{j-1} \leq i < i_j$ with $i_0 = 1$ and $i_{k+1} = n$.

The cohesion structure used in Barry and Hartigan (1992) is

$$c_{ij} = \begin{cases} (j-i)^{-3} & \text{for } 1 < i < j < n \\ (j-i)^{-2} & \text{for } i = 1 \text{ or } j = n \\ n^{-1} & \text{for } i = 1 \text{ and } j = n, \end{cases}$$

while Yao (1984) proposed a geometric dependence. Given $0 < p < 1$,

$$c_{ij} = (1-p)^{j-i-1} p \quad (j < n), \qquad c_{in} = (1-p)^{n-i-1}.$$

The distributions on $X_i$ and $\xi_{i_j}$ are normal $(i_{j-1} \leq i < i_j)$

$$X_i \sim \mathcal{N}(\xi_i, \sigma^2), \qquad \xi_{i_j} \sim \mathcal{N}\left(\mu_0, \frac{\sigma_0^2}{i_j - i_{j-1}}\right),$$

while the prior distributions on the parameters $\sigma$, $\sigma_0$, $\mu_0$, and $p$ are

$$\pi(\sigma^2) = 1/\sigma^2, \qquad \frac{\sigma_0^2}{\sigma_0^2 + \sigma^2}\bigg|\sigma \sim \mathcal{U}_{[0,w_0]}, \qquad \pi(\mu_0) = 1 \text{ and } p \sim \mathcal{U}_{[0,p_0]},$$

with $p_0 \leq 1$ and $w_0 \leq 1$. The equivalent of [A.54] is then deduced from the posterior distribution of $\nu, i_1, \ldots, i_\nu, \xi_{i_1}, \ldots, \xi_{i_\nu}, \xi_n$,

$$\pi(\nu, i_1, \ldots, \xi_n | \sigma, \sigma_0, p, \mu_0, x_1, \ldots, x_n) \propto c_{1 i_1} \ldots c_{i_\nu n}$$

$$\times \exp\left\{ -\frac{(i_1 - 1)(\xi_{i_1} - \mu_0)^2}{2\sigma_0^2} \right\} \cdots \exp\left\{ -\frac{(n - i_\nu)(\xi_n - \mu_0)^2}{2\sigma_0^2} \right\}$$

$$\times \prod_{j=1}^{\nu+1} \prod_{i_{j-1} \leq i < i_j} \exp\left\{ -\frac{(x_i - \xi_{i_j})^2}{2\sigma^2} \right\}$$

$$\times (i_1 - 1)^{1/2} \ldots (i_j - i_{j-1})^{1/2} \ldots (n - i_\nu)^{1/2} .$$

Integrating $\xi_{i_j}$ out leads to

$$\pi(\nu, i_1, \ldots, i_\nu | \sigma, \sigma_0, p, \mu_0, x_1, \ldots, x_n) \propto c_{1 i_1} \ldots c_{i_\nu n}$$

$$(9.7.1) \times \prod_{j=1}^{\nu+1} \exp\left\{ -\sum_{i=i_{j-1}}^{i_j - 1} \frac{(x_i - \overline{x}_{i_{j-1}})^2}{2\sigma^2} - \frac{(i_j - i_{j-1})(\overline{x}_{i_{j-1}} - \mu_0)^2}{2(\sigma^2 + \sigma_0^2)} \right\} .$$

It is impossible to generate directly from (9.7.1). Moreover, the Gibbs sampler does not apply to a division of (9.7.1) in conditional distributions because $\pi(\nu | i_1, \ldots, i_\nu)$ is meaningless. This type of impossibility is common in problems where the dimension of the parameter is a parameter itself.[7] The alternative discussed in §7.3.3 is to use a Metropolis–Hastings step to simulate $(\nu, i_1, \ldots, i_\nu)$ from (9.7.1) (see also §6.5.1). The other steps of the Gibbs sampler are standard and can be easily recovered.

More Markovian alternatives can also be considered. In the first mixture model of (9.3.1), one can assume that

$$\xi_{i+1} | \xi_i \sim p_i g(\xi_{i+1} | \xi_i) + (1 - p_i) \mathbb{I}_{\xi_i}(\xi_{i+1}) .$$

The conditional distribution $f(\xi | \xi_i)$ can correspond to a random walk, $\mathcal{N}(\xi_i, \sigma_0^2)$, as in Chernoff and Zacks (1964). Finally, hidden semi-Markov chains can also be connected with changepoint models, by defining the cohesions of Barry and Hartigan (1993) as

$$c_{ab} = \frac{\lambda^{b-a}}{(b-a)!} e^{-\lambda} .$$

### 9.7.3 Generalized Linear Models

Except in special cases, a Bayesian analysis of the generalized linear models of McCullagh and Nelder (1989) cannot be done analytically. The logit model developed in Problem 9.3 and probit model of Algorithm [A.47] provide illustrations of cases where missing data play an important role in both the completion of the model and the choice of the Gibbs sampler. In general, however, generalized linear models are not naturally associated with missing data, even though they can be processed via Markov chain Monte Carlo algorithms (see, for instance, Clayton 1996 or Goutis and Robert 1998).

By definition, generalized linear models relate observations $Y_i$ to covariates $x_i \in \mathbb{R}^p$ in an exponential family model

$$(9.7.2) \qquad f(y_i | \theta_i) = e^{\{y_i \theta_i - \psi(\theta_i)\}/\varphi} m(y_i | \varphi)$$

---

[7] This is, for instance, the case for the mixtures (9.3.1) when the number of components is unknown (that is, when $k \sim \pi(k)$). A hybrid procedure replacing the Gibbs sampler is then necessary (see Green 1995 and Richardson and Green 1997).

through a linear constraint on the mean $\xi_i = \mathbb{E}_{\theta_i}[y_i] = \psi'(\theta_i)$, in the sense that

(9.7.3)                    $h(\xi_i) = x_i'\beta \,, \qquad \beta \in \mathbb{R}^p \,.$

The *link function* $h$ is generally fixed (but can be estimated as in Mallick and Gelfand 1994), with often a standard choice depending on $f$.

The constraint (9.7.3) often leads to complexities in the posterior distribution of $\beta$, which prevent analytical processing of the model and result in a need for Markov chain Monte Carlo methods. The resulting difficulties are even more evident in the generalization of (9.7.3) to a *random effects* model, also known as a *generalized linear mixed model*. Here, we assume that $\mathbb{E}[Y_{ij}] = \xi_{ij}$, where

$$h(\xi_{ij}) = x_{ij}'\beta + z_{ij}'b_i, \qquad i = 1,\ldots,I \quad j = 1,\ldots n_i,$$

where $x_{ij}$ and $z_{ij}$ are covariates, $b_i$ is a $\mathcal{N}_q(0,\Sigma)$ random variable, and, for simplicity, we assume that $\varphi = 1$.

With Jeffreys' prior as prior distribution $\pi(\beta,\Sigma) = |\Sigma|^{-(q+1)/2}$, the associated posterior distribution is

$$\pi(b,\beta,\Sigma) \propto \exp\left\{\sum_{i,j}(y_{ij}\theta_{ij} - \psi(\theta_{ij}))\right\}$$

$$\exp\left\{-\frac{1}{2}\sum_i b_i'\Sigma^{-1}b_i\right\} |\Sigma|^{-I/2-(q+1)/2}$$

and it is well defined (see §7.4) if $f$ is the normal density. It is then possible to implement a Gibbs sampler (see Problems 9.21 and 9.22).

There has been an enormous amount of work done on generalized linear models. For example, Albert (1988) takes a more Bayesian approach, Dellaportas and Smith (1993) propose another Gibbs sampler and Clayton (1996), Stephens and Dellaportas (1996), and Chib *et al.* (1998) proposes extensions to mixed effect models. McCulloch (1997, Problem 9.23) addresses maximum likelihood estimation in generalized linear mixed models. An alternate approach to estimation in generalized linear mixed models is through the SIMEX algorithm (Cook and Stefanski 1994, Carroll *et al.* 1995, Chapter 4, and Carroll *et al.* 1996).

APPENDIX A

# Probability Distributions

We recall here the density and the two first moments of most of the distributions used in this book. An exhaustive review of probability distributions is provided by Johnson and Kotz (1969–1972), or the more recent Johnson *et al.* (1992, 1994, 1995). The densities are given with respect to Lebesgue or counting measure depending on the context.

## A.1. Normal Distribution, $\mathcal{N}_p(\theta, \Sigma)$

($\theta \in \mathbb{R}^p$ and $\Sigma$ is a $(p \times p)$ symmetric positive definite matrix.)

$$f(\mathbf{x}|\theta, \Sigma) = (\det \Sigma)^{-1/2}(2\pi)^{-p/2}e^{-(\mathbf{x}-\theta)^t \Sigma^{-1}(\mathbf{x}-\theta)/2}.$$

$\mathbb{E}_{\theta,\Sigma}[\mathbf{X}] = \theta$ and $\mathbb{E}_{\theta,\Sigma}[(\mathbf{X} - \theta)(\mathbf{X} - \theta)^t] = \Sigma$.

When $\Sigma$ is not positive definite, the $\mathcal{N}_p(\theta, \Sigma)$ distribution has no density with respect to Lebesgue measure on $\mathbb{R}^p$. For $p = 1$, the *log-normal* distribution is defined as the distribution of $e^X$ when $X \sim \mathcal{N}(\theta, \sigma^2)$.

## A.2. Gamma Distribution, $\mathcal{G}a(\alpha, \beta)$

($\alpha, \beta > 0$.)

$$f(x|\alpha, \beta) = \frac{\beta^\alpha}{\Gamma(\alpha)}x^{\alpha-1}e^{-\beta x}\mathbb{I}_{[0,+\infty)}(x).$$

$\mathbb{E}_{\alpha,\beta}[X] = \alpha/\beta$ and $\text{var}_{\alpha,\beta}(X) = \alpha/\beta^2$.

Particular cases of the Gamma distribution are the *Erlang distribution*, $\mathcal{G}a(\alpha, 1)$, the *exponential distribution* $\mathcal{G}a(1, \beta)$ (denoted by $\mathcal{E}xp(\beta)$), and the *chi squared distribution*, $\mathcal{G}a(\nu/2, 1/2)$ (denoted by $\chi^2_\nu$). (Note also that the opposite convention is sometimes adopted for the parameter, namely that $\mathcal{G}a(\alpha, \beta)$ may also be noted as $\mathcal{G}a(\alpha, 1/\beta)$. See, e.g., Berger 1985.)

## A.3. Beta Distribution, $\mathcal{B}e(\alpha, \beta)$

($\alpha, \beta > 0$.)

$$f(x|\alpha, \beta) = \frac{x^{\alpha-1}(1 - x)^{\beta-1}}{B(\alpha, \beta)}\mathbb{I}_{[0,1]}(x)$$

where

$$B(\alpha, \beta) = \frac{\Gamma(\alpha)\Gamma(\beta)}{\Gamma(\alpha + \beta)}.$$

$\mathbb{E}_{\alpha,\beta}[X] = \alpha/(\alpha + \beta)$ and $\text{var}_{\alpha,\beta}(X) = \alpha\beta/[(\alpha + \beta)^2(\alpha + \beta + 1)]$.

The beta distribution can be obtained as the distribution of $Y_1/(Y_1+Y_2)$ when $Y_1 \sim \mathcal{G}a(\alpha, 1)$ and $Y_2 \sim \mathcal{G}a(\beta, 1)$.

## A.4. Student's $t$ Distribution, $\mathcal{T}_p(\nu, \theta, \Sigma)$

($\nu > 0$, $\theta \in \mathbb{R}^p$, and $\Sigma$ is a $(p \times p)$ symmetric positive-definite matrix.)

$$f(\mathbf{x}|\nu, \theta, \Sigma) = \frac{\Gamma((\nu + p)/2)/\Gamma(\nu/2)}{(\det \Sigma)^{1/2}(\nu\pi)^{p/2}} \left[1 + \frac{(\mathbf{x} - \theta)^t \Sigma^{-1}(\mathbf{x} - \theta)}{\nu}\right]^{-(\nu+p)/2}.$$

$\mathbb{E}_{\nu,\theta,\Sigma}[\mathbf{X}] = \theta$ ($\nu > 1$) and $\mathbb{E}_{\theta,\Sigma}[(\mathbf{X} - \theta)(\mathbf{X} - \theta)^t] = \nu\Sigma/(\nu - 2)$ ($\nu > 2$).
When $p = 1$, a particular case of Student's $t$ distribution is the *Cauchy distribution*, $\mathcal{C}(\theta, \sigma^2)$, which corresponds to $\nu = 1$. Student's $t$ distribution can be derived as the distribution of $\mathbf{X}/Z$ when $\mathbf{X} \sim \mathcal{N}_p(\theta, \Sigma)$ and $\nu Z^2 \sim \chi_\nu^2$.

## A.5. Fisher's $F$ Distribution, $\mathcal{F}(\nu, \rho)$

($\nu, \rho > 0$.)

$$f(x|\nu, \rho) = \frac{\Gamma((\nu + \rho)/2)\nu^{\rho/2}\rho^{\nu/2}}{\Gamma(\nu/2)\Gamma(\rho/2)} \frac{x^{(\nu-2)/2}}{(\nu + \rho x)^{(\nu+\rho)/2}} \mathbb{I}_{[0,+\infty)}(x).$$

$\mathbb{E}_{\nu,\rho}[X] = \rho/(\rho - 2)$ ($\rho > 2$) and $\text{var}_{\nu,\rho}(X) = 2\rho^2(\nu + \rho - 2)/[\nu(\rho - 4)(\rho - 2)^2]$ ($\rho > 4$).
The distribution $\mathcal{F}(p, q)$ is also the distribution of $(\mathbf{X} - \theta)^t \Sigma^{-1}(\mathbf{X} - \theta)/p$ when $\mathbf{X} \sim \mathcal{T}_p(q, \theta, \Sigma)$. Moreover, if $X \sim \mathcal{F}(\nu, \rho)$, $\rho X/(\nu + \rho X) \sim \mathcal{B}e(\nu, \rho)$.

## A.6. Inverse Gamma Distribution, $\mathcal{IG}(\alpha, \beta)$

($\alpha, \beta > 0$.)

$$f(x|\alpha, \beta) = \frac{\beta^\alpha}{\Gamma(\alpha)} \frac{e^{-\beta/x}}{x^{\alpha+1}} \mathbb{I}_{[0,+\infty[}(x).$$

$\mathbb{E}_{\alpha,\beta}[X] = \beta/(\alpha - 1)$ ($\alpha > 1$) and $\text{var}_{\alpha,\beta}(X) = \beta^2/((\alpha - 1)^2(\alpha - 2))$ ($\alpha > 2$).
This distribution is the distribution of $X^{-1}$ when $X \sim \mathcal{G}a(\alpha, \beta)$.

## A.7. Noncentral Chi Squared Distribution, $\chi_\nu^2(\lambda)$

($\lambda \geq 0$.)

$$f(x|\lambda) = \frac{1}{2}(x/\lambda)^{(p-2)/4} I_{(p-2)/2}(\sqrt{\lambda x})e^{-(\lambda+x)/2}.$$

$\mathbb{E}_\lambda[X] = p + \lambda$ and $\text{var}_\lambda(X) = 3p + 4\lambda$.
This distribution can be derived as the distribution of $X_1^2 + \cdots + X_p^2$ when $X_i \sim \mathcal{N}(\theta_i, 1)$ and $\theta_1^2 + \ldots + \theta_p^2 = \lambda$.

## A.8. Dirichlet Distribution, $\mathcal{D}_k(\alpha_1, \ldots, \alpha_k)$

($\alpha_1, \ldots, \alpha_k > 0$ and $\alpha_0 = \alpha_1 + \cdots + \alpha_k$.)

$$f(x|\alpha_1, \ldots, \alpha_k) = \frac{\Gamma(\alpha_0)}{\Gamma(\alpha_1)\ldots\Gamma(\alpha_k)} x_1^{\alpha_1-1} \ldots x_k^{\alpha_k-1} \mathbb{I}_{\{\sum x_i = 1\}}.$$

$\mathbb{E}_\alpha[X_i] = \alpha_i/\alpha_0$, $\text{var}(X_i) = (\alpha_0 - \alpha_i)\alpha_i/[\alpha_0^2(\alpha_0 + 1)]$ and $\text{cov}(X_i, X_j) = -\alpha_i\alpha_j/[\alpha_0^2(\alpha_0 + 1)]$ ($i \neq j$).
As a particular case, note that $(X, 1 - X) \sim \mathcal{D}_2(\alpha_1, \alpha_2)$ is equivalent to $X \sim \mathcal{B}e(\alpha_1, \alpha_2)$.

## A.9. Pareto Distribution, $\mathcal{P}a(\alpha, x_0)$

($\alpha > 0$ and $x_0 > 0$.)

$$f(x|\alpha, x_0) = \alpha\frac{x_0^\alpha}{x^{\alpha+1}} \mathbb{I}_{[x_0,+\infty[}(x).$$

$\mathbb{E}_{\alpha,x_0}[X] = \alpha x_0/(\alpha - 1)$ ($\alpha > 1$) and $\text{var}_{\alpha,x_0}(X) = \alpha x_0^2/[(\alpha - 1)^2(\alpha - 2)]$ ($\alpha > 2$).

### A.10. Binomial Distribution, $\mathcal{B}(n, p)$.

$(0 \leq p \leq 1.)$

$$f(x|p) = \binom{n}{x} p^x (1-p)^{n-x} \mathbb{I}_{\{0,\dots,n\}}(x).$$

$\mathbb{E}_p(X) = np$ and $\mathrm{var}(X) = np(1-p)$.

### A.11. Multinomial Distribution, $\mathcal{M}_k(n; p_1, \dots, p_k)$

$(p_i \geq 0 \ (1 \leq i \leq k)$ and $\sum_i p_i = 1.)$

$$f(x_1, \dots, x_k | p_1, \dots, p_k) = \binom{n}{x_1 \ \dots \ x_k} \prod_{i=1}^{k} p_i^{x_i} \, \mathbb{I}_{\sum x_i = n}.$$

$\mathbb{E}_p(X_i) = np_i$, $\mathrm{var}(X_i) = np_i(1-p_i)$, and $\mathrm{cov}(X_i, X_j) = -np_i p_j \ (i \neq j)$.
Note that, if $X \sim \mathcal{M}_k(n; p_1, \dots, p_k)$, $X_i \sim \mathcal{B}(n, p_i)$, and that the binomial distribution $X \sim \mathcal{B}(n, p)$ corresponds to $(X, n - X) \sim \mathcal{M}_2(n; p, 1 - p)$.

### A.12. Poisson Distribution, $\mathcal{P}(\lambda)$

$(\lambda > 0.)$

$$f(x|\lambda) = e^{-\lambda} \frac{\lambda^x}{x!} \mathbb{I}_{\mathbb{N}}(x).$$

$\mathbb{E}_\lambda[X] = \lambda$ and $\mathrm{var}_\lambda(X) = \lambda$.

### A.13. Negative Binomial Distribution, $\mathcal{N}eg(n, p)$

$(0 \leq p \leq 1.)$

$$f(x|p) = \binom{n + x + 1}{x} p^n (1-p)^x \mathbb{I}_{\mathbb{N}}(x).$$

$\mathbb{E}_p[X] = n(1-p)/p$ and $\mathrm{var}_p(X) = n(1-p)/p^2$.

### A.14. Hypergeometric Distribution, $\mathcal{H}yp(N; n; p)$

$(0 \leq p \leq 1, \ n < N$ and $pN \in \mathbb{N}.)$

$$f(x|p) = \frac{\binom{pn}{x}\binom{(1-p)N}{n-x}}{\binom{N}{n}} \mathbb{I}_{\{n-(1-p)N, \dots, pN\}}(x) \mathbb{I}_{\{0,1,\dots,n\}}(x).$$

$\mathbb{E}_{N,n,p}[X] = np$ and $\mathrm{var}_{N,n,p}(X) = (N - n)np(1-p)/(N-1)$.

# Notation

## B.1 Mathematical

| | |
|---|---|
| $\mathbf{h} = (h_1, \ldots, h_n) = \{h_i\}$ | boldface signifies a vector |
| $H = \{h_{ij}\} = \|h_{ij}\|$ | uppercase signifies matrices |
| $I, \mathbf{1}, J = \mathbf{1}\mathbf{1}'$ | Identity matrix, vector of ones, and matrix of ones |
| $A \prec B$ | $(B - A)$ is a positive definite matrix |
| $|A|$ | determinant of the matrix $A$ |
| $\mathrm{tr}(A)$ | trace of the matrice $A$ |
| $a^+$ | $\max(a, 0)$ |
| $C_n^p, \binom{n}{p}$ | binomial coefficient |
| $D_\alpha$ | logistic function |
| $_1F_1(a; b; z)$ | confluent hypergeometric function |
| $F^-$ | generalized inverse of $F$ |
| $\Gamma(x)$ | gamma function $(x > 0)$ |
| $\Psi(x)$ | digamma function, $(d/dx)\Gamma(x)$ $(x > 0)$ |
| $\mathbb{I}_A(t)$ | indicator function (1 if $t \in A$, 0 otherwise) |
| $I_\nu(z)$ | modified Bessel function $(z > 0)$ |
| $\binom{n}{p_1 \ldots p_n}$ | multinomial coefficient |
| $\nabla f(z)$ | gradient of $f(z)$, the vector with coefficients $(\partial/\partial z_i) f(z)$ $(f(z) \in \mathbb{R}$ and $z \in \mathbb{R}^p)$ |
| $\nabla^t f(z)$ | divergence of $f(z)$, $\sum (\partial/\partial z_i) f(z)$ $(f(z) \in \mathbb{R}^p$ and $z \in \mathbb{R})$ |
| $\Delta f(z)$ | Laplacian of $f(z)$, $\sum (\partial^2/\partial z_i^2) f(z)$ |
| $\| \cdot \|_{TV}$ | total variation norm |
| $|\mathbf{x}| = (\Sigma x_i^2)^{1/2}$ | Euclidean norm |
| $[x]$ or $\lfloor x \rfloor$ | greatest integer less than $x$ |
| $\lceil x \rceil$ | smallest integer larger than $x$ |
| $f(t) \propto g(t)$ | the functions $f$ and $g$ are proportional |
| $\mathrm{supp}(f)$ | support of $f$ |
| $\langle x, y \rangle$ | scalar product of $x$ and $y$ in $\mathbb{R}^p$ |
| $x \vee y$ | maximum of $x$ and $y$ |
| $x \wedge y$ | minimum of $x$ and $y$ |

## B.2 Probability

| | |
|---|---|
| $X, Y$ | random variable (uppercase) |
| $(\mathcal{X}, \mathcal{P}, \mathcal{B})$ | Probability triple: sample space, probability distribution, and $\sigma$-algebra of sets |
| $\beta_n$ | $\beta$-mixing coefficient |

$\delta_{\theta_0}(\theta)$          Dirac mass at $\theta_0$

$E(\theta)$          energy function of a Gibbs distribution

$\mathcal{E}(\pi)$          entropy of the distribution $\pi$

$F(x|\theta)$          cumulative distribution function of $X$,
         conditional on the parameter $\theta$

$f(x|\theta)$          density of $X$, conditional on the parameter $\theta$,
         with respect to Lebesgue or counting measure

$X \sim f(x|\theta)$          $X$ is distributed with density $f(x|\theta)$

$\mathbb{E}_\theta[g(X)]$          expectation of $g(x)$ under the distribution $X \sim f(x|\theta)$

$\mathbb{E}^V[h(V)]$          expectation of $h(v)$ under the distribution of $V$

$\mathbb{E}^\pi[h(\theta)|x]$          expectation of $h(\theta)$ under the distribution of $\theta$
         conditional on $x$, $\pi(\theta|x)$

iid          independent and identically distributed

$\lambda(dx)$          Lebesgue measure, also denoted by $d\lambda(x)$

$P_\theta$          probability distribution, indexed by the parameter $\theta$

$p \star q$          convolution product of the distributions $p$ and $q$,
         that is, distribution of the sum of $X \sim p$ and $Y \sim q$

$p^{n\star}$          convolution $n$th power,
         that is, distribution of the sum of $n$ iid rv's distributed from $p$

$\varphi(t)$          density of the Normal distribution $\mathcal{N}(0,1)$

$\Phi(t)$          cumulative distribution function of the Normal distribution $\mathcal{N}(0$

$O(n), o(n)$          big "Oh", little "oh." As $n \to \infty$, $\frac{O(n)}{n} \to$ constant,

or $O_p(n), o_p(n)$          $\frac{o(n)}{n} \to 0$, and the subscript $p$ denotes *in probability*

## B.3 Distributions

$\mathcal{B}(n,p)$          binomial distribution

$\mathcal{B}e(\alpha, \beta)$          beta distribution

$\mathcal{C}(\theta, \sigma^2)$          Cauchy distribution

$\mathcal{D}_k(\alpha_1, \ldots, \alpha_k)$          Dirichlet distribution

$\mathcal{E}xp(\lambda)$          exponential distribution

$\mathcal{F}(p,q)$          Fisher's $F$ distribution

$\mathcal{G}a(\alpha, \beta)$          gamma distribution

$\mathcal{IG}(\alpha, \beta)$          inverse gamma distribution

$\chi_p^2$          chi squared distribution

$\chi_p^2(\lambda)$          noncentral chi squared distribution
         with noncentrality parameter $\lambda$

$\mathcal{M}_k(n; p_1, .., p_k)$          multinomial distribution

$\mathcal{N}(\theta, \sigma^2)$          univariate normal distribution

$\mathcal{N}_p(\theta, \Sigma)$          multivariate normal distribution

$\mathcal{N}eg(n, p)$          negative binomial distribution

$\mathcal{P}(\lambda)$          Poisson distribution

$\mathcal{P}a(x_0, \alpha)$          Pareto distribution

$\mathcal{T}_p(\nu, \theta, \Sigma)$          multivariate Student's $t$ distribution

$\mathcal{U}_{[a,b]}$          continuous uniform distribution

$\mathcal{W}e(\alpha, c)$          Weibull distribution

$\mathcal{W}_k(p, \Sigma)$          Wishart distribution

## B.4 Markov Chains

| | |
|---|---|
| $\alpha$ | atom |
| $AR(p)$ | autoregressive process of order $p$ |
| $ARMA(p,q)$ | autoregressive moving average process of order $(p,q)$ |
| $C$ | small set |
| $d(\alpha)$ | period of the state or atom $\alpha$ |
| $\dagger$ | "dagger," absorbing state |
| $\Delta V(x)$ | drift of $V$ |
| $\mathbb{E}_\mu[h(X_n)]$ | expectation associated with $P_\mu$ |
| $\mathbb{E}_{x_0}[h(X_n)]$ | expectation associated with $P_{x_0}$ |
| $\eta_A$ | total number of passages in $A$ |
| $Q(x,A)$ | probability that $\eta_A$ is infinite, starting from $x$ |
| $\gamma_g^2$ | variance of $S_N(g)$ for the Central Limit Theorem |
| $K_\epsilon$ | kernel of the resolvent |
| $MA(q)$ | moving average process of order $q$ |
| $L(x,A)$ | probability of return to $A$ starting from $x$ |
| $\nu, \nu_m$ | minorizing measure for an atom or small set |
| $P(x,A)$ | transition kernel |
| $P^m(x,A)$ | transition kernel of the chain $(X_{mn})_n$ |
| $P_\mu(\cdot)$ | probability distribution of the chain $(X_n)$ with initial state $X_0 \sim \mu$ |
| $P_{x_0}(\cdot)$ | probability distribution of the chain $(X_n)$ with initial state $X_0 = x_0$ |
| $\pi$ | invariant measure |
| $S_N(g)$ | empirical average of $g(x_i)$ for $1 \le i \le N$ |
| $T_{p,q}$ | coupling time for the initial distributions $p$ and $q$ |
| $\tau_A$ | return time to $A$ |
| $\tau_A(k)$ | $k$th return time to $A$ |
| $U(x,A)$ | average number of passages in $A$, starting from $x$ |
| $X_t, X^{(t)}$ | generic element of a Markov chain |
| $\check{X}_n$ | augmented or split chain |

## B.5 Statistics

| | |
|---|---|
| $x, y$ | realized values (lowercase) of the random variables $X$ and $Y$ (uppercase) |
| $\mathcal{X}, \mathcal{Y}$ | sample space (uppercase script Roman letters) |
| $\theta, \lambda$ | parameters (lowercase Greek letters) |
| $\Theta, \Omega$ | parameter space (uppercase script Greek letters) |
| $B^\pi(x)$ | Bayes factor |
| $\delta^{JS}(x)$ | James–Stein estimator |
| $\delta^\pi(x)$ | Bayes estimator |
| $\delta^+(x)$ | positive-part James–Stein estimator |
| $H_0$ | null hypothesis |
| $I(\theta)$ | Fisher information |
| $L(\theta, \delta)$ | loss function, loss of estimating $\theta$ with $\delta$ |
| $L(\theta|x)$ | likelihood function, a function of $\theta$ for fixed $x$, mathematically identical to $f(x|\theta)$ |
| $\ell(\theta|x)$ | the logarithm of the likelihood function |

| | |
|---|---|
| $L^P(\theta|x)$, $\ell^P(\theta|x)$ | profile likelihood |
| $m(x)$ | marginal density |
| $\pi(\theta)$ | generic prior density for $\theta$ |
| $\pi^J(\theta)$ | Jeffreys prior density for $\theta$ |
| $\pi(\theta|x)$ | generic posterior density $\theta$ |
| $\bar{x}$ | sample mean |
| $s^2$ | sample variance |
| $X^*, Y^*, x^*, y^*$ | latent or missing variables (data) |

## B.6 Algorithms

| | |
|---|---|
| $[A_n]$ | symbol of the $n$th algorithm |
| $B$ | backward operator |
| $B_T$ | interchain variance after $T$ iterations |
| $W_T$ | intrachain variance after $T$ iterations |
| $D_T^i$ | cumulative sum to $i$, for $T$ iterations |
| $\delta_{rb}, \delta^{RB}$ | Rao–Blackwellized estimators |
| $F$ | forward operator |
| $g_i(x_i|x_j, j \neq i)$ | conditional density for Gibbs sampling |
| $\Gamma(\theta, \xi)$ | regeneration probability |
| $K(x, y)$ | transition kernel |
| $\tilde{K}$ | transition kernel for a mixture algorithm |
| $K^\star$ | transition kernel for a cycle of algorithms |
| $\lambda_k$ | duration of excursion |
| $q(x|y)$ | transition kernel, typically used for |
| | an instrumental variable |
| $\rho(x, y)$ | acceptance probability |
| | for a Metropolis–Hastings algorithm |
| $S_T$ | empirical average |
| $S_T^C$ | conditional version of the empirical average |
| $S_T^{MP}$ | recycled version of the empirical average |
| $S_T^P$ | importance sampling version of the empirical average |
| $S_T^R$ | Riemann version of the empirical average |
| $S_h(\omega)$ | spectral density of the function $h$ |

APPENDIX C

# References

Aarts, E. and Kors, T.J. (1989) *Simulated Annealing and Boltzman Machines: a Stochastic Approach to Combinatorial Optimisation and Neural Computing.* J. Wiley, New York.

Abramowitz, M. and Stegun, I. (1964) *Handbook of Mathematical Functions.* Dover, New York.

Ackley, D.H., Hinton, G.E. and Sejnowski, T.J. (1985) A learning algorithm for Boltzmann machines. *Cognitive Science* **9**, 147–169.

Ahrens, J. and Dieter, U. (1974) Computer methods for sampling from gamma, beta, Poisson and binomial distributions. *Computing* **12**, 223–246.

Albert, J.H. (1988) Computational methods using a Bayesian hierarchical generalized linear model. *J. Amer. Statist. Assoc.* **83**, 1037–1044.

Albert, J.H. and Chib, S. (1993a) Bayes inference via Gibbs sampling of autoregressive time series subject to Markov mean and variance shifts. *J. Business Economic Statistics* **1**, 1–15.

Albert, J.H. and Chib, S. (1993b) Bayesian analysis of binary and polychotomous response data. *J. Amer. Statist. Assoc.* **88**, 669–679.

Aldous, D. (1987) On the Markov chain simulation method for uniform combinatorial distributions and simulated annealing. *Pr. Eng. Inform.* **1**, 33–46.

Aldous, D. (1990) A random walk construction of uniform spanning trees and uniform labelled trees. *SIAM J. Discrete Math.* **3**, 450–465.

Andrieu, C. and Doucet, A. (1998) Simulated annealing for Bayesian estimation of hidden Markov models. Tech. Report TR 317, Department of Engineering, University of Cambridge.

Archer, G.E.B. and Titterington, D.M. (1995) Parameter estimation for hidden Markov chains. Tech. Report, Dept. of Stat., University of Glasgow.

Arnold, B.C. and Press, S.J. (1989) Compatible conditional distributions. *J. Amer. Statist. Assoc.* **84**, 152–156.

Asmussen, S. (1979) *Applied Probability and Queues.* J. Wiley, New York.

Asmussen, S., Glynn, P.W. and Thorisson, H. (1992) Stationarity detection in the initial transient problem. *ACM Trans. Modelling and Computer Simulations* **2**, 130–157.

Athreya, K.B., Doss, H. and Sethuraman, J.(1996) On the convergence of the Markov Chain simulation method. *Ann. Statist.* **24**, 69–100.

Athreya, K.B. and Ney, P. (1978) A new approach to the limit theory of recurrent Markov chains. *Trans. Amer. Math. Soc.* **245**, 493–501.

Atkinson, A. (1979) The computer generation of Poisson random variables. *Appl. Statist.* **28**, 29–35.

Azencott, R. (1988) Simulated annealing. *Séminaire Bourbaki 40ième année, 1987–1988* **697**.

Barbe, P. and Bertail, P. (1995) *The Weighted Bootstrap*. Lecture Notes in Statistics 98. Springer–Verlag, New York.

Barndorff-Nielsen, O. (1983) On a formula for the distribution of the maximum likelihood estimator. *Biometrika* **70**, 343-365.

Barndorff-Nielsen, O. (1991) Modified signed log likelihood ratio. *Biometrika* **78**, 557-563.

Barndorff-Nielsen, O. and Cox, D. R. (1994) *Inference and Asymptotics*. Chapman & Hall, London.

Barone, P. and Frigessi, A. (1989) Improving stochastic relaxation for Gaussian random fields. *Biometrics* **47**, 1473–1487.

Barry, D. and Hartigan, J. (1992) Product partition models for change point problems. *Ann. Statist.* **20**, 260–279.

Barry, D. and Hartigan, J. (1993) A Bayesian analysis of change point problems. *J. Amer. Statist. Assoc.* **88**, 309–319.

Basu, D. (1988) *Statistical Information and Likelihood.* J.K. Ghosh (Ed.). Springer-Verlag, New York.

Baum, L.E. and Petrie, T. (1966) Statistical inference for probabilistic functions of finite state Markov chains. *Ann. Math. Statist.* **37**, 1554–1563.

Baum, L.E., Petrie, T., Soules, G. and Weiss, N. (1970) A maximization technique occurring in the statistical analysis of probabilistic functions of Markov chains. *Ann. Math. Statist.* **41**, 164–171.

Bauwens, L. (1984) *Bayesian Full Information of Simultaneous Equations Models Using Integration by Monte Carlo*. Lecture Notes in Economics and Mathematical Systems 232. Springer–Verlag, New York.

Bauwens, L. (1991) The "pathology" of the natural conjugate prior density in the regression model. *Ann. Econom. Statist.* **23**, 49-64.

Bauwens, L. and Richard, J.F. (1985) A 1-1 Poly-*t* random variable generator with application to Monte Carlo integration. *J. Econometrics* **29**, 19–46.

Bennett, J.E., Racine-Poon, A. and Wakefield, J.C. (1996) MCMC for nonlinear hierarchical models. In *Markov chain Monte Carlo in Practice*. W.R. Gilks, S.T. Richardson and D.J. Spiegelhalter (Eds.). 339–358. Chapman & Hall, London.

Bergé, P., Pommeau, Y. and Vidal, C. (1984) *Order Within Chaos*. J. Wiley, New York.

Berger, J.O. (1985) *Statistical Decision Theory and Bayesian Analysis* (2nd edition). Springer–Verlag, New York.

Berger, J.O. (1990) Robust Bayesian analysis: sensitivity to the prior. *J. Statist. Plann. Inference* **25**, 303–328.

Berger, J.O. (1994) An overview of of robust Bayesian analysis (with discussion). *TEST* **3**, 5–124.

Berger, J.O. and Bernardo, J.M. (1989) Estimating a product of means: Bayesian analysis with reference priors. *J. Amer. Statist. Assoc.* **84**, 200–207.

Berger, J.O. and Bernardo, J.M. (1992) On the development of the reference prior method. In *Bayesian Statistics 4*. J.O. Berger, J.M. Bernardo, A.P. Dawid and A.F.M. Smith (Eds), 35–49. Oxford University Press, London.

Berger, J.O., Philippe, A. and Robert, C.P. (1998) Estimation of quadratuc functions: reference priors for non-centrality parameters. *Statistica Sinica* 8(2), 359–375.

Berger, J.O. and Wolpert, R. (1988) *The Likelihood Principle* (2nd edition). IMS Lecture Notes — Monograph Series 9, Hayward, CA.

Bernardo, J.M. (1979) Reference posterior distributions for Bayesian inference (with discussion). *J. Roy. Statist. Soc. Ser. B* **41**, 113–147.

Bernardo, J.M. and Giròn, F.J. (1986) A Bayesian approach to cluster analysis. In *Second Catalan International Symposium on Statistics*, Barcelona, Spain.

Bernardo, J.M. and Giròn, F.J. (1988) A Bayesian analysis of simple mixture problems. In *Bayesian Statistics 3*. J.M. Bernardo, M.H. DeGroot, D.V. Lindley and A.F.M. Smith (Eds.). 67–78. Oxford University Press, Oxford.

Bernardo, J.M. and Smith, A.F.M. (1994) *Bayesian Theory*. J. Wiley, New York.

Besag, J. (1974) Spatial interaction and the statistical analysis of lattice systems (with discussion). *J. Roy. Statist. Soc. Ser. B* **36**, 192–326.

Besag, J. (1986) On the statistical analysis of dirty pictures. *J. Roy. Statist. Soc. Ser. B* **48**, 259–279.

Besag, J. (1989) Towards Bayesian image analysis. *J. Applied Statistics* **16**, 395–407.

Besag, J.E. (1994) Discussion of "Markov chains for exploring posterior distributions." *Ann. Statist.* **22**, 1734-1741.

Besag, J. and Green, P.J. (1993) Spatial statistics and Bayesian computation (with discussion). *J. Roy. Statist. Soc. Ser. B* **55**, 25–38.

Besag, J., Green, E., Higdon, D. and Mengersen, K.L. (1995) Bayesian computation and stochastic systems (with discussion). *Statistical Science* **10**, 3–66.

Best, D.J. (1978) Letter to the editor. *Applied Statistics* (Ser. C) **27**, 181.

Best, N.G., Cowles, M.K. and Vines, K. (1995) CODA: Convergence diagnosis and output analysis software for Gibbs sampling output, Version 0.30. Tech. Report, MRC Biostatistics Unit, Univ. of Cambridge.

Billingsley, P. (1968) *Convergence of Probability Measures*. J. Wiley, New York.

Billingsley, P. (1986) *Probability and Measure* (2nd edition). J. Wiley, New York.

Billingsley, P. (1995) *Probability and Measure* (3rd edition). J. Wiley, New York.

Billio, M. and Monfort, A. (1995) Switching state space models. Doc. Travail CREST no. 9557, Insee, Paris.

Billio, M., Monfort, A. and Robert, C.P. (1999) Bayesian estimation of wwitching ARMA Models. *J. Econometrics* (to appear).

Billio, M., Monfort, A. and Robert, C.P. (1998b) The simulated likelihood ratio method, Doc. Travail CREST, INSEE, Paris.

Boch, R.D. and Aitkin, M. (1981) Marginal maximum likelihood estimation of item parameters: Application of an EM algorithm. *Psychometrika* **46**, 443–459.

Bollerslev, T., Chou, R.Y. and Kroner, K.F. (1992) ARCH modeling in finance. A review of the theory and empirical evidence. *J. Econometrics* **52**, 5–59.

Booth, J.G., Hobert, J.P. and Ohman, P.A. (1999) On the probable error of the ratio of two gamma means. *Biometrika* **86** (to appear).

Bouleau, N. and Lépingle, D. (1994) *Numerical Methods for Stochastic Processes.* J. Wiley, New York.

Box, G.E.P. and Muller, M. (1958) A note on the generation of random normal variates. *Ann. Math. Statist.* **29**, 610–611.

Box, G.E.P. and Tiao, G.C. (1968). Bayesian estimation of means for the random effect model. *J. Amer. Statist. Assoc.* **63**, 174–181.

Box, G.E.P. and Tiao, G.C. (1973) *Bayesian Inference in Statistical Analysis.* Addison-Wesley, Reading, Massachusetts.

Boyles, R.A. (1983) On the convergence of the EM algorithm. *J. Roy. Statist. Soc. Ser. B* **45**, 47–50.

Bradley, R.C. (1986) Basic properties of strong mixing conditions. In *Dependence in Probability and Statistics*. E. Ebberlein and M. Taqqu (Eds.). 165–192. Birkhäuser, Boston.

Breslow, N.E. and Clayton, D.G. (1993) Approximate inference in generalized linear mixed models. *J. Amer. Statist. Assoc.* **88**, 9–25.

Brockwell, P.J. and Davis, P.A. (1996) *Introduction to Time Series and Forecasting*, Springer Texts in Statistics, Springer–Verlag, New York.

Broniatowski, M., Celeux, G. and Diebolt, J. (1984) Reconnaissance de mélanges de densités par un algorithme d'apprentissage probabiliste. In *Data Analysis and Informatics* **3**. E. Diday (Ed.). North-Holland, Amsterdam, 359–373.

Brooks, S.P. (1998a) Markov chain Monte Carlo and its application, *The Statistician* **47**, 69–100.

Brooks, S.P. (1998b) MCMC convergence diagnosis via multivariate bounds on log-concave densities. *Ann. Statist.* **26**, 398–433.

Brooks, S.P. (1998c) Quantitative convergence diagnosis for MCMC via Cusums. *Statistics and Computing* **8**(3), 267–274.

Brooks, S.P., Dellaportas, P. and Roberts, G.O. (1997) A total variation method for diagnosing convergence of MCMC algorithms. *J. Comput. Graph. Statist.* **6**, 251–265.

Brooks, S.P. and Gelman, A. (1998) General methods for monitoring convergence of iterative simulations. *J. Comput. Graph. Statist.* **7**(4), 434–455.

Brooks, S.P. and Roberts, G.O. (1997) On quantile estimation and MCMC convergence. Tech. Report, University of Bristol.

Brooks, S.P and Roberts, G. (1999) Assessing convergence of Markov chain Monte Carlo algorithms. *Statistics and Computing* (to appear).

Brown, L.D. (1971) Admissible estimators, recurrent diffusions, and insoluble boundary-value problems, *Ann. Math. Statist.* **42**, 855–903.

Brown, L.D. (1986) *Foundations of Exponential Families*. IMS Lecture Notes — Monograph Series 6, Hayward, CA.

Brownie, C., Hines, J.E., Nichols, J.D., Pollock, K.H. and Hestbeck, J.B. (1993) Capture-recapture studies for multiple strata including non-Markovian transition probabilities. *Biometrics* **49**, 1173-1187.

Bucklew, J.A. (1990) *Large Deviation Techniques in Decision, Simulation and Estimation.* J. Wiley, New York.

Burnham, K.D., Anderson, D.R., White, G.C., Brownie C. and Pollock, K.H. (1987) Design and analysis methods for fish survival experiments based on release-recapture. *American Fisheries Society Monograph* **5**.

Carlin, B.P. (1992) State space modeling of non-standard actuarial time series. *Insurance: Mathematics and Economics* **11**, 209–222.

Carlin, B.P. and Chib, S. (1995) Bayesian model choice through Markov-Chain Monte Carlo. *J. Roy. Statist. Soc.* (Ser. B) **57**(3), 473–484.

Carlin, B.P. and Gelfand, A. (1990) Approaches for empirical Bayes confidence intervals. *J. Amer. Statist. Assoc.* **85**, 105–114.

Carlin, B.P. and Gelfand, A. (1991a) A sample reuse method for accurate parametric empirical Bayes confidence intervals. *J. Roy. Statist. Soc. Ser. B* **53**, 189–200.

Carlin, B.P. and Gelfand, A. (1991b) An iterative Monte Carlo method for nonconjugate Bayesian analysis. *Statist. Comput.* **1**, 119–128.

Carlin, B.P., Gelfand, A.E., and Smith, A.F.M. (1992) Hierarchical Bayesian analysis of change point problems. *Applied Statistics* (Ser. C) **41**, 389–405.

Carlin, B.P. and Louis, T.A. (1996) *Bayes and Empirical Bayes Methods for Data Analysis.* Chapman & Hall, London.

Carlin, B.P. and Polson, N.G. (1991) Inference for nonconjugate Bayesian models using the Gibbs sampler. *Canadian Journal of Statistics* **19**, 399–405.

Carlin, B.P. and Gelfand, A.E. (1991) An iterative Monte Carlo method for nonconjugate Bayesian analysis. *Statistics and Computing* **1**, 119-28.

Carlin, J.B. (1992) Meta-Analysis for 2 x 2 tables: a Bayesian approach. *Statistics in Medicine* **11**, 141-59.

Carroll, R.J., Küchenhoff, H., Lombard, F. and Stefanski, L.A. (1996) Asymptotics for the SIMEX estimator in nonlinear measurement error models. *J. Amer. Statist. Assoc.* **91**, 242–250.

Carroll, R.J. and Lombard, F. (1985) A note on $N$ estimators for the binomial distribution. *J. Amer. Statist. Assoc.* **80**, 423–426.

Carroll, R.J., Ruppert, D. and Stefanski, L.A. (1995) *Measurement Error in Nonlinear Models.* Chapman & Hall, London.

Casella, G. (1986) Stabilizing binomial $n$ estimators. *J. Amer. Statist. Assoc.* **81** 171–175.

Casella, G. (1996) Statistical theory and Monte Carlo algorithms (with discussion) *TEST* **5** 249-344.

Casella, G. and Berger, R. (1990) *Statistical Inference.* Wadsworth, Belmont, CA.

Casella, G. and Berger, R. (1994) Estimation with selected binomial information or do you really believe that Dave Winfield is batting .471? *J. Amer. Statist. Assoc.* **89**, 1080–1090.

Casella, G. and George, E.I. (1992) An introduction to Gibbs sampling. *Ann. Math. Statist.* **46**, 167–174.

Casella, G. and Robert, C.P. (1995) Recycling in Accept–Reject algorithms. *Note aux Comptes–Rendus de l'Académie des Sciences* **321**(12), 421–426.

Casella, G. and Robert, C.P. (1996) Rao-Blackwellisation of sampling schemes. *Biometrika* **83**(1), 81–94.

Casella, G. and Robert, C.P. (1998) Post-processing Accept–Reject samples: recycling and rescaling. *J. Comput. Graph. Statist.* **7**(2), 139–157.

Castledine, B. (1981) A Bayesian analysis of multiple-recapture sampling for a closed population. *Biometrika* **67**, 197–210.

Celeux, G., Chauveau, D. and Diebolt, J. (1996) Stochastic versions of the EM algorithm: An experimental study in the mixture case. *J. Statist. Comput. Simul.* **55**(4), 287–314.

Celeux, G. and Clairambault, J. (1992) Estimation de chaînes de Markov cachées: méthodes et problèmes. In *Approches Markoviennes en Signal et Images*, GDR CNRS Traitement du Signal et Images, 5–19. CNRS, Paris.

Celeux, G. and Diebolt, J. (1985) The SEM algorithm: a probabilistic teacher algorithm derived from the EM algorithm for the mixture problem. *Comput. Statist. Quater.* **2**, 73–82.

Celeux, G. and Diebolt, J. (1990) Une version de type recuit simulé de l'algorithme EM. *Notes aux Comptes Rendus de l'Académie des Sciences* **310**, 119–124.

Celeux, G. and Diebolt, J. (1992) A classification type EM algorithm for the mixture problem. *Stochastics and Stochastics Reports* **41**, 119–134.

Cellier, D. (1996) Comptage et méthodes de Monte Carlo par chaînes de Markov. Notes de cours, Université de Rouen.

Chaitin, G. J. (1982) Algorithmic information theory. In *Encyclopedia of Statistical Science* **1**, 38–41.

Chaitin, G. J. (1988) Randomness in arithmetic. *Scient. Amer.* **259**, 80–85.

Chan, K.S. and Geyer, C.J. (1994) Discussion of "Markov chains for exploring posterior distribution." *Ann. Statist.* **22**, 1747–1758.

Chauveau, D., Diebolt, J. and Robert, C.P. (1998) Control by the Central Limit theorem. In *Discretization and MCMC Convergence Assessment*, C.P. Robert (Ed.). Chapter 5, 99–126. Lecture Notes in Statistics 135. Springer–Verlag, New York.

Chen, M.H. and Schmeiser, B.W. (1993) Performances of the Gibbs, hit-and-run, and Metropolis samplers. *J. Comp. Graph Statist.* **2**, 251–272.

Chen, M.H. and Schmeiser, B.W. (1998) Towards black-box sampling. *J. Comp. Graph Statist.* **7**, 1–22.

Chen, M.M. and Shao, Q.M. (1997) On Monte Carlo methods for estimating ratios of normalizing constants. *Ann. Statist.* **25**, 1563–1594.

Cheng, B. and Titterington, D.M. (1994) Neural networks: A review from a statistical point of view (with discussion). *Stat. Science* **9**, 2–54.

Cheng, R.C.H. (1977) The generation of gamma variables with non-integral shape parameter. *Applied Statistics* (Ser. C), **26**, 71–75.

Cheng, R.C.H. and Feast, G. (1979) Some simple gamma variate generators. *Appl. Statist.* **28**, 290–295.

Chernoff, H. and Zacks, S. (1964) Estimating the current mean of a normal distribution which is subjected to changes in time. *Ann. Math. Statist.* **35**, 999–1018.

Chib, S. (1992) Bayes inference in the tobit censored regression model. *J. Econometrics* **51**, 79–99.

Chib, S. (1995) Marginal likelihood from the Gibbs output. *J. Amer. Statist. Assoc.* **90** 1313–1321.

Chib, S. (1996) Calculating posterior distributions and modal estimates in Markov mixture models. *J. Econometrics* **75**, 79–97.

Chib, S. and Greenberg, E. (1995) Understanding the Metropolis–Hastings algorithm. *Ann. Math. Statist.* **49**, 327–335.

Chib, S., Greenberg, E. and Winkelmann, R. (1998) Posterior simulation and Bayes factors in panel count data models. *J. Econometrics* **86**, 33–54.

Chung, K.L. (1960) *Markov Processes with Stationary Transition Probabilities.* Springer–Verlag, Heidelberg.

Churchill, G.A. (1989) Stochastic models for heterogeneous DNA sequences. *Bull. Math. Biol.* **51**, 79–94.

Churchill, G.A. (1995) Accurate restoration of DNA sequences (with discussion). In *Case Studies in Bayesian Statistics.* C. Gatsonis, J.S. Hodges, R.E. Kass and N.D. Singpurwalla (Eds.). Vol. II, 90–148. Springer–Verlag, New York.

Cipra, B.A. (1987) An introduction to the Ising model. *American Mathematical Monthly* **94**, 937–959.

Clarke, B.S. and Barron, A.R. (1990) Information-theoretic asymptotics of Bayes methods. *IEEE Trans. Information Theory* **36**, 453–471.

Clarke, B.S. and Wasserman, L. (1993) Noninformative priors and nuisance parameters. *J. Amer. Statist. Assoc.* **88**, 1427–1432.

Clayton, D.G. (1996) Generalized linear mixed models. In *Markov Chain Monte Carlo in Practice.* W.R. Gilks, S. Richardson, and D.J. Spiegelhalter (Eds.). 275–301. Chapman & Hall, London.

Cocozza-Thivent, C. and Guédon, Y. (1990) Explicit state occupancy modeling by hidden semi-Markov models: application of Derin's scheme. *Comput. Speech Lang.* **4**, 167–192.

Cook, J.R. and Stefanski, L.A. (1994) Simulation-extrapolation estimation in parametric measurement error models. *J. Amer. Statist. Assoc.* **89**, 1314–1328.

Copas, J. B. (1975) On the unimodality of the likelihood for the Cauchy distribution. *Biometrika* **62** 701–704.

Cowles, M.K. and Carlin, B.P. (1996) Markov chain Monte Carlo convergence diagnostics: a comparative study. *J. Amer. Statist. Assoc.* **91**, 883–904.

Cowles, M.K. and Rosenthal, J.S. (1998) A simulation approach to convergence rates for Markov chain Monte Carlo. *Statistics and Computing* **8**, 115–124.

Crawford, S.L., DeGroot, M.H., Kadane, J.B. and Small, M.J. (1992) Modelling lake-chemistry distributions: Approximate Bayesian methods for estimating a finite-mixture model. *Technometrics* **34**, 441–453.

Cressie, N. (1993) *Spatial Statistics.* J. Wiley, New York.

Damien, P. and Walker, S. (1996) Sampling probability densities via uniform random variables and a Gibbs sampler. Tech. Report, Business School, University of Michigan.

Damien, P., Wakefield, J. and Walker, S. (1999) Gibbs sampling for Bayesian non-conjugate and hierarchical models by using auxiliary variables. *J. Roy. Statist. Soc. Ser. B* **61**(2), 331–344.

Daniels, H. E. (1954) Saddlepoint approximations in statistics. *Ann. Math. Statist.* **25**, 631-650.

Daniels, H. E. (1980) Exact saddlepoint approximations. *Biometrika* **67**, 59-63.

Daniels, H. E. (1983) Saddlepoint approximations for estimating equations. *Biometrika* **70**, 89-96.

Daniels, H. E. (1987) Tail probability approximations. *International Statistical Review* **55**, 37–48.

Dawid, A.P. and Lauritzen, S.L. (1993) Hyper Markov laws in the statistical analysis of decomposable graphical models. *Ann. Statist.* **21**, 1272-1317.

Dawid, A.P., Stone, N. and Zidek, J.V. (1973) Marginalization paradoxes in Bayesian and structural inference (with discussion). *J. Roy. Statist. Soc. Ser. B* **35**, 189–233.

Davies, R.B. (1977) Hypothesis testing when a nuisance parameter is present only under the alternative. *Biometrika* **64**, 247–254.

Davydov, Y.A. (1973) Mixing conditions for Markov chains. *Theory Probab. Appl.* **18**, 312–328.

Decoux, C. (1997) Estimation de modèles de semi-chaînes de Markov cachées par échantillonnage de Gibbs. *Revue de Statistique Appliquée* **45**, 71–88.

Dellaportas, P. (1995) Random variate transformations in the Gibbs sampler: Issues of efficiency and convergence. *Statis. Comput.* **5**, 133–140.

Dellaportas, P. and Foster, J.J. (1996) Markov chain Monte Carlo model determination for hierarchical and graphical log-linear models. Technical Report, University of Southampton.

Dellaportas, P. and Smith, A.F.M. (1993) Bayesian inference for generalised linear and proportional hazards models via Gibbs sampling. *Appl. Statist.* **42**, 443–459.

Dempster, A.P., Laird, N.M. and Rubin, D.B. (1977) Maximum likelihood from incomplete data via the EM algorithm (with discussion). *J. Roy. Statist. Soc. Ser. B* **39**, 1–38.

Denison, D.G.T., Mallick, B.K. and Smith, A.F.M. (1998) Automatic Bayesian curve fitting. *J. Roy. Statist. Soc. Ser. B* **60**, 333–350.

Dette, H. and Studden, W.J. (1997) *The Theory of Canonical Moments with Applications in Statistics, Probability and Analysis*. J. Wiley, New York.

Devroye, L. (1981) The computer generation of Poisson random variables. *Computing* **26**, 197–207.

Devroye, L. (1985) *Non-Uniform Random Variate Generation*. Springer–Verlag, New York.

Diaconis, P. and Hanlon, P. (1992) Eigen analysis for some examples of the Metropolis algorithm. *Contemporary Mathematics* **138**, 99-117.

Diaconis, P. and Holmes, S. (1994) Gray codes for randomization procedures. *Statistics and Computing* **4**, 287–302.

Diaconis, P. and Sturmfels, B. (1998) Algebraic algorithms for sampling from conditionnal distributions. *Ann. Statist.* **26** 363–397.

Diaconis, P. and Ylvisaker, D. (1979) Conjugate priors for exponential families. *Ann. Statist.* **7**, 269–281.

DiCiccio, T. J. and Martin, M. A. (1993) Simple modifications for signed roots of likelihood ratio statistics. *J. Roy. Statist. Soc. Ser. B* **55**, 305-316.

Dickey, J.M. (1968) Three multidimensional integral identities with Bayesian applications. *Ann. Statist.* **39**, 1615–1627.

Diebold, D. and Nerlove, M. (1989) The dynamic of exchange rate volatility: a multivariate latent factor ARCH model. *J. Appl. Econom.* **4**, 1–22.

Diebolt, J. and Celeux, G. (1993) Asymptotic properties of a Stochastic EM algorithm for estimating mixture proportions. *Stochastics Models* **9**, 599–613.

Diebolt, J. and Ip, E.H.S. (1996) Stochastic EM: method and application. In *Markov Chain Monte Carlo in Practice.* W.R. Gilks, S. Richardson, and D.J. Spiegelhalter (Eds.). 259–274. Chapman & Hall, London.

Diebolt, J. and Robert, C.P. (1990a) Bayesian estimation of finite mixture distributions, Part I: Theoretical aspects. Rapport tech. # 110, LSTA, Université Paris VI.

Diebolt, J. and Robert, C.P. (1990b) Estimation des paramètres d'un mélange par échantillonnage bayésien. *Notes aux Comptes–Rendus de l'Académie des Sciences I* **311**, 653–658.

Diebolt, J. and Robert, C.P. (1990c) Bayesian estimation of finite mixture distributions, Part II: Sampling implementation. Rapport tech. # 111, LSTA, Université Paris VI.

Diebolt, J. and Robert, C.P. (1993) Discussion of "Bayesian computations via the Gibbs sampler" by A.F.M. Smith and G. Roberts. *J. Roy. Statist. Soc. Ser. B* **55**, 71–72.

Diebolt, J. and Robert, C.P. (1994) Estimation of finite mixture distributions by Bayesian sampling. *J. Roy. Statist. Soc. Ser. B* , **56**, 363–375.

Dobson, A.J. (1983) *An Introduction to Statistical Modelling.* Chapman & Hall, London.

Dobson, A.J. (1990) *An Introduction to Generalized Linear Models.* Chapman & Hall, London.

Dodge, Y. (1996) A natural random number generator. *Int. Statist. Rev.* **64** 329–344.

Doob, J. (1953) *Stochastic Processes.* J. Wiley, New York.

Doukhan, P., Massart, P. and Rio, E. (1994) The functional central limit theorem for strongly mixing processes. *Annales de l'I.H.P.* **30**, 63–82.

Draper, D. (1998) *Bayesian Hierarchical Modeling.* In preparation.

Duflo, M. (1996) *Random Iterative Models.* Applications of Mathematics. I. Karatzas and M. Yor (Eds.). Vol. 34. Springer–Verlag, Berlin.

Dupuis, J.A. (1995a) *Analyse stochastique bayésienne de modèles de capture-recapture.* Thèse de Doctorat de l'Université Paris 6.

Dupuis, J.A. (1995b) Bayesian estimation of movement probabilities in open populations using hidden Markov chains. *Biometrika* **82**(4), 761–772.

Dupuis, J.A. (1999) Estimation bayésienne de modèles multi-états markoviens: une application en dynamique des populations. In *Statistique Bayésienne.* J.J. Droesbeke, J. Fine and G. Saporta (Eds.). Editions Technip, Bruxelles (to appear).

Durrett, R. (1991) *Probability: Theory and Examples.* Wadsworth and Brooks/ Cole, Pacific Grove, CA.

Dykstra, R.L. and Robertson, T. (1982) An algorithm for isotonic regression for two or more independent variables. *Ann. Statist.* **10**, 708–716.

Eaton, M.L. (1992) A statistical dyptich: Admissible inferences – Recurrence of symmetric Markov chains. *Ann. Statist.* **20**, 1147–1179.

Eberly, L. E. (1997) Convergence of interval estimates from the Gibbs sampler. Ph.D. Thesis, Biometrics Unit, Cornell University, Ithaca, NY.

Efron, B. (1979a) Bootstrap methods: another look at the jacknife. *Ann. Statist.* **7**, 1–26.

Efron, B. (1981) Nonparametric standard errors and confidence intervals (with discussion). *Can. J. Statist.* **9** 139-172.

Efron, B. (1982) *The Jacknife, the Bootstrap and Other Resampling Plans.* Regional Conference in Applied Mathematics, **38**. SIAM, Philadelphia.

Efron, B. and Tibshirani, R.J. (1994) *An Introduction to the Bootstrap.* Chapman & Hall, London.

Enders, W. (1994) *Applied Econometric Time Series.* J. Wiley, New York.

D'Epifanio, G. (1989) Un approccio all'inferenza basato sulla ricerca di un punto fisso. *Metron* **47**, 1–4.

D'Epifanio, G. (1996) Notes on a recursive procedure for point estimate. *TEST* **5**, 203–225.

Escobar, M.D. (1994) Estimating normal means with a Dirichlet process prior. *J. Amer. Statist. Assoc.* **89**, 363–386.

Escobar, M.D. and West, M. (1995) Bayesian prediction and density estimation. *J. Amer. Statist. Assoc.* **90**, 577–588.

Ethier, S.N. and Kurtz, T.G. (1986) *Markov processes: Characterization and Convergence.* J. Wiley, New York.

Evans, G. (1993) *Practical Numerical Integration.* J. Wiley, New York.

Evans, M. and Swartz, T. (1995) Methods for approximating integrals in Statistics with special emphasis on Bayesian integration problems. *Statist. Sci.* **10**, 254–272.

Everitt, B.S. (1984) *An Introduction to Latent Variable Models.* Chapman & Hall, London.

Fan, J. and Gijbels, I. (1996) *Local Polynomial Modelling and its Applications* Chapman & Hall, London.

Fang, K.T. and Wang, Y. (1994) *Number-Theoretic Methods in Statistics.* Chapman & Hall, London.

Feller, W. (1970) *An Introduction to Probability Theory and its Applications,* Vol. 1. J. Wiley, New York.

Feller, W. (1971) *An Introduction to Probability Theory and its Applications,* Vol. 2. J. Wiley, New York.

Ferguson, T.S. (1978) Maximum likelihood estimates of the parameters of the Cauchy distribution for samples of size 3 and 4. *J. Amer. Statist. Assoc.* **73**, 211–213.

Ferrenberg, A.M., Landau, D.P. and Wong, Y.J. (1992) Monte Carlo simulations: hidden errors from "good" random number generators. *Physical Review Letters* **69**(23), 3382–3385.

Field, C. and Ronchetti, E. (1990) *Small Sample Asymptotics.* IMS Lecture Notes — Monograph Series, Hayward, CA.

Fieller, E.C. (1954) Some problems in interval estimation. *J. Roy. Statist. Soc. Ser. B* **16**, 175–185.

Fienberg, S.E. (1977) *The Analysis of Cross-Classified Categorical Data.* The MIT Press, Cambridge, MA.

Fill, J.A. (1991) Eigenvalue bounds on convergence to stationarity for non reversible Markov chains with applications to the exclusion process. *Ann. Applied Prob.* **1**, 62–87.

Fill, J.A. (1998a) An interruptible algorithm for exact sampling via Markov chains. *Ann. Applied Prob.* **8**, 131–162.

Fill, J.A. (1998b) The move-to front rule: A case study for two perfect sampling algorithms. *Prob. Eng. Info. Sci.* (to appear).

Finch, S.J, Mendell, N.R. and Thode, H.C. (1989) Probabilistic measures of adequacy of a numerical search for a global maximum. *J. Amer. Statist. Assoc.* **84**, 1020–1023.

Fishman, G.S. (1996) *Monte Carlo.* Springer–Verlag, New York.

Fletcher, R. (1980) *Practical Methods of Optimization.* Vol. 1. J. Wiley, New York.

Gamboa, F. and Gassiat, E. (1997) Blind deconvolution of discrete linear systems. *Ann. Statist.* **24** 1964–1981.

Gamerman, D. (1997) *Markov Chain Monte Carlo.* Chapman & Hall, London.

Gassiat, E. (1994) Déconvolution aveugle de systèmes linéaires discrets bruités. *Notes aux Comptes Rendus de l'Académie des Sciences I* **319**, 489–492.

Gassiat, E. (1995) Déconvolution aveugle d'un système non-causal. *Matapli* **45**, 45–54.

Garren, S.T. and Smith, R.L. (1993) Convergence diagnostics for Markov chain samplers. Tech. Report, Department of Statistics, University of North Carolina.

Gauss, C.F. (1810) *Méthode des Moindres Carrés. Mémoire sur la Combination des Observations.* Transl. J. Bertrand. Mallet-Bachelier, Paris (1955).

Gaver, D.P. and O'Muircheartaigh, I.G. (1987) Robust empirical Bayes analysis of event rates. *Technometrics* **29**, 1–15.

Gelfand, A.E. and Dey, D. (1994) Bayesian model choice: asymptotics and exact calculations. *J. Roy. Statist. Soc.* (Ser. B) **56**, 501–514.

Gelfand, A.E., Hills, S., Racine-Poon, A. and Smith, A.F.M. (1990) Illustration of Bayesian inference in normal data models using Gibbs sampling. *J. Amer. Statist. Assoc.* **85**, 972–982.

Gelfand, A.E. and Sahu, S.K. (1994) On Markov chain Monte Carlo acceleration. *J. Comput. Graph. Statist.* **3**(3), 261–276.

Gelfand, A.E., Sahu, S.K. and Carlin, B.P. (1995) Efficient parametrization for normal linear mixed models. *Biometrika* **82**, 479-488.

Gelfand, A.E., Sahu, S.K. and Carlin, B.P. (1996) Efficient parametrizations for generalised linear mixed models (with discussion). In *Bayesian Statistics 5*. J.M. Bernardo, J.O. Berger, A.P. Dawid and A.F.M. Smith (Eds.). 165–180. Oxford University Press, Oxford.

Gelfand, A.E. and Smith, A.F.M. (1990) Sampling based approaches to calculating marginal densities. *J. Amer. Statist. Assoc.* **85**, 398–409.

Gelfand, A.E., Smith, A.F.M. and Lee, T.M. (1992) Bayesian analysis of constrained parameters and truncated data problems using Gibbs sampling. *J. Amer. Statist. Assoc.* **87**, 523–532.

Gelman, A. (1996) Inference and monitoring convergence. In *Markov Chain Monte Carlo in Practice*. W.R. Gilks, S. Richardson, and D.J. Spiegelhalter (Eds.). 131–143. Chapman & Hall, London.

Gelman, A., Gilks, W.R. and Roberts, G.O. (1996) Efficient Metropolis jumping rules. In *Bayesian Statistics 5*, J.O. Berger, J.M. Bernardo, A.P. Dawid, D.V. Lindley and A.F.M. Smith (Eds.). 599–608. Oxford University Press, Oxford.

Gelman, A. and Meng, X.L. (1998) Simulating normalizing constants: From importance sampling to bridge sampling to path sampling. *Statist. Sci.* **13** 163-185.

Gelman, A. and Rubin, D.B. (1992) Inference from iterative simulation using multiple sequences (with discussion). *Statist. Sci.* **7**, 457–511.

Gelman, A. and Speed, T.P. (1993) Characterizing a joint probability distribution by conditionals. *J. Roy. Statist. Soc. Ser. B* **55**, 185–188.

Geman, S. and Geman, D. (1984) Stochastic relaxation, Gibbs distributions and the Bayesian restoration of images. *IEEE Trans. Pattern Anal. Mach. Intell.* **6**, 721–741.

George, E.I., Makov, U. and Smith, A.F.M. (1993) *Scand. J. Statist.* **20** 147–156.

George, E.I. and Robert, C.P. (1992) Calculating Bayes estimates for capture-recapture models. *Biometrika* **79**(4), 677–683

Geweke, J. (1988) Antithetic acceleration of Monte Carlo integration in Bayesian inference. *J. Econometrics* **38**, 73–90.

Geweke, J. (1989) Bayesian inference in econometric models using Monte Carlo integration. *Econometrica* **57**, 1317–1340.

Geweke, J. (1991) Efficient simulation from the multivariate normal and Student $t$-distributions subject to linear constraints. *Computer Sciences and Statistics: Proc. 23d Symp. Interface*, 571–577.

Geweke, J. (1992) Evaluating the accuracy of sampling-based approaches to the calculation of posterior moments (with discussion). In *Bayesian Statistics 4*, J.M. Bernardo, J.O. Berger, A.P. Dawid and A.F.M. Smith (Eds.). 169–193. Oxford University Press, Oxford.

Geyer, C.J. (1992) Practical Monte Carlo Markov Chain (with discussion). *Statist. Sci.* **7**, 473–511.

Geyer, C.J. (1993) Estimating normalizing constants and reweighting mixtures in Markov chain Monte Carlo. Tech. Report 568, School of Statistics, Univ. of Minnesota.

Geyer, C.J. (1994) On the convergence of Monte Carlo maximum likelihood calculations. *J. R. Statist. Soc. B* **56**, 261-274.

Geyer, C.J. (1995) Conditioning in Markov Chain Monte Carlo. *J. Comput. Graph. Statis.* **4**, 148–154.

Geyer, C.J. (1996) Estimation and optimization of functions. In *Markov chain Monte Carlo in Practice*, W.R. Gilks, S.T. Richardson and D.J. Spiegelhalter (Eds.). 241–258. Chapman & Hall, London.

Geyer, C.J. and Thompson, E.A. (1992) Constrained Monte Carlo maximum likelihood for dependent data (with discussion). *J.Royal Statist. Soc.* (Ser. B) **54**, 657–699.

Geyer, C.J. and Thompson, E.A. (1995) Annealing Markov chain Monte Carlo with applications to ancestral inference. *J. Amer. Statist. Assoc.* **90**, 909-920.

Ghosh, M., Carlin, B. P. and Srivastiva, M. S. (1995) Probability matching priors for linear calibration. *TEST* **4**, 333–357.

Gilks, W.R. (1992) Derivative-free adaptive rejection sampling for Gibbs sampling. In *Bayesian Statistics 4*, J.M. Bernardo, J.O. Berger, A.P. Dawid and A.F.M. Smith (Eds.). 641–649. Oxford University Press, Oxford.

Gilks, W.R., Best, N.G. and Tan, K.K.C. (1995) Adaptive rejection Metropolis sampling within Gibbs sampling. *Applied Statist.* (Ser. C) **44**, 455–472.

Gilks, W.R., Clayton, D.G., Spiegelhalter, D.J., Best, N.G., McNeil, A.J., Sharples, L.D., and Kirby, A.J. (1993) Modelling complexity: applications of Gibbs sampling in medicine. *J. Roy. Statist. Soc. Ser. B* **55**, 39–52.

Gilks, W.R. and Roberts, G.O. (1996) Strategies for improving MCMC. In *Markov Chain Monte Carlo in Practice*. W.R. Gilks, S. Richardson, and D.J. Spiegelhalter (Eds.). 89–114. Chapman & Hall, London.

Gilks, W.R., Roberts G.O. and Sahu S.K. (1998) Adaptive Markov Chain Monte Carlo. *J. Amer. Statist. Assoc.* **93**, 1045–1054.

Gilks, W.R. and Wild, P. (1992) Adaptive rejection sampling for Gibbs sampling. *Appl. Statist.* **41**, 337–348.

Giudici, P. and Green, P.J. (1998) Decomposable graphical Gaussian model determination. Technical Report, Università di Pavia.

Gleick, J. (1987) *Chaos*. Penguin, New York.

Gleser, L.J. (1989) The Gamma distribution as a mixture of exponential distributions. *Am. Statist* **43**, 115-117.

Gleser, L.J. and Hwang, J.T. (1987) The non-existence of $100(1 - \alpha)\%$ confidence sets of finite expected diameters in errors-in-variable and related models. *Ann. Statist.* **15**, 1351–1362.

Goldfeld, S.M. and Quandt, R.E. (1973) A Markov model for switching regressions. *J. Econometrics* **1**, 3–16.

Gouriéroux, C. (1996) *ARCH Models*. Springer–Verlag, New York.

Gouriéroux, C. and Monfort, A. (1995) *Simulation Based Econometric Methods*. CORE Lecture Series, CORE, Louvain.

Gouriéroux, C. and Monfort, A. (1996) *Statistics and Econometric Models*. Cambridge University Press, Cambridge.

Gouriéroux, C., Monfort, A. and Renault, E. (1993) Indirect Inference. *J. Applied Econom.* **8**, 85–118.

Goutis, C. and Casella, G. (1999) Explaining the saddlepoint approximation. *The American Statist.* **53**.

Goutis, C. and Robert, C.P. (1998) Model choice in generalized linear models: a Bayesian approach via Kullback–Leibler projections. *Biometrika* **85** 29–37.

Green, P.J. (1995) Reversible jump MCMC computation and Bayesian model determination. *Biometrika* **82**(4), 711–732.

Green, P.J. and Han, X.L. (1992) Metropolis methods, Gaussian proposals, and antithetic variables. In *Stochastic methods and Algorithms in Image Analysis.* P. Barone, A. Frigessi and M. Piccioni (Eds.). Lecture Notes in Statistics 74. 142–164. Springer–Verlag, Berlin.

Green, P.J. and Murdoch, D. (1998) Exact sampling for Bayesian inference: towards general purpose algorithms. In *Bayesian Statistics 6.* J.O. Berger, J.M. Bernardo, A.P. Dawid, D.V. Lindley and A.F.M. Smith (Eds.). Oxford University Press, Oxford (to appear).

Grenander, U. and Miller, M. (1994) Representations of knowledge in complex systems (with discussion). *J. Roy. Statist. Soc. Ser. B* **56**, 549–603.

Gruet, M.A., Philippe, A. and Robert, C.P. (1998) MCMC control spreadsheets for exponential mixture estimation. *J. Comp. Graph Statist.* (to appear).

Guégan, D. (1994) *Séries Chronologiques non Linéaires à Temps Discret.* Economica, Paris.

Guihenneuc–Jouyaux, C. and Richardson, S. (1996) Modèle de Markov avec erreurs de mesure : approche bayésienne et application au suivi longitudinal des lymphocytes T4. In *Biométrie et Applications Bayésiennes.* Economica, Paris.

Guihenneuc–Jouyaux, C., Richardson, S. and Lasserre, V. (1998) Convergence assessment in latent variable models: apllication to longitudinal modelling of a marker of HIV progression. In *Discretization and MCMC Convergence Assessment.* C.P. Robert (Ed.). Chapter 7, 147–160. Lecture Notes in Statistics 135, Springer–Verlag, New York.

Guihenneuc–Jouyaux, C. and Robert, C.P. (1998) Finite Markov chain convergence results and MCMC convergence assessments. *J. Amer. Statist. Assoc.* **93**, 1055–1067.

Gumbel, E.J. (1940) La direction d'une répartition. *Ann. Univ. Lyon, 3A,* **2**, 39–51.

Gumbel, E.J. (1958) *Statistics of Extremes.* Columbia University Press, New York.

Haario, H. and Sacksman, E. (1991) Simulated annealing in general state space. *Adv. Appl. Probab.* **23**, 866–893.

Hàjek, B. (1988) Cooling schedules for optimal annealing. *Math. Operation. Research* **13**, 311–329.

Hall, P. (1992) *The Bootstrap and Edgeworth Expansion.* Springer–Verlag, New York.

Hall, P. (1994) On the erratic behavior of estimators of $N$ in the binomial $N$, $p$ distribution *J. Amer. Statist. Assoc.* **89** 344–352.

Hammersley, J.M. (1974) Discussion of Besag's paper. *J. Roy. Statist. Soc. Ser. B* **36**, 230–231.

Hammersley, J.M. and Clifford, M.S. (1970) Markov fields on finite graphs and lattices. Unpublished.

Hammersley, J.M. and Handscomb, D.C. (1964) *Monte Carlo Methods*, J. Wiley, New York.

Harris, T. (1956) The existence of stationary measures for certain Markov processes. In *Proc. 3rd Berkeley Symp. Math. Statis. Prob.* **2**. 113–124. University of California Press, Berkeley, CA.

Hastings, W.K. (1970) Monte Carlo sampling methods using Markov chains and their application. *Biometrika* **57**, 97–109.

Heidelberger, P. and Welch, P.D. (1983) A spectral method for confidence interval generation and run length control in simulations. *Comm. Assoc. Comput. Machinery* **24**, 233–245.

Heidelberger, P. and Welch, P.D. (1988) Simulation run length control in the presence of an initial transient. *Operations Research* **31**, 1109–1144.

Heitjan, D.F. and Rubin, D.B. (1991) Ignorability and coarse data. *Ann. Statist.* **19**, 2244–2253.

Hesterberg, T. (1998) Weighted average importance sampling and defensive mixture distributions. *Technometrics* **37**, 185–194.

Higdon, D.M. (1996) Auxiliary variable methods for Markov chain Monte Carlo with applications. Discussion paper, ISDS, Duke University.

Hills, S.E. and Smith, A.F.M. (1992) Parametrization issues in Bayesian inference. In *Bayesian Statistics 4*. J.M. Bernardo, J.O. Berger, A.P. Dawid and A.F.M. Smith (Eds.). 641-649. Oxford University Press, Oxford.

Hills, S.E. and Smith, A.F.M. (1993) Diagnostic plots for improved parameterization in Bayesian inference. *Biometrika* **80**, 61–74.

Hjorth, J.S.U. (1994) *Computer Intensive Statistical Methods*. Chapman & Hall, London.

Hobert, J. P and Casella, G. (1996) The effect of improper priors on Gibbs sampling in hierarchical linear models. *J. Amer. Statist. Assoc.* **91** 1461-1473.

Hobert, J. P. and Casella, G. (1998) Functional compatibility, Markov chains, and Gibbs sampling with improper posteriors. *J. Comp. Graph Statist.* **7**, 42–60.

Hobert, J.P. and Robert, C.P. (1999) Eaton's Markov chain, its conjugate partner and $P$-admissibility. *Ann. Statist.* (to appear).

Hobert, J.P., Robert, C.P. and Goutis, C. (1997) Connectedness conditions for the convergence of the Gibbs sampler. *Statis. Prob. Letters* **33**, 235–240.

Hobert, J.P., Robert, C.P. and Titterington, D.M. (1998) On perfect simulation for some mixtures of distributions. *Statistics and Computing* (to appear).

Holmes, S., Diaconis, P., and Neal, R. M. (1997) Analysis of a non-reversible Markov chain sampler. Tech. Report, Cornell University.

Hougaard, P. (1988) Discussion of the paper by Reid. *Statist. Sci.* **3**, 230-231.

Huber, P.J. (1972) Robust statistics: A review. *Ann. Math. Statist.* **47**, 1041–1067.

Hwang, C.R. (1980) Laplace's method revisited: Weak convergence of probability measures. *Ann. Probab.* **8**, 1177–1182.

Ikeda, M. and Watanabe, Y. (1981) *Stochastic Differential Equations and Diffusion Processes*. North-Holland/Elsevier, Amsterdam.

Jaquier, E., Polson, N.G. and Rossi, P.E. (1994) Bayesian analysis of stochastic volatility models (with discussion). *J. Business Economic Stat.* **12**, 371–417.

Jarrett, R.G. (1979) A note on the intervals between coal-mining disasters. *Biometrika* **66**, 191–193.

Jeffreys, H. (1961) *Theory of Probability* (3rd edition). Oxford University Press, Oxford. [First edition: 1939.]

Jennison, C. (1993) Discussion on the meeting on the Gibbs sampler and other Markov chain Monte Carlo methods. *J. Roy. Statist. Soc. Ser. B* **55**, 54–56.

Jensen, J. L. (1995) *Saddlepoint Approximations*. Clarendon Press, Oxford.

Jensen, S.T., Johansen, S. and Lauritzen, S.L. (1991) Globally convergent algorithms for maximizing a likelihood function. *Biometrika* **78**(4), 867-877.

Jöhnk, M.D. (1964) Erzeugung von Betaverteilten und Gammaverteilten Zufallszahen. *Metrika* **8**, 5–15.

Johnson, N.L. and Kotz, S.V. (1969–1972) *Distributions in Statistics* (4 vols.). J. Wiley, New York.

Johnson, N. L., Kotz. S., and Balakrishnan, N. (1994) *Continuous Univariate Distributions, Volume 1* (2nd edition). J. Wiley, New York.

Johnson, N. L., Kotz. S., and Balakrishnan, N. (1995) *Continuous Univariate Distributions, Volume 2* (2nd edition). J. Wiley, New York.

Johnson, N. L., Kotz. S., and Kemp, A. W. (1992) *Univariate Discrete Distributions* (2nd edition). J. Wiley, New York.

Johnson V.E. (1996) Studying convergence of Markov chain Monte Carlo algorithms using coupled sample paths. *J. Amer. Statist. Assoc.* **91**, 154–166.

Juang, B.H. (1984) On the hidden Markov model and dynamic time warping for speech recognition – A unified view. *BellTech* **63**, 1213–1243.

Juang, B.H. and Rabiner, L.R. (1991) Hidden Markov models for speech recognition. *Technometrics* **33**, 251–272.

Kass, R.E. and Raftery, A.E. (1995) Bayes factors. *J. Amer. Statist. Assoc.* **90**, 773–795.

Kemeny, J.G. and Snell, J.L. (1960) *Finite Markov Chains*. Van Nostrand, Princeton.

Kendall, W. (1998) Perfect simulation for the area-interaction point process. In *Probability Towards 2000*. Heyde, C.C. and Accardi, L. (Eds.). 218–234. Springer–Verlag, New York.

Kennedy, W.J. and Gentle, J.E. (1980) *Statistical Computing*. Marcel Dekker, New York.

Kersting, G. (1987) Some results on the asymptotic behavior of the Robbins-Monro process. *Bull. Int. Statis. Inst.* **47**, 327–335.

Kiefer, J. and Wolfowitz, J. (1952) Stochastic estimation of the maximum of a regression function. *Ann. Math. Statist.* **23**, 462–466.

Kiiveri, H. and Speed, T.P. (1982) Structural analysis of multivariate data: A review. In *Sociological Methodology, 1982*. S. Leinhardt (Ed.). 209–289. Jossey Bass, San Francisco.

Kim, S., Shephard, N. and Chib, S. (1998) Stochastic volatility: Likelihood inference and comparison with ARCH models. *Rev. Econom. Stud.* **65**, 361–393.

Kinderman, A., Monahan, J. and Ramage, J. (1977) Computer methods for sampling from Student's $t$-distribution. *Math. Comput.* **31**, 1009–1018.

Kipnis, C. and Varadhan, S.R. (1986) Central limit theorem for additive functionals of reversible Markov processes and applications to simple exclusions. *Comm. Math. Phys.* **104**, 1–19.

Kirkpatrick, S., Gelatt, C.D. and Vecchi, M.P. (1983) Optimization by simulated annealing. *Science* **220**, 671–680.

Knuth, D. (1981) *The Art of Computer Programing. Volume 2: Seminumerical Algorithms* (2nd edition). Addison-Wesley, Reading, MA.

Kolassa, J. E. (1994) *Series Approximation Methods in Statistics.* Springer-Verlag, New York.

Kong, A., Liu, J.S. and Wong, W.H. (1994) Sequential imputations and Bayesian missing data problems. *J. Amer. Statist. Assoc.* **89**, 278–288.

Kubokawa, T. and Robert, C.P. (1994) New perspectives on linear calibration. *J. Mult. Anal.* **51** 178–200.

Kuehl, R. O. (1994) *Statistical Principles of Research Design and Analysis.* Duxbury.

Kullback, S. (1968) *Information Theory and Statistics.* J. Wiley, New York.

Laird, N., Lange, N. and Stram, D. (1987) Maximum likelihood computations with repeated measures: Application of the EM algorithm. *J. Amer. Statist. Assoc.* **82**, 97–105.

Lauritzen, S.L. (1996) *Graphical Models.* Oxford University Press, Oxford.

Lavielle, M. and Moulines, E. (1997) On a stochastic approximation version of the EM algorithm. *Statist. Comput.* **7**, 229–236.

Lavine, M. and West, M. (1992) A Bayesian method for classification and discrimination. *Canad. J. Statist.* **20**, 451–461.

Le Cun, Y., Boser, D., Denker, J.S., Henderson, D., Howard, R.E., Hubbard, W., and Jackel, L.D. (1989) Handwritten digit recognition with a backpropagation network. In *Advances in Neural Information Processing Systems II.* D.S. Touresky (Ed.). 396–404. Morgan Kaufman, San Mateo, CA.

L'Ecuyer, P. and Cordeau, J.-F. (1996) Tests sur les points rapprochés par des générateurs pseudo-aléatoires. *Proc. 28th ASU Meeting*, 479–482.

Lee, P. (1989) *Bayesian Statistics: an Introduction.* Oxford University Press, Oxford.

Legendre, A. (1805) *Nouvelles Méthodes pour la Détermination des Orbites des Comètes.* Courcier, Paris.

Lehmann, E.L. (1975) *Nonparametrics: Statistical Methods based on Ranks.* McGraw-Hill, New York.

Lehmann, E.L. (1986) *Testing Statistical Hypotheses.* J. Wiley, New York.

Lehmann, E. L. (1998) *Introduction to Large-Sample Theory.* Springer–Verlag, New York.

Lehmann, E.L. and Casella, G. (1998) *Theory of Point Estimation* (revised edition). Springer–Verlag, New York.

Leroux, B.G. and Putterman, M.L. (1992) Maximum-penalised-likelihood estimation for independent and Markov dependent mixture models. *Biometrics* **48**, 545–558.

Levine, R. (1996) Post-processing random variables. Ph.D. Thesis, Biometrics Unit, Cornell University.

Liao, J.G. (1998) Variance reduction in Gibbs sampler using quasi-random numbers. *J. Comp. Graph Statist.* **7**(3), 253–266.

Lindsay, B.G. (1995) *Mixture Models: Theory, Geometry and Applications.* IMS Monographs, Hayward, CA.

Lindvall, T. (1992) *Lectures on the Coupling Theory.* J. Wiley, New York.

Little, R.J.A. and Rubin, D.B. (1983) On jointly estimating parameters and missing data by maximizing the complete-data likelihood. *The American Statistician* **37**(3), 218–220.

Little, R.J.A. and Rubin, D.B. (1987) *Statistical Analysis with Missing Data.* J. Wiley, New York.

Liu, C. and Rubin, D.B. (1994) The ECME algorithm: a simple extension of EM and ECM with faster monotonous convergence. *Biometrika* **81**, 633–648.

Liu, C., Liu, J.S., and Rubin, D.B. (1992) A variational control variable for assessing the convergence of the Gibbs sampler. In *Proceedings of the American Statistical Association*, Statistical Computing Section, 74–78.

Liu, C., Liu, J.S., and Rubin, D.B. (1995) A variational control for assessing the convergence of the Gibbs sampler. In *Proceedings of the American Statistical Association, Statistical Computing Section*, 74–78.

Liu, J.S. (1995) Metropolized Gibbs Sampler: An improvement. Tech. report, Dept. of Statistics, Stanford University, CA.

Liu, J.S. (1996) Peskun's Theorem and a mondified discrete-state Gibbs sampler. *Biometrika* **83** 681–682.

Liu, J.S. (1998) Metropolized independent sampling with comparisons to rejection sampling and importance sampling. *Statistics and Computing* (to appear).

Liu, J.S. and Chen, R. (1995) Blind deconvolution via sequential imputations. *J. Amer. Statist. Assoc.* **90**, 567–576.

Liu, J.S. and Sabbati, C. (1999) Simulated sintering: Markov chain Monte Carlo with spaces of varying dimension. In *Bayesian Statistics 6*, J.O. Berger, J.M. Bernardo, A.P. Dawid, D.V. Lindley and A.F.M. Smith (Eds.). Oxford University Press, Oxford (to appear).

Liu, J.S., Wong, W.H. and Kong, A. (1994) Covariance structure of the Gibbs sampler with applications to the comparisons of estimators and sampling schemes. *Biometrika* **81**, 27–40.

Liu, J.S., Wong, W.H. and Kong, A. (1995) Correlation structure and convergence rate of the Gibbs sampler with various scans. *J. Roy. Statist. Soc. Ser. B* **57**, 157–169.

Loh, W.L. (1996) On Latin hypercube sampling. *Ann. Statist.* **24**, 2058-2080.

Louis, T.A. (1982) Finding the observed information matrix when using the EM algorithm. *J. Roy. Statist. Soc. Ser. B* **44**, 226–233.

Lugannani, R. and Rice, S. (1980) Saddlepoint approximation for the distribution of the sum of independent random variables. *Adv. Appl. Probab.* **12** 475-490.

MacEachern, S.N. and Berliner, L.M. (1994) Subsampling the Gibbs sampler. *The American Statistician* **48**, 188–190.

MacLachlan, G. and Basford, K. (1988) *Mixture Models: Inference and Applications to Clustering.* Marcel Dekker, New York.

MacLachlan, G. and Krishnan, T. (1997) *The EM Algorithm and Extensions.* J. Wiley, New York.

Madigan, D. and York, J. (1995) Bayesian graphical models for discrete data. *International Statistical Review* **63**, 215-232.

Maguire, B.A., Pearson, E.S. and Wynn, A.H.A. (1952) The time interval between industrial accidents. *Biometrika* **38**, 168–180.

Mallick, B. and Gelfand, A.E. (1994) Generalized linear models with unknown link function. *Biometrika* **81**, 237–246.

Maritz, J.S. and Lwin, T. (1989) *Empirical Bayes Methods* (2nd edition). Chapman & Hall, London.

Marsaglia, G. (1964) Generating a variable from the tail of a normal distribution. *Technometrics* **6**, 101–102.

Marsaglia, G. (1977) The squeeze method for generating gamma variables. *Computers and Mathematics with Applications*, **3**, 321–325.

Marsaglia, G. and Zaman, A. (1993) The KISS generator. Tech. Report, Dept. of Statistics, University of Florida.

Martin-Löef, P. (1966) The definition of random sequences. *Information and Control* **9**, 602–619.

Matthews, P. (1993) A slowly mixing Markov chain with implications for Gibbs sampling. *Statist. Prob. Letters* **17**, 231–236.

McCullagh, P. and Nelder, J. (1989) *Generalized Linear Models.* Chapman & Hall, London.

McCulloch, C.E. (1997) Maximum likelihood algorithms for generalized linear mixed models. *J. Amer. Statist. Assoc.* **92**, 162–170.

McCulloch, R. and Tsay, R. (1994) Statistical analysis of macroeconomic time series via Markov switching models. *J. Time Series Analysis* **155**, 523–539.

McKay, M.D., Beckman, R.J., and Conover, W.J. (1979) A comparison of three methods for selecting values of output variables in the analysis of output from a computer code. *Technometrics* **21**, 239–245.

McKeague, I.W. and Wefelmeyer, W. (1996) Markov chain Monte Carlo and Rao–Blackwellisation. Tech. Report, Florida State University.

Mead, R. (1988) *The Design of Experiments.* Cambridge University Press, Cambridge.

Meng, X.L. and Rubin, D.B. (1991) Using EM to obtain asymptotic variance-covariance matrices. *J. Amer. Statist. Assoc.* **86**, 899–909.

Meng, X.L. and Rubin, D.B. (1992) Maximum likelihood estimation via the ECM algorithm: A general framework. *Biometrika* **80**, 267–278.

Meng, X.L. and van Dyk, D. (1997) The EM algorithm–an old folk-song sung to a new tune (with discussion). *J. Roy. Statist. Soc. Ser. B* **59**, 511–568.

Meng, X.L. and Wong, W.H. (1996) Simulating ratios of normalizing constants via a simple identity: a theoretical exploration. *Statist. Sinica* **6**, 831–860.

Mengersen, K.L. and Robert, C.P. (1996) Testing for mixtures: A Bayesian entropic approach (with discussion). In *Bayesian Statistics 5*, J.O. Berger, J.M. Bernardo, A.P. Dawid, D.V. Lindley and A.F.M. Smith (Eds.). 255–276. Oxford University Press, Oxford.

Mengersen, K.L., Robert, C.P. and Guihenneuc-Jouyaux, C. (1999) MCMC convergence diagnostics: a "reviewwww". In *Bayesian Statistics 6*. J.O. Berger, J.M. Bernardo, A.P. Dawid, D.V. Lindley and A.F.M. Smith (Eds.). Oxford University Press, Oxford (to appear).

Mengersen, K.L. and Tweedie, R.L. (1996) Rates of convergence of the Hastings and Metropolis algorithms. *Ann. Statist.* **24** 101–121.

Metropolis, N. and Ulam, S. (1949) The Monte Carlo method. *J. Amer. Statist. Assoc.* **44**, 335–341.

Metropolis, N., Rosenbluth, A.W., Rosenbluth, M.N., Teller, A.H., Teller, E. (1953) Equations of state calculations by fast computing machines. *J. Chem. Phys.* **21**, 1087–1092.

Meyn, S.P. and Tweedie, R.L. (1993) *Markov Chains and Stochastic Stability.* Springer–Verlag, New York.

Meyn, S.P. and Tweedie, R.L. (1994) Computable bounds for convergence rates of Markov chains. *Ann. Appl. Probab.* **4**, 981–1011.

Milshtein, G.N. (1974) Approximate integration of stochastic differential equations. *Theory of Probability and its Applications* **19**, 557-562.

Mira, A. and Geyer, C.J. (1998) Ordering Monte Carlo Markov chains. Tech. Report, University of Minnesota.

Mira, A. and Tierney, L. (1998) On the use of auxiliary variables in Markov chain Monte Carlo methods. Tech. Report, University of Minnesota.

Mitra, D., Romeo, F. and Sangiovanni-Vincentelli, A. (1986) Convergence and finite-time behaviour of simulated annealing. *Advances in Applied Probability* **18**, 747-771.

Moussouris, J. (1974) Gibbs and Markov random systems with constraints. *J. Statist. Phys.* **10**, 11–33.

Müller, P. (1991) A generic approach to posterior integration and Gibbs sampling. Tech. Report # 91-09, Purdue University, West Lafayette, Indiana.

Müller, P. (1993) Alternatives to the Gibbs sampling scheme. Tech. Report, Institute of Statistics and Decision Sciences, Duke University.

Müller, P., Erkanli, A. and West, M. (1996) Curve fitting using Dirichlet process mixtures. *Biometrika* **83**(1), 55-66.

Murray, G.D. (1977) Comments on "Maximum likelihood from incomplete data via the EM algorithm." *J. Roy. Statist. Soc. Ser. B* **39**, 27–28.

Murdoch D.J. and Green, P.J. (1998) Exact sampling for a continuous state *Scandinavian J. Statist.* **25**(3), 483–502.

Muri, F., Chauveau, D. and Cellier, D. (1998) Convergence assessment in latent variable models: DNA applications. In *Discretization and MCMC Convergence Assessment.* C.P. Robert (Ed.). Chapter 6, 127–146. Lecture Notes in Statistics 135, Springer–Verlag, New York.

Mykland, P., Tierney, L. and Yu, B. (1995) Regeneration in Markov chain samplers. *J. Amer. Statist. Assoc.* **90**, 233–241.

Natarajan, R. and McCulloch, C.E. (1998) Gibbs sampling with diffuse proper priors: A valid approach to data-driven inference? *J. Comp. Graph Statist.* **7**, 267–277.

Naylor, J.C. and Smith, A.F.M. (1982) Application of a method for the efficient computation of posterior distributions. *Applied Statistics* **31**, 214–225.

Neal, R.M. (1993) Probabilistic inference using Markov chain Monte Carlo methods. Ph.D. thesis, Dept. of Computer Science, University of Toronto.

Neal, R.M. (1995) Suppressing random walks in MCMC using ordered over-relaxation. Dept. of Statistics, University of Toronto, Canada.

Neal, R.M. (1996) *Bayesian Learning for Neural Networks.* Lecture Notes 118, Springer–Verlag, New York.

Neal, R.M. (1997) Markov chain Monte Carlo methods based on 'slicing' the density function. University of Toronto.

Newton, M.A. and Raftery, A.E. (1994) Approximate Bayesian inference by the weighted likelihood boostrap (with discussion). *J. Roy. Statist. Soc. Ser. B* **56**, 1–48.

Niederreiter, H. (1992) *Random Number Generation and Quasi-Monte Carlo Methods.* SIAM, Philadelphia.

Nobile, A. (1998) A hybrid Markov chain for the Bayesian analysis of the multinomial probit model. *Statistics and Computing* **8**, 229–242.

Norris, J.R.. (1997) *Markov Chains.* Cambridge University Press, Cambridge.

Nummelin, E. (1978) A splitting technique for Harris recurrent chains. *Zeit. Warsch. Verv. Gebiete* **43**, 309–318.

Nummelin, E. (1984) *General Irreducible Markov Chains and Non-Negative Operators.* Cambridge University Press, Cambridge.

Ó Ruanaidh, J.J.K. and Fitzgerald, W.J. (1996) *Numerical Bayesian Methods Applied to Signal Processing.* Springer–Verlag, New York.

Oh, M.S. (1989) Integration of multimodal functions by Monte Carlo importance sampling, using a mixture as an importance function. Tech. Report, Dept. of Statistics, University of California, Berkeley.

Oh, M.S. and Berger, J.O. (1993) Integration of multimodal functions by Monte Carlo importance sampling. *J. Amer. Statist. Assoc.* **88**, 450–456.

Olkin, I., Petkau, A.J. and Zidek, J.V. (1981) A comparison of $n$ estimators for the binomial distribution. *J. Amer. Statist. Assoc.* **76**, 637–642.

Olver, F.W.J. (1974) *Asymptotics and Special Functions.* Academic Press, New York.

Orey, S. (1971) *Limit Theorems for Markov Chain Transition Probabilities.* Van Nostrand, London.

Osborne, C. (1991) Statistical calibration: a review. *International Statistical Review* **59**, 309–336.

Pearl, J. (1988) *Probabilistic Reasoning in Intelligent Systems: Networks of Plausible Inference.* Morgan Kaufmann, San Mateo, CA.

Pearson, K. (1894) Contribution to the mathematical theory of evolution. *Proc. Trans. Roy. Soc. A* **185**, 71–110.

Pearson, K. (1915) On certain types of compound frequency distributions in which the individual components can be individually described by binomial series. *Biometrika* **11**, 139–144.

Perk, W. (1947) Some observations on inverse probabilities including a new indifference rule. *J. Inst. Actuaries* **73**, 285–312.

Peskun, P.H. (1973) Optimum Monte Carlo sampling using Markov chains. *Biometrika* **60**, pp. 607–612.

Peskun, P.H. (1981) Guidelines for chosing the transition matrix in Monte Carlo methods using Markov chains. *Journal of Computational Physics* **40**, pp. 327–344.

Pflug, G.C. (1994) Distributed stochastic approximation: Does cooperation pay? Tech. Report TR-94-07, Institute of Statistics, University of Wien, Wien.

Philippe, A. (1997a) Simulation output by Riemann sums. *J. Statist. Comput. Simul.* **59**(4), 295-314.

Philippe, A. (1997b) Importance sampling and Riemann sums. *Prepub. IRMA* **43**, VI, Université de Lille.

Philippe, A. (1997c) Simulation of right and left truncated Gamma distributions by mixtures. *Stat. Comput.* **7**, 173–181.

Philippe, A. and Robert, C.P. (1998a) A note on the confidence properties of the reference prior for the calibration model. *TEST* **7**(1), 147–160.

Philippe, A. and Robert, C.P. (1998b) Linking discrete and continuous chains. In *Discretization and MCMC Convergence Assessment*, C.P. Robert (Ed.). Chapter 3, 47–66. Lecture Notes in Statistics, Springer–Verlag, New York.

Phillips, D.B. and Smith, A.F.M. (1996) Bayesian model comparison via jump diffusions. In *Markov chain Monte Carlo in Practice*, W.R. Gilks, S.T. Richardson and D.J. Spiegelhalter (Eds.). 215–240. Chapman & Hall, London.

Pike, M.C. (1966) A method of analyis of a certain class of experiments in carcinogenesis. *Biometrics* **22**, 142–161.

Pincus, M. (1968) A closed form solution of certain programming problems. *Oper. Research* **18**, 1225–1228.

Polson, N.G. (1996) Convergence of Markov chain Monte Carlo Algorithms (with discussion). In *Bayesian Statistics 5*. J.M. Bernardo, J.O. Berger, A.P. Dawid and A.F.M. Smith (Eds.). 297–323. Oxford University Press, Oxford.

Propp, J.G. and Wilson, D.B. (1996) Exact sampling with coupled Markov chains and applications to statistical mechanics. *Random Structures and Algorithms* **9**, 223–252.

Qian, W. and Titterington, D.M. (1990) Parameter estimation for hidden Gibbs chains. *Statis. Prob. Letters* **10**, 49-58.

Qian, W. and Titterington, D.M. (1991) Estimation of parameters in hidden Markov models. *Phil. Trans. Roy. Soc. London* A **337**, 407–428.

Qian, W. and Titterington, D.M. (1992) Stochastic relaxations and EM algorithms for Markov random fields. *J. Statist. Comput. Simulation* **40**, 55–69.

Raftery, A.E. and Akman, V.E. (1986) Bayesian analysis of a Poisson process with a change point. *Biometrika* **73**, 85–89.

Raftery, A.E. and Banfield, J.D. (1991) Stopping the Gibbs sampler, the use of morphology, and other issues in spatial statistics. *Ann. Inst. Statist. Math.* **43**, 32–43.

Raftery, A.E. and Lewis, S. (1992a) How many iterations in the Gibbs sampler? In *Bayesian Statistics 4*. J.O. Berger, J.M. Bernardo, A.P. Dawid and A.F.M. Smith (Eds). 763–773. Oxford University Press, Oxford.

Raftery, A.E. and Lewis, S. (1992b) The number of iterations, convergence diagnostics and generic Metropolis algorithms. Tech. Report, Department of Statistics, University of Washington, Seattle.

Raftery, A.E. and Lewis, S. (1996) Implementing MCMC. In *Markov chain Monte Carlo in Practice* W.R. Gilks, S.T. Richardson and D.J. Spiegelhalter (Eds.). 115–130. Chapman & Hall, London.

Raiffa, H. and Schlaifer, R. (1961) *Applied Statistical Decision Theory*. Division of Research, Graduate School of Business Administration, Harvard University.

Randles, R. and Wolfe, D. (1979) *Introduction to the Theory of Nonparametric Statistics*. J. Wiley, New York.

Rao, C.R. (1973) *Linear Statistical Inference and its Applications*. J. Wiley, New York.

Redner, R. and Walker, H. (1984) Mixture densities, maximum likelihood and the EM algorithm. *SIAM Rev.* **26**, 195–239.

Reid, N. (1988) Saddlepoint methods and statistical inference (with discussion). *Statist. Sci.* **3**, 213–238.

Reid, N. (1991) Approximations and asymptotics. In *Statistical Theory and Modelling, Essays in Honor of D. R. Cox.*. 287–334. Chapman & Hall, London..

Resnick, S. (1994) *Adventures in Stochastic Process*. Birkhauser, Basel.

Revuz, D. (1984) *Markov Chains* (2nd edition). North-Holland, Amsterdam.

Richardson, S. and Green, P.J. (1997) On Bayesian analysis of mixtures with an unknown number of components (with discussion). *J. Roy. Statist. Soc. Ser. B* **59**, 731–792.

Ripley, B.D. (1987) *Stochastic Simulation*. J. Wiley, New York.

Ripley, B.D. (1994) Neural networks and related methods for classification (with discussion). *J. Roy. Statist. Soc. Ser. B* **56**, 409–4560.

Ripley, B.D. (1996) *Pattern Recognition and Neural Networks*. Cambridge: Cambridge University Press.

Ritter, C. and Tanner, M.A. (1992) Facilitating the Gibbs sampler: The Gibbs stopper and the Griddy-Gibbs sampler. *J. Amer. Statist. Assoc.* **87**, 861–868.

Robbins, H. (1964) The empirical Bayes approach to statistical decision problems. *Ann. Math. Statist.* **35** 1-20.

Robbins, H. (1983) Some thoughts on empirical Bayes estimation. *Ann. Statist.* **11**, 713-723.

Robbins, H. and Monro, S. (1951) A stochastic approximation method. *Ann. Math. Statist.* **22**, 400–407.

Robert, C. (1991) *Modèles Statistiques pour l'Intelligence Artificielle : l'Exemple du Diagnostic Médical*. Masson, Paris.

Robert, C.P. (1988) An explicit formula for the risk of the positive-part James–Stein estimator. *Canadian J. Statis.* **16**, 161–168.

Robert, C.P. (1990) Modified Bessel functions and their applications in Probability and Statistics. *Statist. Prob. Lett.* **9**, 155–161.

Robert, C.P. (1991) Generalized Inverse Normal distributions. *Statist. Prob. Lett.* **11**, 37–41.

Robert, C.P. (1993) Prior Feedback: A Bayesian approach to maximum likelihood estimation. *Comput. Statist.* **8**, 279–294.

Robert, C.P. (1994a) *The Bayesian Choice.* Springer–Verlag, New York.

Robert, C.P. (1994b) Discussion of "Markov chains for exploring posterior distribution". *Ann. Statist.* **22**, 1742–1747.

Robert, C.P. (1995a) Simulation of truncated normal variables. *Statistics and Computing* **5**, 121–125.

Robert, C.P. (1995b) Convergence control techniques for Markov chain Monte Carlo algorithms. *Statis. Science* **10**(3), 231–253.

Robert, C.P. (1996a) *Méthodes de Monte Carlo par Chaînes de Markov.* Economica, Paris.

Robert, C.P. (1996b) Inference in mixture models. In *Markov Chain Monte Carlo in Practice*, W.R. Gilks, S. Richardson and D.J. Spiegelhalter (Eds.). 441–464. Chapman & Hall, London.

Robert, C.P. (1996c) Intrinsic loss functions. *Theory and Decision* **40**(2), 191–214.

Robert, C.P. (1997) Discussion of Richardson and Green's paper. *J. Roy. Statist. Soc. Ser. B* **59**, 758–764.

Robert, C.P. (1998) *Discretization and MCMC Convergence Assessment.* Lecture Notes in Statistics 135, Springer–Verlag, New York.

Robert, C.P., Celeux, G. and Diebolt, J. (1993) Bayesian estimation of hidden Markov models: A stochastic implementation. *Statistics & Probability Letters* **16**, 77–83.

Robert, C.P. and Hwang, J.T.G. (1996) Maximum likelihood estimation under order constraints. *J. Amer. Statist. Assoc.* **91**, 167–173.

Robert, C.P. and Mengersen, K.L. (1999) Reparametrization issues in mixture estimation and their bearings on the Gibbs sampler. *Comput. Statis. Data Ana.* **29**, 325–343.

Robert, C.P., Rydén, T. and Titterington, D.M. (1998) Convergence controls for MCMC algorithms, with applications to hidden Markov chains. Doc. Travail DT9805, CREST, INSEE.

Robert, C.P. and Soubiran, C. (1993) Estimation of a mixture model through Bayesian sampling and prior feedback. *TEST* **2**, 125–146.

Robert, C.P. and Titterington, M. (1998) Reparameterisation strategies for hidden Markov models and Bayesian approaches to maximum likelihood estimation. *Statistics and Computing* **8**(2), 145–158.

Roberts, G.O. (1992) Convergence diagnostics of the Gibbs sampler. In *Bayesian Statistics 4.* J.M. Bernardo, J.O. Berger, A.P. Dawid and A.F.M. Smith (Eds.). 775–782. Oxford University Press, Oxford.

Roberts, G.O. (1994) Methods for estimating $L^2$ convergence of Markov chain Monte Carlo. In *Bayesian Statistics and Econometrics: Essays in Honor of Arnold Zellner.* D. Barry, K. Chaloner and J. Geweke (Eds.). J. Wiley, New York.

Roberts, G.O. (1998) Introduction to MCMC. Talk at the *TMR Spatial Statistics Network Workshop*, Aussois, France, January 19–21.

Roberts, G.O., Gelman, A. and Gilks, W.R. (1997) Weak convergence and optimal scaling of random walk Metropolis algorithms. *Annals of Applied Probability* **7**, 110–120.

Roberts, G.O. and Polson, N. (1994) A note on the geometric convergence of the Gibbs sampler. *J. Roy. Statist. Soc. Ser. B* **56**, 377–384.

Roberts, G.O. and Rosenthal, J.S. (1998) Markov chain Monte Carlo: Some practical implications of theoretical results (with discussion). *Can. J. Statist.* **26**, 5–32.

Roberts, G.O., Rosenthal, J.S. and Schwartz, P.C. (1995) Convergence properties of perturbated Markov chains. Res. report, Statistics Laboratory, University of Cambridge.

Roberts, G.O. and Sahu, S.K. (1997) Updating schemes, covariance structure, blocking and parametrisation for the Gibbs sampler. *J. Roy. Statist. Soc. Ser. B* **59**, 291–318.

Roberts, G.O. and Tweedie, R.L. (1995) Exponential convergence for Langevin diffusions and their discrete approximations. Res. Report, Statistics Laboratory, University of Cambridge.

Roberts, G.O. and Tweedie, R.L. (1996) Geometric convergence and Central Limit Theorems for multidimensional Hastings and Metropolis algorithms. *Biometrika* **83**, 95–110.

Robertson, T., Wright, F.T. and Dykstra, R.L. (1988) *Order Restricted Statistical Inference.* J. Wiley, New York.

Robinson, G.K. (1982) Behrens-Fisher problem. *Encyclopedia of Statistical Sciences*, **1**, 205–209. S. Kotz and N. Johnson (Eds.). J. Wiley, New York.

Roeder, K. (1992) Density estimation with confidence sets exemplified by superclusters and voids in galaxies. *J. Amer. Statist. Assoc.* **85**, 617–624.

Roeder, K. and Wasserman, L. (1997) Practical Bayesian density estimation using mixtures of Normals. *J. Amer. Statist. Assoc.* **92**, 894–902.

Romano, J.P. and Siegel, A.F. (1986) *Counterexamples in Probability and Statistics.* Wadsworth, Belmont, CA.

Rosenblatt, M. (1971) *Markov Processes: Structure and Asymptotic Behavior.* Springer–Verlag, New York.

Rosenthal, J.S. (1995) Minorization conditions and convergence rates for Markov chain Monte Carlo. *J. Amer. Statist. Assoc.* **90**, 558–566.

Rubin, D.B. (1987) *Multiple Imputation for Nonresponse in Surveys.* J. Wiley, New York.

Rubinstein, R.Y. (1981) *Simulation and the Monte Carlo Method.* J. Wiley, New York.

Rudin, W. (1976) *Principles of Real Analysis.* McGraw-Hill, New York.

Ruelle, D. (1987) *Chaotic Evolution and Strange Attractors.* Cambridge University Press, Cambridge.

Saxena, K. and Alam, K. (1982) Estimation of the non-centrality parameter of a chi-squared distribution. *Ann. Statist.* **10**, 1012–1016.

Scherrer, J. A. (1997) Monte Carlo estimation of transition probabilities in capture-recapture data. Masters Thesis, Biometrics Unit, Cornell University, Ithaca, New York.

Schervish, M.J. (1995) *Theory of Statistics.* Springer–Verlag, New York.

Schervish, M.J. and Carlin, B. (1992) On the convergence of successive substitution sampling. *J. Comput. Graphical Statist.* **1**, 111–127.

Schmeiser, B. (1989) Simulation experiments. Working paper SMS 89-23, School of Industrial Engineering, Purdue University.

Schmeiser, B. and Shalaby, M. (1980) Acceptance/rejection methods for beta variate generation. *J. Amer. Statist. Assoc.* **75**, 673–678.

Schoener, T.W. (1968) The *anolis* lizards of Bimini: Resource partitioning in a complex fauna. *Ecology* **49**, 724–726.

Schruben, L., Singh, H. and Tierney, L. (1983) Optimal tests for initialization bias in simulation output. *Operation. Research* **31**, 1176–1178.

Schukken, Y.H., Casella, G., and van den Broek, J. (1991) Overdispersion in clinical mastitis data from dairy herds: a negative binomial approach. *Preventive Veterinary Medicine* **10**, 239–245.

Searle, S.R., Casella, G. and McCulloch, C.E. (1992) *Variance Components.* J. Wiley, New York.

Seber, G.A.F. (1983) Capture-recapture methods. In *Encyclopedia of Statistical Science.* S. Kotz and N. Johnson (Eds.). J. Wiley, New York.

Seber, G.A.F. (1992) A review of estimating animal abondance (II). *Workshop on Design of Longitudinal Studies and Analysis of Repeated Measure Data* **60**, 129–166.

Seidenfeld, T. and Wasserman, L. (1993) Dilation for sets of probabilities. *Ann. Statist.* **21**, 1139–1154.

Serfling, R.J. (1980) *Approximation Theory of Mathematical Statistics.* J. Wiley, New York.

Shephard, N. (1994) Partial non-Gaussian state space. *Biometrika* **81**, 115–131.

Shephard, N. and Pitt, M.K. (1997) Likelihood analysis of non-Gaussian measurement time series. *Biometrika* **84**, 653-668.

Silverman, B.W. (1986) *Density Estimation for Statistics and Data Analysis.* Chapman & Hall, London.

Smith, A.F.M. and Gelfand A.E. (1992) Bayesian statistics without tears: A sampling-resampling perspective. *The American Statistician* **46**, 84–88.

Smith, A.F.M. and Makov, U.E. (1978) A quasi–Bayes sequential procedure for mixtures. *J. Roy. Statist. Soc. Ser. B* **40**, 106–112.

Smith, A.F.M. and Roberts, G.O. (1993) Bayesian computation via the Gibbs sampler and related Markov chain Monte Carlo methods (with discussion). *J. Roy. Statist. Soc. Ser. B* **55**, 3–24.

Spiegelhalter, D.J., Best, N.G., Gilks, W.R. and Inskip, H. (1996) Hepatitis B: a case study in MCMC methods. In *Markov Chain Monte Carlo in Practice.* W.R. Gilks, S. Richardson and D.J. Spiegelhalter (Eds.). 21–44. Chapman & Hall, London.

Spiegelhalter, D.J., Dawid, A.P., Lauritzen, S.L. and Cowell, R.G. (1993) Bayesian analysis in expert systems (with discussion). *Statist. Science* **8**, 219–283.

Spiegelhalter, D.J. and Lauritzen, S.L. (1990) Sequential updating of conditional probabilities on directed graphical structures. *Networks* **20**, 579-605.

Spiegelhalter, D.J., Thomas, A., Best, N. and Gilks, W.R. (1995a) BUGS: Bayesian Inference Using Gibbs Sampling. Version 0.50, Tech. Report, Medical Research Council Biostatistics Unit, Institute of Public Health, Cambridge University.

Spiegelhalter, D.J., Thomas, A., Best, N. and Gilks, W.R. (1995b) BUGS Examples Volume 1, Version 0.50. MRC Biostatistics Unit, Cambridge University.

Spiegelhalter, D.J., Thomas, A., Best, N. and Gilks, W.R. (1995c) BUGS Examples Volume 2, Version 0.50. MRC Biostatistics Unit, Cambridge University.

Stein, M.L. (1987) Large sample properties of simulations using Latin hypercube sampling. *Technometrics* **29**, 143–151.

Stephens, D. and Dellaportas, P. (1992) Bayesian analysis of generalised linear models with covariate measurement error. In *Bayesian Statistics 4*. J.M. Bernardo, J.O. Berger, A.P. Dawid and A.F.M. Smith (Eds.). Clarendon Press, Oxford.

Stigler, S. (1986) *The History of Statistics*. Belknap, Harvard.

Stramer, O. and Tweedie, R.L. (1997a) Geometric and subgeometric convergence of diffusions with given stationary distributions, and their discretizations. Tech. Report, University of Iowa.

Stramer, O. and Tweedie, R.L. (1997b) Self-targeting candidates for Hastings-Metropolis algorithms. Tech. Report, University of Iowa.

Strawderman, R. (1996) Discussion of Casella's article. *TEST* **5** 325-329.

Stuart, A. (1962) Gamma-distributed products of independent random variables. *Biometrika* **49**, 564–565.

Stuart, A. and Ord, J. K. (1987) *Kendall's Advanced Theory of Statistics*, Vol. I (5th edition). Oxford University Press, New York.

Stuchlik, J.B., Robert, C.P. and Plessis, B. (1994) Character recognition through Bayes theorem. Tech. Report DT9439, CREST, INSEE, Paris.

Swendson, R.H. and Wang, J.S. (1987) Nonuniversal critical dynamics in Monte Carlo simulations. *Physical Review Letters* **58**, 86–88.

Swendson, R.H., Wang, J.S. and Ferrenberg, A.M. (1992) New Monte Carlo methods for improved efficiency of computer simulations in statistical mechanics. In *The Monte Carlo Method in Condensed Matter Physics*. K. Binder (Ed.). Springer–Verlag, Berlin.

Talay, D. (1995) Simulation and numerical analysis of stochastic differential systems: A review. In *Probabilistic Methods in Applied Physics*. P. Krée and W. Wedig (Eds.). Lecture Notes in Physics 451. Chapter 3, 54–96. Springer–Verlag, Berlin.

Tanner, M. (1996) *Tools for Statistical Inference: Observed Data and Data Augmentation Methods* (3rd edition). Springer–Verlag, New York.

Tanner, M. and Wong, W. (1987) The calculation of posterior distributions by data augmentation. *J. Amer. Statist. Assoc.* **82**, 528–550.

Thisted, R.A. (1988) *Elements of Statistical Computing: Numerical Computation*. Chapman & Hall, London.

Tierney, L. (1994) Markov chains for exploring posterior distributions (with discussion). *Ann. Statist.* **22**, 1701–1786.

Tierney, L. (1995) A note on Metropolis–Hastings kernels for general state spaces. Tech. Rep. 606, University of Minnesota.

Tierney, L. and Kadane, J.B. (1986) Accurate approximations for posterior moments and marginal densities. *J. Amer. Statist. Assoc.* **81**, 82–86.

Tierney, L., Kass, R.E. and Kadane, J.B. (1989) Fully exponential Laplace approximations to expectations and variances of non-positive functions. *J. Amer. Statist. Assoc.* **84**, 710–716.

Tierney, L., and Mira, A. (1998) Some adaptive Monte Carlo methods for Bayesian inference. *Statistics in Medicine* (to appear).

Titterington, D.M., Smith, A.F.M. and Makov, U.E. (1985) *Statistical Analysis of Finite Mixture Distributions.* J. Wiley, New York.

Tobin, J. (1958) Estimation of relationships for limited dependent variables. *Econometrics* **26**, 24–36.

Tukey, J.W. (1977) *Exploratory Data Analysis.* Addison-Wesley, New York.

Van Dijk, H.K. and Kloeck, T. (1984) Experiments with some alternatives for simple importance sampling in Monte Carlo integration. In *Bayesian Statistics II.* J.M. Bernardo, M.H. DeGroot, D.V. Lindley and A.F.M. Smith (Eds.). North-Holland, Amsterdam.

Van Laarhoven, P.J. and Aarts, E.H.L. (1987) *Simulated Annealing: Theory and Applications.* CWI Tract 51. Reidel, Amsterdam.

Verdinelli, I. and Wasserman, L. (1992) Bayesian analysis of outliers problems using the Gibbs sampler. *Statist. Comput.* **1**, 105–117.

Vines, S.K. and Gilks, W.R. (1994) Reparameterising random interactions for Gibbs sampling. Tech. Report, MRC Biostatistics Unit, Cambridge University.

Vines, S.K., Gilks, W.R. and Wild, P. (1995) Fitting multiple random effect models. Tech. Report, MRC Biostatistics Unit, Cambridge University.

Von Neumann, J. (1951) Various techniques used in connection with random digits. *J. Resources of the National Bureau of Standards – Applied Mathematics Series* **12**, 36–38.

Wahba, G. (1981) Spline interpolation and smoothing on the sphere. *SIAMSSC* **2**, 5–16.

Wakefield, J.C., Gelfand, A.E. and Smith, A.F.M. (1991) Efficient generation of random variates via the ratio-of-uniforms method. *Statistics and Computing* **1**, 129–33.

Wakefield, J.C., Smith, A.F.M., Racine-Poon, A. and Gelfand, A.E. (1994) Bayesian analysis of linear and non-linear population models using the Gibbs sampler. *Applied Statistics* (Ser. C) **43**, 201–22.

Walker, S. (1997) Robust Bayesian analysis via scale mixtures of Beta distributions. Tech. Report, Dept. of Mathematics, Imperial College, London.

Wand, M.P. and Jones, M.C. (1995) *Kernel Smoothing.* Chapman & Hall, London.

Wasan, M.T. (1969) *Stochastic Approximation.* Cambridge University Press, Cambridge.

Wei, G.C.G. and Tanner, M.A. (1990a) A Monte Carlo implementation of the EM algorithm and the poor man's data augmentation algorithm. *J. Amer. Statist. Assoc.* **85**, 699–704.

Wei, G.C.G. and Tanner, M.A. (1990b) Posterior computations for censored regression data. *J. Amer. Statist. Assoc.* **85**, 829–839.

West, M. (1992) Modelling with mixtures. In *Bayesian Statistics 4*. J.O. Berger, J.M. Bernardo, A.P. Dawid and A.F.M. Smith (Eds). 503–525. Oxford University Press, Oxford.

Whittaker, J. (1990) *Graphical Models in Applied Multivariate Statistics*. Wiley, Chichester.

Winkler, G. (1995) *Image Analysis, Random Fields and Dynamic Monte Carlo Methods*. Springer–Verlag, New York.

Wolff, U. (1989) *Physical Review Letters* **62**, 361.

Wolter, W. (1986) Some coverage error models for census data. *J. Amer. Statist. Assoc.* **81**, 338–346.

Wood, A. T. A., Booth, J. G. and Butler, R. W. (1993) Saddlepoint approximations to the CDF of some statistics with nonnormal limit distributions. *J. Amer. Statist. Assoc.* **88**, 680-686.

Wu, C.F.J. (1983) On the convergence properties of the EM algorithm. *Ann. Statist.* **11**, 95–103.

Yakowitz, S., Krimmel, J.E. and Szidarovszky, F. (1978) Weighted Monte Carlo integration. *SIAM J. Numer. Anal.* **15**(6), 1289–1300.

Yao, Y.C. (1984) Estimation of a noisy discrete-time step function: Bayes and empirical Bayes approaches. *Ann. Statist.* **12**, 1434–1447.

Yu, B. (1994) Monitoring the convergence of Markov samplers based on estimated $L^1$ error. Tech. Report 9409, Department of Statistics, University of California, Berkeley.

Yu, B. (1995) Discussion to Besag *et al.* (1995) *Statistical Science* **10**, 3-66.

Yu, B. and Mykland, P. (1998) Looking at Markov samplers through cusum path plots: A simple diagnostic idea. *Statistics and Computing* **8**(3), 275–286.

Zeger, S. and Karim, R. (1991) Generalized linear models with random effects; a Gibbs sampling approach. *J. Amer. Statist. Assoc.* **86**, 79–86.

Zellner, A. and Min, C.K. (1995) Gibbs sampler convergence criteria. *J. Amer. Statist. Assoc.* **90**, 921–927.

# Author Index

Aarts, E., 201
Abramowitz, M., 8, 21, 29, 44, 129
Ackley, D.H., 199
Ahrens, J., 64
Aitkin, M., 345
Akman, V.E., 434, 444
Alam, K., 74
Albert, J.H., 335, 418, 430, 448, 450
Aldous, D., 199, 343
Anderson, D.R., 420
Andrieu, C., 202
Archer, G.E.B., 448
Arnold, B.C., 341
Asmussen, S., 365, 388
Athreya, K.B., 149, 152, 189, 388
Atkinson, A., 55, 56

Balakrishnan, N., 451
Banfield, J.D., 397
Barbe, P., 33
Barndorff-Nielsen, O.E., 71, 138
Barone, P., 297
Barron, A., 32
Barry, D., 448
Basford, K., 422
Basu, D., 22
Baum, L.E., 342
Bauwens, L., 24, 72
Beckman, R.J., 137
Bennett, J.E., 299
Bergé, P., 38
Berger, J.O., 5, 11, 13, 18, 26, 28, 32,
    75, 80, 125, 167, 222, 328,
    379, 451
Berger, R., 5, 11, 16, 28, 226, 342
Berliner, L.M., 365
Bernardo, J.M., 11, 32, 222, 423
Bertail, P., 33

Besag, J., 254, 265, 279, 286, 290, 294,
    298, 299, 332, 335, 341, 349,
    355
Best, D.J., 54
Best, N.G., 58, 249, 300, 344–347, 355,
    356, 367, 410, 411
Billingsley, P., 33, 35, 70, 139, 163,
    168, 179, 190, 316
Billio, M., 210, 212, 334, 430, 432, 448
Boch, R.D., 345
Bollerslev, T., 211
Booth, J.G., 138, 244
Boser, D., 229
Bouleau, N., 229, 230
Box, G.E.P., 46
Boyles, R.A., 215
Bradley, R.C., 191
Breslow, N.E., 346
Brockwell, P.J., 409
Broniatowski, M., 227, 349, 415
Brooks, S.P., 363, 367, 376, 381,
    397–399, 408, 410
Brown, L.D., 5, 28, 67, 115, 188, 222
Brownie, C., 218, 420
Bucklew, J.A., 37, 132
Burnham, K.D., 420
Butler, R.W., 138

Carlin, B.P., 16, 73, 264, 326, 327,
    340, 344, 345, 350, 352, 353,
    355, 363, 367, 398, 408, 410,
    411, 433, 434, 445
Carroll, R.J., 22, 450
Casella, G., 5, 11, 16, 22, 28, 32, 33,
    66, 73, 74, 79, 85, 93, 94,
    105, 116–119, 127, 226,
    255–258, 279, 300, 311,
    328–332, 334, 341, 342, 425
Castledine, B., 307

Celeux, G., 227, 349, 415, 425, 430, 448
Cellier, D., 405, 413, 448
Chaitin, G., 36
Chan, K.S., 155, 184, 191
Chauveau, D., 228, 393, 413, 448
Chen, M.H., 122, 221, 249, 321, 323, 407
Chen, R., 440
Cheng, B., 55, 228
Cheng, R.C.H., 64
Chernoff, H., 449
Chib, S., 211, 264, 276, 279, 334, 343, 418, 425, 430, 435, 436, 448, 450
Chou, R.Y., 211
Chung, K.L., 139
Churchill, G.A., 201
Cipra, B.A., 201
Clairambault, J., 448
Clarke, B., 32
Clayton, D.G., 300, 326, 346, 450
Clifford, M.S., 298, 349
Cocozza-Thivent, C., 448
Conover, W.J., 137
Cook, J.R., 450
Copas, J. B., 22
Cordeau, J.F., 61
Cowell, R., 357
Cowell, R.G., 2
Cowles, M.K., 363, 367, 398, 402, 408, 410, 411
Cox, D.R., 71
Crawford, S.L., 428
Cressie, N., 201
Cyrano de Bergerac, 33

D'Epifanio, G., 204
Damien, P., 290
Daniels, H.E., 105, 108, 125, 137
Davies, R.B., 79
Davis, P.A., 409
Dawid, A.P., 2, 29, 357, 358
Decoux, C., 448
DeGroot, M.H., 428
Dellaportas, P., 345, 359, 376, 404, 450
Dempster, A.P., 79, 213, 214, 216, 304, 415
Denison, D.G.T., 261
Denker, J.S., 229

Dette, H., 348, 356
Devroye, L., 44, 46, 47, 54–56, 70
Dey, D., 342
Diaconis, P., 34, 264, 277, 343
DiCiccio, T.J., 138
Dickey, J.M., 116
Diebold, D., 334
Diebolt, J., 227, 315, 349, 393, 415, 423–425, 430, 440
Dieter, U., 64
Dobson, A.J., 346
Dodge, Y., 61
Doob, J., 139
Doss, H., 189
Doucet, A., 202
Doukhan, P., 268, 378
Draper, D., 299
Duflo, M., 197, 202, 204, 230, 353
Dupuis, J.A., 59, 217, 339, 422
Durrett, R., 158
Dykstra, R.L., 7, 206

Eaton, M.L., 188, 189, 286
Eberly, L., 300
Efron, B., 33, 107
Enders, W., 211
Erkanli, A., 430
Escobar, M.D., 423, 430
Ethier, S.N., 265
Evans, M., 122
Everitt, B.S., 415, 422
Ewans, G., 17

Fan, J., 314
Fang, K.T., 70, 360
Feast, G., 64
Feller, W., 70, 106, 139, 142, 149, 152, 156, 158, 169, 177
Ferguson, T.S., 22
Ferrenberg, A.M., 38, 202
Field, C., 105, 137
Fieller, E.C., 16
Fienberg, S.E., 417
Fill, J.A., 353, 411, 413
Finch, S.J., 215
Fishman, G.S., 39, 343
Fitzgerald, W.J., 120, 194, 218, 222, 277
Fletcher, R., 17
Foster, J.J., 359

Frigessi, A., 297

Gamboa, F., 440
Garren, S.T., 410, 411
Gassiat, E., 440
Gauss, C.F., 6
Gaver, D.P., 301, 302, 371, 374, 379
Gelatt, C.D., 199
Gelfand, A.E., 116, 121, 255, 279, 290,
    301, 311, 326, 327, 331, 339,
    340, 342, 344, 345, 347, 349,
    368, 433, 434, 445, 450
Gelman, A., 123, 266, 281, 283, 297,
    298, 341, 366, 378, 381,
    394–397, 411, 436
Geman, D., 201, 279, 349
Geman, S., 201, 279, 349
Gentle, J.E., 17, 193
George, E.I., 59, 279, 307, 328
Geweke, J., 53, 84, 89, 112, 409, 411,
    436
Geyer, C.J., 155, 172, 184, 191, 194,
    209, 210, 220, 278, 315, 326,
    355, 366
Ghosh, M., 16
Gijbels, I., 314
Gilks, W.R., 57–59, 249, 266, 281,
    283, 297, 300, 326, 327,
    344–347, 355, 356, 372, 390
Giròn, F.J., 423
Giudici, P., 359
Gleick, J., 38
Gleser, L.J., 16, 49
Glynn, P.W., 365
Goldfeld, S.M., 448
Gouriéroux, C., 11, 32, 72, 79, 80, 211,
    334, 338
Goutis, C., 14, 105, 299, 327, 449
Green, P.J., 172, 254, 255, 259–261,
    276, 290, 332, 335, 355, 359,
    407, 411, 413, 449
Greenberg, E., 279, 450
Grenander, U., 264
Gruet, M.A., 442
Guédon, Y., 448
Guihenneuc–Jouyaux, C., 301, 363,
    401, 436, 448
Gumbel, E.J., 20, 67

Haario, H., 202

Hàjek, B., 200
Hall, P., 22, 33, 34, 106
Hammersley, J.M., 136, 298
Han, X.L., 255
Hancomb, D.C., 136
Hanlon, P., 277
Harris, T., 154
Hartigan, J., 448
Hastings, W.K., 139, 244, 249, 270,
    279, 349, 367, 409
Heidelberger, P., 410, 411, 437
Heitjan, D.F., 416
Henderson, D., 229
Hestbeck, J.B., 218
Hesterberg, T., 91
Higdon, D.M., 290, 332, 335
Hills, S., 318, 327, 331
Hills, S.E., 326, 344
Hines, J.E., 218
Hinton, G.E., 199
Hjorth, J.S.U., 33
Hobert, J.P., 189, 244, 299, 327,
    330–332, 341, 407
Holmes, S., 34, 264
Hougaard, P., 107
Howard, R.E., 229
Hubbard, W., 229
Huber, P.J., 73
Hwang, C.R., 202, 230
Hwang, J.T., 16, 206

Ikeda, M., 133
Inskip, H., 300
Ip, E.H.S., 228

Jöhnk, M.D., 47
Jackel, L.D., 229
Jacquier, E., 435, 436
Jarrett, R.G., 445
Jeffreys, H., 28, 32, 329
Jennison, C., 326
Jensen, J.L., 105
Johnson, N. L., 451
Johnson, V.E., 366, 408, 409
Jones, M.C., 314
Juang, B.H., 448

Kadane, J.B., 103, 204, 342, 428
Karim, R., 445
Kass, R.E., 14, 103, 204, 342

Kemeny, J.G., 175, 180, 182, 183, 278, 397
Kemp, A. W., 451
Kendall, W.S., 411
Kennedy, W.J., 17, 193
Kersting, G., 230
Kiefer, J., 229
Kim, S., 211, 334, 435, 436
Kinderman, A., 63
Kipnis, C., 171
Kirby, A.J., 300
Kirkpatrick, S., 199
Kloeck, T., 85
Knuth, D., 39, 63
Kolassa, J.E., 105, 137
Kong, A., 116, 308, 310, 311, 313, 337, 350, 352–355, 440
Kors, T.J., 201
Kotz, S., 451
Krimmel, J.E., 70, 97, 98, 383
Krishnan, T., 216, 415
Kroner, K.F., 211
Kubokawa, T., 16
Küchenhoff, H., 450
Kuehl, R.O., 137
Kullback, S., 222
Kurz, T.G., 265

Laird, N.M., 79, 213, 214, 216, 304, 415
Landau, D.P., 38
Lange, N., 216
Lauritzen, S.L., 2, 357, 358
Lavielle, M., 228, 415
Lavine, M., 423
Le Cun, Y., 229
L'Ecuyer, P., 61
Lee, T.M., 11, 347
Legendre, A., 6
Lehmann, E.L., 5, 11, 28, 30, 32–34, 37, 73, 74, 79, 80, 256, 311, 329, 370, 397, 425
Lépingle, 229, 230
Leroux, B.G., 432
Levine, R., 313
Lewis, S., 315, 365, 367, 397–399, 401, 404, 405, 410, 411, 437
Liao, J.G., 360
Lindvall, T., 161
Little, R.J.A., 216, 415, 419

Liu, C., 216, 228, 289, 366, 377, 379, 403
Liu, J.S., 116, 242, 276, 297, 308, 310, 311, 313, 324, 326, 337, 340, 350, 352–355, 366, 377, 379, 403, 440
Loh, W.L., 137
Lombard, F., 22, 450
Louis, T.A., 73, 216
Lugannani, R., 138
Lwin, T., 73

MacEachern, S.N., 365
MacLachlan, G., 216, 415, 422
Madigan, D., 358, 359
Maguire, B.A., 445
Makov, U.E., 20, 30, 422, 423
Mallick, B., 261, 450
Maritz, J. S., 73
Marsaglia, G., 37, 39, 42, 53
Martin, M.A., 138
Martin-Löef, P., 37
Massart, P., 268, 378
Matthews, P., 369
McCullagh, P., 449
McCulloch, C.E., 73, 332, 446, 450
McCulloch, R., 430, 448
McKay, M.D., 137
McKeague, I.W., 255
McNeil, A.J., 300
Mead, R., 137
Mendell, N.R., 215
Meng, X.L., 123, 216, 228, 289
Mengersen, K.L., 185, 239, 246, 247, 254, 270, 271, 276, 332, 335, 363, 424, 426, 436
Metropolis, N., 199, 202, 279, 349
Meyn, S.P., 139, 148, 149, 152, 155, 159, 163, 167, 169, 175, 186, 187
Miller, M., 264
Milshtein, G.N., 134
Min, C.K., 408
Mira, A., 278, 290, 291
Mitra, D., 201
Monahan, J., 63
Monfort, A., 11, 32, 72, 79, 80, 210, 212, 334, 338, 430, 432, 448
Monro, S., 229
Moulines, E., 228, 415

Moussouris, J., 298
Muller, B., 46
Müller, P., 282, 322, 323, 430
Murdoch, D.J., 407, 411, 413
Muri, F., 413, 448
Murray, G.D., 222
Mykland, P., 371, 374, 375, 380, 390, 410

Natarajan, R., 332
Naylor, J.C., 18
Neal, R.M., 199, 201, 228, 264, 277, 291
Nelder, J., 449
Nerlove, M., 334
Newton, M.A., 123
Ney, P., 149, 152, 388
Neyman, J., 11
Nichols, J.D., 218
Niederreiter, H., 69, 70
Nobile, A., 321
Norris, J.R., 139
Norris, M., 185, 265
Nummelin, E., 139, 147–149

Oh, M.S., 18
Olkin, I., 22
Oman, P.A., 244
O'Muircheartaigh, I.G., 301, 302, 371, 374, 379
Ord, J.K., 106, 107
Orey, S., 179
Ó Ruanaidh, J.J.K., 120, 194, 218, 222, 277
Osborne, C., 16

Pearl, J., 2
Pearson, E.S., 11, 445
Pearson, K., 20
Peskun, P.H., 278–280, 324, 349
Petkau, A.J., 22
Petrie, T., 342
Pflug, G.C., 230
Philippe, A., 13, 16, 68, 97, 99, 125, 338, 384, 439, 442
Phillips, D.B., 139, 264, 449
Pike, M.C., 20
Pincus, M., 204
Pitt, M.K., 324, 445
Pollock, K.H., 218, 420

Polson, N.G., 280, 350, 354, 355, 435, 436
Pommeau, Y., 38
Press, S.J., 341
Propp, J.G., 364, 406, 409, 411–413
Putterman, M.L., 432

Qian, W., 349, 415
Quant, R.E., 448
Quian, W., 216

Rabiner, C.R., 448
Racine-Poon, A., 300, 331, 344
Raftery, A.E., 14, 123, 315, 365, 367, 397–399, 401, 404, 405, 410, 411, 434, 437, 444
Raiffa, H., 25
Ramage, J., 63
Randles, R., 37
Rao, C.R., 216
Redner, R., 10
Reid, N., 105, 107
Renault, E., 334
Resnick, S.I., 265
Revuz, D., 139
Rice, S., 138
Richard, J.F., 72
Richardson, S., 261, 276, 448, 449
Rio, E., 268, 378
Ripley, B.D., 39–41, 80, 219, 220, 228
Ritter, C., 376, 408
Robbins, H., 73, 229
Robert, C., 2
Robert, C.P., 5, 9, 11, 13–16, 21, 28, 32, 33, 52, 59, 66, 71, 74, 85, 90, 93, 94, 111, 115–119, 125, 127, 167, 189, 191, 204–206, 210, 212, 222, 255–258, 276, 299, 301, 307, 311, 315, 327–329, 334, 338, 357, 363, 365, 371, 382, 384, 401, 404, 407, 422–424, 426, 427, 430, 432, 434, 436, 437, 442, 449
Roberts, G.O., 139, 140, 216, 237, 239, 245, 265, 266, 277, 280, 281, 283, 290, 297, 326, 332, 342, 363, 367, 372, 375, 377, 390, 398, 399, 403, 408, 410
Robertson, T., 7, 206

Roeder, K., 276, 442
Romano, J.P., 22
Romeo, F., 201
Ronchetti, E., 105, 137
Rosenblatt, M., 190
Rosenbluth, A.W., 199, 202, 279, 349
Rosenbluth, M.N., 199, 202, 279, 349
Rosenthal, J.S., 140, 245, 265, 266, 290, 402
Rossi, P.E., 435, 436
Rubin, D.B., 79, 121, 213, 214, 216, 289, 304, 366, 377–379, 381, 394–396, 403, 415, 416, 419, 436
Rubinstein, R.Y., 39, 84, 136, 197
Rudin, W., 97
Ruelle, D., 38
Ruppert, D., 450
Rydén, T., 365, 367, 430

Sabbati, C., 297
Sacksman, E., 202
Sahu, S.K., 255, 326, 327, 332, 340, 344, 372, 390
Sangiovanni-Vincentelli, A., 201
Saxena, M.J., 74
Scherrer, J.A., 218, 225, 339
Schervish, M.J., 103, 204, 350, 352, 353, 355
Schlaifer, R., 25
Schmeiser, B.W., 65, 249, 321, 323, 365
Schoener, T.W., 417
Schruben, L., 410
Schukken, Y.H., 300, 334
Schwartz, P.C., 140, 245
Searle, S.R., 73
Seber, G.A.F., 59, 420
Seidenfeld, T., 178
Sejnowski, T.J., 199
Serfling, R.J., 74
Sethuraman, J., 189
Shalaby, M., 65
Shao, Q.M., 122, 221, 407
Sharples, L.D., 300
Shephard, N., 211, 324, 334, 430, 435, 436, 445, 448
Siegel, A.F., 22
Silverman, B.W., 409
Singh, H., 410

Small, M.J., 428
Smith, A.F.M., 11, 18, 20, 30, 32, 116, 121, 139, 216, 222, 261, 264, 279, 290, 301, 311, 318, 327, 331, 339, 342, 344, 345, 347, 349, 368, 422, 423, 433, 434, 445, 449, 450
Smith, R.L., 410, 411
Snell, J.L., 175, 180, 182, 183, 278, 397
Soubiran, C., 204, 424
Soules, G., 342
Speed, T.P., 298, 341
Spiegelhalter, D.J., 2, 299, 300, 344–347, 355–357
Srivastava, M.S., 16
Stefanski, L.A., 450
Stegun, I., 8, 21, 29, 44, 129
Stein, M.L., 137
Stephens, D., 450
Stigler, S., 6
Stone, M., 29
Stram, D., 216
Stramer, O., 239, 264, 266, 278
Strawderman, R.L., 127
Stuart, A., 48, 106, 107
Studden, W.J., 348, 356
Sturmfels, B., 343
Swartz, T., 122
Swendson, R.H., 202, 279
Szidarovszky, F., 70, 97, 98, 383

Talay, D., 131, 134, 135
Tan, K.K.C., 58, 249
Tanner, M.A., 216, 254, 279, 289, 301, 304, 305, 325, 349, 364, 369, 376, 383, 405, 408, 415, 441
Teller, A.H., 199, 202, 279, 349
Teller, E., 199, 202, 279, 349
Thisted, R.A., 17, 18, 193, 197
Thode, H.C., 215
Thomas, A., 344–347, 355, 356
Thompson, E.A., 210, 221
Thorisson, H., 365
Tibshirani, R.J., 33
Tierney, L., 103, 155, 172, 191, 204, 249, 250, 269, 278, 290, 291, 319, 320, 342, 367, 371, 374, 375, 390, 410
Titterington, D.M., 20, 205, 216, 228, 349, 365, 367, 407, 415, 422,

427, 430, 432, 448
Tobin, J., 338
Tsay, R., 430, 448
Tukey, J.W., 275
Tweedie, R.L., 139, 148, 149, 152, 155,
    159, 163, 167, 169, 175,
    185–187, 237, 239, 246, 247,
    264–266, 270, 271, 278, 280

van den Broek, J., 300, 334
Van Dijk, H.K., 85
van Dyk, D., 216, 228
Van Laarhoven, P.J., 201
Varadhan, S.R., 171
Vecchi, M.P., 199
Verdinelli, I., 423
Vidal, C., 38
Vines, K., 326, 327, 367, 410, 411
Von Neumann, J., 36

Wahba, G., 18
Wakefield, J.C., 290, 300
Walker, H., 10
Walker, S., 65, 290
Wand, M.P., 314
Wang, J.S., 38, 202, 279
Wang, Y., 70, 360
Wasan, M.T., 230
Wasserman, L., 32, 178, 423, 442
Watanabe, Y., 133
Wefelmeyer, W., 255
Wei, G.C.G., 216, 383, 415
Weiss, N., 342
Welch, P.D., 410, 411, 437
West, M., 422, 423, 430
White, G.C., 420
Whittaker, J., 357
Wild, P., 57, 59, 327
Wilson, D.B., 364, 406, 409, 411–413
Winfield, D., 226
Winkelmann, R., 450
Winkler, G., 197, 199, 200, 219, 230,
    279, 280
Wolfe, D., 37
Wolff, U., 38
Wolfowitz, J., 229
Wolpert, R., 5, 11
Wolpert, R.L., 33, 167
Wolter, W., 307, 420

Wong, W.H., 116, 123, 254, 279, 289,
    304, 305, 308, 310, 311, 313,
    325, 337, 349, 350, 352–355,
    364, 408, 440
Wood, A.T.A., 138
Wright, F.T., 7, 206
Wu, C.F.J., 215, 222
Wynn, A.H.A., 445

Yakowitz, S., 70, 97, 98, 383
Yao, Y.C., 448
York, J., 358, 359
Yu, B., 371, 374, 375, 380, 390, 410

Zacks, S., 449
Zaman, A., 39, 42
Zeger, S., 445
Zellner, A., 408
Zidek, J.V., 22, 29

# Subject Index

$\alpha$-mixing, 184, 190, 191
absorbing component, 424
Accept–Reject, 84, 111, 112, 119, 232,
    234, 239, 299, 322, 446
    and Metropolis–Hastings
      algorithm, 241, 242
    and recycling, 53, 92–94, 118, 135
    and wrong bounds, 269
    and importance sampling, 92
    approximative, 250
    bound, 135
    corrected, 270
    criticism, 53
    definition, 49
    for GLMs, 446
    optimization, 51
    sample distribution, 126
    stopping rule, 93
    tight bound, 50
    with envelope, 53
acceptance rate, 252
    and simulated annealing, 202
    expected, 241
    optimal, 266, 281
acidity level, 428
adaptive rejection sampling, 57
aggregation, 416
algorithm
    acceleration of, 44, 109, 117, 251,
      255, 320
    Accept–Reject, 49
    ARMS, 142, 249–251, 285
    ARS, 57–60, 232, 249
    autoregressive
      Metropolis–Hastings, 249
    Best's, 55
    Box–Muller, 46, 62
    CFTP, 406
    comparison, 82, 109, 242

convergence of, 100, 115, 135,
    197, 200
domination, 95, 136
ECM, 228, 289
EM, 2, 79, 213, 289, 304, 415,
    423, 447
Gibbs, 233, 236
hit-and-run, 249, 321, 323
hybrid, 295, 319, 321, 322, 374,
    428, 449
independent
    Metropolis–Hastings, 239,
      241, 256, 321, 373
initial conditions, 227
*Kiss*, 39, 40, 42, 43
*Kiss* (C program), 42
MCEM, 289
MCMC, 146, 164, 172, 231
Metropolis–Hastings, 199, 201,
    232, 233, 299, 328, 379
Newton–Raphson, 17
optimization, 51, 109
performances, 99
pool-adjacent-violators, 7, 23
random walk
    Metropolis–Hastings, 245,
    246
Robbins–Monro, 229, 230
SAEM, 228
SEM, 228, 231
SIMEX, 450
simulated annealing, 199, 326
SIR, 121
stability, 100, 101, 103
stochastic, 193
Swendson–Wang, 279
symmetric Metropolis–Hastings,
    374
algorithmic complexity, 44

allocation
    hidden Markov, 430
    hybrid, 428
    indicator, 423
    map, 437, 438
    probability, 425
    stable, 429
analysis
    conditional, 167
    mortality, 335
    spectral, 410
antithetic variable, 267
aperiodicity, 150, 177, 236
approximation, 245
    first-order, 104
    Laplace, 204
    Normal, 396
    Normal, for the binomial
        distribution, 89
    Riemann, 96, 383
    saddlepoint, 104
    second-order, 104
    third-order, 104
asymptotic normality, 79
asymptotic relative efficiency, 313
atom, 147, 149, 152, 156, 160, 165
    accessible, 147
    ergodic, 160, 162
    geometrically ergodic, 165
    Kendall, 165
    recurrent, 408
augmentation technique, 388, 391
autoexponential model, 286
    uniform ergodicity, 303

$\beta$-mixing, 191
baseball, 225
batch sampling, 314
Bayes estimator, 12
    admissibility of, 189
    linear, 31
Bayes factor, 14
    computation, 122, 342
Bayesian inference, 11
    noninformative, 432
bias, 409
binomial parameter, 22
bits, 41
bootstrap, 33, 71
bridge estimator, 123

Brownian motion, 265
BUGS, 19, 345, 355, 411

C, ix, 42, 355
calibration, 249, 251, 267, 369, 408
    algorithmic, 282, 446
    linear, 16
    step, 282
canonical moments, 347, 348, 356
capture–recapture, 59, 217, 225, 307,
    339, 420
censoring model, 2, 19, 207, 420
    type of, 2
Cesàro average, 179
chains
    parallel, 410
    parallel versus single, 232, 367
changepoint, 432, 448
chaos, 38
Chapman–Kolmogorov equations, 143,
    146, 151, 152
character recognition, 229
classification of Metropolis–Hastings
    algorithms, 239
clique, 358
    perfect, 358
coalescence, 412
CODA, x, 356, 411, 436, 437
combinatorial explosion, 423
comparison of models, 13
completion, 289, 290, 416
    of missing data, 415
    natural, 289, 419
computational difficulties, 74
computer discretization, 245
computing time, 30, 90
    comparison, 45
    excessive, 38
    expensive, 52
condition
    Doeblin's, 166, 280
    drift, 153, 171, 186–188, 247
    geometric drift, 187
    Hilbert–Schmidt, 350, 352, 355,
        410
    minorizing, 141, 147, 149, 152
    positivity, 298
    transition kernel regularity, 332
conditional distributions, 297
conditioning, 116, 119, 310

confidence interval, 34, 77
conjugate prior, 31, 422
    and BUGS, 355
    for mixtures, 404
connected components, 296
connexity of the support, 299
constrained parameter, 7
contingency table, 343, 417
continuous-time process, 264
control
    binary, 399, 400
    of Markov chain Monte-Carlo
        algorithms, 374
    of Metropolis–Hastings
        algorithms, 251
    of simulation methods, 135
    by renewal, 371
control variate, 113, 115
    integration, 114
    optimal constant, 114
convergence, 197
    acceleration, 195
    and duality principle, 315
    and irreducibility, 239, 245
    assessment, 95, 109, 144, 161,
        232, 389
    conservative criterion for, 135
    control, 188, 369, 383
    in distribution, 408
    of the empirical mean, 251, 365,
        377
    ergodic, 280
    evaluation of, 287
    geometric, 164, 239, 280, 301,
        352, 353, 357, 424
    geometric uniform, 280
    Gibbs sampler, 393
    graphical evaluation of, 368, 382
    indicator, 392, 399
    of Markov chain Monte Carlo
        algorithms, 231, 363
    of SEM, 415
    slow, 85, 251, 379
    speed, 100, 101, 164, 205
    test, 75
    to a maximum, 201
    to the stationary distribution,
        364, 366, 377
    towards the mode, 369
    of the variance criterion, 395

convolution, 439
correlation, 109, 111
    between components, 399
    negative, 112
coupling, 161, 270, 409, 412
    from the past, 406, 412
    monotone, 406
    multigamma, 413
    time, 162
covariate, 449
covering, 152
criterion
    conservative, 392
    convergence, 389
    efficiency, 281
    unidimensional, 397
cumulant generating function, 7
cumulative sums, 380
curse of dimensionality, 18, 99, 137,
    383
cycle, 150
    of kernels, 320

dagger (†), 421
data
    censored, 2
    coarsening, 416
    complete, 208
    grouped, 416
    longitudinal, 418, 420, 433
    missing, 196, 208, 415, 417, 418,
        449
    missing at random, 419
data augmentation, 188, 289, 293,
        303, 308, 317, 357, 412, 430
    reversible, 309
    specificities, 293
deconvolution, 439, 440
demarginalization, 207, 213, 288, 290,
    334
    for recursion formulae, 357
density
    instrumental, 49
    log-concave, 56, 58, 59, 67
    multimodal, 254
    spectral, 409
    target, 49
dependence
    on initial values, 409
    logistic, 251

Markovian, 167
detailed balance condition, 235
deviance, 346
diagnostic
    autocorrelation, 411
    coupling, 408
    distance, 376, 377
    multiple estimates, 388
    parallel, 397
*Die Hard*, 37
diffusion, 264
dilation, 178
dimension matching, 260
dimensionality of a problem, 99, 259
discretization, 140, 402, 410
    of a diffusion, 265
    and Laplace approximation, 266
    Euler, 134
    Milshtein, 134
distance
    Kullback–Leibler, 32, 222
    to the stationary distribution,
      377
    unbiased estimator, 375, 403
distribution
    arcsine, 38, 60
    beta, 8, 12
    binomial, 22
    Burr's, 55
    Cauchy, 10, 80
    conditional, 328
    conjugate, 115
    Dirichlet, 98
    double exponential, 51
    exponential, 88
    Gamma, 51, 54, 94
    generalized inverse Gaussian, 67
    generalized inverse normal, 28
    geometric, 144, 152, 247
    Gumbel, 67
    heavy tail, 84
    improper prior, 328, 335
    infinitely divisible, 70
    instrumental, 49, 52, 82, 86, 90,
      97, 205, 233, 239, 323, 425,
      446
    instrumental (choice of), 285, 318
    invariant, 140
    inverse Gaussian, 253
    Laplace, 132

limiting, 234
log-concave, 271
log-normal, 82
logistic, 55, 59, 333
marginal, 13
multinomial, 393
negative binomial, 47
noncentral chi-squared, 8, 21, 24,
    49, 74
nonuniform, 43
Normal, 44, 46, 86, 111, 246, 417
Pareto, 9
Poisson, 46, 47, 55
posterior, 11
prior, 11
stationary, 155, 156, 235
stationary (distance to), 375, 377
Student, 9
truncated Gamma, 439
truncated Normal, 52, 90, 418
uniform, 35, 43, 87, 112
Weibull, 3, 19, 20, 345
Wishart, 28, 65, 418, 434
witch's hat, 368, 369, 376, 380,
    396, 402, 406
divergence, 69, 181, 330
Doeblin's condition, 166, 271, 407
domination of a Markov chain, 324
drift condition, 185, 248
    geometric, 187
dugong, 345

Edgeworth expansion, 106
efficiency, 258
    loss of, 365, 377, 398, 410
eigenvalues, 242, 405, 410
    of a covariance matrix, 327
    of operators, 351, 352, 354, 355
    of the kernel, 350
EM algorithm
    and simulated annealing, 215
    identity, 213
    introduction, 213
    monotonicity, 214
    Monte Carlo, 216
    steps, 214, 227
empirical
    cdf, 71, 79, 80
    mean, 75, 81, 96, 100, 116
empirical Bayes principles, 73

energy function, 199, 201, 230
entropy distance, 222
envelope, 53, 57
  method, 53
  pseudo, 249
ergodicity, 163, 167
  definition of, 160
  geometric, 164, 184, 239, 246, 316
  of Metropolis–Hastings
    algorithms, 236, 237, 251
  uniform, 166, 191, 239
  uniform (sufficient conditions),
    320
  uniform, of Metropolis–Hastings
    algorithms, 246
errors in variables, 11
estimation
  decisional, 78
  nonparametric, 410
  parallel, 135
estimator
  empirical Bayes, 73, 74
  James–Stein, 73
  maximum likelihood, 79, 227, 424
  maximum likelihood (lack of),
    425
  parallel, 109
  Rao–Blackwellized, 126
  truncated James–Stein, 110
expert systems, 2, 357
exponential
  family, 5, 7, 9, 12, 28, 107, 115,
    204, 253, 422
  tilting, 107, 125
  variate generation, 45
extra-binomial variations, 335

Fibonacci series, 61
fixed point, 214
function
  analytical properties of, 193
  confluent hypergeometric, 29
  digamma, 6, 8
  dilog, 129, 135
  empirical cumulative
    distribution, 33
  evaluation, 53
  link, 450
  logistic, 38
  loss, 71–73, 109

modified Bessel, 9, 49
  potential, 153, 185, 186, 188
  regularity of, 193
  risk, 73
  sigmoid, 229
functional compatibility, 341
fundamental matrix, 180

Gamma variate generator, 47, 54, 94
gap, 354
Gauss, 19, 45
generalized inverse, 35
generalized linear model, 251, 446
  Bayesian inference, 449
  completion, 447
  definition, 449
  Gibbs sampler, 445
  link function, 450
  mixed, 450
generator
  Central Limit Theorem, 62
  chaotic, 38
  congruential, 40, 42
  pseudo-random, 19, 35, 36, 245
  pseudo-random (tests for), 37
  RANDU, 40, 61
  shift register, 40
genetic linkage, 216
geometric
  ergodicity, 440
  rate of convergence, 249, 257
Gibbs
  measure, 204, 230
  random field, 349
  sampling, 285
Gibbs sampling, 232, 279, 285
  and ECM, 228
  completion, 288
  definition, 285
  Harris recurrence of, 295
  irreducibility of, 294
  random, 313
  random scan, 310
  random sweep, 310
  reversible, 310
  slow convergence of, 427
  symmetric scan, 310
  symmetric sweep, 310
  uniform ergodicity of, 301
  validity of, 318

versus Metropolis–Hastings
    algorithm, 287, 293, 296, 317
Gibbs stopper, 405, 408
gradient
    methods, 17
    stochastic, 197
graph, 2, 358
    directed, 358
    directed acyclic, 358
    undirected, 358
Gray codes, 34
grouped data, 304, 305

harmonic function, 168, 169, 236, 354
    bounded, 169
Harris positivity, 186
Harris recurrence, 169
    and geometric ergodicity, 354
    importance, 154, 155
Hessian matrix, 266
Hewitt–Savage 0 − 1 Law, 168
hidden
    Markov chains, 413, 429, 430, 448
    Markov Poisson model, 432
    semi-Markov chains, 448, 449
*how long is long enough?*, 411
hybrid strategy, 299

identifiability, 76, 79, 426–428, 430
    constraint, 426–428
image processing, 201, 349
implicit equations, 32, 71
importance, 80
importance sampling, 80–82, 84–88,
        91, 118, 119, 135, 209, 231
    and Accept–Reject, 83–85, 90–92
    and infinite variance, 92, 211, 386
    by defensive mixtures, 92
    for convergence assessment, 382
    implementation, 90
IMSL, 19
independence, 365, 366
inference
    asymptotic, 74
    Bayesian, 5, 11–13, 232, 422
    Bayesian noninformative, 425
    difficulties of, 4, 5
    empirical Bayes, 73
    generalized Bayes, 328
    nonparametric, 5

statistical, 1, 7, 71
information
    Fisher, 329
    Kullback–Leibler, 32
    prior, 13, 300, 330
informative censoring, 19
initial condition, 252, 289, 295, 353
    dependence on, 252, 294, 365
initial distribution, 144, 237
    and parallel chains, 366
    influence of, 294, 366, 396
initial state, 142
integration, 5, 17, 71, 78, 109
    approximative, 75, 97, 103
    Monte Carlo, 74
    numerical, 71
    problems, 11
    recursive, 205, 424, 432
    by Riemann sums, 18
    weighted Monte Carlo, 97
interleaving, 365
    and reversibility, 308
    definition, 308
    property, 308, 312
intrinsic losses, 71
inversion
    constraint, 409
    of Gibbs steps, 375
    of the cdf, 36, 43, 54
irreducibility, 140, 145, 148, 151, 237,
        245
    definition of, 146
Ising model, 201, 220

Jacobian, 260, 261

Kalman filter, 223
$K_\varepsilon$-chain, 144, 146, 153
kernel
    estimation, 314
    transition, 140–142
Kolmogorov–Smirnov test, 37, 69, 370
Kuiper test, 370
Kullback–Leibler
    distance, 222
    information, 32, 222

Lagrangian, 17
Langevin algorithm, 266
    and geometric ergodicity, 266

extensions of, 266
Langevin diffusion, 265
Laplace approximation, 103, 220, 342
large deviations, 37, 80, 132
latent index, 432
Latin hypercubes, 137
Law of Large Numbers, 74, 76, 167,
    168, 391
    strong, 75, 170
lemma
    Pitman–Koopman, 31
likelihood, 11
    function, 5, 6
    integral, 11
    maximum, 73, 74, 206
    multimodal, 10
    profile, 8, 16, 71
    pseudo-, 7
    unbounded, 10
likelihood ratio approximation, 211
likelihood surface approximation, 212
linear model, 26
lizard, 417, 420
local maxima, 195, 318
log-concavity, 246
longitudinal data, 3, 418, 420, 433
    and capture–recapture, 421
loss
    intrinsic, 71
    posterior, 12
    quadratic, 12
low-discrepancy sequence, 69

$M$-estimator, 32
map, 412
Maple, 19
marginalization, 335
    Monte Carlo, 122
Markov chain, 88–90, 139, 145, 293
    $\varphi$-irreducible, 146
    augmented, 149
    average behavior, 167
    covariance, 172
    definition, 141
    divergent, 331
    ergodic, 231
    essentials, 140
    Harris positive, 156, 160
    Harris recurrent, 154, 155, 168,
        186

hidden, 429, 430, 448
homogeneous, 142, 197, 250, 398
instrumental, 308
interleaved, 308
irreducible, 146
limit theorem for, 167, 189
lumpable, 183
nonhomogeneous, 197, 200, 250,
    281, 282, 421
observed, 167
positive, 156
recurrent, 152, 186
reversible, 171, 188
semi-, 448
for simulation, 231
split, 149
stability, 151, 155
strongly aperiodic, 150, 152
strongly irreducible, 146, 304
transient, 185, 422
two state, 397
weakly mixing, 380
Markov chain Monte Carlo method,
    140, 231
Markov property
    strong, 145, 174
    weak, 144
mastitis, 300
Mathematica, 19, 45, 205
Matlab, 19, 45
matrix
    Choleski decomposition of, 65
    regular, 180
    transition, 141, 142, 144, 150, 156
maxima
    global, 201, 204
    local, 197, 199, 230
maximization, 5
maximum likelihood, 5, 7, 9, 74
    constrained, 7, 206
    difficulties of, 9, 11
    estimation, 5, 7
    existence of, 10
    justification, 5
MCEM algorithm, 216, 218
MCMC algorithm
    calibration, 282
    definition, 231
    heterogeneous, 320
    history, 279

implementation, 232
motivation, 231
measure
    counting, 158
    invariant, 155, 157
    Lebesgue, 147, 157
    maximal irreducibility, 146
method
    Accept–Reject, 49, 50, 52, 53
    gradient, 196
    kernel, 409
    least squares, 6, 206, 229
    Markov chain Monte Carlo, 199,
        231
    of moments, 7, 32
    monitoring, 330
    Monte Carlo, 74, 78, 80, 82, 85,
        110, 135
    nonparametric, 397, 410
    numerical, 1, 17, 19, 71, 78, 97,
        193, 394
    quasi-Monte Carlo, 69, 360
    simulated annealing, 199, 200,
        228, 230
    simulation, 17
Metropolis algorithm, 200
Metropolis–Hastings algorithm
    and Accept–Reject, 241
    and importance sampling, 234
    drawbacks of, 319
    efficiency of, 285
    ergodicity of, 246
    independent, 239
    irreducibility of, 236
    random walk, 244
    validity of, 233
Metropolization, 324
mining accidents, 445
missing data, 2, 4, 289, 304, 305
    processing of, 415
    reconstruction, 417
    simulation of, 227, 415
missing data models
    introduction to, 207
mixing, 387
    condition, 190
    speed, 365
mixture, 3, 232, 254, 422, 425
    beta, 450
    defensive, 92

exponential, 20, 221, 442
geometric, 356
indicator, 423
infinite, 429
kernel, 319
negative weight, 391
nonparametric, 356
Normal, 10, 20, 79, 318, 404, 424,
    426, 427
number of components (test on),
    79
Poisson, 20
reparameterization of, 425
representation, 70, 107, 289
for simulation, 48, 65, 70, 116,
    288
stabilization by, 91, 121
Student's $t$, 394
with unknown number of
    components, 259, 261, 449
variance, 105
mode, 195, 369, 381, 396
global, 205
local, 10, 18, 205, 229
local (attraction towards), 318
local (trapping effect of), 320
model
    ANOVA, 346
    AR, 143, 159, 173, 187, 222
    ARCH, 211, 272, 323, 334
    ARMA, 259, 409
    autoexponential, 286, 297, 332
    autoregressive, 27
    Bernoulli–Laplace, 142, 158
    capture–recapture, 59, 217, 225,
        307, 420
    censored, 291
    changepoint, 276, 347, 429, 432
    completed, 418, 423, 438
    decomposable, 358
    embedded, 261
    generalized linear, 251, 446, 449
    graphical, 357
    hierarchical, 299
    Ising, 38, 201, 287, 411
    linear calibration, 289
    logistic, 114, 335
    logit, 14, 201, 418, 419, 438, 449
    MA, 4
    missing data, 415

mixed-effects, 300
multilayer, 228
multinomial, 305, 325
Normal, 329
Normal hierarchical, 26
Normal logistic, 59
overparameterized, 254
Potts, 279
probit, 321, 418
qualitative, 418
random effects, 327, 331
Rasch, 345
stochastic volatility, 435
switching AR, 430, 431
switching state, 429
tobit, 338
model choice, 259
modeling, 1, 422
and reduction, 5
moment generating function, 7
monitoring of Markov chain Monte
          Carlo algorithms, 363, 389
monotonicity
of covariance, 310, 337, 365
for perfect sampling, 412
Monte Carlo
approximation, 208
EM, 216, 217
maximization, 209
optimization, 194
move
merge, 261
split, 261
multimodality, 18, 19
multiple chains, 366

neural network, 228, 448
nodes, 358
nonidentifiability, 335
noninformative prior, 32
nonlinearity, 446
nonparametric mixture, 356
nonresponse, 419
nonstationarity, 368
Normal approximation, 244
Normal cdf
approximation, 78
Normal variate generation, 44, 46, 51
normalization, 383

normalizing constant, 13–15, 50, 100,
          122, 221, 234, 261, 329, 372,
          376, 423
and Rao–Blackwellization, 273
approximation, 85
evaluation, 210, 377
nuclear pump failures, 302, 371, 375
null recurrence and stability, 170
Nummelin's splitting, 149, 157

on-line monitoring, 381
open population, 420, 421
operator
adjoint, 353
associated with the transition
          kernel, 353
transition, 349
optimality, 281
of an algorithm, 44, 59, 82, 85,
          100, 289
of an estimator, 13, 73
of the Metropolis–Hastings
          algorithm, 236
optimization, 17, 52, 71, 252, 253, 279,
          281
exploratory methods, 196
stochastic, 231
order statistics, 47, 64
Orey's inequality, 179, 270
Ornstein–Uhlenbeck equation, 131
outlier, 300
overdispersion, 300
overparameterization, 426

paradox of trapping states, 425
parallel chains, 365, 377
parameter
of interest, 8, 16, 24, 32, 433
location–scale, 426
natural, 7
noncentrality, 108
nuisance, 8, 71, 79
parameterization, 299, 318, 333, 400,
          427
partition product, 448
partitioning, 136
path, 364
properties, 377
perceptron, 277
perfect simulation, 167, 409, 411

performances
    of congruential generators, 40
    of estimators, 73
    of the Gibbs sampler, 318, 326,
        424
    of importance sampling, 86
    of integer generators, 39
    of the Langevin diffusion, 267
    of methods of convergence
        control, 369
    of the Metropolis–Hastings
        algorithm, 252, 255, 282
    of Monte Carlo estimates, 85
    of renewal control, 393
    of simulated annealing, 202
period, 39, 40, 150
Pharmacokinetics, 300
Physics, 37
Pitman–Koopman lemma, 9
Poisson variate generation, 55
polar coordinates, 76
positivity, 156, 236, 298, 299
    condition, 294
potential function, 185, 248
power of a test, 11
principle
    duality, 315, 317, 393, 412, 424,
        430
    likelihood, 33, 167
    squeeze, 53
prior
    conjugate, 12, 31, 73
    Dirichlet, 359
    feedback, 202, 204
    hierarchical, 299, 433
    improper, 328, 329, 427
    influence of, 204
    information, 31
    Jeffreys, 28
    noninformative, 300
    reference, 13, 32
    robust, 75
probability integral transform, 36
probability of regeneration, 374
problem
    Behrens–Fisher, 11
    Fieller's, 16
process
    branching, 151, 152
    forward recurrence, 179

Langevin diffusion, 281
Poisson, 46, 55, 302, 433
renewal, 161
programming, 45
    parallel, 135
properties
    mixing, 411
    sample path, 167
pseudo-likelihood, 7

quadratic loss, 72
quality control, 3
quantile, 397

random effects, 331, 335
random walk, 157, 158, 176, 184, 244,
        245, 247, 251, 252, 254, 280,
        282, 449
    with drift, 266
    multiplicative, 175
randomness, 36
range of proposal, 255
Rao–Blackwellization, 116–119,
        255–258, 273, 310, 312, 337,
        382, 383, 393, 408
    of densities, 314
    improvement, 313
    for mixtures, 424
    parametric, 388
rate
    acceptance, 250, 252, 254,
        281–283, 287, 297, 429
    mixing, 367
    regeneration, 372
    renewal, 392
raw sequence plot, 381
recurrence, 140, 151, 154, 156, 186
    and admissibility, 188
    and positivity, 156
    Harris, 140, 155, 162, 170, 295
    null, 155
    positive, 155, 330
recursion equation, 356
recycling, 310, 382, 397
reference prior, 32
    for the calibration model, 16
referential
    common, 426
    local, 426
reflecting boundary, 248

regeneration, 371, 374
region
    confidence, 15
    highest posterior density, 15, 16
regression
    changepoint, 434
    isotonic, 23, 206
    linear, 6, 26, 27, 396
    logistic, 14, 115
    Poisson, 60
    qualitative, 14
relativity of convergence assessments,
    367
reliability, 20
    of an estimation, 85
renewal, 188, 371
    and coupling, 407
    and regeneration, 371
    control, 392
    probabilities, 374, 391
    rate, 392
    theory, 147, 188, 371, 397, 413
    time, 149, 388
Renyi representation, 64
reparameterization, 195, 422, 427
resampling, 33
resolvant, 144
    chain, 178
reversibility, 171, 259, 264
reversibilization, 353
reversible jump, 259
Riemann integration, 18
Riemann sum
    and Rao–Blackwellization, 338
    control variate, 381
robustness, 9, 80
rounding effect, 416

saddlepoint, 105, 197
    approximation, 207, 243
sample
    independent, 109
    preliminary, 400
    test, 398
    uniform, 110
    uniform common, 408
sampling
    batch, 312
    Gibbs, 285, 299, 328, 449

Gibbs and Metropolis–Hastings
    algorithms, 296
    importance, 258, 394
    stratified, 136
scale, 199
SDE, 133
seeds, 42
semi-Markov chain, 444
    hidden, 449
sensitivity, 80
separators, 358
set
    Kendall, 187
    recurrent, 152
    small, 148, 150, 153, 164, 186,
        238, 371, 372, 388, 392, 409
    transient, 175
    uniformly transient, 152, 175
shift
    left, 42
    right, 42
signal processing, 194
Simpson's rule, 18
simulated annealing, 19, 142, 195, 245,
    281
    and prior feedback, 205
simulation, 19, 71, 74, 231, 232, 328
    in parallel, 366
    motivation for, 1, 4, 5, 72
    numerical, 365
    parallel, 349, 366
    philosophical paradoxes of, 36, 37
    recycling of, 255
    Riemann sum, 97, 99–102
    twisted, 132
    univariate, 286
    versus numerical methods, 17–19,
        77, 193
    waste, 232, 255
size
    convergence, 398
    warm-up, 398
skewness, 106
slice sampler, 208, 288, 290
    difficulties, 291
small dimensions problems, 19
spectral analysis, 409
speed
    convergence, 254, 314, 318, 323,
        387

convergence (for empirical
  means), 377
mixing, 365, 380, 408
splitting, 152
S-Plus, ix, 45, 355, 411
squared error loss, 96
stability, 386, 426
  of a path, 85
stabilizing, 392
start-up table, 40
state
  absorbing, 422, 425
  initial, 142
  period, 150
  recurrent, 151
  transient, 151
  trapping, 425, 427
state-space
  continuous, 143
  discrete, 156, 279, 324
  finite, 140, 158, 200, 201, 244,
    316, 392, 411, 448
stationarity, 167, 381
stationarization, 363
stationary distribution, 140, 411
  and diffusions, 139
  as limiting distribution, 140
statistic
  order, 97, 98
  sufficient, 8, 9, 89, 116
Statistics
  sequential, 397
  spatial, 201
Statlib, 399
stochastic
  approximation, 207, 229
  differential equation, 133, 264
  exploration, 194
  monotonicity, 412
  optimization, 231, 234
  restoration, 304
  volatility, 435
stochastic gradient, 245
  and Langevin algorithm, 267
  equation, 265
stopping rule, 145, 243, 364, 367, 389,
    394, 397, 399, 409
Student's $t$
  approximation, 244
  variate generation, 48, 63

subsampling, 365, 397
  and convergence assessment, 365
  and independence, 370
sufficiency, 25
support
  connected, 238
  restricted, 86, 90
survey, 419
sweeping, 327
switching models, 448

tail area, 108, 243
  approximation, 81, 120, 122, 138
tail event, 168
temperature, 199, 200
  decrease rate of, 201
termwise Rao–Blackwellized
    estimator, 127
test
  $\chi^2$, 410, 417
  Die Hard, 37
  independence between factors,
    417
  Kolmogorov–Smirnov, 37, 69, 410
  likelihood ratio, 79
  for Markov chains, 397
  nonparametric, 370
  optimal, 30
  pivotal quantity for, 30
  power of, 30, 78–80
  randomness, 38
  stationarity, 364
  UMP, 30
  uniform, 37
testing, 78
Theorem
  Cramer, 132
  Kac's, 156
theorem
  Bayes, 11, 13, 50
  Central Limit, 46, 74, 164, 166,
    167, 170, 171, 188, 190, 310,
    388, 389
  ergodic, 168, 169, 232, 364, 365,
    379
  Fubini, 12
  Glivenko–Cantelli, 33
  Hammersley–Clifford, 297, 298,
    333, 382
  Kendall, 165

Rao–Blackwell, 116, 256
theory
    Decision, 13, 73, 74
    Neyman–Pearson, 11
    renewal, 371
time
    computation, 384
    interjump, 282
    renewal, 149, 388
    stopping, 144, 145, 149, 363
time series, 4, 409
    for convergence assessment, 409
total variation norm, 140, 160, 164,
    182
training sample, 228
trajectory (stability of), 151
transience of a discretized diffusion,
    265
transition, 236
    Metropolis–Hastings, 233
    pseudo-reversible, 373
transition kernel, 140, 141, 143, 388
    atom, 147
    choice of, 251
    strictly positive, 190
    symmetric, 141
traveling salesman problem, 341
truncated Normal distribution, 288

unbiased estimator
    density, 383
    $L_2$ distance, 403
uniform ergodicity, 188
uniform random variable, 35
    generation, 35–37
universality, 59, 238, 255
    of Metropolis–Hastings
        algorithms, 235

variable
    antithetic, 109, 111
    auxiliary, 2
    explanatory, 14
    latent, 2, 321, 418, 423, 432, 435,
        437
    selection, 14
variance
    asymptotic, 355, 389, 393
    between- and within-chains, 394
    finite, 84, 86

infinite, 90
variance reduction, 81, 82, 85, 110,
    112–116, 118, 137
    and antithetic variables, 126
variate
    control, 113
vertices, 358
virtual observation, 204
volatility, 435

Wiener process, 131, 133

you've only seen where you've been,
    367, 371, 381

# Springer Texts in Statistics (continued from page ii)

*Noether:* Introduction to Statistics: The Nonparametric Way
*Peters:* Counting for Something: Statistical Principles and Personalities
*Pfeiffer:* Probability for Applications
*Pitman:* Probability
*Rawlings, Pantula and Dickey:* Applied Regression Analysis
*Robert:* The Bayesian Choice: A Decision-Theoretic Motivation
*Robert and Casella:* Monte Carlo Statistical Methods
*Santner and Duffy:* The Statistical Analysis of Discrete Data
*Saville and Wood:* Statistical Methods: The Geometric Approach
*Sen and Srivastava:* Regression Analysis: Theory, Methods, and
    Applications
*Shao:* Mathematical Statistics
*Terrell:* Mathematical Statistics: A Unified Introduction
*Whittle:* Probability via Expectation, Third Edition
*Zacks:* Introduction to Reliability Analysis: Probability Models and Statistical
    Methods